Ninth Edition

Mader's Understanding
Human Anatomy & Physiology

Susannah Nelson Longenbaker
Columbus State Community College, Columbus, OH

Mc
Graw
Hill
Education

MADER'S UNDERSTANDING HUMAN ANATOMY & PHYSIOLOGY, NINTH EDITION

Published by McGraw-Hill Education, 2 Penn Plaza, New York, NY 10121. Copyright © 2017 by McGraw-Hill Education. All rights reserved. Printed in the United States of America. Previous editions © 2014, 2011, and 2008. No part of this publication may be reproduced or distributed in any form or by any means, or stored in a database or retrieval system, without the prior written consent of McGraw-Hill Education, including, but not limited to, in any network or other electronic storage or transmission, or broadcast for distance learning.

Some ancillaries, including electronic and print components, may not be available to customers outside the United States.

This book is printed on acid-free paper.

1 2 3 4 5 6 7 8 9 0 RMN/RMN 1 0 9 8 7 6

ISBN 978-1-259-29643-7
MHID 1-259-29643-1

Senior Vice President, Products & Markets: *Kurt L. Strand*
Vice President, General Manager, Products & Markets: *Marty Lange*
Vice President, Content Design & Delivery: *Kimberly Meriwether David*
Managing Director: *Michael Hackett*
Brand Managers: *Chloe Bouxsein/Amy Reed*
Director, Product Development: *Rose Koos*
Product Developer: *Fran Simon*
Executive Marketing Manager: *James F. Connely*
Marketing Manager: *Jessica Cannavo*
Director of Digital Content: *Michael G. Koot, PhD*
Digital Product Analyst: *Jake Theobald*
Director, Content Design & Delivery: *Linda Avenarius*
Program Manager: *Angela R. FitzPatrick*
Content Project Managers: *April R. Southwood/Christina Nelson*
Buyer: *Sandy Ludovissy*
Design: *Tara McDermott*
Content Licensing Specialists: *Lori Hancock/Lorraine Buczek*
Cover Image: *© ERproductions Ltd/Blend Images LLC*
Compositor: *MPS Limited*
Printer: *R. R. Donnelley*

All credits appearing on page or at the end of the book are considered to be an extension of the copyright page.

Library of Congress Cataloging-in-Publication Data

Longenbaker, Susannah Nelson, author.
  Mader's understanding human anatomy & physiology / Susannah Nelson Longenbaker
  Mader's understanding human anatomy and physiology | Understanding human anatomy & physiology
  Ninth edition. | New York, NY : MHE, 2017. | Includes index.
  LCCN 2015036331 | ISBN 9781259296437 (alk. paper)
  LCSH: Human physiology—Textbooks. | Human anatomy—Textbooks.
  LCC QP34.5 .M353 2017 | DDC 612—dc23
  LC record available at http://lccn.loc.gov/2015036331

The Internet addresses listed in the text were accurate at the time of publication. The inclusion of a website does not indicate an endorsement by the authors or McGraw-Hill Education, and McGraw-Hill Education does not guarantee the accuracy of the information presented at these sites.

mheducation.com/highered

Contents

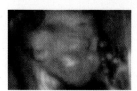

About the Author

After earning a baccalaureate degree in biology from St. Mary's College (Notre Dame, Indiana) and a master's degree in physiology from the Ohio State University, Susannah Nelson Longenbaker began her teaching career at Columbus State Community College in Columbus, Ohio. She continues to teach anatomy and physiology courses there, as she has for over 30 years. During that time, she earned the college's Distinguished Teaching Award and *Ohio Magazine's* Excellence in Education award. She founded and serves as co-coordinator for Columbus State Community College *Fantastic Fridays* and *Fantastic Fridays Thinking Science*. These community outreach programs introduce middle school and high school students to the fun and excitement of laboratory science. In 2015, she was awarded the Columbus City Schools Community Excellence Award in recognition of her work in community outreach and science education.

In 2006, Sue was offered a unique opportunity by Dr. Sylvia Mader: to become the primary author for *Understanding Human Anatomy and Physiology*. Dr. Mader began her long career as a college biology professor, then left the classroom to become one of the most prolific authors of biology and human biology textbooks in the country. Her works are well known for their direct writing style and carefully crafted pedagogy. Dr. Mader's many titles have been published and enjoyed by students worldwide for almost 40 years.

Sue is honored to continue Dr. Mader's legacy to education, as the writer for this ninth edition of the textbook. She looks forward to and appreciates suggestions or comments from instructors and students alike. Feel free to contact her at the following address:

Sue Longenbaker
Department of Biological and Physical Sciences
Columbus State Community College
Columbus, Ohio 43215 (614) 287-2430
slongenb@cscc.edu

Preface

Welcome to the ninth edition of *Mader's Understanding Human Anatomy and Physiology*! It is a joy and a privilege to work on this project, which is so fulfilling to me as a scientist, educator, and creative artist. I am honored to continue the vision of the book's original author, Dr. Sylvia Mader, who introduced the book almost two decades ago. We believe that a book designed to introduce the fascinating workings of the human body should be creative, informative, accurate, and most important, *relevant* to today's students. This book is tailored to appeal to a wide audience, from students in pre-nursing and allied health fields, to non-science majors who want a clear and concise explanation of how their bodies work. As soon as the student opens the book for the very first time, I want to capture that student's interest. Then, I want to keep the reader's attention as he or she learns something new about how we humans work.

Mader's Understanding Human Anatomy and Physiology continues to be the perfect text for a one-semester course. Each chapter begins with a brief introduction, designed to seize attention and stimulate curiosity, while drawing the reader in for a more detailed exploration. For example, the introduction for Chapter 6 on the skeletal system answers an age-old and commonly asked question: Does repeatedly cracking one's knuckles result in arthritis? And who among us has not experienced a "brain-freeze" headache when eating very cold foods too rapidly? The introduction for Chapter 7 on the nervous system explains why it occurs. Historical anecdotes will also intrigue the reader. The brief story of Henrietta Lacks that begins Chapter 4 is a fascinating account of how one woman's cancer cells continue to benefit humanity. The profile of retired astronaut and Senator John Glenn in Chapter 12 gives a fascinating insight to the very beginning of America's space program, and some of the medical issues that arose when humans were put into space for the first time.

Next, each chapter's Learning Outcomes are carefully constructed to be achievable to students with no prior training in anatomy and physiology. These Learning Outcomes are repeated as each new section begins, so that the reader never loses sight of what he or she is expected to learn. At the conclusion of each topic, the Content Check-Up feature allows the reader to test comprehension before continuing. Students who use the wonderful McGraw-Hill Connect® software with this text will be able to use this text's Learning Outcomes to check their progress. McGraw-Hill Learn-Smart® is the most widely used and intelligent adaptive learning resource that is proven to strengthen memory recall, improve course retention, and boost grades.

Throughout the text, the Begin Thinking Clinically feature asks a student to do exactly that: start thinking as though he or she was already working in a clinic or hospital setting. Each question fosters critical thinking skills by requiring the student to conduct further investigation into the chapter's subject matter. A great deal of thought and attention have gone into the conclusion of each chapter.

First, the Learning Outcomes are briefly summarized, and Key Terms and Clinical Key Terms are included, along with a pronunciation guide. Study Questions can be used as a checklist to ensure that important concepts are well understood. Each asks the student to craft a short essay. Learning Outcome Questions allow the student to "take the test" because they replicate the types of short answer questions often used in the classroom (matching, true-false, multiple choice, and the like). Finally, a Medical Terminology Exercise that concludes the chapter helps to build a working vocabulary, thus facilitating comprehension and increasing student confidence.

My own students love to relate examples about anatomy, physiology, and pathophysiology that they've seen in the media or come across on the job. For this reason, the many features in each chapter of this text are tailored toward the varied interests of today's students. The Focus on Forensics articles relate anatomy and physiology principles to the process of solving a crime. Every In Case of Emergency feature will be particularly relevant to those training to be first responders (emergency medical technicians and paramedics, for example), though everyone can benefit from knowing how to respond in a medical crisis situation. Each What's New reading describes a cutting-edge development in medicine and/or biotechnology. For example, you may have read in the popular media about the many uses of 3-D printers—but did you know that they can be used to craft a scaffolding to grow tissues, and may one day make it possible to grow entire organs? You can read about it in Chapter 4! New Medical Focus articles can be found throughout the book as well, and each existing Medical Focus article has been carefully researched and updated for this edition. However, perhaps the most important thing you'll notice throughout the book is the quality of the artwork. The new line drawings are realistic, detailed, and colorful; photographs are fresh and up-to-date. In addition, this ninth edition has been enriched by the incorporation of many fine images from McGraw-Hill Education's outstanding resource, *Anatomy & Physiology* REVEALED®. You'll find some of the best artwork in the industry in this edition of *Mader's Understanding Human Anatomy and Physiology*. Couple this with a completely redesigned layout, and I trust you'll find this book to be visually pleasing as well as accurate and informative.

I have been blessed to have the best job in the world—being a college professor teaching the biological sciences—for over 30 years. Being in the classroom daily helps me to understand the ways my students think, as well as what's happening in their world. Each semester's new batch of students has something to teach me, and I am fortunate to be able to learn something new every day. Further, I am privileged to work with a fine group of colleagues who are generous with both their expertise and their advice. I continue to develop new strategies to describe anatomical and physiological concepts, using more and better examples and analogies. In this book, it's my goal to share the ideas that work for me with both students and teachers. I know that this text will help you, the instructor, to engage and excite your students in the fascinating study of the human body.

Acknowledgements

Every new edition of *Understanding Human Anatomy and Physiology* presents a unique challenge for me. It's my goal to create a work with content that is precisely correct, up-to-date, and worthwhile for an increasingly diverse and rapidly evolving student population. When you have an amazing support team like the one I have at McGraw-Hill Education, the task becomes much easier. I owe a tremendous debt of gratitude to two individuals who directly supported me and with whom I communicated on an almost daily basis: my Product Developer, Fran Simon, and Content Project Manager, April Southwood. Ladies, thanks for your patience, understanding, and good humor. I appreciate everything you've done for this edition.

Further, each of these individuals deserves special recognition for her hard work: Brand Managers, Chloe Bouxsein and Amy Reed; Marketing Managers, Jessica Cannavo and Jim Connely; Photo Researcher, Lori Hancock; Designer, Tara McDermott; and Buyer, Sandy Ludovissy. My copyeditor, Kevin Campbell, and proofreaders, Angie Sigwarth and Carey Lange, helped to ensure accuracy throughout the entire project. Photo researcher Jo Johnson contributed hours of effort to find just the right photo illustrations in each chapter.

I would also like to thank the many others who contributed to the ancillary products associated with this text: Morris Butcher, Jeanette Ferguson, Cindy Hansen, Susan Rohde, and Phillip Snider, Jr.

It's very gratifying to know that one's colleagues will take the time and make the effort to provide comments and suggestions for a new edition. I would like to thank the individuals listed below for the observations and detailed recommendations they shared with me. As an author, it's comforting to know that you have skilled and talented peer educators to review your content and help to improve it.

Finally, I'd like to express my profound thanks to my three coworkers, allies, and buddies at Columbus State Community College: Dr. Jeanette Ferguson, Professor Eric Kenz, and Professor Lyndsy Wolff. Each one generously contributed advice, evaluation and review for this edition. Guys, you are the best, and I'm proud to work with you. And to the folks who always have my back—my husband, Bill, and my family—I can't do anything that I do without your love and support, and I'll always remember that.

— Sue Longenbaker

Dedication:

To the One through whom all things are possible: *Ad majorem dei gloriam.* And for Claire, Molly, Maya, and Julia, and for all future students: may my efforts help you learn.

Reviewers

Melody Bell
Vernon College

Rolfe Bryant
Phillips Community College–Stuttgart

Kim Zahn (Demnicki)
Thomas Nelson Community College

Cindy Hansen
Community College of Rhode Island–Warwick

Jaime Malcore Tjossem
Rochester Community & Technical College

Barry Markillie
Cape Fear Community College

Craig Mauk
KCTCS Gateway Community & Technical College

Payman Nasr
Kent State University–Ashtabula

Jennifer Presley
Ohio Valley University

Susan Rainone
Community College of Rhode Island–Warwick

Rebecca Roush
Sandhills Community College

Deborah Sanderson
Piedmont Virginia Community College

Hope Sasway
Suffolk CCC–BRENTWOOD

Jackie Spencer
Thomas Nelson Community College

Guided Tour Through a Chapter!

McGraw-Hill Education A&P: We have learning down to a science!

At McGraw-Hill Education we work every day to unlock the full potential of each learner. Our mission is to accelerate learning through intuitive, engaging, efficient and effective experiences—grounded in research. MHE Anatomy & Physiology is your trusted, data-driven partner in A&P education. Since 2009, our adaptive programs in A&P have hosted 600,000 unique users who have answered more than 600 million probes, giving us the only data-driven solutions to help your students get from their first college-level course to program readiness.

Learning Outcomes

at the beginning of every chapter will help students understand what they should know after studying the chapter.

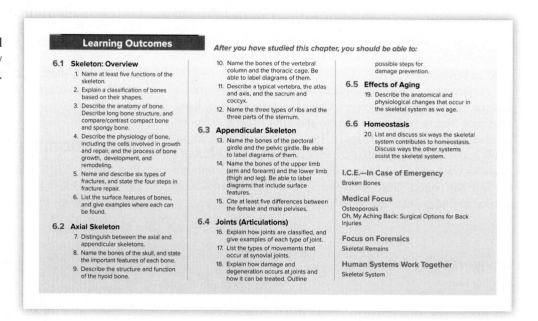

Learning Outcomes

After you have studied this chapter, you should be able to:

6.1 Skeleton: Overview
1. Name at least five functions of the skeleton.
2. Explain a classification of bones based on their shapes.
3. Describe the anatomy of bone. Describe long bone structure, and compare/contrast compact bone and spongy bone.
4. Describe the physiology of bone, including the cells involved in growth and repair, and the process of bone growth, development, and remodeling.
5. Name and describe six types of fractures, and state the four steps in fracture repair.
6. List the surface features of bones, and give examples where each can be found.

6.2 Axial Skeleton
7. Distinguish between the axial and appendicular skeletons.
8. Name the bones of the skull, and state the important features of each bone.
9. Describe the structure and function of the hyoid bone.

10. Name the bones of the vertebral column and the thoracic cage. Be able to label diagrams of them.
11. Describe a typical vertebra, the atlas and axis, and the sacrum and coccyx.
12. Name the three types of ribs and the three parts of the sternum.

6.3 Appendicular Skeleton
13. Name the bones of the pectoral girdle and the pelvic girdle. Be able to label diagrams of them.
14. Name the bones of the upper limb (arm and forearm) and the lower limb (thigh and leg). Be able to label diagrams that include surface features.
15. Cite at least five differences between the female and male pelvises.

6.4 Joints (Articulations)
16. Explain how joints are classified, and give examples of each type of joint.
17. List the types of movements that occur at synovial joints.
18. Explain how damage and degeneration occurs at joints and how it can be treated. Outline

possible steps for damage prevention.

6.5 Effects of Aging
19. Describe the anatomical and physiological changes that occur in the skeletal system as we age.

6.6 Homeostasis
20. List and discuss six ways the skeletal system contributes to homeostasis. Discuss ways the other systems assist the skeletal system.

I.C.E.—In Case of Emergency
Broken Bones

Medical Focus
Osteoporosis
Oh, My Aching Back: Surgical Options for Back Injuries

Focus on Forensics
Skeletal Remains

Human Systems Work Together
Skeletal System

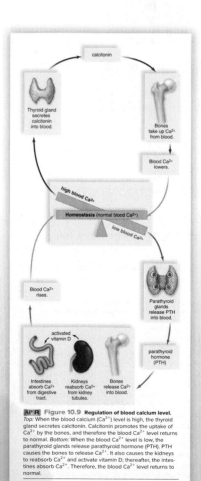

AP|R Figure 10.9 Regulation of blood calcium level. *Top:* When the blood calcium (Ca^{2+}) level is high, the thyroid gland secretes calcitonin. Calcitonin promotes the uptake of Ca^{2+} by the bones, and therefore the blood Ca^{2+} level returns to normal. *Bottom:* When the blood Ca^{2+} level is low, the parathyroid glands release parathyroid hormone (PTH). PTH causes the bones to release Ca^{2+}. It also causes the kidneys to reabsorb Ca^{2+} and activate vitamin D; thereafter, the intestines absorb Ca^{2+}. Therefore, the blood Ca^{2+} level returns to normal.

Accessible Writing Style More important than any other component of a textbook, the writing must be appropriate for the level of the reader. *Mader's Understanding Human Anatomy and Physiology* features the **perfect writing style for the one-semester course.** It has always been written and designed for the one-semester course, not adapted from a two-semester textbook. Paragraph introductions, explanations, comparisons, and relevant, everyday examples are used with these students in mind. The flow of the text is logical and accessible without being overly "chatty" and consistently makes use of relevant examples and analogies.

Easy-to-Understand Art covers what's important but leaves out unnecessary, confusing detail.

Good examples of this are the homeostasis illustrations – instead of lots of various colored arrows and boxes with explanations, these simple visual pieces get the message across beautifully.

Another example is stepped-out art, which shows key stages of an illustration identified by numbered circles. This type of explanation builds comprehension sequentially.

Several Anatomy & Physiology REVEALED® images have been added with icons from APR.

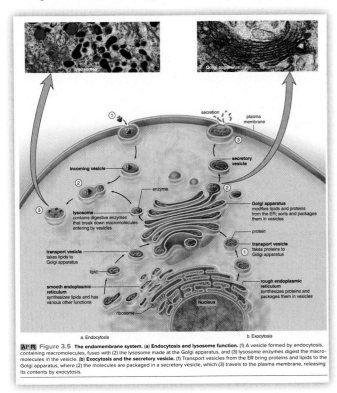

AP|R Figure 3.5 **The endomembrane system. (a)** Endocytosis and lysosome function. (1) A vesicle formed by endocytosis, containing macromolecules, fuses with (2) the lysosome made at the Golgi apparatus, and (3) lysosome enzymes digest the macromolecules in the vesicle. **(b)** Exocytosis and the secretory vesicle. (1) Transport vesicles from the ER bring proteins and lipids to the Golgi apparatus, where (2) the molecules are packaged in a secretory vesicle, which (3) travels to the plasma membrane, releasing its contents by exocytosis.

Macro to micro figures give the students an overall perspective.

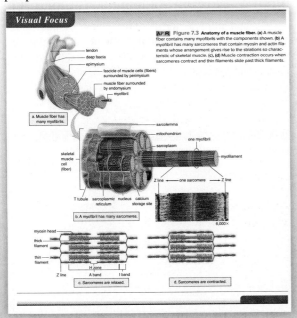

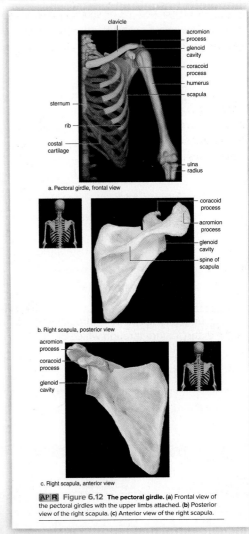

AP|R Figure 6.12 **The pectoral girdle. (a)** Frontal view of the pectoral girdles with the upper limbs attached. **(b)** Posterior view of the right scapula. **(c)** Anterior view of the right scapula.

Learning Outcomes are listed within the chapter! Students will know what that specific section is covering.

6.2 Axial Skeleton

7. Distinguish between the axial and appendicular skeletons.
8. Name the bones of the skull, and state the important features of each bone.
9. Describe the structure and function of the hyoid bone.
10. Name the bones of the vertebral column and the thoracic cage. Be able to label diagrams of them.
11. Describe a typical vertebra, the atlas and axis, and the sacrum and coccyx.
12. Name the three types of ribs and the three parts of the sternum.

The skeleton is divided into the axial skeleton and the appendicular skeleton. The tissues of the axial and appendicular skeletons are bone (both compact and spongy), cartilage (hyaline, fibrocartilage, and elastic cartilage), and dense connective tissue, a type of fibrous connective tissue. (The various types of connective tissues were extensively discussed in Chapter 4.)

In Figure 6.4, the bones of the axial skeleton are colored gray, and the bones of the appendicular skeleton are colored tan for easy distinction. Notice that the **axial skeleton** lies in the midline of the body and contains the bones of the skull, the hyoid bone, the

Guided Tour Through a Chapter!

Built-in Study Aids such as the *Content Check-Up* and the *Begin Thinking Clinically* features allow students to test themselves over major sections of text before continuing. *Content Check-Up* questions will now be found after each major heading.

Content CHECK-UP!

1. The term for the expanded portions at the ends of a long bone is:

    a. diaphysis.

    b. epiphysis.

    c. periosteum.

    d. articular cartilage.

2. Osteons are associated with _____ bone.

3. Which type of bone cell breaks down bone and deposits calcium into the blood?

    a. osteoblast

    b. osteocyte

    c. osteoprogenitor

    d. osteoclast

4. The region in a long bone where growth occurs is the _____.

5. Imagine that an artery has to pass through bone to enter the skull. What is the feature through which the artery will pass? (Refer to Table 6.1.)

Answers in Appendix A.

 Begin Thinking Clinically

You're treating an 11-year-old patient in the emergency room. His right eye was struck by a baseball bat, and he's rapidly developing a nasty black eye. What bones might have been broken by the injury?

Answer and discussion in Appendix A.

End of Chapter

The end-of-chapter material has been newly organized with the Summary updated for every chapter. Key terms are divided into basic and clinical terms. Two levels of additional questions, along with exercises that reinforce medical terminology, are also included with every chapter.

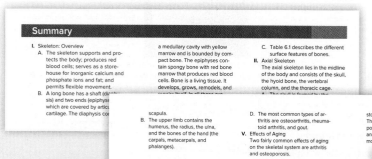

Summary

I. Skeleton: Overview
   A. The skeleton supports and protects the body; produces red blood cells; serves as a storehouse for inorganic calcium and phosphate ions and fat; and permits flexible movement.
   B. A long bone has a shaft (diaphysis) and two ends (epiphyses), which are covered by articular cartilage. The diaphysis con-

a medullary cavity with yellow marrow and is bounded by compact bone. The epiphyses contain spongy bone with red bone marrow that produces red blood cells. Bone is a living tissue. It develops, grows, remodels, and repairs itself. In all these are

C. Table 6.1 describes the different surface features of bones.
II. Axial Skeleton
   The axial skeleton lies in the midline of the body and consists of the skull, the hyoid bone, the vertebral column, and the thoracic cage.

scapula.
   B. The upper limb contains the humerus, the radius, the ulna, and the bones of the hand (the carpals, metacarpals, and phalanges).

D. The most common types of arthritis are osteoarthritis, rheumatoid arthritis, and gout.
V. Effects of Aging
   Two fairly common effects of aging on the skeletal system are arthritis and osteoporosis.

storage and growth of bones. The cardiovascular system transports calcium to the skeleton, and muscles aid the skeleton in movement.

Study Questions

1. What are five functions of the skeleton? (p. XX)
2. What are five major categories of bones based on their shapes? (p. 98)
3. What are the parts of a long bone? What are some differences between compact bone and spongy bone? (pp. XX–XX)
4. How does bone grow in children, and how is it remodeled in all age groups? (pp. XXX–XXX)
5. What are the various types of fractures? Outline the four steps that are required for fracture repair. (p. XXX)
6. List the bones of the axial and appendicular skeletons. (p.103 and Fig. 6.4, p. XXX)
7. What are the bones of the cranium and the face? Describe the special features of the temporal bones, sphenoid bone, and ethmoid bone. (pp. XXX–XXX)
8. What are the parts of the vertebral column, and what are its curvatures? Distinguish between the atlas, axis, sacrum, and coccyx. (pp. XXX–XXX)
9. What are the bones of the rib cage? List several functions of the rib cage. (pp. XXX–XXX)
10. What are girdle? G strate the girdle. W of a scap
11. Name the then outli these bor
12. What are and what
13. What are and what between vises? (p.

Learning Outcome Questions

I. Match the items in the key to the bones listed in questions 1–6.
Key:
   a. forehead
   b. chin
   c. cheekbone
   d. elbow
   e. shoulder blade
   f. hip
   g. leg

   1. temporal and zygomatic bones
   2. tibia and fibula
   3. frontal bone
   4. ulna
   5. coxal bone
   6. scapula

II. Match the items in the key to the bones listed in questions 7–13.
Key:
   a. external acoustic meatus
   b. cribriform plate
   c. xiphoid process
   d. glenoid cavity

   e. olecranon process
   f. acetabulum
   g. greater and lesser trochanters
   7. scapula
   8. sternum
   9. femur
   10. temporal bone
   11. coxal bone
   12. ethmoid bone
   13. ulna

III. Fill in the blanks.
   14. Long bones are _____ than they are wide.
   15. The epiphysis of a long bone contains _____ bone, where red blood cells are produced.
   16. The _____ are the air-filled spaces in the cranium.
   17. The sacrum is a part of the _____, and the sternum is a part of the _____.
   18. The pectoral girdle is specialized for _____, while the pelvic

girdle is specialized for _____.
   19. The term *phalanges* is used for the bones of both the _____ and the _____.
   20. The knee is a freely movable (synovial) joint of the _____ type.

IV. Match the movement with the description in questions 21–25.
   a. extension
   b. circumduction
   c. adduction
   d. flexion
   e. abduction
   21. moving a body part toward the midline
   22. moving a body part away from the midline
   23. moving a body part in a circle
   24. decreasing the angle of a joint
   25. increasing the angle of the joint

Unsurpassed Clinical Coverage is evident all through this text. What's New, Medical Focus, Begin Thinking Clinically, I.C.E.: In Case of Emergency, and Focus on Forensics readings and study aids to relate the very latest research and developments in applied aspects of anatomy and physiology to important concepts in the text. Examples include "Novel Stent for the Severest Strokes," "Brain in a Petri Dish: A Human Model for Alzheimer Research," "Improvements in Transfusion Technology," "Necrotizing Fasciitis," and "Influenza: A Constant Threat of Pandemic." The Focus on Forensics and I.C.E.: In Case of Emergency readings engage students in real-life scenarios that challenge them to use, and expand upon, their recently acquired knowledge.

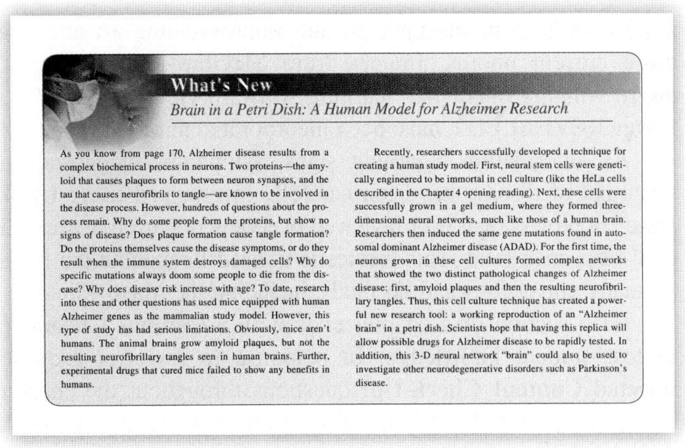

What's New

Brain in a Petri Dish: A Human Model for Alzheimer Research

As you know from page 170, Alzheimer disease results from a complex biochemical process in neurons. Two proteins—the amyloid that causes plaques to form between neuron synapses, and the tau that causes neurofibrils to tangle—are known to be involved in the disease process. However, hundreds of questions about the process remain. Why do some people form the proteins, but show no signs of disease? Does plaque formation cause tangle formation? Do the proteins themselves cause the disease symptoms, or do they result when the immune system destroys damaged cells? Why do specific mutations always doom some people to die from the disease? Why does disease risk increase with age? To date, research into these and other questions has used mice equipped with human Alzheimer genes as the mammalian study model. However, this type of study has had serious limitations. Obviously, mice aren't humans. The animal brains grow amyloid plaques, but not the resulting neurofibrillary tangles seen in human brains. Further, experimental drugs that cured mice failed to show any benefits in humans.

Recently, researchers successfully developed a technique for creating a human study model. First, neural stem cells were genetically engineered to be immortal in cell culture (like the HeLa cells described in the Chapter 4 opening reading). Next, these cells were successfully grown in a gel medium, where they formed three-dimensional neural networks, much like those of a human brain. Researchers then induced the same gene mutations found in autosomal dominant Alzheimer disease (ADAD). For the first time, the neurons grown in these cell cultures formed complex networks that showed the two distinct pathological changes of Alzheimer disease: first, amyloid plaques and then the resulting neurofibrillary tangles. Thus, this cell culture technique has created a powerful new research tool: a working reproduction of an "Alzheimer brain" in a petri dish. Scientists hope that having this replica will allow possible drugs for Alzheimer disease to be rapidly tested. In addition, this 3-D neural network "brain" could also be used to investigate other neurodegenerative disorders such as Parkinson's disease.

MEDICAL FOCUS

Research on Alzheimer Disease: Causes, Treatments, Prevention, and Hop

Alzheimer disease (AD) is an irreversible, fatal disorder characterized by a gradual loss of reason that begins with memory lapses and ends with the inability to perform any activities. Personality changes such agitation and hostility, and memory deficits that affect daily routines often signal the onset of AD. For example, a normal 60- to 70-year-old might forget the name of a friend not seen for years, but someone with AD forgets the name of a neighbor who visits daily. Likewise, a healthy senior might forget where he placed his car keys, while a person with AD will forget what those keys are for. People afflicted with AD become confused and tend to repeat the same question. Signs of mental disturbance eventually appear, and patients gradually become bedridden and die of a complication, such as pneumonia. At the cellular level, AD is characterized by the presence of abnormally structured neurons and a reduced amount of the neurotransmitter acetylcholine (see p. xxx). These defective neurons are especially seen in the portions of the brain involved in reason and memory. The AD neuron has two pathological features. The first consists of bundles of fibrous protein, called neurofibrillary tangles, which surround the nucleus in the cells. The tangles are due to an abnormal form of *tau*, a protein molecule that normally helps to stabilize microtubules that form the cell's cytoskeleton. In addition to these tangles, protein-rich accumulations, called amyloid plaques, envelop the axon branches. Over time, affected neurons will die. The cortex and hippocampus shrivel, the brain shrinks in volume, and the ventricles become enlarged.

Research Regarding Its Causes

As techniques for genetic study continue to improve, several genetic mutations specific to Alzheimer have been identified. One set, designated by the acronyms APP, PS1, and PS2, are termed *deterministic*. People who inherit one of these three mutated genes will always develop the disease, called autosomal dominant Alzheimer disease (ADAD). It's interesting to note that APP, the first of these defective genes to be discovered, is found on chromosome 21. Down syndrome results from the inheritance of three copies of chromosome 21, and people with Down syndrome tend to develop AD. (You will learn more about autosomal dominant disorders in Chapter 19.) Mutation of a fourth gene, designated APO, puts patients at risk but does not always result in disease. Scientists are now studying victims with mutations to try to discover the exact cause for the disease. Recent findings have led researchers to believe that the neuron deterioration seen in Alzheimer disease patients may be caused by the spread of the tau protein from one cell to the next, much as a virus is spread from one infected cell to another. Other studies have implicated a second protein, striatal-enriched tyrosine phosphatase, or STEP, in the cell destruction found in Alzheimer sufferers. Further, other investigators are exploring the role of cell lysosomes in AD, suspecting that these essential organelles may be failing to destroy the abnormal proteins found in diseased cells.

Research into Its Treatment

Each new finding about what causes Alzheimer disease creates new possibilities for its treatment as well. Researchers are now conducting clinical testing on antibodies that block cell-to-cell transmission of the tau protein. (You can read more about antibodies in Chapter 13.) A second treatment might involve the creation of drugs that block formation of the STEP protein. Boosting lysosomal degradation of the abnormal

Alzheimer cell protein is another p
time, only five drugs are accepted f
cholinesterase inhibitors (Aricept®,
works at neuron synapses in the brai
that breaks down acetylcholine (AC
synapses keeps memory pathways
period of time. The newest drug, me
toxicity: the tendency of diseased ne
tion is used only in moderately to s
drug allows neurons involved in me
affected patients. However, it's impo
medication cures AD. Both merely
toms, allowing the patient to functio
of time. Additional research is curre
ness of anticholesterol statin drugs, a
tions, in slowing the progress of the

Research on Prevention

Much of current research on AD fo
have shown that risk factors for cardi
stroke—also contribute to an increa
elevated blood cholesterol and bloo
tary lifestyle, and diabetes mellitus (
caused by gum disease has also been
developing heart disease, and by e
evidence suggests that a lifestyle tai
may also prevent AD. Slight change
developing AD: boosting vitamins
salmon, and drinking coffee. Further
blows to the head. It's been shown that head injuries (such as those experienced by football players) can increase the risk of developing AD in later life 19-fold. Wearing seat belts and helmets and taking steps to prevent falls are commonsense, easy ways to prevent head injury. Finally, staying mentally, physically, and socially active—as long as possible—will help to slow the course of mental impairment for AD sufferers.

Early Detection and Hope for a Cure

Currently, researchers are testing vaccines for AD that would target the patient's immune system to destroy amyloid protein. Early study results show some promising outcomes of this treatment in early-stage patients. However, scientists believe that curing AD will require an early diagnosis because it's thought that the disease may begin in the brain 15 to 20 years before symptoms ever develop. At present, diagnosis can't be made with absolute certainty until the brain is examined at autopsy. In the future, cerebrospinal fluid testing may allow amyloid proteins to be detected before disease symptoms appear. Researchers are also developing ways to tag the amyloid protein with radioactive molecules, which will allow detection of the protein using a PET scan. (You learned about PET and other imaging techniques in Chapter 1.) The Medical Focus reading in Chapter 9 describes an eye scan technique that might allow an earlier diagnosis, and the What's New reading on page XXX describes an exciting breakthrough in cell culture that will create new options for studying neurons and drug therapies in the laboratory.

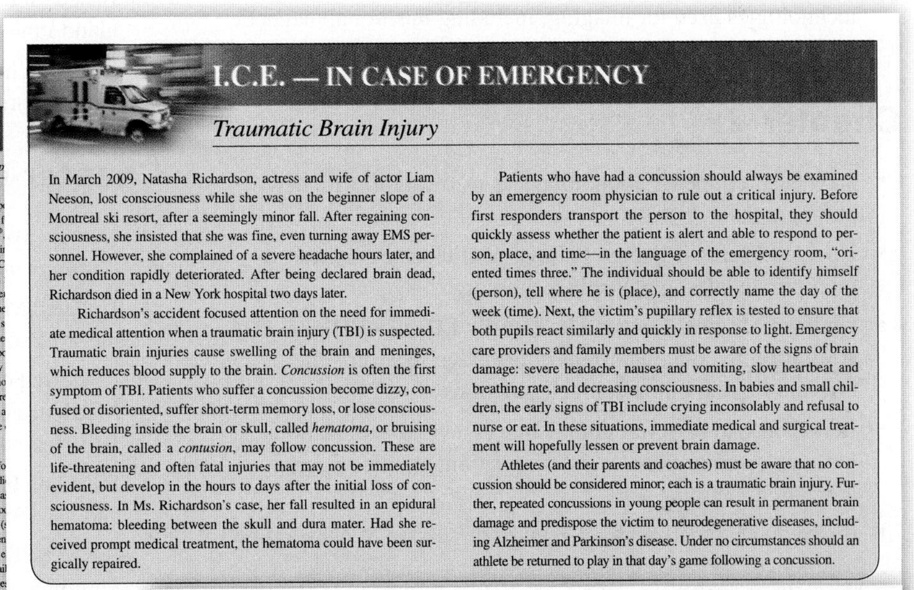

I.C.E. — IN CASE OF EMERGENCY

Traumatic Brain Injury

In March 2009, Natasha Richardson, actress and wife of actor Liam Neeson, lost consciousness while she was on the beginner slope of a Montreal ski resort, after a seemingly minor fall. After regaining consciousness, she insisted that she was fine, even turning away EMS personnel. However, she complained of a severe headache hours later, and her condition rapidly deteriorated. After being declared brain dead, Richardson died in a New York hospital two days later.

Richardson's accident focused attention on the need for immediate medical attention when a traumatic brain injury (TBI) is suspected. Traumatic brain injuries cause swelling of the brain and meninges, which reduces blood supply to the brain. *Concussion* is often the first symptom of TBI. Patients who suffer a concussion become dizzy, confused or disoriented, suffer short-term memory loss, or lose consciousness. Bleeding inside the brain or skull, called *hematoma*, or bruising of the brain, called a *contusion*, may follow concussion. These are life-threatening and often fatal injuries that may not be immediately evident, but develop in the hours to days after the initial loss of consciousness. In Ms. Richardson's case, her fall resulted in an epidural hematoma: bleeding between the skull and dura mater. Had she received prompt medical treatment, the hematoma could have been surgically repaired.

Patients who have had a concussion should always be examined by an emergency room physician to rule out a critical injury. Before first responders transport the person to the hospital, they should quickly assess whether the patient is alert and able to respond to person, place, and time—in the language of the emergency room, "oriented times three." The individual should be able to identify himself (person), tell where he is (place), and correctly name the day of the week (time). Next, the victim's pupillary reflex is tested to ensure that both pupils react similarly and quickly in response to light. Emergency care providers and family members must be aware of the signs of brain damage: severe headache, nausea and vomiting, slow heartbeat and breathing rate, and decreasing consciousness. In babies and small children, the early signs of TBI include crying inconsolably and refusal to nurse or eat. In these situations, immediate medical and surgical treatment will hopefully lessen or prevent brain damage.

Athletes (and their parents and coaches) must be aware that no concussion should be considered minor; each is a traumatic brain injury. Further, repeated concussions in young people can result in permanent brain damage and predispose the victim to neurodegenerative diseases, including Alzheimer and Parkinson's disease. Under no circumstances should an athlete be returned to play in that day's game following a concussion.

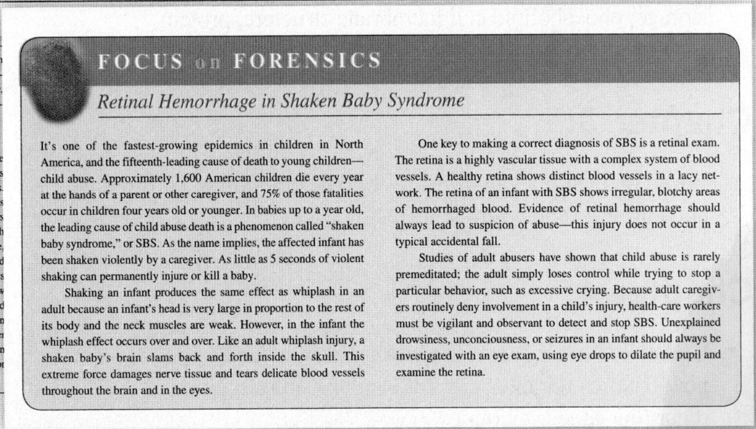

FOCUS on FORENSICS

Retinal Hemorrhage in Shaken Baby Syndrome

It's one of the fastest-growing epidemics in children in North America, and the fifteenth-leading cause of death to young children—child abuse. Approximately 1,600 American children die every year at the hands of a parent or other caregiver, and 75% of those fatalities occur in children four years old or younger. In babies up to a year old, the leading cause of child abuse death is a phenomenon called "shaken baby syndrome," or SBS. As the name implies, the affected infant has been shaken violently by a caregiver. As little as 5 seconds of violent shaking can permanently injure or kill a baby.

Shaking an infant produces the same effect as whiplash in an adult because an infant's head is very large in proportion to the rest of its body and the neck muscles are weak. However, in the infant the whiplash effect occurs over and over. Like an adult whiplash injury, a shaken baby's brain slams back and forth inside the skull. This extreme force damages nerve tissue and tears delicate blood vessels throughout the brain and in the eyes.

One key to making a correct diagnosis of SBS is a retinal exam. The retina is a highly vascular tissue with a complex system of blood vessels. A healthy retina shows distinct blood vessels in a lacy network. The retina of an infant with SBS shows irregular, blotchy areas of hemorrhaged blood. Evidence of retinal hemorrhage should always lead to suspicion of abuse—this injury does not occur in a typical accidental fall.

Studies of adult abusers have shown that child abuse is rarely premeditated; the adult simply loses control while trying to stop a particular behavior, such as excessive crying. Because adult caregivers routinely deny involvement in a child's injury, health-care workers must be vigilant and observant to detect and stop SBS. Unexplained drowsiness, unconciousness, or seizures in an infant should always be investigated with an eye exam, using eye drops to dilate the pupil and examine the retina.

Each chapter contains updated and improved line art and new, more current photos. Images from McGraw-Hill Education's award-winning interactive learning software, *Anatomy and Physiology* REVEALED®, have been incorporated throughout the text.

All information regarding signs, symptoms, diagnosis, and treatment of disease has been carefully investigated using **Up To Date**®, a professional peer-reviewed overview of current research in each respective field. This service is utilized throughout the nation by many universities and hospitals, including the Mayo Clinic.

Each section of a chapter ends with a **Content Check-Up!** to test student knowledge. In response to reviewer requests, selected **Content Check-Up!** questions throughout the chapters have been replaced with higher-level questions requiring critical thinking and assimilation of ideas.

Chapter 1:

- Updated **Medical Focus: Imaging the Body** to include latest technologies used for imaging, including functional magnetic resonance imaging and diffusion tensor imaging.

Chapter 2:

- New chapter opener about toxins as medication, with new photographs.
- Updated **Medical Focus: Prions: Malicious Proteins?** to incorporate latest diagnostic technology.
- Updated **Medical Focus: The Deadly Effects of High-Level Radiation** to contain current information regarding the effects of radiation on cell-cell junctions.
- In response to reviewer commentary: revised explanations for mass number; low levels of radiation; atomic structure, ionization, buffers.
- Based on heat map analysis, revised discussion of dehydration reaction, cation and anion structure, disaccharides, glycogen storage, phospholipid cell membrane structure, protein function.

Chapter 3:

- Based on heat map analysis: revised description of chromatin and chromosomes, simple diffusion.

Chapter 4:

- Updated **What's New: Targeting the Traitor Inside**, which now features the most current available information regarding cancer therapies.

- New Reading: **What's New: 3-D Printing to Create Complex Tissues.**
- Updated the story of Henrietta Lacks, the subject of the chapter introduction. Her gravesite has recently been located, and after many decades has been permanently identified with an appropriate grave marker.
- Based on heat map analysis, revised description of epithelial tissues to make it more complete.

Chapter 5:

- New chapter opener, featuring new photos.
- Per reviewer request, relocated and revised the section on Functions of the Skin.
- Completely revised sections on homeostasis to improve readability.
- Based on heat map analysis, revised description of sebaceous gland function.
- Based on heat map analysis, revised descriptions of temperature regulation.

Chapter 6:

- New artwork throughout, incorporating multiple images from *Anatomy and Physiology* REVEALED®.
- Updated **Medical Focus: Osteoporosis** to reflect state-of-the-art knowledge about medical research in the field.
- Reviewed current findings on causes and therapies to update **Medical Focus: Oh, My Aching Back—Options for Back Injuries.**
- Based on heat map analysis, revised description of lacunae and canaliculi.
- Based on heat map analysis, revised description of transverse foramina.

Chapter 7:

- Completely reworked Figure 7.6.
- Researched current findings and professional recommendations to overhaul **Medical Focus: Benefits of Exercise**. The article features a table of practical, real-world recommendations about incorporating exercise into daily living.
- Based on reviewer feedback, updated discussion of all-or-none law, recruitment, muscle tone.
- Based on reviewer feedback, updated discussion of twitch, tetanus, fatigue.

- Based on heat map analysis, revised description of the neuromuscular junction.
- Based on heat map analysis, revised discussion of myosin power stroke.

Chapter 8:

- Updated articles: **Medical Focus: Research on Alzheimer Disease**, and **In Case of Emergency: Traumatic Brain Injury**. Both readings now feature current research and recommendations from the Alzheimer's Association and the American Heart Association, respectively.
- Two new articles added:
  What's New: Epidural Stimulation in Spinal Cord Injuries: Cause for Hope What's New: Brain in a Petri Dish: A Human Model for Alzheimer Research
- New figures added: Figure 8.4, 8.15.
- Based on heat map analysis, added more detail to description of peripheral nervous system, neuron structure, all-or-none property of the neuron, meninges, pia mater, ventricles, brainstem functions.

Chapter 9:

- Updated statistics for child abuse, hearing loss, and ototoxicity.

Chapter 10:

- Updated all statistics regarding diabetes mellitus.
- New reading: **What's New: Options for Type I Diabetics: The Artificial Pancreas System, Beta Cell Transplants, and the BioHub**
- Updated **Medical Focus: Side Effects of Anabolic Steroids.**
- Based on heat map analysis, added more detail regarding Cushing's syndrome, ketoacidosis.

Chapter 11:

- Updated **What's New: Improvements in Transfusion Technology** to reflect new developments in this field.
- Based on heat map analysis, added more detail to the discussion of hemoglobin breakdown; positive feedback in blood clotting.

Chapter 12:

- In response to reviewer feedback: updated and revised discussion of intercalated disks and gap junctions.
- Updated **Medical Focus: Arteriosclerosis, Atherosclerosis, and Coronary Artery Disease** to contain up-to-date findings in diagnosis and therapy.
- Updated **I.C.E.—IN CASE OF EMERGENCY: Cardiopulmonary Resuscitation and Automated External Defibrillation** to include most current recommendations from the American Heart Association.
- New article: **What's New: Novel Stent for the Severest Strokes.**

Chapter 13:

- Updated **Medical Focus: AIDS Epidemic.**
- Updated **Medical Focus: Immunization: The Great Protector.**
- Updated **Medical Focus: Influenza: A Constant Threat of Pandemic.**
- Updated **What's New: Parasite Prescription for Autoimmune Disease.**
- Based on heat map analysis, added more detail about first, second and third lines of defense against infection, and antibody actions.

Chapter 14:

- Updated **Medical Focus: The Most-Often-Asked Questions About Tobacco and Health** to include current statistics and information about electronic cigarettes.
- Based on heat map analysis, added more detail about paranasal sinuses, muscles of forced expiration, respiratory volumes and capacities.

Chapter 15:

- New article: **Medical Focus: Disorders of the Digestive Tract** presents information about causes, signs and symptoms, and treatment of gastrointestinal disease.
- Revised and updated **Medical Focus: Tips for Effectively Using Nutrition Labels.**
- Researched and incorporated information about the most current pharmaceutical treatments for obesity.

- Researched and incorporated the latest statistics regarding obesity rates in the United States.
- Based on heat map analysis, improved explanations about salivary enzymes and lysozyme and anatomy of the esophagus.

Chapter 16:

- New chapter opener, featuring a fascinating historical account of the first renal dialysis machine and its developer.
- Based on reviewer request, included a section on lifetime renal function.
- In response to reviewer feedback, added greater detail regarding the two types of nephrons and the juxtaglomerular apparatus.

Chapter 17:

- In response to reviewer feedback, added more details to the explanation of the process of meiosis.
- Researched and updated diagnostic criteria for ovarian cancer, incorporating findings and recommendations by the American Cancer Society.
- Thorough review and revision of all information regarding contraceptive methods available in the United States, including statistics about success/failure rates and health precautions for each one.
- Incorporated up-to-date descriptions of the proper techniques for breast and testicular self-examination, using information from the American Cancer Society.

- Investigated newest research regarding endocrine-disrupting contaminants, and included novel recommendations from the United States Environmental Protection Agency (EPA).
- Revised information regarding causes and treatment of infertility to include contemporary findings and recommendations.

Chapter 18:

- Updated information regarding prevention of birth defects, utilizing information from the March of Dimes U.S.A.©
- Based on reviewer recommendation, revised description of implantation and the basis for a positive pregnancy test.

Chapter 19:

- NEW article: **What's New: A Profound Dilemma: Bioengineered Babies.**
- Added information about cell-free DNA analysis to the explanation of karyotyping.
- Revised **Medical Focus: Preimplantation Genetic Studies** to include information regarding polar body testing.
- Updated statistics for **Focus on Forensics: The Innocence Project**.
- Incorporated information about DNA repair into the section on Gene Therapy.

LearnSmart® Prep is an adaptive learning tool that prepares students for college-level work in Anatomy & Physiology. **Prep for Anatomy & Physiology now comes standard to students with Connect.** The tool individually identifies concepts the student does not fully understand and provides learning resources to teach essential concepts so he or she enters the classroom prepared. Data-driven reports highlight areas where students are struggling, helping to accurately identify weak areas.

Ph.I.L.S. 4.0 has been updated! Users have requested and we are providing five new exercises (Respiratory Quotient, Weight & Contraction, Insulin and Glucose Tolerance, Blood Typing, and Anti-Diuretic Hormone). Ph.I.L.S. 4.0 is the perfect way to reinforce key physiology concepts with powerful lab experiments. Created by Dr. Phil Stephens at Villanova University, this program offers 42 laboratory simulations that may be used to supplement or substitute for wet labs. All 42 labs are self-contained experiments—no lengthy instruction manual required. Users can adjust variables, view outcomes, make predictions, draw conclusions, and print lab reports. This easy-to-use software offers the flexibility to change the parameters of the lab experiment. There is no limit!

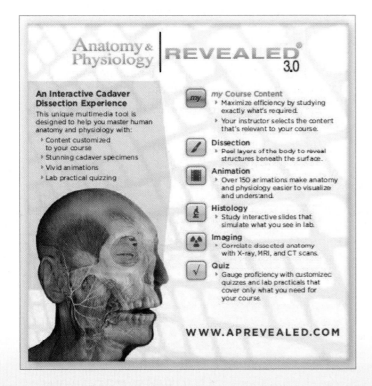

Anatomy & Physiology REVEALED® is now Mobile! Also, available in Cat and Fetal Pig versions.

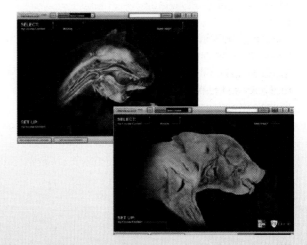

McGraw-Hill Connect®
Learn Without Limits

Connect is a teaching and learning platform that is proven to deliver better results for students and instructors.

Connect empowers students by continually adapting to deliver precisely what they need, when they need it, and how they need it, so your class time is more engaging and effective.

Course outcomes improve with Connect.

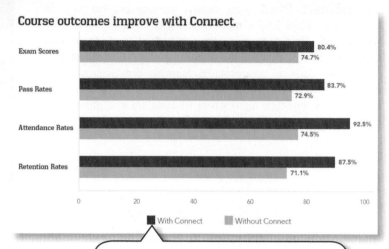

	With Connect	Without Connect
Exam Scores	80.4%	74.7%
Pass Rates	83.7%	72.9%
Attendance Rates	92.5%	74.5%
Retention Rates	87.5%	71.1%

Using **Connect** improves passing rates by **10.8%** and retention by **16.4%**.

88% of instructors who use **Connect** require it; instructor satisfaction **increases** by 38% when **Connect** is required.

Analytics

Connect Insight®

Connect Insight is Connect's new one-of-a-kind visual analytics dashboard—now available for both instructors and students—that provides at-a-glance information regarding student performance, which is immediately actionable. By presenting assignment, assessment, and topical performance results together with a time metric that is easily visible for aggregate or individual results, Connect Insight gives the user the ability to take a just-in-time approach to teaching and learning, which was never before available. Connect Insight presents data that empowers students and helps instructors improve class performance in a way that is efficient and effective.

Connect helps students achieve better grades

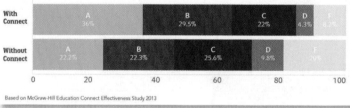

	A	B	C	D	F
With Connect	36%	29.5%	22%	4.3%	8.2%
Without Connect	22.2%	22.3%	25.6%	9.8%	20%

Based on McGraw-Hill Education Connect Effectiveness Study 2013

Students can view their results for any **Connect** course.

Mobile

Connect's new, intuitive mobile interface gives students and instructors flexible and convenient, anytime–anywhere access to all components of the Connect platform.

Adaptive

THE FIRST AND ONLY ADAPTIVE READING EXPERIENCE DESIGNED TO TRANSFORM THE WAY STUDENTS READ

> More students earn **A's** and **B's** when they use McGraw-Hill Education **Adaptive** products.

SmartBook®

Proven to help students improve grades and study more efficiently, SmartBook contains the same content within the print book, but actively tailors that content to the needs of the individual. SmartBook's adaptive technology provides precise, personalized instruction on what the student should do next, guiding the student to master and remember key concepts, targeting gaps in knowledge and offering customized feedback, and driving the student toward comprehension and retention of the subject matter. Available on smartphones and tablets, SmartBook puts learning at the student's fingertips—anywhere, anytime.

> Over **4 billion questions** have been answered, making McGraw-Hill Education products more intelligent, reliable, and precise.

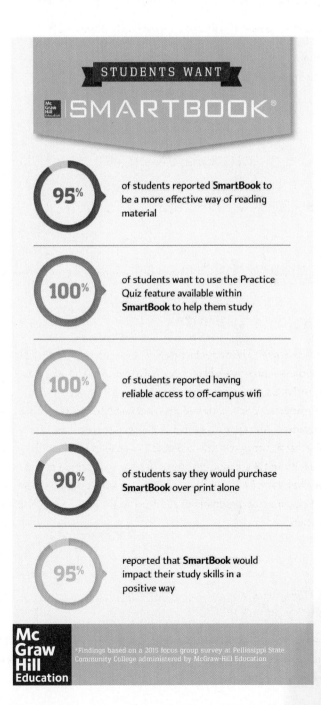

STUDENTS WANT SMARTBOOK®

95% of students reported **SmartBook** to be a more effective way of reading material

100% of students want to use the Practice Quiz feature available within **SmartBook** to help them study

100% of students reported having reliable access to off-campus wifi

90% of students say they would purchase **SmartBook** over print alone

95% reported that **SmartBook** would impact their study skills in a positive way

McGraw Hill Education

*Findings based on a 2015 focus group survey at Pellissippi State Community College administered by McGraw-Hill Education

1 Organization of the Body

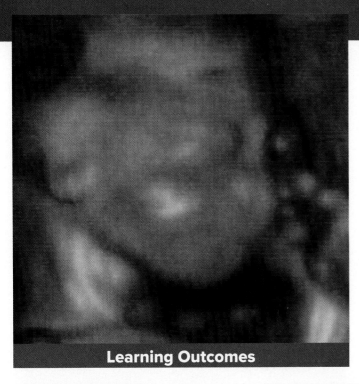

Recognize anything familiar? The little face, chest, and hands shown here belong to a baby in the womb, detailed using ultrasound, one of several imaging techniques described on page 16. All modern imaging methods have evolved from simple two-dimensional X rays, invented in 1895. The three-dimensional ultrasound procedure used to create this photo reveals structures with great detail and clarity. Ultrasound technology can monitor the infant's development and help to ensure a safe delivery for baby and mother. It can even enable physicians to diagnose fetal abnormalities and treat the baby while still in the womb!

Learning Outcomes

After you have studied this chapter, you should be able to:

1.1 The Human Body

1. Define anatomy and physiology, and explain how they are related.
2. Describe and give examples for each level of organization of the body.

1.2 Anatomical Terms

3. Use anatomical terms to describe the relative positions of the body parts, the regions of the body, and the planes that can be used to section the body.

1.3 Body Cavities and Membranes

4. List the cavities of the body, and state their locations.
5. Name the organs located in each of the body cavities.
6. Name the membranes that line each body cavity, and the membranes that cover the organs.

1.4 Organ Systems

7. List the organ systems of the body, and state the major organs associated with each.

8. Describe in general the functions of each organ system.

1.5 Homeostasis

9. Describe how a feedback system maintains homeostasis.
10. Describe the role of each body system in the maintenance of homeostasis.

Medical Focus

Meningitis and Serositis
Imaging the Body

1.1 The Human Body

1. Define anatomy and physiology, and explain how they are related.
2. Describe and give examples for each level of organization of the body.

Anatomy and physiology both involve the study of the human body. **Anatomy** is concerned with the structure of a part, as well as its relationship with other structures. For example, the stomach is a J-shaped, pouchlike organ, found between the esophagus and the small intestine, two other digestive system structures (Fig. 1.1). The stomach wall has thick folds, which disappear as the stomach expands to increase its capacity. **Physiology** is concerned with a body part's function, both individually and as a component of an entire system. For example, the stomach receives food from the esophagus, temporarily stores it and secretes digestive juices, then passes on partially digested food to the small intestine. Signals from the nervous system and the endocrine, or hormone system, direct stomach activities.

Anatomy and physiology are closely connected because the structure of an organ suits its function. For example, the stomach's pouchlike shape and ability to expand are well-suited to its function of storing food. In addition, the microscopic structure of the stomach wall is suitable to its secretion of digestive juices, as we will see in Chapter 15.

The Body's Organization Levels

The structure of the body can be studied at different *levels of organization* (Fig. 1.1). Initially, all substances, including body parts, are composed of chemicals made up of submicroscopic particles called **atoms.** Atoms join to form **molecules,** which can in turn join to form **macromolecules.** For example, molecules called amino acids join to form macromolecules called proteins. Different proteins make up the bulk of our muscles.

Macromolecules compose the cellular **organelles,** which are found within all cells. Organelles are tiny structures that perform cellular functions. For example, the organelle called the nucleus acts as a "control center" by directing cellular activity. Another organelle, called the mitochondrion, supplies the cell with energy. **Cells** are the basic units of living things.

Tissues are the next level of organization. A **tissue** is composed of similar types of cells and performs a specific function. An **organ** is composed of several types of tissues and performs a particular function within an **organ system.** For example, the stomach is an organ that is a part of the digestive system. It has a specific role in this system, whose overall function is to supply the body with the nutrients needed for growth and repair. The other systems of the body (see pages 12–15) also have specific functions.

All of the body systems together make up the **organism**—for example, a human being. Human beings are complex animals, but this complexity can be broken down and studied at ever simpler levels. Each simpler level is organized and constructed in a particular way.

Content CHECK UP!

1. Which would an anatomy student be studying: the structural organization of the skin, or functions of the skin?
2. Groups of organs are organized into _____.
3. Cells contain small structures called _____ that each perform a specific function.

Answers in Appendix A.

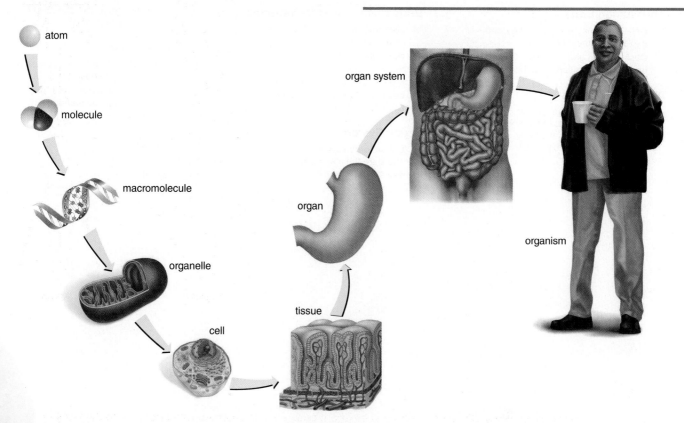

Figure 1.1 Levels of organization of the human body. Each level is more complex than the previous level.

atom · molecule · macromolecule · organelle · cell · tissue · organ · organ system · organism

1.2 Anatomical Terms

3. Use anatomical terms to describe the relative positions of the body parts, the regions of the body, and the planes that can be used to section the body.

Certain terms are used to describe the location of body parts, regions of the body, and imaginary planes that can be used to section the body. You should become familiar with these terms before your study of anatomy and physiology begins. Anatomical terms are useful only if everyone has in mind the same position of the body and is using the same reference points. Therefore, we will assume that the body is in the **anatomical position:** standing erect, with face forward, arms at the sides, and palms and toes directed forward, as illustrated in Figure 1.2.

Directional Terms

Directional terms are used to describe the location of one body part in relation to another (Fig. 1.2):

Anterior (ventral)—a body part is located toward the front. The windpipe (trachea) is *anterior* to the esophagus.

Posterior (dorsal)—a body part is located toward the back. The heart is *posterior* to the sternum (breastbone).

Superior—a body part is located above another part, or toward the head. The face is *superior* to the neck.

Inferior—a body part is below another part, or toward the feet. The navel is *inferior* to the chin.

Medial—a body part is nearer than another part to an imaginary midline of the body. The bridge of the nose is *medial* to the eyes.

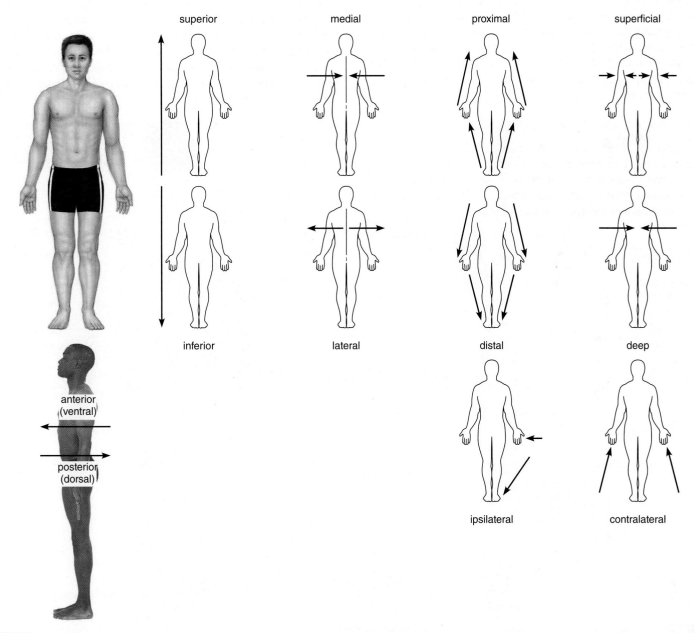

AP|R **Figure 1.2 Directional terms.** Directional terms tell us where body parts are located with reference to the body in anatomical position.

Lateral—a body part is farther away from the midline. The eyes are *lateral* to the nose.

Proximal—a body part is closer to a specific point of origin or attachment, or closer to the trunk of the entire body. For example, if the point of attachment is the shoulder, it is correct to say the elbow is *proximal* to the hand.

Distal—a body part is farther from a specific point of origin or attachment, or farther from the trunk of the entire body. For example, if the point of attachment is the hip, it is correct to say the foot is *distal* to the knee.

Superficial (external)—a body part is located closer to the surface than another. The skin is *superficial* to the muscles.

Deep (internal)—a body part is located farther from the surface than another. The intestines are *deep* to the spine.

Central—a body part is situated at the center of the body or an organ. The *central* nervous system is *centrally* located along the main axis of the body.

Peripheral—a body part is situated away from the center of the body or an organ. The *peripheral* nervous system is located outside the central nervous system.

Ipsilateral—a body part is on the same side of the body as another body part. The right hand is *ipsilateral* to the right foot.

Contralateral—a body part is on the opposite side of the body from another body part. The right hand is *contralateral* to the left hand.

Regions of the Body

The human body can be divided into axial and appendicular portions. The **axial portion** includes the head, neck, spinal column, and ribs. The trunk can be divided into the thorax, abdomen, and pelvis. The pelvis is that part of the trunk associated with the hips. The **appendicular portion** of the human body includes the limbs—that is, the upper limbs and the lower limbs.

The human body is further divided as shown in Figure 1.3. The labels in Figure 1.3 don't include the word "region." It is understood that you will supply the word *region* in each case. The anatomical term for each region is followed by the common name for that region. For example, the cephalic region is commonly called the head.

Notice that the upper limb includes (among other parts) the brachial region (arm) and the antebrachial region (forearm). Similarly, the lower limb includes the femoral region (thigh), the crural

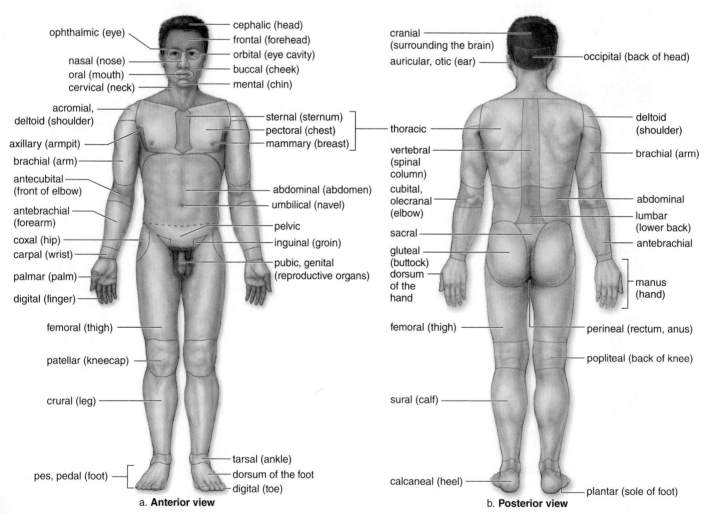

a. **Anterior view**

b. **Posterior view**

AP|R Figure 1.3 **Terms for body parts and areas. (a)** Anterior. **(b)** Posterior.

region (leg), and the pedal region (foot). In other words, contrary to common usage, the terms *arm* and *leg* refer only to a part of the upper limb and lower limb, respectively.

Most likely, it will take practice to learn the terms in Figure 1.3, but you'll be glad you did. Try pointing to various regions of your own body and see if you can give the scientific name for that region. Check your answer against the figure.

Planes and Sections of the Body

To observe the structure of an internal body part, it is customary to section (cut) the body along a plane. A plane is an imaginary flat surface passing through the body. The body is customarily sectioned along the following planes (Fig. 1.4):

A **sagittal** (median) **plane** extends lengthwise and divides the body into right and left portions. A midsagittal plane passes exactly through the midline of the body. The head and neck are shown in a midsagittal section (Fig. 1.4*a*). Sagittal cuts that are not along the midline are called parasagittal (paramedian) sections.

A **frontal** (coronal) **plane** also extends lengthwise, but it is perpendicular to a sagittal plane and divides the body or an organ into anterior and posterior portions. Here, the knee joint is shown in frontal section (Fig. 1.4*b*).

A **transverse** (horizontal) **plane** is perpendicular to the body's long axis and therefore divides the body horizontally to produce a cross section. A transverse cut divides the body or an organ into superior and inferior portions. Figure 1.4*c* is a transverse section of abdomen at the level of the umbilicus (navel).

The terms *longitudinal section* and *cross section* are often applied to individual body parts that have been removed and cut either lengthwise or straight across, respectively.

Content **CHECK-UP!**

4. Choose the correct directional term and finish the sentence: The chin is _____ to the navel.

5. If you point to your cheek, what region of the body are you identifying?

6. Suppose a CT scan creates images showing transverse sections of the head in a migraine headache patient. Are these horizontal or vertical images?

Answers in Appendix A.

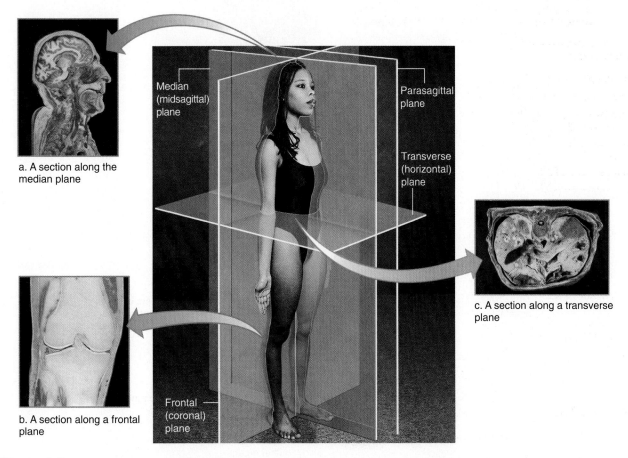

a. A section along the median plane

b. A section along a frontal plane

Median (midsagittal) plane

Parasagittal plane

Transverse (horizontal) plane

Frontal (coronal) plane

c. A section along a transverse plane

AP|R **Figure 1.4** **Body planes and sections.** Observation of internal parts requires sectioning the body along various planes.

1.3 Body Cavities and Membranes

4. List the cavities of the body, and state their locations.
5. Name the organs located in each of the body cavities.
6. Name the membranes that line each body cavity, and the membranes that cover the organs.

During embryonic development, the body is first divided into two internal cavities: the posterior (dorsal) body cavity and the anterior (ventral) body cavity. Each of these major cavities is then subdivided into smaller cavities. The cavities, as well as the organs in the cavities (called the **viscera**), are lined by membranes.

Posterior (Dorsal) Body Cavity

The posterior body cavity is subdivided into two parts: (1) The **cranial cavity,** enclosed by the bony cranium, contains the brain; (2) the **vertebral canal,** enclosed by vertebrae, contains the spinal cord (Fig. 1.5*a*).

The posterior body cavity is lined by three membranous layers collectively called the **meninges** (sing., *men'inx*). The innermost, or deepest, of the meninges is tightly bound to the surface of the brain and the spinal cord. The space between this layer and the next layer is filled with cerebrospinal fluid. Cerebrospinal fluid supports and nourishes the brain and the spinal cord and enables their cells to transmit electrical signals. In the skull, the most superficial of the three meninges lies directly under the skull bone. In the vertebral column, the outermost meninx is deep to a layer of fat and connective tissue, as you'll see in Chapter 8.

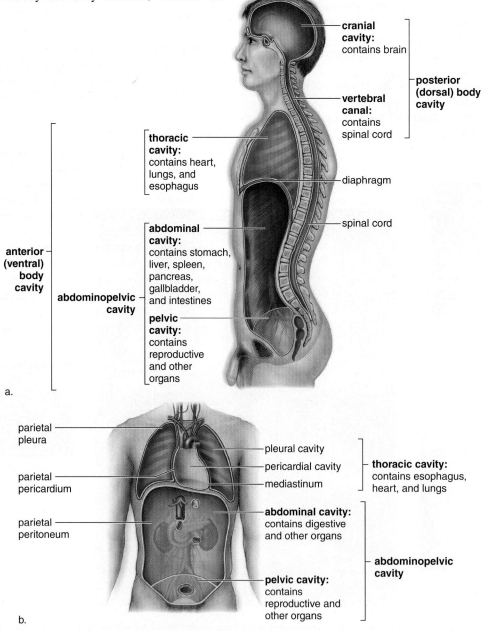

cranial cavity: contains brain

vertebral canal: contains spinal cord

posterior (dorsal) body cavity

thoracic cavity: contains heart, lungs, and esophagus

diaphragm

spinal cord

abdominal cavity: contains stomach, liver, spleen, pancreas, gallbladder, and intestines

pelvic cavity: contains reproductive and other organs

anterior (ventral) body cavity

abdominopelvic cavity

a.

parietal pleura

parietal pericardium

parietal peritoneum

pleural cavity

pericardial cavity

mediastinum

thoracic cavity: contains esophagus, heart, and lungs

abdominal cavity: contains digestive and other organs

abdominopelvic cavity

pelvic cavity: contains reproductive and other organs

b.

AP|R Figure 1.5 **The two major body cavities and their subdivisions. (a)** Left lateral view **(b)** Frontal view.

Anterior (Ventral) Body Cavity

The large anterior body cavity is subdivided into the superior **thoracic cavity** and the inferior **abdominopelvic cavity** (Fig. 1.5a). A muscular partition called the **diaphragm** separates the two cavities. Membranes that line these cavities are called **serous membranes** because they secrete a fluid that is similar to blood **serum.** Serum is the fluid that remains if all of the clotting proteins are removed from the blood. **Serous fluid** between the smooth serous membranes reduces friction when the viscera rub against each other or against the body wall.

To understand the relationship among serous membranes, the outer body wall, and an organ, consider the following example: Imagine a soft, pliable balloon (the serous membrane) filled with a small amount of fluid (serous fluid). The balloon sits inside a container (the inner body wall), tightly pressed to all sides of the container. An organ (the lung, for example) is pushed into this balloon and is then covered by the balloon (Fig. 1.6). As a result, two layers of serous membrane are created, separated from each other by the serous fluid. The balloon's outermost layer (lining the inner body wall) is termed the **parietal serous membrane.** The inner layer covering the organ is the **visceral serous membrane.** Thus, the parietal membrane is a cavity lining, and the visceral membrane is an organ covering. Inflammation of the serous membrane or infection of the serous fluid in the body cavities causes serious and potentially fatal illness (see Medical Focus, p. 9).

Thoracic Cavity

The thoracic cavity is enclosed by the rib cage and has three portions: the left, right, and medial portions. The medial portion, called the **mediastinum,** contains the heart, thymus gland, trachea, esophagus, and other structures (Fig. 1.5b).

The right and left portions of the thoracic cavity contain the lungs. The lung tissue is covered by a serous membrane—the **visceral pleura.** The **parietal pleura** lines the thoracic cavity. In between these two **pleurae** is the pleural cavity, which contains a small amount of pleural fluid. Similarly, in the medial thoracic cavity, the

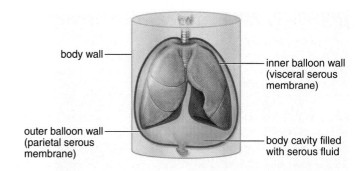

AP|R **Figure 1.6** **Relationship between the body wall, serous membranes, and organs.**

heart is covered by the **visceral pericardium.** The visceral pericardium contributes to the outermost connective tissue layer of the heart. Forming a tough connective tissue sac around the heart is the **fibrous pericardium,** whose inner lining is the **parietal pericardium**. Together, these structures create the **pericardial cavity.** The heart, inside its visceral pericardial sac, is separated from the outer parietal pericardium by a small amount of pericardial fluid.

Abdominopelvic Cavity

The abdominopelvic cavity has two portions: the superior **abdominal cavity** and the inferior **pelvic cavity.** The stomach, liver, spleen, gallbladder, and most of the small and large intestines are in the abdominal cavity. The pelvic cavity contains the rectum, the urinary bladder, the internal reproductive organs, and the rest of the large intestine. Males have an external extension of the abdominal wall, called the **scrotum,** where the testes are located.

Many of the organs of the abdominopelvic cavity are covered by the **visceral peritoneum,** whereas the wall of the abdominal cavity is lined with the **parietal peritoneum.** Peritoneal fluid fills the cavity between the visceral and parietal peritoneum. Table 1.1 summarizes our discussion of body cavities and membranes.

TABLE 1.1 Body Cavities and Membranes			
Name of Cavity	**Contents of Cavity**	**Membranes**	
POSTERIOR BODY CAVITY			
Cranial cavity	Brain	Meninges	
Vertebral canal	Spinal cord	Meninges	
ANTERIOR BODY CAVITY			
Thoracic Cavity		*Parietal Membrane*	*Visceral Membrane*
Pleural cavity	Lungs, serous fluid	Parietal pleura	Visceral pleura
Pericardial cavity	Heart, serous fluid	Fibrous pericardium and parietal pericardium	Visceral pericardium (epicardium)
Abdominopelvic Cavity			
Abdominal cavity	Stomach, intestines, liver	Parietal peritoneum	Visceral peritoneum
Pelvic cavity	Reproductive organs, urinary bladder, rectum	Parietal peritoneum	Visceral peritoneum

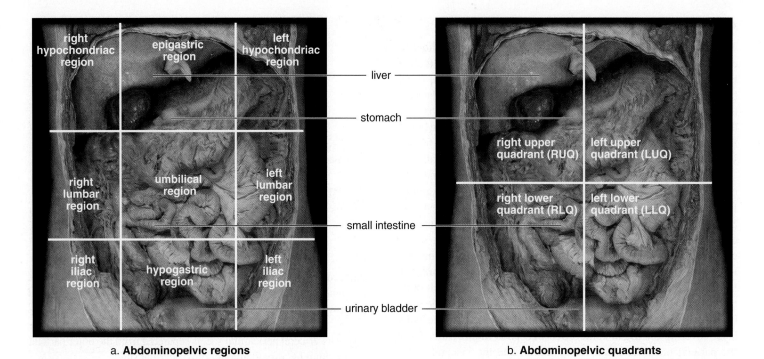

a. **Abdominopelvic regions**

b. **Abdominopelvic quadrants**

AP|R **Figure 1.7** **The abdominopelvic cavity.** The abdominopelvic cavity can be subdivided into **(a)** nine regions or **(b)** four quadrants.

It's important that clinicians use the same terminology to reference various regions of the abdominopelvic cavity. Either of two systems can be used. The first uses nine regions (imagine a "tic-tac-toe" grid, with the umbilicus [navel] in the center square). The upper regions are right hypochondriac, epigastric, and left hypochondriac. The center regions are right lumbar, umbilical, and left lumbar. The lower regions are right inguinal (iliac), pubic, and left inguinal (iliac) (Fig. 1.7a). Note that the terms used are those for each body area, as illustrated in Figure 1.3.

Alternatively, the abdominopelvic cavity can be divided into four quadrants by running a horizontal plane across the median plane at the point of the navel (Fig. 1.7b).

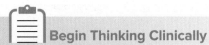

Begin Thinking Clinically

Imagine that you are caring for a small child, who tells you that his stomach hurts. However, he points to his umbilical region, immediately below his navel. What organ is more likely the source of his pain?

Answer and discussion in Appendix A.

Physicians commonly use these quadrants to identify the locations of patients' symptoms. The four quadrants are: (1) right upper quadrant, (2) left upper quadrant, (3) right lower quadrant, and (4) left lower quadrant.

Figure 1.7 compares the two methods of referencing the abdominopelvic region and shows the organs within each region.

Content **CHECK-UP!**

7. Match each of the serous membranes to its function.

parietal pleura _____   A. covers the heart

visceral pericardium _____   B. lines walls of right and left portions of thoracic cavity

visceral peritoneum _____   C. covers the abdominal organs

8. Pleurisy refers to infection or inflammation of the pleurae. You might expect a common symptom of pleurisy to be _____.

Answers in Appendix A.

1.4 Organ Systems

7. List the organ systems of the body, and state the major organs associated with each.

8. Describe in general the functions of each organ system.

The organs of the body work together in systems. The reference figures found on pages 14 and 15 summarize the functions of each of these systems. Corresponding figures that complete each organ system chapter show how that particular system interacts with all the other systems. In this text, the organ systems of the body have been divided into four categories, as discussed next.

Support, Movement, and Protection

The **integumentary system,** discussed in Chapter 5, includes the skin and accessory organs, such as the hair, nails, sweat glands, and sebaceous glands. The skin protects underlying tissues, prevents

Meningitis and Serositis

The anterior and posterior body cavities are enclosed areas that are protected by bone, muscle, connective tissues, and skin. Inflammation of the membranes lining these cavities is a fairly rare, but serious, illness. If body defenses are overcome by bacteria, viruses, or other microbes, the result is a serious, potentially fatal infection and inflammation of the meninges (meningitis) or the serous membranes (**serositis**). Pleurisy, pericarditis, and peritonitis are all forms of serositis.

Meningitis is the term for inflammation of the meninges—linings of the posterior body cavity that cover the brain and spinal cord. The most dangerous form is caused by bacteria that commonly inhabit the nose. In the bacterial meningitis patient, a previous viral infection (which may be a simple common cold) allows these bacteria to enter the bloodstream and infect the meninges. Symptoms of bacterial meningitis include a severe headache and stiff neck, sensitivity to light, fever, weakness, and fatigue. Even with aggressive antibiotic treatment, bacterial meningitis is fatal in 25% of adults. The best treatment is prevention by immunization—especially important for young college students living in the close quarters of a college dorm.

Pleurisy is an inflammation of the pleurae—linings of the thoracic cavity that also cover the lungs. It is often caused by a cold virus, although it can signal the presence of more serious infections or even lung cancer. Its symptoms include chest pain that worsens with deep breathing, and *pleural friction rub*—a rough, grating sound in the chest that can be heard with a stethoscope placed over the painful area. Treatment for pleurisy depends on its cause; most often, pleurisy that results from a common cold requires only pain medication such as aspirin or ibuprofen. Treatment for bacterial infection requires antibiotics.

Pericarditis affects the linings surrounding the heart. Like meningitis, it often results from previous infections and can be extremely dangerous. It is a common complication in drug abusers who use dirty needles for injections. Symptoms include severe chest pain (which may be mistaken for a heart attack), fever, and weakness. Physicians can hear *pericardial friction rub* by placing a stethoscope over the patient's heart. Fluid accumulation inside the pericardial sac surrounding the heart may interfere with blood flow to and from the heart. Bacterial pericarditis is treated with antibiotics, pain medications, and drugs that reduce swelling.

Peritonitis affects the lining of the abdominopelvic cavity. It usually results from bacterial infection; a common cause of infection is a ruptured appendix from appendicitis. Severe pain, fever, elevated white blood cell counts, and tenderness are common symptoms. Aggressive treatment with antibiotics is necessary to prevent bacteria from invading the blood.

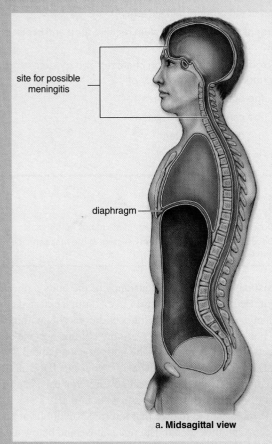

site for possible meningitis

diaphragm

a. **Midsagittal view**

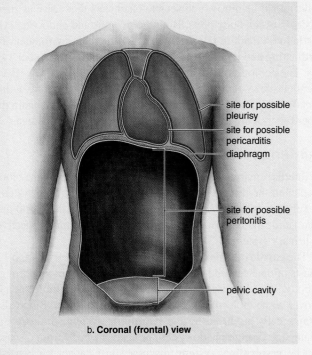

site for possible pleurisy

site for possible pericarditis

diaphragm

site for possible peritonitis

pelvic cavity

b. **Coronal (frontal) view**

Figure 1A **Meningitis and serositis. (a)** Meningitis is infection or inflammation of the linings of the cranial cavity and vertebral canal. **(b)** Serositis is infection or inflammation of the ventral body cavities. Pleurisy affects the pleural cavities, pericarditis affects the pericardial cavity, and peritonitis affects the abdominopelvic cavities.

infection and water loss, helps regulate body temperature, contains sense organs, and even manufactures certain chemicals that affect the rest of the body.

The **skeletal system** and the **muscular system** give the body support and the ability to move. The skeletal system, discussed in Chapter 6, consists of the bones of the skeleton and associated cartilage, as well as the ligaments that bind these structures together. The skeleton protects body parts. For example, the skull forms a protective case for the brain, as does the rib cage for the heart and lungs. Some bones produce blood cells, and all bones are a storage area for calcium and phosphorus compounds. The skeleton as a whole serves as a place of attachment for the muscles.

Contraction of *skeletal muscles,* discussed in Chapter 7, accounts for our ability to move voluntarily and to respond to outside stimuli. These muscles also maintain posture and are responsible for the production of body heat. *Cardiac muscle* and *smooth muscle* are called involuntary muscles because they contract automatically. Cardiac muscle makes up the heart, and smooth muscle is found within the walls of internal organs and blood vessels.

Integration and Coordination

The **nervous system,** discussed in Chapter 8, consists of the brain, spinal cord, and associated nerves. The nerves conduct sensory nerve signals to the brain and spinal cord. They also conduct nerve signals from the brain and spinal cord to the muscles, glands, and organs.

The sense organs (discussed in Chapter 9) provide us with information about our internal and external environments. The brain and spinal cord then process this information, and the individual responds to environmental stimuli using organs, glands, and/or muscles.

The **endocrine system,** discussed in Chapter 10, consists of the hormonal glands, which secrete chemicals that serve as messengers between body parts. Both the nervous and endocrine systems help maintain a relatively constant internal environment by coordinating and regulating the functions of the body's other systems. The nervous system acts quickly but has a short-lived effect; the endocrine system acts more slowly but has a more long-acting effect on body parts. The endocrine system also helps maintain the proper functioning of the male and female reproductive organs.

Maintenance of the Body

The internal environment of the body is the blood within the blood vessels and the tissue fluid that surrounds the cells. Five systems add substances to and/or remove substances from the blood and tissue fluid: the cardiovascular, lymphatic, respiratory, digestive, and urinary systems.

The **cardiovascular system,** discussed in Chapters 11 and 12, consists of the heart and the blood vessels that carry blood through the body. Blood transports nutrients and oxygen to the cells and removes waste molecules to be excreted from the body. Blood also contains cells produced by the **lymphatic system,** discussed in Chapter 13. The lymphatic system protects the body from disease.

The **respiratory system,** discussed in Chapter 14, consists of the lungs and the tubes that take air to and from the lungs. The respiratory system brings oxygen into the lungs and takes carbon dioxide out of the lungs.

The **digestive system** (see Fig. 1.1), discussed in Chapter 15, consists of the mouth, esophagus, stomach, small intestine, and large intestine, along with the accessory organs: teeth, tongue, salivary glands, liver, gallbladder, and pancreas. This system receives food and digests it into nutrient molecules, which can enter the cells of the body.

The **urinary system,** discussed in Chapter 16, contains the kidneys and the urinary bladder. This system rids the body of nitrogenous (nitrogen containing) wastes and helps regulate the fluid level and chemical content of the blood.

Reproduction and Development

The male and female **reproductive systems,** discussed in Chapter 17, contain different organs. The *male reproductive system* consists of the testes, other glands, and various ducts that conduct semen to and through the penis. The *female reproductive system* consists of the ovaries, uterine tubes, uterus, vagina, and external genitalia. Both systems produce sex cells, but in addition, the female system receives the sex cells of the male and also nourishes and protects the fetus until the time of birth. Development before birth, the birth process, and the process of genetic inheritance are discussed in Chapters 18 and 19.

Content **CHECK UP!**

9. Which two organ systems function mainly to control the activities of the other organ systems?

10. Organs of the _____ system receive food and break it down into nutrient molecules.

11. Acromegaly is a condition characterized by excess growth hormone. This is a disorder of:

   a. the endocrine system.

   b. the cardiovascular system.

   c. the digestive system.

Answers in Appendix A.

1.5 **Homeostasis**

9. Describe how a feedback system maintains homeostasis.

10. Describe the role of each body system in the maintenance of homeostasis.

Homeostasis is the relative constancy of the body's internal environment. Because of homeostasis, even though external conditions may change dramatically, internal conditions stay within a narrow range. For example, regardless of how cold or hot one's environment gets, the temperature of the body stays around 37°C (97° to 99°F). Likewise, no matter how acidic your meal, the blood's acidity remains relatively constant, and even if you eat a candy bar, the amount of sugar in your blood is just about 0.1%.

It is important to realize that internal conditions are not absolutely constant; they tend to fluctuate above and below a particular value. Therefore, the internal state of the body is often described as one of *dynamic* equilibrium. If internal conditions change to any great degree, illness results. This makes the study of homeostatic mechanisms medically important.

Negative Feedback

Negative feedback is the primary homeostatic mechanism that keeps a variable close to a particular value, or set point. A homeostatic mechanism has three components: a sensor, a control center, and an effector (Fig. 1.8a). As you study processes controlled by negative feedback, note that all follow the same basic pattern of events: (1) The sensor detects a change in the external or internal environment and alerts the control center; (2) the control center activates the effector; and (3) the effector reverses the initial change and brings conditions back to normal again. Once the initial change is corrected, the sensor is no longer activated and the correction mechanism is turned off. These steps are illustrated in Figures 1.8 and 1.9.

Mechanical Example

A home heating system illustrates how a negative feedback mechanism works (Fig. 1.8b). You set the thermostat at, say, 68°F. This is the set point. The thermostat contains a thermometer, a sensor that detects when the room temperature falls below the set point. The thermostat is also the control center; it turns the furnace on. The furnace plays the role of the effector. On a chilly day, a fall in room temperature activates the thermostat, which starts up the furnace. The heat given off by the furnace raises the temperature of the room to 68°F. Once the room is warmed, the furnace turns off because the sensor is no longer activated.

Notice that a negative feedback mechanism prevents change in the same direction; the room does not get warmer and warmer because warmth inactivates the system.

Human Examples for Negative Feedback

The thermostat for body temperature is located in a part of the brain called the hypothalamus. When the body temperature falls below normal, the control center directs (via nerve signals) the blood vessels of the skin to constrict (Fig. 1.9). This conserves heat. If body temperature falls even lower, the control center sends nerve signals to the skeletal muscles, and shivering occurs. Shivering generates heat, and gradually body temperature rises to 37°C. When the temperature rises to normal, the regulatory center is inactivated.

When the body temperature is higher than normal, the control center directs the blood vessels of the skin to dilate. This allows more blood to flow near the surface of the body, where heat can be lost to the environment. In addition, the nervous system activates the sweat glands, and the evaporation of sweat helps lower body temperature. Gradually, body temperature decreases to 37°C.

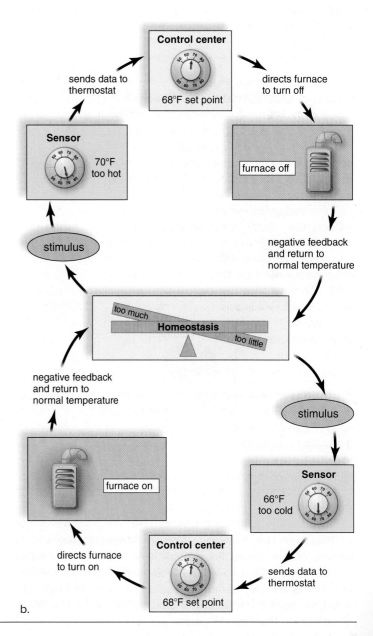

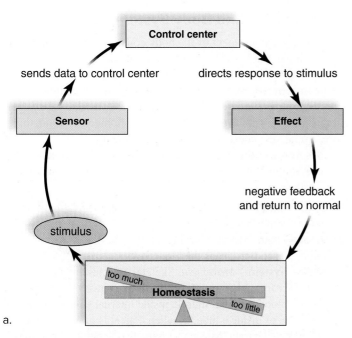

Figure 1.8 Negative feedback. In both examples, an internal or external stimulus (pink) activates a sensor (green). In turn, the sensor signals a control center (tan) which causes an effect (blue). The effect reverses the starting stimulus, and the system returns to homeostasis. **(a)** The general pattern. **(b)** A mechanical example.

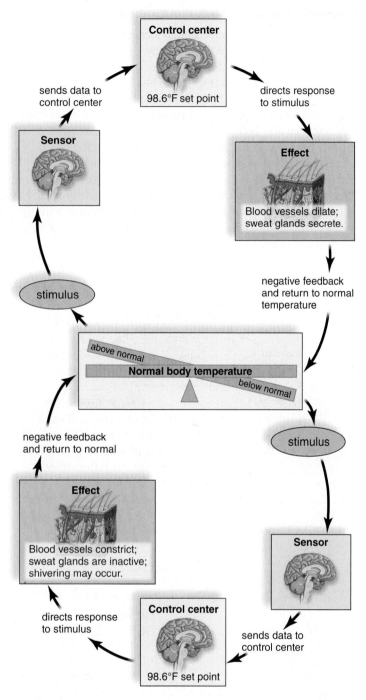

Figure 1.9 Body temperature regulation, a human example of negative feedback. Temperature remains relatively stable (37°C) and is returned to normal when it fluctuates above or below the set point.

Numerous other examples of negative feedback control help to maintain homeostasis. For example, when blood pressure falls, sensory receptors signal a control center in the brain. This center sends out nerve signals to the heart and artery walls. The heart contracts faster and more forcefully, arterial walls constrict, and blood pressure rises. Once the blood pressure is within a homeostatic range, the signals from the brain diminish.

Positive Feedback

Like negative feedback, in a **positive feedback** mechanism an initial stimulus will cause the sensor to trigger the control center. The control center in turn activates the effector. However, the effector in this type of feedback mechanism produces a response that continues to stimulate the sensor. Thus, positive feedback causes the initial stimulus to continue to increase (instead of resolving the stimulus, as in negative feedback). A positive feedback mechanism can be harmful, as when a fever causes metabolic changes that push the fever still higher. Death occurs at a body temperature of 45°C because cellular proteins are destroyed at this temperature and metabolism stops.

Still, positive feedback loops such as those involved in blood clotting, the stomach's digestion of protein, and childbirth assist the body in completing a process that has a definite cutoff point.

Consider that when a woman is giving birth, the head of the baby begins to press against the uterine cervix, which is the opening into the womb. Pressure on the cervix stimulates sensory receptors built into its walls (the sensor in this example). When nerve signals reach the brain (control center), the brain causes the pituitary gland (effector) to secrete the hormone oxytocin. Oxytocin travels in the blood and causes the uterus (also an effector) to contract. As labor continues, the cervical sensory receptors are increasingly stimulated as the baby's head descends in the birth canal, and uterine contractions become increasingly stronger until birth occurs (Fig. 1.10). Once the birth process is complete, the sensory receptors of the cervix are no longer stimulated. However, continued secretion of oxytocin is necessary to safeguard the mother from excessive bleeding after her baby is born. Stimulating the nipples by allowing the baby to suckle signals the brain and continues oxytocin secretion. This process is also controlled by positive feedback.

Homeostasis and Body Systems

The internal environment of the body consists of blood and tissue fluid. Tissue fluid bathes all the cells of the body. Oxygen and nutrients move from blood to tissue fluid, and wastes move from tissue fluid into the blood (Fig. 1.11). Tissue fluid remains constant only as long as blood composition remains constant. All systems of the body contribute toward maintaining homeostasis and, therefore, a relatively constant internal environment.

As you use the text, you will be introduced to each of the organ systems. Remember the job of body systems in maintaining homeostasis as you study each:

Integumentary System (skin)—*Support and protection* of delicate internal structures.

Skeletal System—*Movement.* In addition, the skeleton stores minerals and produces the blood cells.

Muscular System—*Movement.* Skeletal muscles and bones work together for movement. Cardiac muscle causes the heart to pump blood. Smooth muscle inside internal structures moves substances inside a tube, as when a meal is moved through the digestive tract. Muscle activity also generates heat, thus helping to maintain body temperature.

Nervous and Sensory Systems—*Control.* We have already seen that in negative and positive feedback mechanisms, sensory receptors send nerve signals to control centers in the brain, which then activate effectors: muscles, glands, or organs. The nervous system can cause rapid change, if needed, to maintain homeostasis.

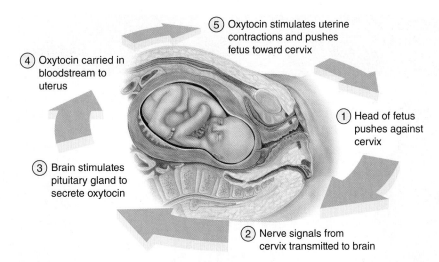

⑤ Oxytocin stimulates uterine contractions and pushes fetus toward cervix

④ Oxytocin carried in bloodstream to uterus

① Head of fetus pushes against cervix

③ Brain stimulates pituitary gland to secrete oxytocin

② Nerve signals from cervix transmitted to brain

Figure 1.10 Positive feedback. Positive feedback loops help the body to complete a process with a definite cutoff point, like labor and delivery.

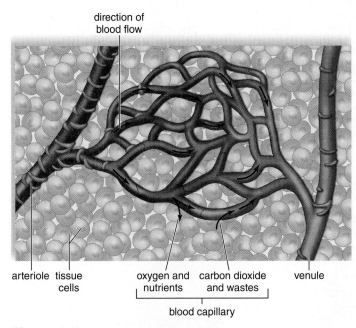

direction of blood flow

arteriole tissue cells

oxygen and nutrients

carbon dioxide and wastes

venule

blood capillary

Figure 1.11 Regulation of tissue fluid composition. Cells are surrounded by tissue fluid (blue), which is continually refreshed because oxygen and nutrient molecules constantly exit the bloodstream, and carbon dioxide and waste molecules continually enter the bloodstream.

Endocrine System—*Control.* Endocrine glands secrete hormones into the blood. Endocrine hormones bring about a slower, more lasting change that keeps the internal environment relatively stable.

Blood and Cardiovascular System—*Transportation and defense.* Red blood cells and blood plasma (the liquid fraction of blood) transport oxygen, carbon dioxide, nutrients, and wastes. Platelets in blood participate in the clotting process, preventing excess blood loss. White blood cells defend against infection. The cardiovascular system conducts blood to and away from capillaries, where exchange occurs. The heart pumps the blood and thereby keeps it moving toward the capillaries.

Lymphatic System—*Transportation and defense.* The lymphatic system assists the cardiovascular system. Lymphatic capillaries collect excess tissue fluid, which is returned via lymphatic vessels to the cardiovascular veins. Lymph nodes help to purify lymph and keep it free of pathogens (disease-causing agents such as bacteria and viruses). This action is assisted by the white blood cells that are housed within lymph nodes.

Respiratory System—*Gas exchange.* The respiratory system adds oxygen to and removes carbon dioxide from the blood. It also plays a role in regulating acid-base balance in blood and tissue fluid. Removal of CO_2 helps to prevent excessive acidity of the blood.

Digestive System—*Nourishment and waste removal.* The digestive system takes in and digests food, providing nutrient molecules to replace the nutrients that are constantly being used by the body cells. Substances that cannot be digested are eliminated. The liver, an accessory digestive organ, also manufactures urea, a waste product of protein digestion. The liver also removes toxic chemicals such as alcohol and other drugs. Additionally, the liver regulates blood glucose (sugar). As glucose enters the blood after a meal, any excess is removed by the liver and stored as glycogen. Later, the glycogen can be broken down to replace the glucose used by the body cells. In this way, the glucose composition of blood remains constant.

Urinary System—*Waste removal.* Urea and other metabolic waste molecules are excreted by the kidneys, which are a part of the urinary system. Urine formation by the kidneys is extremely critical to the body, not only because it rids the body of unwanted substances but also because urine formation offers an opportunity to carefully regulate blood volume, salt balance, and acid-base balance.

Reproductive System—*Survival of the species.* Although individuals can survive and thrive without reproducing, the human species cannot continue without this vital system.

The contributions of each of the body's systems are summarized in the Human Systems Work Together illustration on pages 14 and 15.

Human Systems Work Together

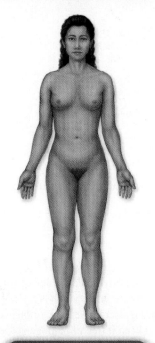

Integumentary System

External support and protection of body; helps maintain body temperature.

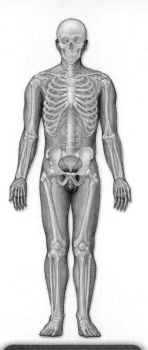

Skeletal System

Internal support and protection; body movement; production of blood cells.

Muscular System

Body movement; production of heat that maintains body temperature.

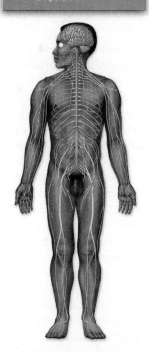

Nervous System

Regulatory centers for control of all body systems; learning and memory.

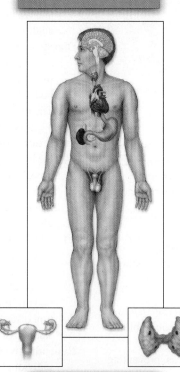

Endocrine System

Secretion of hormones for chemical regulation of all body systems.

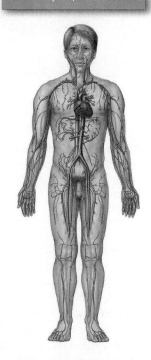

Blood/Cardiovascular System

Transport of nutrients to body cells and transport of wastes away from cells

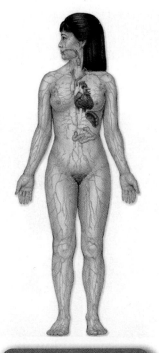

Lymphatic System/Immunity

Drainage of tissue
fluid; purifies tissue
fluid and keeps it free
of pathogens.

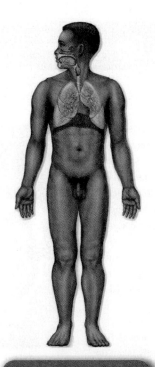

Respiratory System

Rids the blood of carbon
dioxide and supplies the
blood with oxygen; helps
maintain the pH of the blood.

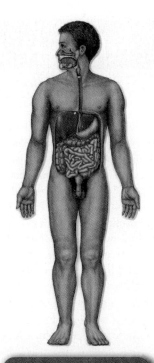

Digestive System

Breakdown of food
and absorption of
nutrients into blood.

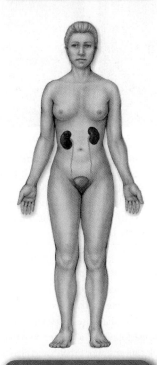

Urinary System

Maintenance of
volume and
chemical composition
of blood.

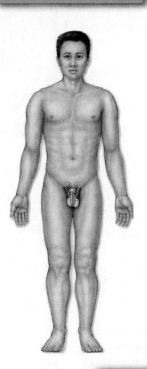

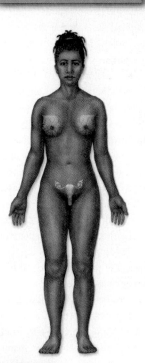

Reproductive System

Production of sperm and
egg; transfer of sperm
to female system where
development occurs.

Imaging the body for diagnosis of disease has certainly changed since the accidental invention of the X ray by Wilhelm Roentgen in 1895. Many new techniques allow clinicians to visualize internal structures with great accuracy. The most widely used imaging technique remains the X ray, which is produced when high-speed electrons strike a heavy metal. Dense structures (such as bone) absorb X rays well, showing up as light areas. Soft tissues absorb to a lesser extent and show up as dark areas. Injecting opaque dye into blood vessels allows blood vessel visualization, as in a coronary angiogram (an X ray of the heart's arteries). Digestive tract imaging is possible on patients who have first consumed opaque dye solutions. In the past, X-ray images had to be developed on photographic film for study. Now images can be digitized and projected on a computer, making it possible to easily store and share an X ray.

A bone density scan is a specialized X ray. The patient is first injected with a harmless radioactive tracer, which is rapidly taken up by bone. Subsequent X rays can reveal areas of increased bone metabolism (as in a cancerous tumor) or decreased metabolism (as in osteoporosis). Similarly, radioactive iodine is used to study the thyroid gland because the thyroid is the only tissue to use the element iodine.

During computed tomography, or CT scan, a computer uses the X ray information taken from various angles to form a series of cross sections. CT scanning has reduced the need for exploratory surgery and can guide the surgeon in visualizing complex body structures during surgical procedures.

PET (positron emission tomography) and SPECT (single-photon emission computerized tomography) are both variations on CT scanning. Radioactively labeled compounds are injected into the body, where they are taken up by metabolically active tissues. The tissues then emit gamma rays. Again, a computer generates cross-sectional images of the body, but this time the image indicates metabolic activity in addition to structure. PET scanning is used to diagnose brain disorders, such as a brain tumor, Alzheimer disease, epilepsy, or stroke.

During MRI (magnetic resonance imaging), the patient lies in a massive, hollow, cylindrical magnet and is exposed to short bursts of a powerful magnetic field. This causes the protons in the nuclei of the body's billions of hydrogen atoms to align. Then, when exposed to strong radio waves, the protons move out of alignment and produce signals. A computer changes these signals into an image. Tissues with many hydrogen atoms (such as fat) show up as bright areas, while tissues with few hydrogen atoms (bone, for example) appear black. This is the opposite of an X ray, which is why MRI is more useful than an X ray for imaging soft tissues. However, many people cannot undergo MRI, because the magnetic field can actually pull a metal object (like a tooth filling or an artificial hip) out of the body!

MRI technology continues to evolve, enabling scientists to study the active brain. In functional magnetic resonance imaging (fMRI), as with traditional MRI, the patient is placed inside a magnetic field. He or she is then asked to do small tasks (for example, touching thumb to fingers, or answering simple questions). fMRI highlights metabolically

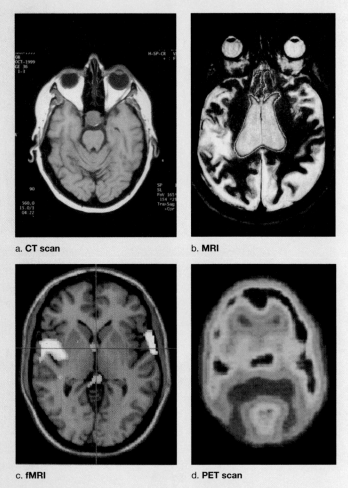

a. **CT scan** b. **MRI**

c. **fMRI** d. **PET scan**

Figure 1B Techniques for imaging the body. (a) a CT scan highlights bony structures in white, while soft tissue is dark. A contrast medium has to be used to see brain tissue. **(b)** In an MRI, dense bone is dark and soft tissues are illuminated in lighter shades. **(c)** and **(d)** Both fMRI and PET scan show areas of highest metabolic activity.

active brain tissue, which has a greater blood supply than neighboring inactive regions. The newest MRI technology application, called diffusion tensor imaging, traces water movement in brain cells. Abnormal cells have different water movement patterns than normal cells. Both imaging techniques are used to diagnose and treat brain injury and disease (such as brain damage following a stroke, or Alzheimer disease), and can also help to track cancer tumor growth and plan brain surgeries.

The least expensive method of creating tissue images is sonography, or ultrasound. High-frequency sound waves are transmitted into tissues, which reflect the sound waves to create an image. Sonography avoids both radiation exposure and rare allergic reactions to dyes. Ultrasound is safe for imaging the fetus in a pregnant woman, and can show an amazing amount of detail, as you can see in the chapter introduction.

Disease

Disease occurs when homeostasis fails and the body (or part of the body) no longer functions properly. The effects may be limited or widespread. A **local disease** is more or less restricted to a specific part of the body. On the other hand, a **systemic disease** affects the entire body or involves several organ systems. For example, streptococcal tonsillitis, or "strep throat," is a local disease. If not effectively treated with antibiotics, strep throat can progress to become a dangerous systemic disease—rheumatic fever. Diseases may also be classified on the basis of their severity and duration. **Acute diseases** occur suddenly and generally last a short time. **Chronic diseases** tend to be less severe, develop slowly, and are long term. Acute bronchitis is a bacterial infection of lung airways and a frequent complication of the common cold; chronic bronchitis is typically caused by years of smoking.

The medical profession has many ways of diagnosing disease, including imaging internal body parts (see Medical Focus, page 16).

Content CHECK-UP!

12. The two organ systems responsible for eliminating wastes are the _____ and _____ systems.

13. The two organ systems that defend the body against infection are:

   a. cardiovascular and urinary

   b. cardiovascular and respiratory

   c. cardiovascular and skeletal

   d. cardiovascular and lymphatic

14. A person with osteoarthritis, a degenerative disorder that causes stiff, painful joints, developed this condition at age 60 after decades of "wear and tear." Osteoarthritis can best be described as:

   a. acute

   b. chronic

Answers in Appendix A.

Selected New Terms

Basic Key Terms

abdominal cavity (ăb-dŏm′ĭ-nŭhl kăv′ĭ-tē), p. 7

abdominopelvic cavity (ăb-dŏm″ĭ-nō-pĕl′vĭk kăv′ĭ-tē), p. 7

anatomical position (ăn″ŭh-tŏm′ĭ-kŭhl pō-zĭsh′ŭn), p. 3

anatomy (ŭh-năt′ō-mē), p. 2

anterior (ăn-tēr′ē-e˘r), p. 3

appendicular portion (ăp″ŭhn-dĭk′yŭh-lĕr pŏr′shŭhn), p. 4

atoms (ăt′ŭhms), p. 2

axial portion (ăk′sē-ŭhl pŏr′shŭhn), p. 4

cells (se˘ls), p. 2

central (cĕn′trŭhl), p. 4

contralateral (kŏn′trŭh-lăt′ŭhr-ŭhl), p. 4

cranial cavity (krā′nē-ŭhl kăv′ĭ-tē), p. 6

deep (dēp), p. 4

diaphragm (dī′ŭh-frăm″), p. 7

distal (dĭs′tŭhl), p. 4

fibrous pericardium (fī′-brŭs pĕr′ĭ-kar′dē-ŭm), p. 7

frontal plane (frŭn′tl plăn), p. 5

homeostasis (hō″mē-ō-stā′sĭs), p. 10

inferior (ĭn-fēr′ē-ŭhr), p. 3

ipsilateral (ĭp′sŭh-lăt′ĕr-ŭhl), p. 4

lateral (lăt′ĕr-ŭhl), p. 4

macromolecules (măk′rō-mŏl′ĭ-kyūlz), p. 2

medial (mē′dē-ŭhl), p. 3

mediastinum (mē″dē-ŭh-stī′nŭm), p. 7

meninges (mŭh-nĭn′jēz), p. 6

molecules (mŏl′ĭ-kyūlz), p. 2

negative feedback (nĕg′ŭh-tĭv fēd′băk), p. 11

organ (ŏr′gŭhn), p. 2

organelles (ŏr″gŭh-nĕlz′), p. 2

organism (ŏr′gŭh-nĭz′ŭhm), p. 2

organ system (ŏr′gŭhn sĭs′tŭhm), p. 2

parietal pericardium (pŭh-rī′-ŭh-tŭhl pĕr″ĭ-kar′dē-ŭhm), p. 7

parietal peritoneum (pŭh-rī′-ŭh-tŭhl pĕr″ĭ-tŭh-nē′ŭhm), p. 7

parietal pleurae (pŭh-rī′-ŭh-tŭhl plŭr′ē), p. 7

parietal serous membrane (pŭh-rī′ŭh-tŭhl sēr′ŭhs mĕm′brăn), p. 7

pelvic cavity (pĕl′vĭk kăv′ĭ-tē), p. 7

pericardial cavity (pĕr″ĭ-kăr′dē-ŭhl kăv′ĭ-tē), p. 7

peripheral (pŭh-rĭf′ŭhr-ŭhl), p. 4

physiology (fĭz″ē-ŏl′ō-jē), p. 2

pleurae (plŭr′ē), p. 7

positive feedback (pŏz′ĭ-tĭv fēd′băk), p. 12

posterior (pŏs-tēr′ē-ŭhr), p. 3

proximal (prŏk′sĭ-mŭhl), p. 4

sagittal plane (săj′ĭ-tŭhl plăn), p. 5

scrotum (skrō′tŭm), p. 7

serous fluid (sēr′ŭs flū′ĭd), p. 7

serous membrane (sēr′ŭs mĕm′brăn), p. 7

serum (sēr′ŭm), p. 7

superficial (sū″păr-fĭsh′ŭhl), p. 4

superior (sū-pēr′ē-ŭhr), p. 3

thoracic cavity (thō-răs′ĭk kăv′ĭ-tē), p. 7

tissue (tĭsh′ū), p. 2

transverse plane (trăns-vĕrs′ plăn), p. 5

vertebral canal (vŭr′tŭh-brŭhl kŭh-năl), p. 6

viscera, (vĭs′ĕr-ŭh), p. 6

visceral pericardium (vĭs′ĕr-ŭhl pĕr″ĭ-kar′dē-ŭm), p. 7

visceral peritoneum (vĭs′ĕr-ŭhl pĕr″ĭ-tŭh-nē′ŭm), p. 7

visceral pleura, (vĭs′ĕr-ŭhl plŭ′rŭh), p.7

visceral serous membrane (vĭs′ĕr-ŭhl sēr′ŭs mĕm′brăn), p. 7

Clinical Key Terms

acute disease (ŭh-kyūt′ dĭ-zēz′), p. 17

chronic disease (krŏn′-ĭk dĭ-zēz′), p. 17

disease (dĭ-zēz′), p. 17

local disease (lō′kŭhl dĭ-zēz′), p. 17

meningitis (mĕn″ĭn-jī′tĭs), p. 9

pericarditis (pĕr″ĭ-kăr-dī′tĭs), p. 9

peritonitis (pĕr″ĭ-tō-nī′tĭs), p. 9

pleurisy (plū′rĭ-sē), p. 9

serositis (sē″rō-sī′tĭs), p. 9

systemic disease (sĭs-tĕm′ĭk dĭ-zēz′), p. 17

Summary

1.1 The Human Body

   A. Anatomy is the study of the structure of body parts, and physiology is the study of the function of these parts. Structure is suited to the function of a part. The body has levels of organization that progress from atoms to molecules, macromolecules, organelles, cells, tissues, organs, organ systems, and finally, the organism.

1.2 Anatomical Terms
   Various terms are used to describe the location of body organs when the body is in the anatomical position (standing erect, with face forward, arms at the sides, and palms and toes directed forward).
   A. The terms *anterior/posterior, superior/inferior, medial/lateral, proximal/distal, superficial/deep, central/peripheral, and contralateral/ipsilateral* describe the relative positions of body parts.
   B. The body can be divided into axial and appendicular portions, each of which can be further subdivided into specific regions. For example, *brachial* refers to the arm, and *pedal* refers to the foot.
   C. The body or its parts may be sectioned (cut) along certain planes. A median or midsagittal (vertical) cut divides the body into equal right and left portions. A parasagittal or paramedian section is a sagittal section parallel to the midline. A frontal (coronal) cut divides the body into anterior and posterior parts. A transverse (horizontal) cut is a cross section.

1.3 Body Cavities and Membranes
   A. The posterior (dorsal) body cavity contains the cranial cavity and vertebral canal.
   B. The anterior (ventral) body cavity contains the thoracic and abdominopelvic cavities, which are separated by the diaphragm. Specific serous membranes line these cavities (as well as the posterior cavity) and adhere to the organs within them.

1.4 Organ Systems
   The body has a number of organ systems. These systems have been characterized as follows:
   A. Support, movement, and protection. The integumentary system, which includes the skin, not only protects the body, but also has other functions. The skeletal system contains the bones, and the muscular system contains the three types of muscles. The primary function of the skeletal and muscular systems is support and movement, but they have other functions as well.
   B. Integration and coordination. The nervous system contains the brain, spinal cord, and nerves. Because the nervous system communicates with both the sense organs and the muscles, it allows us to respond to outside stimuli. The endocrine system consists of the hormonal glands. The nervous and endocrine systems coordinate and regulate the activities of the body's other systems.
   C. Maintenance of the body. The cardiovascular system (heart and vessels), lymphatic system (lymphatic vessels and nodes, spleen, and thymus), respiratory system (lungs and conducting tubes), digestive system (mouth, esophagus, stomach, small and large intestines, and associated organs), and urinary system (kidneys and bladder) all perform specific processing and transporting functions to maintain the normal conditions of the body.
   D. Reproduction and development. The reproductive system in males (testes, other glands, ducts, and penis) and in females (ovaries, uterine tubes, uterus, vagina, and external genitalia) carries out those functions that give humans the ability to reproduce.

1.5 Homeostasis
   Homeostasis is the relative constancy of the body's internal environment, which is composed of blood and the tissue fluid that bathes the cells.
   A. Negative feedback mechanisms help maintain homeostasis by stopping a stimulus when the response of its effector reaches the normal range.
   B. Positive feedback occurs in processes with a definite cutoff point.
   C. All of the body's organ systems contribute to homeostasis. The respiratory, digestive, and urinary systems remove and/or add substances to the blood. Blood is then pumped through the body by the cardiovascular system. In this way, all tissues in the body have access to substances in the blood. The nervous and endocrine systems regulate the activities of other systems.
   D. Disease is a failure of homeostasis. Whether local, systemic, acute, or chronic, disease often requires intervention by a medical professional.

Study Questions

1. Distinguish between the study of anatomy and the study of physiology. (p. 2)
2. Give an example that shows the relationship between the structure and the function of body parts. (p. 2)
3. List the levels of organization within the human body in reference to a specific organ. (p. 2)
4. What purpose is served by directional terms as long as the body is in anatomical position? (pp. 3–4)
5. Distinguish between the axial and appendicular portions of the body.
6. State at least two anatomical terms that pertain to the head, thorax, abdomen, and limbs. (pp. 4–5)
7. Distinguish between a midsagittal section, a parasagittal section, a transverse section, and a coronal section. (p. 5)
8. Distinguish between the posterior and anterior body cavities, and name two smaller cavities that occur within each. (pp. 6–7)
9. Name the four quadrants and the nine regions of the abdominopelvic cavity. (p. 8)
10. Name the major organ systems, and describe the general functions of each. (pp. 8, 10)
11. List the major organs found within each organ system. (pp. 8, 10)
12. Define homeostasis, and give examples of negative feedback and positive feedback mechanisms. (pp. 10–13)
13. Discuss the contribution of each body system to homeostasis. (pp. 12–15)
14. Define the term disease. Compare and contrast local vs. systemic disease, and acute vs. chronic disease. (p. 17)

Learning Outcome Questions

I. Match the terms in the key to the relationships listed in questions 1–5.

Key:

   a. anterior
   b. posterior
   c. superior
   d. inferior
   e. medial
   f. lateral
   g. proximal
   h. distal

1. the esophagus in relation to the stomach
2. the ears in relation to the nose
3. the shoulder in relation to the hand
4. the intestines in relation to the vertebrae
5. the large intestine in relation to the mouth

II. Match the terms in the key to the body regions listed in questions 6–12.

Key:

   a. oral
   b. occipital
   c. gluteal
   d. carpal
   e. palmar
   f. cervical
   g. axillary

6. buttocks
7. palm
8. back of head
9. mouth
10. wrist
11. armpit
12. neck

III. Match the terms in the key to the organs listed in questions 13–18.

Key:

   a. cranial cavity
   b. vertebral canal
   c. thoracic cavity
   d. abdominal cavity
   e. pelvic cavity

13. stomach
14. heart
15. urinary bladder
16. brain
17. liver
18. spinal cord

IV. Match the organ systems in the key to the organs listed in questions 19–25.

Key:

   a. digestive system
   b. urinary system
   c. respiratory system
   d. cardiovascular system
   e. reproductive system
   f. nervous system
   g. endocrine system

19. thyroid gland
20. lungs
21. heart
22. ovaries
23. brain
24. stomach
25. kidneys

V. Fill in the blanks.

26. Learning the body regions and body cavities is the first step in learning _____.
27. A(n) _____ is composed of several types of tissues and performs a particular function.
28. The imaginary vertical plane that passes through the midline of the body is called the _____ plane.
29. All the organ systems of the body together function to maintain _____, a relative constancy of the internal environment.
30. When blood estrogen levels fall below normal, the ovaries make more estrogen. When blood levels return to normal, the ovaries decrease their production of estrogen. This is an example of _____ feedback to maintain estrogen homeostasis.

Medical Terminology Exercise

After studying this chapter, see if you can derive the definitions for the medical terms listed below. Many of the prefixes and suffixes used to create these terms can be found throughout the chapter. For additional help, use McGraw-Hill Connect™ at www.mcgrawhillconnect.com and consult Appendix B.

1. Suprapubic (sū″prŭh-pyū′bĭk) means _____ the pubis.
2. Infraorbital (ĭn″frŭh-ŏr′bĭ-tŭhl) means _____ the eye orbit.
3. Gastrectomy (găs-trĕk′tō-mē) means excision of the _____.
4. Celiotomy (sē″lē-ŏt′ō-mē) means incision (cut into) of the _____.
5. Macrocephalus (măk″rō-sĕf′ŭh-lŭs) means large _____.
6. Transthoracic (trăns″thō-răs′ĭk) means across the _____.
7. Bilateral (bī-lăt′ŭhr-ŭhl) means two or both _____.
8. Ophthalmoscope (ŏf-thăl′mō-skōp) is an instrument to view inside the _____.
9. Dorsalgia (dŏr-săl′jē-ŭh) means pain in the _____.
10. Endocrinology (ĕn″dō-krĭ-nŏl′ō-jē) is the _____ of the endocrine system.
11. The pectoralis (pĕk-tō-răl′ĭs) muscle can be found on the _____.
   a. chest   b. head   c. buttocks
   d. thigh
12. The sacral (sā′krŭhl) nerves are located in the _____.
   a. lower back   b. neck   c. upper back
   d. head
13. Hematuria (hē-mŭh-tū′rē-ŭh) means _____ in the urine.
14. Nephritis (nĕf-rī′tĭs) is inflammation of the _____.
   a. lungs   b. heart   c. liver   d. kidneys
15. Tachypnea (tăk-ĭp-nē′ŭh) is a breathing rate that is _____.
   a. faster than normal
   b. slower than normal

Online Study Tools

LEARNSMART®   **connect®**

APR

Anatomy & Physiology REVEALED includes cadaver photos that allow you to peel away layers of the human body to reveal structures beneath the surface. This program also includes animations, radiologic imaging, audio pronunciations, and practice quizzing. To learn more visit www.aprevealed.com

Anatomy & Physiology REVEALED
aprevealed.com

2 Chemistry of Life

Clockwise from the upper left, the critters you see here are an Israeli death stalker scorpion, a Chilean rose tarantula, a Brazilian pit viper, and a vampire bat. Besides causing goose bumps, these animals share another common trait: all are venomous. Venoms are proteins that are directly injected into the victim, and all are exquisitely tailored to disable, paralyze, and/or ultimately kill the animal's prey. However, scientists are also crafting modified venom proteins into unique medicines. By combining it with a fluorescent dye, scorpion protein has been bioengineered into "tumor paint," which specifically targets cancer cells in the brain, breast, and colon. Even small cancer cell clusters can be identified and surgically removed when tagged with this protein. Tarantula toxin has been shown to have unique pain relieving ability, and drugs to fight high blood pressure have been created from pit viper venom. Further, if that vampire bat wants a good blood feast, his venom must keep his "dinner's" blood from clotting. Researchers are currently perfecting a drug that actually dissolves blood clots in stroke victims from this protein. You can learn more about proteins on page 31. And if you want to find out why you get goose bumps from something you find scary, check out the autonomic nervous system in Chapter 8.

Learning Outcomes

After you have studied this chapter, you should be able to:

2.1 Basic Chemistry

1. Describe how an atom is organized, and tell why atoms interact.
2. Define *radioactive isotopes*, and describe how they can be used in the diagnosis and treatment of disease.
3. Distinguish between an ionic bond and a covalent bond.

2.2 Water, Acids, and Bases

4. Describe the characteristics of water and three functions of water in the human body.
5. Explain the difference between an acid and a base, using examples of each.
6. Use and understand the pH scale.

2.3 Molecules of Life

7. List the four classes of macromolecules in cells, and distinguish between a dehydration reaction and a hydrolysis reaction.
8. Name the individual subunits that link to form carbohydrates, lipids, proteins, and nucleic acids.

2.4 Carbohydrates

9. Give some examples of different types of carbohydrates and their specific functions in cells.

2.5 Lipids

10. Describe the composition of a neutral fat, and give examples of how lipids function in the body.

2.6 Proteins

11. State the major functions of proteins, and tell how globular proteins are organized.

2.7 Nucleic Acids

12. Describe the structure and function of DNA and RNA in cells.
13. Explain the importance of ATP in the body.

Medical Focus

Prions: Malicious Proteins?
The Deadly Effects of High-Level Radiation.

2.1 Basic Chemistry

1. Describe how an atom is organized, and tell why atoms interact.
2. Define *radioactive isotopes*, and describe how they can be used in the diagnosis and treatment of disease.
3. Distinguish between an ionic bond and a covalent bond.

Matter is anything that takes up space and has mass. It can be a solid, a liquid, or a gas. Therefore, we humans are matter, just like the water we drink and the air we breathe.

Elements and Atoms

All matter is composed of basic substances called **elements.** It's quite remarkable that there are only 92 naturally occurring elements. It is even more surprising that over 90% of the human body is composed of just four elements: carbon, nitrogen, oxygen, and hydrogen.

Every element has a name and a symbol consisting of one or two letters. For example, carbon has been assigned the atomic symbol C (Fig. 2.1*a*). When the symbol consists of two letters, only the first letter is capitalized. Some of the symbols we use for elements are derived from Latin. For example, the symbol for sodium is Na because the Latin word *natrium* means sodium.

Elements are composed of tiny particles called atoms. The same name is given to both an element and its atoms.

Atoms

An **atom** is the smallest unit of an element that still retains the chemical and physical properties of the element. Although it is possible to split an atom by physical means, an atom is the smallest unit to enter into chemical reactions. For our purposes, it's satisfactory to think of each atom as having a central nucleus and pathways about the nucleus called *shells* (sometimes called *energy levels*). The subatomic particles called **protons** and **neutrons** are located in the nucleus, and **electrons** orbit about the nucleus in the shells (Fig. 2.1*b*). Most of an atom is empty space. If we could draw an atom the size of a football stadium, the nucleus would be like a gumball in the center of the field, and the electrons would be tiny specks whirling about in the upper stands.

The **atomic number** of an atom tells you how many protons an atom has. All atoms of any particular element have the same number of protons. Protons carry a positive ($+$) charge, and electrons have a negative ($-$) charge. Neutrons have no charge. Atoms are electrically neutral because the numbers of protons and electrons are equal. For example, the atomic number of carbon is 6. Therefore, carbon has six protons and six electrons. How many electrons are in each shell of an atom? The inner shell is the lowest energy level and can hold only two electrons; after that, each shell, for the atoms noted in Figure 2.1*a*, can hold up to eight electrons. Using this information, we can calculate that carbon has two shells and that the outer shell has four electrons.

The number of electrons in the outer shell determines the chemical properties of an atom, including how readily it enters into chemical reactions. As we will see, an atom is most stable when the outer shell has eight electrons. (Hydrogen and helium, with only one shell, are exceptions to this statement. Atoms with only one shell are stable when this shell contains two electrons.) If an atom gains or loses one or more electrons, it becomes an **ion.** An atom that acquires electrons becomes negatively charged, while an atom that loses electrons becomes positively charged.

The **mass number** of an atom is generally equal to the sum of its protons and neutrons. Both protons and neutrons are so light that their weight is indicated by a special designation called an *atomic mass unit* (amu). An atom's protons and neutrons each weigh one atomic mass unit, and its electrons have almost no mass. Using this knowledge, you can determine the number of neutrons in the nucleus of an atom. For example, here is how you calculate the number of neutrons in carbon (C): Carbon's mass number is 12, and you know from its atomic number that it has six protons. Thus, its mass number (12) minus its protons (6) equals its neutrons (6). Likewise, the mass number of sodium (23) minus its protons (11) equals its neutrons (12). (Fig. 2.1*b*)

Further, as shown in Figure 2.1*b*, the atomic number of an atom is often written as a subscript to the lower left of the atomic symbol. The mass number is often written as a superscript to the upper left of the atomic symbol. Thus, determining neutron number is as easy as subtracting the bottom number from the top number.

Common Elements in Living Things

Element	Atomic Symbol	Atomic Number	Mass Number	Comment
hydrogen	H	1	1	These
carbon	C	6	12	elements
nitrogen	N	7	14	make up
oxygen	O	8	16	most
phosphorus	P	15	31	biological
sulfur	S	16	32	molecules.
sodium	Na	11	23	These
magnesium	Mg	12	24	elements
chlorine	Cl	17	35	occur mainly
potassium	K	19	39	as dissolved
calcium	Ca	20	40	salts.

a.

p = protons
n = neutrons
= electrons
= nucleus

6p
6n

Carbon

mass number
$^{12}_{6}C$
b.   atomic number

AP|R Figure 2.1 Elements and atoms. (a) The atomic symbol, number, and mass number are given for common elements in the body. **(b)** The structure of carbon shows that an atom contains the subatomic particles called protons (p) and neutrons (n) in the nucleus (colored pink) and electrons (colored blue) in shells about the nucleus.

In everyday life, length is measured in units such as centimeters, meters, inches, and feet. Mass, or weight, is measured in grams or pounds. Chemists have their own terminology for measuring the number of atoms of an element or molecules of a compound. The unit is a **mole,** and one mole is a specific, very large number: 6.02×10^{23}. (This number, termed *Avogadro's number,* is the number of carbon atoms in exactly 12 grams of carbon 12). Thus, when a chemist refers to a mole of helium, he or she is referring to 6.02×10^{23} helium atoms. Likewise, a mole of oxygen is 6.02×10^{23} molecules of oxygen.

Isotopes

Isotopes of the same element differ in the number of neutrons and therefore differ in their weight. For example, the element carbon has three common isotopes:

$$^{12}_{6}C \qquad ^{13}_{6}C \qquad ^{14}_{6}C*$$
$$*\text{radioactive}$$

Carbon 12 has six neutrons, carbon 13 has seven neutrons, and carbon 14 has eight neutrons. (Note that the mass number minus the atomic number equals the number of neutrons.) Unlike the other two isotopes of carbon, carbon 14 is unstable and breaks down over time. As carbon 14 decays, it releases various types of energy in the form of rays and subatomic particles, and therefore it is a **radioactive isotope.** The radiation given off by radioactive isotopes can be detected in various ways. You may be familiar with the use of a Geiger counter to detect radiation.

Low Levels of Radiation

The importance of chemistry to biology and medicine can easily be appreciated by understanding the many uses of radioactive isotopes. A radioactive isotope behaves the same as do the stable isotopes of an element. This means that you can put a small amount of radioactive isotope in a sample, and it becomes a **tracer** by which to detect molecular and cellular changes. For example, as mentioned in the Medical Focus on page 16, radioactive iodine can be absorbed by the thyroid gland, which uses the element to make thyroid hormone. Abnormal thyroid cells (such as cancer cells) will absorb iodine differently than their normal counterparts, and the changes can be detected using an X ray.

High Levels of Radiation

Radioactive substances in the environment can harm or kill cells, damage DNA, and cause cancer (see the Medical Focus on page 38). The harmful effects of radiation can also be put to good use, however. Radiation from radioactive isotopes has been used for many years to sterilize medical and dental products. Radiation has even been used to sterilize the U.S. mail to free it of possible pathogens, such as anthrax spores.

The ability of radiation to kill cells is often applied to cancer cells. Radioisotopes can be introduced into the body in a way that allows radiation to destroy only the cancerous cells. For example, radioactive iodine is not used just for diagnosis, but also for treatment of thyroid cancer. High levels of radioactive iodine can poison thyroid cancer cells.

Molecules and Compounds

Atoms often bond with each other to form a chemical unit called a **molecule.** A molecule can contain atoms of the same kind, as when an oxygen atom joins with another oxygen atom to form oxygen gas. The atoms could also be different, as when an oxygen atom joins with two hydrogen atoms to form water. Combining different atoms forms a **compound.**

Two types of bonds join atoms: the **ionic bond** and the **covalent bond.** Ionic bonds are created by electrical attraction between ions—atoms that have become electrically charged by gaining or losing electrons. Covalent bonds are created by the sharing of electrons between atoms.

Ionic Bonds

Remember that atoms with more than one shell are most stable when the outer shell contains eight electrons. Sometimes during a reaction, atoms give up or take on electron(s) in order to achieve a stable outer shell.

Figure 2.2*a* depicts a reaction between a sodium (Na) atom and a chlorine (Cl) atom. Sodium, with one electron in the outer shell, reacts with a single chlorine atom. Why? Because once the reaction is finished and sodium loses one electron to chlorine, its outer shell will have eight electrons. Similarly, a chlorine atom, which already has seven electrons, needs only to acquire one more electron to have a stable outer shell.

As you know, ions are atoms that have lost electrons and carry a positive charge, or atoms that have gained electrons and carry a negative charge. Positively charged ions are called **cations,** and negatively charged ions are called **anions.** When the reaction between sodium and chlorine is finished, the sodium cation carries a positive charge because it now has 11 positively charged protons and only 10 negatively charged electrons (11^{+}, 10^{-}; net charge: 1^{+}). Meanwhile, the negatively charged chloride anion has 17 protons and 18 electrons (17^{+}, 18^{-}; net charge: 1^{-}). The attraction between oppositely charged sodium ions and chloride ions forms an ionic bond. The resulting compound, sodium chloride, is table **salt,** which we use to improve the taste of foods. Salts characteristically form an **ionic lattice** that *dissociates* (separates into ions) after dissolving in water (Fig. 2.2*b*). When dissolved in water, salts are referred to as *electrolytes.*

In contrast to sodium, why would calcium, with two electrons in the outer shell, react with two chlorine atoms? Because whereas calcium needs to lose two electrons, each chlorine, with seven electrons already, requires only one more electron to have a stable outer shell. The resulting salt ($CaCl_2$) is called calcium chloride.

The balance of various ions in the body is important to our health. Too much sodium in the blood can contribute to **hypertension** (high blood pressure); not enough calcium leads to **rickets** (a bowing of the legs) in children; too much or too little potassium results in **arrhythmia** (heartbeat irregularities). Bicarbonate, hydrogen, and hydroxide ions are all involved in maintaining the acid-base balance of the body (see pages 25–27).

Covalent Bonds

In a covalent bond, atoms share electrons instead of losing or gaining them. The overlapping outermost shells in Figure 2.3 indicate

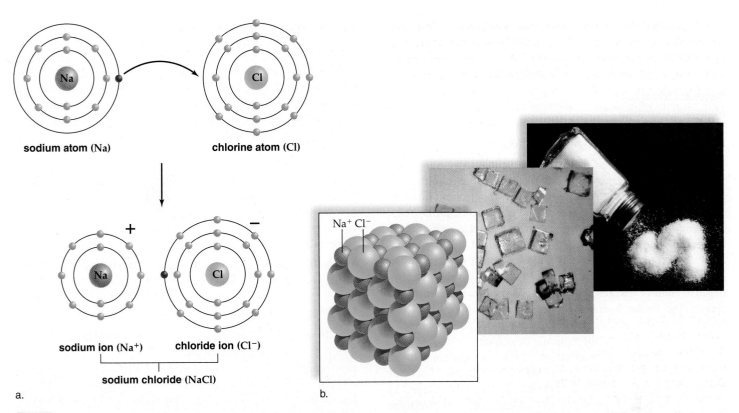

sodium atom (Na)   chlorine atom (Cl)

sodium ion (Na⁺)   chloride ion (Cl⁻)

sodium chloride (NaCl)

a.   b.

Na⁺ Cl⁻

AP|R **Figure 2.2** **Ionic reaction. (a)** During the formation of sodium chloride, an electron is transferred from the sodium atom to the chlorine atom. At the completion of the reaction, each ion formed has eight electrons in the outer shell, but each also carries a charge as shown. **(b)** In a sodium chloride crystal, bonding between ions creates a three-dimensional lattice in which each Na^+ ion is surrounded by six Cl^- ions, and each Cl^- is surrounded by six Na^+.

that the atoms are sharing electrons. Just as two people's hands participate in a handshake, each atom contributes one electron to the pair that is shared. These electrons spend part of their time in the outer shell of each atom; therefore, they are counted as belonging to both bonded atoms.

Covalent bonds can be represented in a number of ways. In contrast to the diagrams in Figure 2.3, structural formulas use straight lines to show the covalent bonds between the atoms. Each line represents a pair of shared electrons. A molecular formula describes the molecule's composition, but molecular formulas indicate only the number of each type of atom making up a molecule. A comparison follows:

Structural formula: Cl—Cl
Molecular formula: Cl_2

Double and Triple Bonds

In addition to a single bond, in which atoms share only a pair of electrons, a double or a triple bond can form. In a double bond, atoms share two pairs of electrons, and in a triple bond, atoms share three pairs of electrons between them. For example, in Figure 2.3, each nitrogen atom (N) requires three electrons to achieve a total of eight electrons in the outer shell. Notice that six electrons are placed in the outer overlapping shells in the diagram and that three straight lines are in the structural formula for nitrogen gas (N_2).

What are the structural and molecular formulas for carbon dioxide? Carbon, with four electrons in the outer shell, requires

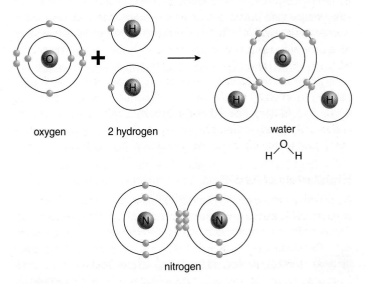

oxygen   2 hydrogen   water

$$\underset{H}{O}\diagdown H$$

nitrogen

AP|R **Figure 2.3** **Covalent reactions.** After a covalent reaction, each atom will have filled its outer shell by sharing electrons. To determine this, it is necessary to count the shared electrons as belonging to both bonded atoms. Oxygen and nitrogen are most stable with eight electrons in the outer shell. Hydrogen is most stable with two electrons in the outer shell.

four more electrons to complete its outer shell. Each oxygen, with six electrons in the outer shell, needs only two electrons to complete its outer shell. Therefore, carbon shares two pairs of electrons with each oxygen atom, and the formulas are as follows:

Structural formula: O=C=O
Molecular formula: CO_2

Polar and Nonpolar Covalent Bonds

In atoms that form a covalent molecule, electrons are shared in single, double, or triple pairs. Custody of the shared electrons is not always equal. One of the elements often keeps the electrons for a longer period of time. Such elements are called **electronegative** (think "electron-grabbers"). When these electronegative elements such as oxygen are teamed with smaller, weaker elements like hydrogen, the resulting compound (water, H_2O) forms a **polar covalent bond.** Because the electrons spend most of their time with the electronegative element (oxygen), this element acquires a partial negative charge. The weaker hydrogen atoms gain a partial positive charge because they surrender their electrons for a longer period of time.

What if the atoms in the molecule share their electrons equally, so that each atom keeps the electrons for the same amount of time? This arrangement results in a **nonpolar covalent bond.** Molecules linked by this type of bond do not have an electrical charge. As you'll see, the bonding properties of covalent molecules determine that molecule's characteristic role in living organisms.

Content CHECK-UP!

1. One particular isotope of the element potassium has the following symbol: ^{39}K. The atomic number (number of protons) for potassium is 19. How many neutrons are contained in this isotope?

2. Fluorine gas contains two atoms of fluorine that share two electrons. Draw the structural and molecular formulas for fluorine gas.

Answers in Appendix A.

Begin Thinking Clinically

One of the world's worst nuclear accidents occurred at Chernobyl, Ukraine, in 1986. The damaged nuclear reactor released radioactive strontium into the air. Strontium is chemically similar to calcium. What body tissue might be damaged by radioactive strontium?

Answer and discussion in Appendix A.

2.2 Water, Acids, and Bases

4. Describe the characteristics of water and three functions of water in the human body.
5. Explain the difference between an acid and a base, using examples of each.
6. Use and understand the pH scale.

Water is the most abundant molecule in living organisms, usually making up about 60–70% of the total body weight. Even so, water

is an **inorganic molecule** because it does not contain carbon atoms. Carbon atoms are common to **organic molecules.**

As mentioned previously, in water the electrons spend more time circling the larger oxygen (O) atom than the smaller hydrogen (H) atoms. This gives water both a slight negative charge (symbolized as δ^-) to the oxygen and a slight positive charge (symbolized as δ^+) to the hydrogen atoms. Therefore, water is a **polar molecule** with negative and positive ends:

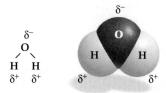

The diagram on the left shows the structural formula of water, and the one on the right is called a space-filling model.

Hydrogen Bonds

A **hydrogen bond** occurs whenever a covalently bonded hydrogen has a slight positive charge, and is attracted to a slightly negatively charged atom nearby. A hydrogen bond is represented by a dotted line because it is relatively weak and can be broken rather easily. In Figure 2.4, you can see that each hydrogen atom, being slightly positive, bonds to the slightly negative oxygen atom of another water molecule nearby.

Properties of Water

Polarity and hydrogen bonding cause water to have many properties beneficial to life, including the three to be mentioned here.

1. Water is a **solvent** for polar (charged) molecules and ionic compounds, and thereby facilitates chemical reactions both outside and within our bodies.

When ions and molecules disperse in water, they move about and collide, allowing reactions to occur. Therefore, water is a

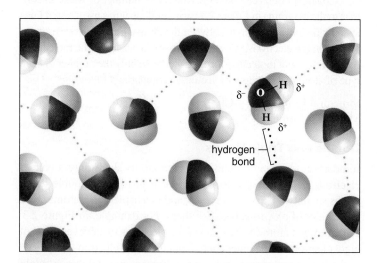

AP|R **Figure 2.4 Hydrogen bonding between water molecules.** The polarity of the water molecules causes hydrogen bonds (dotted lines) to form between the molecules.

solvent that facilitates chemical reactions. For example, when a salt such as sodium chloride (NaCl) is put into water, the negative ends of the water molecules are attracted to the sodium ions, and the positive ends of the water molecules are attracted to the chloride ions. This causes the sodium ions and the chloride ions to dissolve and to separate in water:

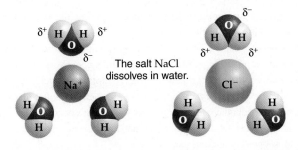

The salt NaCl dissolves in water.

Ions and molecules that are water-soluble are said to be **hydrophilic** ("water-loving"). Nonionized and nonpolar molecules that do not interact with water are said to be **hydrophobic** ("water-fearing").

2. Water molecules are both cohesive and adhesive.

Water molecules cling together because of hydrogen bonding, a property called **cohesion.** Water can also cling to other surfaces by hydrogen bonding, and this behavior is **adhesion.** Thus, water flows freely and water-based solutions fill vessels, such as blood vessels. These characteristics of water make it an excellent transport medium. Within our bodies, the blood that fills our arteries and veins is 92% water. Blood transports oxygen and nutrients to the cells and removes wastes such as carbon dioxide from the cells.

3. Water has a high **specific heat capacity** and a high **heat of vaporization.**

Specific heat capacity is the amount of heat energy needed to change an object's temperature by exactly 1°C. Water has the highest specific heat of almost any substance in nature. Thus, water can absorb a tremendous amount of heat energy without changing its temperature. This is due to the hydrogen bonds that link water molecules together. Maintaining a fairly constant body temperature is possible, in part, because the body is roughly 60% water. High water content helps to insulate the body from extreme heat and cold in the environment. Hydrogen bonding also accounts for water's high heat of vaporization—the amount of energy needed to turn water into steam. As water evaporates during sweating, the body can be cooled rapidly and effectively.

Acids and Bases

When water molecules dissociate (break up), they release an equal number of hydrogen ions (H^+) and hydroxide ions (OH^-):

$$H-O-H \rightleftharpoons H^+ + OH^-$$
water   hydrogen   hydroxide
                  ion   ion

Only a few water molecules at a time dissociate, and the actual number of H^+ and OH^- is very small (1×10^{-7} moles/liter).

Acids are substances that dissociate in water, releasing hydrogen ions (H^+). For example, an important inorganic acid is hydrochloric acid (HCl), which dissociates in this manner:

$$HCl \longrightarrow H^+ + Cl^-$$

Dissociation is almost complete; therefore, HCl is called a *strong acid.* If hydrochloric acid is added to a beaker of water, the number of hydrogen ions (H^+) increases greatly. Lemon juice, vinegar, tomatoes, and coffee are all acidic solutions.

Bases are substances that either take up hydrogen ions (H^+) or release hydroxide ions (OH^-). For example, an important inorganic base is sodium hydroxide (NaOH), which dissociates in this manner:

$$NaOH \longrightarrow Na^+ + OH^-$$

Dissociation is almost complete; therefore, sodium hydroxide is called a *strong base.* If sodium hydroxide is added to a beaker of water, the number of hydroxide ions increases. Milk of magnesia and ammonia are common basic solutions.

When an acid and a base are combined, the reaction produces a salt and water. For example, if hydrochloric acid and sodium hydroxide are combined:

$$HCl + NaOH \longrightarrow NaCl + H_2O$$

It's interesting to note that many compounds formed by reacting acids and bases are water-soluble alkaline salts that can have direct effects on our brains. You will recognize many of these: nicotine, caffeine, morphine, codeine, and cocaine. When taken in excess, all can be poisonous. Emergency care in a poisoning crisis is decribed in the In Case of Emergency reading on the following page.

pH Scale

In chemistry, pH is a chemist's "shorthand." For practical purposes, it's useful to think of pH as a measure of hydrogen ion concentration. Technically, pH is defined as the negative of the base 10 logarithm of free hydrogen ion concentration. Suppose a chemist stirs up a solution containing 0.00001 moles per liter of free hydrogen ions. The solution's concentration can be written (using scientific notation) as 1×10^{-5} moles per liter. The base 10 logarithm is the exponent, -5. Using the pH system, concentration is expressed as negative times (-5), or pH 5. (Recall that a negative number multiplied by a negative number equals a positive number.)

It's important to remember that as pH *decreases,* H^+ concentration exponentially *increases.* For example, a solution of pH 2 (0.01 moles/liter) has 100 times *more* H^+ than a solution of pH 4 (0.0001 moles/liter).

The **pH scale,** which ranges from 0 to 14, is used to indicate the acidity and basicity (alkalinity) of a solution. A solution of exactly pH 7, which is the pH of water, is at a neutral pH because water releases an equal number of hydrogen ions (H^+) and hydroxide ions (OH^-). Notice in Figure 2.5 that any solution with a pH above 7 is basic, with more hydroxide ions than hydrogen ions. Any solution with a pH below 7 is acidic, with more hydrogen ions

I.C.E. — IN CASE OF EMERGENCY

When There's a Poison Involved

The world can be a pretty dangerous place for a child. Toddlers will examine a new object visually at first, and then it usually goes right into their mouths, often regardless of how awful the taste or smell might be. Small children also mimic the older people in their lives, watching as adults consume drinks or medication. Sometimes, curiosity leads to indiscriminate tasting of substances such as lipstick, crayons, hand lotion, shaving cream, modeling clay, and chalk (all are not toxic in small doses, but ingesting any of them still warrants a call to a poison control center). Unfortunately, accidental poisoning in children happens all too frequently when toxins are accessible.

Accidental poisoning occurs in adults as well, especially to seniors who may be confused about prescription medications. Obviously, poisoning is intentional when an older child, teen, or adult attempts suicide. When date rape or murder is the motive, poisoning is also deliberate (see **Focus on Forensics: Rape** in Chapter 17 for further discussion of date rape).

First responders must quickly take action: **first, call 911.** Then, continue with the **A-B-Cs** of emergency care: A → establish an **A**irway and make sure the person is **B**reathing, and maintain **C**irculation. (The techniques of cardiopulmonary resuscitation are described in Chapter 12). Other effects of the poison demand a quick reaction as well. Alkaloid poisons like nicotine often increase pulse rate and blood pressure to dangerous levels, while other toxins will cause the opposite effect. Further, many poisons cause delirium, seizures, and coma. In these cases, rapid transport to the emergency room can mean the difference between life and death.

Whenever possible, the patient's history should be recorded to give clues about the ingested substance. The victim's caregivers, relatives, friends, and co-workers can help. Partially empty containers, spilled material, and evidence of recreational drug use such as needles and syringes provide valuable information about the poison's possible identity. Once the poison is identified, local and national poison control centers can help, suggesting antidotes and therapies.*

EMS personnel shouldn't attempt to cause vomiting (unless a physician orders an *emetic*, a drug that causes vomiting). In the emergency room, clinicians often have the patient drink a solution of ground-up charcoal, called *activated charcoal*, which combines with many different types of chemicals. Stomach pumping—repeated filling and emptying of the stomach to wash out the poison, with or without charcoal solution—may also be used. However, if the poison is a chemical like gasoline or kerosene, or a burning chemical like drain cleaner, these procedures can't be used because they might further damage the nose, throat, and esophagus. In very serious poisonings, the patient may need dialysis, a technique for cleansing blood artificially.

*The U.S. National Poison Control Center Hotline telephone number is 1-800-222-1222.

Figure 2.5 The pH scale. The proportionate amount of hydrogen ions to hydroxide ions is indicated by the numerical values on the dial. Any solution with a pH above 7 is basic, while any solution with a pH below 7 is acidic.

than hydroxide ions. As we move toward a higher pH, each unit has 10 times the basicity of the previous unit, and as we move toward a lower pH, each unit has 10 times the acidity of the previous unit. This means that even a small change in pH represents a large change in the proportional number of hydrogen and hydroxide ions in the body.

For us to remain healthy, the pH of body fluids needs to be maintained within a narrow range. The normal pH of our blood is always about 7.4—that is, just slightly basic (alkaline). If the pH value drops below 7.35, the person is said to have **acidosis.** If it rises above 7.45, the condition is called **alkalosis.** The pH stability is normally possible because the body has built-in mechanisms to prevent pH changes. The respiratory and urinary systems cooperate to maintain normal pH (see Chapters 14 and 16). Additionally, **buffers** in blood and body fluids help to keep pH within its normal range. Buffers are chemicals, or pairs of chemicals, that take up excess hydrogen ions (H^+) or hydroxide ions (OH^-). One of the most important buffer pairs in the body is the combination of carbonic acid (H_2CO_3) and bicarbonate ions (HCO_3^-), which helps to keep the pH of the blood within normal range. Carbonic acid is a weak acid—so called because not all H^+ separates from the molecules. Carbonic acid can separate and re-form in the following manner:

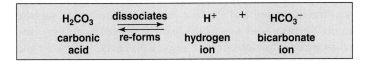

Note that the arrows shown in this chemical equation go forward and backward. This indicates that the reaction is reversible. When hydrogen ions are added to blood (let's say, because you drink a can of pop) the following reaction occurs:

$$H^+ + HCO_3^- \longrightarrow H_2CO_3$$

What happens if a person takes magnesium hydroxide (often called milk of magnesia) for an upset stomach? When hydroxide ions (OH^-) are added to blood, this reaction occurs:

$$OH^- + H_2CO_3 \longrightarrow HCO_3^- + H_2O$$

In this way, the carbonic acid–bicarbonate buffer pair can temporarily prevent significant changes in the pH of blood.

Electrolytes

As we have seen, salts, acids, and bases are molecules that dissociate; that is, they ionize in water. For example, when a salt such as sodium chloride is put in water, the Na^+ ion separates from the Cl^- ion.

Substances that release ions when put into water are called **electrolytes** because the ions can conduct an electrical current. The electrolyte balance in the blood and body tissues is important for good health because it affects the functioning of vital organs such as the brain and the heart.

Content CHECK-UP!

3. A soda pop has a pH of 4. How many free hydrogen ions are in the soda in moles per liter?

4. Depending on what you've eaten recently (as well as other factors), the pH of your urine can vary from 5 to 8.
   a. Which is more acidic—urine at pH 5 or pH 8?
   b. How many more free hydrogen ions are in a solution of pH 5, compared to one of pH 8?

Answers in Appendix A.

2.3 Molecules of Life

7. List the four classes of macromolecules in cells, and distinguish between a dehydration reaction and a hydrolysis reaction.
8. Name the individual subunits that link to form carbohydrates, lipids, proteins, and nucleic acids.

Four categories of molecules called carbohydrates, lipids, proteins, and nucleic acids are the building blocks of cells. As you can see

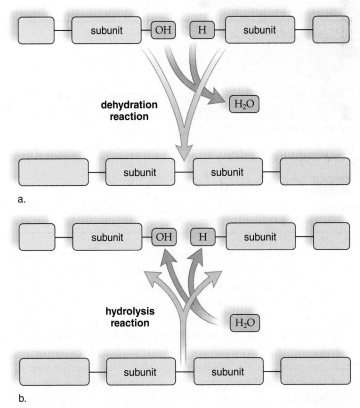

Figure 2.6 Synthesis and decomposition of macromolecules. (a) In cells, synthesis often occurs when subunits bond following a dehydration reaction (removal of H_2O). **(b)** Decomposition occurs when the subunits in a macromolecule separate after a hydrolysis reaction (addition of H_2O).

from the table below, they are called *macromolecules,* or *polymers,* because each is formed by linking together many smaller subunits, or *monomers.*

Category	Polymer	Monomer(s) (subunits)
Carbohydrates	Polysaccharide	Monosaccharide
Lipids	Fats, oils, steroids	Glycerol and fatty acids, steroid nucleus
Proteins	Polypeptide	Amino acid
Nucleic acids	DNA, RNA	Nucleotide

To link monomers into polymers, the cell uses a **dehydration reaction,** so called because an —OH (hydroxyl group) and an —H (hydrogen atom)—the equivalent of a water molecule—are removed as the molecule forms (Fig. 2.6*a*). With each dehydration reaction, another monomer can be added to create a larger polymer, just like a train can be lengthened by adding boxcars. To break up macromolecules, the cell reverses dehydration, using a **hydrolysis reaction.** In hydrolysis ("splitting with water"), the components of water are added (Fig. 2.6*b*).

Content CHECK-UP!

5. List the four major macromolecules found in the human body.

6. Based on what you know about how macromolecules are broken up, why is it important to drink liquids with a meal?

Answers in Appendix A.

2.4 Carbohydrates

9. Give some examples of different types of carbohydrates and their specific functions in cells.

Carbohydrates are organic molecules that always contain carbon (C), hydrogen (H), and oxygen (O) atoms. Carbohydrate molecules are characterized by the presence of the atomic grouping H—C—OH, in which the ratio of hydrogen atoms (H) to oxygen atoms (O) is approximately 2:1. Because this ratio is the same as the ratio in water, the name "hydrates of carbon" works well. **Carbohydrates** first and foremost function for quick, short-term cellular energy in all organisms, including humans. The role of carbohydrates and other nutrients will be discussed completely in Chapter 15.

Simple Carbohydrates

If the number of carbon atoms in a carbohydrate is low (between three and seven), it is called a simple sugar, or **monosaccharide.** The designation **pentose** means a 5-carbon sugar, and the designation **hexose** means a 6-carbon sugar. **Glucose,** the hexose our

bodies use as an immediate source of energy, can be written in any one of these ways:

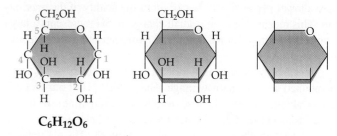

$$C_6H_{12}O_6$$

Other common hexoses are fructose, found in fruits, and galactose, a constituent of milk.

A **disaccharide** (*di,* two; *saccharide,* sugar) is made by joining only two monosaccharides together by a dehydration reaction (Fig. 2.6*a*). **Maltose** is a disaccharide that contains two glucose molecules:

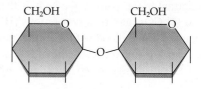

maltose $C_{12}H_{22}O_{11}$

When glucose and fructose join, the disaccharide **sucrose** forms. Sucrose—ordinary table sugar—is derived from sugar cane or sugar beets. **Lactose,** or milk sugar, is formed from glucose and galactose. (Individuals who are *lactose intolerant* cannot digest this sugar. If they consume dairy products, they experience intestinal cramping and diarrhea.)

Complex Carbohydrates (Polysaccharides)

Macromolecules such as starch, glycogen, and cellulose are **polysaccharides** that contain many glucose units. Although polysaccharides can contain other sugars, we'll study the ones that use glucose.

Starch and Glycogen

Starch and **glycogen** are ready storage forms of glucose in plants and animals, respectively. Some of the macromolecules in starch are long chains of up to 4,000 glucose units. Starch has fewer side branches, or chains of glucose that branch off from the main chain, than does glycogen, as shown in Figures 2.7 and 2.8. Wheat flour and potatoes are examples of high-starch foods.

After we eat starchy foods such as potatoes, bread, and pasta, glucose enters the bloodstream. The liver and muscles store excess glucose as glycogen. In between meals, the liver releases glucose from glycogen, so that the blood glucose concentration is always about 0.1%. Maintaining a fairly constant blood glucose concentration is important because the brain and nerve tissues work best when they use glucose as their energy source.

Cellulose

The polysaccharide **cellulose** is found in plant cell walls. In cellulose, the glucose units are joined by a slightly different type of

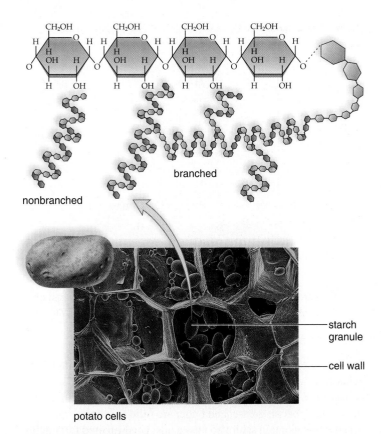

Figure 2.7 **Starch structure and function.** Starch has straight chains of glucose molecules. Some chains are also branched, as indicated. The electron micrograph shows starch granules in potato cells. Starch is the storage form of glucose in plants.

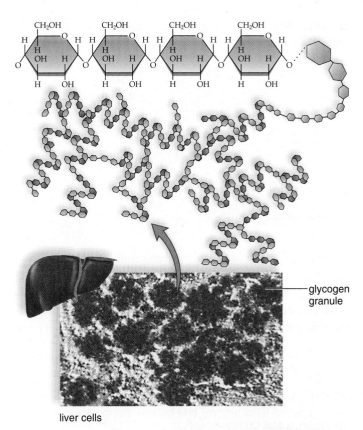

Figure 2.8 **Glycogen structure and function.** Glycogen is more branched than starch. The electron micrograph shows glycogen granules in liver cells. Glycogen is the storage form of glucose in humans.

linkage from that in starch or glycogen. Although this might seem to be a technicality, actually it is important because humans cannot digest foods containing this type of linkage. Cellulose largely passes through our digestive tract as fiber, or roughage. Dietary fiber is essential to good health, and its many functions will be discussed in Chapter 15.

Content **CHECK-UP!**

7. Categorize each carbohydrate as a monosaccharide, disaccharide, or polysaccharide:

a. glucose

b. fructose

c. sucrose

d. starch

e. glycogen

f. cellulose

8. The carbohydrate that humans eat, but cannot digest, is _____.

Answers in Appendix A.

2.5 Lipids

10. Describe the composition of a neutral fat, and give examples of how lipids function in the body.

Lipids contain more energy per gram than carbohydrates and proteins, and some function as long-term energy storage molecules in organisms. Other lipids are part of cell membranes that enclose individual cells, as well as membranes that surround the organelles found inside cells. **Steroids** are a large class of lipids that includes, among other molecules, the sex hormones.

Lipids are diverse in structure and function, but they have a common characteristic: They do not dissolve in water. Their low solubility in water is due to an absence of polar groups. They contain little oxygen and consist mostly of carbon and hydrogen atoms.

Fats and Oils

The most familiar lipids are those found in fats and oils. **Fats,** which are usually of animal origin (e.g., lard and butter), are solid at room temperature. **Oils,** which are usually of plant origin (e.g., corn oil and soybean oil), are liquid at room temperature. Fat has several functions in the body: It is used for long-term energy storage, it insulates against heat loss, and it forms a protective cushion around major organs.

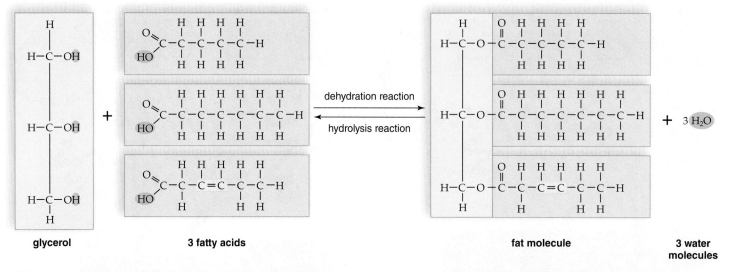

Figure 2.9 **Synthesis and degradation of a fat molecule.** Fatty acids can be saturated (no double bonds between carbon atoms) or unsaturated (have double bonds, colored yellow, between carbon atoms). When a fat molecule forms, three fatty acids combine with glycerol, and three water molecules are produced.

A fat or an oil forms when one **glycerol** molecule reacts with three fatty acid molecules (Fig. 2.9). Fats and oils are sometimes called **triglycerides** because of their three-part structure. They can also be referred to as **neutral fats** because the molecules are nonpolar and have no electrical charge.

Emulsification

Emulsifiers can cause fats to mix with water. They contain molecules with a nonpolar end and a polar end. The molecules position themselves about an oil droplet so that their nonpolar ends project. Now the droplet disperses in water, which means that **emulsification** has occurred.

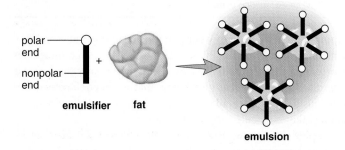

Soaps and detergents are emulsifiers that can remove fats and oils from dirty clothes or dishes. In our bodies, fats and oils from food must be emulsified before they can be digested. **Bile** is the emulsifier produced by the liver, stored in the gallbladder, and released when you eat a fat-containing meal.

Saturated and Unsaturated Fatty Acids

A **fatty acid** is a carbon–hydrogen chain that ends with the acidic group —COOH (Fig. 2.9). Most of the fatty acids in cells contain 16 or 18 carbon atoms per molecule, although smaller ones with fewer carbons are also known.

Fatty acids are either saturated or unsaturated. **Saturated fatty acids** have only single covalent bonds because the carbon chain is *saturated,* or combined with all the hydrogens it can hold. Fats such as lard, bacon grease, and butter are solid at room temperature because they contain saturated fatty acids. **Unsaturated fatty acids** have double bonds between carbon atoms wherever fewer than two hydrogens are bonded to a carbon atom. Oils such as safflower, corn, and peanut oil are liquid at room temperature because of their unsaturated fatty acids. Hydrogenation of vegetable oils can convert them to margarine and products such as Crisco®. For more about the role of fats and oils in nutrition, health, and wellness, check out Chapter 15.

Phospholipids

Phospholipids, as their name implies, contain a phosphate group (Fig. 2.10a). Essentially, they are constructed like fats, except that in place of the third fatty acid, there is a phosphate group or a grouping that contains both phosphate and nitrogen. Unlike fats and oils, phospholipid molecules are not electrically neutral because the phosphate and nitrogen-containing groups are ionized. They form the so-called hydrophilic head of the molecule, while the rest of the molecule becomes the hydrophobic tails. Because of their unique chemistry, phospholipids form the backbone of cellular membranes. When surrounded by water (as body cells are), their hydrophilic phosphate heads face outward while the hydrophobic fatty acid tails are sandwiched in between. Thus, phospholipids spontaneously create a double layer, called a bilayer (Fig. 2.10b).

Steroids

Steroids are lipids that have an entirely different structure from that of fats. Steroid molecules have a backbone of four fused carbon rings, called the steroid nucleus. Each one differs primarily by the side-chain molecules, called **functional groups,** attached to the rings.

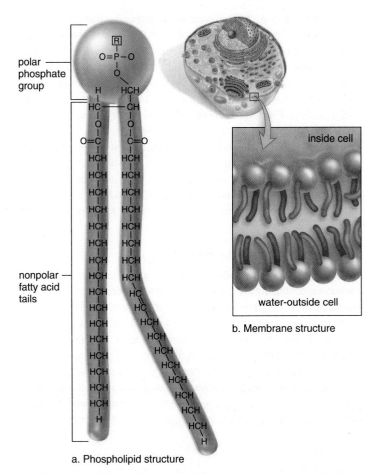

a. Phospholipid structure

b. Membrane structure

inside cell

water-outside cell

polar phosphate group

nonpolar fatty acid tails

Figure 2.10 **Phospholipid structure and function. (a)** Phospholipids are structured like fats, but one fatty acid is replaced by a polar phosphate group. Therefore, the head is polar, while the tails are nonpolar. **(b)** This causes the molecule to arrange itself as shown when exposed to the water-based environment outside the cell.

Cholesterol is a component of a human or an animal cell's outer membrane and is the precursor of several other steroids, such as the male sex hormone testosterone, and a group of female sex hormones that are collectively termed estrogens. Testosterone is formed primarily in the testes, a male's sex organs. Estrogens, such as the estradiol molecule pictured in Fig. 2.11, are formed primarily in a woman's ovaries. Testosterone and estradiol differ only by the functional groups attached to the same carbon backbone, yet they have a profound effect on the body and on our sexuality (Fig. 2.11a, b). Further, testosterone is a steroid that causes males to have greater muscle strength than females. However, taking synthetic testosterone for this purpose is dangerous to your health, as will be discussed in Chapter 10.

We know that a diet high in saturated fats and cholesterol can cause fatty material to accumulate inside the lining of blood vessels, thereby reducing blood flow. As discussed in the Medical Focus on page 364, nutrition labels are now required to list the calories from fat per serving and the percent daily value from saturated fat and cholesterol.

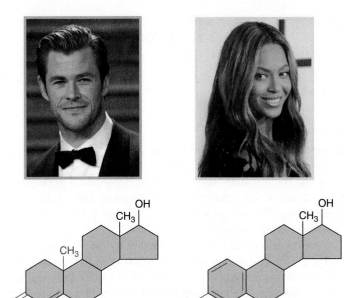

a. Testosterone

b. Estradiol (a form of estrogen)

Figure 2.11 **Steroids.** All steroids have four rings, but they differ by attached groups. Seemingly small chemical differences between **(a)** testosterone and **(b)** estradiol help to produce two distinct genders.

Content **CHECK-UP!**

9. Which fatty acid, saturated or unsaturated, contains more hydrogen atoms? Which has double bonds between some of the carbons?

10. Are phospholipids water- or oil-soluble? Explain.

11. Which lipid acts as a precursor to steroid hormones such as testosterone?

Answers in Appendix A.

2.6 Proteins

11. State the major functions of proteins, and tell how globular proteins are organized.

Proteins perform a wide range of functions, including the following:

- Collagen (found in connective tissue and skin) and keratin (found in skin, hair, and nails) are fibrous structural proteins.
- Many hormones, which are messengers that influence cellular metabolism, are proteins.
- The proteins actin and myosin account for the movement of cells and the ability of our muscles to contract.
- Some proteins transport molecules in the blood; for example, hemoglobin is a complex protein in our blood that transports oxygen and carbon dioxide.
- Antibodies in blood and other body fluids are proteins that combine with pathogens or their toxins.

- Enzymes are globular (round) proteins that *catalyze*, or speed up chemical reactions, allowing them to happen at body temperature.

As you know from the chapter opener, proteins can have medicinal applications as well.

Structure of Proteins

Proteins are macromolecules composed of amino acid subunits. An **amino acid** has a central carbon atom bonded to a hydrogen atom and three groups. The name of the molecule is appropriate because one of these groups is an amino group and another is an acidic group. The third group is called an *R* group because it is the *remainder* of the molecule (Fig. 2.12*a*).

Amino acids differ from one another by the *R* group, which varies in structure from a single hydrogen atom to a complicated ring. When two amino acids join, a **dipeptide** results. **Polypeptides** contain three or more amino acids (Fig. 2.12*b*). Proteins are large polypeptides. All human polypeptides are created from just 20 different amino acids. The sequence of amino acids in a polypeptide is called its **primary structure** (Fig. 2.12*b*).

The bond between amino acids is termed a **peptide bond.** The atoms of a peptide bond share electrons unevenly, making hydrogen bonding possible in a polypeptide. Thus, polypeptide molecules are polar. Due to hydrogen bonding, the polypeptide often twists to form a coil or folds into a fan shape. Coiling or folding creates the **secondary structure** of the protein (Fig. 2.12*c*) Finally, the coil bends and twists into a particular shape because of bonding between *R* groups. Hydrogen, ionic, and covalent bonding all occur in polypeptides, creating the **tertiary structure** (Fig. 2.12*d*). Hydrophobic portions of a polypeptide are found inside the molecule, while the hydrophilic portions are outside where they can contact water.

Some proteins have just one polypeptide, while others have more than one polypeptide, each with its own primary, secondary, and tertiary structures. If a protein has more than one polypeptide, the arrangement of individual polypeptides gives a protein its **quaternary structure** (Fig. 2.12*e*). The oxygen-carrying protein found in red blood cells, called *hemoglobin*, is an example of a protein with a quaternary structure. Hemoglobin molecules contain four complete protein subunits.

The final three-dimensional shape of a protein is very important to its function. When proteins are exposed to extremes in heat and pH, they undergo an irreversible change in shape called **denaturation.** For example, addition of acid to milk (as when milk goes "sour") causes curdling. Heating causes egg white protein, called *albumin,* to coagulate (form clumps). Denaturation occurs because the normal bonding between the *R* groups has been disturbed. Once a protein loses its normal shape, it is no longer able to function normally. Researchers hypothesize that an alteration in protein organization may be the cause of Alzheimer disease (see the Medical Focus in Chapter 8) and variant Creutzfeldt-Jacob disease (see the Medical Focus on page 34).

Enzymatic Reactions

Metabolism is the sum of all of the chemical reactions that occur in a cell. Most cellular reactions cannot happen unless an **enzyme** is present. Enzymes are protein *catalysts*—molecules that enable a particular metabolic reaction to occur at the body's normal temperature. Atoms and molecules often do not react unless they are first activated in some way. *Energy of activation* is the energy needed to start a reaction. In the lab, heat is often used to supply the energy needed for a reaction. For example, sugar can be broken down to form carbon dioxide and water, but doing so in a laboratory requires tremendous heat. In the body, enzymes lower the energy of activation by forming a complex with particular molecules. Just as a mutual friend can cause particular people to interact at a crowded party, enzymes in cells bring together atoms or molecules and cause reactions. Our bodies use enzymes to convert sugar into CO_2 and water (just like in a lab setting), but the reaction can occur at body temperature.

Enzyme-Substrate Complex

In any reaction, the molecules that interact are called *reactants,* while the substances that form as a result of the reaction are the *products.* The reactants in an enzymatic reaction are its **substrate(s).** Enzymes are often named for their substrate(s); for example, maltase is the enzyme that digests maltose. Enzymes are highly specific for their particular reactions. This specificity is due to the shape and chemical composition of the enzyme's *active site*, the region where the reaction occurs. There, the enzyme and its substrate(s) fit together, much like pieces of a jigsaw puzzle (Fig. 2.13). After a reaction is complete and the products are released, the enzyme is ready to catalyze its reaction again:

$$E + S \longrightarrow ES \longrightarrow E + P$$

(where E = enzyme, S = substrate, ES = enzyme-substrate complex, and P = product). The enzyme is neither changed nor used up during the chemical reaction.

Many enzymes require **cofactors.** Cofactors assist an enzyme and may even accept or contribute atoms to the reaction. Some cofactors are inorganic metals, such as copper, zinc, and iron. Other cofactors are organic, nonprotein molecules called coenzymes. Vitamins are often components of coenzymes.

Types of Reactions

Certain types of chemical reactions are common to metabolism.

Synthesis Reactions During **synthesis reactions,** two or more reactants combine to form a larger and more complex product (Fig. 2.13*a*). The dehydration synthesis reaction we have already studied (i.e., the joining of subunits to form a macromolecule) is an example of a synthesis reaction (see Fig. 2.8). When glucose molecules join in the liver, forming glycogen, dehydration synthesis has occurred. Notice that synthesis reactions always involve bond formation and require energy.

Degradation (Decomposition) Reactions During **degradation reactions** (also called decomposition reactions), a larger and more complex molecule breaks down into smaller, simpler products (Fig. 2.13*b*). The hydrolysis reactions that break down macromolecules into their subunits are degradation reactions. For example, digestion of a protein to amino acids is a degradation reaction that occurs in the stomach and small intestine.

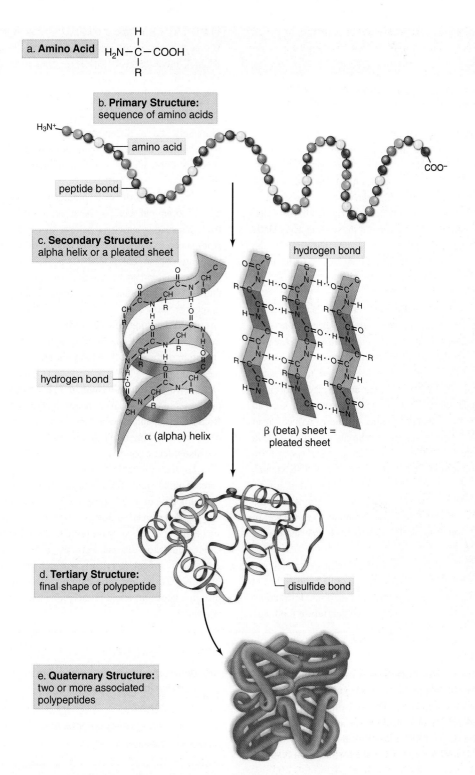

a. Amino Acid

$$H_2N-C-COOH$$

H
|
C
|
R

b. Primary Structure:
sequence of amino acids

H_3N^+

amino acid

peptide bond

COO^-

c. Secondary Structure:
alpha helix or a pleated sheet

hydrogen bond

hydrogen bond

α (alpha) helix

β (beta) sheet =
pleated sheet

d. Tertiary Structure:
final shape of polypeptide

disulfide bond

e. Quaternary Structure:
two or more associated
polypeptides

AP|R **Figure 2.12** **Levels of polypeptide structure.** (**a**) Amino acids are the subunits of polypeptides. Note that an amino acid contains nitrogen. (**b**) Polypeptides differ by the sequence of their amino acids, which are joined by peptide bonds. (**c**) A polypeptide often twists to form a coil or folds into a sheet due to hydrogen bonding between amino acids in the primary sequence. (**d**) The tertiary structure of polypeptide structure is due to various types of bonding between the *R* groups of the amino acids. (**e**) The grouping of two or more polypeptides creates the quaternary structure of a large protein molecule.

Prions: Malicious Proteins?

Infectious diseases are known to be caused by minute organisms that successfully dodge the body's defense mechanisms. These invaders include parasites such as liver flukes and parasitic worms; smaller, one-celled parasites like disease-causing strains of *Amoeba;* and the smallest unicellular organisms, the bacteria. Viruses are disease-causing particles that are even smaller than bacteria. All of these disease agents share a common trait: All contain the nucleic acids DNA or RNA, or both. Nucleic acids are large molecules that comprise the genetic material of parasites, bacteria, and viruses. At one time, scientists believed that infectious organisms had to contain genetic material.

However, disease outbreaks in England, Europe, Canada, and the United States have demonstrated that even smaller agents, consisting only of a protein molecule, are capable of causing disease. These proteinaceous infectious particles, called **prions** (pronounced pree-ahns), are responsible for a set of diseases termed transmissible spongiform encephalopathies (TSEs). There are numerous forms of TSEs: Creutzfeld-Jakob (croits-feld yay-kob) disease and kuru in humans, bovine spongiform encephalopathy (formerly termed "mad cow disease") in cattle, scrapie in sheep, and chronic wasting disease in wild deer and elk, among others. (In addition, some spongiform encephalopathies are inherited.) All human and animal TSE victims show similar findings at autopsy: holes pockmark the brain, giving the tissue a spongy, Swiss-cheese appearance. Currently, TSEs affect approximately 1:1,000,000 in a given population each year. In the U.S., this currently translates to 250 to 300 newly infected people per year.

Scientists theorize that prions cause normal brain cell proteins to change into prion proteins. Accumulating prions kill brain cells and spread throughout brain tissue. Symptoms indicate brain destruction: "Mad" cows (deer and elk as well) wobble, stagger, refuse food, and ultimately starve. Sheep with scrapie will frantically rub wool off their bodies. Infected humans lose memory, muscle control, and finally stop breathing. At this time, there is no treatment—all TSEs are fatal.

The addition of waste brain/spinal cord tissue from slaughterhouses to animal feed has been outlawed in all TSE-affected countries, because researchers believe that prions were initially spread from sheep to cattle, and later among infected cattle, when the animals ate contaminated feed. (It is interesting to note that kuru in humans was spread by cannibalism among the Fore' tribe of New Guinea, who honored dead relatives by eating their brains!) Further, hunters are now being warned to protect against chronic wasting disease by discarding meat from deer or elk that appear ill, to strictly avoid contact with any brain or spinal cord tissue (especially consumption of the tissue!) and to wear gloves if a carcass is butchered in the field.

However, human-to-human TSE transmission can occur if certain types of brain-derived matter, including brain coverings, corneas from the eyes, and certain hormones are transferred from an infected donor into a recipient. Prion-contaminated neurosurgery instruments can also infect a patient.

To date, probable diagnosis of TSEs has relied on analysis of cerebrospinal fluid (fluid obtained from a spinal tap) and electroencephalogram, or EEG. Absolute confirmation of the disease could only be determined at autopsy. Recently developed blood tests have successfully detected prion proteins in individuals with one form of Creutzfeld-Jakob disease. Perhaps early discovery could lead to an effective treatment. Very early research has shown success in treating prion disease in test tube cultures of nerve cells. However, many puzzling questions about prions remain, and much additional research is needed.

Replacement (Exchange) Reactions Replacement **reactions** involve both degradation (decomposition) and synthesis. Reactions between acids and bases are replacement reactions. For example, when hydrochloric acid (HCl) is reacted with sodium hydroxide (NaOH), the chloride ion (Cl^-) trades places with the hydroxide ion (OH^-). Table salt (NaCl) and water (H_2O) are formed as a result.

Content **CHECK-UP!**

12. The sequence of amino acids found in a protein is that protein's _____ structure.

   a. primary   c. tertiary

   b. secondary   d. quaternary

13. What happens to a protein when it is denatured? How would this affect biologically active proteins?

14. What type of reaction is the digestion of glucose to form carbon dioxide and water?

   a. synthesis   c. replacement

   b. degradation/decomposition

Answers in Appendix A.

2.7 Nucleic Acids

12. Describe the structure and function of DNA and RNA in cells.

13. Explain the importance of ATP in the body.

Nucleic acids are huge macromolecules composed of nucleotides. Every **nucleotide** is a molecular complex of three types of subunit

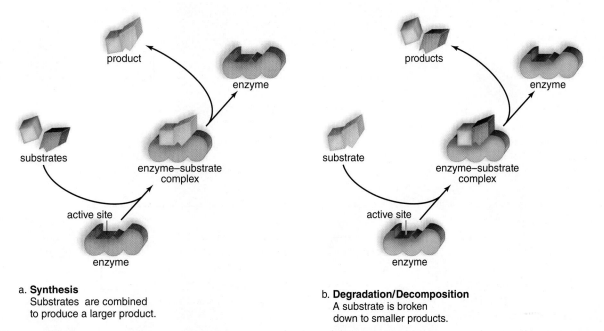

a. **Synthesis**
Substrates are combined
to produce a larger product.

b. **Degradation/Decomposition**
A substrate is broken
down to smaller products.

AP|R **Figure 2.13** **Enzymatic action.** An enzyme has an active site where the substrates come together and react. The products are released, and the enzyme is free to act again. (**a**) In synthesis, the substrates join to produce a larger product. (**b**) In degradation/decomposition, the substrate breaks down to smaller products.

molecules—a phosphate (phosphoric acid), a pentose sugar, and a nitrogen-containing base:

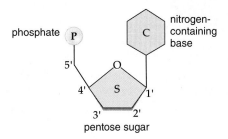

Nucleic acids contain hereditary information that determines which proteins a cell will have. Two classes of nucleic acids are in cells: **DNA (deoxyribonucleic acid)** and **RNA (ribonucleic acid).** DNA makes up the hereditary units called **genes.** Genes pass on from generation to generation the instructions for replicating DNA, making RNA, and joining amino acids to form the proteins of a cell. RNA is an intermediary in the process of protein synthesis, conveying information from DNA regarding the amino acid sequence in proteins. As an analogy, think of DNA as a "blueprint"—a set of instructions for assembling an entire building. Using this example, RNA would be the directions, made from the blueprint, describing how to put beams together to make a wall.

The nucleotides in DNA contain the 5-carbon sugar deoxyribose; the nucleotides in RNA contain the sugar ribose. This difference accounts for their respective names. As indicated in Figure 2.14, there are four different types of bases in DNA: A = adenine, T = thymine, G = guanine, and C = cytosine. The base can

have two rings (like adenine or guanine) or one ring (like thymine or cytosine). In RNA, the base uracil replaces the base thymine.

The bases in DNA and RNA are nitrogen-containing bases—that is, a nitrogen atom is a part of the ring. Like other bases, the presence of the nitrogen-containing base in DNA and RNA raises the pH of a solution.

The nucleotides in DNA and RNA form a linear molecule called a *strand.* A strand has a backbone made up of phosphate-sugar-phosphate-sugar, with the bases projecting to one side of the backbone. Because the nucleotides occur in a definite order, so do the bases. Any particular DNA or RNA has a definite sequence of bases, although the sequence can vary between molecules. RNA is usually single-stranded, while DNA is usually double-stranded, with the two strands twisted about each other in the form of a double helix (like a spiral staircase). The molecular differences between DNA and RNA are listed in Table 2.1.

TABLE 2.1	DNA Structure Compared to RNA Structure	
	DNA	**RNA**
Sugar	Deoxyribose	Ribose
Bases	Adenine, guanine, thymine, cytosine	Adenine, guanine, uracil, cytosine
Strands	Double-stranded	Single-stranded
Helix	Yes	No

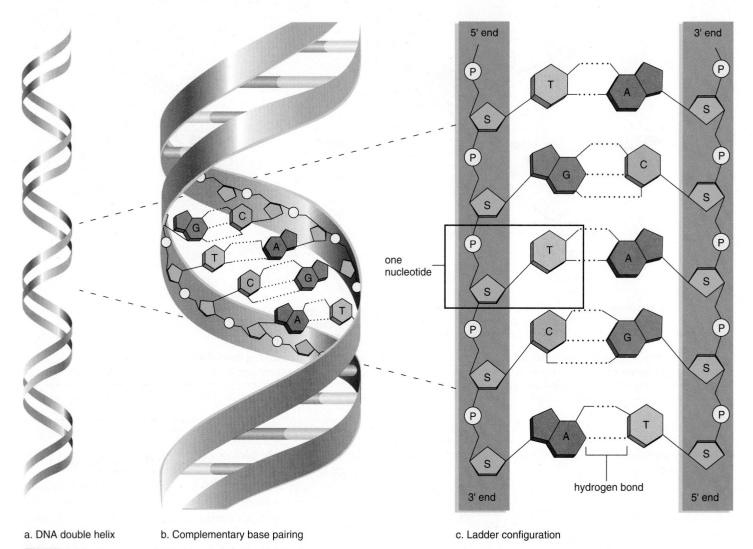

a. DNA double helix   b. Complementary base pairing   c. Ladder configuration

AP|R **Figure 2.14** **Overview of DNA structure.** (**a**) Double helix. (**b**) Complementary base pairing between strands. (**c**) Ladder configuration. Notice that the uprights are composed of phosphate and sugar molecules and that the rungs are complementary paired bases.

In DNA, the two strands are held together by hydrogen bonds between the bases (Fig. 2.14). When unwound, DNA resembles a stepladder. The sides of the ladder are made entirely of phosphate and sugar molecules, and the rungs of the ladder are made only of complementary paired bases. Thymine (T) always pairs with adenine (A), and guanine (G) always pairs with cytosine (C) (Fig. 2.14). This is called **complementary base pairing.**

Complementary bases pair because they have shapes that fit together. We'll see that complementary base pairing allows DNA to replicate in a way that ensures the sequence of bases will remain the same. When RNA is produced, complementary base pairing occurs between DNA and RNA, but uracil takes the place of thymine. Then, the sequence of the bases in RNA determines the sequence of amino acids in a protein because every three bases code for a particular amino acid (see Chapter 3, pp. 56–58). The code is nearly universal and is the same in other organisms as it is in humans.

ATP (Adenosine Triphosphate)

Individual nucleotides can have metabolic functions in cells. Some nucleotides are important in energy transfer. When adenosine (adenine plus ribose) is modified by the addition of three phosphate groups, it becomes **ATP (adenosine triphosphate),** the primary energy carrier in cells.

Cells require a constant supply of ATP. To obtain it, they continually break down glucose (and other food molecules as well) and convert the energy that is released into ATP molecules. To give an analogy, the energy in glucose is like a $100 bill, and the energy in ATP is like a $10 bill. The larger bill (glucose energy), when broken down, will yield ten smaller bills (ATP energy). Cells "spend" ATP energy when cellular reactions require energy. Therefore, ATP is called the *energy currency* of cells.

However, converting food energy into ATP energy is not 100% complete. To continue the comparison, imagine that a store

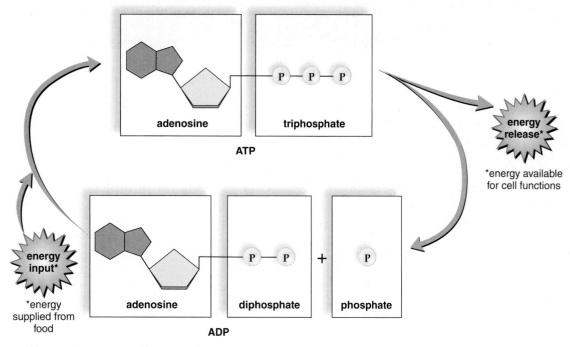

Figure 2.15 **Breakdown and formation of ATP.**

won't accept your $100 bill because it's too large. You go to a bank and exchange your $100 bill for $10 bills. Naturally, you receive ten $10 bills from the bank—but you lose six $10 bills on the way home.

A similar situation occurs for body cells. Energy in foods must be released, but during the conversion about 60% of food energy is not converted to ATP. In the body, this energy is given off as heat. Thus, energy released from the breakdown of food fuels cellular reactions and warms the body at the same time.

Cells use ATP energy when macromolecules such as carbohydrates and proteins are synthesized. In muscle cells, ATP is used for muscle contraction. In nerve cells, it is used for the conduction of nerve signals. Likewise, all other body cells use ATP energy to fuel their various functions.

ATP is a high-energy molecule because the last two phosphate bonds are unstable and easily broken. In cells, the terminal phosphate bond is usually hydrolyzed, releasing energy for cell reactions. The terminal bond is sometimes called a high-energy bond, symbolized by a wavy line. But this terminology is misleading—the breakdown of ATP *releases* energy because the products of hydrolysis are *more* stable than ATP. Breakdown of ATP leaves one molecule of ADP (adenosine diphosphate) and a molecule of inorganic phosphate, Ⓟ (Fig. 2.15).

After ATP breaks down and the energy is used for a cellular purpose, ATP is rebuilt by the addition of Ⓟ to ADP again (Fig. 2.15). There is enough energy in one glucose molecule to build roughly 36 to 40 ATP molecules in this way. Homeostasis is only possible because cells continually produce and use ATP molecules. The use of ATP as the energy currency of cells also occurs in other organisms, ranging from bacteria to humans.

Content **CHECK-UP!**

15. For the following terms, indicate whether it is characteristic of RNA, DNA, or both:

   a. uracil

   b. single-stranded

   c. cytosine

   d. thymine

   e. double-stranded

   f. ribose

16. Suppose someone presented data from DNA analysis of a newly discovered species showing that the relative amounts of the DNA bases were 30% guanine, 20% thymine, 30% adenine, and 20% cytosine. Based on what you know about DNA structure, why doesn't this data make sense?

17. The complimentary base sequence for TTAGC is _____.

18. The breakdown of ATP yields _____, a phosphate group, and energy.

Answers in Appendix A.

Begin Thinking Clinically

The disease sickle-cell anemia results from improper substitution of a single DNA base (in this case, in part of the DNA sequence that ultimately results in hemoglobin, blood's oxygen-carrying protein). If DNA's structure is defective, what are the other two types of molecules that will be affected?

Answer and discussion in Appendix A.

On August 6, 1945, and then again on August 9, 1945, the world saw the effects of massive doses of ionizing radiation as two atomic bombs were dropped: first on Hiroshima, Japan, and subsequently on Nagasaki. Fast-forward to March 2011: a massive earthquake and subsequent tsunami in Japan damaged nuclear reactors in Fukushima, Japan, causing release of radioactive steam into the air. Three people were killed in this accident, and radioactive fallout will likely contaminate soil and water for centuries. Small-scale radiation accidents have also repeatedly happened in the decades following the discovery of radiation in the early twentieth century. For example, there are many people who have been exposed to high-level radiation by accidental contact with radioactive waste.

What is ionizing radiation? As the name suggests, it's an extremely high-powered form of energy that causes ions—charged subatomic particles—to form when unstable radioactive chemical elements fall apart. There are three forms of ionizing radiation: alpha particles, beta particles, and gamma rays. Alpha particles consist of two protons and two neutrons. When a radioactive element spits out an alpha particle, it actually changes into a new element. For example, if radium (atomic number 88) releases its alpha particle, it changes into radon gas (atomic number 86) because it loses two protons. Beta particles are formed when an excess neutron converts to a proton (which remains in the element) and an electron (which is given off as a very high-speed particle). As with an alpha-particle emitter, a radioactive element that gives off a beta particle becomes a new element—this time, by adding a proton. For example, radioactive strontium (atomic number 38) becomes yttrium (atomic number 39) by releasing a beta particle. Gamma rays are pure energy.

Throughout our lives, we're all exposed to many forms of low-level ionizing radiation: from cosmic and solar energy, radioactive material in soil and water, and radiation used for medical procedures such as dental X rays and mammograms. Absorbed radiation is measured in units called grays, abbreviated Gy. The typical person's average annual dosage of radiation from all sources is 0.0062 Gy. As you know, there are many medical and commercial uses for radioactive elements. For example, radioactive material selectively destroys cancer tissue during radiation therapy, and gamma rays are used to sterilize delicate surgical equipment.

However, high levels of environmental ionizing radiation can be deadly. By reacting with molecules in the body, each damages cells and tissues. Acute radiation poisoning occurs when a person's entire body is exposed to 1 Gy or more, from radioactive alpha or beta particles touching the skin, consumed in contaminated food or drink, or inhaled from the air. Because gamma rays aren't particles, they directly penetrate the body. Recent research has shown that radiation breaks down cell-to-cell connections in the small intestine, allowing bacteria to enter the bloodstream. The first signs of acute radiation poisoning are nausea, vomiting, and diarrhea, and the faster these symptoms appear, the greater the dosage. Bone marrow that produces blood cells is destroyed, and the person becomes anemic (due to red blood cell death) and at risk for infection (due to white blood cell destruction). In the days or weeks following exposure, any tissue with a rapid growth rate—skin and digestive tract lining, for example—may be destroyed. A dose of 6 Gy or greater is almost always fatal. Over time, radiation exposure predisposes a person to many forms of cancer.

Acute radiation poisoning caused by alpha or beta particles is treated by having the person consume very high doses of stable forms of the radioactive element. For example, exposure to radioactive iodine (following a nuclear reactor incident) is treated by taking pills containing stable iodine compounds. However, once radiation poisoning injury has occurred, the patient must receive intensive supportive treatment if he or she is to live. Long-term therapy using intravenous fluids, blood transfusions, antibiotics to fight infections, drugs to stimulate bone marrow growth, and pain medication will all be essential to keep the person alive until damaged organs or tissues can regrow.

There are practical ways to prevent radiation poisoning in our day-to-day lives. In homes, screen for radon gas, which can be released from soil surrounding a basement. If you're preparing to receive an X ray, CT scan, or other medical procedure, ask to have other body areas shielded from ionizing radiation. Avoid contact with radioactive waste. Contact the authorities if you discover radioactive material (for example, discarded medical or industrial waste). Most important, seek immediate medical attention if you're accidentally exposed.

Basic Key Terms

acid (ăs'id), p. 25

adhesion (ăd-hē'zhŭn), p. 25

amino acid (ŭh-mē'nō ăs'īd), p. 32

anion (ăn' ī-ŭhn), p. 22

atom (ă' tŭhm), p. 21

atomic number (ă-tăw'mĭk nŭhm'běr), p. 21

ATP (adenosine triphosphate) (ă-děn'ō-sēn trī-fŏs'fāt), p. 36

base (bās), p. 25

bile (bīl), p. 30

buffer (bŭf'ěr), p. 27

carbohydrate (kăr"bō-hī'drāt), p. 28

cation (cătă ī-ŭhn), p. 22

cellulose (sěl'yŭh-lōs"), p. 28

cholesterol (kō-lěs'tŭh-răwl), p. 31

cofactors (kō'făk-tŏrz), p. 32

cohesion (kō-hē'zhŭn), p. 25

complementary base pairing (kŏm'plŭh-měn'tŭh-rē bās pār-ĭng), p. 36

compound (kŏm-pŏwnd'), p. 22

covalent bond (kō-vā'lěnt bond), p. 22

degradation decomposition reactions (děg'rŭh-dā'shŭhn dē-kŏm"pŭh-zĭsh'ŭhn rē-ăk' shŭhnz), p. 32

dehydration reaction (dē'hī-drā'shŭhn rē-ăk'shŭhnz), p. 28

denaturation (dē-nă'chŭr-ā'shŭn), p. 32

dipeptide (dī-pěp'tīd), p. 32

disaccharide (dī-săk'ŭh-rīd), p. 28

DNA (deoxyribonucleic acid) (dē-ŏks'ē-rī"-bō-nū-klā"ĭk ăs'īd), p. 35

electrolyte (ē-lěk'trō-līt), p. 27

electronegative (ē-lěk'trō-něg'ŭh-tĭv), p. 24

electrons (ē-lek'tronz), p. 21

elements (ěl'ŭh-měntz), p. 21

emulsification (ē-mŭl'sŭh-fŭh-cā'shŭhn), p. 30

emulsifier (ē-mŭl' sŭh-fī-ŭhr), p. 30

enzyme (ěn'zīm), p. 32

fats (fătz), p. 29

fatty acid (făt'ē ăs'īd), p. 30

functional groups (fŭngk'shŭh-nŭhl grūpz), p. 30

gene (jēn), p. 35

glucose (glū'kōs), p. 28

glycerol (glĭs'ěr-ŏl), p. 30

glycogen (glī'kō-jěn), p. 28

heat of vaporization (hēt ŭv vā"pŭh-rī-zā'shŭhn), p. 25

hexose (hěk'sōs), p. 28

hydrogen bond (hī'drō-jěn bŏnd), p. 24

hydrolysis reaction (hī-drŏl'ĭ-sĭs rē-ăk'shŭn), p. 28

hydrophilic (hī'drō-fĭl'ĭk), p. 25

hydrophobic (hī'drō-fō'bĭk), p. 25

inorganic molecule (ĭn-ŏr-găn'ĭk mŏl'ě kyūl), p. 24

ion (ī'ŏn), p. 21

ionic bond (ī-ŏn'ĭk bŏnd), p. 22

ionic lattice (ī-ŏn'-ĭk lă'-tĭs), p. 22

isotope (ī'sō-tōp), p. 22

lactose (lăk'tōs), p. 28

lipid (lĭp'ĭd), p. 29

maltose (măwl'tōs), p. 28

mass number (măs nŭhm'bŭr), p. 21

metabolism (mŭh-tăb'ō-lĭzm), p. 32

mole (mōl), p. 22

molecule (mŏl'ě kyūl), p. 22

monosaccharide (mŏn"ō-săk'ŭh-rīd), p. 28

neutral fats (nū'trŭhl făts), p. 30

neutrons (nū'trŏnz), p. 21

nonpolar covalent bond (nŏn-pō'lěr kō'vā'lěnt bŏnd), p. 24

nucleic acid (nū-klā'ĭk ăs'īd), p. 34

nucleotide (nū'klē-ŭh-tīd), p. 34

oils (ŏylz), p. 29

organic molecule (ŏr-găn'ĭk mŏl'ě-kyūl), p. 24

pentose (pěn'tōs), p. 28

peptide bond (pěp'tīd bŏnd), p. 32

pH scale (pē āch skāl), p. 25

phospholipids (fŏs'fō-lĭp'ĭdz), p. 30

polar covalent bond (pō'lěr kō-vā'lěnt bŏnd), p. 24

polar molecule (pō'lěr mŏl'ě-kyūl), p. 24

polypeptide (pŏl"ē-pěp'tīd), p. 32

polysaccharide (pol"ē-săk'ŭh-rīd), p. 28

primary structure (prī'mār-ē strŭk'chŭr), p. 32

protein (prō'tēn), p. 32

protons (prō'tŏnz), p. 21

quaternary structure (kwă'těr-năr-ē strŭk'chŭr), p. 32

radioactive isotope (rā"dē-ō-ăk'tĭv ī'sō-tōp), p. 22

replacement reactions (rē-plās'měnt rē-ăk'shŭhnz), p. 34

RNA (ribonucleic acid) (rī"bō-nū-klā'ĭk ăs'īd), p. 35

salt (săwlt), p. 22

saturated fatty acids (sătch'ŭ-rāt-ěd făt'ē ăs'īdz), p. 30

secondary structure (sěk'ŏn-dār-ē strŭk'chŭr), p. 32

solvent (sŏl'věnt), p. 24

specific heat capacity (spĭ-sĭf'ĭk hēt kŭh-păs'ĭ-tē), p. 25

starch (stărch), p. 28

steroids (stěr'ŏydz), p. 29

substrate (sŭb'strāt), p. 32

sucrose (sū'krōs), p. 28

synthesis reactions (sĭn'thĭ-sĭs rē-ăk'shŭhnz), p. 32

tertiary structure (těr'shē-ā-rē strŭk'chěr), p. 32

triglycerides (trī-glĭs'ŭh-rīdz), p. 30

unsaturated fatty acids (ŭn-sătch'ŭ-rāt-ěd făt'ē ăs'īdz), p. 30

Clinical Key Terms

acidosis (ăs"ĭ-dō'sĭs), p. 27

alkalosis (ăl"kŭh-lō'sĭs), p. 27

arrhythmia (ā-rĭth'mē-ŭh), p. 22

hypertension (hī"pěr-těn'shŭn), p. 22

prion (prē'ŏn), p. 34

rickets (rĭk'ěts), p. 22

tracer (trā'sěr), p. 22

Summary

2.1 Basic Chemistry
  A. All matter is composed of elements, each made up of just one type of atom. An atom has an atomic symbol, atomic number (number of protons and, therefore, electrons when neutral), and mass number (number of protons and neutrons). The isotopes of some atoms are radioactive and have biological and medical applications.
  B. Isotopes of the same element have a different number of neutrons. Radioisotopes are used in medicine in a variety of ways, from imaging the body to killing cancer cells.
  c. Atoms react with one another to form molecules. Following an ionic reaction, charged ions are

attracted to one another. Following a covalent reaction, atoms share electrons. Polar covalent bonding results from unequal sharing of electrons. Nonpolar covalent bonds occur when electrons are shared equally.

2.2 Water, Acids, and Bases
  A. In water, the electrons are shared unequally, and the result is a polar molecule. Hydrogen bonding can occur between polar molecules.
  B. Water is a polar molecule and acts as a solvent; it dissolves various chemical substances and facilitates chemical reactions. Because of hydrogen bonding, water molecules are cohesive and adhesive. Water heats up and cools down slowly. This helps keep body temperature within normal limits.
  C. Substances such as salts, acids, and bases that dissociate in water are called electrolytes. The electrolyte balance in the blood and body tissues is important for good health. Acids have a pH less than 7, and bases have a pH greater than 7. The actions of the lungs and kidneys, and the presence of buffers, help to keep the pH of body fluids around pH 7.

2.3 Molecules of Life
  Carbohydrates, lipids, proteins, and nucleic acids are the molecules of life. A monosaccharide, such as glucose, is a subunit (monomer) for larger carbohydrates. Glycerol and fatty acids are monomers for fat. Amino acids are monomers for proteins, and nucleotides are monomers for nucleic acids.

2.4 Carbohydrates
  A. Monosaccharides are carbohydrate monomers. The monosaccharide glucose is an immediate source of energy in cells. Disaccharides are made by joining two monosaccharides together.
  B. Glycogen, starch and cellulose are complex carbohydrates, made of many monosaccharides joined together. Glycogen stores energy in the body, starch is a dietary source of energy, and cellulose is a fiber in the diet.

2.5 Lipids
  A. Fats and oils are lipids that function in long-term energy storage. They consist of glycerol and three fatty acids, which may be saturated or unsaturated.
  B. Phospholipids have hydrophilic heads and hydrophobic tails.
  C. Cholesterol is a type of steroid that has important functions in the body, including serving as a precursor to steroid hormones like testosterone.

2.6 Proteins
  A. Proteins, which are composed of one or more long polypeptides, have both structural and physiological functions. Polypeptides have several levels of structure. Their three-dimensional shape is necessary to their function.
  B. Enzymes are proteins necessary to metabolism. The reaction occurs at the active site of an enzyme.

2.7 Nucleic Acids
  Both DNA and RNA are polymers of nucleotides; DNA is usually double-stranded. DNA makes up the genes, and along with RNA, specifies protein synthesis. ATP is the energy "currency" of cells because its breakdown supplies energy for many cellular processes.

Study Questions

1. Name the subatomic particles of an atom. Using the carbon atom as an example, describe their charge, atomic mass unit, and location. (p. 22)

2. What is an isotope? A radioactive isotope? Discuss the clinical uses of radioactive isotopes. (p. 23)

3. Give an example of an ionic reaction, and explain it. (p. 23)

4. Give an example of a covalent reaction, and explain it. (pp. 23–25)

5. Relate three characteristics of water to its polarity and hydrogen bonding between water molecules. (pp. 25–26)

6. What is an acid? A base? (p. 26)

7. On the pH scale, which numbers indicate a basic solution? An acidic solution? Why? (pp. 26–28)

8. What are buffers, and how do they function? (p. 28)

9. Name the four categories of macromolecules (polymers) in cells; give an example for each category, and name the subunits (monomers) of each. (pp. 28–29)

10. Tell how macromolecules are built up and broken down. (p. 28)

11. Name some monosaccharides, disaccharides, and polysaccharides, and give the functions for each. (pp. 29–30)

12. What is a lipid? A saturated fatty acid? An unsaturated fatty acid? What is the function of fats? (pp. 30–31)

13. Relate the structure of a phospholipid to that of a neutral fat. What is the function of a phospholipid? (p. 31)

14. Name two steroids that function as sex hormones in humans. (p. 32)

15. What are some functions of proteins? Why do proteins stop functioning if exposed to the wrong pH or high temperature? (pp. 32–33)

16. Discuss the levels of protein structure. (pp. 33–34)

17. How do enzymes function? Name three types of metabolic reactions. (pp. 33, 35)

18. Discuss the structure and function of the nucleic acids DNA and RNA. (pp. 36–37)

19. What is ATP? How does the cell use ATP as the energy source for cell reactions? (pp. 37–39)

Learning Outcome Questions

Fill in the blanks.

1. _____ are the smallest units of matter nondivisible by chemical means.
2. Isotopes differ by the number of _____ in the nucleus.
3. The two primary types of reactions and bonds are _____ and _____.
4. A type of weak bond, called a _____ bond, exists between water molecules.
5. Acidic solutions contain more _____ ions than basic solutions, but they have a _____ pH.
6. Glycogen is a polymer of _____ molecules that serve to give the body immediate _____.
7. A fat hydrolyzes to give one _____ molecule and three _____ molecules.
8. A polypeptide has levels of structure. The first level is the sequence of _____; the second level is very often a _____; the third level is its final _____.
9. _____ speed chemical reactions in cells.
10. Genes are composed of _____, a nucleic acid formed from _____ molecules that are joined together.

Medical Terminology Exercise

After studying this chapter, see if you can derive the definitions for the medical terms listed below and at right. Many of the prefixes and suffixes used to create these terms can be found throughout the chapter. For additional help, use McGraw-Hill Connect™ at www.mcgrawhillconnect.com and consult Appendix B.

1. anisotonic (ăn-ī″sō-tăwn′ĭk)
2. dehydration (dē″hī-drā′shŭn)
3. hypokalemia (hī″pō-kā-lē′mē-ŭh)
4. hypovolemia (hī″pō-vō-lē′mē-ŭh)
5. nonelectrolyte (nŏn″ē-lĕk′trō-līt)
6. lipometabolism (līp″ō-mŭh-tăb′ō-līzm)
7. hyperlipoproteinemia (hī″pĕr-līp″ō-prō″tēn-ē′mē-ŭh)
8. hyperglycemia (hī″pĕr-glī-sē′mē-ŭh)
9. hypoxemia (hī″pŏk-sē′mē-ŭh)
10. hydrostatic pressure (hī″drō-stăt′-ĭk-prĕsh′ĕr)
11. galactosemia (gŭh-lăk-tō-sē′mē-ŭh)
12. hypercalcemia (hī″pĕr-kăl-sē′mē-ŭh)
13. hyponatremia (hī″pō-nŭh-trē′mē-ŭh)
14. gluconeogenesis (glū″kō-nē″ō-jĕn′ŭh-sĭs)
15. edema (ē-dē′mŭh)

Online Study Tools

LEARNSMART® **connect®**

APR

Anatomy & Physiology REVEALED includes cadaver photos that allow you to peel away layers of the human body to reveal structures beneath the surface. This program also includes animations, radiologic imaging, audio pronunciations, and practice quizzing. To learn more visit www.aprevealed.com

Anatomy & Physiology REVEALED
aprevealed.com

3 Cell Structure and Function

Need another reason to quit using tobacco? The fine, hairlike cilia you see on these cells from the trachea, or windpipe, are exquisitely tailored organelles with an important protective function. Sticky mucus covering the tracheal walls traps harmful pollutants like dust and mold spores before they can reach the lungs. Cilia push the mucus upward toward the throat, and you can either spit it out or swallow it. In either case, the mucus and trapped pollutants are usually harmless. Now consider this: nicotine, the addictive drug found in tobacco, temporarily poisons delicate cilia. As a result, the airways become covered with a sticky layer in which fine particles and toxic residue from the burnt tobacco are trapped. A deep, hacking *smoker's cough* becomes the only way to clear mucus from the airways. You'll find tips to help you stop smoking in Chapter 14.

Learning Outcomes

After you have studied this chapter, you should be able to:

3.1 Cellular Organization

1. Name the three main parts of a human cell.
2. Describe the structure and function of the plasma membrane.
3. Explain the structure and function of the nucleus.
4. Describe the structures and roles of the endoplasmic reticulum and the Golgi apparatus in the cytoplasm.
5. Detail the structures of lysosomes and the role of these organelles in the breakdown of molecules.
6. Describe the structure of mitochondria and their role in producing ATP.

7. Describe the structures of centrioles, cilia, and flagella and their roles in cellular movement.
8. Describe the structures and function of the cytoskeleton.

3.2 Crossing the Plasma Membrane

9. Describe how substances move across the plasma membrane, and distinguish between passive and active transport.

3.3 The Cell Cycle

10. Describe the phases of the cell cycle.
11. As a part of interphase, describe the process of DNA replication.

12. As a part of interphase, also describe how cells carry out protein synthesis.
13. Describe the phases of mitosis, and explain the function of mitosis.

Visual Focus
The Cell

Medical Focus
Dehydration and Water Intoxication

Focus on Forensics
DNA Fingerprinting

3.1 Cellular Organization

1. Name the three main parts of a human cell.
2. Describe the structure and function of the plasma membrane.
3. Explain the structure and function of the nucleus.
4. Describe the structures and roles of the endoplasmic reticulum and the Golgi apparatus in the cytoplasm.
5. Detail the structures of lysosomes and the role of these organelles in the breakdown of molecules.
6. Describe the structure of mitochondria and their role in producing ATP.
7. Describe the structures of centrioles, cilia, and flagella and their roles in cellular movement.
8. Describe the structures and function of the cytoskeleton.

Every human cell has a plasma membrane, a single nucleus, and cytoplasm. (There are some exceptions to this rule. A mature erythrocyte, or red blood cell, ejects its nucleus as it develops. Thus, erythrocytes are *anucleate*. Cells of skeletal muscle, liver, and other tissues may have up to 50 nuclei and are *multinucleate*.) The **plasma membrane,** which surrounds the cell and keeps it intact, regulates what enters and exits a cell. The plasma membrane is a phospholipid bilayer that is said to be *semipermeable* (or *selectively* permeable) because it allows certain molecules but not others to enter the cell. Proteins present in the plasma membrane play important roles in allowing substances to enter the cell.

The **nucleus** is a large, centrally located structure that contains the chromosomes and is the control center of the cell. It controls the metabolic functioning and structural characteristics of the cell. The **nucleolus** is a region inside the nucleus.

The **cytoplasm** is the portion of the cell between the nucleus and the plasma membrane. The matrix of cytoplasm is a gelatinous, semifluid medium that contains water and various types of molecules suspended or dissolved in the medium. The presence of proteins accounts for the semifluid nature of cytoplasm.

Along with its nucleus, the cell cytoplasm contains numerous **organelles,** which are usually membranous structures (Table 3.1 and Fig. 3.1). The largest organelle, the nucleus, can often be seen with a light microscope. Others are best seen with an electron microscope.[1] Each type of organelle has a specific function. As you know, the nucleus directs cellular activities. Another type of organelle transports substances, and a different type produces ATP for the cell. Organelles compartmentalize the cell, keeping the various cellular activities separated from one another. Just as the rooms in your house

---

[1] Electron microscopes are high-powered instruments that are used to generate detailed photographs of cellular contents. The photographs are called electron micrographs. Scanning electron micrographs have depth, while transmission electron micrographs are flat (see Figs. 3.1*a* and 3.3). Light microscopes are used to generate photomicrographs that are often simply called micrographs.

TABLE 3.1 Structures in Human Cells

Name	Composition	Function
MEMBRANOUS STRUCTURES		
Plasma membrane	Phospholipid bilayer with embedded proteins	Cell border; selective passage of molecules into and out of cell; location of cell markers, cell receptors
Nucleus	Nuclear membrane (envelope) surrounding nucleoplasm, chromatin, and nucleolus	Storage of genetic information; control center of cell; cell replication
Nucleolus	Concentrated area of chromatin, RNA, and proteins; found in the nucleus	Ribosomal formation
Ribosome	Two subunits composed of protein and RNA	Protein synthesis
Endoplasmic reticulum	Complex system of tubules, vesicles, and sacs	Synthesis and/or modification of proteins and other substances; transport by vesicle formation
Rough endoplasmic reticulum	Endoplasmic reticulum studded with ribosomes	Protein synthesis for export
Smooth endoplasmic reticulum	Endoplasmic reticulum without ribosomes	Varies: lipid and/or steroid synthesis; calcium storage
Golgi apparatus	Stacked, concentrically folded membranes	Processing, packaging, and distribution of molecules
Vacuole	Small membranous sac	Isolates substances inside cell
Vesicle	Small membranous sac	Storage and transport of substances into/out of cell
Lysosome	Vesicle containing digesting enzymes	Intracellular digestion; self-destruction of the cell
Peroxisome	Vesicle containing oxidative enzymes	Detoxifies drugs, alcohol, etc.; breaks down fatty acids
Mitochondrion	Inner membrane within outer membrane	Cellular respiration
Cytoskeleton	Microtubules, actin filaments	Shape of cell and movement of its parts
Cilia and flagella	9 + 2 pattern of microtubules	Movement by cell; movement of substances inside a tube
Centriole	9 + 0 pattern of microtubules	Formation of basal bodies for cilia and flagella; formation of spindle in cell division

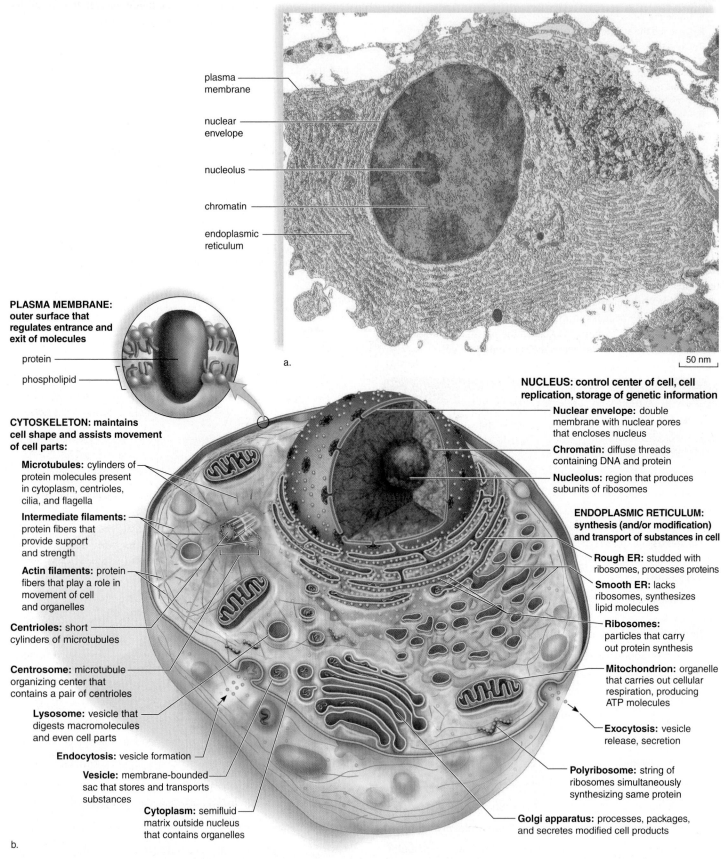

PLASMA MEMBRANE: outer surface that regulates entrance and exit of molecules

protein

phospholipid

CYTOSKELETON: maintains cell shape and assists movement of cell parts:

Microtubules: cylinders of protein molecules present in cytoplasm, centrioles, cilia, and flagella

Intermediate filaments: protein fibers that provide support and strength

Actin filaments: protein fibers that play a role in movement of cell and organelles

Centrioles: short cylinders of microtubules

Centrosome: microtubule organizing center that contains a pair of centrioles

Lysosome: vesicle that digests macromolecules and even cell parts

Endocytosis: vesicle formation

Vesicle: membrane-bounded sac that stores and transports substances

Cytoplasm: semifluid matrix outside nucleus that contains organelles

plasma membrane

nuclear envelope

nucleolus

chromatin

endoplasmic reticulum

a.

50 nm

NUCLEUS: control center of cell, cell replication, storage of genetic information

Nuclear envelope: double membrane with nuclear pores that encloses nucleus

Chromatin: diffuse threads containing DNA and protein

Nucleolus: region that produces subunits of ribosomes

ENDOPLASMIC RETICULUM: synthesis (and/or modification) and transport of substances in cell

Rough ER: studded with ribosomes, processes proteins

Smooth ER: lacks ribosomes, synthesizes lipid molecules

Ribosomes: particles that carry out protein synthesis

Mitochondrion: organelle that carries out cellular respiration, producing ATP molecules

Exocytosis: vesicle release, secretion

Polyribosome: string of ribosomes simultaneously synthesizing same protein

Golgi apparatus: processes, packages, and secretes modified cell products

b.

AP|R **Figure 3.1** **The Cell.** (**a**) Transmission electron micrograph of the cell. (**b**) The interior of the cell, with its organelles.

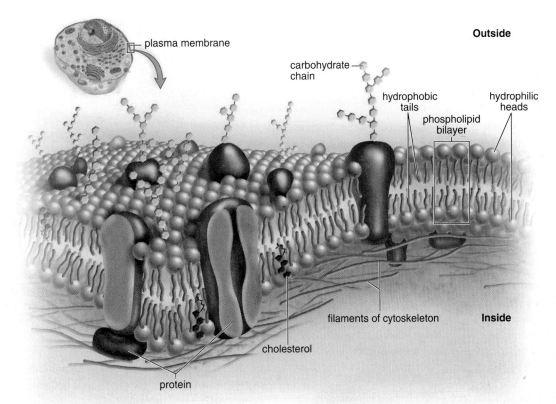

Outside

plasma membrane

carbohydrate chain

hydrophobic tails

hydrophilic heads

phospholipid bilayer

filaments of cytoskeleton

Inside

cholesterol

protein

AP|R Figure 3.2 **Fluid-mosaic model of the plasma membrane.** In the phospholipid bilayer, the hydrophilic (polar) heads are on the inner and outer surfaces of the bilayer, and the hydrophobic (nonpolar) tails are sandwiched in between the two hydrophilic layers. Embedded proteins are scattered throughout the bilayer. Carbohydrate molecules attach to lipids to form glycolipids. Likewise, carbohydrates and proteins form glycoproteins.

have particular pieces of furniture that serve a particular purpose, organelles have a structure that suits their function.

Cells also have a **cytoskeleton,** a network of interconnected filaments and tiny hollow tubes called microtubules in the cytoplasm. The name *cytoskeleton* is convenient because it allows us to compare the cytoskeleton to our bones and muscles. Bones and muscles give us structure and produce movement. Similarly, the elements of the cytoskeleton maintain cell shape and allow the cell and its contents to move. Some cells move by using cilia and flagella, which are made up of microtubules.

The Plasma Membrane

Our cells are surrounded by an outer plasma membrane. The plasma membrane separates the cytoplasm and organelles inside the cell from the outside of the cell. Plasma membrane integrity is necessary for the cell to live.

The plasma membrane is a phospholipid bilayer with attached (also called *peripheral*) or embedded (also called *integral*) proteins. As you discovered in Chapter 2, each phospholipid molecule has a polar head and nonpolar tails (Fig. 3.2). Because the polar heads are charged, they are *hydrophilic* (water-loving) and face outward, where they are likely to encounter a watery environment. The nonpolar tails are *hydrophobic* (water-fearing) and face inward, where there is no water. When phospholipids are placed in

water, they naturally form a spherical bilayer because of the chemical properties of the heads and the tails.

At body temperature, the phospholipid bilayer is a liquid; it has the consistency of olive oil, and the proteins are able to change their positions by moving laterally. The *fluid-mosaic model,* a working description of membrane structure, suggests that the protein molecules have a changing pattern (form a mosaic[2]) within the fluid phospholipid bilayer (Fig. 3.2). Our plasma membranes also contain a substantial number of cholesterol molecules. These molecules stabilize the phospholipid bilayer and prevent a drastic decrease in fluidity at low temperatures.

Short chains of sugars are attached to the outer surfaces of some proteins and lipids within the plasma membrane. This creates molecules termed *glycoproteins* and *glycolipids,* respectively. Unique glycoprotein molecules, specific to each cell, mark the cell as belonging to a particular individual. These cell markers account for characteristics such as blood type or why a patient's system sometimes rejects an organ transplant. Some glycoproteins have a special configuration that allows them to act as receptors for

[2] A mosaic is a picture or decorative design that's created when pieces of colored glass, tile, or paper are placed in or on a surface to create a pattern. In the fluid-mosaic analogy, proteins are placed into the phospholipid layer.

a chemical messenger such as a hormone. Some integral plasma membrane proteins form channels through which certain substances can enter cells, while others are carriers involved in the passage of molecules through the membrane.

The Nucleus

The nucleus is a prominent structure in human cells. The nucleus is of primary importance because it stores the genetic information that determines the characteristics of the body's cells and their metabolic functioning. The unique chemical composition of each person's DNA forms the basis for DNA fingerprinting (see p. 61). Every cell contains a copy of genetic information, but each cell type has certain genes turned on and others turned off. Activated DNA, with messenger RNA (mRNA) acting as an intermediary, controls protein synthesis (see pp. 56–58). The proteins of a cell determine its structure and the functions it can perform.

When you look at the nucleus, even in an electron micrograph, you cannot see DNA molecules, but you can see chromatin (Fig. 3.3). Chemical analysis shows that **chromatin** contains DNA and much protein, as well as some RNA. Chromatin coils into individual rodlike structures called **chromosomes** just before the cell divides. Chromatin is immersed in a semifluid medium called nucleoplasm.

Most likely, too, when you look at an electron micrograph of a nucleus (Fig. 3.3), you'll see one or more regions that look darker than the rest of the chromatin. These are nucleoli (sing., nucleolus) where another type of RNA, called ribosomal RNA (rRNA), is

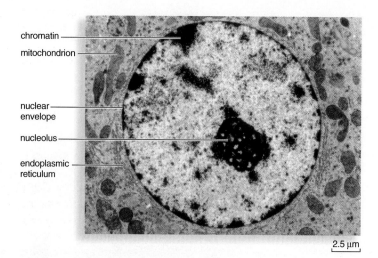

2.5 μm

AP|R **Figure 3.3** **The nucleus.** The double membrane nuclear envelope surrounds the chromatin. The nucleus has a special region called the nucleolus where rRNA is produced and ribosomal subunits are assembled.

produced and where rRNA joins with proteins to form the subunits of ribosomes. (Ribosomes are small bodies in the cytoplasm that contain rRNA and proteins.)

The nucleus is separated from the cytoplasm by a double membrane known as the **nuclear envelope,** which is continuous with the endoplasmic reticulum (Fig. 3.4) discussed on page 47.

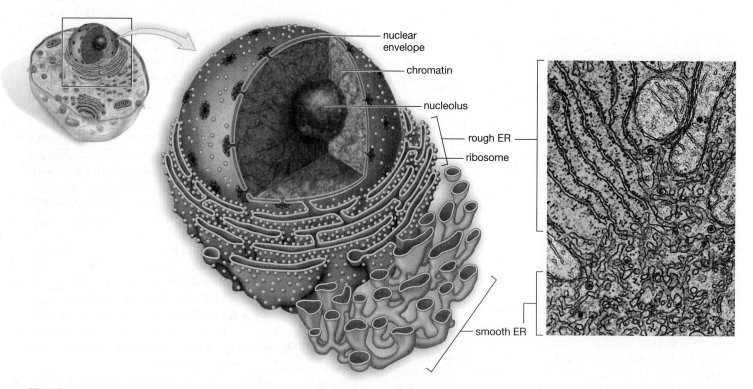

AP|R **Figure 3.4** **Endoplasmic reticulum.** Rough endoplasmic reticulum is studded with ribosomes where protein synthesis occurs. Smooth endoplasmic reticulum, which has no attached ribosomes, produces lipids and often has other functions as well in particular cells.

The nuclear envelope has **nuclear pores** that are large enough to permit the passage of proteins into the nucleus and ribosomal subunits out of the nucleus. Once again, remember the most important function of the nuclear envelope: to surround and contain DNA.

Ribosomes

Ribosomes are composed of two subunits, one large and one small. Each subunit has its own mix of proteins and rRNA. Protein synthesis occurs at the ribosomes.

Ribosomes are found free within the cytoplasm either singly or in groups called **polyribosomes** (polysomes). Ribosomes are often attached to the endoplasmic reticulum, a membranous system of saccules and channels discussed next (Fig. 3.4).

Proteins synthesized by cytoplasmic ribosomes are used inside the cell for various purposes. Those produced by ribosomes attached to endoplasmic reticulum may eventually be secreted from the cell.

Endomembrane System

The **endomembrane system** consists of the nuclear envelope, the endoplasmic reticulum, the Golgi apparatus, lysosomes, and **vesicles** (tiny membranous sacs) (Fig. 3.5). All four of these components of the cell work together to produce and secrete a product.

The Endoplasmic Reticulum

The **endoplasmic reticulum (ER),** a complicated system of membranous channels and saccules (flattened vesicles), is physically continuous with the outer membrane of the nuclear envelope. Rough ER is studded with ribosomes on the side of the membrane that faces the cytoplasm. Here proteins are synthesized and enter the ER interior, where processing and modification begin. Some of these proteins are incorporated into the membrane, and some are for export. Smooth ER, which is continuous with rough ER, does not have attached ribosomes. Smooth ER synthesizes the phospholipids that occur in membranes and has various other functions, depending on the particular cell. For example, in the testes, smooth ER produces testosterone, and in the liver it helps detoxify drugs.

Regardless of any specialized function, ER also forms vesicles in which large molecules are transported to other parts of the cell. Often these vesicles are on their way to the plasma membrane or the Golgi apparatus.

The Golgi Apparatus

The **Golgi apparatus** is named for Camillo Golgi, who discovered its presence in cells in 1898. The Golgi apparatus consists of a stack of 3 to 20 slightly curved saccules that resemble a stack of pancakes (Fig. 3.5). In human and animal cells, one side of the stack (the inner face) is directed toward the ER, and the other side of the stack (the outer face) is directed toward the plasma membrane. Vesicles can frequently be seen at the edges of the saccules.

The Golgi apparatus receives protein and/or lipid-filled vesicles that bud from the ER. Some biologists believe that these fuse to form a saccule at the inner face and that this saccule remains a part of the Golgi apparatus until the molecules are repackaged in new vesicles at the outer face. Others believe that the vesicles from the ER proceed directly to the outer face of the Golgi apparatus, where processing and packaging occur within its saccules. The Golgi apparatus contains enzymes that modify proteins and lipids. For example, it can add a chain of sugars to proteins and lipids, thereby making them into glycoproteins and glycolipids, which are molecules found in the plasma membrane.

The vesicles that leave the Golgi apparatus move to other parts of the cell. Some vesicles proceed to the plasma membrane where they discharge their contents. Because this is secretion, note that the Golgi apparatus is involved in processing, packaging, and secretion. Other vesicles that leave the Golgi apparatus are lysosomes.

Lysosomes

Lysosomes, membranous sacs produced by the Golgi apparatus, contain hydrolytic digestive enzymes. Sometimes macromolecules are brought into a cell by vesicle formation at the plasma membrane (Fig. 3.5). When a lysosome fuses with such a vesicle, the lysosome's contents are digested by lysosomal enzymes into simpler subunits that then enter the cytoplasm. Even parts of a cell are digested by its own lysosomes (called *autodigestion*). Normal cell rejuvenation most likely takes place in this manner, but autodigestion is also important during development. For example, when a tadpole becomes a frog, lysosomes digest away the cells of the tail. The fingers of a human embryo are at first webbed, but they are freed from one another as a result of lysosomal action.

Occasionally, a child is born with **Tay-Sachs disease,** a metabolic disorder involving a missing or inactive lysosomal enzyme in nerve cells. In these cases, the lysosomes fill to capacity with macromolecules that cannot be broken down. The nerve cells become so crowded with these nonfunctional lysosomes that they can no longer function. Nerve signals in the brain and spinal cord are no longer transmitted. The brain stops developing and the child dies. Someday soon, it may be possible to provide the missing enzyme for these children.

Peroxisomes and Vacuoles

Although **peroxisomes** are not part of the endomembranous system, they are similar to lysosomes in structure. Like lysosomes, they are vesicles that contain enzymes. Peroxisome enzymes function to detoxify drugs, alcohol, and other potential toxins. The liver and kidneys contain large numbers of peroxisomes because these organs help to cleanse the blood. Peroxisomes also break down fatty acids from fats so that fats can be metabolized by the cell.

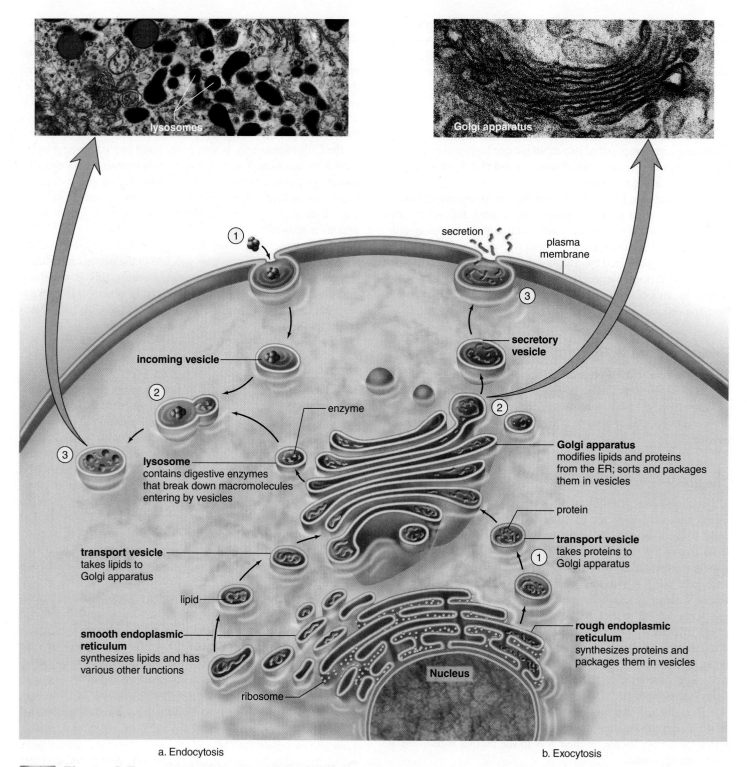

lysosomes

Golgi apparatus

secretion

plasma membrane

①

③

secretory vesicle

incoming vesicle

②

enzyme

③

lysosome
contains digestive enzymes that break down macromolecules entering by vesicles

Golgi apparatus
modifies lipids and proteins from the ER; sorts and packages them in vesicles

protein

②

transport vesicle
takes proteins to Golgi apparatus

①

transport vesicle
takes lipids to Golgi apparatus

lipid

smooth endoplasmic reticulum
synthesizes lipids and has various other functions

rough endoplasmic reticulum
synthesizes proteins and packages them in vesicles

Nucleus

ribosome

a. Endocytosis

b. Exocytosis

AP|R **Figure 3.5** **The endomembrane system.** (a) Endocytosis and lysosome function. (1) A vesicle formed by endocytosis, containing macromolecules, fuses with (2) the lysosome made at the Golgi apparatus, and (3) lysosome enzymes digest the macro-molecules in the vesicle. (b) Exocytosis and the secretory vesicle. (1) Transport vesicles from the ER bring proteins and lipids to the Golgi apparatus, where (2) the molecules are packaged in a secretory vesicle, which (3) travels to the plasma membrane, releasing its contents by exocytosis.

Vacuoles are occasionally found within human cells, where they isolate substances captured inside the cell. For example, vacuoles may contain parasites that are awaiting digestion by lysosomes.

Mitochondria

Although the size and shape of mitochondria (sing., **mitochondrion**) can vary, all are bounded by a double membrane. The inner membrane is folded to form little shelves called *cristae,* which project into the *matrix,* an inner space filled with a gel-like fluid (Fig. 3.6). The matrix also contains mitochondrial DNA, which is distinct from nuclear DNA.

Mitochondria are the site of ATP (adenosine triphosphate) production involving complex metabolic pathways. As described in section 2.7, ATP molecules are the common carrier of energy in cells. A shorthand way to indicate the chemical transformation that involves mitochondria is as follows:

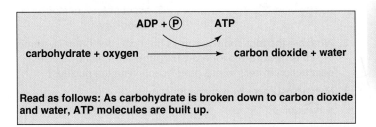

Read as follows: As carbohydrate is broken down to carbon dioxide and water, ATP molecules are built up.

Mitochondria are often called the powerhouses of the cell: Just as a powerhouse burns fuel to produce electricity, the mitochondria convert the chemical energy of carbohydrate molecules into the chemical energy of ATP molecules. In the process, mitochondria use up oxygen and give off carbon dioxide and water. The oxygen you breathe in enters cells and then mitochondria; the carbon dioxide you breathe out is released by mitochondria. Because oxygen is used up and carbon dioxide is released, we say that mitochondria carry on **cellular respiration.**

Fragments of digested carbohydrate, protein, and lipid enter the mitochondrial matrix from the cytoplasm. The matrix contains enzymes for metabolizing these fragments to carbon dioxide and water. Energy released from metabolism is used for ATP production, which occurs at the cristae. The protein complexes that aid in the conversion of energy are located in an assembly-line fashion on these membranous shelves.

Every cell uses a certain amount of ATP energy to synthesize molecules, but many cells use ATP to carry out their specialized functions. For example, muscle cells use ATP for muscle contraction, which produces movement, and nerve cells use it for the conduction of nerve signals, which make us aware of our environment.

The Cytoskeleton

Several types of filamentous protein structures form a cytoskeleton that helps maintain the cell's shape and either anchors the organelles or assists their movement as appropriate. The cytoskeleton

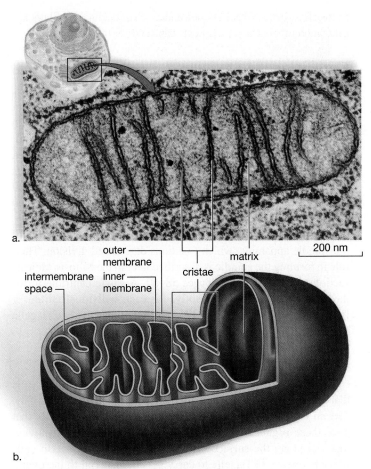

AP|R **Figure 3.6 Mitochondrion structure.** Mitochondria are involved in cellular respiration. **(a)** Electron micrograph of a mitochondrion. **(b)** Generalized drawing in which the outer membrane and portions of the inner membrane have been cut away to reveal the cristae.

includes microtubules, intermediate filaments, and actin filaments (see Fig. 3.1).

Microtubules are hollow cylinders whose wall is made up of 13 longitudinal rows of the globular protein tubulin. Remarkably, microtubules can assemble and disassemble. Microtubule assembly is regulated by the **centrosome,** which lies near the nucleus. The centrosome is the region of the cell that contains the centrioles. (As noted in Table 3.1, centrioles help to form cilia, flagella, and the spindle apparatus.) Microtubules radiate from the centrosome, helping to maintain the shape of the cell and acting as tracks along which organelles move. It is well known that during cell division, microtubules form spindle fibers, which assist the movement of chromosomes.

Intermediate filaments differ in structure and function. Because they are tough and resist stress, intermediate filaments often form cell-to-cell junctions. Intermediate filaments join skin cells in the outermost skin layer, the epidermis. **Actin filaments** are long, extremely thin fibers that usually occur in bundles or other groupings. Actin filaments have been isolated from various types of cells, especially those in which movement occurs. Microvilli, which project from certain cells and can shorten and extend, contain actin filaments. Like microtubules, actin filaments can assemble and disassemble.

Centrioles

Centrioles are short cylinders with a 9 + 0 pattern of microtubules, meaning that there are nine outer microtubule triplets and no center microtubules (see Fig. 3.8a). Each cell has a pair of centrioles in the centrosome, a region located near the cell nucleus. The members of each pair of centrioles are at right angles to one another. Before a cell divides, the centrioles duplicate, and the members of the new pair are also at right angles to one another. During cell division, the pairs of centrioles separate so that each daughter cell gets one centrosome.

Centrioles are directed by the centrosome in the formation of the spindle apparatus, which functions during cell division. Their role in forming the spindle is described later in the Mitosis/Meiosis section of this chapter on pages 58–60. A single centriole forms the anchor point, or **basal body,** for each individual cilium or flagellum. Basal bodies direct the formation of cilia and flagella as well.

Cilia and Flagella

Cilia and flagella (sing., **cilium, flagellum**) are projections of cells that can move either in a wavelike fashion, like a whip, or stiffly, like an oar. Cilia are shorter than flagella (Fig. 3.7). Cells that have these organelles are capable of self-movement or moving material along the surface of the cell. For example, sperm cells move by means of flagella to carry genetic material to the ovum, or egg. As you know from the chapter opener, the cells that line our respiratory tract are ciliated. These cilia sweep debris trapped within mucus back up the throat, and this action helps keep the lungs clean. Within a woman's uterine tubes, ciliated cells move

the ovum toward the uterus (womb), where a fertilized ovum grows and develops.

Recall that each cilium and flagellum is anchored by its basal body, which lies in the cytoplasm of the cell. Basal bodies, like centrioles, have a 9 + 0 pattern of microtubule triplets (Fig. 3.8a). They are believed to organize the structure of cilia and flagella even though cilia and flagella have a 9 + 2 pattern of microtubules. In cilia and flagella, nine microtubule doublets surround two central microtubules. This arrangement is believed to be necessary for their ability to move (Fig. 3.8b).

Content **CHECK-UP!**

1. Match each part of the plasma membrane to its function:

   a. phospholipid molecules _____
   b. glycoproteins _____
   c. cholesterol _____

   1. stabilize the phospholipid bilayer
   2. form a bilayer
   3. account for a person's blood type

2. What are the four components of the endomembrane system? Which is responsible for protein synthesis? forming vesicles? destruction of cells or cell parts? containing the nucleus?

3. Match the organelle with the function:

   a. nucleolus
   b. mitochondrion
   c. peroxisome

   d. cytoskeleton

   1. break down toxins
   2. ribosome production
   3. allow movement within the cell
   4. ATP production

   Answers in Appendix A.

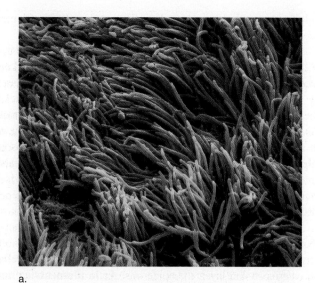

a.

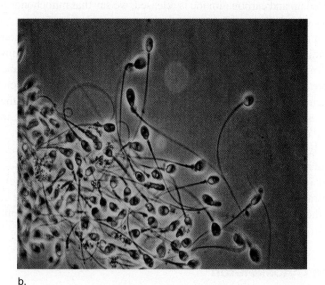

b.

AP|R **Figure 3.7** **Cilia and flagella. (a)** Cilia are common on the surfaces of certain tissues, such as the one that forms the inner lining of the respiratory tract. **(b)** Flagella form the tails of human sperm cells.

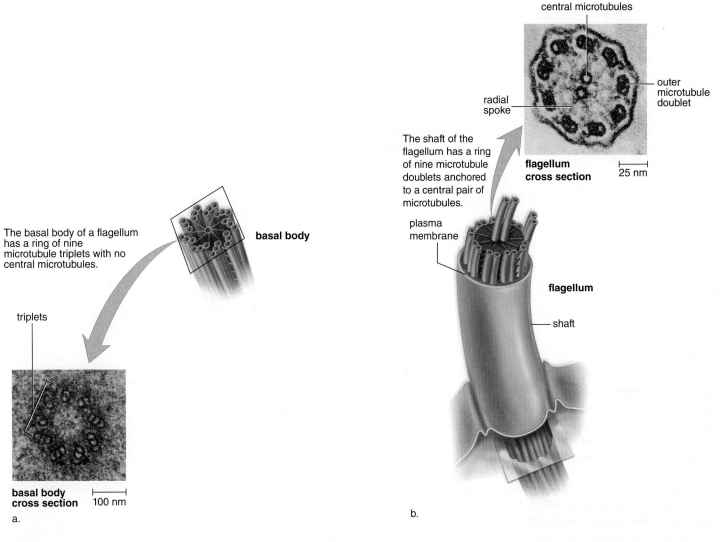

central microtubules

radial spoke

outer microtubule doublet

flagellum cross section

25 nm

The shaft of the flagellum has a ring of nine microtubule doublets anchored to a central pair of microtubules.

The basal body of a flagellum has a ring of nine microtubule triplets with no central microtubules.

basal body

plasma membrane

flagellum

shaft

triplets

basal body cross section

100 nm

a.

b.

AP|R **Figure 3.8** **Structure of basal bodies and flagella.** (**a**) Basal bodies have a 9 + 0 pattern of microtubule triplets. (**b**) A flagellum has a 9 + 2 pattern of microtubules.

3.2 Crossing the Plasma Membrane

9. Describe how substances move across the plasma membrane, and distinguish between passive and active transport.

The plasma membrane keeps a cell intact. It allows only certain molecules and ions to enter and exit the cytoplasm freely; therefore, the plasma membrane is said to be **selectively permeable** (semipermeable). Both passive and active methods are used to cross the plasma membrane (Table 3.2). A passive transport mechanism is one that doesn't require cellular energy. By contrast, active transport must be fueled by cellular energy (usually in the form of ATP).

Simple Diffusion

Simple diffusion is the random movement of simple atoms or molecules from an area of higher concentration to an area of lower concentration until they are equally distributed. (When atoms or molecules move from high to low concentration, they are said to be moving *down* their concentration gradient.) To illustrate diffusion, imagine putting a tablet of dye into water. The water eventually takes on the color of the dye as the dye molecules diffuse (Fig. 3.9).

The chemical and physical properties of the plasma membrane allow only a few types of molecules to enter and exit a cell by simple diffusion. Lipid-soluble molecules such as alcohols can diffuse through the membrane because lipids are the membrane's main structural components. Gases can also diffuse through the lipid bilayer; this is the mechanism by which oxygen enters cells and carbon dioxide exits cells. For example, consider the movement of oxygen from the lungs to the bloodstream. When you inhale, oxygen fills the tiny air sacs, or *alveoli,* within your lungs. Neighboring lung capillaries contain red blood cells with a very low oxygen concentration. Oxygen diffuses from the area of highest concentration to the area of lowest concentration: first through alveolar cells, then lung capillary cells, and finally into the red blood cells.

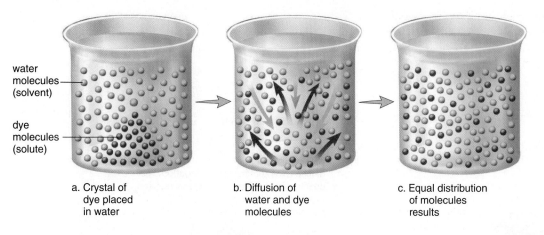

water
molecules
(solvent)

dye
molecules
(solute)

a. Crystal of
dye placed
in water

b. Diffusion of
water and dye
molecules

c. Equal distribution
of molecules
results

AP|R **Figure 3.9** **Diffusion.** Kinetic (thermal) energy causes constant, random movement of both water and dye molecules in a solution. As dye molecules move from areas of higher to lower concentration, the water takes on the color of the dye.

When atoms or molecules diffuse from areas of higher to lower concentration across plasma membranes, no cellular energy is involved. Instead, kinetic or thermal energy of matter is the energy source for diffusion.

Osmosis

Osmosis is the diffusion of water across a plasma membrane. Osmosis occurs whenever an unequal concentration of water exists on either side of a selectively permeable membrane. (Recall that a selectively permeable membrane allows water to pass freely, but not most dissolved substances.) In a solution, water is *more* concentrated when it contains fewer dissolved substances, or **solutes** (and thus is closest to pure water). Water is *less* concentrated as solute concentration increases. When water moves from the area of higher water concentration to the area of lower water concentration (i.e., higher concentration of solute) it exerts a force on the selectively permeable membrane called **osmotic pressure.**

Tonicity is the degree to which a solution's concentration of solute-versus-water causes water to move into or out of cells. Normally, body fluids are **isotonic** to cells (Fig. 3.10a)—that is, there is an equal concentration of solutes (dissolved substances) and solvent (water) on both sides of the plasma membrane, and cells maintain their usual size and shape. Medically administered intravenous solutions usually have this tonicity. Body fluids that are *not* isotonic to body cells are the result of dehydration or water intoxication (see Medical Focus, p. 54).

Solutions (solute plus solvent) that cause cells to swell or even to burst due to an intake of water are said to be **hypotonic** solutions. If red blood cells are placed in a hypotonic solution, which has a higher concentration of water (i.e., lower concentration of solute) than do the cells, water enters the cells and they swell to bursting (Fig. 3.10b). The term **lysis** refers to breaking or disintegration of cells. **Hemolysis** is the disintegration of red blood cells.

Solutions that cause cells to shrink or to shrivel due to a loss of water are said to be **hypertonic** solutions. If red blood cells are placed in a hypertonic solution, which has a lower concentration of water (i.e., higher concentration of solute) than do the cells, water leaves the cells and they shrink (Fig. 3.10c). The term **crenation** refers to red blood cells in this condition.

 Begin Thinking Clinically

An intravenous hypertonic solution may help relieve swelling in the brain in patients with head trauma. How does this work?

Answer and discussion in Appendix A.

Filtration

Capillaries (described in detail in Chapter 12) are the smallest blood vessels. Because capillary walls are only one cell thick, small molecules (e.g., water or small solutes) tend to passively diffuse across these walls, from areas of higher concentration

 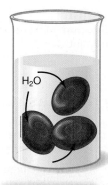 

a. Isotonic solution
(same solute concentration as in cell)

b. Hypotonic solution
(lower solute concentration than in cell)

c. Hypertonic solution
(higher solute concentration than in cell)

AP|R **Figure 3.10** **Tonicity.** The arrows indicate the *net* movement of water from high to low concentration.

to those of lower concentration. However, blood pressure aids matters by pushing water and dissolved solutes out of the capillary, through tiny pores between capillary cells. This process, called **filtration,** is the movement of liquid from high pressure to low pressure.

You can observe filtration in a drip coffeemaker. Water moves from the area of high pressure (the water reservoir) to the area of low pressure (the coffee pot). Large substances (the coffee grounds) remain behind in the coffee filter, but small molecules (caffeine, flavor molecules) and water pass through.

Transport by Carriers

Most solutes do not simply diffuse across a plasma membrane; rather, they are transported by means of protein carriers within the membrane. During **facilitated diffusion** (facilitated transport), a molecule (e.g., an amino acid or glucose) is transported across the plasma membrane from the side of higher concentration to the side of lower concentration. The cell doesn't need to expend energy for this type of transport because the molecules are moving down their concentration gradient.

Begin Thinking Clinically

If the disease diabetes isn't well controlled, the concentration of glucose found in blood soars after meals. The protein carriers can't transport it all into cells. What happens to that extra glucose?

Answer and discussion in Appendix A.

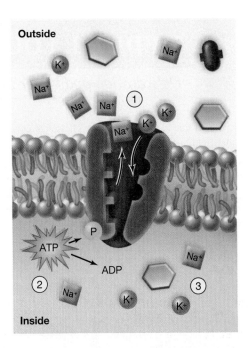

AP|R **Figure 3.11** **Active transport through a plasma membrane.** Active transport allows a molecule to cross the membrane from lower concentration to higher concentration. ① Ion (or molecule) enters carrier. ② Breakdown of ATP induces a change in shape that ③ drives the molecule across the membrane.

During **active transport,** a molecule is moving contrary to the normal direction—that is, from lower to higher concentration (Fig. 3.11). For example, iodine collects in the cells of the thyroid gland; sugar is completely absorbed from the gut by cells that line the digestive tract; and sodium (Na$^+$) is sometimes almost completely

TABLE 3.2 Crossing the Plasma Membrane			
Name	**Direction**	**Requirement**	**Examples**
PASSIVE METHODS			
Simple diffusion	High to low concentration	Concentration gradient	Lipid-soluble molecules, gases
Osmosis	High to low concentration	Semipermeable membrane, water concentration gradient	Absorption of water from digestive tract to bloodstream
Facilitated diffusion	High to low concentration	Carrier molecule plus concentration gradient	Sugars and amino acids
Filtration	High to low pressure	Pressure gradient	Water and dissolved solutes out of capillaries
ACTIVE METHODS			
Active transport	Low to high concentration	Carrier molecule plus cell energy	Ions, sugars, amino acids
Endocytosis			
Phagocytosis	Into the cell ("cell eating")	Vesicle formation	Bacterial cells, viruses, cell debris
Pinocytosis	Into the cell ("cell drinking")	Vesicle formation	Breast milk absorption in infants
Exocytosis	Out of the cell	Vesicle fuses with plasma membrane	Hormones, messenger chemicals, other macromolecules

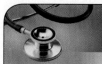

Dehydration and Water Intoxication

Dehydration is due to a loss of water. The solute concentration in extracellular fluid increases—that is, tissue fluid becomes hypertonic to cells, and water leaves the cells, so that they crenate. A common cause of dehydration is excessive sweating, perhaps during exercise, without any replacement of the water lost. Dehydration can also be a side effect of any illness that causes prolonged vomiting or diarrhea.

The signs of moderate dehydration are a dry mouth, sunken eyes, and skin that will not bounce back after light pinching. If dehydration becomes severe, the pulse and breathing rate are rapid, the hands and feet are cold, and the lips are blue. Although dehydration leads to weight loss, deliberately dehydrating to lose weight is extremely dangerous and can be fatal.

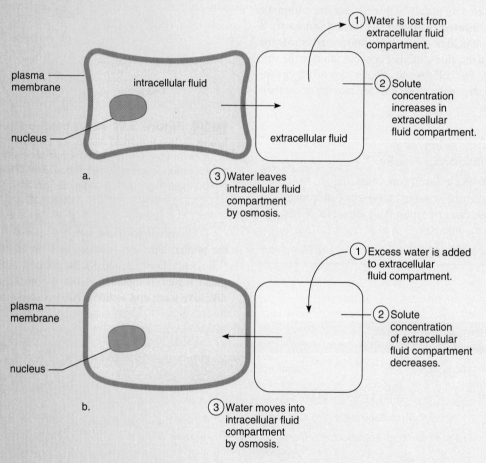

Figure 3A **Dehydration versus water intoxication. (a)** If extracellular fluid loses too much water, cells lose water by osmosis and become dehydrated. **(b)** If extracellular fluid gains too much water, cells gain water by osmosis and water intoxication occurs.

Water intoxication may be caused by excessive consumption of pure water. The tissue fluid becomes hypotonic to the cells, and water enters the cells. Water intoxication can lead to pulmonary edema (excess tissue fluid in the lungs) and swelling in the brain. In extreme cases, it is fatal. Water intoxication is not nearly as common in adults as is dehydration. It can result from a mental disorder termed *psychogenic polydipsia*. Another cause can be the intake of too much pure water during vigorous exercise: for example, a marathon race. Marathoners who collapse and have nausea and vomiting after a race

may be suffering from water intoxication. The cure, an intravenous solution containing high amounts of sodium, is the opposite of that for dehydration. Therefore, it is important that physicians be able to diagnose water intoxication in athletes who have had an opportunity to drink fluids over a period of a few hours. To prevent both dehydration and water intoxication, athletes should replace lost fluids continuously. Pure water is a good choice if the exercise period is short. Low-sodium solutions, such as sports drinks, are a good choice for longer-duration events like marathons.

withdrawn from urine by cells lining kidney tubules. Active transport requires a protein carrier and the use of cellular energy obtained from the breakdown of ATP. When ATP is broken down, energy is released, and in this case the energy is used by a carrier to carry out active transport. Therefore, it is not surprising that cells involved in active transport have a large number of mitochondria near the plasma membrane at which active transport is occurring.

Proteins involved in active transport often are called pumps because just as a water pump uses energy to move water against the force of gravity, proteins use energy to move substances against their concentration gradients. One type of pump that is active in all cells but is especially associated with nerve and muscle cells moves sodium ions (Na^+) to the outside of the cell and potassium ions (K^+) to the inside of the cell.

The passage of salt (NaCl) across a plasma membrane is of primary importance in cells. First, sodium ions are pumped across a membrane; then, chloride ions simply diffuse through channels that allow their passage. Chloride ion channels malfunction in persons with **cystic fibrosis,** and this leads to the symptoms of this inherited (genetic) disorder.

Endocytosis and Exocytosis

During **endocytosis,** a portion of the plasma membrane forms an inner pocket to envelop a substance, and then the membrane pinches off to form an intracellular vesicle (see Fig. 3.5, *left*). Two forms of endocytosis exist: **phagocytosis,** or "cell eating," is a mechanism that allows the cell to ingest solid particles. White blood cells consume bacterial cells by phagocytosis. Once inside the cell, the bacterial cell can be destroyed. **Pinocytosis,** or "cell drinking," allows the cell to consume solutions. An infant's intestinal lining ingests breast milk by pinocytosis, allowing the mother's protective antibodies to enter the baby's bloodstream.

During **exocytosis,** a vesicle fuses with the plasma membrane as secretion occurs (see Fig. 3.5, *right*). This is the way insulin leaves insulin-secreting cells, for instance. Table 3.2 summarizes the various ways molecules cross the plasma membrane.

Content **CHECK-UP!**

4. Which process requires cellular ATP energy?

   a. osmosis

   b. facilitated diffusion (facilitated transport)

   c. active transport

   d. simple diffusion

5. A researcher studying the white blood cells of a patient infected with tuberculosis (TB) bacteria notices the bacteria are in *vesicles* in the cytoplasm. How did the bacteria come to be inside the cell?

6. A cell will swell when placed in a(n) _____ solution.

   a. hypotonic

   b. isotonic

   c. hypertonic

Answers in Appendix A.

3.3 The Cell Cycle

10. Describe the phases of the cell cycle.
11. As a part of interphase, describe the process of DNA replication.
12. As a part of interphase, also describe how cells carry out protein synthesis.
13. Describe the phases of mitosis, and explain the function of mitosis.

The **cell cycle** is an orderly set of stages that take place between the time a cell divides and the time the resulting daughter cells also divide. The cell cycle is controlled by internal and external signals. A signal is a molecule that stimulates or inhibits a metabolic event. For example, *growth factors* are external signals received at the plasma membrane that cause a resting cell to undergo the cell cycle. For example, when you skin your knee, a growth factor found in blood causes skin fibroblasts (immature skin cells) in the injured area to complete their cell cycle. New skin cells are produced, and the injury heals. Other signals ensure that the stages follow one another in the normal sequence and that each stage is properly completed before the next stage begins.

The cell cycle has a number of checkpoints, which are places where the cell cycle stops if it is not proceeding normally. Any cell that did not successfully complete mitosis and is abnormal undergoes apoptosis at the *restriction checkpoint*. **Apoptosis** is often defined as *programmed cell death* because the cell progresses through a series of events that bring about its destruction. The cell rounds up and loses contact with its neighbors. The nucleus fragments and the plasma membrane develops blisters. Finally, the cell fragments and its bits and pieces are engulfed by white blood cells and/or neighboring cells. The enzymes that bring about apoptosis are ordinarily held in check by inhibitors, but are unleashed by either internal or external signals.

Following a certain number of cell cycle revolutions, cells are apt to become specialized and no longer go through the cell cycle. For example, once they have matured, muscle and nerve cells rarely, if ever, go through the cell cycle. At the other extreme, some cells in the body, called stem cells, are always immature and go through the cell cycle repeatedly. There is a great deal of interest in stem cells today because it may be possible to control their future development into particular tissues and organs.

Cell Cycle Stages

The cell cycle has two major portions: **interphase** and the **mitotic stage** (Fig. 3.12).

Interphase

The cell in Figure 3.1 is in interphase because it is not dividing. During interphase, the cell carries on its regular activities, and it also gets ready to divide if it is about to complete the cell cycle. For these cells, interphase has three stages, called G_1 phase, S phase, and G_2 phase.

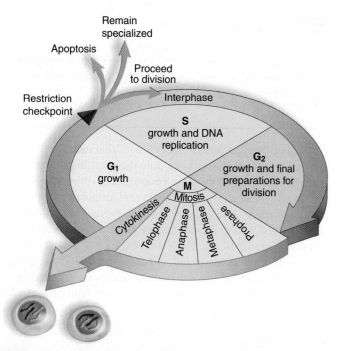

AP|R **Figure 3.12** **The cell cycle.** The cell cycle consists of interphase, during which cellular components duplicate, and a mitotic stage, during which the cell divides. Interphase consists of two so-called growth phases (G₁ and G₂) and a DNA synthesis (S) phase. The mitotic stage consists of the phases noted plus cytokinesis.

G₁ Phase Early microscopists named the phase before DNA replication G₁, and they named the phase after DNA replication G₂. G stood for "gap." Now that we know how metabolically active the cell is, it is correct to think of G as standing for "growth." During the G₁ phase, cells return to their normal, pre-mitotic condition. Each cell enlarges and doubles its number of organelles, such as mitochondria and ribosomes. Proteins are synthesized and other materials are accumulated in preparation for DNA synthesis. Most important, it is during the G₁ phase that the cell performs all its "tasks"—endocrine cells make and secrete hormones, white blood cells defend against invaders, and so on. As you read about the functions of cells in a particular tissue, remember that these are cells in the G₁ phase.

S Phase Following G₁, the cell enters the S (for "synthesis") phase. During the S phase, DNA replication occurs. At the beginning of the S phase, each chromosome is composed of one DNA double helix, which is called a **chromatid.** At the end of this phase, each chromosome has two identical DNA double helix molecules, called *sister chromatids.* Another way of describing these events is to say that DNA replication has resulted in duplicated chromosomes.

G₂ Phase During this phase, the cell synthesizes proteins that will assist cell division, such as the protein found in microtubules.

The cell also completes replication of the centrioles. The role of centrioles and microtubules in cell division is described later in this section.

Events During Interphase

Two significant events during interphase are replication of DNA and protein synthesis.

Replication of DNA

As described previously, during **replication,** an exact copy of a DNA helix is produced. (DNA and RNA structure are described on pp. 34–36.) The double-stranded structure of DNA aids replication because each strand serves as a template for the formation of a complementary strand. (In everyday terminology, a **template** is most often a mold used to produce a shape opposite to itself.) In the case of DNA, each old (parental) strand is a template for each new (daughter) strand.

Figures 3.13 and 3.14 show how replication is carried out. Figure 3.14 uses the ladder configuration of DNA for easy viewing.

1. Before replication begins, the two strands that make up parental DNA are hydrogen-bonded to one another.
2. During replication, the old (parental) DNA strands unwind and "unzip" (i.e., the weak hydrogen bonds between the two strands break).
3. New complementary nucleotides, which are always present in the nucleus, pair with the nucleotides in the old strands. A pairs with T and C pairs with G. The enzyme DNA polymerase joins the new nucleotides forming new (daughter) complementary strands.
4. When replication is complete, the two resulting double helix DNA molecules are identical.

Thus, when replication is complete, each chromosome has an exact copy, just as duplicating your course notes on a copy machine gives you two copies of the same written text. The identical DNA strands, or chromatids (remember, each is a complete double helix) are called sister chromatids. The chromosome is called a **duplicated chromosome** (see Fig. 3.16).

Protein Synthesis

DNA not only serves as a template for its own replication, but is also a template for RNA formation and for the construction of proteins by the cell. Protein synthesis requires two steps: transcription and translation. During **transcription,** a messenger RNA (mRNA) molecule is produced, using DNA as the template. During **translation,** this mRNA specifies the order of amino acids in a particular polypeptide (Fig. 3.15). A gene (i.e., a piece of DNA in a chromosome) contains coded information for the sequence of amino acids in a particular polypeptide. The code is a **triplet code:** Every three bases in DNA (and therefore in mRNA) stand for a particular amino acid.

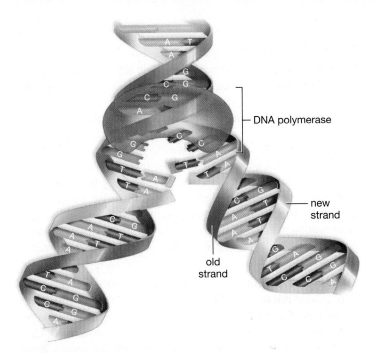

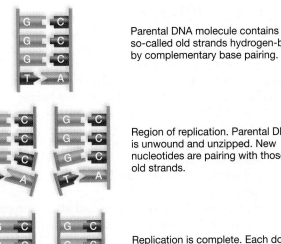

Parental DNA molecule contains so-called old strands hydrogen-bonded by complementary base pairing.

Region of replication. Parental DNA is unwound and unzipped. New nucleotides are pairing with those in old strands.

Replication is complete. Each double helix is composed of an old (parental) strand and a new (daughter) strand.

AP|R **Figure 3.13** **Overview of DNA replication.** During replication, an old strand serves as a template for a new strand. The new double helix is composed of an old (parental) strand and a new (daughter) strand.

AP|R **Figure 3.14** **Ladder configuration and DNA replication.** Use of the ladder configuration better illustrates how complementary nucleotides available in the cell pair with those of each old strand before they are joined together to form a daughter strand.

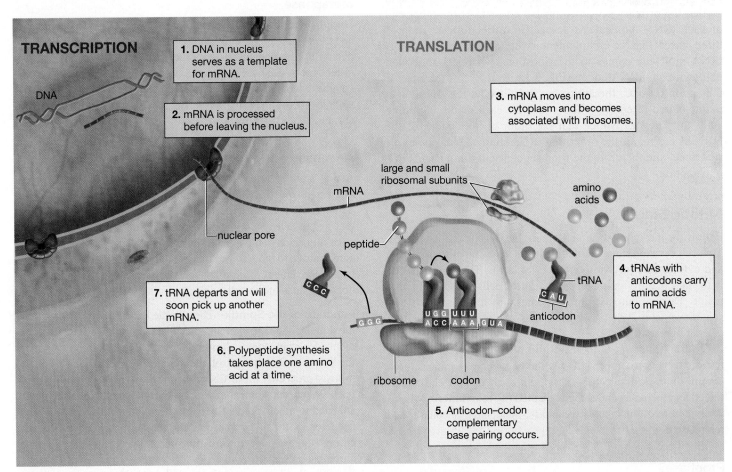

TRANSCRIPTION

1. DNA in nucleus serves as a template for mRNA.

DNA

2. mRNA is processed before leaving the nucleus.

TRANSLATION

3. mRNA moves into cytoplasm and becomes associated with ribosomes.

mRNA

nuclear pore

large and small ribosomal subunits

amino acids

peptide

7. tRNA departs and will soon pick up another mRNA.

tRNA

anticodon

4. tRNAs with anticodons carry amino acids to mRNA.

6. Polypeptide synthesis takes place one amino acid at a time.

ribosome   codon

5. Anticodon–codon complementary base pairing occurs.

AP|R **Figure 3.15** **Protein synthesis.** The two steps required for protein synthesis are transcription, which occurs in the nucleus, and translation, which occurs in the cytoplasm at the ribosomes.

Transcription and Translation

During transcription, DNA unwinds and unzips, as though it is preparing for replication. Complementary RNA nucleotides from an RNA nucleotide pool in the nucleus pair with the DNA nucleotides of one strand. The RNA nucleotides are joined by an enzyme called RNA polymerase, and an RNA molecule results (Fig. 3.15, *step 1*). There are three forms of RNA: messenger, transfer, and ribosomal; however, only messenger RNA determines amino acid sequence. When mRNA forms, it has a sequence of bases complementary to that of DNA. A sequence of three bases in mRNA that is complementary to the DNA triplet code is a **codon** (Fig. 3.15).

Translation requires several enzymes and all three types of RNA. Messenger RNA, containing the polypeptide's code, is sandwiched between the two ribosome subunits (Fig. 3.15, *step 3*). **Transfer RNA (tRNA)** molecules deliver amino acids to the ribosomes, which are composed of **ribosomal RNA (rRNA)** and protein (Fig. 3.15, *step 4*). There is at least one tRNA molecule for each of the 20 amino acids found in proteins. The amino acid binds to one end of the molecule, and the entire complex is designated as tRNA-amino acid.

Remember that the primary sequence of a protein is determined by the ordering of its amino acids (see Chapter 2, page 32). Transfer RNA must deliver the correct amino acid so that the primary sequence is also correct. At the other end of each tRNA molecule is a specific **anticodon,** a group of three bases that is complementary to an mRNA codon. A tRNA molecule comes to the ribosome, where its anticodon pairs with an mRNA codon (Fig. 3.15, *step 5*). For example, if the codon is ACC, the anticodon is UGG and the amino acid is one called threonine. (The codes for each of the 20 amino acids are known.) Notice that the order of the codons of the mRNA determines the order that tRNA-amino acids come to a ribosome, and therefore the final sequence of amino acids in a polypeptide.

Mitotic Stage

Following interphase, the cell enters the M (for mitotic) stage. This cell division stage includes **mitosis** (division of the nucleus) and **cytokinesis** (division of the cytoplasm). The **parental cell** is the cell that divides, and the **daughter cells** are the cells that result. During mitosis, chromosomes are distributed to two separate nuclei. When cytokinesis is complete, two daughter cells are present.

During interphase, the centrioles double in preparation for mitosis (Fig. 3.16, upper left). At the beginning of mitosis, each chromosome is duplicated—it is composed of two chromatids joined together at a central region, called the **centromere.** Mitosis is divided into four phases: prophase, metaphase, anaphase, and telophase (Fig. 3.16).

Prophase

Several events occur during **prophase** that visibly indicate the cell is about to divide. Each of the two pairs of centrioles is now part of a structure called a centrosome. The two centrosomes outside the nucleus begin moving away from each other, toward opposite ends of the nucleus. Long, thread-like structures called **spindle fibers** appear between the separating centrosomes. The nuclear envelope begins to fragment, and the nucleolus begins to disappear (Fig. 3.16).

As mentioned previously, the disorganized chromatin DNA found in the nucleus now coils into distinct duplicated chromosomes (humans have 46 chromosomes, but only four are shown in Figure 3.16, so that the phases of mitosis are easier to understand). The duplicated chromosomes are now fully visible and randomly scattered in the nucleus. Spindle fibers attach to the chromosome centromeres as the chromosomes continue to shorten and thicken.

Structure of the Spindle At the end of prophase, a cell has a fully formed **mitotic spindle.** The spindle has two **poles** located at opposite ends of the dividing cell. Each pole is formed by a centrosome. Short microtubule bundles called **asters** radiate away from the poles, and spindle fibers stretch between the poles. When complete, each half of the mitotic spindle resembles a lopsided bicycle wheel, with a centrosome as a "hub" and asters and fibers as the "spokes."

Metaphase

During **metaphase,** the nuclear envelope is completely fragmented, and the completed mitotic spindle occupies the region formerly occupied by the nucleus. The paired chromosomes (each with two sister chromatids) are now lined up at the equator (center) of the spindle. Metaphase is characterized by a fully formed spindle, and spindle fibers can easily be seen as they stretch toward the line of chromosomes (Fig. 3.17).

Anaphase

At the start of **anaphase,** the sister chromatids separate. *Once separated, the chromatids are once again called chromosomes.* Separation of the sister chromatids ensures that each cell receives a copy of each type of chromosome and has a complete gene set once mitosis is complete. During anaphase, the daughter chromosomes move to the poles of the spindle. Anaphase is characterized by the movement of chromosomes toward each pole, and thus to opposite sides of the cell.

Function of the Spindle The spindle brings about chromosome movement. Two types of spindle fibers are involved in the movement of chromosomes during anaphase. One type extends from the poles to the equator of the spindle; there, they overlap. As mitosis proceeds, these fibers increase in length, and this helps *push* the chromosomes apart. The chromosomes themselves are attached to other spindle fibers that simply extend from their centromeres to the poles. These fibers get shorter and shorter as the chromosomes move toward the poles. Therefore, they *pull* the chromosomes apart.

Spindle fibers, as stated earlier, are composed of microtubules. Microtubules can assemble and disassemble by the addition or subtraction of tubulin (protein) subunits. This is what enables spindle

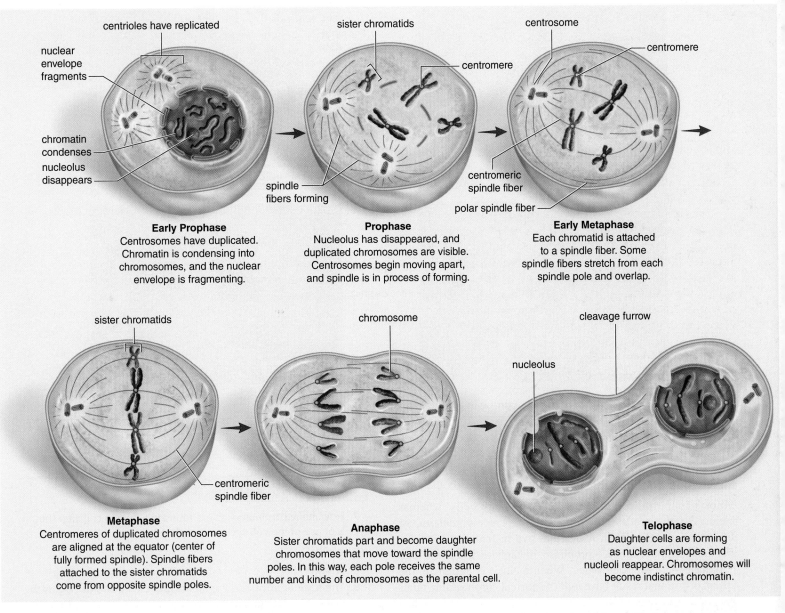

Early Prophase
Centrosomes have duplicated. Chromatin is condensing into chromosomes, and the nuclear envelope is fragmenting.

Prophase
Nucleolus has disappeared, and duplicated chromosomes are visible. Centrosomes begin moving apart, and spindle is in process of forming.

Early Metaphase
Each chromatid is attached to a spindle fiber. Some spindle fibers stretch from each spindle pole and overlap.

Metaphase
Centromeres of duplicated chromosomes are aligned at the equator (center of fully formed spindle). Spindle fibers attached to the sister chromatids come from opposite spindle poles.

Anaphase
Sister chromatids part and become daughter chromosomes that move toward the spindle poles. In this way, each pole receives the same number and kinds of chromosomes as the parental cell.

Telophase
Daughter cells are forming as nuclear envelopes and nucleoli reappear. Chromosomes will become indistinct chromatin.

AP|R **Figure 3.16** **The mitotic stage of the cell cycle.** Humans have 46 chromosomes; four are shown here. The blue chromosomes were originally inherited from the father, and the red were originally inherited from the mother.

fibers to lengthen and shorten, and it ultimately causes the movement of the chromosomes.

Telophase and Cytokinesis

Telophase begins when the chromosomes arrive at the poles. During telophase, the chromosomes become indistinct chromatin again. The spindle disappears as nucleoli appear, and nuclear envelope components reassemble in each cell. Telophase is characterized by the presence of two distinct and separate daughter nuclei.

Cytokinesis is division of the cytoplasm and organelles. In human cells, a slight indentation called a **cleavage furrow** passes around the circumference of the cell. Actin filaments form a contractile ring, and as the ring gets smaller and smaller, the cleavage furrow pinches the cell in half. As a result, each cell becomes enclosed by its own plasma membrane.

Importance of Mitosis

Because of mitosis, each cell in our bodies is genetically identical, meaning that it has the same number and kinds of chromosomes. Mitosis is important to the growth and repair of multicellular organisms. When a baby develops in the mother's womb, mitosis occurs as a component of growth. As a wound heals, mitosis occurs, and the damage is repaired.

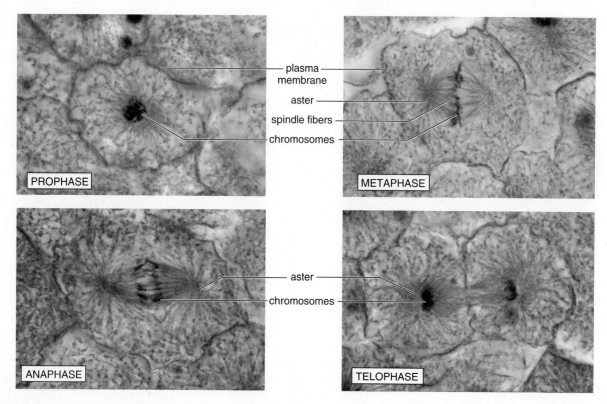

Figure 3.17 Micrographs of mitosis occurring in a whitefish embryo.

Meiosis: Reduction-Division

Mitosis is the process for growing new body cells, whereas meiosis is the process for producing a person's *gametes,* or sex cells: sperm cells in males and ova in females. In meiosis, the stages of mitosis—prophase, metaphase, anaphase, and telophase—are repeated twice. When meiosis is complete, the sperm or ova that result have *half* the normal chromosome number, or 23 (instead of 46 chromosomes in a body cell). When the ovum is joined by a sperm at conception, 23 ovum chromosomes join with 23 sperm chromosomes to form the genetic makeup of the new zygote. (See Chapter 17 for a complete discussion of meiosis.)

Content **CHECK-UP!**

7. How many chromosomes are found in a human cell after the S phase of the cell cycle is complete?

   a. 46   c. 92 (46 × 2)

   b. 23

8. The nuclear envelope fragments and the nucleolus begins to disappear during:

   a. anaphase.   c. telophase.

   b. prophase.   d. metaphase.

9. Formation of a cleavage furrow signals the beginning of which stage?

   a. anaphase   c. telophase

   b. metaphase   d. cytokinesis

Answers in Appendix A.

DNA Fingerprinting

It's almost a cliché of modern life: Look no further than the evening news if you want to view evidence of human cruelty. Tragic scenarios are played out daily in emergency rooms and morgues across North America, as increasing numbers of innocent people become victims of violent crimes—beatings, shootings, stabbings, sexual assault. However, recent advancements in biotechnology make the likelihood of catching and convicting criminals greater than ever before. DNA identification technology, also called DNA fingerprinting, identifies the DNA samples recovered from a crime scene. Like actual fingerprints, the composition of DNA is unique to each individual. Only identical twins share the same DNA.

The process of identifying human DNA begins with sample collection at the crime scene. Blood, hair, bone, tissue, nail clippings, saliva, even vomit—all contain human cells. An individual's DNA is contained within the nucleus and the mitochondrion of the cell, although investigators use nuclear DNA whenever possible because of its larger sample size. Exceedingly small samples of only a nanogram (one-billionth of a gram, approximately 160 human cells) are sufficient, making it possible to identify human DNA from saliva on a cigarette butt or a single human hair.

DNA is first extracted from the sample; if necessary, it can be reproduced from a very small sample to create adequate amounts for testing. The DNA is digested into fragments of differing sizes using restriction enzymes, which act as a molecular "scissors." Each fragment is polar—possessing either a positive or negative charge. A technique called *gel electrophoresis* separates the fragments by size and electrical charge, spreading the fragments throughout a gel medium.

Identifying the exact nucleotide sequence for each fragment is never completed—to do so would be extremely expensive and time consuming. Instead, the best DNA analysis technique identifies STRs, or "short tandem repeats." STRs are segments of repeating nucleotides (for example, A-A-A-A or T-T-T-T). They can be found in all human DNA, but their exact location in the molecule is unique to each individual. Each STR is tagged with radioactive molecules. Finally, the radioactively labeled gel is exposed to X-ray film. The result is a pattern similar to a grocery store "bar code," with alternating light and dark bands.

If a suspect's DNA matches the crime scene DNA, the probability of finding another person on Earth with the same DNA pattern is 1 in 300 billion (a number roughly equal to the number of human beings found on Earth!). Thus, DNA fingerprinting is recognized as a very powerful tool in the courtroom for convicting a guilty individual. However, the same technique can also be used to clear the innocent. If there is no match between the crime scene DNA and that of the suspect, the suspect can be declared innocent of the crime (see Chapter 19, "The Innocence Project").

Crime labs at the state, national, and international levels now maintain computerized databases of DNA fingerprints. These databases allow rapid electronic comparison of DNA samples from crime scenes to the DNA fingerprints of known and suspected felons. Laws currently exist in all 50 states that require felons convicted of murder, assault, sexual offenses, etc., to submit DNA samples on demand to the Combined DNA Index System (CODIS), which is maintained by the FBI.

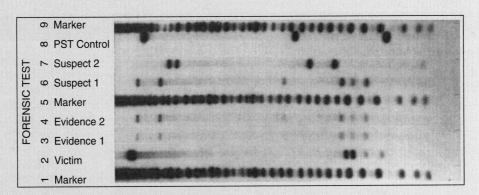

Figure 3B **Actual DNA fingerprint, from a rape trial.** Lanes 1, 2, 5, 8, and 9 are controls. Lane 2 is the victim's DNA, taken from a blood sample. Lanes 3 and 4 were created using semen samples taken from the victim's body after the rape was committed. Study lanes 6 and 7 carefully—who committed the crime—suspect 1 or suspect 2?

Basic Key Terms

actin filaments (ăk′tĭn fĭl′ŭh-mŭhntz), p. 49

active transport (ăk′tĭv trăns′pōrt), p. 53

anaphase (ăn′ŭh-fāz), p. 58

anticodon (ăn″tĭ-kō′dŏn), p. 58

apoptosis (ăp″ō-tō′-sĭs), p. 55

asters (ăs′tŭhrz), p. 58

basal body (bā′sŭhl bŏd′ē), p. 50

cell cycle (sĕl sī′-kŭhl), p. 55

cellular respiration (sĕl′yŭh-lŭr rĕs″pŭh-rā′shŭn), p. 49

centriole (sĕn′trē-ōl), p. 50

centromere (sĕn′trō-mēr), p. 58

centrosome (sĕn′trō-sōm), p. 49

chromatid (krō′mŭh-tĭd), p. 56

chromatin (krō′mŭh-tĭn), p. 46

chromosome (krō′mō-sōm), p. 46

cilium (sĭl′ē-ŭhm), p. 50

cleavage furrow (klēv′ĭj fŭr′ō), p. 59

codon (kō′dŏn), p. 58

crenation (krē-nā′shŭn), p. 52

cytokinesis (sī″tō-kĭ-nē′sĭs), p. 58

cytoplasm (sī′tō-plăzm), p. 43

cytoskeleton (sī′tō-skĕl′ē-tŭn), p. 45

daughter cells (dăh′tŭr sĕlz), p. 58

duplicated chromosome (dū′plŭh-kā-tĕd krō′mō-sōm), p. 56

endocytosis (ĕn″dō-sī′tō-sĭs), p. 55

endomembrane system (ĕn″dō-mĕm′brān sĭs′tĕm), p. 47

endoplasmic reticulum (ĕn-dō-plăz′mĭc rĕ-tĭk′yū-lŭm), p. 47

exocytosis (ĕks′ō-sī-tō′sĭs), p. 55

facilitated diffusion (fŭh-sĭl′-ĕ-tāt″ĭd dĭ-fyū′zhŭn), p. 53

filtration (fĭl-trā′shŭn), p. 53

flagellum (flŭh-jĕl′ŭhm), p. 50

Golgi apparatus (gōl′jē ăp″ŭh-ră′tŭs), p. 47

hemolysis (hē-mŏl′ĭ-sĭs), p. 52

hypertonic (hī′pĕr-tŏn-ĭk), p. 52

hypotonic (hī′pō-tŏn-ĭk), p. 52

intermediate filaments (ĭn″tŭr-mē′dē-ĭt fĭl′ŭh-mĕntz), p. 49

interphase (ĭn′tŭr-fāz), p. 55

isotonic (ī′sō-tŏn-ĭk), p. 52

lysis (lī′sĭs), p. 52

lysosome (lī′sō-sōm), p. 47

metaphase (mĕt′ŭh-fāz), p. 58

microtubule (mī″krō-tū′byŭl), p. 49

mitochondrion (mī″tō-kon′drē-on), p. 49

mitosis (mī-tō′sis), p. 58

mitotic spindle (mī-tŏt′ĭc spĭn′dŭl), p. 58

mitotic stage (mī-tŏt′ĭc stāj), p. 55

nuclear envelope (nū′klē-ĕr ĕn′vĕ-lōp), p. 46

nuclear pore (nū′klē-ĕr pŏr), p. 47

nucleolus (nū-klē′ō-lŭs), p. 43

nucleus (nū′klē-ŭs), p. 43

organelle (or′gŭh-nĕl), p. 43

osmosis (ŏz-mō′sis), p. 52

osmotic pressure (ŏz-mŏt′ĭk prĕsh′ŭr), p. 52

parental cell (pŭh-rĕn′tŭl sĕl), p. 58

peroxisome (pĕr-ŏk′sē-sōm), p. 47

phagocytosis (făg″ō-sī-tō′sĭs), p. 55

pinocytosis (pĭn″ō-sī-tō′sĭs), p. 55

plasma membrane (plăz′mŭh mĕm′brān), p. 43

poles (pōlz), p. 58

polyribosomes (pŏl-ē-rī′bō-sōmz), p. 47

prophase (prō′fāz), p. 58

replication (rĕp″lĭ-kā′shŭn), p. 56

ribosomal RNA (rī″bō-sōm′ŭl RNA), p. 58

ribosome (rī′bō-sōm), p. 47

selectively permeable (sĕ-lĕk′tĭv-lē pĕr′mē-ŭh-bŭl), p. 51

simple diffusion (sĭm′pŭl dĭ-fyū′zhŭn), p. 51

solute (sŏl′yūt), p. 52

spindle fibers (spĭn′dl fī′bĕrz), p. 58

telophase (tĕl′ŭh-fāz), p. 59

template (tĕm′plĭt), p. 56

tonicity (tō-nĭs′ĭ-tē), p. 52

transcription (trăns-krĭp′shŭn), p. 56

transfer RNA (trăns′fŭr RNA), p. 58

translation (trăns-lā′shŭn), p. 56

triplet code (trĭp′lĕt cōd), p. 56

vacuole (văk′yū-ōl), p. 49

vesicle (vĕs′ĭ-kŭl), p. 47

Clinical Key Terms

cystic fibrosis (sĭs′tĭk fī-brō′sĭs), p. 55

dehydration (dē″hī-drā′shŭn), p. 54

Tay-Sachs disease (tā săks dī-zēz′), p. 47

water intoxication (wŏh′-tĕr ĭn-tŏk″sĭ-kā′shŭn), p. 54

Summary

Cells differ in shape and function, but even so, a generalized cell can be described.

3.1 Cellular Organization

All human cells, despite varied shapes and sizes, have a plasma membrane. The plasma membrane is said to be selectively permeable because it regulates what goes through it to enter or leave the cell.

A. The plasma membrane, composed of phospholipid and protein molecules, regulates the entrance and exit of other molecules into and out of the cell.

B. The nucleus contains chromatin, which condenses into chromosomes just prior to cell division. Genes, composed of DNA, make up the chromosomes, and they code for the production of proteins in the cytoplasm. The nucleolus is involved in ribosome formation.

C. Ribosomes are small organelles where protein synthesis occurs. Ribosomes occur in the cytoplasm, both singly and in groups. Numerous ribosomes are attached to the endoplasmic reticulum.

D. The endomembrane system consists of the nuclear envelope, endoplasmic reticulum (ER), the Golgi apparatus, the lysosomes, and various transport vesicles. The ER is involved in protein synthesis (rough ER) and various other processes such as lipid synthesis (smooth ER). Molecules produced or modified in the ER are eventually enclosed in vesicles that take them to the Golgi apparatus. The Golgi apparatus processes and packages molecules, distributes them within the cell, and transports them out of the cell. Lysosomes are produced by the Golgi apparatus, and their hydrolytic enzymes digest macromolecules from various sources.

E. Peroxisomes digest toxins and fatty acids. Vacuoles isolate substances captured by the cell so that lysosomes can digest them.

F. Mitochondria are involved in cellular respiration, a metabolic pathway that provides ATP molecules to cells.

G. Notable among the contents of the cytoskeleton are microtubules, intermediate filaments, and actin filaments. The cytoskeleton maintains the shape of the cell and also directs the movement of cell parts.

H. Centrioles lie near the nucleus and are involved in the production of the spindle during cell division and in the formation of cilia and flagella.

3.2 Crossing the Plasma Membrane
When energy is required for a substance to cross the plasma membrane, the process is called active transport. If no energy is required, the process is passive transport.

A. Some substances can simply diffuse across a plasma membrane, moving from high concentration to low concentration.

B. The diffusion of water is called osmosis. When water diffuses by osmosis, it creates an osmotic pressure. In an isotonic solution, cells neither gain nor lose water.

In a hypotonic solution, cells swell. In a hypertonic solution, cells shrink.

C. During filtration, movement of water and small molecules out of a blood vessel is aided by blood pressure.

D. During facilitated diffusion (facilitated transport), a carrier is required, but energy is not because the substance is moving from higher to lower concentration. Active transport, which requires a carrier and ATP energy, moves substances from lower to higher concentration.

E. Endocytosis involves the uptake of substances by a cell through vesicle formation. Phagocytosis, a type of endocytosis, takes up solids, whereas pinocytosis is ingestion of a liquid. Exocytosis involves the release of substances from a cell as vesicles within the cytoplasm fuse with the plasma membrane.

3.3 The Cell Cycle
The cell cycle, or life span of a cell, consists of a series of stages. It is controlled by internal and external signals such as growth factors. Stem cells are cells that can undergo cell division multiple times without becoming further differentiated.

A. The cell cycle consists of interphase (G_1 phase, S phase, G_2 phase) and the mitotic stage, which includes mitosis and cytokinesis. During G_1, the cell prepares for DNA replication but also performs its normal activities. During S, DNA is replicated. Finally, during G_2, the cell prepares for cell division.

B. During the S phase of interphase, DNA serves as a template for its own replication: The DNA parental molecule unwinds and unzips, and new (daughter) strands form by complementary base pairing.

C. Protein synthesis consists of transcription and translation. During transcription, DNA serves as a template for the formation of RNA. During translation, mRNA, rRNA, and tRNA are involved in polypeptide synthesis.

D. Mitosis consists of a number of phases (prophase, metaphase, anaphase, and telophase) during which each newly formed cell receives a copy of each chromosome. Later, the cytoplasm divides by furrowing. Mitosis occurs during tissue growth and repair.

Study Questions

1. What are the three main parts to any human cell? (p. 43)
2. Describe the fluid-mosaic model of membrane structure. (pp. 45–46)
3. Describe the nucleus and its contents, and include the terms *DNA* and *RNA* in your description. (pp. 46–47)
4. Describe the structure and function of ribosomes. (p. 47)
5. What is the endomembrane system? What organelles belong to this system? (pp. 47–49)
6. Describe the structure and function of endoplasmic reticulum (ER). Include the terms *smooth ER, rough ER,* and *ribosomes* in your description. (pp. 47–48)
7. Describe the structure and function of the Golgi apparatus. Mention vesicles

and lysosomes in your description. (pp. 47–48)
8. Describe the structure and function of mitochondria. Mention the energy molecule ATP in your description. (p. 49)
9. What is the cytoskeleton? What role does the cytoskeleton play in cells? (pp. 49–50)
10. Describe the structure and function of centrioles. Mention the mitotic spindle in your description. (pp. 50–51)
11. Contrast passive transport (diffusion, osmosis, filtration) with active transport of molecules across the plasma membrane. (pp. 51–53, 55)
12. Define osmosis, and discuss the effects of placing red blood cells in

isotonic, hypotonic, and hypertonic solutions. (p. 52)
13. What is the cell cycle? What stages occur during interphase? What happens during the mitotic stage? (pp. 55–60)
14. Review the structure of DNA. How does this structure contribute to the process of DNA replication? (p. 56)
15. Briefly describe the events of protein synthesis. (pp. 56–58)
16. List the phases of mitosis. What happens during each phase? (pp. 58–60)
17. Discuss the importance of mitosis in humans. (p. 59)

Learning Outcome Questions

I. Match the organelles in the key to the functions listed in questions 1–5.

Key:
- a. mitochondria
- b. nucleus
- c. Golgi apparatus
- d. rough ER
- e. centrioles

1. packaging and secretion
2. cell division
3. powerhouse of the cell
4. protein synthesis
5. control center for the cell

II. Fill in the blanks.

6. The _____, which is the substance outside the nucleus of a cell, contains bodies called _____, each with a specific structure and function.
7. The fluid-mosaic model of membrane structure says that _____ molecules drift about within a double layer of _____ molecules.

8. Rough ER has _____, but smooth ER does not.
9. Basal bodies that organize the microtubules within cilia and flagella are believed to be derived from _____.
10. Water will enter a cell when it is placed in a _____ solution.
11. Active transport requires a protein _____ and _____ for energy.
12. Vesicle formation occurs when a cell takes in material by _____.
13. At the conclusion of mitosis, each newly formed cell in humans contains _____ chromosomes.

III. Match the molecules in the key to the functions listed in questions 14–17.

Key:
- a. DNA
- b. mRNA
- c. tRNA
- d. rRNA

14. Joins with proteins to form subunits of a ribosome.

15. Contains codons that determine the sequence of amino acids in a polypeptide.
16. Contains a code and serves as a template for the production of RNA.
17. Brings amino acids to the ribosomes during the process of transcription.

IV. Match the events in the key with the cell cycle stage in questions 18–23.

Key:
- a. DNA replication
- b. nuclear envelope disappears
- c. chromosomes line up in the middle of the cell
- d. sister chromatids move to opposite poles
- e. spindle disappears
- f. division of cytoplasm

18. anaphase
19. cytokinesis
20. interphase
21. metaphase
22. prophase
23. telophase

Medical Terminology Exercise

After studying this chapter, see if you can derive the definitions for the medical terms listed at right. Many of the prefixes and suffixes used to create these terms can be found throughout the chapter. For additional help, use McGraw-Hill Connect™ at www.mcgrawhillconnect.com and consult Appendix B.

1. hemolysis (hē″mŏl′ĭ-sĭs)
2. cytology (sī-tŏl′ō-jē)
3. cytometer (sī-tŏm′ĕ-tĕr)
4. nucleoplasm (nu′klē-ō-plăzm)
5. pancytopenia (păn″sī-tō-pē′nē-ŭh)
6. cytogenic (sī-tō-jĕn′ĭk)
7. erythrocyte (ĕ-rĭth′rō-sīt)
8. atrophy (ăt′rō-fē)

9. hypertrophy (hī-pĕr′trō-fē)
10. oncotic pressure, colloid osmotic pressure (ŏng-kŏt′ĭk prĕsh′ĕr) (kŏl′ŏyd ŏz-măh′-tĭk prĕsh′ĕr)
11. hyperplasia (hī-pĕr-plā′zhē-ŭh)

Online Study Tools

LEARNSMART **connect**

APR

Anatomy & Physiology REVEALED includes cadaver photos that allow you to peel away layers of the human body to reveal structures beneath the surface. This program also includes animations, radiologic imaging, audio pronunciations, and practice quizzing. To learn more visit www.aprevealed.com

4 Body Tissues and Membranes

How long can human cells survive? In the case of cancer, some types of cells might, in fact, be "immortal." Consider the cells of Henrietta Lacks, a young and very poor African American woman who died of cervical cancer in 1951. Cells taken from her original tumor, designated *HeLa* (from *He*nrietta *La*cks) became the first human cells to easily grow in a laboratory. In their 60-plus years of survival, thousands of research projects have used these sturdy cells. Developing vaccines, studying drug effects, investigating virus behavior, developing tests for genetic disorders, and of course, research into cancer—these are only a few of their uses. They can be found in tissue culture laboratories all over the world, including those on the International Space Station. If you're interested in cell research, *HeLa* cells can even be purchased from catalogs. The message on Henrietta's tombstone is a fitting eulogy for this remarkable woman:

In loving memory of a phenomenal woman, wife and mother who touched the lives of many. Here lies Henrietta Lacks (HeLa). Her immortal cells will continue to help mankind forever.

You can read more about cancer in Medical Focus: *Cancer: The Traitor Inside* on page 79.

Learning Outcomes

After you have studied this chapter, you should be able to:

4.1 Epithelial Tissue

1. Describe the general characteristics and functions of epithelial tissue.
2. Name the major types of epithelial tissue, and relate each one to a particular organ.

4.2 Connective Tissue

3. Describe the general characteristics and functions of connective tissue.
4. Name the major types of connective tissue, and relate each one to a particular organ.

4.3 Muscular Tissue

5. Describe the general characteristics and functions of muscular tissue.

6. Name the major types of muscular tissue, and relate each one to a particular organ.

4.4 Nervous Tissue

7. Describe the general characteristics and functions of nervous tissue.

4.5 Extracellular Junctions, Glands, and Membranes

8. Describe the structure and function of three types of extracellular junctions.
9. Describe the difference between an exocrine and an endocrine gland, and be able to give examples of each one.

10. Describe the way the body's membranes are organized.
11. Name and describe the major types of membranes in the body.

Medical Focus

Necrotizing Fasciitis
Cancer: The Traitor Inside

What's New

Targeting the Traitor Inside
3-D Printing to Create Complex Tissues

A tissue is composed of specialized cells of similar structure that perform a common function in the body. There are four major types of tissues: (1) Epithelial tissue, also called epithelium, covers body surfaces and organs. It also lines body cavities and hollow body structures such as the heart, blood vessels, and digestive tract; (2) connective tissue binds and supports body parts; (3) muscular tissue contracts; (4) nervous tissue responds to stimuli and transmits signals from one body part to another. You can use the "4Cs" mnemonic to remember these functions: *Tissues Cover-Connect-Contract-Communicate*. Body organs are typically composed of all four tissue types. For example, the heart is covered by *epithelium;* its valves are supported by *connective tissue;* cardiac *muscle* pumps blood, and *nerves* regulate how rapidly the heart beats.

4.1 Epithelial Tissue

1. Describe the general characteristics and functions of epithelial tissue.
2. Name the major types of epithelial tissue, and relate each one to a particular organ.

Epithelial tissues share a set of common characteristics. In all epithelial tissues, the cells are tightly packed, with little space between them. Externally, this tissue protects the body from drying out, injury, and bacterial invasion. On internal surfaces, epithelial tissue protects, but it also may have an additional function. For example, epithelial tissue in the respiratory tract sweeps up impurities by means of cilia. Along the digestive tract, it secretes mucus, which protects the lining from digestive enzymes. In kidney tubules, its absorptive function is enhanced by the presence of fine, cellular extensions called microvilli.

Epithelial cells readily divide to produce new cells that replace lost or damaged ones. Skin cells as well as those that line the stomach and intestines are continually being replaced. To support its very high rate of reproduction, epithelial tissue must get its nutrients from underlying connective tissues. Epithelial tissues are *avascular*—they lack blood vessels. Thus, they can be shed safely without the risk of bleeding.

Because epithelial tissue covers surfaces and lines cavities, it always has a *free surface*. The other surface is attached to underlying tissue by a nonliving layer of carbohydrates and proteins called the *basement membrane.*

Epithelial tissues are classified according to the shape of the cells and the number of cell layers (Table 4.1). A **simple epithelium** is composed of a single layer, and **stratified epithelium** is composed of two or more layers. **Squamous** epithelium has flattened cells; **cuboidal** epithelium has cube-shaped cells; and **columnar** epithelium has elongated cells. In addition, there are two specialized epithelial tissues: transitional epithelium and pseudostratified columnar epithelium.

Squamous Epithelium

Simple squamous epithelium is composed of a single layer of flattened cells, and therefore its protective function is not as significant as that of other epithelial tissues (Fig. 4.1). It is

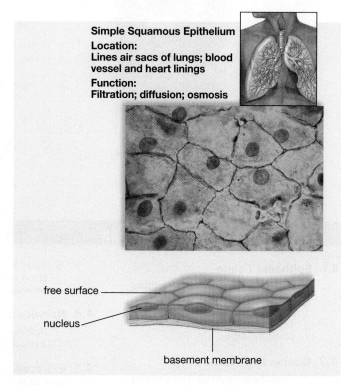

Simple Squamous Epithelium
Location:
Lines air sacs of lungs; blood vessel and heart linings
Function:
Filtration; diffusion; osmosis

free surface
nucleus
basement membrane

AP|R **Figure 4.1 Simple squamous epithelium.** The thin and flat cells are tightly joined. The nuclei tend to be broad and thin.

TABLE 4.1 Epithelial Tissues

Type of Tissue	Description	Location
Simple squamous	One layer of flattened cells	Blood capillaries; air sacs (alveoli) of lungs
Stratified squamous	Many layers; cells flattened at free surface	Skin and body orifices
Simple cuboidal	One layer of cube-shaped cells	Secreting glands, ovaries, linings of kidney tubules
Stratified cuboidal	Two or more layers of cube-shaped cells	Linings of salivary gland and mammary gland ducts
Simple columnar	One layer of elongated cells	Lining of digestive organs; lining of uterine tubes
Stratified columnar	Two or more layers of elongated cells	Pharynx (back of throat); male urethra (tube that carries urine out of body)
Pseudostratified columnar	One layer of elongated, tapered cells; appear stratified	Air passages of the respiratory system
Transitional	Many layers; when tissue stretches, layers become fewer	Urinary bladder (stores urine); ureters and urethra (tubes carrying urine)

Stratified Squamous Non-Keratinized

Location:
Lines mouth, esophagus, anal canal, vagina

Function:
Protection from abrasion and dessication (drying out)

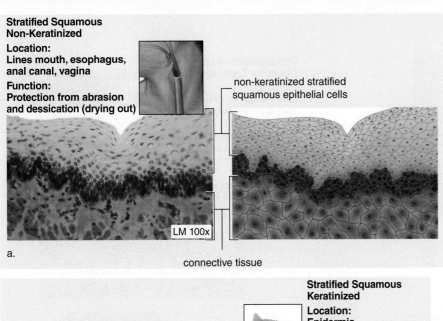

non-keratinized stratified squamous epithelial cells

LM 100x

a.

connective tissue

AP|R **Figure 4.2** **Stratified squamous epithelial tissue.** (**a**) non-keratinized. (**b**) keratinized.

found in areas where simple diffusion occurs. (Remember that simple diffusion is movement of a substance from high to low concentration.) For example, simple squamous epithelium forms the tiny air sacs (called alveoli) of the lungs, where oxygen and carbon dioxide are exchanged. Capillaries (the smallest blood vessels) are tubes of simple squamous epithelium. Nutrients and wastes are exchanged between capillaries and neighboring cells by diffusion.

Stratified squamous epithelium has many cell layers and does play a protective role (Fig. 4.2). While the deeper cells may be cuboidal or columnar, the outer layer is composed of squamous-shaped cells. (Note that stratified epithelial tissues are named according to the cells on their outer layer.) There are two forms of stratified squamous epithelium. The first is stratified squamous non-keratinized epithelium (Fig. 4.2*a*). This tissue acts as a lining for moist surfaces near the openings of body orifices, such as the mouth, vagina and anus. There, it protects against abrasion and drying out, or *dessication*. The second form of stratified squamous epithelium is the superficial (outer) portion of skin, which is keratinized (Fig. 4.2*b*).

Stratified Squamous Keratinized

Location:
Epidermis

Function:
Protection from injury, dessication (drying out)

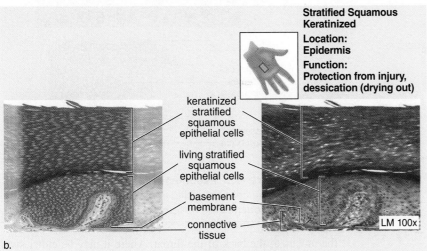

keratinized stratified squamous epithelial cells

living stratified squamous epithelial cells

basement membrane

connective tissue

LM 100x

b.

New skin cells produced in the *basal* (bottom) layer become reinforced by keratin, a protein that waterproofs and provides strength. As they move toward the skin's surface, the cells accumulate more and more keratin, and ultimately die. Thus, the outermost layer of skin is composed of dead cells, as you'll discover in Chapter 5.

Cuboidal Epithelium

Simple cuboidal epithelium (Fig. 4.3) consists of a single layer of cube-shaped cells attached to a basement membrane. This type of epithelium is frequently found in glands, such as salivary glands, the thyroid gland, and the pancreas, where its function is secretion. Simple cuboidal epithelium also covers the ovaries and lines most of the kidney tubules, the portion of the kidney where urine is formed. In one part of the kidney tubule, it absorbs substances from the tubule, and in another part it secretes substances into the tubule. Tubular absorption and secretion are both forms of active transport. Thus, the cuboidal epithelial cells contain many mitochondria, which supply the ATP needed for active transport. Additionally, where the cells function in reabsorption, microvilli (tiny, fanlike folds in the plasma membrane) increase the surface area of the cells.

Stratified cuboidal epithelium is mostly found lining the larger ducts of certain glands, such as the mammary glands and the salivary glands. Often this tissue has only two layers.

Simple Cuboidal Epithelium

Location:
Lines kidney tubules; ducts of many glands; covers surface of ovaries

Function:
Secretion; absorption

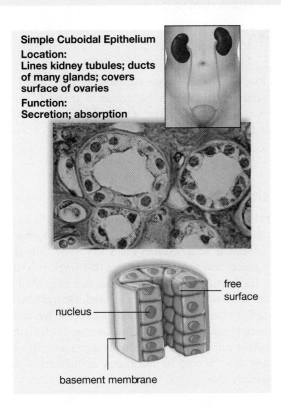

free surface

nucleus

basement membrane

AP|R **Figure 4.3** **Simple cuboidal epithelium.** The cells are cube-shaped. Spherical nuclei tend to be centrally located.

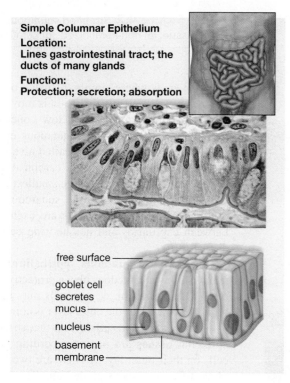

Simple Columnar Epithelium

Location:
Lines gastrointestinal tract; the ducts of many glands

Function:
Protection; secretion; absorption

free surface
goblet cell secretes mucus
nucleus
basement membrane

AP|R **Figure 4.4** **Simple columnar epithelium.** The cells are longer than they are wide. The nuclei are in the lower half of the cells.

Columnar Epithelium

Simple columnar epithelium (Fig. 4.4) has cells that are longer than they are wide. They are modified to perform particular functions. Some of these cells are **goblet cells** that secrete mucus onto the free surface of the epithelium.

This tissue is well known for lining digestive organs, including the small intestine, where microvilli expand the surface area and aid in absorbing the products of digestion. Simple columnar epithelium also lines the uterine tubes. Here, many cilia project from the cells and propel the egg toward the uterus, or womb.

Stratified columnar epithelium is not very common but does exist in parts of the pharynx (back of the throat) and the male urethra.

Pseudostratified Columnar Epithelium

Pseudostratified columnar epithelium is so named because it appears to be layered (*pseudo,* false; *stratified,* layers). However, true layers do not exist because each cell touches the basement membrane. Each cell is tapered and narrow at one end; the opposite end contains the nucleus. The irregular placement of the nuclei creates the appearance of several layers where only one really exists.

Pseudostratified ciliated columnar epithelium (Fig. 4.5) lines parts of the reproductive tract as well as the air passages of the respiratory system, including the nasal cavities and the trachea (windpipe) and its branches. Mucus-secreting goblet cells are scattered among the ciliated epithelial cells. A surface covering of mucus traps foreign particles, and upward ciliary motion carries

Pseudostratified Ciliated Columnar Epithelium

Location:
Lines respiratory tract; parts of the reproductive tracts

Function:
Protection; secretion; movement of mucus and sex cells

cilia
goblet cell secretes mucus
basement membrane

AP|R **Figure 4.5** **Pseudostratified ciliated columnar epithelium.** The cells have cilia and appear to be stratified, but each actually touches the basement membrane.

the mucus to the back of the throat, where it may be either swallowed or expectorated (spit out).

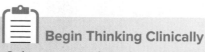
Begin Thinking Clinically

In long-term smokers, the pseudostratified ciliated columnar epithelium in the bronchi is gradually converted to stratified squamous epithelium. Why do you think this change occurs?

Answer and discussion in Appendix A.

Transitional Epithelium

The term **transitional epithelium** implies changeability, and this tissue changes in response to tension. It forms the lining of the urinary bladder, the ureters (tubes that carry urine from the kidneys to the bladder), and part of the urethra (the single tube that carries urine to the outside). All are structures that may need to stretch. When the walls of the bladder are relaxed, the transitional epithelium consists of several layers of cuboidal cells. When the bladder is distended because it is filled with urine, the epithelium stretches, and the outer cells take on a squamous appearance. It's interesting to observe that the cells in transitional epithelium of the bladder are physically able to slide in relation to one another, while at the same time forming a barrier that prevents any part of urine from diffusing into the internal environment.

1. Which of the following are characteristics of epithelial tissues?

    a. possess a blood supply

    b. rapid rate of mitosis

    c. very tightly packed

    d. b and c are correct

    e. All are correct

2. A single layer of tall, skinny epithelial cells would be best described as:

    a. stratified columnar epithelium

    b. simple cuboidal epithelium

    c. simple squamous epithelium

    d. simple columnar epithelium

3. Imagine that while studying cells under a microscope, you see microvilli. How would you describe the function of these cells?

Answers in Appendix A.

4.2 Connective Tissue

3. Describe the general characteristics and functions of connective tissue.
4. Name the major types of connective tissue, and relate each one to a particular organ.

Connective tissue binds structures together, provides support and protection, fills spaces, produces blood cells, and stores fat. As a general rule, connective tissue cells are widely separated by a nonliving, extracellular **matrix** composed of an *organic ground substance* that contains *fibers* and varies in consistency from solid to semifluid to completely fluid. Whereas the functional and physical properties of epithelial tissues stem from their cells, connective tissue properties are largely derived from the characteristics of the matrix (Table 4.2).

The fibers within the matrix are of three types. **Collagen fibers** contain the fibrous protein *collagen,* a substance that gives the fibers flexibility and tremendous strength. **Elastic fibers** contain the protein *elastin,* which is not as strong as collagen but is more elastic. **Reticular fibers** are very thin, highly branched, collagenous fibers that form delicate supporting networks.

TABLE 4.2 Classification of Connective Tissue

Type	Structure of Matrix	Types of Cells	Example of Location
FIBROUS CONNECTIVE TISSUE			
Loose connective tissue			
Areolar	Collagen, elastin, and reticular fibers	Primarily fibroblasts; white blood cells, etc.	Between tissues and organs
Adipose	Fibroblasts enlarge and store fat; very little matrix	Fibroblasts	Beneath skin; around organs
Dense connective tissue			
Regular	Bundles of parallel collagen fibers	Fibroblasts	Tendons; ligaments; aponeuroses
Irregular	Bundles of nonparallel collagen fibers	Fibroblasts	Dermis of skin; joint capsules
Reticular connective tissue	Reticular fibers	Primarily fibroblasts; many white blood cells	Lymphatic organs and liver
CARTILAGE			
Hyaline cartilage	Fine collagen fibers		Ends of long bones; rib cartilages; nose
Elastic cartilage	Many elastin fibers	Chondroblasts and chondrocytes	External ear
Fibrocartilage	Strong collagen fibers		Between vertebrae of spine
BONE			
Compact	Collagen plus calcium salts; arranged in *osteons*		Skeleton
Spongy	Collagen plus calcium salts; arranged in *trabeculae*	Osteoblasts and osteocytes	Ends of long bones
BLOOD	Plasma plus cells, platelets	Red and white blood cells	Inside blood vessels

Fibrous Connective Tissue

Fibrous connective tissue includes both loose connective tissue and dense connective tissue. The body's membranes are composed of an epithelium and fibrous connective tissue (see pp. 80, 82).

Loose connective tissue (also called *areolar connective tissue*) commonly lies between other tissues or between organs, binding them together. Areolar tissue has a fine, spider-web appearance. The cells of this tissue are mainly **fibroblasts**—large, star-shaped cells that produce extracellular fibers (Fig. 4.6). The cells are located some distance from one another because they are separated by a matrix with a jellylike ground substance containing many collagen and elastin fibers. The collagen fibers occur in bundles and are strong and flexible. The elastin fibers form a highly elastic network that returns to its original length after stretching. **Adipose tissue** (Fig. 4.7) is a type of loose connective tissue in which the fibroblasts enlarge and store fat, and there is limited extracellular

matrix. Adipose protects and cushions many organs, including the eye and kidney. In addition, adipose stores energy and insulates the body against the cold.

Dense connective tissue (Fig. 4.8) has a matrix produced by fibroblasts that contains thick bundles of collagen fibers. In *dense regular connective tissue,* the bundles are parallel as in **tendons** (which connect muscles to bones), **ligaments** (which connect bones to other bones at joints), and **aponeuroses** (sing., *aponeurosis;* which join muscle to muscle). In *dense irregular connective tissue,* the bundles run in different directions. This stretchy tissue is found in the inner portion of the skin (called the *dermis*) and in joint capsules (Fig. 4.9).

The fibroblasts of **reticular connective tissue** are called reticular cells, and the matrix contains only reticular fibers. This tissue, also called **lymphatic tissue,** is found in lymph nodes, the spleen, thymus, and red bone marrow. These organs are a part of the immune system because they store and/or produce white blood cells, particularly lymphocytes. All types of blood cells are produced in red bone marrow.

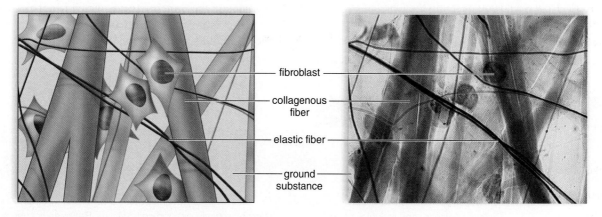

- fibroblast
- collagenous fiber
- elastic fiber
- ground substance

AP|R **Figure 4.6** **Loose (areolar) connective tissue.** This tissue has a loose network of fibers.

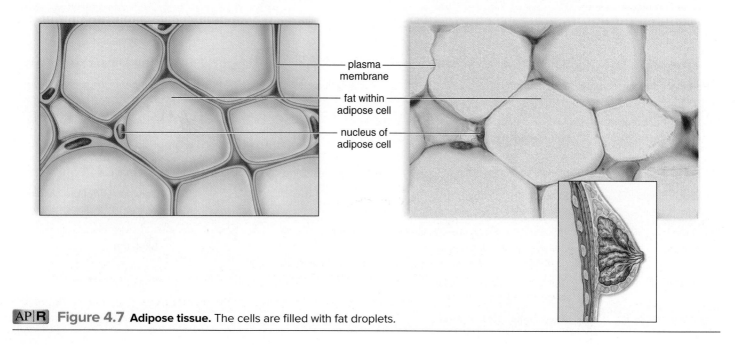

- plasma membrane
- fat within adipose cell
- nucleus of adipose cell

AP|R **Figure 4.7** **Adipose tissue.** The cells are filled with fat droplets.

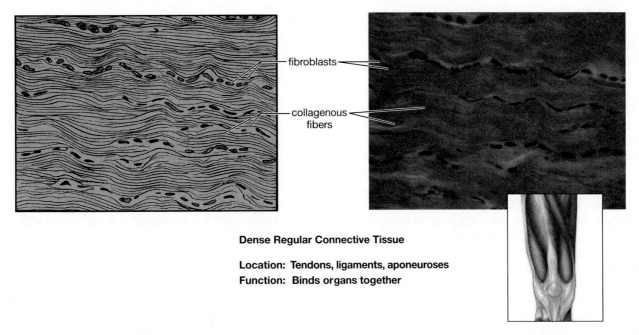

Dense Regular Connective Tissue

Location: Tendons, ligaments, aponeuroses
Function: Binds organs together

AP|R **Figure 4.8** **Dense regular connective tissue.** Parallel bundles of collagenous fibers are closely packed.

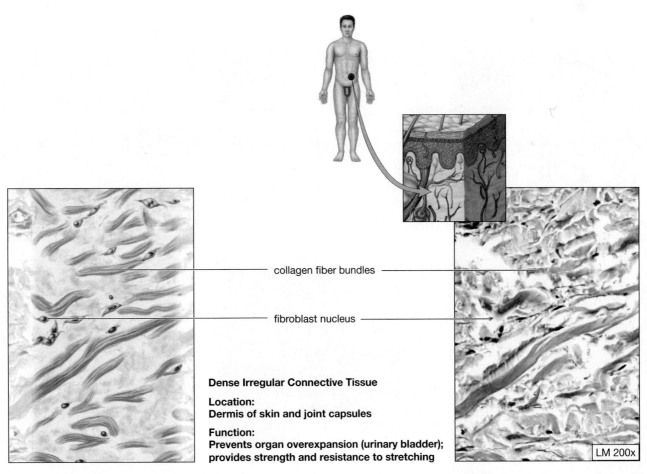

Dense Irregular Connective Tissue

Location:
Dermis of skin and joint capsules

Function:
Prevents organ overexpansion (urinary bladder); provides strength and resistance to stretching

AP|R **Figure 4.9** **Dense irregular connective tissue.** Bundles of collagen fibers run in different directions, enabling the tissue to resist or withstand multidirectional forces.

Cartilage

In **cartilage,** the cells (*chondrocytes*), which lie in small chambers called **lacunae,** are separated by a matrix that is solid yet flexible. Immature chondrocytes, called *chondroblasts,* help cartilage to grow. Unfortunately, because this tissue lacks a direct blood supply, it heals very slowly if injured. The three types of cartilage are classified according to the type of fiber in the matrix.

Hyaline cartilage (Fig. 4.10*a*) is the most common type of cartilage. It is strong and durable, yet flexible. The matrix, which contains only very fine collagen fibers, has a glassy, white, opaque appearance. This type of cartilage is found in the nose, at the ends of the long bones and ribs, and in the supporting rings of the trachea. The fetal skeleton is also made of this type of cartilage, although the cartilage is later replaced by bone.

Elastic cartilage (Fig. 4.10*b*) has a matrix containing many elastic fibers, in addition to collagen fibers. For this reason, elastic cartilage is more flexible than hyaline cartilage. For example, elastic cartilage is found in the framework of the outer ear.

Fibrocartilage (Fig. 4.10*c*) has a matrix containing strong collagen fibers. This type of cartilage absorbs shock, and is found in structures that withstand tension and pressure, such as the disks between the vertebrae in the backbone and the pads in the knee joint.

Bone

Bone is the most rigid of the connective tissues. Two types of cells, the osteoblasts and osteocytes, form an extremely hard matrix of mineral salts, notably calcium salts, deposited around collagen fibers. The minerals give bone rigidity. The collagen fibers provide elasticity and strength, much like steel rods reinforce concrete.

The outer portion of a long bone contains compact bone. **Compact bone** consists of many cylindrical-shaped units called *osteons,* or *Haversian systems* (Fig. 4.11). In an osteon, matrix is deposited in thin layers called *lamellae* that form a concentric pattern around tiny tubes called *central canals.* The canals contain nerve fibers and blood vessels. The blood vessels bring nutrients to bone cells (called *osteocytes*) that are located in small hollows called *lacunae* between the lamellae. The nutrients can reach all of the cells because minute canals (*canaliculi*) containing thin extensions of the osteocytes connect the osteocytes with one another and with the central canals.

The ends of a long bone contain spongy bone, which has an entirely different structure. **Spongy bone** contains numerous bony bars and plates called **trabeculae** separated by irregular spaces (Fig. 4.11). Although lighter than compact bone, spongy bone is still designed for strength. Like braces used for support in buildings, the solid plates of spongy bone follow lines of stress. Blood cells are formed within red marrow found in spongy bone at the ends of certain long bones.

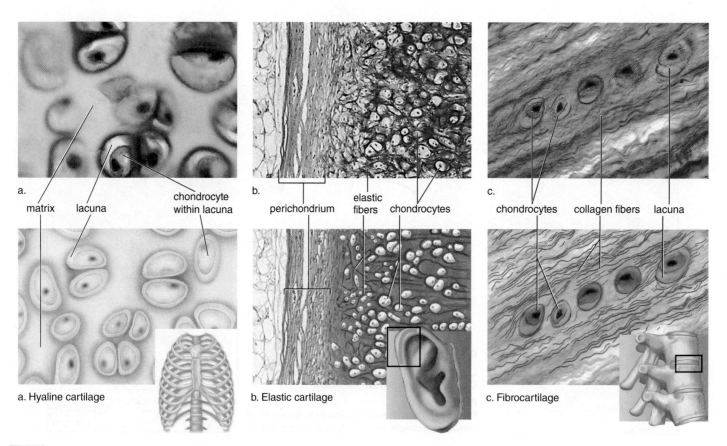

a.
matrix lacuna chondrocyte within lacuna

b.
perichondrium elastic fibers chondrocytes

c.
chondrocytes collagen fibers lacuna

a. Hyaline cartilage

b. Elastic cartilage

c. Fibrocartilage

AP|R **Figure 4.10 Three types of cartilage. (a)** Hyaline cartilage is durable and flexible. **(b)** Elastic cartilage has the greatest flexibility. **(c)** Fibrocartilage is the strongest type.

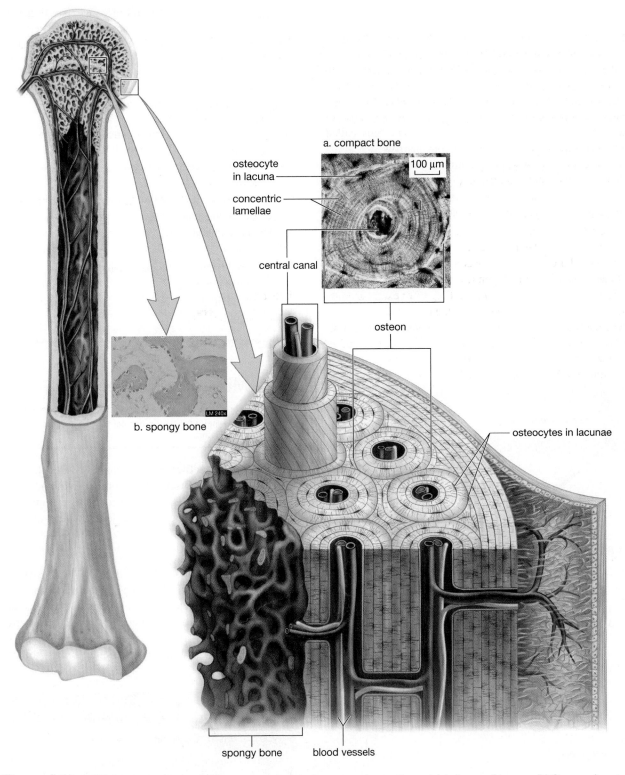

a. compact bone

osteocyte in lacuna

concentric lamellae

central canal

100 μm

osteon

LM 240x

b. spongy bone

osteocytes in lacunae

spongy bone

blood vessels

AP|R **Figure 4.11** **Bone tissue. (a)** Compact bone, composed of osteons, forms the outside layer of bones. **(b)** Spongy bone, composed of trabeculae, is found inside the bone.

Blood

Blood (Fig. 4.12) is a connective tissue composed of **formed elements** suspended in a liquid matrix called **plasma.** There are three types of formed elements: **red blood cells (erythrocytes),** which carry oxygen; **white blood cells (leukocytes),** which aid in fighting infection; and **platelets (thrombocytes),** which are important to the initiation of blood clotting. Platelets are not complete cells; rather, they are fragments of giant cells called **megakaryocytes,** which are found in the bone marrow.

In red bone marrow, stem cells continually divide to produce new cells that mature into the different types of blood cells. The rate of cell division is high because blood cells have a relatively short life span and must be replaced constantly.

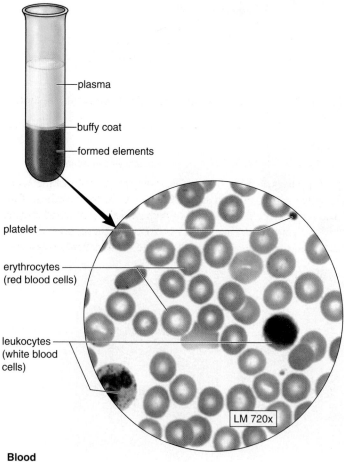

plasma

buffy coat

formed elements

platelet

erythrocytes
(red blood cells)

leukocytes
(white blood
cells)

LM 720x

Blood
Location:
In the blood vessels

Function:
Supplies cells with nutrients and oxygen and takes away their wastes; fights infection

AP|R **Figure 4.12** **Blood.** When a blood sample is centrifuged, the formed elements settle out below the plasma. White blood cells form a layer termed the "buffy coat" on top of red blood cells. Plasma is the liquid portion of the blood. Red blood cells, white blood cells, and platelets are called the formed elements.

Blood is unlike other types of connective tissue because the extracellular matrix *(plasma)* is not made by the cells of the tissue. Plasma is a mixture of different types of molecules that enter blood at various organs.

4.3 Muscular Tissue

5. Describe the general characteristics and functions of muscular tissue.
6. Name the major types of muscular tissue, and relate each one to a particular organ.

Muscular (contractile) tissue is composed of cells called muscle fibers (Table 4.3). Muscle fibers contain actin and myosin, which are protein filaments that interact to cause movement. The three types of vertebrate muscles are skeletal, smooth, and cardiac.

Skeletal Muscle

Skeletal muscle, also called *voluntary muscle* (Fig. 4.13), is attached by tendons to the bones of the skeleton, or directly to the skin. When skeletal muscle contracts, the muscle shortens, and body parts such as arms and legs move. Contraction of skeletal muscle is generally under one's conscious control. Skeletal muscle fibers are cylindrical and quite long—sometimes they

TABLE 4.3   Classification of Muscular Tissue

Type	Fiber Appearance	Location	Control
Skeletal	Striated and cylindrical	Attached to skeleton or skin	Voluntary
Smooth	Nonstriated, spindle-shaped	Wall of hollow organs (e.g., intestine, urinary bladder, uterus, and blood vessels)	Involuntary
Cardiac	Striated, cylindrical and branched	Heart	Involuntary

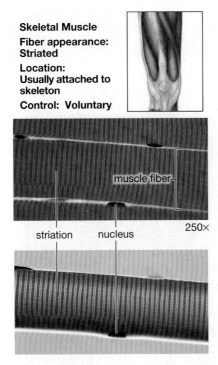

Skeletal Muscle

Fiber appearance: Striated

Location: Usually attached to skeleton

Control: Voluntary

muscle fiber

striation   nucleus   250×

AP|R **Figure 4.13** **Skeletal muscle.** The cells are long, cylindrical, and multinucleated.

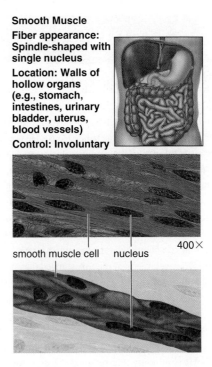

Smooth Muscle

Fiber appearance: Spindle-shaped with single nucleus

Location: Walls of hollow organs (e.g., stomach, intestines, urinary bladder, uterus, blood vessels)

Control: Involuntary

smooth muscle cell   nucleus   400×

AP|R **Figure 4.14** **Smooth muscle.** The cells are spindle-shaped.

run the length of the muscle. They arise during development when several cells fuse, resulting in one fiber with multiple nuclei. The nuclei are located at the periphery of the cell, just inside the plasma membrane. The fibers have alternating light and dark bands that give them a *striated* (striped) appearance. These bands are due to the placement of actin filaments and myosin filaments in the fiber, and they give skeletal muscle its strength.

Smooth Muscle

Smooth (visceral) muscle is so named because the arrangement of actin and myosin does not give the appearance of cross-striations. The *spindle-shaped cells* form layers in which the thick middle portion of one cell is opposite the thin ends of adjacent cells. (A *spindle* is a long, pointed, oval structure.) Consequently, the nuclei form an irregular pattern in the tissue (Fig. 4.14).

Smooth muscle is not under conscious control and therefore is said to be *involuntary*. Smooth muscle is found in the walls of hollow structures and organs, such as the blood vessels, intestines, stomach, uterus, and urinary bladder. This muscle type contracts more slowly than skeletal muscle but can remain contracted for a longer time. Contractility is an important characteristic of smooth muscle, and it contracts rhythmically on its own. However, its contraction can be modified by the nervous and endocrine systems. Intestinal smooth muscle contracts in waves, thereby moving food along its *lumen* (central cavity). When the smooth muscle of blood vessels contracts, blood vessels constrict and their diameter decreases. This helps to regulate blood flow and blood pressure.

Cardiac Muscle

Cardiac muscle (Fig. 4.15) is found only in the walls of the *heart.* Its contraction pumps blood and accounts for the heartbeat. Cardiac muscle combines features of both smooth muscle and skeletal muscle. Like skeletal muscle, it has striations, but the contraction of the heart is *involuntary* (although the use of relaxation therapy does

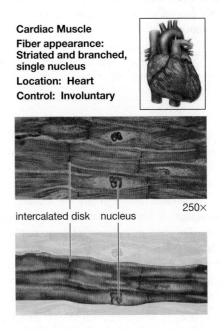

Cardiac Muscle

Fiber appearance: Striated and branched, single nucleus

Location: Heart

Control: Involuntary

intercalated disk   nucleus   250×

AP|R **Figure 4.15** **Cardiac muscle.** The cells are cylindrical but branched.

Necrotizing Fasciitis

On June 30, 1924, 16-year-old Calvin Coolidge, Jr., son of the president, strode out onto the tennis court at the White House. He was anxious to face off against his older brother John, and was in a hurry. He'd forgotten his socks, and chose to play the match without them. As happens to so many of us, Cal left the tennis court with a blister on his foot. One week later, he was dead, from what physicians of the day termed "blood poisoning."

The precise cause of his death will never be known. Calvin Jr. was never autopsied, and it would have been unlikely that testing methods of that era would determine the bacterial organism that infected him. However, medical records showed that surgery had been performed before his death to try to stop the infection. Based on this history, it's quite possible that young Coolidge may have been killed by a condition called *necrotizing fasciitis.* This condition is most commonly caused by *Streptococcus pyogenes,* a group of dangerous pathogenic bacteria that also cause "strep" throat. This strain of *Streptococcus* is commonly referred to as "flesh-eating bacteria."

As you know, fascia, or areolar connective tissue, binds tissues to one another. Tissues are organized into groups by the fascia as well. For example, muscles in the upper and lower limbs are divided into compartments by fascia. In necrotizing fasciitis, the *Streptococcus* bacteria enter the body through a break in the skin—a small abrasion, a cut, an insect bite, or a broken blister. The person will notice that the skin around the wound quickly becomes inflamed and painful. Once infection starts, the bacteria travel extremely rapidly, using fascia channels to rapidly invade tissues all over the body. Poisons produced by the bacteria dissolve and kill connective tissues; thus the term *necrotizing,* which means "causing death." Victims experience a high fever, tremendous pain, and dehydration. If the condition progresses, the person goes into **shock** (you'll read more about shock in Chapter 12). The body's organ systems shut down, resulting in death.

Today, victims of necrotizing fasciitis are aggressively treated with antibiotics and the patient is carefully monitored for signs of

Calvin Coolidge, Jr. (1908–1924).

shock. Surgery to remove dead tissue is frequently necessary to stop the spread of the bacteria. Limbs may need to be amputated. Plastic surgery often must be performed after recovery to restore function, as well as for cosmetic reasons.

Taking a very cautious approach to cleanliness is the best way to stop the disease from occurring. Frequent hand-washing is an obvious—but often neglected—way to prevent all forms of bacterial infection. Further, any damage to the skin must be taken seriously. Even small cuts should be washed with soap and water, and bandaged for a short time until a scab begins to form. Resist the urge to "pop" a blister. Once it opens on its own, disinfect the area thoroughly. If any symptoms of infection develop (high fever, inflammation, and pain around the wound area), seek medical attention as quickly as possible.

---

enable some people to consciously slow the heart). Further, like skeletal muscle, its contractions are strong, but like smooth muscle, the contraction of the heart is inherent and rhythmical. Cardiac muscle contraction can be modified by the nervous and endocrine systems.

Even though cardiac muscle fibers are *striated,* the cells differ from skeletal muscle fibers in that they have a single, centrally placed nucleus. The cells are branched and seemingly fused to one another, and the heart appears to be composed of one large, interconnecting mass of muscle cells. Actually, cardiac muscle cells are separate and individual, but they are bound end-to-end at **intercalated disks,** areas where folded plasma membranes between two cells contain adhesion junctions and gap junctions (see page 75). These permit extremely rapid spread of contractile stimuli so that the fibers contract almost simultaneously.

Content CHECK-UP!

8. Match the muscle type with the characteristic:

   a. skeletal   1. striated and branched cells

   b. cardiac   2. no striations and spindle-shaped

   c. smooth   3. striated and multinucleate

9. Which muscle type or types contract on their own, without outside stimulation?

10. What feature binds cardiac muscle cells to one another so they contract as a unit?

Answers in Appendix A.

4.4 Nervous Tissue

7. Describe the general characteristics and functions of nervous tissue.

Nervous tissue, found in the brain and spinal cord, contains specialized cells called neurons that generate and conduct nerve signals. A **neuron** (Fig. 4.16) has three parts: (1) A *dendrite* receives signals that may result in a nerve signal; (2) the *cell body* contains the nucleus and most of the cytoplasm of the neuron; and (3) the *axon* conducts nerve impulses.

Long axons are called *fibers.* In the brain and spinal cord, fibers form *tracts.* Outside the brain and spinal cord, fibers are bound together by connective tissue to form **nerves.** Nerves conduct signals from sense organs to the spinal cord and brain, where the phenomenon called *sensation* occurs. They also conduct nerve signals away from the spinal cord and brain to muscles, glands, and organs.

In addition to neurons, nervous tissue contains neuroglia.

Neuroglia

Neuroglia are cells that outnumber neurons nine to one and take up more than half the volume of the brain. The primary function of neuroglia is to nourish and support neurons. For example, types of neuroglia found in the brain are microglia, astrocytes, oligodendrocytes, and ependymal cells. *Microglia,* in addition to supporting neurons, engulf bacterial and cellular debris. *Astrocytes* provide nutrients to neurons and produce a hormone known as glial-derived growth factor, which someday might be used as a cure for Parkinson disease and other diseases caused by neuron degeneration. *Oligodendrocytes* form **myelin,** a protective layer of fatty insulation. *Ependymal cells* line the hollow cavities, or ventricles, of the brain.

Schwann cells are the neuroglia that enclose all long nerve fibers located outside the brain or spinal cord. Each Schwann cell covers only a small section of a nerve fiber. The gaps between Schwann cells are called **nodes of Ranvier.** Collectively, the Schwann cells provide nerve fibers with an insulating myelin layer, the **myelin sheath,** separated by the nodes. The myelin sheath speeds conduction because

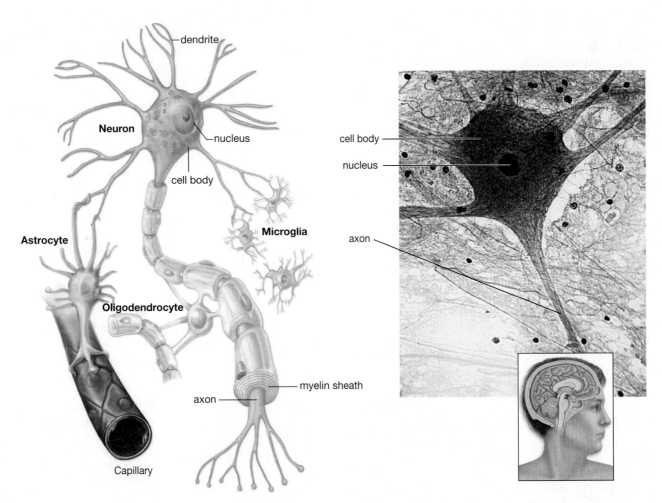

AP|R **Figure 4.16 Nervous tissue.** Neurons are surrounded by neuroglia, such as Schwann cells, which envelop axons. Only neurons conduct nerve signals.

the nerve impulse jumps from node to node. Because the myelin sheath is white, all myelinated nerve fibers appear white.

Content CHECK-UP!

11. Complete this statement: In a nerve cell, a(n) _____ receives signals while a(n) _____ conducts signals.

    a. axon, dendrite

    b. dendrite, axon

    c. dendrite, cell body

12. What two types of neuroglial cells produce myelin?

Answers in Appendix A.

4.5 Extracellular Junctions, Glands, and Membranes

8. Describe the structure and function of three types of extracellular junctions.

9. Describe the difference between an exocrine and an endocrine gland, and be able to give examples of each one.

10. Describe the way the body's membranes are organized.

11. Name and describe the major types of membranes in the body.

Extracellular Junctions

The cells of a tissue can function in a coordinated manner when the plasma membranes of adjoining cells interact. The junctions that occur between cells help cells function as a tissue.

A **tight junction** forms an impermeable barrier because adjacent plasma membrane proteins actually join, producing a zipperlike fastening (Fig. 4.17a). In the stomach, digestive secretions are contained, and in the kidneys, the urine stays within kidney tubules because epithelial cells are joined by tight junctions. A **gap junction** forms when two adjacent plasma membrane channels join (Fig. 4.17b). This lends strength, but it also allows ions, sugars, and small molecules to pass between the two cells. Gap junctions in heart and smooth muscle ensure synchronized contraction. In an **adhesion junction** (desmosome), the adjacent plasma membranes do not touch but are held together by extracellular filaments firmly attached to cytoplasmic plaques, composed of dense protein material (Fig. 4.17c). Desmosomes that join heart muscle cells prevent the cells from tearing apart during contraction. Similarly, desmosomes in the *cervix*, the opening to the uterus (womb), prevent the cervix from ripping when a woman gives birth.

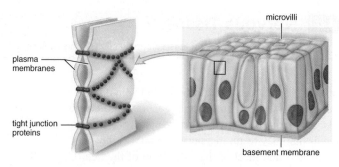

a. Tight junction in digestive tract epithelium

b. Gap junction in smooth muscle

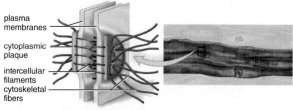

c. Adhesion junction in cardiac muscle

Figure 4.17 Extracellular junctions. Tissues are held together by **(a)** tight junctions that are impermeable; **(b)** gap junctions that allow materials to pass from cell to cell; and **(c)** adhesion junctions that allow tissues to stretch.

Glands

A **gland** consists of one or more cells that produce and secrete a product. Most glands are composed primarily of epithelium in which the cells secrete their product by exocytosis. During secretion, the contents of a vesicle are released when the vesicle fuses with the plasma membrane.

The mucus-secreting goblet cells within the columnar epithelium lining the digestive tract are single cells (see Fig. 4.4). Glands with ducts that secrete their product onto the outer surface (e.g., sweat glands and mammary glands) or into a cavity (e.g., pancreas) are called *exocrine glands.* Ducts can be simple or compound, as illustrated in Figure 4.18.

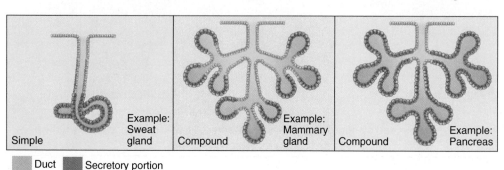

Simple   Example: Sweat gland   Compound   Example: Mammary gland   Compound   Example: Pancreas

Duct   Secretory portion

AP|R Figure 4.18
Multicellular exocrine glands. Exocrine glands have ducts that can be simple or compound. Compound glands vary according to the placement of secretory portions.

Cancer: The Traitor Inside

The life of almost every person has been touched, either directly or indirectly, by the specter of cancer. **Cancer** is not one disease, but perhaps several hundred diseases, all sharing a common characteristic: rapid, uncontrolled, and disorganized growth of tissue cells. Thus, any cell in any of the body's tissues can be the starting point for cancer. Cancers are classified according to the type of tissue from which they arise. **Carcinomas,** the most common type, are cancers of epithelial tissues (skin and linings); **sarcomas** are cancers that arise in connective tissue (muscle, bone, and cartilage); **leukemias** are cancers of the blood; and **lymphomas** are cancers of reticular connective tissue. The chance of cancer occurring in a particular tissue is related to the rate of cell division. As you know, epithelial cells reproduce at a high rate. That's why carcinomas account for 90% of all human cancers.

In the body, a cancer cell divides to form a **malignant neoplasm** ("new tissue"), or a malignant **tumor,** that invades and destroys neighboring tissue. Cancer cells can also detach and spread to other sites by invading the blood vessels or the lymphatic vessels. Through this process, called **metastasis,** cancer tumors colonize healthy tissue elsewhere in the body. By contrast, noncancerous, or **benign tumors** are encapsulated (surrounded by a connective tissue capsule) and stay in one place. To support their growth, cancer cells release a growth factor that causes neighboring blood vessels to branch into the cancerous tissue, a process called **vascularization.** Cancer development seems to occur by a two-step process involving (1) initiation and (2) promotion. Cancer initiation is caused by a change, or a **mutation,** in the DNA (genes) of a cell, which results in runaway cell growth. Some mutations, such as those that result in certain forms of breast cancer, are genetically inherited. Agents that are known to actually cause DNA mutations are called **carcinogens.** Known carcinogens include viruses, excessive radiation, and certain chemicals. For example, cigarette smoke contains chemical carcinogens that may initiate cancers of the lung, throat, mouth, and urinary bladder. A cancer **promoter** is any influence that causes a mutated cell to start growing in an uncontrolled manner. A promoter might cause a second mutation or provide the environment for cells to form a tumor. For example, evidence suggests that a diet rich in saturated fats and cholesterol promotes colon cancer. Considerable time may elapse between initiation and promotion, and this is one reason why cancer is seen more often in older people.

Cancer can be detected by physical examination, assisted by various means of viewing the internal organs. Mammograms can detect early breast cancer using low-level X ray, and thyroid cancer is diagnosed using radioactive iodine (see the Medical Focus on page 16). Specific blood tests exist for tumors that secrete a particular chemical in the blood. For example, the level of prostate-specific antigen (PSA) appears to increase in the blood according to the size of a prostate tumor. Tissue samples can also detect early malignancy. During a **Pap smear** (named for George Papanicolaou, the Greek doctor who first described the test), a small sample of epithelial tissue lining the cervix at the opening of the uterus is obtained using a cotton swab, then examined for cervical cancer cells. A **biopsy** is the removal of a suspect sample of tissue using a plunger-like device. A **pathologist** is a physician who is skilled at recognizing the abnormal characteristics that allow for cancer **diagnosis.** If cancer is found and treated before metastasis occurs, chances for a complete cure are greatly increased.

Tumors can often be removed surgically, but there is always the danger that they have metastasized. For this reason, surgery is often preceded or followed by radiation therapy and/or chemotherapy to destroy rapidly dividing cancer cells. Radiation therapy using radioactive protons is preferred over X ray because proton beams can be aimed directly at the tumor. Chemotherapy drugs kill actively growing cancer cells, but sometimes cancer cells become resistant to chemotherapy (even when several drugs are used in combination). The plasma membrane in resistant cells contains a carrier that pumps toxic chemicals out of the cell. Researchers are testing drugs known to poison the pump in an effort to restore sensitivity to chemotherapy.

Unfortunately, both chemotherapy and radiation kill normal cells as well as the cancer. The patient will suffer the negative side effects of therapy: nausea, vomiting, hair loss, weight loss, anemia, etc. Thus, the use of chemotherapy and radiation must be balanced carefully: strong enough to kill cancer, but not so strong as to cause the person's death. The What's New reading on page 80 describes emerging technologies that will specifically target cancer cells while sparing healthy cells.

Individuals should be aware of the seven danger signals for cancer (Table 4A) and inform their doctor when any one of these are observed. Further, the evidence is clear that the risk of certain types of cancer can be reduced by lifestyle changes. For example, avoiding excessive sunlight reduces the risk of skin cancer, and abstaining from cigarettes, cigars, and chewing tobacco reduces the risk of oral, throat, and lung cancers, as well as other types of cancer. Exercise and a healthy diet are also believed to be important. Recommendations include:

1. Lowering total fat intake
2. Eating more high-fiber foods
3. Increasing consumption of foods rich in vitamins A and C
4. Reducing consumption of salt-cured and smoked foods
5. Including vegetables of the cabbage family in the diet
6. Consuming only moderate amounts of alcohol

TABLE 4.A Danger Signals for Cancer

C	hange in bowel or bladder habits
A	sore that does not heal
U	nusual bleeding or discharge
T	hickening or lump in breast or elsewhere
I	ndigestion or difficulty in swallowing
O	bvious change in wart or mole
N	agging cough or hoarseness

Glands that no longer have a duct are appropriately known as the ductless glands, or endocrine glands. *Endocrine glands* (e.g., pituitary gland and thyroid) secrete their products internally so they are transported by the bloodstream. Endocrine glands produce hormones that help promote homeostasis. Each type of hormone influences the metabolism of a particular target organ or cells.

Glands are composed of epithelial tissue, but they are supported by connective tissue, as are other epithelial tissues.

Membranes

Membranes line the internal spaces of organs and tubes that open to the outside, and they also line the body cavities discussed on pages 6–8.

Mucous Membranes

Mucous membranes line the interior walls of the organs and tubes that open to the outside of the body, such as those of the digestive, respiratory, urinary, and reproductive systems. These membranes consist of an epithelium overlying a layer of loose connective tissue. The epithelium contains goblet cells that secrete mucus.

The mucus secreted by mucous membranes ordinarily protects interior walls from invasion by bacteria and viruses. For example, more mucus is secreted when a person has a cold, resulting in a "runny nose." In addition, mucus usually protects the walls of the stomach and small intestine from digestive juices, but this protection breaks down when a person develops an *ulcer.*

Serous Membranes

You'll remember from Chapter 1 (pages 6–7) that **serous membranes** line cavities, including the thoracic and abdomino-pelvic cavities, and cover internal organs such as the intestines. The term *parietal* refers to the wall of the body cavity, while the term *visceral* pertains to the internal organs. Therefore, parietal membranes line the interior of the thoracic and abdominopelvic cavities, and visceral membranes cover the organs.

Serous membranes consist of a layer of simple squamous epithelium overlying a layer of loose connective tissue. They secrete serous fluid, which keeps the membranes lubricated. Serous membranes support the internal organs and tend to compartmentalize the large thoracic and abdominopelvic cavities. This helps to slow the spread of any infection.

In the thorax, the **pleurae** are serous membranes that form a double layer around the lungs. The parietal pleura lines the inside of the thoracic wall, while the visceral pleura adheres to the surface of the lungs. Similarly, a double-layered serous membrane is a part of the **pericardium,** a covering for the heart (see Fig. 12.2).

The **peritoneum** is the serous membrane within the abdomen. The parietal peritoneum lines the abdominopelvic wall, and the visceral peritoneum covers the organs. In between the organs, the visceral peritoneum comes together to form a double-layered **mesentery** that supports these organs (see Fig. 15.5).

What's New

Targeting the Traitor Inside

"When you get into a tight place and everything goes against you, till it seems as though you could not hang on a minute longer, never give up then, for that is just the place and time that the tide will turn."
—Harriet Beecher Stowe, novelist

A diagnosis of cancer is a terrifying event for anyone. Suddenly, life is turned upside-down, and decisions must quickly be made about treatment options. Radiation therapy and chemotherapy have existed for decades and continue to improve in effectiveness. However, these techniques are comparable to "carpet-bombing" in wartime—throwing many deadly bombs to blanket large areas and destroy as much as possible. As in a real-world conflict, chemotherapy and radiation therapy generally hit their cancer target, but they cause a lot of collateral damage. Frequently, these types of treatments cause extensive damage to other cells and tissues, which may be fatal. Furthermore, cancer cells from the original tumor that survive the original attack have also been observed to mutate over time, becoming increasingly stronger and resistant to both chemotherapy and radiation. When this occurs, these older techniques don't work, and the cancer returns.

Discovering the Enemy

Increasingly, oncologists (doctors who specialize in cancer treatment) have new options to offer their patients. One rapidly improving technique is to begin the treatment process by identifying the exact genetics of the patient's cancer cells. As you know from Focus on Forensics in Chapter 3, cellular DNA can be studied by creating a "fingerprint." Once the cancer is precisely identified, targeted therapies can be developed. Targeted therapies are sometimes referred to as the result of "rational drug design," because the treatments are tailored to damage or destroy only one type of cells—the cancer cells. Normal cells are largely unharmed, survival rates increase, side effects are reduced, and the patient's quality of life is improved.　　　*—Continued*

Continued—

Targeted therapies work by directly interfering with cancer growth and progression. These treatments may function externally (directly on the plasma membrane, for example) or by obstructing internal metabolism. One of the first targeted therapies to be developed is directed at the cell membrane estrogen receptors of breast cancer cells. You discovered in Chapter 2 that estrogens stimulate the growth of all female structures. Unfortunately, breast cancer cells also increase their growth rate in response to estrogen. Selective estrogen receptor modulator (SERM) drugs block cancer cell estrogen receptors by binding to them in place of estrogen. Without estrogen as a promoter, cancer cell growth slows and sometimes completely stops. You may have heard of Tamoxifen, a commonly used SERM. Similar drugs are used for prostate, thyroid, and uterine cancers, and new receptor blockers aimed at other cancers are in development.

Strengthening the Defense

Many cancer researchers believe that cancer develops as a result of a patient's sluggish, underactive immune (defensive) system. Thus, stimulating a stronger immune response, called *immunotherapy*, is the strategy used in yet another type of targeted therapy. One existing immunotherapy involves using small immune system proteins called *antibodies*. Antibodies can be tailored to only bind on a specific set of cancer cells. One type simply "marks" the cancer cells when it attaches. Leukocytes then can destroy the cancer cell. (A good analogy for this process would be pinning a bull's-eye target to a wall—you'd know exactly where to aim!) Different antibodies are used to deliver toxic chemotherapy molecules precisely to specific cells, and others directly kill the cancer cell all by themselves.

A currently used antibody (bevacizumab, marketed as Avastin®) blocks the action of VEGF, or vascular endothelial growth factor. VEGF is produced in huge quantities by cancer cells and stimulates nearby blood vessels to sprout new capillaries. Without VEGF, tumor cells starve and have no route to spread to other body areas.

Bioengineers are now conducting clinical trials of a small, aspirin-sized sponge tablet that is implanted under the skin to fight melanoma, the deadliest form of skin cancer. Inside the tablet, the researchers place proteins obtained from the patient's tumor cells, along with a substance called granulocyte macrophage colony-stimulating factor (GMC-SF). This chemical attracts leukocytes, which quickly recognize the cancer proteins as foreign. These activated leukocytes then cause a potent immune response, which has actually stopped the melanoma in experimental animals. If it works as effectively on human melanoma, this approach could be used to fight this cancer and other cancer forms as well.

Now imagine vaccinating people for cancer, using tumor cell vaccines that would work in a way similar to commonly used vaccines. For example, to immunize a person for the viral diseases *hepatitis* (a liver disease) and *influenza* (flu), a person is injected with pieces of virus. The virus segment can't cause disease, but it does alert the immune system to fight the virus if you're ever exposed. Recently, scientists developed a type of mouse cancer vaccine by mixing lab-grown leukocytes with tumor cell cultures. Exposing the two cell types to each other trained the leukocytes to recognize tumor cells, and the activated defense cells were then injected into the mice. After a second injection with more cancer cells, the resulting tumors in vaccinated mice were ten times smaller than those in unvaccinated mice. Much additional research and testing is needed, but human cancer vaccines created using similar techniques might be possible in the future.

Destroying the Enemy

Receptor-blocking drugs and antibodies can't enter the cancer cell, but modern small-molecule drugs are typically able to diffuse into the cell. Once inside, these drugs attack enzymes involved in cancer cell DNA replication, RNA transcription, or protein translation. (As you'll recall from Chapter 3, these three interphase processes are essential for the cell to reproduce.) Likewise, ongoing research is investigating *nanoparticles* to attack the cancer cell from inside. Nanoparticles are tiny, water-soluble shells roughly half the size of the smallest bacterial cell. These particles are filled with a chemotherapy drug, which spills out and kills the cancer cell once the cell takes up the nanoparticle by phagocytosis. (You also studied phagocytosis in Chapter 3.)

Normally we think of viruses as enemies, because they are naturally occurring invaders that kill cells. However, cancer investigators are now using knowledge of their biology to develop strains that exclusively invade tumor cells. One strain under investigation delivers an inactive drug, called a prodrug, into colon cancer cells. Once inside, the prodrug activates and destroys the cell. Another genetically engineered virus that is currently in clinical trials selectively infects tumor cells of the liver, kidney, ovary, and skin. It then multiplies and destroys the cell while simultaneously triggering additional immune system activity.

Winning the Fight

Rational design therapies such as these may be used alone or combined with traditional chemotherapy and/or radiation based on the patient's own tumor cells. In the future, cancer scientists envision targeted viruses, small-molecule drugs, antibodies, and other new strategies to use in the fight against cancer. Hopefully, as research progresses, we will continue to see even more success stories: patients who have been completely cured.

Synovial Membranes

Synovial membranes line freely movable joint cavities and are composed of connective tissues. They secrete synovial fluid into the joint cavity. This fluid lubricates the ends of the bones so that they can move freely (see Fig. 6.21). In *rheumatoid arthritis,* the synovial membrane becomes inflamed and grows thicker. Fibrous tissue then invades the joint. The invading tissue may eventually become bony so that the bones of the joint are no longer capable of moving.

Meninges

The **meninges** are membranes found within the posterior cavity (see Fig. 1.5). They are composed only of connective tissue and serve as a protective covering for the brain and spinal cord. *Meningitis* is a life-threatening infection of the meninges (see the Medical Focus on page 9).

Cutaneous Membrane

The **cutaneous membrane,** or skin, forms the outer covering of the body. It consists of an outer portion of keratinized stratified squamous epithelium attached to a thick underlying layer of dense irregular connective tissue. The skin is discussed in detail in Chapter 5.

Content **CHECK-UP!**

13. Match each type of extracellular junction to its function:

a. desmosome _____	1. allows ions and small molecules to pass between cells
b. tight junction _____	2. prevents adjacent cells from tearing apart
c. gap junction _____	3. forms an impermeable barrier

14. Explain the difference between exocrine and endocrine glands. Give examples of each gland type.

15. Match each type of membrane to its function:

a. synovial membrane _____	1. forms outer covering of the body
b. cutaneous membrane _____	2. lines walls of organs that open to the outside of the body
c. mucous membrane _____	3. lines the interior of a joint capsule

Answers in Appendix A.

What's New

3-D Printing to Create Complex Tissues

How would you construct a building? Start with a framework, or scaffold, formed from a strong, rigid material such as wood or steel. Around it, add layers: wood, brick, or metal, for example. Inside, insulate and add some form of finishing material (for example, wood or plaster). Now imagine a printer that will spray liquid materials—concrete, strong plastic materials, plaster—and build up that same house layer by layer. That's how a 3-D printer works, and many products such as automobile parts, tools, rocket components and household items are currently being manufactured in this way. Now, bioengineers are now exploring this new technology of 3-D printing to create functional tissues, layer by layer, using stem cells. Stem cells are so-called undifferentiated cells that can "change their minds"—that is, they can form multiple types of different cells.

For several decades, scientists have been able to grow cells in simple sheets, in a process called tissue culture. (As you know from the chapter introduction, HeLa cells were the first to be easily grown in tissue culture.) Multiple applications are now widely used. For example, cultured cells can be grown from a burn patient's remaining healthy skin, then placed over the burned area as a skin graft. Bladder-shaped epithelial linings can be created using a connective tissue framework coated with the patient's own epithelial cells. The linings have been successfully transplanted into a patient's existing bladder. More recently, researchers used a framework created from a mouse heart that was stripped of its cells, and then coated with human heart cells. Not only did the cells grow, they beat in a coordinated way, much like the actual heart. Likewise, skeletal muscle cells grown on a framework were able to contract.

Now imagine that tissue layers could be crafted into organs. As you know from Chapter 1, organs are created from layers of tissues. However, to grow an actual organ from tissue culture layers, a complex network of blood vessels would have to be simultaneously created to provide the organ with oxygen and nutrients, and to carry away waste. Bioengineers have now begun this organ creation process using 3-D printers and tissue-friendly inks called "bio-inks." Each ink performed a different function: one created a scaffold, a second contained living cells, and when complete, the third created a weblike network of tiny hollow tubes. Each bio-ink created its layers in a 3-D process. When the printing was complete, a three-layer tissue was created. Most important, the network of tubes was lined with simple squamous epithelium, just like a capillary (the smallest blood vessel) or a blood vessel lining.

Scientists continue to develop and improve stem cell culture techniques, and refinements in 3-D printing of tissue scaffolds will likely happen as well. In the future, it may soon be possible to completely grow entire organs (such as a heart or a kidney) from the patient's own stem cells.

Selected New Terms

Basic Key Terms

adhesion junction (ăd-hē′zhŭn jŭngk′shŭn), p. 78

adipose tissue (ăd′ĭ-pōs tĭsh′ū), p. 70

aponeuroses (ăp″ō-nū-rō′sĭs), p. 70

blood (blŭd), p. 74

bone (bōn), p. 72

cartilage (kăr′tĭl-ĭj), p. 72

collagen fibers (kŏl′ŭh-jŭn fī′bŭrz), p. 69

columnar epithelium (kŭh-lŭm′ nĕr ĕp-ŭh-thē′lē-ŭhm), p. 66

connective tissue (kŭh-nĕk′tiv tĭsh′ū), p. 69

cuboidal epithelium (kyū-bōyd′ ŭl), p. 66

cutaneous membrane (kyū-tā′nē-ŭs mĕm′brān), p. 82

dense connective tissue (dĕns kŭh-nĕk′ tĭv tĭsh′ū), p. 70

elastic cartilage (ĭ-lăs′tĭk kăr′tĭl-ĭj), p. 72

elastic fibers (ĭ-lăs′tĭk fī′bŭrz), p. 69

epithelial tissue (ĕp-ĭ-thē′lē-ŭhl tĭsh′ū), p. 66

fibroblasts (fī′brō-blăst″), p. 70

fibrocartilage (fī′brō-kăr″tĭl-ĭj), p. 72

gap junction (găp jŭngk′shŭn), p. 78

goblet cell (gŏb′lĕt sĕl), p. 68

hyaline cartilage (hī′ŭh-lĭn kăr′tĭl-ĭj), p. 72

intercalated disk (ĭn-tĕr′kĕ-lāt-ĕd dĭsk), p. 76

lacunae (lŭh-kū′nā), p. 72

ligaments (lĭg′ŭh-mĕntz), p. 70

loose connective tissue (lūs kŭh-nĕk′ tiv tĭsh′ ū), p. 70

lymphatic tissue (lĭm-făt′ĭk tĭsh′ū), p. 70

matrix (mā′trĭks), p. 69

meninges (mĕ-nĭn′jēz), p. 82

mesentery (mĕs′ĕn-tĕr″ē), p. 80

mucous membrane (myū′kŭs mĕm′brān), p. 80

muscular tissue (mŭs′kyū-lĕr tĭsh′ū), p. 74

myelin (mīĕ-lĭn), p. 77

myelin sheath (mī′ĕ-lĭn shēth), p. 77

nervous tissue (nĕr′vŭs tĭsh′ū), p. 77

neuroglia (nū-rō-glē′ŭh), p. 77

neuron (nū′rŏn), p. 77

nodes of Ranvier (nōdz ŭv răhn′vē-ā), p. 77

pericardium (pĕr″ĭ-kăr′dē-ŭhm), p. 80

peritoneum (pĕr″ĭ-tō-nē′ŭhm), p. 80

pleurae (plū′rā), p. 80

pseudostratified columnar epithelium (sū″dō-străt′ ĭ-fīd kŭh-lŭm′nĕr ĕp-ŭh-thē′lē-ŭhm), p. 68

reticular connective tissue (rĕ-tĭk′ū-lăr kŭh-nĕk′ tiv tĭsh′ū), p. 70

reticular fibers (rĕ-tĭk′ū-lăr fī′bŭrz), p. 69

serous membrane (sēr′ŭs mĕm′brān), p. 80

Schwann cell (shwăn sĕl), p. 77

simple epithelium (sĭm′pŭl ĕp-ŭh-thē′lē-ŭhm), p. 66

simple columnar epithelium (sĭm′ pŭl kŭh lŭm′ nĕr ĕp-ŭh-thē′lē-ŭhm), p. 68

simple cuboidal epithelium (sĭm′ pŭl kyū bōyd′ ŭl ĕp-ŭh-thē′ lē-ŭhm), p. 67

simple squamous epithelium (sĭm′ pŭl skwā′ mŭhs ĕp-ŭh-thē′ lē-ŭhm), p. 66

squamous epithelium (skwā′mŭhs ĕp-ŭh-thē′lē-ŭhm), p. 66

stratified columnar epithelium (străt′ĭ-fīd kŭh-lŭm′ nĕr ĕp-ŭh-thē′lē-ŭhm), p. 68

stratified cuboidal epithelium (străt′ĭ-fīd kyū-bōyd′ŭl ĕp-ŭh-thē′lē-ŭhm), p. 67

stratified epithelium (străt′ĭ-fīd ĕp-ŭh-thē′lē-ŭhm), p. 66

stratified squamous epithelium (străt′ ĭ-fīd skwā′ mŭhs ĕp-ŭh-thē′ lē-ŭhm), p. 67

synovial membrane (sĭ-nō′vē-ŭl mĕm′brān), p. 82

tendons (tĕn′dŭhnz), p. 70

tight junction (tīt jŭnk′shŭn), p. 78

trabeculae (trŭh-bĕk′yŭh-lē), p. 72

transitional epithelium (trăn-zĭsh′ŭhn-ŭl ĕp-ŭh-thē′lē-ŭhm), p. 68

Clinical Key Terms

benign (bē-nīn′) tumor, p. 79

biopsy (bī′ŏp-sē), p. 79

cancer (kăn′sĕr), p. 79

carcinogen (kăr-sĭn′ō-jĕn), p. 79

carcinoma (kăr-sĭn′ō-mŭh), p. 79

diagnosis (dī-ŭhg-nō′sĭs), p. 79

leukemia (lū-kē′mē-ŭh), p. 79

lymphoma (lĭm-fō′mŭh), p. 79

malignant (mă-lĭg′nŭnt), p. 79

metastasis (mĕ-tăs′tă-sĭs), p. 79

mutation (myū-tā′shŭn), p. 79

neoplasm (nē′ŭh-plăz″ŭhm), p. 79

Pap smear (păp smēr), p. 79

pathologist (pŭh-thŏl′ŭh-jĭst), p. 79

promoter (prŭh-mō′tĕr), p. 79

sarcoma (săr-kō′mŭh), p. 79

shock (shăwk) p. 76

tumor (tū′ mĕr), p. 79

vascularization (văs″kyū-lŭr-ĭ-zā′shŭn), p. 79

Summary

4.1 Epithelial Tissue

Body tissues are categorized into four types: epithelial, connective, muscular, and nervous. Epithelial tissue is classified according to cell shape and the number of cell layers. The cell shape can be squamous, cuboidal, or columnar. Simple tissues have one layer of cells, and stratified tissues have several layers.

A. Simple squamous epithelium allows substances to diffuse, while stratified squamous is for protection.

B. Simple cuboidal epithelium functions in absorption and secretion. Stratified cuboidal epithelium forms ducts in glands.

C. Simple columnar epithelium is found in mucous membranes. Stratified columnar epithelium is not common in the body.

D. Pseudostratified columnar epithelium appears stratified, but is actually one cell layer thick.

E. Transitional epithelial cells change shape from cuboidal to squamous in response to tension placed on the tissue.

4.2 Connective Tissue

In connective tissue, cells are separated by a matrix (organic ground substance plus fibers).

A. Fibrous connective tissue can be loose connective tissue, in which fibroblasts are separated by a semisolid ground substance, or dense connective tissue, which contains bundles of collagenous fibers. Adipose tissue is a type of loose connective tissue in which the fibroblasts enlarge and store fat. Reticular connective tissue is found in lymphatic tissues.

B. There are three types of cartilage, based on their fiber composition: hyaline, elastic and fibrocartilage.
C. There are two types of bone tissue: compact and spongy. The matrix of bone is mineral salts crystallized on collagen fibers.
D. Blood is a connective tissue in which the matrix is plasma. The three formed elements of blood are red blood cells, white blood cells, and platelets.

4.3 Muscular Tissue
A. Muscular tissue contains actin and myosin protein filaments that interact to cause movement.
B. Skeletal muscle tissue is voluntary, striated, and contains many nuclei in each cell.
C. Smooth muscle has spindle-shaped cells with one nucleus.

They are involuntary and do not have striations.
D. Cardiac muscle is striated, involuntary, and contains unique intercalated disks.

4.4 Nervous Tissue
Nervous tissue contains conducting cells called neurons. Neurons have processes called axons and dendrites. In the brain and spinal cord, axons are organized into tracts. Outside the brain and spinal cord, axons (fibers) are found in nerves.
A. Neuroglia support and nourish neurons. There are five different kinds, each with their own role in the nervous system.

4.5 Extracellular Junctions, Glands, and Membranes
A. In a tissue, cells can be joined by tight junctions, gap junctions, or adhesion junctions (desmosomes).

B. Glands are composed of epithelial tissue that produces and secretes a product, usually by exocytosis. Glands can be unicellular or multicellular. Multicellular exocrine glands have ducts and secrete onto surfaces; endocrine glands are ductless and secrete into the bloodstream.
C. Mucous membranes line the interior of organs and tubes that open to the outside. Serous membranes line the thoracic and abdominopelvic cavities, and cover the organs within these cavities. Synovial membranes line certain joint cavities. Meninges are membranes that cover the brain and spinal cord. The skin forms a cutaneous membrane.

Study Questions

1. What is a tissue? (p. 66)
2. Name the four major types of tissues. (p. 66)
3. What are the functions of epithelial tissue? Name the different kinds of epithelial tissue, and give a location for each. (pp. 66–68)
4. What are the functions of connective tissue? Name the different kinds of connective tissue, and give a location for each type. (pp. 69–71)
5. Contrast the structure of cartilage with that of bone, using the words *lacunae*

and *central canal* in your description. Describe the difference between compact and spongy bone. (pp. 72–73)
6. Describe the composition of blood, and give a function for each type of blood cell. (p. 74)
7. What are the functions of muscular tissue? Name the different kinds of muscular tissue, and give a location for each. (pp. 74–76)
8. What types of cells does nervous tissue contain? Which organs in the

body are made up of nervous tissue? (pp. 77–78)
9. Name three types of junctions, and state the function of each with examples. (p. 78)
10. Describe the structure of a gland. What is the difference between an exocrine gland and an endocrine gland? (pp. 78, 80)
11. Name the different types of body membranes, and associate each type with a particular location in the body. (pp. 80, 82)

Learning Outcome Questions

I. **Fill in the blanks.**
1. Most organs contain several different types of _____.
2. Pseudostratified ciliated columnar epithelium contains cells that appear to be _____, have projections called _____, and are _____ in shape.
3. Connective tissue cells are widely separated by a _____ that usually contains _____.

4. Both cartilage and blood are classified as _____ tissue.
5. Which is a type of loose connective tissue?
   a. areolar
   b. hyaline cartilage
   c. compact bone
6. A mucous membrane contains _____ tissue overlying _____ tissue.

II. **Match the organs in the key to the epithelial tissues listed in questions 7–10.**

Key:
   a. kidney tubules
   b. small intestine
   c. air sacs of lungs
   d. trachea (windpipe)

7. simple squamous
8. simple cuboidal

9. simple columnar
10. pseudostratified ciliated columnar

III. **Match the muscle tissues in the key to the descriptions listed in questions 11–13.**

Key:
   a. skeletal muscle
   b. smooth muscle
   c. cardiac muscle

11. striated and branched, involuntary
12. striated and voluntary
13. visceral and involuntary

IV. **Fill in the blanks.**

14. The cell type that conducts electrical signals is the _____. Supporting cells are called _____.

15. The three types of cell junctions are _____, _____ and _____.
16. The type of gland that secretes into the bloodstream is a(n)_____ gland. The type of gland that secretes into ducts is a(n) _____ gland.
17. Membranes that line cavities open to the environment are _____ membranes.

Medical Terminology Exercise

After studying this chapter, see if you can derive the definitions for the medical terms listed below and at right. Many of the prefixes and suffixes used to create these terms can be found throughout the chapter. For additional help, use McGraw-Hill Connect™ at www.mcgrawhillconnect.com and consult Appendix B.

1. epithelioma (ĕp-ĭ-thē″lē-ō′mŭh)
2. fibrodysplasia (fī″brō-dĭs-plā′sē-ŭh)
3. meningoencephalopathy (mĕ-nĭng″gō-ĕn-sĕf″ŭl-lŏp′ŭh-thē)
4. pericardiocentesis (pĕr″ĭ-kăr″dē-ō-sĕn-tē′sĭs)
5. peritonitis (pĕr″ĭ-tō-nī′tĭs)
6. intrapleural (ĭn″trŭ-plūr′ŭl)
7. neurofibromatosis (nū″rō-fī″brō′mŭh-tō′sĭs)
8. submucosa (sŭb″myū-kō′sŭh)
9. polyarthritis (pŏl″ē-ăr-thrī′tĭs)

10. cardiomyopathy (kăr′dē-ō-mī-ăh′pŭh-thē)
11. encephalitis (ĕn-sĕf″ŭh-lī′-tĭs)
12. glioma (glē-ō′mŭh)
13. pleurisy (plūr′ĭ-sē)
14. chondroblast (kŏn′drō-blăst)
15. osteology (ŏs′tē-ŏl′ō-jē)

Online Study Tools

LEARNSMART® **Connect**

APR

Anatomy & Physiology REVEALED includes cadaver photos that allow you to peel away layers of the human body to reveal structures beneath the surface. This program also includes animations, radiologic imaging, audio pronunciations, and practice quizzing. To learn more visit www.aprevealed.com

Anatomy & Physiology REVEALED®
aprevealed.com

PART II

5 The Integumentary System

The photograph you see at left is a microscopic image of a human hair growing in a *hair follicle*, a tube-shaped accessory organ found in skin. All human hair strands have the same basic chemistry—they're formed from keratin, the fibrous protein you first studied in Chapter 2. So what accounts for the distinct hair colors and textures that you see here? Fundamentally, they're determined by genetic inheritance. Each color is made by combining the two basic forms of *melanin,* the pigment produced by skin cells that also creates skin color. Different blends of these two melanins create the infinite variations of red, blonde, brown, and black color shades that are found in human hair. As melanin production slows with age, a person develops gray hair; a complete lack of melanin causes white hair. Your own hair's texture is created by the shape of your hair follicles. Rounded follicles result in straight hair, such as that seen in many people of Asian descent. If the follicle is flatter, the hair becomes more and more curly. So if you're African, or African American, with very curly hair, your flattened hair follicles are the reason for your curls. Don't like your hair the way it naturally is? By changing its chemistry, you can temporarily straighten it, curl it, or change its color. You can read more about hair follicles, hair growth, and hair loss on pages 89 and 90.

Learning Outcomes

After you have studied this chapter, you should be able to:

5.1 Structure of the Skin

1. Describe the regions of the skin and the hypodermis.
2. Name the epidermal layers, and describe their structure and function.
3. Describe the structure and function of the dermis.

5.2 Accessory Structures of the Skin

4. Describe the structure and growth of hair and nails.
5. List three glands of the skin, and describe their structure and function.

5.3 Functions of the Skin

6. List and describe the six major homeostatic functions of the skin.

7. Explain what occurs in hyperthermia and hypothermia.

5.4 Disorders of the Skin

8. Name the three types of skin cancer, and state their risk factors.
9. Describe several skin diseases, and outline the steps by which a skin wound heals.
10. Name and describe four types of burns with regard to depth.
11. Describe how the "rule of nines" may be used to estimate the extent of a burn.

5.5 Effects of Aging

12. Describe the anatomical and physiological changes that occur in the integumentary system as we age.

5.6 Homeostasis

13. Explain how the integumentary system interacts with other body systems to ensure homeostasis.

Medical Focus
Decubitus Ulcers

I.C.E—In Case of Emergency
Burns

Medical Focus
Body Art: Buyer Beware!

Human Systems Work Together
Integumentary System

5.1 Structure of the Skin

1. Describe the regions of the skin and the hypodermis.
2. Name the epidermal layers, and describe their structure and function.
3. Describe the structure and function of the dermis.

The skin, sometimes called the **cutaneous membrane** or the **integument,** covers the entire surface of the human body. In an adult, the skin has a surface area of about 1.8 square meters (20.83 square feet). Thus, it is the largest organ in the human body. It is important to note that the skin is an organ, comprised of all four tissue types: epithelial, connective, muscle, and nervous tissue. Because the skin has several accessory organs, it is also technically an organ system, sometimes referred to as the **integumentary system.**

The skin (Fig. 5.1) has two regions: the epidermis and the dermis. These layers are joined together by a basement membrane. The **hypodermis,** a **subcutaneous** ("under the skin") **tissue,** is found between the skin and any underlying structures, such as muscle. Usually, the hypodermis is only loosely attached to underlying muscle tissue, but where no muscles are present, the hypodermis attaches directly to bone. For example, there are *flexion creases* where the skin attaches directly to the joints of the fingers.

Epidermis

The **epidermis** is the outer and thinner region of the skin. It is made up of stratified squamous epithelium divided into five separate layers, or *strata* (sing., *stratum*). From deepest to most superficial, the layers are *stratum basale, stratum spinosum, stratum granulosum, stratum lucidum,* and *stratum corneum.* Like all epithelial tissues, epidermis lacks blood vessels and has tightly packed cells.

Stratum Basale

The basal cells of the **stratum basale** lie just superficial to the dermis. Here, epidermal stem cells constantly divide and produce new cells that are pushed to the surface of the epidermis in two to four weeks. As the cells move away from the dermis, they get progressively farther away from the blood vessels in the dermis. Because these cells are not being supplied with nutrients and oxygen (since epidermis lacks blood vessels), they eventually die and are sloughed off.

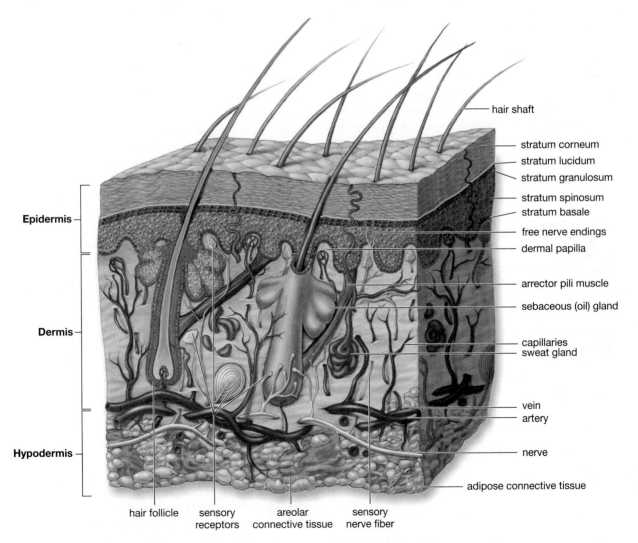

AP|R **Figure 5.1 Skin anatomy.** Skin is composed of two regions: the epidermis and the dermis. The hypodermis, or subcutaneous layer, is located beneath the skin.

Keratinocytes are the most numerous cells in the stratum basale, comprising approximately 95% of the cells in this layer. As their name suggests, keratinocytes produce and store a thick, waterproof protein called **keratin.** These sturdy cells form a barrier between the outer environment and the body. As keratinocytes move superficially into the stratum spinosum layer, they enlarge by packing in additional keratin and other types of protein.

Langerhans cells are macrophages found deep in the epidermis. Macrophages are a type of white blood cell (leukocyte). These cells phagocytize microbes and then travel to lymphatic organs, where they stimulate the immune system to react.

Melanocytes are another type of specialized cell located in the deeper epidermis. Melanocytes produce **melanin,** the pigment primarily responsible for skin and hair color. Because the number of melanocytes is about the same in all individuals, variation in skin color is due to the amount of melanin produced and its distribution. When skin is exposed to the sun, melanocytes produce more melanin to protect the skin from the damaging effects of the ultraviolet (UV) radiation in sunlight. The melanin is passed to other epidermal cells, and the result is tanning, or in some people, the formation of patches of melanin called freckles. A hereditary trait characterized by the lack of ability to produce melanin is known as **albinism.** Individuals with this disorder lack pigment not only in the skin, but also in the hair and eyes. Another pigment, called carotene, is present in epidermal cells and in the dermis and gives the skin of certain Asians its yellowish hue. The pinkish color of fair-skinned people is due to the pigment hemoglobin in the red blood cells in the capillaries of the dermis.

Sensory nerves also supply the stratum basale. **Free nerve endings** supply pain and temperature sensations to the brain (Fig. 5.1). **Tactile cells** (also called **Merkel cells**) signal the brain that an object has touched the skin.

Stratum Spinosum and Stratum Granulosum

Immediately superior to the stratum basale are two additional layers of cells—the **stratum spinosum** and **stratum granulosum.** Like the stratum basale, stratum spinosum cells can reproduce by mitosis. Their name is derived from their spiny appearance, which is created by keratin filaments. Stratum granulosum cells are flattened cells that get their name from the dark-staining protein granules found in their cytoplasm. These cells contain still more keratin than stratum spinosum cells. Stratum granulosum cells are tightly sealed together and form an effective barrier.

Stratum Lucidum

As you examine your skin, you can likely identify areas where constant abrasion has created calluses. These are areas where the epidermis has formed **stratum lucidum,** just deep to the stratum corneum. This additional layer is found only in thick skin: the palms of the hands, soles of the feet, elbows, etc. In these areas, both the stratum lucidum and an extra-thick layer of stratum corneum provide protection from constant friction.

Stratum Corneum

As cells are pushed toward the surface of the skin, they become flat and hard, forming the tough, uppermost layer of the epidermis, the **stratum corneum.** Hardening is caused by keratinization of the entire cell, which causes the uppermost cell layers of the epidermis

to die. We constantly shed these dead cells throughout our environment. Over much of the body, keratinization is minimal. However, in areas containing an underlying stratum lucidum, a particularly thick layer of dead, keratinized cells affords extra protection.

The waterproof nature of keratin protects the body from water loss and water gain. The stratum corneum allows us to live in a desert or a tropical rain forest without damaging our inner cells. The stratum corneum also serves as a mechanical barrier against microbe invasion. This protective function of skin is assisted by the secretions of *sebaceous glands* (discussed in section 5.2).

Begin Thinking Clinically

Psoriasis is a disorder of the immune system that causes immune cells to target the epidermis. The cells of the epidermis reproduce much more rapidly than normal. What do you think will happen to the skin's appearance?

Answer and discussion in Appendix A.

Dermis

The **dermis,** a deeper and thicker region than the epidermis, is composed of dense irregular connective tissue. The upper layer of the dermis has fingerlike projections called *dermal papillae.* Dermal papillae project into and anchor the epidermis. In addition, dermal papillae cause ridges in the overlying epidermis. Epidermal ridges, commonly known as "fingerprints," are spiral and concentric skin patterns that increase friction and provide a better gripping surface. Because they are unique to each person, fingerprints and footprints can be used for identification purposes.

The dermis contains collagen and elastic fibers. The *collagen fibers* are flexible but offer great resistance to overstretching; they prevent the skin from being torn. The *elastic fibers* stretch to allow movement of underlying muscles and joints, but they maintain normal skin tension. The dermis also contains blood vessels that supply oxygen and nutrients to its cells, and those of the epidermis as well. Blood rushes into these vessels when a person blushes; *pallor* (pale skin) develops when blood flow to dermal vessels is reduced. If blood is not adequately supplied with oxygen (perhaps because of lung disease), the person turns *cyanotic,* or "blue."

Extended periods of diminished blood flow to the dermis can cause the formation of decubitus ulcers, or bedsores (see Medical Focus, p. 89).

There are also numerous sensory nerve fibers in the dermis that take nerve signals to and from the accessory structures of the skin, which are discussed in section 5.2.

Hypodermis

As mentioned previously, the hypodermis, or subcutaneous tissue, lies below the dermis. (From the names for this layer, we get the terms **subcutaneous injection,** performed with a **hypodermic needle.**)

The hypodermis is composed of loose connective tissue, including adipose (fat) tissue. Fat is an energy storage form that can be called upon when necessary to supply the body with molecules for cellular respiration. Adipose tissue also helps insulate

Decubitus Ulcers

Decubitus ulcers, or *bedsores,* are a critical problem that should concern *anyone* who is a care provider for patients requiring long-term or extended care. Bedsores develop when there is constant, unrelieved pressure on a single area of the skin. Blood supply to the dermis is blocked by the continuous pressure. Because the epidermis relies on diffusion of oxygen from underlying dermis, epidermal cells will begin to die. Areas of the body where bone is close to the skin are common areas for the development of bedsores. These include the sacral and coxal areas (see Fig. 1.3), as well as the ankle, heel, shoulder, and elbow.

Populations at risk for bedsores include elderly patients, quadriplegics and paraplegics, and brain-injured individuals. Additional risk factors include anemia (decreased red blood cell count), urinary/fecal incontinence, and malnutrition. Signs of injury to the skin begin with a sunburn-like redness; as injury worsens, blisters develop. Progressive injury will cause skin loss, first involving the epidermis and subsequently spreading to the dermis and hypodermis (Fig. 5A). Full-thickness bedsores may penetrate to underlying muscle tissue or even to the bone itself. Patients will have a fever and increased white blood cell count. If the patient is able to feel pain, the sore is very painful; however, many quadriplegics and paraplegics have no pain sensation. Infection of the bedsore can slow the healing process, and in severe cases will be fatal.

Once formed, decubitus ulcers are very difficult to cure. Keeping pressure off the affected area is critical. Moist, sterile dressings are applied to the wound to protect growing skin. Therapy using high-pressure oxygen is sometimes helpful. Severe cases may require a skin graft to replace lost tissue.

The best treatment for bedsores is prevention. Bedridden patients must be turned and repositioned hourly. Wet, soiled clothing must be

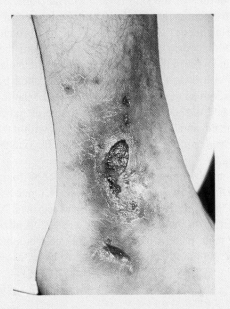

Figure 5A Decubitus ulcer (bedsore). The most frequent sites for bedsores are in the skin overlying a bony projection, such as on the hip, ankle, heel, shoulder, or elbow.

changed immediately to prevent further irritation to the skin. Bony areas at risk for bedsore development should be well padded. Massaging the skin gently to stimulate blood flow is helpful. Proper nutrition is important to provide the nutrients needed by rapidly growing skin cells.

---

the body. A well-developed hypodermis gives the body a rounded appearance and provides protective padding against external assaults. Excessive development of adipose tissue in the hypodermis layer results in obesity.

Content **CHECK-UP!**

1. Blood vessels can be found in:
   a. the epidermis, dermis, and hypodermis.
   b. the dermis only.
   c. the hypodermis and the dermis.

2. Which of the following white blood cells found in the epidermis phagocytize microbes and stimulate the immune system?
   a. Merkel cells
   b. Langerhans cells
   c. melanocytes

3. What is the deepest layer of the epidermis? What is the most superficial layer?

Answers in Appendix A.

5.2 Accessory Structures of the Skin

4. Describe the structure and growth of hair and nails.
5. List three glands of the skin, and describe their structure and function.

Hair, nails, and glands are structures of epidermal origin, even though some parts of hair and glands are located largely in the dermis.

Hair and Nails

Hair is found on all body parts except the palms, soles, lips, nipples, and portions of the external reproductive organs. Most of this hair is fine and downy, but the hair on the head includes stronger types as well. After puberty, when sex hormones are made in quantity, there is noticeable hair in the axillary and pubic regions of both sexes. In males, a beard develops, and other parts of the body may also become quite hairy. When women produce more male sex hormone than usual, they can develop **hirsutism,** a condition characterized by excessive body and facial hair. Hormonal injections and procedures to kill hair roots are possible treatments for this condition.

Hairs project from complex structures called **hair follicles.** These hair follicles are formed from epidermal cells but are located in the dermis of the skin (Fig. 5.2). Hair follicle cells are found in a structure called the **hair matrix,** located at the base of the hair. Hair matrix cells continually divide, producing new keratinocytes that form a hair. As you know from the chapter opener, melanocytes in the hair matrix produce the melanins that give hair its color. At first, the cells are nourished by dermal blood vessels, but as the hair grows up and out of the follicle, its cells are pushed farther away from this source of nutrients, become keratinized, and die. The portion of a hair within the follicle is called the *root,* and the portion that extends beyond the skin is called the *shaft.*

The life span of an eyelash is usually three to four months, while a scalp hair survives for three to four years. Once its life span is complete, the eyelash or hair is shed and regrows. **Alopecia,** meaning hair loss, can have many causes. **Androgenetic alopecia,** or male pattern baldness, is the most common form. It is a genetically inherited condition caused by an excess of the male hormone dihydrotestosterone. Lotions containing minoxidil (Rogaine®) may promote blood flow to the hair follicles and cause hair regrowth in some men with male pattern baldness. *Alopecia areata* is characterized by the sudden onset of patchy hair loss. It is most common among children and young adults, and can affect either sex.

Each hair has one or more oil, or sebaceous, glands, whose ducts empty into the follicle. A smooth muscle, the **arrector pili,** attaches to the follicle in such a way that contraction of the muscle causes the hair to stand on end. If a person has had a scare or is cold, "goose bumps" develop due to contraction of these muscles.

Nails grow from special epithelial cells at the base of the nail in the region called the *nail root* (Fig 5.3). These cells become keratinized as they grow out over the nail bed. The visible portion of the nail is called the *nail body.* The cuticle is a fold of skin that hides the nail root. Ordinarily, nails grow only about 1 millimeter per week.

The pink color of nails is due to the vascularized dermal tissue beneath the nail. The whitish color of the half-moon-shaped base, or **lunula,** results from the thicker layer of rapidly reproducing cells in this area.

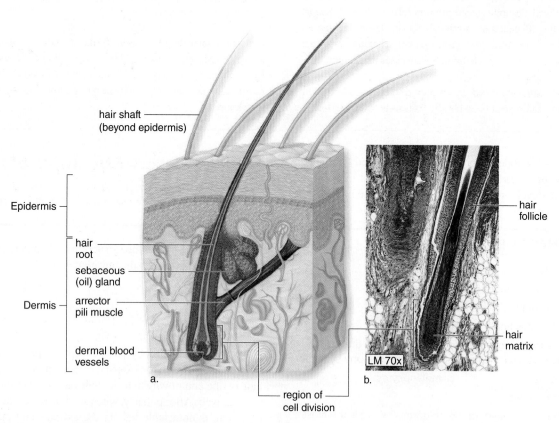

AP|R Figure 5.2 **Hair follicle and hair shaft.** (a) A hair grows from the base of a hair follicle where epidermal cells produce new cells as older cells move outward and become keratinized. (b) A hair follicle and matrix magnified 70x.

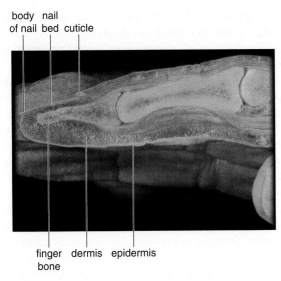

body nail
of nail bed cuticle

finger dermis epidermis
bone

AP|R **Figure 5.3** **Sagittal section of a nail.** Cells produced by the nail root become keratinized, forming the nail body.

Glands

The *glands* in the skin are groups of cells specialized to produce and secrete a substance into ducts.

Sweat Glands

Sweat glands, or sudoriferous glands, are present in all regions of the skin. There can be as many as 90 glands per square centimeter on the leg, 400 glands per square centimeter on the palms and soles, and an even greater number on the fingertips. A sweat gland is tubular. The tubule is coiled, particularly at its origin within the dermis. These glands become especially active when a person is under stress.

Two types of sweat glands are shown in Figure 5.4. Both types secrete their products by exocytosis (see Chapter 3). *Apocrine glands* open into hair follicles in the anal region, groin, and armpits. These glands begin to secrete at puberty, and a component of their secretion may act as a sex attractant. *Eccrine glands* open onto the surface of the skin. They become active when a person is hot, helping to lower body temperature as sweat evaporates. The sweat (perspiration) produced by these glands

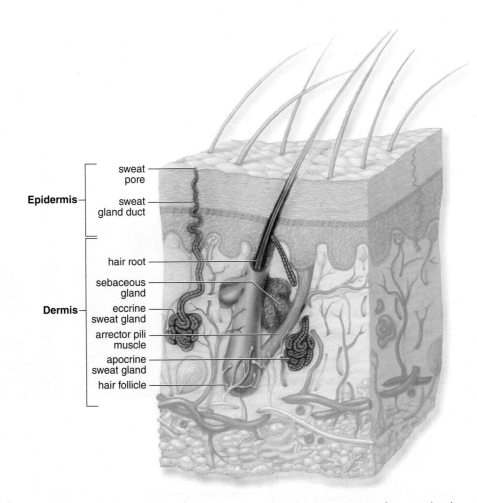

Epidermis
- sweat pore
- sweat gland duct

Dermis
- hair root
- sebaceous gland
- eccrine sweat gland
- arrector pili muscle
- apocrine sweat gland
- hair follicle

AP|R **Figure 5.4** **Types of skin glands.** Apocrine glands and eccrine glands are types of sweat glands.

is mostly water, but it also contains salts and some urea, a waste substance. Therefore, sweat is a form of excretion. The opening into the ear, called the external ear canal, is lined with *ceruminous glands,* which are modified sweat glands. These glands produce cerumen, or earwax.

Sebaceous Glands

Most **sebaceous glands** are associated with a hair follicle. These glands secrete an oily substance called **sebum** that flows into the follicle and then out onto the skin surface. This secretion softens and lubricates the hair and skin and helps to waterproof them both. Sebum also weakens or kills bacteria on the skin surface.

Particularly on the face and back, the sebaceous glands may fail to discharge sebum, and the secretions collect, forming whiteheads or blackheads. If pus-inducing bacteria are also present, a boil or pimple may result.

Acne vulgaris, the most common form of acne, is an inflammation of the sebaceous glands that most often occurs during adolescence. Hormonal changes during puberty cause the sebaceous glands to become more active at this time.

Mammary Glands

Mammary glands are modified apocrine sweat glands that produce milk only after childbirth. The anatomy and physiology of mammary glands are discussed in Chapter 17.

Content CHECK-UP!

4. Which part of the hair contains actively growing cells?

  a. hair matrix   c. hair root

  b. hair shaft

5. Contrast hirsutism and alopecia. Give reasons why each might occur.

6. Which glands begin secretion at puberty and contain a sex attractant?

  a. eccrine glands   c. ceruminous glands

  b. apocrine glands   d. sebaceous glands

Answers in Appendix A.

5.3 Functions of the Skin

6. List and describe the six major homeostatic functions of the skin.
7. Explain what occurs in hyperthermia and hypothermia.

Now that you understand the structure of the skin, consider the many functions of this essential organ system.

Skin protects. First and foremost, the skin forms a defensive shield over the entire body, safeguarding delicate underlying structures from physical trauma. The melanocytes in the stratum basale protect the skin itself from UV radiation. Further, sebaceous gland secretions, phagocytic Langerhans cells, and the outermost layer of dead cells all help to protect against invasion by pathogens.

Skin helps regulate water loss. Since outer skin cells are dead and keratinized, the skin is waterproof, thereby preventing water loss. The skin's waterproofing also prevents water from entering the body when the skin is immersed.

Skin helps to eliminate excess water and waste. Sweat glands secrete water from the body through sensible and insensible perspiration. Sensible perspiration can be felt; insensible perspiration occurs without one's awareness as water evaporates from the body. This perspiration contains small amounts of salt, ammonia, urea, and other wastes. Thus, skin plays a minor role in waste elimination.

Skin produces vitamin D. When keratinocyte cells are exposed to sunlight, the ultraviolet (UV) rays assist them in producing vitamin D from a precursor molecule. Only a small amount of UV radiation is needed for this process to occur. Vitamin D leaves the skin and enters the liver and kidneys, where it is converted to a hormone called calcitriol. Calcitriol regulates calcium uptake by the digestive system, as well as cellular metabolism of both calcium and phosphorus. Calcium–phosphorus compounds are essential for bone development and mineralization. They are deposited in bone, making it strong, yet flexible. In addition, calcium is essential both for nervous activity and muscular contraction. Most milk today is fortified with vitamin D, which helps prevent rickets. **Rickets** is especially characterized by soft and deformed bones (Fig. 5.5*b*).

Skin gathers sensory information from one's surroundings. The sensory receptors in the epidermis and dermis specialized for touch, pressure, pain, hot, and cold supply the central nervous system with information about the external environment. The fingertips contain the greatest number of touch receptors, allowing the fingers to be used for delicate tasks. The sensory receptors also facilitate communication between people. In particular, touch is important for sexual arousal.

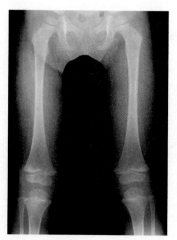 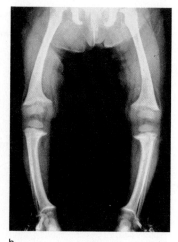

a.   b.

Figure 5.5 **(a) X ray of a normal child. (b) X ray of a child with rickets.** Rickets develops from an improper diet and also from a lack of ultraviolet (UV) light (sunlight). Under these conditions, vitamin D does not form in the skin, which leads to a reduction in calcium uptake. Low calcium in bones weakens them and can cause bowing.

Skin helps regulate body temperature. When muscles contract and ATP is broken down, heat is released. If body temperature starts to rise above normal, the smooth muscle layer found in skin blood vessels will relax. This will cause the diameter of the blood vessel to increase, a process called *vasodilation.* When skin blood vessels vasodilate, more blood can move to the surface of the body, and excess body heat can be released to the environment. Further, the sweat glands are also activated as blood moves to the skin's surface. Sweat absorbs body heat, and this heat is carried away as sweat evaporates. If the weather is humid, evaporation can be assisted by a cool breeze. However, if the outer temperature is cool, the sweat glands remain inactive. At the same time, *vasoconstriction* occurs in the skin's blood vessels: smooth muscle cells in the vessels contract, causing the diameter of the blood vessel to decrease. As a result, less blood is brought to the skin's surface and less body heat is lost to the environment. In addition, whenever the body's temperature falls below normal, the skeletal muscles start *shivering:* rhythmic muscular contractions that don't cause movement, but produce heat instead. Further, arrector pili muscles attached to hair follicles cause goose bumps to form when a person is cold. The arrector pili response in animals (birds, cats, and dogs, for example) provides extra insulation when hair or feathers stand upright, but the insulating effect is absent in humans because body hair is sparse. If the outside temperature is extremely cold and blood flow to the skin is severely restricted for an extended period, a portion of the skin will die, resulting in frostbite.

Hyperthermia and Hypothermia

Hyperthermia, a body temperature above normal, and **hypothermia,** a body temperature below normal, indicate that the body's regulatory mechanisms have been overcome. In *heat exhaustion*, blood pressure may be low, and salts may have been lost due to profuse sweating. Even so, body temperature remains high. *Heat stroke* is characterized by an elevated temperature, up to 43°C (110°F), with no sweating. *Fever* is a special case of hyperthermia that can be brought on by immune system response and/or by a bacterial infection. When the fever "breaks," sweating occurs as the normal set point for body temperature returns.

At first, hypothermia is characterized by uncontrollable shivering, incoherent speech, and lack of coordination (body temperature 90°–95°F). Then the pulse rate slows, and hallucinations occur as unconsciousness develops (body temperature 80°–90°F). Breathing becomes shallow, and shivering diminishes as rigidity sets in. This degree of hypothermia is associated with a 50% mortality rate.

Content **CHECK-UP!**

7. How does the skin help the urinary system eliminate wastes from the body?

8. When body temperature drops below normal, blood vessels in the dermis _____. Explain how this prevents body temperature from dropping further.

Answers in Appendix A.

5.4 Disorders of the Skin

8. Name the three types of skin cancer, and state their risk factors.

9. Describe several skin diseases, and outline the steps by which a skin wound heals.

10. Name and describe four types of burns with regard to depth.

11. Describe how the "rule of nines" may be used to estimate the extent of a burn.

The skin is subject to many disorders, some of which are more annoying than life-threatening. For example, **athlete's foot** is caused by a fungal infection that usually involves the skin of the toes and soles. **Impetigo** is a highly contagious disease occurring most often in young children. It is caused by a bacterial infection that results in pustules that crust over. **Candidiasis** is caused by a yeast organism, which develops in moist areas such as the diaper area in infants. **Eczema,** an inflammation of the skin, is caused by sensitivity to various chemicals (e.g., soaps or detergents), to certain fabrics, or even to heat or dryness. **Dandruff** is a skin disorder not caused by a dry scalp, as is commonly thought, but by an accelerated rate of keratinization in certain areas of the scalp, producing flaking and itching. **Urticaria,** or hives, is an allergic reaction characterized by the appearance of reddish, elevated patches and often by itching.

Individuals who choose to pierce or tattoo their skin must take care to avoid skin damage or life-threatening infection (see Medical Focus, p. 98).

Skin Cancer

Skin cancer is categorized as either melanoma or nonmelanoma. Like all cancers, it begins with mutation of the skin cell DNA.

Nonmelanoma cancers, which include basal cell carcinoma and squamous cell carcinoma, are much less likely to metastasize than melanoma cancer. **Basal cell carcinoma** (Fig. 5.6*a*), the most common type of skin cancer, begins when ultraviolet (UV) radiation causes epidermal basal cells to form a tumor, while at the same time suppressing the immune system's ability to detect the tumor. The signs of a tumor are varied. They include an open sore that will not heal; a recurring reddish patch; a smooth, circular growth with a raised edge; a shiny bump; or a pale mark. About 95% of patients are easily cured by surgical removal of the tumor, but recurrence is common.

Squamous cell carcinoma (Fig. 5.6*b*) begins in the superficial cells of the epidermis. While five times less common than basal cell carcinoma, it is more likely to spread to nearby organs, and death occurs in about 1% of cases. It, too, is triggered by excessive UV exposure. The signs of squamous cell carcinoma are the same as those for basal cell carcinoma, except that it may also appear as a wartlike or scaly growth that bleeds and scabs over, but refuses to heal.

Melanoma (Fig. 5.6*c*), the type that is more likely to be malignant, starts in the melanocytes and looks like an unusual **mole** (Table 5.1). Unlike a normal mole, which is dark, circular, and confined, a melanoma mole looks like a spilled ink spot, and may display a variety of shades. A melanoma mole can also itch, hurt, or feel numb. The melanoma mole may be elevated above the skin surface, and the skin around it turns gray, white, or red. Melanoma is most common in fair-skinned persons, particularly if they have suffered occasional severe burns as children. Melanoma risk increases with

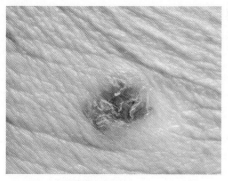

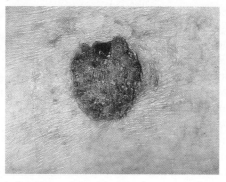

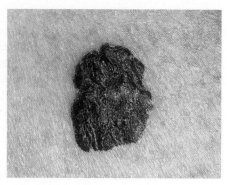

a. Basal cell carcinoma b. Squamous cell carcinoma c. Melanoma

Figure 5.6 **Skin cancer.** In each of the three types shown, the skin clearly has an abnormal appearance.

TABLE 5.1	Warning Signs for Melanoma: The ABCDE Rule

WHEN EXAMINING A MOLE ON THE BODY, NOTE THE FOLLOWING SIGNS OF MELANOMA:

A. **Asymmetrical**—instead of being perfectly round, the mole has an oval or irregular shape.

B. **Borders** of the mole are irregular and have notches or indentation in them.

C. **Color** is uneven, and several colors may be present: black, brown, tan, red, or blue.

D. **Diameter** of the mole is greater than 6 mm (larger than a pencil eraser).

E. **Evolving**—the mole has changed in size, shape or color.

the number of moles a person has. Most moles appear before the age of 14, and their appearance is linked to sun exposure. Melanoma rates have doubled in the last decade. Each year, an estimated 76,000 new cases of melanoma are diagnosed in the United States, and melanoma is responsible for approximately 9,000 fatalities annually.

Protecting the skin from the harmful effects of ultraviolet radiation is the best strategy to prevent all forms of skin cancer:

• Use a broad-spectrum sunscreen with a sun-protective factor (SPF) of at least 15, and make sure it protects against both UV-A and UV-B radiation.

• Wear protective clothing: long-sleeved shirts made from fabrics with a tight weave, for example. Choose a wide-brimmed hat over a baseball cap, which doesn't protect the ears from sunburn. At high altitude or in areas of the Earth where the ozone layer is thin, extra care must be taken. For example, in Australia (where ozone is thin due to the Earth's rotation), children should wear long-sleeved shirts and wide-brimmed hats when they play outside.

• Wear sunglasses that reflect both forms of UV radiation. Children's sunglasses should also protect, so avoid sunglasses that are cute but not protective.

• Stay out of the sun between 10 A.M. and 3 P.M. Doing so will reduce annual exposure to the sun's rays by as much as 60%.

• Avoid tanning machines, unless prescribed by a physician for Seasonal Affective Disorder (SAD).

Kaposi's sarcoma is a form of skin cancer that is most commonly seen in patients with AIDS, and in others whose immune system defenses are weakened or nonfunctional. The tumors of Kaposi's sarcoma appear as red, blue, or black spots on the skin. The tumors respond to treatment with a combination of drugs commonly referred to as the "AIDS cocktail."

Raised growths on the skin, such as moles and **warts,** usually are not cancerous. Moles are due to an overgrowth of melanocytes, and warts are due to a viral infection.

Wound Healing

Injury to the skin will cause an inflammatory response, characterized by redness, swelling, heat, and pain (inflammation is fully discussed in Chapter 13). A wound that punctures a blood vessel will fill the surrounding area with blood. Chemicals released by damaged tissue cells will cause the blood to clot. The clot effectively plugs the wound, preventing pathogens and toxins from spreading to other tissues (Fig. 5.7a). The part of the clot exposed to air will dry and harden, gradually becoming a scab. At the same time, inflammatory chemicals draw white blood cells and fibroblasts to the injured area. There, white blood cells help fight infection, and fibroblasts are able to pull the margins of the wound together (Fig. 5.7b). Fibroblasts promote tissue regeneration, and the basal layer of the epidermis begins to produce new cells at a faster-than-usual rate. The proliferating fibroblasts bring about scar formation; the scar may or may not be visible from the surface (Fig. 5.7c). A scar is a tissue composed of many collagen fibers arranged to provide maximum strength. A scar doesn't contain the accessory organs of the skin, and sensations usually can't be felt on the scarred area. In any case, epidermis and dermis have now healed (Fig. 5.7d).

Content **CHECK-UP!**

9. Which skin cancer type is the most common? The most dangerous? Which causes all forms of skin cancer?

10. What cell type is responsible for scar formation?

Answers in Appendix A.

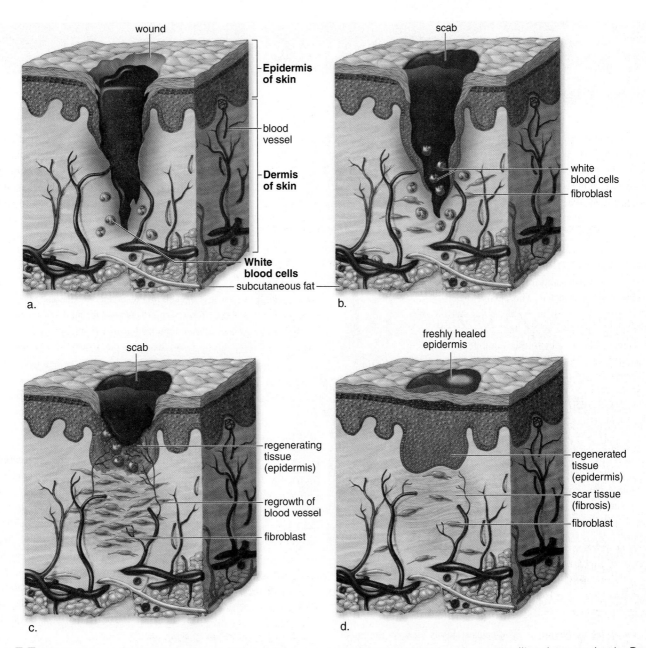

Figure 5.7 **The process of wound healing.** (**a**) Tissue injury causes inflammation, with redness, swelling, heat, and pain. Deep wounds rupture blood vessels, and blood fills the wound. (**b**) After a blood clot forms, a protective scab develops. Fibroblasts and white blood cells migrate to the wound site. (**c**) New epidermis forms, and fibroblasts promote tissue regeneration. (**d**) Freshly healed skin.

Burns

Burns are skin injuries that are usually caused by heat but can also be caused by radioactive, chemical, and electrical agents. Burns account for approximately 1 million emergency department visits yearly in the United States. Emergency caregivers must quickly assess burn severity and take appropriate supportive action.

Two criteria are used to estimate burn severity: depth and thickness. In *first-degree burns* (a moderate sunburn, for example), only the epidermis is affected. The person experiences redness and pain, but no blisters or swelling. The pain subsides within 48–72 hours, and the injury heals without further complications or scarring. Treatment involves pain management. The damaged skin peels off in about a week.

Second-degree burns extend through the entire epidermis and part of the dermis. The person experiences not only redness and pain, but also blistering. The deeper the burn, the more prevalent the blisters, which can enlarge after the injury. Unless they become infected, superficial second-degree burns heal in 10–14 days without complications and with little scarring. If the burn extends deep into the dermis, it heals more slowly over a period of 30–105 days. The healing epidermis is extremely fragile, and scarring is common. First- and second-degree burns are sometimes referred to as partial-thickness burns.

Third-degree burns, or full-thickness burns, destroy the entire thickness of the skin. The surface of the wound is leathery and may be brown or black. Pain receptors, blood vessels, sweat glands, sebaceous glands, and hair follicles are destroyed. *Fourth-degree burns* involve tissues down to the bone. Unless a very limited area of the body is affected, the patient will likely not survive.

The extent of the burn must also be estimated when judging a burn's severity. The "rule of nines" is used to divide the body into regions: the head and neck, 9% of the total body surface; each upper limb, 9%; each lower limb, 18%; the front and back portions of the trunk, 18% each; and the perineum, which includes the anal and urogenital regions, 1% (Fig. 5B). Physicians use the Lund-Browder chart to estimate the extent of burns in children. This system adjusts for the fact that a child's head is proportionally larger than an adult's.

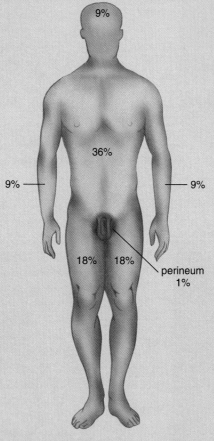

Figure 5B The "rule of nines" for estimating the extent of burns.

A burn is a critical injury if: (1) Second-degree burns cover 25% or more of the patient's body; (2) third-degree burns cover 10% or more of the patient's body; (3) any portion of the body has a fourth-degree burn; or (4) third-degree burns occur on the face, hands, or feet. Facial burns may be accompanied by lung damage from smoke inhalation. Burns to the hands or feet can scar, resulting in loss of joint movement.

The first responder's reactions when caring for a burn patient (and for all other patients) can be summarized using the letters ABC:

A—establish the *airway.* A tube may need to be placed into the trachea (windpipe), especially if the patient might have suffered smoke inhalation. Smoke inhalation should be suspected whenever there are facial burns, or if the patient has trouble breathing.

B—make sure the person is *breathing,* using artificial respiration when necessary. If smoke inhalation is suspected, a paramedic will administer 100% oxygen.

C—*circulation* must be maintained. The patient's pulse is taken, and CPR is begun when needed. The patient must be carefully observed for signs of **shock**—a dangerously low drop in blood pressure (see Chapter 12 for discussion of shock). A paramedic may begin an IV (intravenous administration of a balanced salt solution) under a physician's direction.

First responders must also act quickly to stop the burn. All clothing must be carefully removed from the body and any smoldering must be extinguished. In the case of a chemical burn, the affected area is flushed continuously with water.

The major complications resulting from severe burns are fluid loss, heat loss, and bacterial infection. Fluid loss is counteracted by continuous IV therapy. Heat loss is minimized by placing the burn patient in a warm environment. Application of antibacterial dressings and isolation help prevent infection. As soon as possible, the damaged tissue is removed and skin grafting is begun. The skin needed for grafting is usually taken from other parts of the patient's body. However, if the burned area is quite extensive, skin can be grown in the laboratory from only a few cells taken from the patient.

5.5 Effects of Aging

12. Describe the anatomical and physiological changes that occur in the integumentary system as we age.

As aging occurs, the epidermis maintains its thickness, but the rate of cell mitosis decreases. The dermis becomes thinner, the dermal papillae flatten, and the dermis is held less tightly to the underlying fascia, so that the skin is looser. Adipose tissue in the hypodermis of the face and hands also decreases. These cellular changes have important implications for the overall health of the older patient. Decreased adipose tissue insulation means that older people are more likely to feel cold. In addition, older individuals will bruise more easily as skin becomes thinner. It's important for caregivers to watch for any signs of bedsores in elderly patients (see Medical Focus, Decubitus Ulcers, p. 89).

The fibers within the dermis change with age. The collagenous fibers become coarser, thicker, and farther apart; therefore, there is less collagen than before. Elastic fibers in the upper layer of the dermis are lost, and those in the lower dermis become thicker, less elastic, and disorganized. The skin wrinkles because (1) the epidermis is loose, (2) the dermal fibers are fewer and those remaining are disorganized, and (3) the hypodermis has less padding.

With aging, homeostatic adjustment to heat is limited due to less vasculature (fewer blood vessels) and fewer sweat glands. Thus, an older person may be more sensitive to the effects of overheating. The number of hair follicles decreases, causing the hair on the scalp and extremities to thin. Because of a reduced number of sebaceous glands, the skin tends to crack.

As a person ages, the number of melanocytes decreases. This causes the hair to turn gray and the skin to become paler. In contrast, some of the remaining pigment cells are larger, and pigmented blotches appear on the skin.

Many of the changes that occur in the skin as a person ages appear to be due to sun damage. Ultraviolet radiation causes rough skin, mottled pigmentation, fine lines and wrinkles, deep furrows, numerous benign skin growths, and the various types of skin cancer discussed in section 5.4.

Content CHECK-UP!

11. Why do older people feel colder as they age?

Answers in Appendix A.

5.6 Homeostasis

13. Explain how the integumentary system interacts with other body systems to ensure homeostasis.

The illustration on page 99, called Human Systems Work Together, tells how the functions of the skin assist the other systems of the body and how the other systems help the skin carry out these functions.

Skin Interacts with All Organ Systems

Skin assists the lymphatic system in protecting the organs of all body systems. Healthy skin aids the lymphatic system by safeguarding all underlying structures from physical damage, infection by pathogens, and radiation.

Skin helps the kidneys regulate water balance and waste excretion. Waterproof skin protects the body from water loss and dehydration, as well as water gain and hypotonic swelling when the skin is surrounded by water. The skin's sweat glands secrete water from the body through sensible and insensible perspiration. Because sweat contains a small amount of waste, skin plays a minor role in waste elimination. In turn, if a person sweats excessively, the kidneys correct for water loss by returning filtered water to the bloodstream.

Skin aids the skeletal system in storing calcium. The vitamin D produced by skin cells that are exposed to UV light is converted to the hormone calcitriol. In the digestive system, calcitriol enables calcium to be absorbed from the foods we eat. Calcium compounds are deposited in bone, giving bone both rigidity and flexibility. The skeletal system interacts with skin by providing support and sites for skin attachment.

Skin serves the nervous system by obtaining sensory information from the environment. Skin contains receptors that can respond to many different types of sensory stimuli: temperature, pain, light touch and deep pressure, itching, tickling, etc. Sensations from these receptors provide constant, real-time information to the nervous system. The nervous system contributes to skin homeostasis by regulating the function of blood vessels, glands, and accessory structures such as arrector pili muscles.

Skin helps the muscular system regulate body temperature. When muscles contract and ATP is broken down, heat is released. If body temperature starts to rise above normal, the skin blood vessels vasodilate, blood moves to the surface of the body, and heat is released to the surroundings. Cooler temperatures cause vasoconstriction, and less body heat is lost to the environment. Muscle shivering also helps to raise body temperature to normal. Skin sweat glands activate when one is too warm, and remain inactive if body temperature falls. Though skin arrector pili muscles merely cause goose bumps in humans, in other animals their contraction creates an insulating layer of fur or feathers. Muscles themselves serve as attachment points for skin, especially in the face.

Skin survives thanks to the cardiovascular, respiratory, digestive, and renal systems. The lungs supply oxygen to the blood, while simultaneously releasing waste carbon dioxide to the environment. Cardiovascular blood vessels deliver oxygen to muscles and remove their wastes, which are then eliminated in urine formed by the kidneys. Finally, the digestive system provides essential nutrients.

Skin responds to endocrine hormones and assists reproduction. Androgens (male hormones) stimulate hair growth and

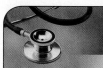

MEDICAL FOCUS

Body Art: Buyer Beware!

There's no doubt about it: Body art is definitely mainstream. Tattooing (the process of injecting inks into the dermal layer of the skin) and piercing the body are practices that have existed for thousands of years in many different human cultures. In the past decade, both practices have become commonplace in Western society as well. Actors, sports figures, and other celebrities can routinely be seen with visible tattoos, pierced navels, multiple earrings, or other forms of wearable art. Moreover, tattoos keep getting larger. It's no longer uncommon to see so-called sleeves—tattoos covering the entire arm—and facial and full-body tattoos as well. The most common sites for piercing remain the earlobes and ear cartilage, but noses, lips, eyebrows, navels, tongues, and genitals are also common sites.

As these trends continue, more and more people are tempted to get their own tattoo or piercing. But as recent reality shows like *Bad Ink, Tattoo Nightmares,* and *America's Worst Tattoos* have shown, a tattoo can create big problems for its wearer. Individuals must consider the decision to tattoo or pierce very carefully, and health-care providers must be prepared to give accurate information about both practices. With the increase in both tattooing and piercing, reports of complications have increased as well.

Tattooing and piercing can result in infections because both practices can potentially introduce bacteria into the skin, blood vessels, or lymphatic vessels. Complications from infection can range from minor skin irritations to life-threatening blood poisoning. Several cases on record have involved infection with *resistant* bacteria, which are extremely dangerous because they are not killed by common antibiotics. Fatalities have been reported from infections caused by both piercing and tattooing.

Further, the reaction of a person's skin to tattoo ink is unpredictable. Tattoo inks are not regulated by any form of federal agency, and many color additives have not been approved for contact with the skin. Some are inks used in paper printing or automobile paint. It is hardly surprising that many tattooed individuals develop allergies to ingredients in tattoo dyes. Allergies to tattoo inks can even develop years after the procedure. Further, tattooed skin may develop unattractive granulomas: hard connective tissue nodules under the skin. Keloids—scars that grow beyond normal boundaries—may also develop as the body reacts to chemicals in the ink. (African Americans are particularly susceptible to keloid formation, for reasons that are not clear.)

Metals (especially nickel) used in body jewelry are another common cause for allergy. Once an allergic reaction develops, it typically persists for life. In such cases, the jewelry must be permanently removed if at all possible. Additionally, the jewelry may damage nerves or blood vessels as piercing occurs. Earlobes can be torn completely and require surgical repair if an earring is snagged. Jewelry in and around the mouth may damage gums, break teeth or dentures, and interfere with chewing or swallowing. Mouth jewelry that becomes loose may be aspirated (sucked into the airways), where it can obstruct an airway and interfere with breathing.

Anyone considering body art must proceed with caution. First, make sure you're not under the influence of drugs or alcohol when making such an important decision. If local, state, or provincial law regulates tattoo and piercing artists, use an approved artist or shop. Never try a "do-it-yourself" approach. Check the cleanliness of the shop and make sure that all equipment is steam-sterilized using a device called an *autoclave*, and that the tattoo inks are also sterile. Insist that the artist or piercer use sterile gloves at all times. Don't let the artist tattoo around or near a mole or other skin growth, as it may make it very difficult to diagnose a skin cancer in the future. For a piercing, make sure that jewelry is appropriate: not too heavy, and made of inert metals such as gold, titanium, niobium, or surgical stainless steel, which are less likely to cause allergic reactions. Keep the affected area scrupulously clean after piercing or tattooing, and touch jewelry as little as possible. Allow adequate time for healing. For example, pierced navels may require up to a year to heal completely. Finally, if complications develop (excessive pain, redness, swelling, fever, etc.), seek medical attention promptly.

Most important, anyone considering body art must remember that the decision may very likely be *permanent*. Piercings may never reclose completely and they may leave a large scar. Tattoos can change their appearance over time as the inks fade and the skin stretches. And though your tattoo might possibly be removed by laser therapy, surgery, or dermabrasion (using a rotating wire brush to scrape off surface skin), such treatment is expensive. It may also leave large scars—and often, it cannot completely remove the tattoo.

---

sebaceous gland secretion, while estrogens stimulate fat deposition in the hypodermis. Skin sensations are an important part of sexual arousal and the sex response. Skin protects the fetus in the womb, and mammary glands support a newborn created by sexual activity.

Content **CHECK-UP!**

12. How does the skin help to make bones stronger?

Answers in Appendix A.

Human Systems Work Together

Skeletal System

Skin protects bones; helps provide vitamin D for Ca^{2+} absorption.

Bones provide support for skin.

Muscular System

Skin protects muscles; rids the body of or conserves heat produced by muscle contraction.

Muscle contraction provides heat to warm skin.

Nervous System

Skin protects nerves, helps regulate body temperature; skin receptors send sensory input to brain.

Brain transmits signals by nerves to regulate size of cutaneous blood vessels, activate sweat glands and arrector pili muscles.

Endocrine System

Skin helps protect endocrine glands.

Androgens activate sebaceous glands and help regulate hair growth.

Cardiovascular System

Skin prevents water loss; helps regulate body temperature; protects heart and blood vessels.

Blood vessels deliver nutrients and oxygen to skin, carry away wastes; blood clots if skin is broken.

How the Integumentary System works with other body systems

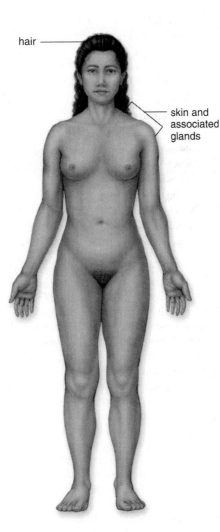

hair

skin and associated glands

Lymphatic System/Immunity

Skin serves as a barrier to pathogen invasion; Langerhans cells phagocytize pathogens; protects lymphatic vessels.

Lymphatic vessels pick up excess tissue fluid; immune system protects against skin infections.

Respiratory System

Skin helps protect respiratory organs.

Gas exchange in lungs provides oxygen to skin and rids body of carbon dioxide from skin.

Digestive System

Skin helps to protect digestive organs; helps provide vitamin D for Ca^{2+} absorption.

Digestive tract provides nutrients needed by skin.

Urinary System

Skin helps regulate water loss; sweat glands carry on some excretion.

Kidneys compensate for water loss due to sweating; activate vitamin D precursor made by skin.

Reproductive System

Skin receptors respond to touch; mammary glands produce milk; skin stretches to accommodate growing fetus.

Androgens activate oil glands; sex hormones stimulate fat deposition, affect hair distribution in males and females.

Selected New Terms

Basic Key Terms

arrector pili (ăh-rĕk′tŏr pī-lī′), p. 90

cutaneous membrane (kyū-tān′ē-ŭs mĕmbrān) p. 87

dermis (dĕr′mĭs), p. 88

epidermis (ĕp″ĭ-dĕr′mĭs), p. 87

free nerve endings (frē nŭrv ĕn′dĭngs), p. 88

hair follicle (hār fŏl′ĭ-kĕl), p. 90

hair matrix ((hār mā′trĭks)) p. 90

hypodermis (hī″pō-dĕr′mĭs), p. 87

integument (ĭn-tĕg′yū-mĕnt), p. 87

integumentary system (ĭn-tĕg″yū-mĕn′tăr-ē sĭs′tĕm), p. 87

keratin (kĕr′ŭh-tĭn), p. 88

keratinocyte (kĕr′ŭh-tĭn-ō-sīt), p. 88

Langerhans cells (lähng′ĕr-hähnz sĕlz), p. 88

lunula (lū′nyū-lŭh), p. 90

mammary glands (mă′mă-rē glănds), p. 92

melanin (mĕl′ŭh-nĭn), p. 88

melanocyte (mĕl′ŭh-nō-sīt), p. 88

Merkel cells (mĕr′kĕl sĕlz), p. 88

sebaceous gland (sĕ-bā′shŭs glănd), p. 92

sebum (sē′bŭm), p. 92

stratum basale (strā′tŭm bā′să-lē), p. 87

stratum corneum (strā′tŭm kŏr′nē-ŭm), p. 88

stratum granulosum (strā′tŭm grăn-yū-lō′sŭm), p. 88

stratum lucidum (strā′tŭm lū′sĭd-ŭm), p. 88

stratum spinosum (strā′tŭm spĭ-nō′sŭm), p. 88

subcutaneous tissue (sŭb′kyū-tā′nē-ŭs tĭsh′ū), p. 87

sweat glands (swĕt glănds), p. 91

tactile cells (tăk′tĭl sĕlz), p. 88

Clinical Key Terms

acne vulgaris (ăk′nē vŭl-gă′rĭs), p. 92

albinism (ăl′bĭ-nĭzm), p. 88

alopecia (ăl-ō-pē′shē-ŭh), p. 90

androgenetic alopecia (ăn″drō-jŭh-nĕ′tĭk ăl-ō-pē′shē-ŭh), p. 90

athlete's foot (ăth′lēts fŭt), p. 93

basal cell carcinoma (bās′ăl sĕl kăr-sĭ-nō′mŭh), p. 93

candidiasis (kăn′dĭd-ī′ŭh-sĭs), p. 93

dandruff (dăn′drŭf), p. 93

decubitus ulcer (dē-kyū′bĭ-tŭs ŭl′sĕr), p. 89

eczema (ĕk′zĕ-mŭh), p. 93

hirsutism (hĕr′sŭh-tĭzm), p. 89

hyperthermia (hī″pĕr-thĕr′mē-ŭh), p. 93

hypodermic needle (hī-pō-dĕr′mĭk nē′dŭl), p. 88

hypothermia (hī″pō-thĕr′mē-ŭh), p. 93

impetigo (ĭm″pĕ-tī′gō), p. 93

Kaposi's sarcoma (kă-pō′sēz săr-kō′mŭh), p. 94

melanoma (mĕl-ŭh-nō′mŭh), p. 93

mole (mōl), p. 93

psoriasis (sō-rī′ŭh-sĭs), p. 88

rickets (rĭk′ĕts), p. 92

shock (shăwk), p. 96

squamous cell carcinoma (skwā′mŭs sĕl kăr-sĭ-nō′mŭh), p. 93

subcutaneous injection (sŭb″kyū-tā′nē-ŭs ĭn-jĕk′shŭn), p. 88

urticaria (ŭr″tĭ-kār′ē-ŭh), p. 93

warts (wărts), p. 94

Summary

5.1 Structure of the Skin
The skin has two regions, the epidermis and the dermis. The hypodermis lies below the skin.
A. The epidermis, the outer region of the skin, is made up of stratified squamous epithelium. New cells continually produced in the stratum basale of the epidermis are pushed outward and become the keratinized cells of the stratum corneum. The stratum lucidum is found only in the thick skin of the palms and soles.
B. The dermis, which is composed of dense irregular connective tissue, lies beneath the epidermis. It contains collagenous and elastic fibers, blood vessels, and nerve fibers.
C. The hypodermis is made up of loose connective tissue and adipose tissue, which insulates the body from heat and cold.

5.2 Accessory Structures of the Skin
Accessory structures of the skin include hair, nails, and glands.
A. Both hair and nails are produced by the division of epidermal cells and consist of keratinized cells.
B. Sweat glands are numerous and present in all regions of the skin. Sweating helps lower the body temperature. Sebaceous glands are associated with a hair follicle and secrete sebum, which lubricates the hair and skin. Mammary glands located in the breasts produce milk after childbirth.

5.3 Functions of the Skin
Skin protects the body from physical trauma and bacterial invasion. Skin helps regulate water loss and gain, which helps the urinary system. Also, sweat glands excrete some urea and other wastes. The skin produces vitamin D following exposure to UV radiation. A hormone derived from vitamin D helps regulate calcium and phosphorus metabolism involved in bone development. The skin contains sensory receptors for touch, pressure, pain, hot, and cold, which help people to be aware of their surroundings. These receptors send information to the nervous system.

The skin helps regulate body temperature. When the body is too hot, dermal blood vessels dilate, and the sweat glands are active. When the body is cold, dermal blood vessels constrict, and the sweat glands are inactive. Hyperthermia and hypothermia are two conditions that can result when the body's temperature regulatory mechanism is overcome.

5.4 Disorders of the Skin
A. Skin cancer, which is associated with ultraviolet radiation, occurs in three forms. Basal cell carcinoma and squamous cell carcinoma can usually be removed surgically. Melanoma is the most dangerous form of skin cancer.
B. The skin has regenerative powers and can grow back on its own if a wound is not too extensive.

5.5 Effects of Aging
Skin wrinkles with age because the epidermis is held less tightly, fibers in the dermis are fewer, and the hypodermis has less padding. The skin has fewer blood vessels, sweat glands, and hair follicles. Although pigment cells are fewer and the hair

turns gray, pigmented blotches appear on the skin. Exposure to the sun results in many of the skin changes we associate with aging.

5.6 Homeostasis
The skin interacts with various organ systems in many ways. For example, the skin helps the skeletal system by producing vitamin D, and the urinary system by eliminating wastes through sweat.

Study Questions

1. In general, describe the two regions of the skin. (p. 87)
2. Describe the process by which epidermal tissue continually renews itself. (pp. 87–88)
3. What function does the dermis have in relation to the epidermis? (p. 88)
4. What primary role does adipose tissue play in the hypodermis? (pp. 88–89)
5. Describe in general the structure of a hair follicle and a nail. How do hair follicles and nails grow? (pp. 88–91)
6. Describe the structure and function of sweat glands and sebaceous glands. (pp. 91–92)
7. Name five functions of the skin, and tell what system of the body is assisted by these functions and how they contribute to homeostasis. (pp. 92–93, 97)
8. Name the three types of skin cancer, and cite the most frequent cause of skin cancer. (pp. 93–94)
9. Describe how a wound heals and how a scar forms. (pp. 94–95)
10. Explain how to determine the severity of a burn. Describe the proper treatment for burns. (p. 96)
11. Explain three changes that happen to the skin with age. (p. 97)

Learning Outcome Questions

I. Match the terms in the key to the items listed in questions 1–5.

Key:
   a. epidermis
   b. dermis
   c. hypodermis

1. blood vessels and nerve fibers
2. fat cells
3. basal cells
4. location of sweat glands
5. many collagenous and elastic fibers

II. Fill in the blanks.

6. Sweat glands are involved in body _____ regulation.
7. Sebaceous glands are associated with _____ in the dermis, and they secrete an oily substance called _____.
8. Skin protects against _____ trauma, _____ invasion, and _____ gain or loss.
9. Skin cells produce vitamin _____, which is needed for strong bones.
10. The type of skin cancer with the highest death rate is _____, while the most common form is _____.
11. The severity of a burn is determined by _____ and _____.

Medical Terminology Exercise

After studying this chapter, see if you can derive the definitions for the medical terms listed at right. Many of the prefixes and suffixes used to create these terms can be found throughout the chapter. For additional help, use McGraw-Hill Connect™ at www.mcgrawhillconnect.com and consult Appendix B.

1. epidermomycosis (ĕp″ĭ-dĕr″mō-mī-kō′sĭs)
2. melanogenesis (mĕl″ŭh-nō-jĕn′ĕ-sĭs)
3. mammoplasty (măm′ō-plăs″tē)
4. antipyretic (an″tĭ-pī-rĕt′ĭk)
5. dermatome (dĕr′mŭh-tōm)
6. hypodermic (hy″pō-dĕr′mĭk)
7. hyperhydrosis (hī″pĕr-hī-drō′sĭs)
8. scleroderma (sklĕr-ō-dĕr′mŭh)
9. piloerection (pī′lō-ĕ-rĕk′shŭn)
10. cellulitis (sĕl′yū-lī′ tĭs)
11. dermatitis (dĕr-mŭh-tī′tĭs)
12. trichopathy (trī-kŏp′ŭh-thē)

Online Study Tools

■LEARNSMART® Mc Graw Hill Education **connect**®

APR

Anatomy & Physiology REVEALED includes cadaver photos that allow you to peel away layers of the human body to reveal structures beneath the surface. This program also includes animations, radiologic imaging, audio pronunciations, and practice quizzing. To learn more visit www.aprevealed.com

Anatomy & Physiology REVEALED®
aprevealed.com

6 The Skeletal System

"Will you please quit doing that?! Keep it up and you'll get arthritis!" Habitual knuckle-crackers might stop their annoying habit when confronted with this threat, but is there any truth to it? Knuckles are freely moving *synovial joints* (see page 123). In a synovial joint, ligaments join bones and create a fluid-filled capsule. Stretching the ligaments suddenly creates an air bubble in the fluid. Popping the bubble causes the "crack" noise. Any synovial joint—toes, ankles, fingers—can crack, but numerous studies have shown that arthritis doesn't result from it. However, ligaments will weaken with habitual cracking, and the bones at a joint can then be dislocated more easily. If you're tempted to crack those stiff joints, massage and gently stretch them instead.

Learning Outcomes

After you have studied this chapter, you should be able to:

6.1 Skeleton: Overview

1. Name at least five functions of the skeleton.
2. Explain a classification of bones based on their shapes.
3. Describe the anatomy of bone. Describe long bone structure, and compare/contrast compact bone and spongy bone.
4. Describe the physiology of bone, including the cells involved in growth and repair, and the process of bone growth, development, and remodeling.
5. Name and describe six types of fractures, and state the four steps in fracture repair.
6. List the surface features of bones, and give examples where each can be found.

6.2 Axial Skeleton

7. Distinguish between the axial and appendicular skeletons.
8. Name the bones of the skull, and state the important features of each bone.
9. Describe the structure and function of the hyoid bone.

10. Name the bones of the vertebral column and the thoracic cage. Be able to label diagrams of them.
11. Describe a typical vertebra, the atlas and axis, and the sacrum and coccyx.
12. Name the three types of ribs and the three parts of the sternum.

6.3 Appendicular Skeleton

13. Name the bones of the pectoral girdle and the pelvic girdle. Be able to label diagrams of them.
14. Name the bones of the upper limb (arm and forearm) and the lower limb (thigh and leg). Be able to label diagrams that include surface features.
15. Cite at least five differences between the female and male pelvises.

6.4 Joints (Articulations)

16. Explain how joints are classified, and give examples of each type of joint.
17. List the types of movements that occur at synovial joints.
18. Explain how damage and degeneration occurs at joints and how it can be treated. Outline

possible steps for damage prevention.

6.5 Effects of Aging

19. Describe the anatomical and physiological changes that occur in the skeletal system as we age.

6.6 Homeostasis

20. List and discuss six ways the skeletal system contributes to homeostasis. Discuss ways the other systems assist the skeletal system.

I.C.E.—In Case of Emergency

Broken Bones

Medical Focus

Osteoporosis
Oh, My Aching Back: Surgical Options for Back Injuries

Focus on Forensics

Skeletal Remains

Human Systems Work Together

Skeletal System

6.1 Skeleton: Overview

1. Name at least five functions of the skeleton.
2. Explain a classification of bones based on their shapes.
3. Describe the anatomy of bone. Describe long bone structure, and compare/contrast compact bone and spongy bone.
4. Describe the physiology of bone, including the cells involved in growth and repair, and the process of bone growth, development, and remodeling.
5. Name and describe six types of fractures, and state the four steps in fracture repair.
6. List the surface features of bones, and give examples where each can be found.

The skeletal system consists of the bones (206 in adults) and joints, along with the cartilage and ligaments that occur at the joints.

Functions of the Skeleton

The skeleton carries out a number of vital functions in the body:

The skeleton supports the body. The bones of the lower limbs support the entire body when we're standing, and the pelvic girdle supports the abdominal cavity.

The skeleton protects soft body parts. The bones of the skull protect the brain; the rib cage protects the heart and lungs.

The skeleton produces blood cells. All bones in the fetus have red bone marrow that produces blood cells. In the adult, only certain bones produce blood cells.

The skeleton stores minerals and fat. All bones have a matrix that contains calcium phosphate, a source of calcium ions and phosphate ions in the blood. Blood calcium is essential for nerves and muscles to function properly. Fat is stored in yellow bone marrow.

The skeleton, along with the muscles, permits flexible body movement. While articulations (joints) occur between all the bones, we associate body movement in particular with the bones of the limbs.

Anatomy of a Long Bone

Bones are classified according to their shape. As the name implies, long bones are longer than they are wide. Short bones are cube shaped—that is, their lengths and widths are about equal. Flat bones, such as those of the skull, are platelike with broad surfaces. Irregular bones have varied shapes that permit connections with other bones. Round bones are circular in shape (Fig. 6.1).

A long bone, such as the one in Figure 6.2a, can be used to illustrate certain principles of bone anatomy. The bone is enclosed in a tough, fibrous, connective tissue covering called the **periosteum,** which is continuous with the ligaments that join bones and the tendons that anchor muscles to bones. The periosteum contains blood vessels that enter the bone and supply its cells. At both ends of a long bone is an expanded portion called an **epiphysis** (*pl., epiphyses*); the portion between the epiphyses is called the **diaphysis.**

As shown in the section of an adult bone in Figure 6.2, the diaphysis, or shaft, of a long bone, is not solid but has a **medullary cavity** containing yellow marrow. Yellow marrow contains large amounts of fat. The medullary cavity is bounded at the sides by

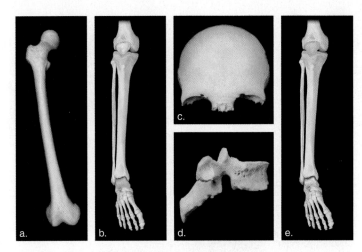

Figure 6.1 Classification of bones. (a) Long bones are longer than they are wide. **(b)** Short bones are cube shaped; their lengths and widths are about equal. **(c)** Flat bones are platelike and have broad surfaces. **(d)** Irregular bones have varied shapes with many places for connections with other bones. **(e)** Round bones are circular.

compact bone. The epiphyses contain spongy bone. Beyond the spongy bone is a thin shell of compact bone and, finally, a layer of hyaline cartilage called the **articular cartilage.** Articular cartilage is so named because it occurs where bones articulate (come together to form a joint). **Articulation** is the joining together of bones at a joint. The medullary cavity and the spaces of spongy bone are lined with *endosteum,* a thin, fibrous membrane.

In infants, **red bone marrow,** a specialized tissue that produces blood cells, is found in the medullary cavities of most bones. In adults, red blood cell formation, called **hematopoiesis,** occurs in the spongy bone of the skull, ribs, sternum (breastbone), and vertebrae, and in the ends of the long bones.

Compact Bone

Compact bone, or dense bone, contains many cylinder-shaped units called osteons. **Osteons** are formed by concentric layers of matrix called *lamellae.* Between lamellae are *lacunae,* tiny chambers where osteocytes (bone cells) can be found. *Canaliculi* are small canals that connect lacunae. Oxygen and nutrients can pass through canaliculi to supply the osteocytes in each lacuna. The matrix contains collagenous protein fibers and mineral deposits, primarily of calcium and phosphorus salts.

In each osteon, the lamellae and lacunae surround a single *central canal.* Canaliculi from adjacent lacunae open into the central canal, and blood vessels and nerves from the periosteum travel within it. These same blood vessels and nerves can travel from one central canal to another by way of *perforating canals* (Fig. 6.2). Because osteocytes send cell extensions into the canaliculi, the osteocytes are connected to each other and also to the central canal.

Spongy Bone

Spongy bone, or cancellous bone, contains numerous bony bars and plates, called *trabeculae.* Although lighter than compact bone,

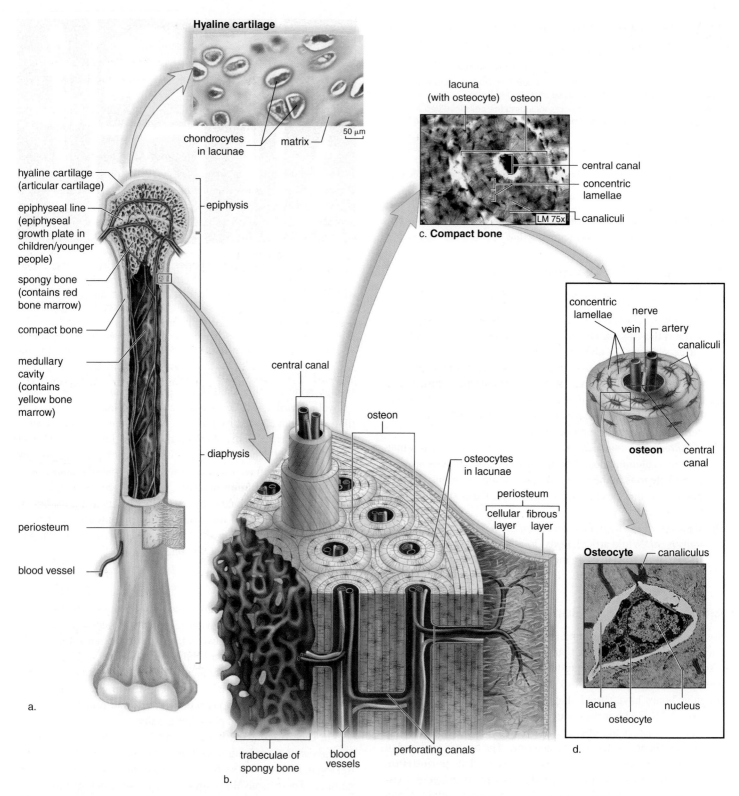

Hyaline cartilage

chondrocytes in lacunae — matrix
50 μm

lacuna (with osteocyte) — osteon
central canal
concentric lamellae
canaliculi
LM 75x

c. **Compact bone**

hyaline cartilage (articular cartilage)
epiphyseal line (epiphyseal growth plate in children/younger people)
spongy bone (contains red bone marrow)
compact bone
medullary cavity (contains yellow bone marrow)
periosteum
blood vessel

epiphysis

diaphysis

a.

central canal

osteon

osteocytes in lacunae

periosteum
cellular layer | fibrous layer

trabeculae of spongy bone
blood vessels
perforating canals

b.

concentric lamellae
nerve
vein — artery
canaliculi
osteon — central canal

Osteocyte — canaliculus
lacuna
osteocyte — nucleus

d.

Figure 6.2 Anatomy of a long bone. (a) The center shaft is the diaphysis, and the ends are epiphyses. The entire bone is covered in periosteum, except at the epiphyses. Hyaline cartilage caps each epiphysis. Spongy bone of the epiphyses contains red bone marrow. The diaphysis contains yellow bone marrow surrounded by compact bone. **(b)** The detailed anatomy of spongy bone and compact bone is shown in the enlargement. **(c)** Photomicrograph of compact bone. **(d)** An osteon is formed by concentric rings called lamellae, and has a central canal with a nervous and blood supply. Bone cells, called osteocytes, live in lacunae.

spongy bone is still designed for strength. Like braces used for support in buildings, the trabeculae of spongy bone follow lines of stress.

Cells for Bone Growth and Repair

Bones are composed of living tissues, as exemplified by their ability to grow and undergo repair. Several different types of cells are involved in bone growth and repair:

Osteoprogenitor cells are unspecialized cells present in the inner portion of the periosteum, in the endosteum, and in the central canal of compact bone.

Osteoblasts are bone-forming cells formed from osteoprogenitor cells. They are responsible for secreting the matrix characteristic of bone.

Osteocytes are mature bone cells derived from osteoblasts. Once the osteoblasts are surrounded by matrix, they become the osteocytes in bone.

Osteoclasts are thought to be derived from monocytes, a type of white blood cell present in red bone marrow.
Osteoclasts perform bone resorption; that is, they break down bone and assist in depositing calcium and phosphate in the blood. The work of osteoclasts is important to the growth and repair of bone.

Bone Development and Growth

The term **ossification** refers to the formation of bone. The bones of the skeleton form during embryonic development in two distinctive ways—intramembranous ossification and endochondral ossification.

In **intramembranous ossification,** bone develops between sheets of fibrous connective tissue. Cells derived from connective tissue become osteoblasts that form a matrix resembling the trabeculae of spongy bone. Other osteoblasts associated with a periosteum lay down compact bone over the surface of the spongy bone. The osteoblasts become osteocytes when they are surrounded by a mineralized matrix. The bones of the skull develop in this manner.

Most of the bones of the human skeleton form by **endochondral ossification.** Hyaline cartilage models, which appear during fetal development, are replaced by bone as development continues. During endochondral ossification of a long bone, the cartilage begins to break down in the center of the diaphysis, which is now covered by a periosteum (Fig. 6.3). Osteoblasts and blood vessels enter the central region. There, these osteoblasts begin to lay down spongy bone in what is called a *primary ossification center.* Other osteoblasts lay down compact bone beneath the periosteum. As the compact bone thickens, the spongy bone of the diaphysis is broken down by osteoclasts, and the cavity created becomes the medullary cavity.

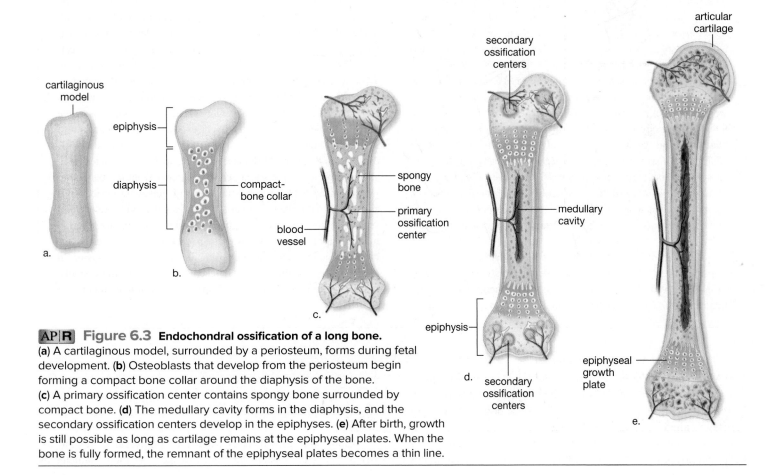

AP|R Figure 6.3 **Endochondral ossification of a long bone.**
(a) A cartilaginous model, surrounded by a periosteum, forms during fetal development. (b) Osteoblasts that develop from the periosteum begin forming a compact bone collar around the diaphysis of the bone. (c) A primary ossification center contains spongy bone surrounded by compact bone. (d) The medullary cavity forms in the diaphysis, and the secondary ossification centers develop in the epiphyses. (e) After birth, growth is still possible as long as cartilage remains at the epiphyseal plates. When the bone is fully formed, the remnant of the epiphyseal plates becomes a thin line.

Imagine how your world would change if you lived with severe back pain. Not the kind that can be fixed with an Icy Hot® patch; rather, this variety only responds to drugs whose side effects might include dizziness and falling. And you're terrified of falling, because you've seen your friends and loved ones break bones after what should have been a trivial misstep. For those with osteoporosis, this pain and fear could be a daily reality. Osteoporosis is a condition caused by a reduction in density of individual bones that make up the skeleton. These weakened bones are particularly susceptible to painful and debilitating fractures, especially at the hip, vertebrae, long bones, and pelvis. Complications of these fractures can be very dangerous and potentially fatal to an older person. Simply managing pain often requires medication with serious side effects; some drugs can be addicting as well. Further, as you learned in Chapter 5, an older person immobilized by a fracture is prone to decubitus ulcers and may contract pneumonia while hospitalized. Moreover, any bedridden person is at increased risk for forming a *thromboembolism*, an abnormal blood clot inside blood vessels. These clots can block arteries to the heart or brain, resulting in heart attack or stroke.

Although osteoporosis can result from various disease processes, it's essentially a disease of aging. Bones are continuously remodeled—built up, broken down, built up again—throughout life. In childhood, bone formation is greater than bone breakdown, and skeletal density increases until approximately age 25. Afterward, rates of bone formation and breakdown are roughly equal for the next two decades, until age 45 to 50. Then, reabsorption begins to exceed formation, and bone density slowly decreases. Over time, men are apt to lose 25%, and women 35%, of bone mass. Male bones are generally denser than female bones because testosterone (the male sex hormone) promotes bone formation and doesn't significantly decline until after about age 65. In contrast, estrogen (female sex hormone), which promotes women's bone formation, begins to decline at about age 45 with the onset of menopause. These differences in hormone levels mean that women are more likely than men to suffer osteoporosis. Women with a slight build are at greatest risk, especially those of Caucasian and Asian descent.

If osteoporosis is detected early enough, treatment can slow or stop bone density decrease. Older at-risk individuals, especially postmenopausal women, should be tested using dual-energy X-ray absorptiometry (DEXA), a specialized exam used to measure bone density.

DEXA can be followed by blood and urine tests to detect high levels of calcium and biochemicals that are associated with bone loss. Early bone thinning, called *osteopenia*, should be aggressively treated to restore bone density and reduce fracture risk. The most commonly used drugs, called bisphosphonates (Fosamax" Actonel") inhibit bone-resorbing osteoclast cells. Hormone therapy is another option, but it is used less often simply because bisphosphonates are so effective. Calcitonin and parathyroid hormone are the body's two naturally occurring hormones for calcium homeostasis. Calcitonin can be administered as a nasal spray or an injection to inhibit osteoclasts and to slow bone thinning. Parathyroid hormone is given by injection to high-risk patients to stimulate osteoblast cells to build new bone. To slow bone loss, estrogen is used for postmenopausal women and testosterone can be given to men. However, sex hormone therapy must be carefully monitored because these hormones may trigger the growth of certain reproductive tissue cancers. The breast cancer drugs tamoxifen and raloxifene are also used occasionally to stimulate the growth of new bone tissue.

Whatever your age, race, or gender, there are steps you can take to avoid having osteoporosis when you get older. The most important thing you can do to protect your skeleton is to make sure your diet contains enough calcium. The U.S. National Institutes of Health advises that adults take in 1,000 mg of dietary calcium daily, accompanied by 600 IU (international units) of vitamin D to promote calcium absorption. After age 65, your calcium intake should increase to 1,200 mg. In addition, because older people have fewer vitamin D receptors in the intestinal tract, you'll need additional vitamin D as well: 800 IU are recommended. Get outside if you can, because the ultraviolet energy from mild sunlight exposure allows your skin to synthesize vitamin D. If you live on or north of an imaginary line drawn from Boston to Milwaukee, to Minneapolis, and then to Boise, chances are you're not getting enough vitamin D during the winter months. If this is the case, look for vitamin D found in fortified foods such as low-fat milk and cereal. Combine regular moderate exercise such as walking, cycling or jogging with weight training to restore and maintain bone strength. And if you're a smoker, quit! Cigarette toxins damage blood vessels, thus decreasing the blood supplied to bone. In addition, the nicotine and other chemicals from cigarette smoke destroy osteoblasts. You can find tips to help you quit smoking in Chapter 14.

After birth, the epiphyses of a long bone continue to grow, but soon secondary ossification centers appear in these regions. Here spongy bone forms and doesn't break down. A band of cartilage called an **epiphyseal plate** remains between the primary ossification center and each secondary center. The limbs keep increasing in length and width as long as epiphyseal plates are still present. The rate of growth is controlled by hormones, such as growth hormones and the sex hormones. Eventually, the epiphyseal plates become ossified, and the bone stops growing in length.

Long bone growth ends when the epiphyseal plates are ossified, but it is possible for bones to increase their diameter by

appositional growth. In this process, osteoprogenitor cells in the inner periosteum convert to osteoblast cells, which in turn add more matrix to the outer surface of the bone. Osteoblasts in the matrix are then converted to osteocytes. Appositional growth causes the bone to become thicker and stronger.

Remodeling of Bones

In the adult, bone is continually being broken down and built up again. Osteoclasts derived from monocytes in red bone marrow break down bone, remove worn cells, and assist in depositing calcium in the blood. After a period of about three weeks, the osteoclasts disappear, and the bone is repaired by the work of osteoblasts. As they form new bone, osteoblasts take calcium from the blood. Eventually some of these cells get caught in the matrix they secrete and are converted to osteocytes, the cells found within the lacunae of osteons.

Though it might seem strange, adults apparently require at least as much calcium in the diet (about 1,000 to 1,200 mg daily) as do actively growing children. Calcium promotes the work of osteoblasts in adults and children. Although adults no longer experience growth in the long bones, high levels of calcium are necessary to prevent **osteoporosis.** In osteoporosis, bones are thin, weak, and fracture easily (see the Medical Focus on osteoporosis, previous page).

Growth of bone is a complex process involving over 20 different known hormones and other messenger chemicals. Three of the most important hormones that regulate bone growth are parathyroid hormone, calcitonin, and growth hormone. Their effects on bone are discussed in Chapter 10.

Surface Features of Bones

As we study the various bones of the skeleton, refer to Table 6.1, which lists and explains the surface features of bones.

TABLE 6.1 Surface Features of Bones

ARTICULATIONS AND PROJECTIONS

Term	Definition	Example
ARTICULATING SURFACES		
Condyle (kon-dile)	A rounded knob of bone	Occipital condyles at the base of the skull (Fig. 6.7)
Epicondyle	Smaller knob above condyle	Epicondyles of the femur (Fig. 6.17)
Head	Large knob of bone that creates a joint proximally	Head of humerus (Fig. 6.13)
Process	Creates a bar of bone or forms a joint with a fossa	Zygomatic and temporal processes of the skull (Figs. 6.6, 6.7) Olecranon process of the ulna (Fig. 6.14)
Suture	Immovable joint of the skull	Coronal, sagittal, lambdoidal, squamosal sutures (Figs. 6.5, 6.6, 6.7)
PROJECTIONS FOR MUSCLE ATTACHMENT		
Crest	A slender ridge	Anterior crest of tibia (Fig. 6.18)
Malleolus	Thickened, triangular knob	Medial malleolus of tibia (Fig. 6.18)
Spine	Slightly thickened ridge	Spine of the scapula (Fig. 6.12)
Trochanter (tro-kan ter)	Large, oval knob found only on the femur	Greater and lesser trochanters (Fig. 6.17)
Tubercle	Small, round knob	Greater tubercle of humerus (Fig. 6.13)
Tuberosity	Large, irregular knob	Tibial tuberosity (Fig. 6.18)
DEPRESSIONS AND OPENINGS		
Foramen	Hole in a bone	Foramen magnum of the skull (Fig. 6.7)
Fossa	Depression in bone	Olecranon fossa of humerus (Fig. 6.13)
Meatus (me-ay tus)	External opening in a canal	External acoustic meatus of temporal bone (Fig. 6.6)
Sinus	Hollow cavity in bone	Frontal sinus in the frontal bone (Fig. 6.5)

I.C.E. — IN CASE OF EMERGENCY

Broken Bones

Raising a child is always an adventure, but having an active, busy child can bring its share of traumas. Wise parents don't want to limit their children's activities unless it's necessary for safety. Lively children often require emergency care for bone fractures. When energetic children grow into adolescence, they often suffer sports-related fractures as well.

A **fracture** is *complete* if the bone is broken through and *incomplete* if the bone isn't separated into two parts. A fracture is *simple* if bone ends don't pierce the skin and *compound* if skin is torn open by bone. When the broken ends are wedged into each other, the fracture is *impacted.* A *spiral* fracture occurs when the break is ragged due to bone twisting. Repair of a fracture is called **reduction.** Closed reduction involves realigning the bone fragments into their normal position without surgery. Open reduction requires surgical repair of the bone using plates, screws, or pins.

Parents or caregivers should always suspect a fracture if a child feels pain in a limb, or if the limb is swollen or bruised. If the child can't move the limb normally, or the limb appears deformed, a fracture is also likely. Emergency care of a fracture involves immobilization of the limb. A temporary splint can be created using rolled-up newspapers or magazines. Caregivers should constantly monitor the affected limb because nerves and blood vessels may be damaged by the injury. If tissues begin turning blue and/or a pulse can't be felt, blood vessel damage might be occurring. Tingling or numbness indicate possible nerve damage. Treatment must begin immediately in these situations.

Pain management should begin as soon as possible—fractures are very painful! Fractures are typically diagnosed with X rays, but a CT scan or MRI is sometimes necessary. The fracture is permanently immobilized using a cast or splint. Bone repair occurs in a series of four steps (Fig. 6A):

1. *Hematoma*—Within six to eight hours after a fracture, blood escapes from ruptured blood vessels and forms a hematoma (mass of clotted blood) in the space between the broken bones.
2. *Fibrocartilaginous callus*—Tissue repair begins, and fibrocartilage fills the space between the ends of the broken bone.
3. *Bony callus*—Osteoblasts produce trabeculae of spongy bone and convert the fibrocartilaginous callus to a bony callus that joins the broken bones together and lasts about three to four months.
4. *Remodeling*—Osteoblasts build new compact bone at the periphery, and osteoclasts reabsorb the spongy bone, creating a new medullary cavity.

In some ways, bone repair parallels the development of a bone. However, a hematoma indicates that injury has occurred. Fibrocartilage precedes the production of compact bone (instead of hyaline cartilage, as in growing bone).

Parents and caregivers should also be aware that bone fractures may sometimes indicate child/elder abuse. In cases where abuse is suspected, health-care professionals are required by law to investigate the circumstances of the injury.

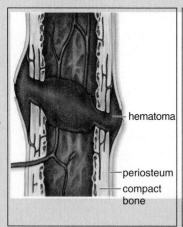 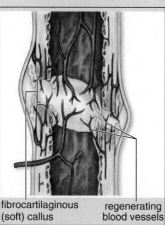 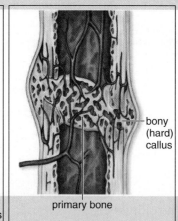 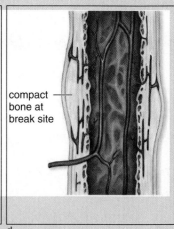

a. b. c. d.

Figure 6A Repair of a broken bone. (a) A hematoma forms between the broken sections of the bone. **(b)** Fibrocartilage fills the space for about three weeks. **(c)** Bony callus is formed by osteoblast cells. **(d)** Osteoclasts reabsorb the callus and create a new medullary cavity.

6.2 Axial Skeleton

7. Distinguish between the axial and appendicular skeletons.
8. Name the bones of the skull, and state the important features of each bone.
9. Describe the structure and function of the hyoid bone.
10. Name the bones of the vertebral column and the thoracic cage. Be able to label diagrams of them.
11. Describe a typical vertebra, the atlas and axis, and the sacrum and coccyx.
12. Name the three types of ribs and the three parts of the sternum.

The skeleton is divided into the axial skeleton and the appendicular skeleton. The tissues of the axial and appendicular skeletons are bone (both compact and spongy), cartilage (hyaline, fibrocartilage, and elastic cartilage), and dense connective tissue, a type of fibrous connective tissue. (The various types of connective tissues were extensively discussed in Chapter 4.)

In Figure 6.4, the bones of the axial skeleton are colored gray, and the bones of the appendicular skeleton are colored tan for easy distinction. Notice that the **axial skeleton** lies in the midline of the body and contains the bones of the skull, the hyoid bone, the

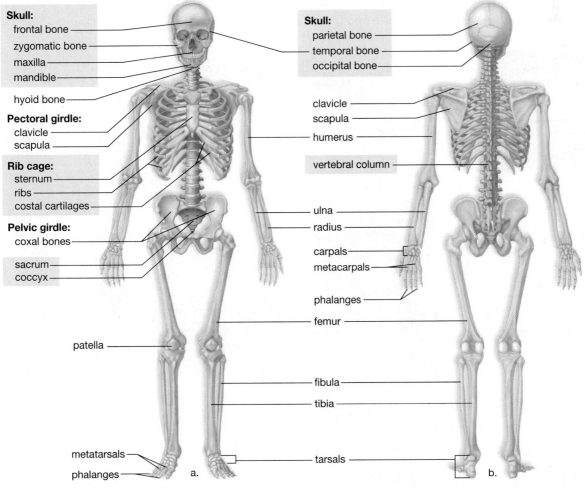

AP|R **Figure 6.4** **Major bones of the skeleton.** (**a**) Anterior view. (**b**) Posterior view. The bones of the axial skeleton are shown in blue-gray, and those of the appendicular skeleton are shown in tan.

vertebral column, and the thoracic cage (composed of the ribs and the sternum, or breastbone). Six tiny middle ear bones (three in each ear) are also in the axial skeleton; we will study them in Chapter 9 in connection with the ear. The **appendicular skeleton** includes the pectoral girdle, upper limbs, pelvic girdle, and lower limbs.

Skull

The skull is formed by the cranium and the facial bones. These bones contain **sinuses,** air spaces lined by mucous membranes that reduce the weight of the skull and give the voice a resonant sound. The paranasal sinuses empty into the nose and are named for their locations. They include the maxillary, frontal, sphenoidal, and ethmoidal sinuses (the frontal and sphenoid sinuses are shown in Fig. 6.5). **Sinusitis** is infection or inflammation of the paranasal sinuses. The two mastoid sinuses drain into the middle ear. **Mastoiditis,** a condition that can lead to deafness, is an inflammation of these sinuses.

Bones of the Cranium

The cranium protects the brain and is composed of eight bones. These bones are separated from each other by immovable joints called **sutures.** Newborns have membranous regions called **fontanels,** where the bones of the cranial vault have not yet fused together. The fontanels permit the bones of the skull to shift during birth as the head passes through the birth canal. The largest

fontanel is the anterior fontanel (often called the "soft spot"), which is located where the two parietal bones meet the two unfused parts of the frontal bone. The anterior fontanel usually closes by the age of two years. Besides the frontal bone, the cranium is composed of two parietal bones, one occipital bone, two temporal bones, one sphenoid bone, and one ethmoid bone (Figs. 6.6 and 6.7).

Frontal Bone One frontal bone forms the forehead, a portion of the nose, and the superior portions of the orbits (bony sockets of the eyes).

Parietal Bones Two parietal bones are just posterior to the frontal bone. They form the roof of the cranium and also help form its sides.

Occipital Bone One occipital bone forms the most posterior part of the skull and the base of the cranium. The spinal cord joins the brain by passing through a large opening in the occipital bone called the **foramen magnum.** The **occipital condyles** are rounded processes on either side of the foramen magnum that articulate with the first vertebra of the spinal column.

Temporal Bones Two temporal bones are just inferior to the parietal bones on the sides of the cranium. They also help form the base of the cranium (Figs. 6.5 and 6.6a). Each temporal bone has the following special features:

- **external acoustic meatus,** a canal that leads to the middle ear;
- **mandibular fossa,** which articulates with the mandible;

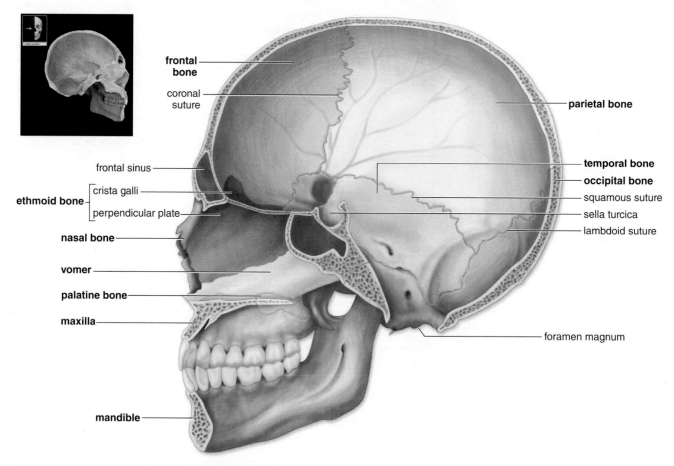

APR Figure 6.5 **Sagittal section of the skull.**

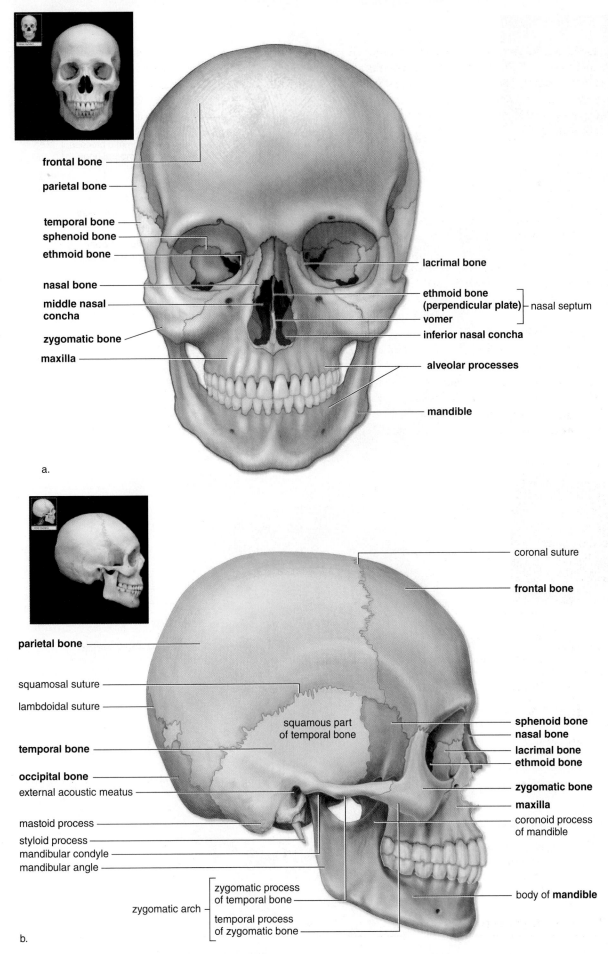

frontal bone

parietal bone

temporal bone
sphenoid bone
ethmoid bone

nasal bone
middle nasal concha

zygomatic bone

maxilla

lacrimal bone

ethmoid bone
(perpendicular plate) ⎤
vomer ⎦ nasal septum

inferior nasal concha

alveolar processes

mandible

a.

coronal suture

frontal bone

parietal bone

squamosal suture

lambdoidal suture

squamous part
of temporal bone

sphenoid bone
nasal bone

temporal bone

lacrimal bone
ethmoid bone

occipital bone

zygomatic bone

external acoustic meatus

maxilla

mastoid process

coronoid process
of mandible

styloid process
mandibular condyle
mandibular angle

zygomatic arch ⎤
⎦

zygomatic process
of temporal bone

temporal process
of zygomatic bone

body of **mandible**

b.

AP|R **Figure 6.6** **Skull anatomy.** (**a**) Anterior view. (**b**) Lateral view.

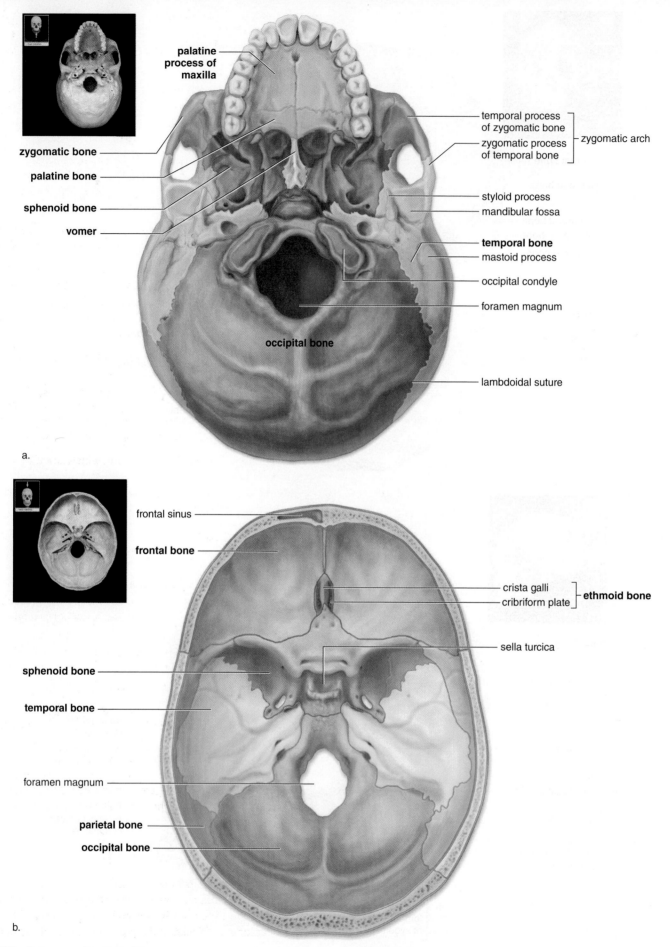

palatine process of maxilla

zygomatic bone

palatine bone

sphenoid bone

vomer

temporal process of zygomatic bone

zygomatic process of temporal bone

} zygomatic arch

styloid process

mandibular fossa

temporal bone

mastoid process

occipital condyle

foramen magnum

occipital bone

lambdoidal suture

a.

frontal sinus

frontal bone

crista galli

cribriform plate

} **ethmoid bone**

sella turcica

sphenoid bone

temporal bone

foramen magnum

parietal bone

occipital bone

b.

APR **Figure 6.7** **Skull anatomy.** (**a**) Inferior view. (**b**) Superior cross-sectional view.

- **mastoid process,** which provides a place of attachment for certain neck muscles;
- **styloid process,** which provides a place of attachment for muscles associated with the tongue and larynx;
- **zygomatic process,** which projects anteriorly and helps form the cheekbone.

Sphenoid Bone The sphenoid bone helps form the sides and floor of the cranium and the rear wall of the orbits. The sphenoid bone is shaped like a butterfly. Its complex shape allows it to articulate with and hold together the other cranial bones (Fig. 6.7). Within the cranial cavity, the sphenoid bone has a saddle-shaped midportion called the **sella turcica** (Fig. 6.7*b*), which houses the pituitary gland in a depression.

Ethmoid Bone The ethmoid bone is anterior to the sphenoid bone and helps form the floor of the cranium. It contributes to the medial sides of the orbits and forms the roof and sides of the nasal cavity (Figs. 6.5, 6.6, and 6.7). Important components of the ethmoid bone include:

- **crista galli** (cock's comb, Fig. 6.7), a triangular process that serves as an attachment for membranes that enclose the brain;
- **cribriform plate** (Fig. 6.7), with tiny holes that serve as passageways for nerve fibers from the olfactory receptors (nerve endings that give us our sense of smell);
- **perpendicular plate** (Fig. 6.5), which projects downward to form the superior part of the nasal septum;
- **superior** and **middle nasal conchae,** which project toward the perpendicular plate. These increase the surface area of the nasal cavity. Projections support mucous membranes that line the nasal cavity.

Bones of the Face

Maxillae The two maxillae form the upper jaw. Aside from contributing to the floors of the orbits and to the sides of the floor of the nasal cavity, each maxilla has the following processes:

- **alveolar process** (Fig. 6.6*a*). The alveolar processes contain the tooth sockets for teeth: incisors, canines, premolars, and molars.
- **palatine process** (Fig. 6.7*a*). The left and right palatine processes form the anterior portion of the **hard palate** (roof of the mouth).

Palatine Bones The two palatine bones contribute to the floor and lateral wall of the nasal cavity (Fig. 6.5). The horizontal plates of the palatine bones form the posterior portion of the hard palate (Fig. 6.7*a*).

Notice that the hard palate consists of (1) portions of the maxillae (i.e., the palatine processes) and (2) horizontal plates of the palatine bones. A cleft palate results when either (1) or (2) have failed to fuse.

Zygomatic Bones The two zygomatic bones form the sides of the orbits (Fig. 6.7*a*). They also contribute to the "cheekbones." Each zygomatic bone has a **temporal process.** A **zygomatic arch,** the most prominent feature of a cheekbone, consists of a temporal process connected to a zygomatic process (a portion of the temporal bone).

Lacrimal Bones The two small, thin lacrimal bones are located on the medial walls of the orbits (Fig. 6.6). A small opening between the orbit and the nasal cavity serves as a pathway for a duct that carries tears from the eyes to the nose.

Nasal Bones The two nasal bones are small, rectangular bones that form the bridge of the nose (see Fig. 6.5). The anterior distal portion of the nose is cartilage, which explains why the nose is not seen on a skull.

Vomer Bone The vomer bone joins with the perpendicular plate of the ethmoid bone to form the nasal septum (see Figs. 6.5 and 6.6*a*).

Inferior Nasal Conchae The two inferior nasal conchae are thin, curved bones that form a part of the inferior lateral wall of the nasal cavity (see Fig. 6.6*a*). Like the superior and middle nasal conchae, they project into the nasal cavity and support the mucous membranes that line the nasal cavity.

Mandible The mandible, or lower jaw, is the only movable portion of the skull. The horseshoe-shaped front and horizontal sides of the mandible, referred to as the *body,* form the chin. The body has an **alveolar process** (see Fig. 6.6*a*), which contains tooth sockets. Superior to the left and right angle of the mandible are upright projections called *rami.* Each ramus has detailed features:

- **mandibular condyle** (see Fig. 6.6*b*), which articulates with a temporal bone;
- **coronoid process** (see Fig. 6.6*b*), which serves as a place of attachment for the muscles used for chewing.

Begin Thinking Clinically

You're treating an 11-year-old patient in the emergency room. His right eye was struck by a baseball bat, and he's rapidly developing a nasty black eye. What bones might have been broken by the injury?

Answer and discussion in Appendix A.

Hyoid Bone

The U-shaped hyoid bone (see Fig. 6.4) is located superior to the larynx (voice box) in the neck. It Is the only bone in the body that does not articulate (form a joint) with another bone. Instead, it is suspended from the styloid processes of the temporal bones by the stylohyoid muscles and ligaments. It anchors the tongue and serves as the site for the attachment of several muscles associated with swallowing.

Vertebral Column (Spine)

The **vertebral column** extends from the skull to the pelvis. It consists of a series of separate bones, the **vertebrae,** separated by pads of fibrocartilage called the **intervertebral disks** (Fig. 6.8). The vertebral column is located in the posterior region of the body at

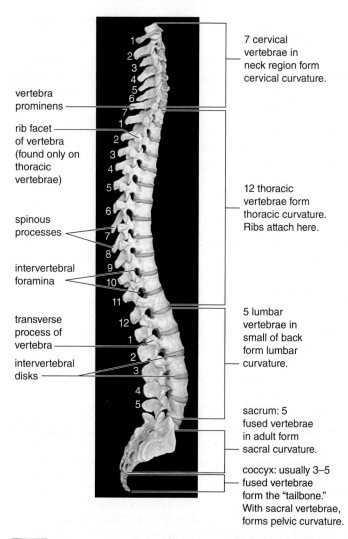

7 cervical vertebrae in neck region form cervical curvature.

vertebra prominens

rib facet of vertebra (found only on thoracic vertebrae)

spinous processes

intervertebral foramina

transverse process of vertebra

intervertebral disks

12 thoracic vertebrae form thoracic curvature. Ribs attach here.

5 lumbar vertebrae in small of back form lumbar curvature.

sacrum: 5 fused vertebrae in adult form sacral curvature.

coccyx: usually 3–5 fused vertebrae form the "tailbone." With sacral vertebrae, forms pelvic curvature.

AP|R **Figure 6.8** **Curvatures of the spine.** The vertebrae are named for their location in the body. Note the presence of the coccyx, also called the tailbone.

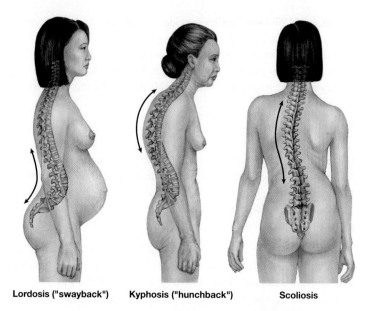

Lordosis ("swayback")   Kyphosis ("hunchback")   Scoliosis

Figure 6.9 **Abnormal curvatures of the vertebral column.**

the midline, where it forms the body's vertical axis. The skull rests on the superior end of the vertebral column, which also supports the rib cage and serves as a point of attachment for the pelvic girdle. The vertebral column also protects the spinal cord, which passes through a vertebral canal formed by the vertebrae. The vertebrae are named according to their location: seven *cervical* (neck) *vertebrae,* twelve *thoracic* (chest) *vertebrae,* five *lumbar* (lower back) *vertebrae,* five *sacral vertebrae* fused to form the sacrum, and three to five *coccygeal vertebrae* fused into one coccyx (tailbone).

When viewed from the side, the vertebral column has four normal curvatures, named for their location (Fig. 6.8). The cervical and lumbar curvatures are concave posteriorly, and the thoracic and sacral curvatures are convex posteriorly. In the fetus, the vertebral column has only one curve, and it is convex posteriorly (the curved "fetal position"). The cervical curve develops three to four months after birth, when the child begins to hold his or her head up. The lumbar curvature develops when a child begins to stand and walk, around one year of age. The curvatures of the vertebral column

provide more support than a straight column would, and they also provide the balance needed to walk upright.

The curvatures of the vertebral column are subject to abnormalities (Fig. 6.9). An abnormally exaggerated lumbar curvature is called **lordosis,** or "swayback." People who are balancing a heavy midsection, such as pregnant women or men with "potbellies," may have swayback. An increased roundness of the thoracic curvature is **kyphosis,** or "hunchback." This abnormality sometimes develops in older people as the center sections of thoracic vertebrae become compressed. An abnormal lateral (side-to-side) curvature is called **scoliosis.** Occurring most often in the thoracic region, scoliosis is usually first seen during late childhood.

Intervertebral Disks

The fibrocartilaginous intervertebral disks located between the vertebrae act as cushions. The disks are filled with gelatinous material, which prevents the vertebrae from grinding against one another and absorbs shock caused by such movements as running, jumping, and even walking. The disks also allow motion between the vertebrae so that a person can bend forward, backward, and from side to side. Unfortunately, these disks become weakened with age, and they can slip or even rupture (called a **herniated disk**). A damaged disk pressing against the spinal cord or the spinal nerves causes pain. Such a disk may need to be removed surgically. If a disk is removed, the vertebrae are fused together, limiting the body's flexibility.

Vertebrae

Figure 6.10*a* shows that a typical vertebra has an anteriorly placed *body* and a posteriorly placed vertebral *arch.* The vertebral arch forms the wall of a *vertebral foramen* (pl., *foramina*). When stacked on top of one another, the foramina become a canal through which the spinal cord passes.

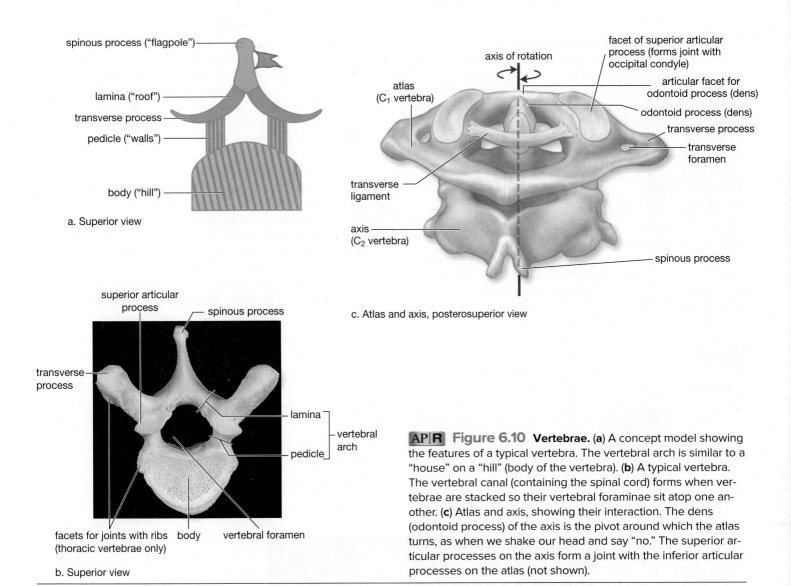

a. Superior view

spinous process ("flagpole")

lamina ("roof")

transverse process

pedicle ("walls")

body ("hill")

axis of rotation

atlas
(C₁ vertebra)

facet of superior articular
process (forms joint with
occipital condyle)

articular facet for
odontoid process (dens)

odontoid process (dens)

transverse process

transverse
foramen

transverse
ligament

axis
(C₂ vertebra)

spinous process

c. Atlas and axis, posterosuperior view

superior articular
process

spinous process

transverse
process

lamina

vertebral
arch

pedicle

facets for joints with ribs
(thoracic vertebrae only)

body

vertebral foramen

b. Superior view

AP|R **Figure 6.10** **Vertebrae. (a)** A concept model showing the features of a typical vertebra. The vertebral arch is similar to a "house" on a "hill" (body of the vertebra). **(b)** A typical vertebra. The vertebral canal (containing the spinal cord) forms when vertebrae are stacked so their vertebral foraminae sit atop one another. **(c)** Atlas and axis, showing their interaction. The dens (odontoid process) of the axis is the pivot around which the atlas turns, as when we shake our head and say "no." The superior articular processes on the axis form a joint with the inferior articular processes on the atlas (not shown).

The structure of a single vertebra can be likened to a house (the vertebral arch) sitting on a hill (the body of the vertebra). *Pedicles* are the upright walls of the house and the *laminae* form a slanting roof. The single *spinous process*, or *spine,* is like a flagpole, and the *transverse processes* are gutters projecting sideways at the corners where pedicles join laminae. Each of these bony projections on an actual vertebra serves as a site for muscle attachment. In addition, *superior* and *inferior articular processes* form joints between an upper and a lower vertebra. When stacked this way, the articular processes create paired openings called *intervertebral foramina* on both sides of the vertebral column (see Fig. 6.8). Spinal nerves exit from the spinal cord through these openings and travel to both sides of the body.

The vertebrae have regional differences. For example, as the vertebral column descends, the bodies get bigger and are better able to carry more weight. In the cervical region, the spines are short and tend to have a split, or bifurcation. The exception is the C7 vertebra, whose long spinous process is clearly seen posteriorly when a person touches his or her chin to the chest (try it!). Thus,

C7 is the prominent vertebra, or in Latin, *vertebra prominens.* It's an important feature that allows health-care providers to determine the transition between cervical and thoracic vertebrae. In addition, cervical vertebrae all have an opening in the transverse process, called a transverse foramen. The vertebral arteries and veins pass through the first six transverse foramina (C1 to C6). The vertebral arteries help to supply blood to the brain, and the veins return blood to the heart.

The thoracic and lumbar vertebrae also have their own unique structural features. The thoracic spines are long and slender and project downward. Further, the bodies and transverse processes of thoracic vertebrae have articular facets, called *costal facets,* for connecting to ribs. The lumbar spines are massive and square and project posteriorly.

Atlas and Axis The first two cervical vertebrae are not typical (Fig. 6.10*c*). The **atlas** supports and balances the head. It has two depressions that articulate with the occipital condyles, allowing movement of the head up and down (as though nodding "yes"). The **axis** has an *odontoid process* (also called the *dens*) that projects into the ring of the atlas. When the head moves from side to side, the atlas pivots around the odontoid process (as though shaking the head "no").

Sacrum and Coccyx The five sacral vertebrae are fused to form the **sacrum.** The sacrum articulates with the pelvic girdle and forms the posterior wall of the pelvic cavity (see Fig. 6.16). The **coccyx,** or tailbone, is the last part of the vertebral column. It is formed from a fusion of three to five vertebrae.

The Rib Cage

The **rib cage** (Fig. 6.11), sometimes called the thoracic cage, is composed of the thoracic vertebrae, the ribs and their costal cartilages, and the sternum.

The rib cage demonstrates how the skeleton is protective but also flexible. The rib cage protects the heart and lungs; yet it swings outward and upward upon inspiration and then downward and inward upon expiration. The rib cage also provides support for the bones of the pectoral girdle (see page 118).

The Ribs

There are 12 pairs of ribs. All 12 pairs connect directly to the thoracic vertebrae in the back, by attaching to costal facets built into the thoracic vertebrae. After connecting with thoracic vertebrae, each rib first curves outward and then forward and downward. The first pair of ribs attaches to the body of the first thoracic vertebra, or T1. It also attaches to the transverse process of T1, at the facet for the joint with the rib. The next eight pairs of ribs (ribs 2–9) attach to the vertebrae at three places (see Fig. 6.10a). First, each

attaches to the body of the same numbered vertebra (rib 4, for example, is attached to the body of the 4th thoracic vertebra). Next, each attaches to the body of the vertebra immediately superior (so, rib 4 also attaches to the body of the 3rd thoracic vertebra). Finally, each rib attaches to the transverse process of the same numbered vertebra (rib 4 attaches to the transverse process of the 4th thoracic vertebra). Rib pairs 10 through 12 attach only to their respective vertebrae.

The upper seven pairs of ribs connect directly to the sternum by means of costal cartilages. These are called the "true ribs," or the *vertebrosternal* ribs. The next five pairs of ribs are called the "false ribs" because they attach indirectly to the sternum or are not attached at all. Ribs 8, 9, and 10 are called *vertebrochondral* ribs; each attaches its costal cartilage to the cartilage of the rib superior to it. All three ribs attach indirectly to the sternum using the costal cartilage of rib 7. Ribs 11 and 12 are *vertebral,* or "floating," ribs. These are short ribs with no attachment to the sternum (Fig. 6.11).

The Sternum

The **sternum,** or breastbone, is a flat bone that has the shape of a blade. The sternum, along with the ribs, helps protect the heart and lungs. During surgery the sternum may be split to allow access to the organs of the thoracic cavity. The sternum is also where a first responder will place his or her hands when performing CPR (see Chapter 14).

The sternum is composed of three bones that fuse during fetal development (Fig. 6.11). These bones are the manubrium,

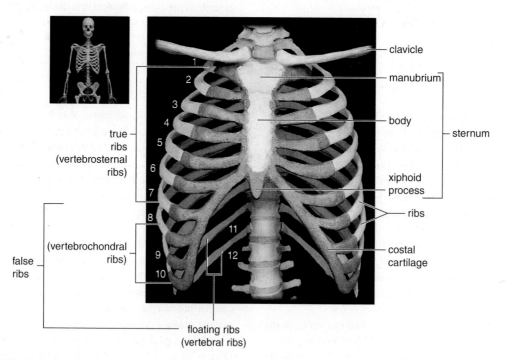

AP|R **Figure 6.11 The rib cage.** This structure includes the thoracic vertebrae, the ribs, and the sternum. The three bones that make up the sternum are the manubrium, body, and xiphoid process. The ribs numbered 1–7 are true ribs; ribs 8–12 are false ribs. Ribs 11 and 12 are also called "floating ribs."

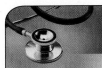

MEDICAL FOCUS

Oh, My Aching Back—Options for Back Injuries

Back pain is going to be an inevitable part of almost everyone's life at some point. It might begin when you bend over to pick up a heavy load and stand up suddenly. Perhaps it comes on gradually with those extra 30 pounds you've put on over the last 10 years. Maybe it was caused by the fall when you were ice skating and landed abruptly on your posterior. Regardless of its cause, in 90% of cases, back pain will slowly improve over time with minimal treatment. Pain management, weight control, gentle massage, physical therapy, and exercise all help to make the patient comfortable and restore mobility. In all but about 5% of back pain cases, the pain is resolved within three months or less.

Many back pain sufferers lack the patience and persistence to allow the back to heal itself normally. These folks often inquire about surgical options to alleviate their back pain. However, back and neck surgery is recommended only in the most extreme cases. For example, excessive movement of the vertebrae or compression of the spinal cord (as in a herniated disk, for example) are reasons to attempt back surgery. Surgery may also be required for those with nerve damage caused by kyphosis, scoliosis, or fractured vertebrae. Loss of bowel and/or bladder control, tingling in the arms or legs, or pain that spreads down arms or legs are all signs of nerve damage.

Traditional surgeries include *laminectomy*, where the bony lamina is cut from the vertebra and surgically removed (the lamina is shown in Fig. 6.10*b*). This procedure can relieve pressure on a pinched spinal cord or spinal nerve. Removal of a herniated disk, called a *diskectomy*, is usually followed by fusion of vertebrae. However, once

vertebrae are fused together, motion and flexibility between vertebrae are limited. Other available options seek to avoid removing the intervertebral disk, and thus avoid having to fuse vertebrae. *Intradiscal electrothermal therapy* (IDET) involves inserting a needle into a ruptured disk. The needle is heated to high temperatures for approximately 20 minutes. The tissues of the disk wall become thickened in response to the heat, and the ruptured area is sealed. A diskectomy and vertebral fusion won't be necessary because the disk will be able to heal.

Vertebroplasty and kyphoplasty are techniques that allow compressed vertebrae to be lifted and separated from one another. In vertebroplasty, bone cement is directly injected into the space between two compressed vertebrae, while in kyphoplasty, a small balloon is inserted between the vertebrae and then inflated. Bone cement is injected into the space created by the balloon, expanding the existing vertebrae. Both of these techniques relieve pressure on the trapped spinal nerve. If vertebrae are fractured (in a car accident, for example), the same technique can be used. Artificial vertebral disks can completely replace an intervertebral disk for some patients. The artificial disk is similar to implants used for hip and knee implants. To install the artificial disk, the patient's own intervertebral disk is first removed, then two metal plates are surgically inserted onto the bodies of the superior and inferior vertebrae. In between, a polyethylene-covered titanium disk re-creates the space originally occupied by the intervertebral disk and allows normal motion of the spinal column.

---

the body, and the xiphoid process. The *manubrium* is the superior portion of the sternum. The *body* is the middle and largest part of the sternum, and the *xiphoid process* is the inferior and smallest portion of the sternum. The manubrium joins with the body of the sternum at an angle. This joint is an important anatomical landmark because it occurs at the level of the second rib, and therefore allows the ribs to be counted. Counting the ribs is sometimes done to determine where the apex (most inferior portion) of the heart is located—usually between the fifth and sixth ribs.

The manubrium articulates with the costal cartilages of the first and second ribs; the body articulates with costal cartilages of the second through seventh ribs; and the xiphoid process doesn't articulate with any ribs.

The xiphoid process is the third part of the sternum. Composed of hyaline cartilage in the child, it becomes ossified (converted into bone) in the adult. The variably shaped xiphoid process serves as an attachment site for the diaphragm, which separates the thoracic cavity from the abdominal cavity.

Content CHECK-UP!

6. Which of the following are cranial bones?
   a. frontal
   b. sphenoid
   c. maxilla
   d. a and b
   e. all of the above

7. When you shake your head "no," you're moving a joint between two bones. What are they?

8. The opening in the center of a vertebra where the spinal cord is located is called the _____ _____.

9. The most superior part of the sternum is the:
   a. xiphoid process.   c. manubrium.
   b. body.   d. costal cartilage.

Answers in Appendix A.

6.3 Appendicular Skeleton

13. Name the bones of the pectoral girdle and the pelvic girdle. Be able to label diagrams of them.
14. Name the bones of the upper limb (arm and forearm) and the lower limb (thigh and leg). Be able to label diagrams that include surface features.
15. Cite at least five differences between the female and male pelvises.

The appendicular skeleton contains the bones of the pectoral girdle, upper limbs, pelvic girdle, and lower limbs.

Pectoral Girdle

The **pectoral girdle** (shoulder girdle) contains four bones: two clavicles and two scapulae (Fig. 6.12). It supports the arms and serves as a place of attachment for muscles that move the arms. The bones of this girdle are held in place by ligaments and muscles. This arrangement allows great flexibility but means that the arm is prone to dislocation at the shoulder joint.

Clavicles

The **clavicles** (collarbones) are slender and S-shaped (Fig. 6.12a). Each clavicle articulates medially with the manubrium of the sternum. This is the only place where the pectoral girdle is attached to the axial skeleton.

Each clavicle also articulates with a scapula. The clavicle serves as a brace for the scapula and helps stabilize the shoulder. It is structurally weak, however, and if undue force is applied to the shoulder, the clavicle will fracture.

Scapulae

The **scapulae** (sing., *scapula*), also called the shoulder blades, are broad bones that somewhat resemble triangles (Figs. 6.12b and c). One reason for the pectoral girdle's flexibility is that the scapulae are not joined to each other (see Fig. 6.4).

Each scapula has a spine, a thick ridge of bone found on its posterior surface. Note the following features as well:

- **acromion process,** which articulates with a clavicle and provides a place of attachment for arm and upper back muscles;
- **coracoid process,** which serves as a place of attachment for arm and chest muscles;
- **glenoid cavity,** which articulates with the head of the arm bone (humerus). The arm joint's flexibility is also a result of the glenoid cavity being smaller than the head of the humerus.

Upper Limb

The upper limb includes the bones of the arm (humerus), the forearm (radius and ulna), and the hand (carpals, metacarpals, and phalanges).[1]

---

[1] The term *upper extremity* is used to include a clavicle and scapula (of the pectoral girdle), an arm, forearm, wrist, and hand.

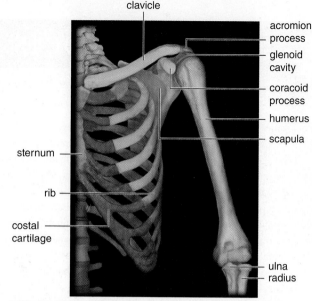

a. Pectoral girdle, frontal view

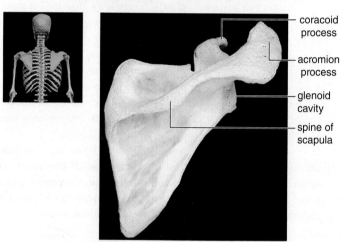

b. Right scapula, posterior view

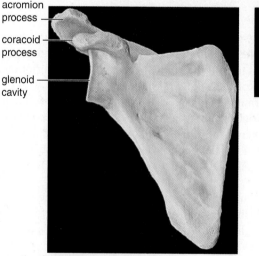

c. Right scapula, anterior view

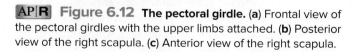

AP|R **Figure 6.12** **The pectoral girdle.** **(a)** Frontal view of the pectoral girdles with the upper limbs attached. **(b)** Posterior view of the right scapula. **(c)** Anterior view of the right scapula.

Humerus

The **humerus** (Fig. 6.13) is the bone of the arm. It is a long bone with the following structural details at the proximal end:

- **head,** which articulates with the glenoid cavity of the scapula;
- **greater** and **lesser tubercles,** which provide attachments for muscles that move the arm and shoulder;
- **intertubercular groove,** which holds a tendon from the biceps brachii, a muscle of the arm;
- **deltoid tuberosity,** which provides an attachment for the deltoid, a muscle that covers the shoulder joint.

At the distal end, these features can be seen on the humerus:

- **capitulum,** a lateral condyle that articulates with the head of the radius;
- **trochlea,** a spool-shaped condyle that articulates with the ulna;
- **coronoid fossa,** a depression for a process of the ulna when the elbow is flexed;
- **olecranon fossa,** a depression for a process of the ulna when the elbow is extended.

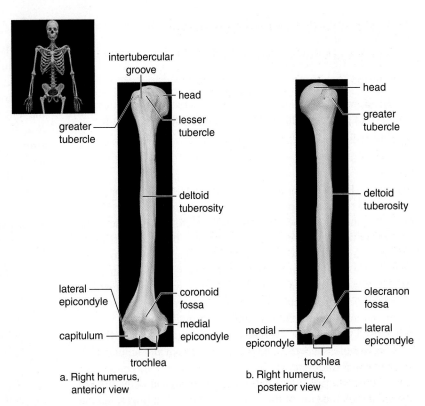

a. Right humerus, anterior view

b. Right humerus, posterior view

AP|R **Figure 6.13** **Right humerus.** (a) Anterior surface view. (b) Posterior surface view.

Radius

The **radius** and **ulna** (see Figs. 6.12*a* and 6.14) are the bones of the forearm. The radius is on the lateral side of the forearm (the thumb side). When you turn your hand from the "palms up" position to the "palms down" position, the radius crosses over the ulna, so the two bones are criss-crossed. Proximally, the radius has the following features:

- **head,** which articulates with the capitulum of the humerus and fits into the radial notch of the ulna;
- **radial tuberosity,** which serves as a place of attachment for the distal tendon from the biceps brachii;

Distally, the radius has the following features:

- **ulnar notch,** which articulates with the head of the ulna;
- **styloid process,** which serves as a place of attachment for ligaments that run to the wrist.

Ulna

The ulna is the longer bone of the forearm. Proximally, the ulna has the following features:

- **coronoid process,** which articulates with the coronoid fossa of the humerus when the elbow is flexed;
- **olecranon process,** the point of the elbow, articulates with the olecranon fossa of the humerus when the elbow is extended;

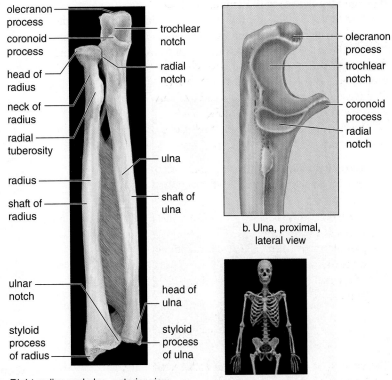

a. Right radius and ulna, anterior view

b. Ulna, proximal, lateral view

AP|R **Figure 6.14** **Right radius and ulna.** (a) The head of the radius articulates with the radial notch of the ulna. The head of the ulna articulates with the ulnar notch of the radius. (b) Lateral view of the proximal end of the ulna.

- **trochlear notch,** which articulates with the trochlea of the humerus at the elbow joint;
- **radial notch,** which articulates with head of the radius.

Distally, the ulna has the following features:

- **head,** which articulates with the ulnar notch of the radius;
- **styloid process,** which serves as a place of attachment for ligaments that run to the wrist.

Hand

Each hand (Fig. 6.15) has a wrist, a palm, and five fingers, or digits.

The wrist, or carpus, contains eight small carpal bones, tightly bound by ligaments in two rows of four each. The distal forearm is where we wear a "wrist watch"—the true wrist is the proximal part of what we generally call the hand. Only two of the carpals (the scaphoid and lunate) articulate with the radius. Anteriorly, the concave region of the wrist is covered by a ligament called the flexor retinaculum, forming the so-called *carpal tunnel.* Inflammation of the tendons running through this area is usually caused by abuse or overuse of the wrist area. The inflamed tendons compress a nerve, and the resulting pain and numbness is *carpal tunnel syndrome.*

Five metacarpal bones, numbered with Roman numerals I to V from the thumb side of the hand toward the little finger, fan out to form the palm. When the fist is clenched, the heads of the metacarpals, which articulate with the phalanges, become obvious. The first metacarpal is more anterior than the others, and this allows the thumb to touch each of the other fingers.

The fingers, including the thumb, contain bones called the *phalanges.* The thumb has only two phalanges (proximal and distal), but the other fingers have three each (proximal, middle, and distal).

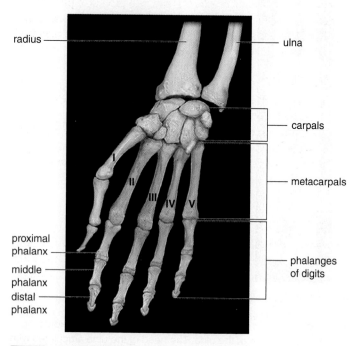

radius — ulna

carpals

metacarpals

proximal phalanx —

phalanges of digits

middle phalanx —
distal phalanx —

AP|R **Figure 6.15** **Right wrist and hand.**

Pelvic Girdle and Pelvis

The **pelvic girdle** is formed by two coxal bones (hip bones). These bones meet anteriorly at the **pubic symphysis,** a tough fibrocartilage joint. Posteriorly, the coxal bones join the sacrum. Inferiorly, the sacrum is joined to the small, triangular coccyx. Combined, the four bones form the **bony pelvis,** also simply called the **pelvis** (Fig. 6.16*a,b;* see also Fig. 6.8). The strong bones of the pelvis are firmly attached to one another and bear the weight of the body. The pelvis also serves as the place of attachment for the lower limbs and protects the urinary bladder, the internal reproductive organs, and a portion of the large intestine.

Coxal Bones

Each **coxal bone** (in Latin, *os coxa*) has the following three parts:

1. **ilium** (Fig. 6.16). The ilium, the largest part of a coxal bone, flares outward to give the hip prominence. The margin of the ilium is called the **iliac crest** (when you put your hands on your hips, you are resting them on the iliac crests). The end points of the iliac crest are called the **anterior superior iliac spine** and the **posterior superior iliac spine** (Fig. 6.16*c*). The **greater sciatic notch** is the site where blood vessels and the large sciatic nerve pass posteriorly into the leg. Each ilium connects posteriorly with the sacrum at a **sacroiliac joint.**
2. **ischium** (Fig. 6.16*c*). The ischium is the most inferior part of a coxal bone. Its posterior region, the **ischial tuberosity,** allows a person to sit. Near the junction of the ilium and ischium is the **ischial spine.** The distance between the ischial spines helps to determine the size of the pelvic cavity.
3. **pubis** (Fig. 6.16*c*). The pubis is the anterior part of a coxal bone. The two pubic bones are joined anteriorly by a fibrocartilage disk at the pubic symphysis. Posterior to where the pubis and the ischium join together is a large opening, the **obturator foramen,** through which blood vessels and nerves pass anteriorly into the leg.

Where the three parts of each coxal bone meet is a depression called the **acetabulum,** which receives the rounded head of the femur.

False and True Pelvises

The false pelvis is the portion of the trunk bounded laterally by the flared parts of the ilium. In Figure 6.16, it is the portion of the pelvis that is shaded green. This space is much larger than that of the true pelvis. The true pelvis (see pink shading on Fig. 6.16), which is inferior to the false pelvis, is the bony ring formed by the sacrum, lower ilium, ischium, and pubic bones. The true pelvis is said to have an upper inlet (also called the *pelvic brim*), and a lower outlet. The dimensions of these outlets are important for females because the outlets must be large enough to allow a baby to pass through during the birth process.

Gender Differences

Female and male pelvises (see Fig. 6.16) usually differ in several ways, including the following:

1. Female iliac bones are more flared than those of the male; therefore, the female has broader hips.

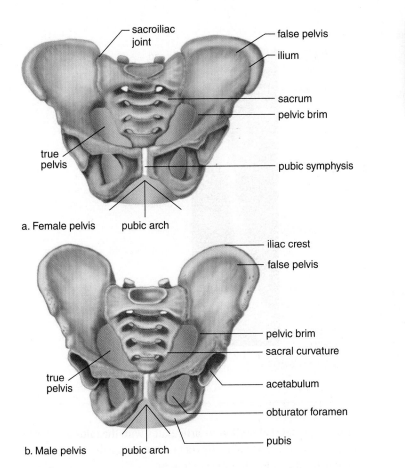

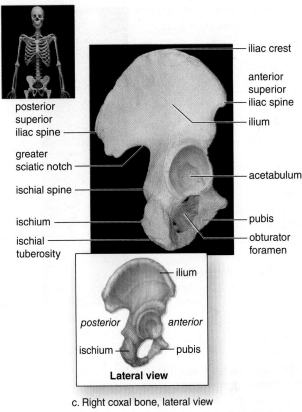

a. Female pelvis

b. Male pelvis

c. Right coxal bone, lateral view

Lateral view

AP|R **Figure 6.16** **The pelvis.** The female pelvis is usually wider in all diameters than that of the male. (**a**) Female pelvis. (**b**) Male pelvis. (**c**) Right coxal bone, lateral view.

2. The female pelvis is wider between the ischial spines and the ischial tuberosities.
3. The female inlet and outlet of the true pelvis are wider.
4. The female pelvic cavity is more shallow, while the male pelvic cavity is more funnel shaped.
5. Female bones are lighter and thinner.
6. The female pubic arch (angle at the pubic symphysis) is wider. In females, the pubic arch resembles an inverted letter U; in males, the pubic arch resembles an inverted V.

In addition to these differences in pelvic structure, male pelvic bones are larger and heavier, the articular ends are thicker, and the points of muscle attachment may be larger.

Lower Limb

The lower limb includes the bones of the thigh (femur), the kneecap (patella), the leg (tibia and fibula), and the foot (tarsals, metatarsals, and phalanges).[2]

Femur

The **femur** (Fig. 6.17), or thighbone, is the longest and strongest bone in the body. Proximally, note these structures on the femur:

- **head,** which fits into the acetabulum of the coxal bone;
- **greater** and **lesser trochanters,** which provide a place of attachment for the muscles of the thighs and buttocks;
- **linea aspera,** a crest that serves as a place of attachment for several muscles.

Distally, the femur has the following components:

- **medial** and **lateral epicondyles,** which serve as sites of attachment for muscles and ligaments;
- **lateral** and **medial condyles,** which articulate with the tibia;
- **patellar surface,** which is located between the condyles on the anterior surface, articulates with the **patella,** a small triangular bone that protects the knee joint.

Tibia

The **tibia** and **fibula** (Fig. 6.18) are the bones of the leg. The tibia, or shinbone, is medial to the fibula. It is thicker than the fibula and bears the weight from the femur, with which it articulates. Observe the structural details of the tibia:

- **medial** and **lateral condyles,** which articulate with the femur;
- **tibial tuberosity,** where the patellar (kneecap) ligaments attach;

[2] The term *lower extremity* is used to include a coxal bone (of the pelvic girdle), the thigh, kneecap, leg, ankle, and foot.

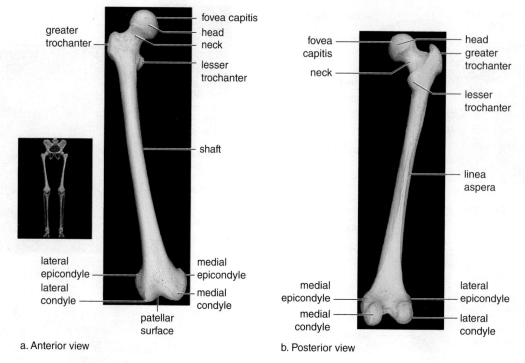

a. Anterior view

b. Posterior view

AP|R **Figure 6.17** **Right femur.** (**a**) Anterior view. (**b**) Posterior view.

- **anterior crest,** commonly called the shin;
- **medial malleolus,** the bulge of the inner ankle, which articulates with the talus in the foot.

Fibula

The fibula is lateral to the tibia and is more slender. It has a head that articulates with the tibia just below the lateral condyle.

Distally, the **lateral malleolus** articulates with the talus and forms the outer bulge of the ankle. Its role is to stabilize the ankle; it does not participate in forming the knee joint.

Foot

Each foot (Fig. 6.19) has an ankle, an instep, and five toes (also called phalanges or digits).

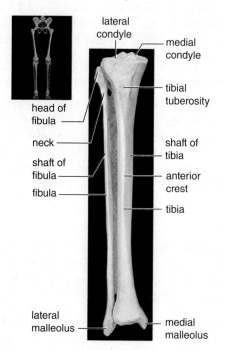

Figure 6.18 **Bones of the right leg, viewed anteriorly.**

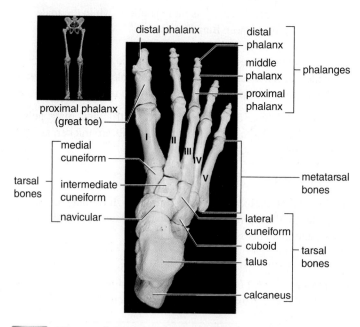

AP|R **Figure 6.19** **The right foot, viewed superiorly.**

The ankle has seven **tarsal bones;** together, they are called the tarsus. Only one of the seven bones, the **talus,** can move freely where it joins the tibia and fibula. The largest of the ankle bones is the **calcaneus,** or heel bone. Along with the talus, it supports the weight of the body.

The instep has five elongated **metatarsal bones.** The distal ends of the metatarsals form the ball of the foot. Along with the tarsals, these bones form the arches of the foot (longitudinal and transverse), which give spring to a person's step. If the ligaments and tendons holding these bones together weaken, fallen arches, or "flat feet," can result.

The toes contain the **phalanges.** The big toe has only two phalanges, but the other toes have three each.

Content CHECK-UP!

10. Name the specific features on the scapula and humerus that form the shoulder joint. Why is this joint relatively unstable?

11. The acetabulum is a part of which bone?

   a. coxal bone   c. humerus

   b. scapula   d. femur

12. Which bone forms the heel?

   a. talus   c. calcaneus

   b. fibula   d. tibia

Answers in Appendix A.

6.4 Joints (Articulations)

16. Explain how joints are classified, and give examples of each type of joint.

17. List the types of movements that occur at synovial joints.

18. Explain how damage and degeneration occur at joints and how they can be treated. Outline possible steps for damage prevention.

Bones articulate at the joints. There are two systems for classifying joints. First, joints can be classified according to the amount of movement they allow. A joint called a **synarthrosis** is immovable, while an **amphiarthrosis** allows slight movement. A **diarthrosis** joint is freely movable. The second classification system (and the convention followed here) is to categorize joints according to their structure:

Fibrous joints occur where fibrous connective tissue joins bone to bone. These joints are typically immovable (thus, synarthrosis joints), although exceptions exist.

Cartilaginous joints occur where fibrocartilage or hyaline cartilage joins bones. These are generally slightly movable (amphiarthrosis joints), with a few exceptions.

Synovial joints are formed when bone ends do not contact each other, but are enclosed in a capsule. These are usually freely movable (diarthrosis joints). Again, there are a few exceptions.

Fibrous Joints

Some bones, such as those that make up the adult cranium, are sutured together by a thin layer of fibrous connective tissue and are immovable. Review Figures 6.5, 6.6, and 6.7, and note the following immovable *sutures:*

coronal suture, between the parietal bones and the frontal bone;

lambdoidal suture, between the parietal bones and the occipital bone;

squamosal suture, between each parietal bone and each temporal bone;

sagittal suture, between the parietal bones (not shown).

The joints formed by each tooth in its tooth socket are also fibrous joints.

Cartilaginous Joints

Where bones are joined by hyaline cartilage or fibrocartilage, the joint that forms is usually slightly movable. The ribs are joined to the sternum by costal cartilages, which are hyaline cartilage (see Fig. 6.11). The epiphyseal plate that separates the diaphysis and epiphyses of growing bones is also a hyaline cartilage joint. The *bodies* of adjacent vertebrae are separated by fibrocartilage intervertebral disks, which increase vertebral flexibility. The pubic symphysis, the joint between the two pubic bones (see Fig. 6.16), consists largely of fibrocartilage. Due to hormonal changes, this joint becomes more flexible during late pregnancy, allowing the pelvis to expand during childbirth.

Synovial Joints

Synovial joints are generally freely movable because, unlike the joints discussed so far, the two bones are separated by a *joint cavity* (Figs. 6.20 and 6.21). The joint cavity is lined by a **synovial membrane,** which produces **synovial fluid,** a lubricant for the joint. The absence of tissue between the articulating bones allows them to be freely movable but means that the joint has to be stabilized in some way.

The joint is stabilized by the joint capsule, a sleeve-like extension of the periosteum of each articulating bone. **Ligaments,** which are composed of dense regular connective tissue, bind the two bones to one another and add even more stability. Tendons, which are cords of dense regular connective tissue that connect muscle to bone, also help stabilize a synovial joint.

The articulating surfaces of the bones are protected in several ways. The bones are covered by a layer of articular (hyaline) cartilage. In addition, the joint, such as the knee, contains **menisci** (sing., *meniscus*), crescent-shaped pieces of cartilage, and fluid-filled sacs called **bursae,** which ease friction between all parts of the joint. Inflammation of the bursae is called **bursitis.** Tennis elbow is a form of bursitis. Articular cartilage can be worn away by years of constant use, such as when performing a sport.

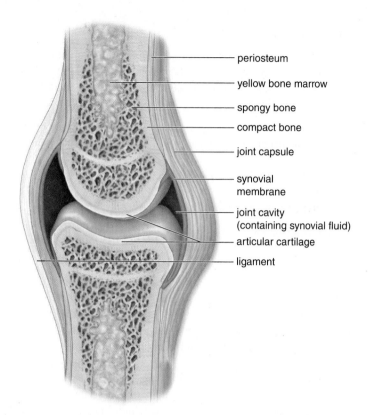

periosteum
yellow bone marrow
spongy bone
compact bone
joint capsule
synovial membrane
joint cavity (containing synovial fluid)
articular cartilage
ligament

AP|R **Figure 6.20** **Generalized anatomy of a synovial joint.**

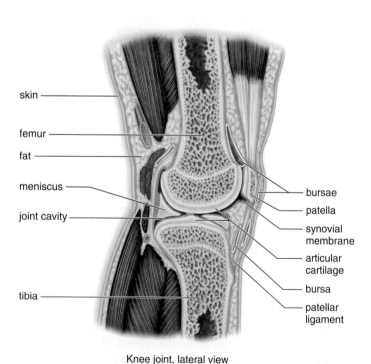

skin
femur
fat
meniscus
joint cavity
tibia

bursae
patella
synovial membrane
articular cartilage
bursa
patellar ligament

Knee joint, lateral view

AP|R **Figure 6.21** **The knee joint.** Notice the menisci and bursae associated with the knee joint.

Types of Synovial Joints

Different types of freely movable joints are listed here and depicted in Figure 6.22.

Saddle joint. Each bone is saddle-shaped and fits into the complementary regions of the other. A variety of movements are possible. *Example*: the joint between the carpal and metacarpal bones of the thumb.

Ball-and-socket joint. The ball-shaped head of one bone fits into the cup-shaped socket of another. Movement in all planes, as well as rotation, are possible. *Examples*: the shoulder and hip joints.

Pivot joint. A small, cylindrical projection of one bone pivots within the ring formed of bone and ligament of another bone. Only rotation is possible. *Examples*: the joint between the proximal ends of the radius and ulna, and the joint between the atlas and axis (see Fig. 6.10).

Hinge joint. The convex surface of one bone articulates with the concave surface of another. Up-and-down motion in one plane is possible. *Examples*: the elbow and knee joints.

Gliding joint. Flat or slightly curved surfaces of bones articulate. Sliding or twisting in various planes is possible. *Examples*: the joints between the bones of the wrist and between the bones of the ankle.

Condyloid joint. The oval-shaped condyle of one bone fits into the elliptical cavity of another. Movement in different planes is possible, but rotation is not. *Examples*: the joints between the metacarpals and phalanges.

Movements Permitted by Synovial Joints

Skeletal muscles are attached to bones by tendons that cross joints. When a muscle contracts, one bone moves in relation to another bone. The more common types of movements are described here.

Angular Movements (Fig. 6.23*a*):

Flexion decreases the joint angle. Flexion of the elbow moves the forearm toward the arm; flexion of the knee moves the leg toward the thigh. *Dorsiflexion* is flexion of the foot upward, as when you stand on your heels; *plantar flexion* is flexion of the foot downward, as when you stand on your toes.

Extension increases the joint angle. Extension of the flexed elbow straightens the upper limb. *Hyperextension* occurs when a portion of the body part is extended beyond 180°. It is possible to hyperextend the head and the trunk of the body, and also the shoulder and wrist (arm and hand).

Adduction is the movement of a body part toward the midline. For example, adduction of the arms or legs moves them back to the sides, toward the body.

Abduction is the movement of a body part laterally, away from the midline. Abduction of the arms or legs moves them laterally, away from the body.

Circular Movements (Fig. 6.23*b*):

Circumduction is the movement of a body part in a wide circle, as when a person makes arm circles. Careful observation of the motion reveals that, because the proximal end of the arm is stationary, the shape outlined by the arm is actually a cone.

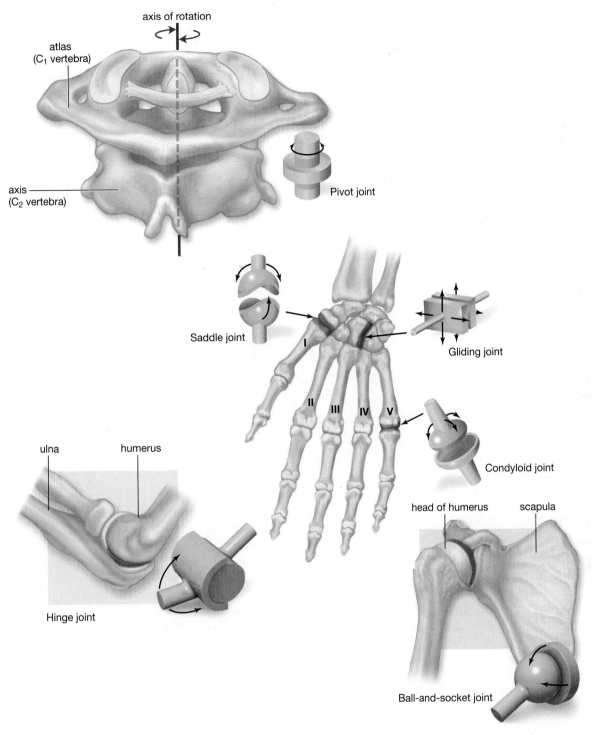

Rotation is the movement of a body part around its own axis, as when the head is turned to answer "no" or when the arm is twisted toward the trunk (medial rotation) and away from the trunk (lateral rotation).

Supination is the rotation of the forearm so that the palm is upward; **pronation** is the opposite—the movement of the forearm so that the palm is downward.

Special Movements (Fig. 6.23c):

Inversion and **eversion** apply only to the feet. Inversion is turning the foot so that the sole faces inward, and eversion is turning the foot so that the sole faces outward.

Elevation and **depression** refer to the lifting up and down, respectively, of a body part, as when you shrug your shoulders or move your jaw up and down.

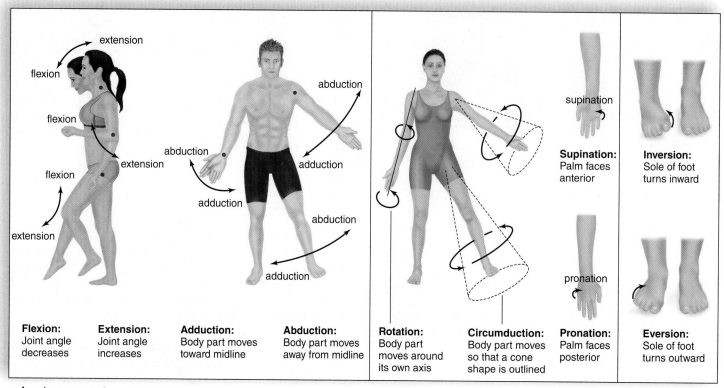

a. Angular movements   b. Circular movements   c. Special movements

Flexion:
Joint angle decreases

Extension:
Joint angle increases

Adduction:
Body part moves toward midline

Abduction:
Body part moves away from midline

Rotation:
Body part moves around its own axis

Circumduction:
Body part moves so that a cone shape is outlined

Pronation:
Palm faces posterior

Eversion:
Sole of foot turns outward

Supination:
Palm faces anterior

Inversion:
Sole of foot turns inward

AP|R Figure 6.23 Joint movements. (a) Angular movements increase or decrease the angle between the bones of a joint. **(b)** Circular movements describe a circle or part of a circle. **(c)** Special movements are unique to certain joints.

Joint Damage and Repair

Joints can be damaged, and even destroyed, by overuse or chronic inflammation. Joint inflammation and destruction is termed **arthritis.** The most common form, **osteoarthritis,** is caused by deterioration of the articular cartilage of a joint. The damage might be caused by chronic overuse. Typical cases of osteoarthritis are seen in joints of older persons after decades of heavy use, or in knee joints of football players after constant misuse and/or abuse. **Rheumatoid arthritis** (RA) occurs when the synovial membrane becomes inflamed and grows thicker cartilage. RA is caused by an autoimmune reaction in which the body's immune system mistakenly attacks the synovial membrane. RA is more common with age, but it can occur in children and younger adults. **Gout,** or gouty arthritis, results from excessive buildup of uric acid (a metabolic waste) in the blood. Crystals of uric acid are deposited in the joints, causing inflammation.

Arthritis causes joint pain and stiffness. The afflicted joint is often swollen and may feel warm to the touch. Without the protective hyaline cartilage, exposed bone ends can grate against each other and cause bone destruction. On an X ray, the joint space is thinner and narrower than normal.

Treatment of arthritis should begin immediately to preserve function. Pain management is an important first step, followed by physical therapy and exercise. It's important that arthritis sufferers keep moving. Without regular exercise, the muscles around the joint will atrophy (shrink in size). Tendons and ligaments around a joint weaken, and the bones at a joint can be dislocated more easily.

Tissue culture (growing cells outside of the patient's body in a special medium) can help younger people and athletes with knee or ankle injuries to regenerate their own hyaline cartilage. In autologous chondrocyte implantation (ACI) surgery, a piece of healthy hyaline cartilage from the patient's joint is first removed surgically. The chondrocyte cells are grown outside the body in tissue culture medium, and then injected into the joint and left to grow. However, ACI isn't always successful, and it can't be used for elderly or overweight patients.

When other treatments for arthritis fail, surgical replacement of a joint can restore movement and relieve pain (Fig. 6.24). Knees and hips are the most common joints to be replaced, but shoulders, elbows, ankles, and even finger joints can be replaced with surgical implants.

After ACI or joint replacement, the patient faces a lengthy rehabilitation. Physical therapy after ACI will stimulate cartilage growth without overstressing the area being repaired. In joint replacement patients, physical therapy will prevent muscle degeneration and promote bone growth around the new implant.

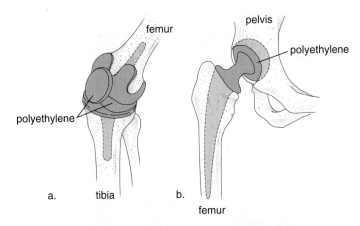

Figure 6.24 **Artificial joints in which polyethylene replaces articular cartilage.** (a) Knee. (b) Hip.

> ## Content CHECK-UP!
>
> **13.** What are the major unfused synarthrosis and amphiarthrosis joints in a newborn? What might happen if these closed prematurely?
>
> **14.** The joint found at the elbow is what type of synovial joint?
>
>   a. hinge   c. saddle
>
>   b. ball-and-socket   d. condyloid
>
> **15.** The most common type of arthritis is _____, and involves the breakdown of _____.
>
> *Answers in Appendix A.*

6.5 Effects of Aging

> **19.** Describe the anatomical and physiological changes that occur in the skeletal system as we age.

Both cartilage and bone tend to gradually deteriorate as a person ages. The chemical nature of cartilage changes, and the bluish color typical of young cartilage changes to an opaque, yellowish color. The chondrocytes die, and reabsorption occurs as the cartilage undergoes calcification, becoming hard and brittle. Calcification interferes with the ready diffusion of nutrients and waste products through the matrix. The articular cartilage may no longer function properly, and symptoms of osteoarthritis can appear. As you know, decades of constant heavy use of a joint can speed the development of arthritis. Osteoporosis, discussed in the Medical Focus on page 106, is present when weak and thin bones lead to fractures. However, it's important to remember that many of the degenerative changes seen in cartilage and bone can be slowed or stopped by regular weight-bearing exercise.

> ## Content CHECK-UP!
>
> **16.** Imagine that you have a patient whose knee joint has calcified with age. What motions are most likely to be affected by this change?
>
> *Answer in Appendix A.*

6.6 Homeostasis

> **20.** List and discuss six ways the skeletal system contributes to homeostasis. Discuss ways the other systems assist the skeletal system.

The illustration in Human Systems Work Together on page 129 tells how the skeletal system assists other systems and how other systems assist the skeletal system. Let's review again the functions of the skeletal system, but this time as they relate to the other systems of the body.

Skeletal System Interacts with All Organ Systems

The bones protect the internal organs. The rib cage protects the heart, lungs, and kidneys; the skull protects the brain; and the vertebrae protect the spinal cord. Certain endocrine organs such as the pituitary gland, pineal gland, and thymus, are also protected by bone. The pelvic bones safeguard the pelvic reproductive structures, thus helping to make reproduction possible.

The bones assist all phases of respiration. The rib cage assists the breathing process, enabling oxygen to enter the blood. Red bone marrow produces the blood cells, including the red blood cells that transport oxygen to the tissues. Without a supply of oxygen, the cells of the body could not efficiently produce ATP, the primary cellular energy source.

The bones store and release calcium needed by the muscular and nervous systems. Hormones control the balance between the concentrations of calcium in the bones and in the blood. Calcium ions play a major role in muscle contraction and nerve conduction. Calcium ions also help regulate cellular metabolism. Protein hormones, which cannot enter cells, are called the first messenger, and a second messenger such as calcium ions jump-starts cellular metabolism, directing it to proceed in a particular way.

The bones assist the lymphatic system and immunity. Red bone marrow produces the white cells, which congregate in the lymphatic organs. White blood cells defend the body against pathogens and cancerous cells. Without the ability to withstand foreign invasion, the body may quickly succumb to disease and die.

The bones assist digestion. The bones used for chewing break food into pieces small enough to be swallowed and chemically digested. Without digestion, needed nutrients would not enter the body.

The skeleton is necessary to locomotion. Our jointed skeleton forms a framework for muscle attachment, and muscle contraction causes joint movement. Thus, we can seek out and move to a more suitable external environment in order to maintain homeostasis.

Other Body Systems Interact with the Skeletal System

How do the other systems of the body help the skeletal system function? The integumentary system and the muscles help the skeletal system protect internal organs. The digestive system absorbs the calcium from food, and the plasma portion of blood transports calcium from the digestive system to the bones. The endocrine system regulates the storage of calcium in the bones, as well as

"JOHN/JANE DOE, SKELETAL REMAINS, AGE UNKNOWN" is the initial identification given by law enforcement officials to the bones of an unidentified human being. The bones may have been found in the woods by a hiker or a hunter, or in a field after a farmer harvests his crops. Bones may be uncovered when a building is demolished, or if natural events such as floods or earthquakes disrupt the soil. Regardless of how human bones are found, questions must be answered. Who was this person? Was this a male or a female, and how old? What was the person's ethnicity? How did the person die, and how long ago? Was this person murdered, or did death come from natural causes?

It's the job of a *forensic anthropologist* to collect, analyze, and ultimately identify the remains. Forensic anthropologists typically have extensive training in the structure of the human skeleton and are able to examine the features of the recovered bones. These scientists rely on a national forensic analysis data bank that contains measurements and observations from thousands of skeletons. In addition, forensic anthropologists are routinely called upon to testify in criminal cases as to a victim's time and cause of death.

Clues about the identity and history of a deceased person can be found throughout the skeleton. Age is approximated by *dentition*—the structure of the teeth in the upper jaw (maxilla) and lower jaw (mandible). For example, infants aged 0–4 months have no teeth present; children aged approximately 6 through 10 have missing deciduous, or "baby" teeth; young adults acquire their last molars, or "wisdom teeth," around age 20. The age of older adults can be approximated by the number and location of missing or broken teeth.

In addition, ossification of bones—that is, replacement of a baby/child's incomplete cartilage skeleton with bone—continues in an orderly fashion until about the age of 20. Studying areas of bone ossification also gives clues to the age of the deceased at the time of death. In older adults, signs of joint breakdown provide additional information about age. Hyaline cartilage becomes worn, yellowed, and brittle with age, and the hyaline cartilages covering bone ends wear down over time. The amount of yellowed, brittle, or missing cartilage helps scientists to estimate the person's age.

If skeletal remains include the individual's pelvic bones, these provide the best method for determining an adult's gender (see pages 120–121). The long bones, particularly the humerus and femur, give information about gender as well. Long bones are thicker and denser in males, and points of muscle attachment are bigger and more prominent. The skull of a male has a square chin and more prominent ridges above the eye sockets or orbits (Figure 6B).

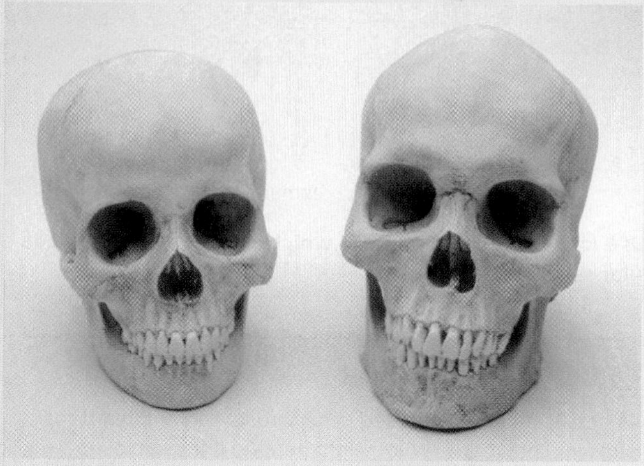

a. Female adult skull   b. Male adult skull

Figure 6B Gender differences of the skull. (a) Note that the female skull is smaller, more delicate, and has a pointed chin. **(b)** The male skull is large, bulky, and has a squared-off chin.

Determining the ethnic origin of skeletal remains can be difficult because so many people have a mixed racial heritage. Forensic anthropologists rely on observed racial characteristics of the skull. In general, individuals of African or African American descent have a greater distance between the eyes, eye sockets that are roughly rectangular, and a jaw that is large and prominent. Skulls of Native Americans typically have round eye sockets, prominent cheek (zygomatic) bones, and a rounded palate. Caucasian skulls usually have a U-shaped palate and a visible suture line between the frontal bones. Additionally, the external ear canals in Caucasians are long and straight, so that the auditory ossicles (tiny bones built inside the temporal bone and used for hearing) are visible.

Once the identity of the individual has been determined, the skeletal remains can be returned to the victim's family for proper burial. Although this can be a sorrowful event, the return of physical remains provides closure and solace to many families. For this reason, special teams of forensic anthropologists employed by the U.S. military are currently researching the identities of bones from soldiers who fought in World War II, as well as the Korean, Vietnam, Gulf, and Iraq/Afghanistan wars. Ancestral remains from Native Americans are protected by the Native American Grave Protection and Repatriation Act, and must be returned to the leadership of the tribe.

the growth of bone and other tissues. The cardiovascular system transports oxygen and nutrients to bone, and wastes from bone. The urinary and digestive systems excrete bone wastes. Movement of the bones would be impossible without contraction of the muscles. In these and other ways, the systems of the body help the skeletal system carry out its functions.

Content CHECK-UP!

17. Which organ systems must cooperate in order to store calcium in the skeleton?

Answer in Appendix A.

Human Systems Work Together

Integumentary System

Bones provide support for skin.

Skin protects bones; helps provide vitamin D for Ca^{2+} absorption.

Muscular System

Bones provide attachment sites for muscles; store Ca^{2+} for muscle function.

Muscular contraction causes bones to move joints; muscles help protect bones.

Nervous System

Bones protect sense organs, brain, and spinal cord; store Ca^{2+} for nerve function.

Receptors send sensory input from bones to central nervous system.

Endocrine System

Bones provide protection for glands; store Ca^{2+} used as second messenger.

Growth hormone regulates bone development; parathyroid hormone and calcitonin regulate Ca^{2+} content.

Cardiovascular System

Rib cage protects heart; red bone marrow produces blood cells; bones store Ca^{2+} for blood clotting.

Blood vessels deliver nutrients and oxygen to bones, carry away wastes.

How the Skeletal System works with other body systems

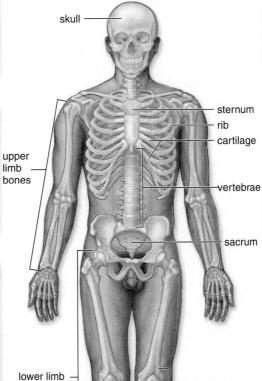

skull

sternum
rib
cartilage

upper limb bones

vertebrae

sacrum

lower limb bones

knee joint

Lymphatic System/Immunity

Red bone marrow produces white blood cells involved in immunity.

Lymphatic vessels pick up excess tissue fluid; immune system protects against infections.

Respiratory System

Rib cage protects lungs and assists breathing; bones provide attachment sites for muscles involved in breathing.

Gas exchange in lungs provides oxygen and rids body of carbon dioxide.

Digestive System

Jaws contain teeth that chew food; hyoid bone assists swallowing.

Digestive tract provides Ca^{2+} and other nutrients for bone growth and repair.

Urinary System

Bones provide support and protection.

Kidneys provide active vitamin D for Ca^{2+} absorption and help maintain blood level of Ca^{2+} needed for bone growth and repair.

Reproductive System

Bones provide support and protection of reproductive organs.

Sex hormones influence bone growth and density in males and females.

Basic Key Terms

abduction (ăb-dŭk'shŭn), p. 124

adduction (ăd-dŭk'shŭn), p. 124

amphiarthrosis (ăm"fē-ăr-thrō'sĭs), p. 123

appendicular skeleton (ăp"ĕn-dĭk'yū-lĕr skĕl'ĕ-tŭn), p. 110

appositional growth (ăp"ŭh-zĭsh'ŭn-ŭl grōth), p. 107

articular cartilage (ăr-tĭk'yū-lĕr kăr'tĭ-lĭj), p. 103

articulation (ăr-tĭk"yū-lā'shŭn), p. 103

atlas (ăt-lŭhs), p. 115

axial skeleton (ăk'sē-ăl skĕl'ĕ-tŭn), p. 109

axis (ăk'sĭs), p. 115

ball-and-socket joint (băwl and sŏk'ĭt jōynt), p. 124

bony pelvis (bō'nē pĕl'vĭs), p. 120

bursae (bŭr'sā), p. 123

calcaneus (kăl-kā'nē-ŭhs), p. 123

cartilaginous joints (kăr'tŭl-ăj'ŭh-nŭs jōyntz), p. 123

circumduction (sĕr"kŭm-dŭk'shŭn), p. 124

clavicles (klăv'ĭ-kŭls), p. 118

coccyx (kŏk'sĭks), p. 116

compact bone (kŏm'păkt bōn), p. 103

condyloid joint (kŏn'dĭl-ōyd jōynt), p. 124

coxal bone (kŏk'săl bōn), p. 120

depression (dĭ-prĕsh'ŭn), p. 125

diaphysis (dī-ăf'ĭ-sĭs), p. 103

diarthrosis (dī-ăr-thrō'sĭs), p. 123

elevation (ĕl'ŭh-vā'shŭn), p. 125

endochondral ossification (ĕn"dō-kŏn'drŭl ŏs"ŭh-fĭ-kā'shŭn), p. 105

epiphyseal plate (ĕp"ĭ-fĭz'ē-ăl plāt), p. 106

epiphysis (ĕ-pĭf'ĭ-sĭs), p. 103

eversion (ē-vĕr'zhŭn), p. 125

extension (ĕk-stĕn'shŭn), p. 124

femur (fē'mŭr), p. 121

fibrous joints (fī'brŭs jōyntz), p. 123

fibula (fĭb'yū-lŭh), p. 121

flexion (flĕk'shŭn), p. 124

fontanel (fŏnt"ŭh-nĕl'), p. 110

foramen magnum (fŏr-ā'-mĕn măg'nŭm), p. 110

gliding joint (glīd'ĭng jōynt), p. 124

hard palate (hărd păl'ut), p. 113

hematopoiesis (hēm"ăh-tō-pōy-ē'sĭs), p. 103

hinge joint (hĭnj jōynt), p. 124

humerus (hyū'mŭr-ŭs), p. 119

ilium (ĭl'ē-ŭm), p. 120

intervertebral disk (ĭn"tĕr-vĕr'tĕ-brŭl dĭsk), p. 113

intramembranous ossification (ĭn"trŭh-mĕm'brăn-ŭs ŏs"ŭ-fĭ-kā'shŭn), p. 105

inversion (ĭn-vĕr'zhŭn), p. 125

ischium (ĭs'kē-ŭm), p. 120

ligament (lĭg'ŭh-mĕnt), p. 123

medullary cavity (mĕd'ū-lār"ē kăv'ĭ-tē), p. 103

menisci (mĕ-nĭs'kī), p. 123

metatarsal bones (mĕt'ŭh-tăr'sŭl bōnz), p. 123

occipital condyle (ŏk'sĭp-ĭ-tŭl kŏn-dīl), p. 110

ossification (ŏs''ĭ-fĭ-kā'shŭn), p. 105

osteoblast (ŏs'tē-ō-blăst"), p. 105

osteoclast (ŏs'tē-ō-klăst"), p. 105

osteocyte (ŏs'tē-ō-sīt), p. 105

osteon (ŏ'stē-ŏn), p. 103

osteoprogenitor cells (ŏs"tē-ō-prō-jĕn'ĭ-tŭr sĕlz), p. 105

patella (pŭh-tĕl' ŭh), p. 121

pectoral girdle (pĕk'tō-rŭl gĕr'dŭl), p. 118

pelvic girdle (pĕl'vĭk gĕr'dŭl), p. 120

pelvis (pĕl'-vĭs), p. 120

periosteum (pĕr"ē-ŏs'tē-ŭm), p. 103

phalanges (fūh-lăn'jēz), p. 123

pivot joint (pĭv'ŭt jōynt), p. 124

pronation (prō-nā'shŭn), p. 125

pubic symphysis (pyū'bĭk sĭm'fĭ-sĭs), p. 120

pubis (pyū'bĭs), p. 120

radius (rā'dē-ŭs), p. 119

red bone marrow (rĕd bōn măh'-rō), p. 103

rib cage (rĭb kāj), p. 116

rotation (rō-tā'shŭn), p. 125

sacrum (sā'krŭm), p. 116

saddle joint (săd'ĕl jōynt), p. 124

scapulae (skăp'yŭh-lā), p. 118

sella turcica (sĕl'ŭh tŭr'sĭk-ŭh), p. 113

sinus (sī'nŭs), p. 110

spongy bone (spŭnj'ē bōn), p. 103

sternum (stŭr'nŭm), p. 116

supination (sū"pĭ-nā'shŭn), p. 125

suture (sū'chĕr), p. 110

synarthrosis (sĭn-ăr-thrō'sĭs), p. 123

synovial fluid (sĭ-nō'vē-ăl flū'ĭd), p. 123

synovial joint (sĭ-nō'vē-ăl jōynt), p. 123

synovial membrane (sĭ-nō'vē-ăl mĕm'brān), p. 123

talus (tă'lŭs), p. 123

tarsal bones (tăr'sŭl bōns), p. 123

temporal process (tĕm'pō-rŭl prŏh'sĕs), p. 113

tibia (tĭb'ē-ŭh), p. 121

ulna (ŭl'nŭh), p. 119

vertebrae (vĕr'tŭh-brā), p. 113

vertebral column (vĕr'tĕ-brăl kŏh'lŭm), p. 113

Clinical Key Terms

arthritis (ăr-thri'tĭs), p. 126

bursitis (bŭr-sī'tĭs), p. 123

fracture (frăk'chĕr), p. 108

gout (gŏwt), p. 126

herniated disk (hĕr'nē-ā-tĕd dĭsk), p. 114

kyphosis (kī-fō'sĭs), p. 114

lordosis (lŏr-dō'sĭs), p. 114

mastoiditis (măs"tōy-dī'tĭs), p. 110

osteoarthritis (ŏs"tē-ō-ăr-thri' tĭs), p. 126

osteoporosis (ŏs"tē-ō-pō-rō'sĭs), p. 107

reduction (rĭ-dŭk'shŭn), p. 108

rheumatoid arthritis (RA), p. 126

scoliosis (skō"lē-ō'sĭs), p. 114

sinusitis (sī"nŭh-sī'tĭs) p. 110

Summary

6.1 Skeleton: Overview

A. The skeleton supports and protects the body; produces red blood cells; serves as a storehouse for inorganic calcium and phosphate ions and fat; and permits flexible movement.

B. A long bone has a shaft (diaphysis) and two ends (epiphyses), which are covered by articular cartilage. The diaphysis contains a medullary cavity with yellow marrow and is bounded by compact bone. The epiphyses contain spongy bone with red bone marrow that produces red blood cells. Bone is a living tissue. It develops, grows, remodels, and repairs itself. In all these processes, osteoclasts break down bone, and osteoblasts build bone.

C. Table 6.1 describes the different surface features of bones.

6.2 Axial Skeleton

The axial skeleton lies in the midline of the body and consists of the skull, the hyoid bone, the vertebral column, and the thoracic cage.

A. The skull is formed by the cranium and the facial bones. The cranium includes the frontal bone, two parietal bones, one occipital bone, two temporal bones, one sphenoid bone, and one ethmoid bone. The facial bones include two maxillae, two palatine bones, two zygomatic bones, two lacrimal bones, two nasal bones, the vomer bone, two inferior nasal conchae, and the mandible.

B. The U-shaped hyoid bone is located in the neck. It anchors the tongue and does not articulate with any other bone.

C. The typical vertebra has a body, a vertebral arch surrounding the vertebral foramen, and a spinous process. The first two vertebrae are the atlas and axis. The vertebral column has four curvatures and contains the cervical, thoracic, lumbar, sacral, and coccygeal vertebrae. Cervical, thoracic, and lumbar vertebrae are separated by intervertebral disks.

D. The rib cage contains the thoracic vertebrae, ribs and associated cartilages, and the sternum.

6.3 Appendicular Skeleton
The appendicular skeleton consists of the bones of the pectoral girdle, upper limbs, pelvic girdle, and lower limbs.

A. Each pectoral girdle (shoulder) is formed by the clavicle and the scapula.

B. The upper limb contains the humerus, the radius, the ulna, and the bones of the hand (the carpals, metacarpals, and phalanges).

C. The pelvic girdle contains two coxal bones, and forms the pelvis when joined to the sacrum and coccyx. The female pelvis is generally wider and more shallow than the male pelvis.

D. The lower limb contains the femur, the patella, the tibia, the fibula, and the bones of the foot (the tarsals, metatarsals, and phalanges).

6.4 Joints (Articulations)
Joints are regions of articulation between bones. They are classified according to their structure and/or degree of movement. Fibrous joints are typically immovable, cartilaginous joints are generally slightly movable, and synovial joints usually are freely movable.

A. Fibrous joints include sutures and the joints between teeth and their sockets.

B. Cartilaginous joints occur where bones are connected by hyaline or fibrocartilage.

C. The different kinds of synovial joints are ball and socket, hinge, condyloid, pivot, gliding, and saddle. Movements at joints are broadly classified as angular (flexion, extension, adduction, abduction), circular (circumduction, rotation, supinatin, pronation), and special (inversion, eversion, elevation, and depression).

D. The most common types of arthritis are osteoarthritis, rheumatoid arthritis, and gout.

6.5 Effects of Aging
Two fairly common effects of aging on the skeletal system are arthritis and osteoporosis.

6.6 Homeostasis

A. The bones protect the internal organs: The rib cage protects the heart, lungs, and kidneys; the skull protects the brain; and the vertebrae protect the spinal cord. The bones assist all phases of respiration. The rib cage assists the breathing process, and red bone marrow produces the red blood cells that transport oxygen. The bones store and release calcium. Calcium ions play a major role in muscle contraction and nerve conduction. Calcium ions also help regulate cellular metabolism. The bones assist the lymphatic system and immunity. Red bone marrow produces not only the red blood cells but also the white blood cells. The bones assist digestion. The bones used for chewing break food into pieces small enough for chemical digestion. The skeleton is necessary for locomotion. Humans have a jointed skeleton for the attachment of muscles that move the bones.

B. The integumentary and muscular systems help the skeletal system protect organs. The digestive system absorbs calcium that is stored in the skeleton. The endocrine system regulates calcium storage and growth of bones. The cardiovascular system transports calcium to the skeleton, and muscles aid the skeleton in movement.

Study Questions

1. What are five functions of the skeleton? (p. 103)
2. What are five major categories of bones based on their shapes? (p. 103)
3. What are the parts of a long bone? What are some differences between compact bone and spongy bone? (pp. 103–105)
4. How does bone grow in children, and how is it remodeled in all age groups? (pp. 105–107)
5. What are the various types of fractures? Outline the four steps that are required for fracture repair. (p. 108)
6. List the bones of the axial and appendicular skeletons. (Fig. 6.4, pp. 109–110)
7. What are the bones of the cranium and the face? Describe the special features of the temporal bones, sphenoid bone, and ethmoid bone. (pp. 110–113)
8. What are the parts of the vertebral column, and what are its curvatures? Distinguish between the atlas, axis, sacrum, and coccyx. (pp. 113–116)
9. What are the bones of the rib cage? List several functions of the rib cage. (pp. 116–117)
10. What are the bones of the pectoral girdle? Give examples to demonstrate the flexibility of the pectoral girdle. What are the special features of a scapula? (p. 118)
11. Name the bones of the upper limb, then outline the special features of these bones. (pp. 118–120)
12. What are the bones of the pelvic girdle, and what are their functions? (p. 120)
13. What are the false and true pelvises, and what are several differences between the male and female pelvises? (pp. 120–121)

14. What are the bones of the lower limb? Describe the special features of these bones. (pp. 121–123)
15. How are joints classified? Give examples of each type of joint. (p. 123)
16. How can joint movements permitted by synovial joints be categorized? Give an example of each category. (pp. 124–125)
17. How does aging affect the skeletal system? (p. 127)
18. What functions of the skeletal system are particularly helpful in maintaining homeostasis? (pp. 127–129)

Learning Outcome Questions

I. **Match the items in the key to the bones listed in questions 1–6.**
   Key:
   a. forehead
   b. chin
   c. cheekbone
   d. elbow
   e. shoulder blade
   f. hip
   g. leg

   1. temporal and zygomatic bones
   2. tibia and fibula
   3. frontal bone
   4. ulna
   5. coxal bone
   6. scapula

II. **Match the items in the key to the bones listed in questions 7–13.**
   Key:
   a. external acoustic meatus
   b. cribriform plate
   c. xiphoid process
   d. glenoid cavity
   e. olecranon process
   f. acetabulum
   g. greater and lesser trochanters

   7. scapula
   8. sternum
   9. femur
   10. temporal bone
   11. coxal bone
   12. ethmoid bone
   13. ulna

III. **Fill in the blanks.**
   14. Long bones are _____ than they are wide.
   15. The epiphysis of a long bone contains _____ bone, where red blood cells are produced.
   16. The _____ are the air-filled spaces in the cranium.
   17. The sacrum is a part of the _____, and the sternum is a part of the _____.
   18. The pectoral girdle is specialized for _____, while the pelvic girdle is specialized for _____.
   19. The term *phalanges* is used for the bones of both the _____ and the _____.
   20. The knee is a freely movable (synovial) joint of the _____ type.

IV. **Match the movement with the description in questions 21–25.**
   a. extension
   b. circumduction
   c. adduction
   d. flexion
   e. abduction
   21. moving a body part toward the midline
   22. moving a body part away from the midline
   23. moving a body part in a circle
   24. decreasing the angle of a joint
   25. increasing the angle of the joint

Medical Terminology Exercise

After studying this chapter, see if you can derive the definitions for the medical terms listed at right. Many of the prefixes and suffixes used to create these terms can be found throughout the chapter. For additional help, use McGraw-Hill Connect™ at www.mcgrawhillconnect.com and consult Appendix B.

1. chondromalacia (kŏn″drō-mŭh-lā′shē-ŭh)
2. osteomyelitis (ŏs″tē-ō-mī″ŭh-lī′tĭs)
3. craniosynostosis (krā″nē-ō-sīn″ōs-tō′sĭs)
4. myelography (mī″ĕ-lŏg′rŭh-fē)
5. acrocyanosis (ăk″rō-sī″ŭh-nō′sĭs)
6. syndactylism (sĭn-dăk′tĭ-lĭzm)
7. orthopedist (ŏr″thō-pē′dĭst)
8. prognathism (prŏg′năh-thĭzm)
9. micropodia (mī″krō-pō′dē-ŭh)
10. arthroscopic (ăr″thrō-skŏp′ĭk)
11. bursectomy (bĕr-sĕk′tō-mē)
12. synovitis (sīn-ō-vī′tĭs)
13. acephaly (ā-sĕf′ŭh-lē)
14. sphenoidostomy (sfē-nōy-dŏs′tō-mē)
15. acetabuloplasty (ăs-ĕ-tăb′yū-lō-plăs-tē)

Online Study Tools

LEARNSMART® **connect**

APR

Anatomy & Physiology REVEALED includes cadaver photos that allow you to peel away layers of the human body to reveal structures beneath the surface. This program also includes animations, radiologic imaging, audio pronunciations, and practice quizzing. To learn more visit www.aprevealed.com

Anatomy & Physiology | REVEALED®
aprevealed.com

7 The Muscular System

If you were asked to describe what skeletal muscles do, you'd probably talk about the conscious limb movements you need for everyday activities like walking, running, lifting, bending, twisting from side to side, performing sports activities, and so on. But here's another skeletal muscle function you might not have thought of: communication through facial expression. We use over 20 muscles just for facial expression (anatomists disagree on the exact number), and each performs a very precise movement. It's interesting to note that human infants mimic the expressions of their caregivers almost from birth. Expressing emotions—joy, fear, frustration, unhappiness, loneliness—without speaking is essential for the baby's survival, for without this ability the child couldn't communicate its needs. Further, facial expressions convey nonverbal messages between people of all races and cultures. Check out the expressions of the people pictured here—what do you think each is feeling? Then, study Figure 7.10 and see if you can determine which facial muscles are used to make each of these expressions.

Learning Outcomes

After you have studied this chapter, you should be able to:

7.1 Functions and Types of Muscles

1. Distinguish between the three types of muscles, and tell where they are located in the body.
2. Describe the connective tissues of a skeletal muscle.
3. Name and discuss five functions of skeletal muscles.

7.2 Microscopic Anatomy and Contraction of Skeletal Muscle

4. Name the components of a skeletal muscle fiber, and describe the function of each.
5. Explain how skeletal muscle fibers are innervated and how they contract.
6. Describe how ATP is made available for muscle contraction.

7.3 Muscle Responses

7. Compare and contrast the behavior of an isolated muscle in the laboratory with that of an intact muscle in the body.
8. Contrast slow-twitch, intermediate-twitch, and fast-twitch muscle fibers.

7.4 Skeletal Muscles of the Body

9. Discuss how muscles work together to achieve the movement of a bone.
10. Give examples to show how muscles are named.

7.5 Skeletal Muscle Groups

11. Describe the locations and actions of the major skeletal muscles of each body region.

7.6 Effects of Aging

12. Describe the anatomical and physiological changes that occur in the muscular system as we age.

7.7 Homeostasis

13. Describe how the muscular system works with other systems of the body to maintain homeostasis.

Visual Focus
Anatomy of a Muscle Fiber

Focus on Forensics
Rigor Mortis

Medical Focus
Benefits of Exercise
Muscular Disorders and Neuromuscular Disease

Human Systems Work Together
Muscular System

7.1 Functions and Types of Muscles

1. Distinguish between the three types of muscles, and tell where they are located in the body.
2. Describe the connective tissues of a skeletal muscle.
3. Name and discuss five functions of skeletal muscles.

All muscles, regardless of type, can contract—that is, shorten. When muscles contract, some part of the body or the entire body moves. As you learned in Chapter 4, humans have three types of muscles: skeletal, cardiac, and smooth (Fig. 7.1). The contractile cells of these tissues are elongated and therefore are called **muscle fibers.**

Skeletal Muscle

Skeletal muscle fibers are cylindrical and have multiple nuclei around the periphery of the cell, just inside the plasma membrane. The fibers have alternating light and dark bands, and you will remember (from Chapter 4) that these are called **striations.** Skeletal muscles attach to the skeleton or directly to the skin (in the case of facial muscles). Skeletal muscle fibers can run the length of a muscle and therefore can be quite long. Skeletal muscle is voluntary because its contraction is always stimulated and controlled by the nervous system. (It's important to note, however, that though skeletal muscle is termed voluntary, it is not always consciously

controlled. In Chapter 8, you'll learn about muscle reflexes, which happen without any conscious thought.) In this chapter, we will explore why skeletal muscle (and cardiac muscle) is striated.

Cardiac Muscle

Cardiac muscle forms the heart wall. Its fibers are striated and cylindrical and have one or two central nuclei. Because they branch, cardiac muscle fibers are able to interlock at intercalated disks. Intercalated disks contain gap junctions and adhesion junctions, which permit contractions to spread quickly throughout the heart. Cardiac fibers relax completely between contractions, which prevents fatigue. Contraction of cardiac muscle fibers is rhythmical and occurs without requiring outside nervous stimulation. Thus, cardiac muscle contraction is involuntary. However, keep in mind that in order to maintain homeostasis, the nerves that supply the heart can increase or decrease both heart rate and strength of contraction. You'll learn more about cardiac muscle in Chapter 12.

Smooth Muscle

Smooth muscle is located in the walls of hollow internal organs and blood vessels. Its involuntary contractions regulate blood flow in blood vessels and move materials through hollow organs such as the digestive organs, uterus, and urinary bladder. Smooth muscle

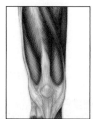

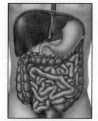

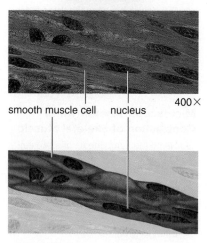

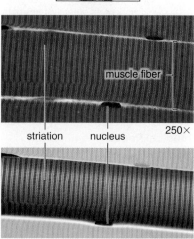

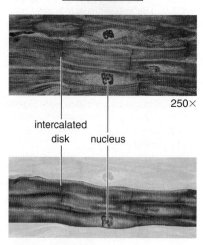

Skeletal muscle

Fiber appearance: striated, cylindrical, multinucleated

Location: attached to skeleton or skin

Control: voluntary

Cardiac muscle

Fiber appearance: striated, cylindrical and branched, one or two nuclei

Location: heart walls

Control: involuntary

Smooth muscle

Fiber appearance: spindle-shaped with single nucleus

Location: walls of hollow organs (e.g., stomach, intestines, urinary bladder, uterus, blood vessels)

Control: involuntary

AP|R **Figure 7.1** **Types of muscles.** The three types of muscles in the body have the appearance and characteristics shown above.

cells are *spindle-shaped*: narrow, tapered cylindrical cells with pointed ends. Each cell is *uninucleate* (has a single nucleus). The cells are usually arranged in parallel lines, forming sheets. Smooth muscle does not have the striations (bands of light and dark) seen in cardiac and skeletal muscle. Although smooth muscle is slower to contract than skeletal muscle, it can sustain prolonged contractions and does not fatigue easily.

Characteristics of Skeletal Muscle

Now that you have had a chance to study the similarities and differences of the three muscle types, we will return to the chapter's primary focus, which is the study of skeletal muscle.

Connective Tissue Coverings

Muscles are organs, and as such they contain other types of tissues, such as nervous tissue, blood vessels, and connective tissue. Connective tissue is essential to the organization of the fibers within a muscle (Fig. 7.2). First, each fiber is surrounded by a thin layer of areolar (loose) connective tissue called the *endomysium*. Blood capillaries and nerve fibers reach each muscle fiber by way of the endomysium. Second, the muscle fibers are grouped into bundles called *fascicles*. The fascicles have a sheath of connective

tissue called the *perimysium*. Finally, the muscle itself is covered by a connective tissue layer called the *epimysium*. The epimysium blends with the *deep fascia*, a layer of fibrous tissue that surrounds a set of muscles. Deep fascia separates muscles from each other and from the *superficial fascia*, also called the hypodermis. Remember (from Chapter 5) that the hypodermis is the layer that lies just deep to the dermis of the skin. Collagen fibers of the epimysium continue as a strong, fibrous **tendon** that attaches the muscle to a bone. The epimysium merges with the periosteum of the bone.

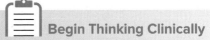

Begin Thinking Clinically

The deep fascia that surrounds a group of several muscles is called a ***muscle compartment.*** It functions as a tight sleeve containing the muscles, nerves, and blood vessels. In a **compartment syndrome,** swelling inside a compartment increases, choking off the blood supply to the muscle. This condition can cause complete muscle destruction and may be fatal. What type of injury could cause a compartment syndrome? How could it be corrected?

Answer and discussion in Appendix A.

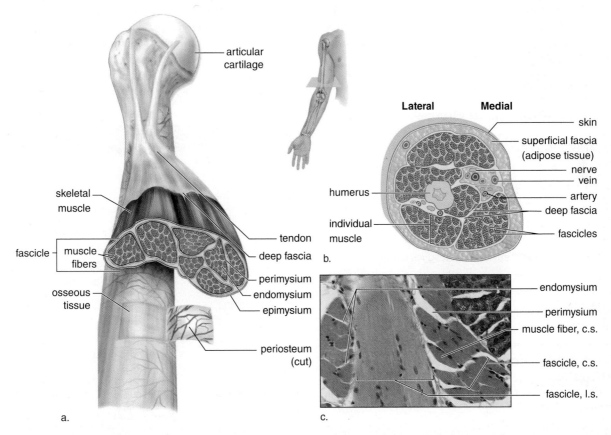

Figure 7.2 Connective tissue of a skeletal muscle. (a) Trace the connective tissue of a muscle from the endomysium to the perimysium to the epimysium, which becomes a part of the deep fascia and from which the tendon extends to attach a muscle to the periosteum of a bone. **(b)** Cross section of the arm showing the arrangement of the muscles, which are separated from the skin by fascia. The superficial fascia contains adipose tissue. **(c)** Photomicrograph of muscle fascicles from the tongue where the fascicles run in different directions. (c.s. = cross section; l.s. = longitudinal section.)

Functions of Skeletal Muscles

This chapter details skeletal muscles, so it's important to independently consider their functions (cardiac and smooth muscle function will be detailed in later chapters). The role of skeletal muscles in homeostasis can be summarized:

Skeletal muscles support the body. Skeletal muscle contraction opposes the force of gravity, allowing us to remain upright. Some skeletal muscles are serving this purpose even when you think you're relaxed.

Skeletal muscles make bones and other body parts move. Muscle contraction accounts not only for the movement of limbs but also for eye movements, facial expressions, and breathing.

Skeletal muscles help maintain a constant body temperature. Skeletal muscle is *thermogenic*; that is, its contractions generate heat. Muscle breaks down ATP as it contracts, and heat energy is distributed around the body. Involuntary muscle contractions that cause shivering help to maintain body temperature in the cold.

Skeletal muscle contraction assists fluid movement in cardiovascular and lymphatic vessels. The pressure of skeletal muscle contraction keeps blood moving in cardiovascular veins and lymph moving in lymphatic vessels.

Skeletal muscles help protect bones and internal organs and stabilize joints. Muscles pad the bones, and the muscular wall in the abdominal region protects the internal organs. Muscle tendons help hold bones together at the joints.

Content CHECK-UP!

1. Choose the phrase you could use when describing skeletal muscle.
   a. has uninucleate cells
   b. can be very long
   c. is under voluntary control
   d. both b. and c.
   e. all of the above

2. Choose the phrase you could use when describing smooth muscle.
   a. is multinucleate
   b. has cells arranged in sheets
   c. is striated
   d. fatigues easily

3. Put the three connective tissue coverings for a muscle (perimysium, endomysium, epimysium) in order from superficial to deep.

Answers in Appendix A.

7.2 Microscopic Anatomy and Contraction of Skeletal Muscle

4. Name the components of a skeletal muscle fiber, and describe the function of each.
5. Explain how skeletal muscle fibers are innervated and how they contract.
6. Describe how ATP is made available for muscle contraction.

We've already examined the structure of skeletal muscle as seen with the light microscope. As you know, skeletal muscle tissue has alternating light and dark bands, giving it a striated appearance. The electron microscope shows that these bands are due to the arrangement of protein filaments, called *myofilaments,* in a muscle fiber.

Structure of a Muscle Fiber

A muscle fiber contains the usual cellular components, but special names have been assigned to some of these components (note that the terms used to describe muscle start with the prefixes *myo-* and *sarco-;* Table 7.1 and Fig. 7.3). Thus, the plasma membrane is called the *sarcolemma;* cytoplasm is the *sarcoplasm;* and the endoplasmic reticulum is the *sarcoplasmic reticulum.* A muscle fiber also has some unique anatomical characteristics. One feature is its T (for transverse) system; the sarcolemma forms **T (transverse) tubules** that penetrate, or dip down, into the cell so that they come into contact—but do not fuse with—expanded portions of the sarcoplasmic reticulum. The expanded portions of the sarcoplasmic reticulum are calcium storage sites. Calcium ions (Ca^{2+}), as we will see, are essential for muscle contraction.

The sarcoplasmic reticulum is a fine endomembrane network that surrounds hundreds and sometimes even thousands of **myofibrils,** which are bundles of myofilaments. Each myofibril is about 1 micrometer in diameter. Myofibrils are the parts of muscle fibers that contract. Any other organelles, such as mitochondria, are located in the sarcoplasm between the myofibrils. The sarcoplasm also contains glycogen, which provides stored energy for muscle contraction, and the red pigment **myoglobin,** which binds oxygen until it is needed for muscle contraction. Muscle is the only tissue that has both myoglobin for a supplemental oxygen supply and glycogen as a nutrient source. Thus, muscle is well-suited to produce the enormous amounts of ATP energy needed for muscle contraction.

TABLE 7.1	Microscopic Anatomy of a Muscle
Name	**Function**
Sarcolemma	Plasma membrane of a muscle fiber that forms T tubules
Sarcoplasm	Cytoplasm of a muscle fiber that contains organelles, including myofibrils
Glycogen	A polysaccharide that stores energy for muscle contraction
Myoglobin	A red pigment that stores oxygen for muscle contraction
T tubule	Extension of the sarcolemma that extends into the muscle fiber and conveys nerve signals that cause Ca^{2+} to be released from the sarcoplasmic reticulum into the sarcoplasm
Sarcoplasmic reticulum	The smooth ER of a muscle fiber that stores Ca^{2+}
Myofibril	A bundle of myofilaments that contracts
Myofilament	Thick and thin filaments whose structure and functions account for muscle striations and contractions

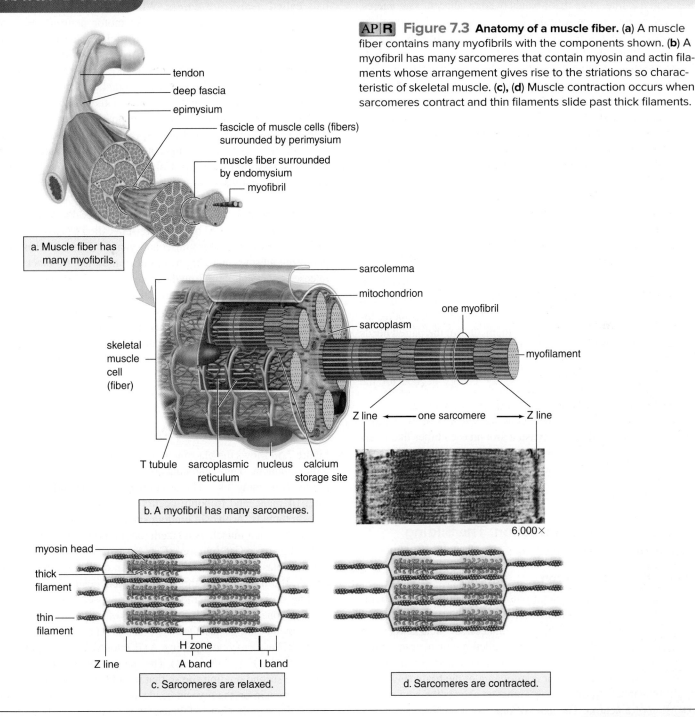

AP|R **Figure 7.3** **Anatomy of a muscle fiber.** (a) A muscle fiber contains many myofibrils with the components shown. (b) A myofibril has many sarcomeres that contain myosin and actin filaments whose arrangement gives rise to the striations so characteristic of skeletal muscle. (c), (d) Muscle contraction occurs when sarcomeres contract and thin filaments slide past thick filaments.

a. Muscle fiber has many myofibrils.

b. A myofibril has many sarcomeres.

6,000×

c. Sarcomeres are relaxed.

d. Sarcomeres are contracted.

Myofibrils and Sarcomeres

Myofibrils are cylindrical and run the length of the muscle fiber. Each myofibril is composed of numerous **sarcomeres,** which are microscopic repeating units (Fig. 7.3). Each sarcomere extends between two dark, vertical lines called **Z lines.** The horizontal stripes, or *striations,* of skeletal muscle fibers are formed by the placement of myofilaments within the sarcomeres. A sarcomere contains two types of protein myofilaments: The **thick filaments** are made up of a single protein called **myosin. Thin filaments** are made up of three proteins: a globular protein called **actin,** plus

tropomyosin and **troponin.** At both ends of each sarcomere, the **I band** is light-colored because it contains only thin filaments attached to a Z line. Note that I bands overlap adjacent sarcomeres. The dark regions of the **A band,** found between the I bands of the sarcomere, contain overlapping thick and thin filaments. A good way to recall the difference between the *I* band and A band is to remember that the letter *I* is found in the word *light*. I bands are light because they contain only thin filaments. Likewise, the letter *A* is part of the word *dark*. A bands are dark because they contain both thick and thin filaments. Directly in the center of the dark A band is the lighter **H zone,** which contains only myosin filaments (Fig. 7.3*b* and *c*).

Myofilaments

The thick and thin filaments differ in the following ways:

Thick Filaments A thick filament is composed of several hundred molecules of the protein myosin. Each myosin molecule is composed of two protein strands, each shaped like a golf club. The straight portions of each strand coil around each other. Each myosin molecule ends in a double globular head. The paired myosin heads are slanted away from the middle of a sarcomere, toward the thin filaments surrounding them. One of the paired heads has an actin binding site that will link to actin in thin filaments during muscle contraction. The second is an enzyme that will break down cellular ATP to release the energy needed for the contraction.

Thin Filaments A thin filament consists primarily of two strands of the globular protein actin, twisted around each other like intertwined bead necklaces. Double strands of tropomyosin coil over each actin strand. Troponin occurs at intervals on the tropomyosin strand (see Fig. 7.5*a*).

Skeletal Muscle Contraction: The Sliding Filament Theory

Role of the Motor Neuron

The mechanism behind skeletal muscle contraction is a phenomenon called the **sliding filament theory,** and it begins with nervous stimulation. Muscle fibers are innervated—that is, they are stimulated to contract by nerve cells called motor neurons. Multiple motor neurons are organized into motor nerves, which are controlled by a specific motor control area of the brain (Fig. 7.4). You'll remember (from Chapter 4) that an axon is the single, long extension of a neuron. The axon of one motor neuron has several branches and can stimulate from a few to several hundred muscle fibers of a particular muscle. This entire region is called a **motor unit,** and it consists of the single motor nerve axon and the entire collection of muscle fibers it innervates. Each branch of the axon ends in an axon terminal, an expanded area that lies in close proximity to the sarcolemma of the muscle fiber. This region is called the **neuromuscular junction.** Because a small gap, called a *synaptic cleft,* separates the axon terminal from the sarcolemma, the two do not physically touch (Fig. 7.4).

Axon terminals contain synaptic vesicles that are filled with **acetylcholine (ACh),** one of a large category of biochemicals called *neurotransmitters*. When signals from the brain's motor control area travel down the motor neuron and arrive at the axon terminal, the synaptic vesicles release neurotransmitter into the synaptic cleft. The ACh quickly diffuses across the cleft and binds to receptors on the sarcolemma. Now the sarcolemma generates an electrical signal, called an action potential, that spreads over the surface of the sarcolemma. (You will learn much more about action potentials in Chapter 8.) The action potential signal next spreads down T tubules to the sarcoplasmic reticulum, triggering the release of calcium from the sarcoplasmic reticulum. Calcium enables the next phase of muscle fiber contraction, involving interaction between myosin and actin myofilaments.

It's interesting to note that one of the world's deadliest poisons, a protein produced by the bacterium called *Clostridium botulinum*, works at the neuromuscular synapse. The **botulism toxin** paralyzes muscle, including respiratory muscles, by blocking the release of acetylcholine from the motor neuron. Without acetylcholine, skeletal muscles cannot generate their action potential signals, and poisoning victims die from suffocation. However, the diluted toxin, called Botox, is routinely used for therapeutic and cosmetic reasons. Painful muscle spasms called *contractures* can be relaxed with Botox, and it has also been used to treat migraine headaches. Cosmetically, the appearance of facial wrinkles can be reduced by paralyzing the muscles of facial expression.

The Role of Actin and Myosin Myofilaments

Figure 7.5*a* shows the placement of the three proteins that make up a thin filament. Thin filaments are composed of a double row of twisted actin molecules. Threads of tropomyosin wind around each actin filament, and troponin occurs at intervals along the tropomyosin threads. A myosin binding site can be found on each actin molecule; when muscle is relaxed, these binding sites are covered by tropomyosin. Calcium ions (Ca^{2+}) that have been released from the sarcoplasmic reticulum combine with troponin. After binding occurs, the tropomyosin threads shift their position, and myosin binding sites on the actin molecules are exposed.

Each one of the paired globular heads of a myosin thick filament has its own binding sites. One site binds to ATP and then functions as an ATPase enzyme, splitting ATP into ADP and a phosphate group $\textcircled{P}$ (see Fig. 7.5*b, step 1*). The energy from this reaction activates the second binding site so that it will bind to actin. The ADP and $\textcircled{P}$ remain on the myosin heads until the heads attach to actin (Fig. 7.5*b, step 2*). Now, ADP and $\textcircled{P}$ are released, and this causes the myosin head to bend sharply toward the center of the sarcomere (Fig. 7.5*b, step 3*). This action of myosin is called the **power stroke,** and it pulls the thin filaments toward the middle of the sarcomere. When another ATP molecule binds to a myosin head, the head detaches from actin (Fig. 7.5*b, step 4*). The cycle begins again, and the thin filaments move nearer the center of the sarcomere each time the cycle is repeated. Some of the myosin heads remain attached to actin during the cycle while others form new bonds. Thus, the thin filaments don't slide back to their resting position while a

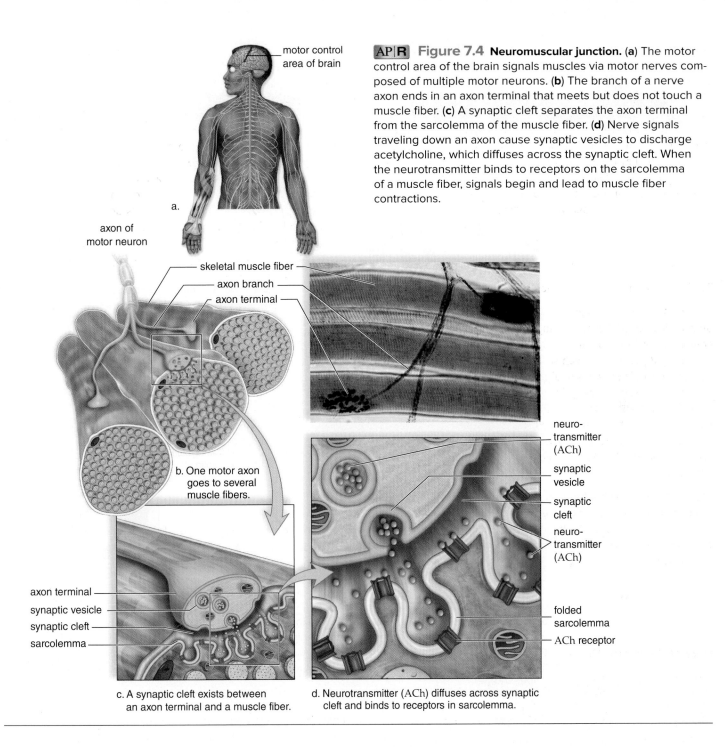

AP|R **Figure 7.4** **Neuromuscular junction.** **(a)** The motor control area of the brain signals muscles via motor nerves composed of multiple motor neurons. **(b)** The branch of a nerve axon ends in an axon terminal that meets but does not touch a muscle fiber. **(c)** A synaptic cleft separates the axon terminal from the sarcolemma of the muscle fiber. **(d)** Nerve signals traveling down an axon cause synaptic vesicles to discharge acetylcholine, which diffuses across the synaptic cleft. When the neurotransmitter binds to receptors on the sarcolemma of a muscle fiber, signals begin and lead to muscle fiber contractions.

a.

axon of motor neuron

skeletal muscle fiber

axon branch

axon terminal

b. One motor axon goes to several muscle fibers.

neuro-transmitter (ACh)

synaptic vesicle

synaptic cleft

neuro-transmitter (ACh)

axon terminal

synaptic vesicle

synaptic cleft

sarcolemma

folded sarcolemma

ACh receptor

c. A synaptic cleft exists between an axon terminal and a muscle fiber.

d. Neurotransmitter (ACh) diffuses across synaptic cleft and binds to receptors in sarcolemma.

contraction is occurring. Note that ATP has two roles in this process: first to energize myosin, and then to break the link between myosin and actin. (A good analogy for the sliding filament action of the myosin myofilaments in a sarcomere is a game of tug-of-war. The players on either side can be likened to myosin molecules. As their hands grab, release, and grab the rope again, each side is pulled toward the other and the distance between them is shortened.)

As the thin filaments slide past the thick filaments toward the sarcomere's center, the entire sarcomere shortens (though the thick and thin filaments themselves remain the same length). This causes the I band to shorten and the H zone to almost or completely disappear (Fig. 7.3c and d). It's important to note that although the thin filaments slide past the thick filaments, the myosin filaments do the work with their power stroke. As sarcomeres shorten, the myofibril and ultimately the entire muscle fiber are shortened.

Contraction continues until nerve signals stop and calcium ions are returned to their storage sites. The membranes of the sarcoplasmic reticulum contain active transport proteins that pump calcium ions back into the interior of the sarcoplasmic reticulum. Of course, this active transport process also requires ATP energy.

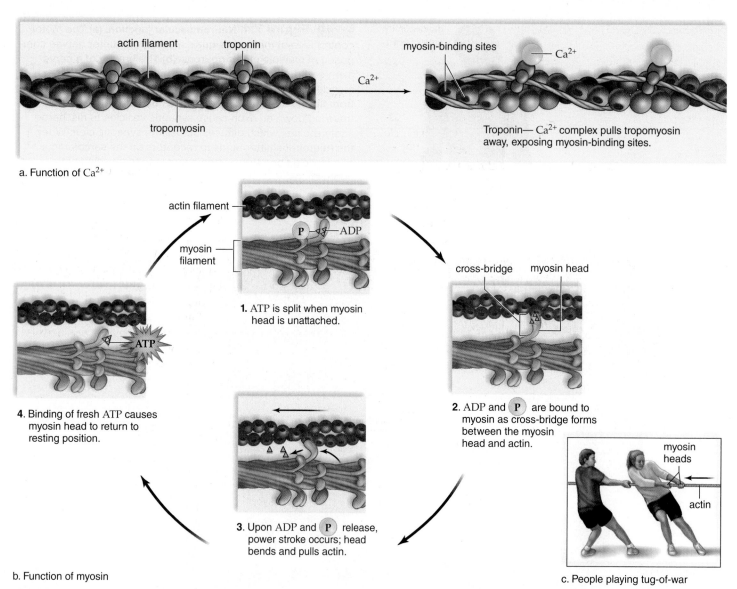

a. Function of Ca^{2+}

1. ATP is split when myosin head is unattached.

2. ADP and (P) are bound to myosin as cross-bridge forms between the myosin head and actin.

3. Upon ADP and (P) release, power stroke occurs; head bends and pulls actin.

4. Binding of fresh ATP causes myosin head to return to resting position.

b. Function of myosin

c. People playing tug-of-war

AP|R **Figure 7.5** **The role of actin and myosin in muscle contraction.** (**a**) In relaxed muscle, tropomyosin covers myosin binding sites on actin. Upon its release, calcium binds to troponin, exposing myosin binding sites. (**b**) After breaking down ATP into ADP and a phosphate group (P), myosin heads bind to actin filaments, forming cross-bridges. Then, a power stroke causes actin to move. (**c**) A tug-of-war game is a good analogy for cross-bridges between myosin and actin. Cross-bridges form, then break and re-form. The sarcomere, myofibril, and muscle fiber shorten.

Contraction of Cardiac Muscle

The events of contraction in cardiac muscle are very similar to those in skeletal muscle. However, in cardiac muscle, the calcium needed to bind to troponin comes from outside the cell as well as inside. After the action potential signal in cardiac muscle, calcium diffuses into the sarcoplasm and triggers the release of more calcium from the sarcoplasmic reticulum. Once calcium has bound to troponin, cross-bridges can form between activated myosin and actin, just as in skeletal muscle.

Contraction of Smooth Muscle

You'll recall that smooth muscle cells are spindle-shaped, that is, uninucleate, cylindrical cells with pointed ends, and that smooth muscle control is involuntary. Like skeletal muscle, smooth muscle contains thick and thin filaments. However, in smooth muscle these filaments are not arranged into myofibrils that create visible striations. Instead, thin filaments in smooth muscle are anchored directly to the sarcolemma or to protein molecules called *dense bodies.* Dense bodies are scattered through the sarcoplasm. When a smooth muscle cell contracts, its fibers shorten in all directions, causing the cylindrical cell to become more oval in shape. In turn, the entire smooth muscle shortens. Smooth muscle contraction occurs very slowly but can last for long periods of time without fatigue.

Energy for Muscle Contraction

ATP produced before strenuous exercise and found in the muscle cell sarcoplasm lasts a few seconds. Then, muscles must manufacture new ATP in three different ways: creatine phosphate breakdown, cellular respiration, and fermentation (Fig. 7.6). Creatine phosphate breakdown and fermentation are **anaerobic,** meaning that they do not require oxygen.

Creatine Phosphate Breakdown

Creatine phosphate is a high-energy compound built up when a muscle is resting. Creatine phosphate cannot participate directly in muscle contraction. Instead, it can recycle ADP inside the cell into ATP by transferring its phosphate to ADP, using the following reaction:

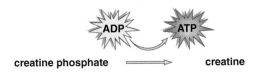

This reaction occurs in the midst of sliding filaments, and therefore is the speediest way to make ATP available to muscles. Creatine phosphate provides enough energy for only about eight seconds of intense activity, and then it is used up. Creatine phosphate is rebuilt when a muscle is resting by transferring a phosphate group from ATP back to creatine (Fig. 7.6a).

Cellular Respiration

Cellular respiration completed in mitochondria usually provides most of a muscle's ATP. Glycogen and fat are stored in muscle cells. Therefore, a muscle cell can use glucose from glycogen and fatty acids from fat as fuel to produce ATP if oxygen is available:

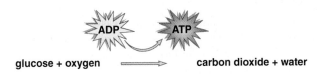

glucose + oxygen $\longrightarrow$ carbon dioxide + water

The red pigment myoglobin is an oxygen carrier manufactured by muscle cells, and it is similar to the hemoglobin protein of red blood cells. Its presence accounts for the reddish-brown color of skeletal muscle fibers. Myoglobin has a higher affinity for oxygen than hemoglobin. Therefore, myoglobin can pull oxygen out of blood and make it available to muscle mitochondria that are carrying on cellular respiration. Then, too, the ability of myoglobin to temporarily store oxygen reduces a muscle's immediate need for oxygen when cellular respiration begins. The end products of cellular respiration (carbon dioxide and water)

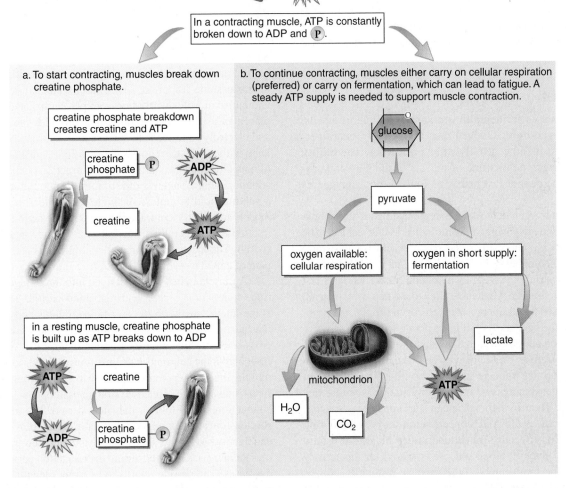

Figure 7.6 Energy sources for muscle contraction. (a) Creatine phosphate can transfer a phosphate group to ADP to form ATP. **(b)** Cellular respiration and fermentation also form ATP in muscle tissue.

are usually no problem in a healthy person's body. Carbon dioxide is exhaled by the lungs. Water simply enters the extracellular space, and any excess can be excreted by the kidneys. The by-product of muscular contraction—heat—keeps the entire body warm.

Fermentation

Fermentation, like creatine phosphate breakdown, supplies ATP without consuming oxygen. Like creatine phosphate breakdown, fermentation is an anaerobic process, and it occurs in the sarcoplasm. Fermentation produces the ATP necessary for short bursts of exercise—for example, a 50-yard dash or a run around the bases in a baseball game. During fermentation, glucose is broken down to lactate (lactic acid):

The accumulation of lactate in a muscle fiber makes the sarcoplasm more acidic, and eventually enzymes cease to function well. If fermentation continues longer than two or three minutes, cramping and fatigue set in. Cramping seems to be due to lack of ATP. As you recall, that is because ATP is needed to pump calcium ions back into the sarcoplasmic reticulum and to break the linkages between the actin and myosin filaments so that muscle fibers can relax.

Oxygen Debt

When a muscle uses fermentation to supply its energy needs, it incurs an **oxygen debt.** Oxygen debt is obvious when a person continues to breathe heavily after exercising. The ability to run up an oxygen debt is one of muscle tissue's greatest assets. Brain tissue cannot last nearly as long without oxygen as muscles can.

Repaying an oxygen debt requires replenishing creatine phosphate supplies and disposing of lactic acid. Lactic acid can be changed back to a compound called pyruvic acid (or pyruvate) and metabolized completely in mitochondria, or it can be sent to the liver to reconstruct glycogen. A marathon runner who has just crossed the finish line is not exhausted due to oxygen debt. Instead, the runner has used up all the muscles' glycogen supply, and probably the liver's glycogen as well. It takes about two days to replace glycogen stores on a high-carbohydrate diet.

People who train rely more heavily on cellular respiration than do people who do not train. In people who train, the number of muscle mitochondria increases, and so fermentation is not needed to produce ATP. Their mitochondria can start consuming glucose and oxygen as soon as the ADP concentration starts rising during muscle contraction. Because mitochondria can break down fatty acids, instead of glucose, blood glucose is spared for the activity of the brain. (The brain, unlike other organs, ordinarily utilizes only glucose to produce ATP.) Because less lactate is produced in people who train, the pH of the blood remains steady, and there is less oxygen debt.

| Content **CHECK-UP!** |

4. The portion of the muscle cell membrane that extends into the muscle and conveys nerve signals is called the:

  a. sarcoplasmic reticulum.   c. transverse (T) tubules.

  b. sarcolemma.   d. sarcoplasm.

5. Thin filaments are composed of which three proteins?

6. Which of the following is the process in a muscle cell that requires oxygen to produce ATP energy?

  a. fermentation   c. cellular respiration

  b. creatine phosphate breakdown

Answers in Appendix A.

7.3 Muscle Responses

7. Compare and contrast the behavior of an isolated muscle in the laboratory with that of an intact muscle in the body.

8. Contrast slow-twitch, intermediate-twitch, and fast-twitch muscle fibers.

Muscles can be studied in the laboratory in an effort to understand how they respond when in the body.

In the Laboratory

Samples of intact muscle can be taken out of the body and studied in a laboratory, and these experiments have allowed scientists to make some important observations about how muscle fibers and entire muscles work. For example, an isolated muscle can be placed in a laboratory solution that provides the ATP, nutrients, and electrolytes it needs to survive and contract. Next, it is stimulated with an electric shock, whose voltage must be strong enough to make the muscle contract. If a contraction occurs, the electrical stimulus is called a threshold stimulus; if not, the stimulus is called a subthreshold stimulus. A single threshold stimulus causes the muscle to quickly contract and relax. This action—a single contraction that lasts only a fraction of a second—is called a **muscle twitch.** The mechanical force of contraction is recorded as a visual pattern called a myogram (Fig. 7.7a).

A muscle twitch can be divided into three stages: the latent period, contraction period, and relaxation period. The latent period is the interval between the threshold stimulus and the onset of the muscle contraction. During this time, all of the events that lead up to cross-bridge formation are occurring in the cell. Acetylcholine diffuses across the neuromuscular junction and binds to receptors on the muscle, the action potential spreads over and throughout the muscle fibers, and calcium is released from the sarcoplasmic reticulum. The contraction period follows the latent period. The muscle physically shortens during this period as cross-bridges are formed, and thick and thin myofilaments slide past one another. Force generated by the muscle increases. Finally, during the relaxation period, muscle force decreases. The muscle returns to its former length as cross-bridges break and calcium is returned to the sarcoplasmic reticulum.

If a muscle twitches and then is allowed to relax in between threshold stimuli, the resulting myogram shows a series of twitches with identical force (Fig. 7.7*b*). However, if the muscle is stimulated rapidly, the contraction force gradually increases, even if the stimulus voltage is exactly the same. The more rapidly the muscle is stimulated, the greater the contraction force becomes. This effect is called **summation** (Fig. 7.7*c*). Scientists believe that summation occurs because when the muscle is stimulated quickly enough, there is not enough time between stimuli to return all the calcium to the sarcoplasmic reticulum. With extra calcium in the sarcoplasm, more crossbridges can be formed after each threshold stimulus. Further, repetitive muscle contraction also generates heat as ATP is broken down to release energy, and scientists believe that muscle enzymes may work more efficiently to cause contraction when muscle fibers are warmer.

During summation, the muscle has less time to relax as the rate of stimulation increases, and the muscle force becomes more and more constant. When the muscle is stimulated very rapidly, the muscle has no time to relax at all. This effect is called **tetanus,**[1] or a **tetanic contraction.** It can be seen on the myogram as a horizontal line, because muscle force is constant during this period. If stimulation continues at the same rate, tetanus will continue until the muscle fatigues. **Fatigue** occurs when the muscle relaxes even though stimulation continues, and on the myogram it can be seen when muscle force falls. There are several reasons why isolated muscles become fatigued. First, ATP is depleted during constant use of a muscle; the muscle essentially "runs out of energy." At the same time, repetitive use causes production of lactic acid by fermentation, which lowers the pH of the sarcoplasm and inhibits muscle function. In addition, the motor nerves that supply muscle can run out of their neurotransmitter, acetylcholine.

However, intact muscles in the body rarely fatigue completely like an isolated muscle in the laboratory does. Muscles are well supplied with blood vessels to transport nutrients and remove lactic acid. Instead, in the body, fatigue is a gradual weakening that occurs after repetitive use. In addition, the brain itself may signal a person to stop exercising, even if the muscles are not truly fatigued. The mechanisms that cause this muscle fatigue are not well understood. People who train can exercise for longer periods without experiencing fatigue.

In the Body

As you know, muscles in the body are stimulated to contract by motor nerves composed of motor neurons. The combination of the neuron and all of the muscle fibers it innervates is called a motor unit. Motor units function according to a property called the **all-or-none law:** since all the muscle fibers in a motor unit are stimulated at once by the same neuron, they all contract simultaneously in response to the neuron, or do not contract. It's interesting to note that the number of muscle fibers within a motor unit can be quite different. For example, in the ocular muscles that move the eyes, the innervation ratio is one motor axon per 23 muscle fibers, while in the gastrocnemius muscle of the leg, the ratio is about one motor axon per 1,000 muscle fibers. No doubt, moving the eyes requires finer control than moving the lower limbs.

Changing the strength of a contraction occurs consciously or unconsciously throughout one's day. For example, think about the

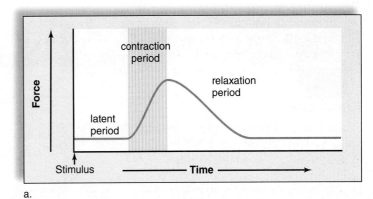

a.

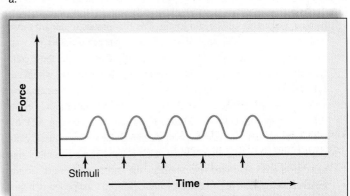

b.

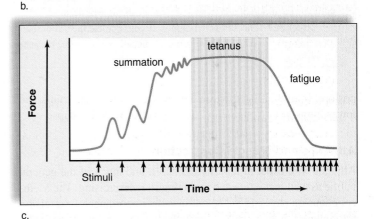

c.

Figure 7.7. Myograms showing (a) a single twitch, (b) a series of twitches, (c) summation of twitches that produces tetanus, or tetanic contraction. Muscle fatigue is caused by pH decrease and depletion of muscle ATP and acetylcholine.

strength it might take to lift a pencil. Now imagine the muscular effort needed to lift a book bag loaded with an entire day's books and supplies. Increasing a muscle's contraction strength can be accomplished by a phenomenon known as **recruitment.** By increasing the intensity of nervous stimulation, the nervous system can activate, or *recruit*, more and more motor units, resulting in stronger and stronger muscle contraction. However, while some muscle fibers within a muscle are contracting, others in the same muscle are relaxing. This allows a muscle to sustain a contraction for a long period of time. In life, even when muscles appear to be at rest, they exhibit **tone,** in which some of their fibers are always contracting. Muscle tone is particularly important in maintaining

---

[1] It's important not to confuse normal muscle tetanus with the disease of the same name. The disease is caused by a bacterial toxin (see Medical Focus, p. 157).

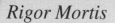
For everything there is a season, and a time for every matter under heaven: a time to be born, and a time to die . . .
—Ecclesiastes 3:1,2 KJV

When a person dies, the physiologic events that accompany death occur in an orderly progression. Respiration ceases, the heart ultimately stops beating, and tissue cells begin to die. The first tissues to die are those with the highest oxygen requirement. Brain and nervous tissues have an extremely high requirement for oxygen. Deprived of oxygen, these cells typically die after only six minutes, as the ATP energy reserve within the cell is used up. However, tissues that can produce ATP by fermentation (which does not require oxygen) can "live" for an hour or more before ATP is completely depleted. Muscle is capable of generating ATP by both fermentation and creatine phosphate breakdown; thus, muscle cells can survive for a time after clinical death occurs. Muscle death is signaled by a process termed *rigor mortis*—the "stiffness of death." Rigor mortis develops and reverses itself within a known time period of hours to days after death occurs. Forensic pathologists use this information to study the condition of a body to approximate the time of death.

Rigor mortis occurs as dying muscle cells deplete the last of their ATP energy reserve. Relaxing the muscle becomes impossible because ATP energy is needed to break the bond between an actin binding site and the myosin cross-bridge. In addition, ATP energy is required for muscle relaxation to return calcium ions from the sarcoplasm to the sarcoplasmic reticulum. Thus, without ATP energy, the muscle remains fixed in the same state of contraction that preceded that person's death. If, for example, a murder victim dies while sitting at a desk, the body in rigor mortis will be frozen in the sitting position.

Forensic pathologists use body temperature and the presence or absence of rigor mortis to estimate time of death. For example, the body of someone who has been dead for 3 hours or less will still be warm (close to normal body temperature, 98.6°F or 37°C) and rigor mortis will be absent. After approximately 3 hours, the body will be significantly cooler than normal, and rigor mortis will begin to develop. The corpse of an individual who has been dead at least 8 hours will be in full rigor mortis, and the temperature of the body will be the same as the surroundings. Rigor mortis resolves approximately 24–36 hours after death. Muscles lose their stiffness because lysosomes inside the cell eventually rupture, releasing enzymes that break the myosin-actin bonds. Forensic pathologists know that a person has been dead for more than 24 hours if the body temperature is the same as the environment and there is no longer a trace of rigor mortis.

posture. If all the fibers within the muscles of the neck, trunk, and lower limbs were to suddenly relax, the body would collapse.

Athletics and Muscle Contraction

Athletes who excel in a particular sport, and much of the general public as well, are interested in staying fit by exercising. The Medical Focus on pages 155–156 outlines the importance of exercising throughout life, and gives suggestions for starting and staying with an exercise program.

Exercise and Size of Muscles Muscles that are not used or that are used for only very weak contractions decrease in size, or atrophy. **Atrophy** can occur when a limb is placed in a cast or when the nerve serving a muscle is damaged. If nerve stimulation is not restored, muscle fibers are gradually replaced by fat and fibrous tissue. Unfortunately, atrophy can cause muscle fibers to shorten progressively, leaving body parts contracted in contorted positions called *contractures*.

Forceful muscular activity over a prolonged period causes muscle to increase in size as the size of the individual myofibrils within the muscle fibers increases. Increase in muscle size, called **hypertrophy,** occurs only if the muscle contracts to at least 75% of its maximum tension.

Some athletes take anabolic steroids, either testosterone or related chemicals, to promote muscle growth. This practice is quite dangerous and has many undesirable side effects, as discussed in the Medical Focus on pages 234–235.

Slow-Twitch and Fast-Twitch Muscle Fibers We've seen that all skeletal muscle fibers metabolize both aerobically (using oxygen during cellular respiration) and anaerobically (without oxygen, using fermentation or creatine phosphate breakdown). However, some muscle fibers utilize one method more than the other to provide myofibrils with ATP. Slow-twitch and intermediate-twitch muscle fibers tend to be aerobic, and fast-twitch fibers tend to be anaerobic. Slow-twitch fibers are also referred to as type I fibers, intermediate-twitch fibers are type IIa, and fast-twitch fibers are called type IIb fibers.

Slow-twitch fibers have motor units with a smaller number of fibers. These muscle fibers are most helpful in endurance sports such as long-distance running or swimming in a triathlon. Because they produce most of their energy aerobically, they tire only when their fuel supply is gone. Slow-twitch fibers have many mitochondria that can maintain a steady, prolonged production of ATP when oxygen is available. An abundant supply of myoglobin, the respiratory pigment found in muscles, gives these fibers a dark color. They are also surrounded by dense capillary beds for a continuous supply of blood and oxygen. Slow-twitch fibers have a low maximum tension, which develops slowly, but these muscle fibers are highly resistant to fatigue.

Like slow-twitch fibers, *intermediate-twitch fibers* are well supplied with myoglobin and contain many mitochondria. Intermediate-twitch fibers have an extensive blood supply as well. However, they contract much more quickly and can be described as fast aerobic fibers. Activities that require moderate strength for shorter periods (such as walking, jogging, or biking) will employ these muscles.

Fast-twitch fibers have fewer capillaries to supply blood and oxygen, and they tend to function anaerobically. These fibers are light in color because they have fewer mitochondria and little myoglobin. They provide explosions of energy and are most helpful in sports activities such as sprinting, weight lifting, swinging a golf club, or pitching a baseball. Fast-twitch fibers can develop maximum tension more rapidly, and their maximum tension is greater. However, their dependence on anaerobic energy leaves them vulnerable to an accumulation of lactic acid that causes them to fatigue quickly.

Content CHECK-UP!

7. A motor unit is:

    a. a group of myofibrils.

    b. a group of skeletal muscles that perform a specific function.

    c. a group of motor nerves in the brain.

    d. a motor nerve and all of the muscle fibers it innervates.

8. An increase in muscle size, caused by increasing the number of myofibrils in individual muscle cells, is called:

    a. recruitment.   c. atrophy.

    b. hypertrophy.   d. tetanic contraction.

9. A certain muscle cell produces most of its energy aerobically, contains many mitochondria, and contracts quickly. What kind is it? What kinds of movements is it specialized for, and why?

Answers in Appendix A.

7.4 Skeletal Muscles of the Body

9. Discuss how muscles work together to achieve the movement of a bone.

10. Give examples to show how muscles are named.

The human body has some 600 skeletal muscles, but this text will discuss only some of the most significant of these. First, let us consider certain basic principles of muscle contraction.

Basic Principles

When a muscle contracts at a joint, one bone remains fairly stationary, and the other one moves. The **origin** of a muscle is on the stationary bone, and the **insertion** of a muscle is on the bone that moves.

Frequently, a body part is moved by a group of muscles working together. Even so, one muscle does most of the work, and this muscle is called the **prime mover** or the **agonist.** For example, in flexing the forearm at the elbow, the prime mover is the brachialis muscle (Fig. 7.8). The assisting muscles are called the **synergists.** The biceps brachii (see Fig. 7.13) is a synergist that helps the brachialis flex the elbow. A prime mover can have several synergists.

When muscles contract, they shorten. Therefore, muscles can only pull; they cannot push. However, agonist muscles are paired with **antagonists,** and antagonistic pairs work opposite one another to bring about movement in opposite directions. For example, the brachialis and biceps brachii and the triceps brachii are antagonists; the first pair flexes the forearm, and the other extends the forearm (Fig. 7.8). Later on in our discussion, we'll encounter other antagonistic pairs.

tendon — origin

biceps brachii (contracted)

triceps brachii (relaxed)

radius

ulna

humerus

insertion

a.

b.

biceps brachii (relaxed)

triceps brachii (contracted)

c.

AP|R **Figure 7.8 Origin and insertion.** The origin of a muscle is on a bone that remains stationary, and the insertion of a muscle is on a bone that moves when a muscle contracts. The two muscles shown here are antagonistic. **(a)** When the biceps brachii and brachialis muscles contract, they are agonists that flex the forearm. **(b)** Strengthening one's arm muscles using hand weights requires forearm flexion and extension. **(c)** When the triceps brachii contracts, it is the antagonist muscle that extends the forearm.

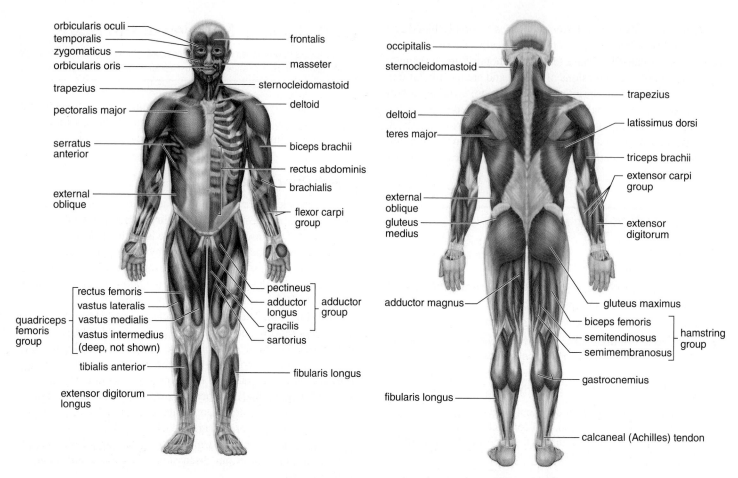

Figure 7.9 (a) Anterior view of the body's superficial skeletal muscles. (b) Posterior view of the body's superficial skeletal muscles.

Naming Muscles

It's easier to learn and remember muscle names if you consider what each muscle's name means. The names of the various skeletal muscles are often combinations of the following terms used to characterize muscles:

1. **Size.** For example, the *gluteus maximus* is the largest muscle that makes up the buttocks. The *gluteus minimus* is the smallest of the gluteal muscles. Other terms used to indicate size are *vastus* (huge), *longus* (long), and *brevis* (short).

2. **Shape.** For example, the *deltoid* is shaped like a delta, or triangle, while the *trapezius* is shaped like a trapezoid. Other terms used to indicate shape are *latissimus* (wide) and *teres* (round).

3. **Direction of fibers.** For example, the *rectus abdominis* is a longitudinal muscle of the abdomen (*rectus* means straight). The *orbicularis oculi* is a circular muscle around the eye. Other terms used to indicate direction are *transverse* (across) and *oblique* (diagonal).

4. **Location.** For example, the *frontalis* overlies the frontal bone. The *external obliques* are located outside the internal obliques. Other terms used to indicate location are *pectoralis* (chest), *gluteus* (buttock), *brachii* (arm), and *sub* (beneath). You should also review these directional terms: *anterior, posterior, lateral, medial, proximal, distal, superficial,* and *deep*.

5. **Attachment.** For example, the *sternocleidomastoid* is attached to the sternum, clavicle, and mastoid process. The *brachioradialis* is attached to the brachium (arm) and the radius.

6. **Number of attachments.** For example, the *biceps brachii* has two attachments, or origins (and is located on the arm). The *quadriceps femoris* has four origins (and is located on the anterior femur).

7. **Action.** For example, the *extensor digitorum* extends the fingers or digits. The *adductor magnus* is a large muscle that adducts the thigh. Other terms used to indicate action are *flexor* (to flex), *masseter* (to chew), and *levator* (to lift).

Try to understand the name of every muscle you learn.

Content **CHECK-UP!**

10. Which muscle is named for its shape?

 a. latissimus dorsi   c. gluteus maximus

 b. triceps brachii   d. frontalis

Answer in Appendix A.

7.5 **Skeletal Muscle Groups**

11. Describe the locations and actions of the major skeletal muscles of each body region.

In our discussion, the muscles of the body (Fig. 7.9) are grouped according to their location and their action. To better understand muscle groups, recall from Chapter 6 that the term *arm* refers to

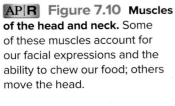

Figure 7.10 Muscles of the head and neck. Some of these muscles account for our facial expressions and the ability to chew our food; others move the head.

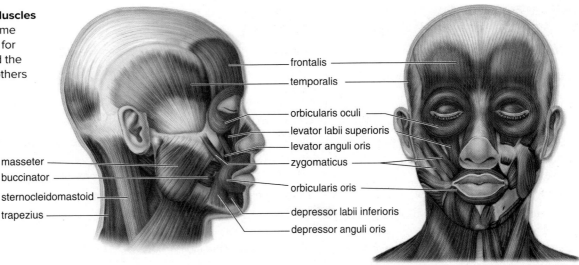

the humeral area, whereas the radius/ulna area is referred to as the *forearm*. Likewise, the femur is the *thigh* area, whereas the tibia/fibular area is referred to as the *leg*.

After you understand the meaning of a muscle's name, try to correlate its name with the muscle's location and the action it performs. Knowing the origin and insertion will also help you remember what the muscle does. Why? Because the insertion is on the bone that moves. You should review the various body movements listed and illustrated in Figure 6.23. Once you're finished, you'll be able to understand the actions of the muscles listed in Tables 7.2 to 7.5. Scientific terminology is necessary because it allows all persons to know the exact action being described for that muscle.

Muscles of the Head

The muscles of the head and neck, the first group of muscles we'll study, are illustrated in Figure 7.10 and listed in Table 7.2. The muscles of the head are responsible for facial expression and mastication (chewing). One muscle of the head and several muscles of the neck allow us to swallow. The muscles of the neck also move the head.

Muscles of Facial Expression

The muscles of facial expression are located on the scalp and face. These muscles are unusual in that they insert into and move the skin. Therefore, we expect them to move the skin and not a bone. As you know from the chapter introduction, these muscles communicate whether we are surprised, angry, fearful, happy, and so forth.

Frontalis lies over the frontal bone; it raises the eyebrows and wrinkles the brow. Frequent use results in furrowing of the forehead.

Orbicularis oculi is a ringlike band of muscle that encircles (forms an orbit about) the eye. It causes the eye to close or blink, and is responsible for "crow's feet" at the eye corners.

Orbicularis oris encircles the mouth and is used to pucker the lips, as in forming a kiss. Frequent use results in lines about the mouth.

Buccinator muscles are located in the cheek areas. When a buccinator contracts, the cheek is compressed, as when a person whistles or blows out air. Therefore, this muscle is called the "trumpeter's muscle." Important to everyday life, the buccinator helps hold food in contact with the teeth during chewing. Babies use this muscle for suckling. It is also used in swallowing, as discussed next.

Zygomaticus extends from each zygomatic arch (cheekbone) to the corners of the mouth. It raises the corners of the mouth when a person smiles.

Levator anguli oris and **levator labii superioris** muscles lift the upper edge and corners of the lip. Simultaneously contracting these muscles on both sides of the mouth helps produce a smile. However, a person will sneer if he uses only the set on one side of his mouth.

Depressor anguli oris and **depressor labii inferioris** pull the lower edge and corners of the lip down, as when a person is frowning or a child is pouting.

Muscles of Mastication

We use the muscles of mastication when we chew food or bite something. Although there are four pairs of muscles for chewing, only two pairs are superficial and shown in Figure 7.10. As you might expect, both pairs insert on the mandible.

Each **masseter** has its origin on the zygomatic arch and its insertion on the mandible. The masseter is a muscle of mastication (chewing) because it is a prime mover for elevating the mandible.

Each **temporalis** is a fan-shaped muscle that overlies the temporal bone. It is also a prime mover for elevating the mandible. The masseter and temporalis are synergists.

Muscles of the Neck

Deep muscles of the neck (not illustrated) are responsible for swallowing. Superficial muscles of the neck move the head (Table 7.2 and Fig. 7.10).

TABLE 7.2　Muscles of the Head and Neck

Name	Function	Origin/Insertion
MUSCLES OF FACIAL EXPRESSION		
Frontalis (frŭn-tă′lĭs)	Raises eyebrows	Cranial fascia/skin and muscles around eye
Orbicularis oculi (ōr-bĭk′yū-lā-rĭs ŏk′yū-lī)	Closes eye	Maxillary and frontal bones/skin around eye
Orbicularis oris (ōr-bĭk′yū-lā-rĭs ō′rĭs)	Closes and protrudes lips	Muscles near the mouth/skin around mouth
Buccinator (bŭk′sĭ-nā″tōr)	Compresses cheeks inward	Outer surfaces of maxilla and mandible/orbicularis oris
Zygomaticus (zī″gō-măt′ĭk-ŭs)	Raises corner of mouth	Zygomatic bone/skin and muscle around mouth
Levator labii superioris (lŭh-vā′ těr lā′bē-ī sū′pē-rē-ō-rĭs)	Lifts the lip	Zygomatic bone/skin and muscles of upper lip
Levator anguli oris (lŭh-vā′těr ăn′ gū-lĭ ō′rĭs)	Lifts the corner of the mouth	Maxilla/corner of upper lip
Depressor labii inferioris (dŭh-prĕs′ōr lā′bē-ī ĭn-fē′rē-ō-rĭs)	Pulls the lip inferiorly	Mandible/skin of inferior lip
Depressor anguli oris (dŭh-prĕs′ōr ăn′gū-lĭ ō′rĭs)	Pulls the corner of the mouth inferiorly	Mandible/skin of lower corner of the mouth
MUSCLES OF MASTICATION		
Masseter (măs-sē′tĕr)	Closes jaw	Zygomatic arch/mandible
Temporalis (tĕm-pō-rā′lĭs)	Closes jaw	Temporal bone/mandibular coronoid process
MUSCLES THAT MOVE THE HEAD		
Sternocleidomastoid (stĕr″nō-klī″dō-măs′tōyd)	Flexes head and rotates head	Sternum and clavicle/mastoid process of temporal bone
Trapezius (trŭh-pē′zē-ŭs)	Extends head and adducts scapula	Occipital bone, C_7 vertebra, all thoracic vertebrae/spine of scapula and clavicle

Swallowing

Swallowing is an important activity that begins after we chew our food. First, the tongue (a muscle) and the buccinators squeeze the food back along the roof of the mouth toward the pharynx. An important bone that functions in swallowing is the hyoid (see Figure 6.4). As you know, the hyoid is the only bone in the body that does not articulate (form a joint) with another bone.

Muscles that lie superior to the hyoid, called the *suprahyoid muscles,* and muscles that lie inferior to the hyoid, called the *infrahyoid muscles,* move the hyoid. Because these muscles lie deep in the neck, they are not illustrated in Figure 7.10. The suprahyoid muscles pull the hyoid forward and upward toward the mandible. Because the hyoid is attached to the larynx, this pulls the larynx upward and forward. The epiglottis now lies over the glottis and closes the respiratory passages. Small palatini muscles (not illustrated) pull the soft palate backward, closing off the nasal passages. Pharyngeal constrictor muscles (not illustrated) push the bolus of food into the pharynx, which widens when the suprahyoid muscles move the hyoid. The hyoid bone and larynx are returned to their original positions by the infrahyoid muscles. Notice that the suprahyoid and infrahyoid muscles are antagonists.

Muscles That Move the Head

Two muscles in the neck are of particular interest: The sternocleidomastoid and the trapezius are listed in Table 7.2 and illustrated in Figure 7.10. Recall that *flexion* is a movement that closes the angle at a joint and *extension* is a movement that increases the angle at a joint. Recall that *abduction* is a movement away from the midline of the body, whereas *adduction* is a movement toward the midline. Also, *rotation* is the movement of a part around its own axis.

Sternocleidomastoid muscles ascend obliquely from their origin on the sternum and clavicle to their insertion on the mastoid process of the temporal bone. Which part of the body do you expect them to move? When both sternocleidomastoid muscles contract, flexion of the head occurs. When only one contracts, the head turns to the opposite side. If you turn your head to the right, you can see how the left sternocleidomastoid shortens, pulling the head to the right.

Each side of the **trapezius** muscle is triangular, but together, they take on a diamond or trapezoid shape. The origin of the trapezius is at the base of the skull. Its insertion is on the clavicles and scapula. You would expect the trapezius muscles to move the scapulae, and they do. They adduct the scapulae when the shoulders are shrugged or pulled back. The trapezius muscles also help extend the head, however. The prime movers for head extension are actually deep to the trapezius and not illustrated in Figure 7.10.

Muscles of the Trunk

The muscles of the trunk are listed in Table 7.3 and illustrated in Figure 7.11. The muscles of the thoracic wall are primarily

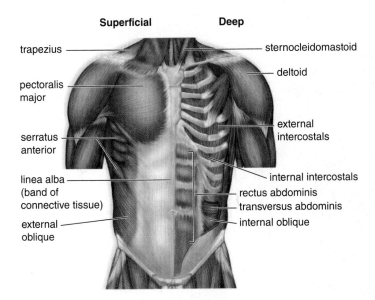

Superficial Deep

trapezius —
sternocleidomastoid
pectoralis major —
deltoid
serratus anterior —
external intercostals
linea alba (band of connective tissue) —
internal intercostals
rectus abdominis
transversus abdominis
external oblique —
internal oblique

Figure 7.11 Muscles of the anterior shoulder and trunk. The right pectoralis major is removed to show the deep muscles of the chest.

involved in breathing. The muscles of the abdominal wall protect and support the organs within the abdominal cavity.

Muscles of the Thoracic Wall

External intercostal muscles occur between the ribs; they originate on a superior rib and insert on an inferior rib. These muscles elevate the rib cage during the inspiration phase of breathing.

The **diaphragm** is a dome-shaped muscle that, as you know, separates the thoracic cavity from the abdominal cavity (see Fig. 1.5). The diaphragm is the primary muscle for respiration, and it is the only muscle used during normal, quiet breathing.

Internal intercostal muscles originate on an inferior rib and insert on a superior rib. These muscles depress the rib cage and contract only during a forced expiration. Normal expiration does not require muscular action.

Muscles of the Abdominal Wall

The abdominal wall has no bony reinforcement (Fig. 7.11). The wall is strengthened by four pairs of muscles that run at angles to one another. The external and internal obliques and the transversus abdominis occur laterally, but the fasciae of these muscle pairs meet at the midline of the body, forming a tendinous area called the *linea alba*. The rectus abdominis is a superficial medial pair of muscles.

All of the muscle pairs of the abdominal wall compress the abdominal cavity and support and protect the organs within the abdominal cavity.

External and **internal obliques** occur on a slant and are at right angles to one another. They are located between the lower ribs and the pelvic girdle. The internal obliques are deep to the external obliques. These muscles also aid trunk rotation and lateral flexion.

Transversus abdominis, deep to the obliques, extends horizontally across the abdomen. The obliques and the transversus abdominis are synergistic muscles.

Rectus abdominis has a straplike appearance but takes its name from the fact that it runs straight (*rectus* means straight) up from the pubic bones to the ribs and sternum. These muscles also help flex and rotate the lumbar portion of the vertebral column.

Muscles of the Shoulder

Muscles of the shoulder are shown in Figures 7.11 and 7.12. They are also listed in Table 7.4. The muscles of the shoulder attach the scapula to the thorax and move the scapula; they also attach the humerus to the scapula and move the arm.

Muscles That Move the Scapula

Of the muscles that move the scapula, you know (from page 148) that the trapezius adducts the scapulae.

Serratus anterior is located below the axilla (armpit) on the lateral chest. It runs between the upper ribs and the scapula. It holds the scapula near the thorax, pulling it forward (as when we're pushing on something in front of us). Because this muscle causes a fast-forward jab of the arm, it is often called the boxer's muscle. It also helps to elevate the arm above the horizontal level.

TABLE 7.3 Muscles of the Trunk		
Name	**Function**	**Origin/Insertion**
External intercostals	Elevate rib cage for inspiration	Superior rib/inferior rib
Internal intercostals	Depress rib cage for forced expiration	Inferior rib/superior rib
External oblique	Tenses abdominal wall; lateral rotation of trunk	Lower eight ribs/iliac crest
Internal oblique	Tenses abdominal wall; lateral rotation of trunk	Iliac crest/lower three ribs
Transversus abdominis	Tenses abdominal wall	Lower six ribs/pubis
Rectus abdominis	Flexes and rotates the vertebral column	Pubis, pubic symphysis/xiphoid process of sternum, fifth to seventh costal cartilages

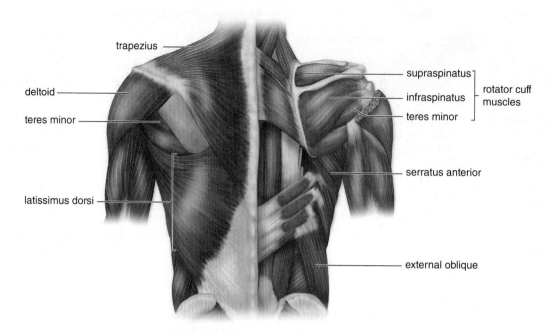

AP|R **Figure 7.12** **Muscles of the posterior shoulder.** The right trapezius and latissimus dorsi are removed to show the deep muscles that move the scapula, as well as three of the four rotator cuff muscles.

TABLE 7.4	Muscles of the Shoulder and Upper Limb	
Name	**Function**	**Origin/Insertion**
MUSCLES THAT MOVE THE SCAPULA AND ARM		
Serratus anterior	Depresses scapula and pulls it forward; elevates arm above horizontal	Upper nine ribs/vertebral border of scapula
Deltoid	Abducts arm to horizontal	Acromion process, spine of scapula, and clavicle/deltoid tuberosity of humerus
Pectoralis major	Flexes, medially rotates, and adducts arm	Clavicle, sternum, second to sixth costal cartilages/intertubular groove of humerus
Latissimus dorsi	Extends, adducts, and medially rotates arm	Iliac crest/intertubular groove of humerus
Rotator cuff (supraspinatus, infraspinatus, subscapularis, teres minor)	Angular and rotational movements of arm	Scapula/humerus
MUSCLES THAT MOVE THE FOREARM		
Biceps brachii	Flexes arm and forearm; supinates forearm	Scapula/radial tuberosity
Triceps brachii	Extends arm and forearm	Scapula, proximal humerus/olecranon process of ulna
Brachialis	Flexes forearm	Anterior humerus/coronoid process of ulna
MUSCLES THAT MOVE THE HAND AND FINGERS		
Flexor carpi and extensor carpi	Move wrist and hand	Humerus/carpals and metacarpals
Flexor digitorum and extensor digitorum	Move fingers	Humerus, radius, ulna/phalanges

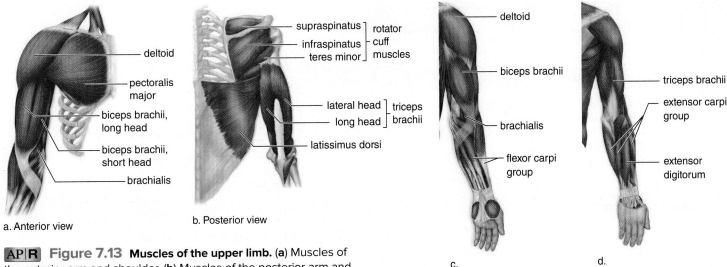

a. Anterior view

b. Posterior view

c.

d.

AP|R **Figure 7.13** **Muscles of the upper limb.** (a) Muscles of the anterior arm and shoulder. (b) Muscles of the posterior arm and shoulder. (c) Muscles of the anterior forearm. (d) Muscles of the posterior forearm.

Muscles That Move the Arm

Deltoid is a large, fleshy, triangular muscle (*deltoid* in Greek means triangular) that covers the shoulder and causes a bulge in the arm where it meets the shoulder. It runs from both the clavicle and the scapula of the pectoral girdle to the humerus. This muscle abducts the arm to the horizontal position.

Pectoralis major (Fig. 7.11) is a large anterior muscle of the upper chest. It originates from a clavicle, but also from the sternum and ribs. It inserts on the humerus. The pectoralis major flexes the arm (raises it anteriorly). It also medially rotates and adducts the arm, pulling it toward the chest.

Latissimus dorsi (Fig. 7.12) is a large, wide, triangular muscle of the back. This muscle originates from the lower spine and sweeps upward to insert on the humerus. The latissimus dorsi extends, medially rotates, and adducts the arm (brings it down from a raised position). This muscle is very important for swimming, rowing, and climbing a rope.

Rotator cuff (Figs. 7.12 and 7.13). This group of muscles is so named because their tendons help form a cuff over the proximal humerus. There are four rotator cuff muscles. Three are located on the posterior scapula: supraspinatus, infraspinatus, and teres minor. The last rotator cuff muscle is the subscapularis muscle located on the anterior surface of the scapula. These muscles lie deep to those already mentioned, and they are synergists to them.

Muscles of the Arm

The muscles of the arm move the forearm. They are illustrated in Figure 7.13 and listed in Table 7.4.

Biceps brachii is a muscle of the proximal anterior arm (Fig. 7.13*a*) that is familiar because it bulges when the forearm is flexed. It also assists in flexing the arm at the shoulder, and supinates the hand when a doorknob is turned or the cap of a jar is unscrewed. The name of the muscle

refers to its two heads that attach to the scapula, where it originates. The biceps brachii inserts on the radius.

Brachialis originates on the humerus and inserts on the ulna. It is a muscle of the distal anterior humerus and lies deep to the biceps brachii. It is the strongest forearm flexor muscle, and the biceps brachii is its synergist.

Triceps brachii is the only muscle of the posterior arm (Fig. 7.13*b*). As its name implies, it has three heads. The long head originates from the scapula and humerus, while the medial and lateral heads only originate from the humerus. All three heads join in a common tendon that inserts on the ulna. The triceps extends the arm and forearm. The triceps is also used in tennis to do a backhand volley.

Muscles of the Forearm

The muscles of the forearm move the hand and fingers. They are illustrated in Figure 7.13*c,d* and listed in Table 7.4. Note that in anatomical position, extensors of the wrists and fingers are on the posterior and lateral forearm and flexors are on the anterior and medial forearm.

Flexor carpi and **extensor carpi** muscles primarily originate on the humerus and insert on the bones of the hand. The flexor carpi flex the wrists and hands, and the extensor carpi extend the wrists and hands.

Flexor digitorum and **extensor digitorum** muscles also primarily originate on the humerus and insert on the bones of the hand. The flexor digitorum (not shown) flexes the wrist and fingers, and the extensor digitorum extends the wrist and fingers (i.e., the digits).

Muscles of the Hip and Lower Limb

The muscles of the hip and lower limb are listed in Table 7.5 and shown in Figures 7.14 to 7.17. These muscles, particularly those

TABLE 7.5 Muscles of the Hip and Lower Limb

Name	Function	Origin/Insertion
MUSCLES THAT MOVE THE THIGH		
Iliopsoas (ĭl′ē-ō-sō′ŭs)	Flexes thigh	Lumbar vertebrae, ilium/lesser trochanter of femur
Gluteus maximus	Extends thigh	Posterior ilium, sacrum/proximal femur
Gluteus medius	Abducts thigh	Ilium/greater trochanter of femur
Adductor group	Adducts thigh	Pubis, ischium/femur and tibia
MUSCLES THAT MOVE THE LEG		
Quadriceps femoris group	Extends leg, steadies hip joint, and assists in thigh flexion	Ilium, femur/patellar tendon that continues as a ligament to tibial tuberosity
Sartorius	Flexes leg; flexes, abducts and laterally rotates thigh	Ilium/medial tibia
Hamstring group	Flexes and rotates leg medially, and extends thigh	Ischial tuberosity/lateral and medial tibia
MUSCLES THAT MOVE THE ANKLE AND FOOT		
Gastrocnemius (găs″trŏk-nē′mē-ŭs)	Flexes leg; plantar flexion and eversion of foot	Condyles of femur/calcaneus by way of Achilles tendon
Tibialis anterior (tĭb″ē-ă′lĭs ăn-tē′rē-ōr)	Dorsiflexion and inversion of foot	Condyles of tibia/tarsal and metatarsal bones
Fibularis group	Plantar flexion and eversion of foot	Fibula/tarsal and metatarsal bones
Flexor and extensor digitorum longus	Moves toes	Tibia, fibula/phalanges

of the hips and thigh, tend to be large and heavy because they are used to move the entire weight of the body and to resist the force of gravity. Therefore, they are important for movement and balance.

Muscles That Move the Thigh

The muscles that move the thigh have at least one origin on the pelvic girdle and insert on the femur. Notice that the iliopsoas is an anterior muscle that moves the thigh, while the gluteal muscles ("glutes") are posterior muscles that move the thigh. The adductor muscles are medial muscles (Figs. 7.14 and 7.15). Before studying the action of these muscles, review the movement of the hip joint when the thigh flexes, extends, abducts, and adducts (see Chapter 6, pp. 124–126).

Iliopsoas (includes psoas major and iliacus) originates at the ilium and the bodies of the lumbar vertebrae, and inserts on the femur medially (Fig. 7.14). This muscle is the prime mover for flexing the thigh and also the trunk, as when we bow. As the major flexor of the thigh, the iliopsoas is important to the process of walking. It also helps prevent the trunk from falling backward when a person is standing erect.

The gluteal muscles form the buttocks. We will consider only the gluteus maximus and the gluteus medius, both of which are illustrated in Figure 7.15. (The third gluteal muscle, gluteus minimus, is deep to both and therefore not shown in the figure.)

Gluteus maximus is the largest muscle in the body and covers a large part of the buttock (*gluteus* means buttocks in Greek). It originates at the ilium and sacrum, and inserts on the femur. The gluteus maximus is a prime mover of thigh extension, as when a person is walking, climbing stairs, or jumping from a crouched position. Notice that the iliopsoas and the gluteus maximus are antagonistic muscles.

Gluteus medius lies partly behind the gluteus maximus (Fig. 7.15). It runs between the ilium and the femur, and functions to abduct the thigh. The gluteus maximus assists the gluteus medius in this function. Therefore, they are synergistic muscles.

Adductor group muscles (pectineus, adductor longus, adductor magnus, gracilis) are located on the medial thigh (Fig. 7.14). All of these muscles originate from the pubis and ischium, and insert on the femur; the deep adductor magnus is shown in Figure 7.14. Adductor muscles adduct the thigh—that is, they lower the thigh sideways from a horizontal position. Because they squeeze the thighs together, these are the muscles that keep a rider on a horse. Notice that the glutes and the adductor group are antagonistic muscles.

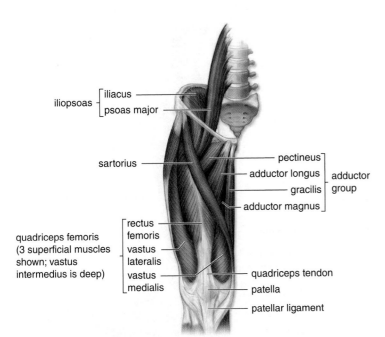

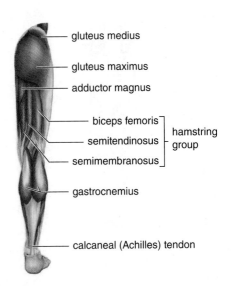

iliopsoas
  iliacus
  psoas major

sartorius

pectineus
adductor longus
gracilis
adductor magnus
} adductor group

quadriceps femoris
(3 superficial muscles
shown; vastus
intermedius is deep)
  rectus femoris
  vastus lateralis
  vastus medialis

quadriceps tendon
patella
patellar ligament

AP|R Figure 7.14 **Muscles of the anterior right hip and thigh.**

gluteus medius
gluteus maximus
adductor magnus

biceps femoris
semitendinosus
semimembranosus
} hamstring group

gastrocnemius

calcaneal (Achilles) tendon

AP|R Figure 7.15 **Muscles of the posterior right hip and thigh.**

Muscles That Move the Leg

The muscles that move the leg originate from the pelvic girdle or femur and insert on the tibia. They are listed in Table 7.5 and illustrated in Figures 7.14 and 7.15. Before studying these muscles, review the movement of the knee when the leg extends and when it flexes (see Chapter 6, pp. 124–126).

Quadriceps femoris group (rectus femoris, vastus lateralis, vastus medialis, vastus intermedius), also known as the "quads," is found on the anterior and medial thigh. The rectus femoris, which originates from the ilium, is external to the vastus intermedius, and therefore the vastus intermedius is not shown in Figure 7.14. These muscles are the primary extensors of the leg, as when you kick a ball by straightening your knee. They also stabilize the hip and assist in thigh flexion.

Sartorius is a long, straplike muscle that has its origin on the iliac spine and then goes across the anterior thigh to insert on the medial side of the knee (Fig. 7.14). Because this muscle crosses both the hip and knee joint, it acts on the thigh in addition to the leg. The insertion of the sartorius is such that it flexes both the leg and the thigh. It also abducts and laterally rotates the thigh, enabling us to sit cross-legged, as tailors were accustomed to do in another era. Therefore, it is sometimes called the "tailor's muscle," and in fact, *sartor* means tailor in Latin.

Hamstring group (biceps femoris, semimembranosus, semitendinosus) is located on the posterior thigh (Fig. 7.15). Notice that these muscles also cross the hip and knee joint because they have origins on the ischium and insert on the tibia. They flex and rotate the leg medially, but they also extend the

thigh. Their strong tendons can be felt behind the knee. These same tendons are present in hogs and were used by butchers as strings to hang up hams for smoking—hence, the name. Notice that the quadriceps femoris group and the hamstring group are antagonistic muscles in that the quads extend the leg and the hamstrings flex the leg. Likewise, the quads assist in flexing the thigh, while the hamstrings extend the thigh.

Muscles That Move the Ankle and Foot

Muscles that move the ankle and foot are shown in Figures 7.16 and 7.17.

Gastrocnemius is a muscle of the posterior leg, where it forms a large part of the calf. It arises from the femur; distally, the muscle joins the strong calcaneal tendon, which attaches to the calcaneus bone (heel). The gastrocnemius is a leg flexor, but its most important function is to act as a powerful plantar flexor of the foot that aids in pushing the body forward during walking or running. It is sometimes called the "toe dancer's muscle" because it allows a person to stand on tiptoe.

Tibialis anterior is a long, spindle-shaped muscle of the anterior leg. It arises from the surface of the tibia and attaches to the bones of the ankle and foot. Contraction of this muscle causes dorsiflexion and inversion of the foot.

Fibularis muscles (fibularis longus, fibularis brevis) are found on the lateral side of the leg, connecting the fibula to the metatarsal bones of the foot. These muscles evert the foot and also help bring about plantar flexion.

Flexor (not shown) and **extensor digitorum longus** muscles primarily originate from the tibia and insert on the toes. They flex and extend the toes, respectively, and assist in other movements of the feet.

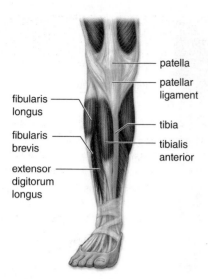

AP|R **Figure 7.16** Muscles of the anterior right leg.

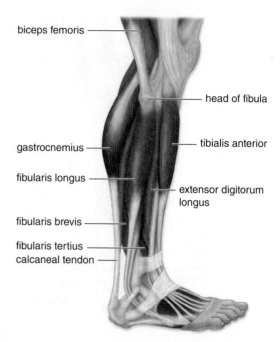

AP|R **Figure 7.17** Muscles of the lateral right leg.

Begin Thinking Clinically

Have you ever sat cross-legged for a long period of time? When you unfolded your legs to stand, you experienced an uncomfortable tingling sensation and you were unable to lift your foot (a condition called a foot-drop). The foot-drop resulted from compressing the nerve supplying the muscles that lift your foot. Which muscles are these?

Answer and discussion in Appendix A.

Content CHECK-UP!

11. Which muscle of facial expression is also used for swallowing?

  a. frontalis   c. zygomaticus

  b. masseter   d. buccinator

12. Imagine you've heard some great news and a big smile spreads over your face. What muscles might you use to create it?

Answers in Appendix A.

7.6 Effects of Aging

12. Describe the anatomical and physiological changes that occur in the muscular system as we age.

Muscle mass and strength tend to decrease as people age. How much of this is due to lack of exercise and a poor diet is under careful study. Deteriorated muscle elements are replaced initially by connective tissue and, eventually, by fat. With age, degenerative changes take place in the mitochondria, and endurance decreases. Also, changes in the nervous and cardiovascular systems adversely affect the structure and function of muscles.

Muscle mass and strength can improve remarkably if elderly people undergo a training program. Exercise at any age appears

to stimulate muscle buildup. As discussed in the Medical Focus, Benefits of Exercise on pages 155–156, exercise has many other benefits as well. For example, exercise improves the cardiovascular system and reduces the risk of elevated blood sugar, metabolic syndrome, diabetes, and glycation. During glycation, excess glucose molecules stick to body proteins so that the proteins no longer have their normal structure and cannot function properly. Exercise burns glucose and, in this way, helps prevent muscle deterioration.

Content CHECK-UP!

13. A scientist is comparing muscle cells obtained from a 20-year old man and a 85-year old man. What differences does this researcher observe in these cells?

Answer in Appendix A.

7.7 Homeostasis

13. Describe how the muscular system works with other systems of the body to maintain homeostasis.

The illustration in Human Systems Work Together on page 158 tells how the muscular system works with other systems of the body to maintain homeostasis.

Cardiac muscle contraction accounts for the heartbeat, which creates blood pressure, the force that propels blood in the arteries and arterioles. The walls of the arteries and arterioles contain smooth muscle. Constriction of arteriole walls is regulated to help maintain blood pressure. Arterioles branch into the capillaries, where exchange takes place that creates and cleanses tissue fluid. Blood and tissue fluid are the internal environment of the body, and

MEDICAL FOCUS

Benefits of Exercise

What's at the top of many New Year's resolution lists every January 1? You guessed it: get more exercise! Have you ever made that resolution, only to find yourself floundering and failing? Keep reading to learn why it's best to stick with your New Year's exercise plan, instead of admitting defeat when February 1 rolls around.

Exercise builds muscles and saves your skeleton. Exercise programs improve muscular strength, endurance, and flexibility. Muscular strength is the force a muscle (or muscle group) can exert against resistance. Muscle endurance is the muscle's ability to contract repeatedly or to sustain a contraction for an extended time. Muscle flexibility is tested by measuring a joint's range of motion.

As muscular strength improves, the muscle's overall size increases. Muscle fibers synthesize more thick and thin myofilaments, and myofibrils increase in size. Though muscle fibers don't reproduce by mitosis, they do enlarge by creating these new myofibrils. Simultaneously, the total protein, numbers of capillaries, and the amounts of connective tissue, including tissue found in tendons and ligaments, also increase. Physical training with weights can improve muscular strength and endurance in all adults, regardless of age.

Over time, increased muscle strength creates stronger bones and increases joint stability. Exercise also helps prevent osteoporosis, the condition described in Chapter 6 in which the bones are weak and easily broken. Exercise stimulates the activity of osteoblasts (the bone building cells) in young and old alike, and even those with joint diseases such as osteoarthritis can benefit. Patients who regularly exercise report much less pain, swelling, fatigue, and depression.

Exercise helps control weight and keeps blood sugar concentration in the normal range. Increased activity can help to take off unwanted pounds and to keep them off (and that's no surprise, right?). Moreover, glucose moves into muscle cells during contraction, which reduces the amount of glucose in the blood. As a result, blood glucose homeostasis is maintained. This is extremely important for your overall health, not just your waistline. Excess body weight and elevated blood glucose levels contribute to metabolic syndrome and type II diabetes, two disorders of glucose metabolism (see pages 230–231). Long-term complications of diabetes include kidney failure, blindness, and cardiovascular disease. However, type II

diabetics who exercise regularly can often reduce or even eliminate the need for insulin and other medications.

Exercise protects your heart, blood vessels, *and* your brain. The life-long benefits of exercise are most apparent with regard to cardiovascular health. Regular exercise raises the blood levels of high-density lipoprotein (HDL, the so-called "good cholesterol"; see Chapter 12), and lowers blood levels of low-density lipoprotein (LDL, or "bad cholesterol"). These effects protect both the heart and all blood vessels—including those supplying the brain—from long-term damage. Thus, regular exercise reduces the risk of heart attack and stroke.

Further, it's important to get moving to protect your brain. Exercise can help to moderate the effects of depression because it leads to an increase in the level of brain neurotransmitters that naturally elevate mood. Both chronic depression and poor cardiovascular health are strongly linked to the development of dementia, including Alzheimer disease.

Exercise helps prevent cancer. You know (from Chapter 4) that cancer prevention requires eating properly, not smoking, and avoiding exposure to radiation and cancer-causing chemicals. To detect cancer early, when it's most curable, you'll need to undergo appropriate medical screening tests and know the early warning signs of cancer. But did you know that exercise also leads to a reduced risk of certain kinds of cancer? Evidence shows that people who exercise are less likely to develop colon, breast, cervical, uterine, and ovarian cancer.

So how do you keep that promise you made to yourself to get more exercise? Remember to use the acronym SMART as you plan your exercise routine. Studies show that successful people choose *specific* and *measurable* goals, for starters. So instead of "exercise more," maybe it's "attend three workout classes at the gym each week" or "take a 30-minute walk around the block five times a week." Activity plans must be *attainable* and *realistic* as well. If you can't fit those three workout classes into your frantically busy life, why not schedule two instead, and walk in between? Realistic exercisers who stick with their programs also choose activities they already enjoy and will be more likely to keep doing. Finally, any exercise goal must be *time-limited*—have a definite end time—or you just might get burned out. Check out Table 7A for ways to include exercise in your lifestyle.

—Continued

Continued—

TABLE 7.A   A Variety of Ways to Stay Fit

	Children, 7–12	Teenagers, 13–18	Adults, 19–65	Seniors, 65 and Older
How much time should I spend exercising?	Vigorous activity for 1–2 hours daily	Vigorous activity for 1 hour 3–5 days a week; otherwise, 30–45 minutes of moderate activity. Can be broken up into three 10–15 minute segments if needed.	Moderate exercise for 1 hour daily, 3 days a week; otherwise, 30–45 minutes of daily moderate activity. Can be divided into three 10–15 minute segments.	
Exercise at School or Work	Physical education classes. Walk or cycle to school whenever possible		Taking the stairs, parking at a distant space, walking or cycling to work.	Taking the stairs, parking at a distant space, walking or cycling to work or to volunteer activities.
Exercising at Home	Free play	Build muscle with resistance exercise. Aerobic exercise to control buildup of fat cells	Build muscle with resistance exercise. Aerobic exercise to prevent lower back pain: stretching, yoga	Do daily stretching exercises. Take a daily walk with a dog. Gardening.
Exercising Socially	Build motor skills through team sports, dance, swimming.	Continue team sports, dancing, swimming. Try a new sport or activity.	Find exercise partners: join a running club, bicycle club, outing group	Try low-impact aerobics; before undertaking new exercises, consult with your doctor.
Innovating and Keeping it Fun	Initiate family outings: golfing, boating, camping, hiking	Try tennis, swimming, horseback riding—sports that can be enjoyed for a lifetime.	Take active vacations: hike, bicycle, cross-country ski.	Learn a new sport or activity: golf, fishing, doubles tennis, ballroom dancing

without cardiac and smooth muscle contraction, blood would never reach the capillaries for exchange to take place. Blood is returned to the heart in cardiovascular veins, and excess tissue fluid is returned to the cardiovascular system within lymphatic vessels. In turn, skeletal muscle contraction presses on the cardiovascular veins and lymphatic vessels, and this creates the pressure that moves fluids in both types of vessels. Without the return of blood to the heart, circulation would stop, and without the return of lymph to the blood vessels, normal blood pressure could not be maintained.

The contraction of sphincters composed of smooth muscle fibers temporarily prevents the flow of blood into a capillary. This is an important homeostatic mechanism because in times of emergency it is more important, for example, for blood to be directed to the skeletal muscles than to the tissues of the digestive tract.

Skeletal muscle in the abdominopelvic region protects all the internal organs it covers. In the digestive system, smooth muscle contraction accounts for peristalsis, the process that moves food along the digestive tract. Without this action, food would never reach all the organs of the digestive tract where digestion releases nutrients that enter the bloodstream. As part of the urinary system, smooth muscle contracts to assist in the voiding of urine, which is necessary for ridding the body of metabolic wastes and for regulating the blood volume, salt concentration, and pH of internal fluids.

Contraction of the skeletal muscles that are part of the respiratory system raises and lowers the rib cage and diaphragm during the active phases of breathing. As we breathe, oxygen enters the blood and is delivered to the tissues, including the muscles, where ATP is produced in mitochondria with heat as a by-product. The heat produced by skeletal muscle contraction allows the body temperature to remain within the normal range for human beings.

The muscular system maintains the integrity of the skeletal system. Repetitive skeletal muscle contraction helps build bone,

Muscular Disorders and Neuromuscular Disease

During the course of a lifetime, nearly everyone will suffer from some type of muscular disorder. Muscular disorders cover a wide spectrum in terms of severity. Minor muscle irritation, inflammation, or injury may resolve without any medical care. However, many diseases affecting the neuromuscular system are extremely serious and eventually prove to be fatal.

Spasms are sudden, involuntary muscular contractions, most often accompanied by pain. Spasms can occur in both smooth and skeletal muscles. A spasm of the intestinal tract is a "bellyache"; most such spasms are not serious. Multiple spasms of skeletal muscles are called **convulsions. Cramps** are strong, painful spasms, especially of the leg and foot, usually due to strenuous athletic activity. Cramps typically occur after a strenuous workout, and may even occur when sleeping. Facial **tics,** such as periodic eye blinking or grimacing, are spasms that can be controlled voluntarily but only with great effort.

Muscles, joints, and their connective tissues are often subject to overuse injuries: strains, sprains, and tendinitis. A **strain** is caused by stretching or tearing of a muscle. A **sprain** is the twisting of a joint, leading to swelling and injury not only of muscles but also of ligaments, tendons, blood vessels, and nerves. The ankle and knee are two areas often subject to sprains. **Tendinitis** is inflammation of a tendon due to repeated athletic activity. The tendons most commonly affected are those associated with the shoulder, elbow, hip, and knee.

Overuse injuries are often minor and can be treated with pain medication and rest. However, an individual should seek medical attention if the injured area is extremely painful, hot, or swollen, or if accompanied by a fever.

Neuromuscular Diseases

Neuromuscular disease can result from pathologic changes to the muscle itself. It can also result from excessive motor nerve stimulation, or from damage or destruction of the motor neurons that supply the muscle.

The disease **tetanus** develops in persons who have not been properly immunized against the toxin of the tetanus bacterium. Tetanus toxin shuts down brain areas that normally inhibit unnecessary muscle contractions. As a result, excessive brain stimulation causes muscles to lock in a tetanic contraction (from which the disease gets its name). A rigidly locked jaw is one of the first signs of bacterial infection and toxin production. Though antibiotics will kill the bacteria, once the toxin is circulating in the bloodstream, it cannot be removed or neutralized. Because muscles can't relax, the patient cannot breathe or swallow, and death may occur due to respiratory failure. Immunization and periodic booster shots will prevent the toxin's effects (see Immunization: The Great Protector, pages 309–310).

Fibromyalgia is a chronic condition whose symptoms include achy pain, tenderness, and stiffness of muscles. Its precise cause is not known, though 80–90% of sufferers are women. Substance P, a neurotransmitter (messenger chemical) of pain pathways in the brain, has been found in the bloodstream of affected individuals. Exercise seems to decrease blood levels of substance P. Therapeutic massage, over-the-counter pain medication, and muscle relaxants are also recommended.

Muscular dystrophy is a broad term applied to a group of disorders that causes progressive degeneration and weakening of muscles. As muscle fibers die, fat and connective tissue take their place. **Duchenne muscular dystrophy,** the most common type, is inherited through a flawed gene carried by the mother. It is now known that the lack of a protein called *dystrophin* causes the condition. When dystrophin is absent, calcium leaks into the cell and activates an enzyme that dissolves muscle fibers. Treatment includes muscle injections with immature muscle cells that do produce dystrophin.

Myasthenia gravis is an autoimmune disease characterized by weakness that especially affects the muscles of the eyelids, face, neck, and extremities. Muscle contraction is impaired because the immune system mistakenly produces antibodies that destroy acetylcholine receptors on the sarcolemma. (Recall that acetylcholine is the neurotransmitter released by motor neurons.) In many cases, the first signs of the disease are drooping eyelids and double vision. Treatment includes drugs that inhibit the enzyme that digests acetylcholine, thus allowing it to accumulate.

Amyotrophic lateral sclerosis (ALS) is often called Lou Gehrig's disease, after its most famous victim, the 1930s-era baseball player. ALS sufferers experience the gradual death of their motor neurons, thus losing the ability to walk, talk, chew, swallow, etc. Intellect and sensation are not affected, however. Drugs can slow the disease's progression, but ALS is always fatal.

and it strengthens joints by stabilizing their movements. Finally, body movements allow us to perform those daily activities necessary to our health and benefit. Although it may seem as if movement of our limbs does not affect homeostasis, it does so by allowing us to relocate our bodies to keep the external environment within favorable limits for our existence.

Content CHECK-UP!

14. How does skeletal muscle assist the heart and blood vessels in maintaining blood pressure in the normal range?

Answer in Appendix A.

Integumentary System

Muscle contraction provides heat to warm skin. Muscle moves skin of face.

Skin protects muscles; rids the body of heat produced by muscle contraction.

Skeletal System

Muscle contraction causes bones to move joints; muscles help protect bones.

Bones provide attachment sites for muscles; store Ca^{2+} for muscle function.

Nervous System

Muscle contraction moves eyes, permits speech, creates facial expressions.

Brain controls nerves that innervate muscles; receptors send sensory input from muscles to brain.

Endocrine System

Muscles help protect glands.

Androgens promote growth of skeletal muscle; epinephrine stimulates heart and constricts blood vessels.

Cardiovascular System

Muscle contraction keeps blood moving in heart and blood vessels.

Blood vessels deliver nutrients and oxygen to muscles, carry away wastes.

How the Muscular System works with other body systems

Lymphatic System/Immunity

Skeletal muscle contraction moves lymph; physical exercise enhances immunity.

Lymphatic vessels pick up excess tissue fluid; immune system protects against infections.

Respiratory System

Muscle contraction assists breathing; physical exercise increases respiratory capacity.

Lungs provide oxygen for, and rid the body of, carbon dioxide from contracting muscles.

Digestive System

Smooth muscle contraction accounts for peristalsis; skeletal muscles support and help protect abdominal organs.

Digestive tract provides glucose for muscle activity; liver metabolizes lactic acid following anaerobic muscle activity.

Urinary System

Smooth muscle contraction assists voiding of urine; skeletal muscles support and help protect urinary organs.

Kidneys maintain blood levels of Na^+, K^+, and Ca^{2+}, which are needed for muscle innervation, and eliminate creatinine.

Reproductive System

Muscle contraction occurs during orgasm and moves gametes; abdominal and uterine muscle contraction occurs during childbirth.

Androgens promote growth of skeletal muscle.

Selected New Terms

Basic Key Terms

A band (ā bănd), p. 138

acetylcholine (ŭh-sē' tŭhl-kō'' lēn) p. 138

actin (ăk'tĭn), p. 137

agonist (ăg' ō-nĭst), p. 145

all-or-none law (āwl-ōr-nŭn lāw), p. 143

anaerobic (ăn' ŭh rō'' bĭk), p. 141

antagonist (ăn-tăg'ō-nĭst), p. 145

cardiac muscle (kăr'dē-ăk mŭs'ĕl), p. 134

creatine phosphate (krē'ŭh-tĭn fŏs'fāt), p. 141

H zone (h zōn), p. 138

I band (ī bănd), p. 138

insertion (ĭn-sĕr'shŭn), p. 145

motor unit (mō'tōr yū'nĭt), p. 138

muscle fiber (mŭs'ŭl fī'bĕr), p. 134

muscle twitch (mŭs'ŭl twĭch), p. 142

myofibril (mī''ō-fī'brĭl), p. 136

myoglobin (mī''ō-glō'bĭn), p. 136

myosin (mī'ō-sĭn), p. 137

neuromuscular junction (nū''rō-mŭs'kyū-lĕr jŭnk'shŭn), p. 138

origin (ōr'ĭ-jĭn), p. 145

oxygen debt (ŏk'sĭ-jĕn dĕbt), p. 142

power stroke (pŏw'ĕr strōk), p. 138

prime mover (prīm mū'vĕr), p. 145

recruitment (rē-krūt'mĕnt), p. 143

sarcomere (săr'kō-mēr), p. 137

skeletal muscle (skĕl'ĕ-tŭl mŭs'ŭl), p. 134

sliding filament theory (slī'dĭng fĭl'ŭh-mĕnt thē'ō-rē), p. 138

smooth muscle (smūth mŭs'ŭl), p. 134

striations (strī ā shŭnz), p. 134

summation (sŭh-mā'shŭn), p. 143

synergist (sĭn'ĕr-jĭst), p. 145

tendon (tĕn'dŭn), p. 135

tetanic contraction (tĕ-tăn'ĭk kŭn-trăk'shŭn), p. 143

tetanus (tĕ' tăn-ŭs), p. 143

thick filaments (thĭk fĭl' uh-ments), p. 137

thin filaments (thĭn fĭl' uh-ments), p. 137

tone (tōn), p. 143

T (transverse) tubules p. 136

tropomyosin (trō'pō-mī'ŭh-sĭn), p. 138

troponin (trō'pō-nĭn), p. 138

Z line (z līn), p. 137

Clinical Key Terms

amyotrophic lateral sclerosis (ALS) p. 157

atrophy (ăt'rō-fē), p. 144

botulism toxin (bŏt' yū-lĭzm tŏks' ĭn), p. 138

compartment syndrome (kŭm-părt'mĕnt sĭn-drōm), p. 135

convulsion (kŭn-vŭl'shŭn), p. 157

cramps (krămpz), p. 157

fatigue (fŭh-tēg'), p. 143

fibromyalgia (fī''brō-mī-ăl'jē-ŭh), p. 157

hypertrophy (hī-pĕr'trō-fē), p. 144

muscular dystrophy (mŭs'kyū-lĕr dĭs'trŭh-fē), p. 157

myasthenia gravis (mī''ăs-thē'nē-ŭh grăh'vĭs), p. 157

spasm (spăzm), p. 157

sprain (sprān), p. 157

strain (strān), p. 157

tendinitis (tĕn''dĕ-nī'tĭs), p. 157

tetanus [disease] (tĕt'ŭh-nŭs), p. 157

tics (tĭks), p. 157

Summary

7.1 Functions and types of muscles
Three types of muscle tissue are found in the body.

A. Skeletal muscle is striated and largely voluntary, and attaches to the skeleton and skin. Cardiac muscle, found in the heart wall, is striated, involuntary, and doesn't fatigue. Smooth muscle is involuntary, not striated, and is located in the walls of internal organs and blood vessels. Though it contracts slowly, it can sustain longer contractions.

B. Individual skeletal muscle cells, called muscle fibers, are surrounded by endomysium. Bundles of muscle fibers, called fascicles, are surrounded by perimysium. Epimysium envelops the entire muscle and is continuous with the muscle tendon. Skeletal muscles support the body, make bones move, help maintain a constant body temperature, assist movement in cardiovascular and lymphatic vessels, help protect internal organs, and stabilize joints.

7.2 Microscopic Anatomy and Contraction of Skeletal Muscle

A. The muscle cell plasma membrane, or sarcolemma, extends into the muscle fiber to form T tubules. The sarcoplasmic reticulum stores calcium. The arrangement of actin and myosin myofilaments in a myofibril creates the striations of skeletal muscle.

B. Skeletal muscle innervation occurs at neuromuscular junctions. Electrical signals called action potentials travel down the T tubules and cause the release of calcium from calcium storage in the sarcoplasmic reticulum.

C. When calcium binds to troponin, myosin myofilaments bind to actin myofilaments. The myosin power stroke causes actin to slide past myosin, shortening sarcomere length. The muscle fiber and ultimately the entire muscle then shorten.

D. Since cardiac muscle is also striated, the contraction mechanism is very similar to that of skeletal muscle.

E. Smooth muscle has thick and thin myofilaments, but they aren't arranged in a way that creates striations.

F. ATP, required for muscle contraction, can be generated by way of creatine phosphate breakdown and fermentation.

G. Lactic acid from fermentation causes an oxygen debt because oxygen is required to metabolize this product. Cellular respiration, an aerobic process, is the best source of ATP.

7.3 Muscle Responses

A. In the laboratory, muscles that are stimulated with a single threshold stimulus will contract in a twitch. The occurrence of a muscle twitch or tetanic contraction depends on the frequency with which a muscle is stimulated. Muscle fatigue occurs when ATP or acetylcholine are depleted, or if the muscle pH decreases because of lactic acid produced by fermentation.

B. In the body, muscle fibers belong to motor units. The strength of muscle contraction depends on the recruitment of motor units. A muscle has tone because some fibers are always contracting.

7.4 Skeletal Muscles of the Body
When muscles cooperate to achieve movement, some act as prime movers or agonists, others as synergists, and still others as antagonists.

7.5 Skeletal Muscle Groups
The skeletal muscles of the body are divided into those that move: the head and neck (see Table 7.2); the trunk (see Table 7.3); the shoulder and arm (see Table 7.4); the forearm (see Table 7.4); the hand and fingers (see Table 7.4); the thigh (see Table 7.5); the leg (see Table 7.5); and the ankle and foot (see Table 7.5).

7.6 Effects of Aging
As we age, muscles become weaker, but exercise can help retain vigor.

7.7 Homeostasis
Smooth muscle contraction helps move blood in the circulatory system and food in the digestive system, and allows the urinary bladder to empty. Cardiac muscle contraction pumps the blood. Skeletal muscle contraction produces heat and is needed for breathing.

Study Questions

1. Name and describe the three types of muscles, and give a general location for each type. (pp. 134–135)
2. List and discuss five functions of skeletal muscles. (p. 136)
3. Describe the anatomy of a muscle, from the whole muscle to the myofilaments within a sarcomere. Name the layers of fascia that cover a skeletal muscle and divide the muscle interior. (pp. 135–138)
4. List the sequential events that occur after a nerve signal reaches a muscle. (pp. 138–140)
5. How is ATP supplied to muscles? What is oxygen debt? (pp. 140–142)

6. What is the difference between a single muscle twitch, summation, and a tetanic contraction? (pp. 142–143)
7. What is the all-or-none law? What is muscle tone? How does muscle contraction affect muscle size? (pp. 143–145)
8. Describe how muscles are attached to bones. Define the terms *prime mover, synergist, agonist,* and *antagonist.* (p. 145)
9. How do muscles get their names? Give an example for each characteristic used in naming muscles. (p. 146)
10. Which of the muscles of the head are used for facial expression? Which are used for chewing? (pp. 147–148)

11. Which muscles of the neck flex and extend the head? (pp. 147–148)
12. What are the muscles of the thoracic wall? What are the muscles of the abdominal wall? (pp. 148–149)
13. Which of the muscles of the shoulder and upper limb move the arm and forearm, and what are their actions? Name the muscles that move the hand and fingers. (pp. 149–151)
14. Which of the muscles of the hip move the thigh, and what are their actions? Which of the muscles of the thigh move the leg, and what are their actions? Which of the muscles of the leg move the feet? (pp. 151–154)

Learning Outcome Questions

I. **Fill in the blanks.**
1. _____ muscle is uninucleated, nonstriated, and located in the walls of internal organs.
2. The fascia called _____ separates muscle fibers from one another within a fascicle.
3. When a muscle fiber contracts, an _____ myofilament slides past a myosin myofilament.
4. The energy molecule _____ is needed for muscle fiber contraction.
5. Whole muscles have _____, a condition in which some fibers are always contracted.
6. When muscles contract, the _____ does most of the work, but the _____ help.

7. The _____ is a muscle in the arm that has two origins.
8. The _____ acts as the origin of the latissimus dorsi, and the _____ acts as the insertion during most activities.

II. **For questions 9–12, name the muscle indicated by the combination of origin and insertion shown.**

Origin	Insertion
9. temporal bone	mandibular coronoid process
10. scapula, clavicle	humerus
11. scapula, proximal humerus	olecranon process of ulna
12. posterior ilium, sacrum	proximal femur

III. **Match the muscles in the key to the actions listed in questions 13–18.**

Key:
 a. orbicularis oculi
 b. zygomaticus
 c. deltoid
 d. serratus anterior
 e. rectus abdominis
 f. iliopsoas
 g. gluteus maximus
 h. gastrocnemius

13. Allows a person to stand on tiptoe
14. Tenses abdominal wall
15. Abducts arm
16. Flexes thigh
17. Raises corner of mouth
18. Closes eyes

Medical Terminology Exercise

After studying this chapter, see if you can derive the definitions for the medical terms listed at right. Many of the prefixes and suffixes used to create these terms can be found throughout the chapter. For additional help, use McGraw-Hill Connect™ at www.mcgrawhillconnect.com and consult Appendix B.

1. hyperkinesis (hī″pĕr-kĭ-nē′sĭs)
2. electromyogram (ē-lĕk″trō-mī′ō-grăm)
3. meniscectomy (mĕn″ĭ-sĕk′tō-mē)
4. tenorrhaphy (tĕ-nōr′ă-fē)
5. myatrophy (mī-ăt′rō-fē)
6. leiomyoma (lī″ō-mī-ō′mŭh)
7. kinesiotherapy (kĭ-nē″sē-ō-thĕr′ŭh-pē)
8. myocardiopathy (mī″ō-kăr″dē-ŏp′ŭh-thē)

Online Study Tools

APR

Anatomy & Physiology REVEALED includes cadaver photos that allow you to peel away layers of the human body to reveal structures beneath the surface. This program also includes animations, radiologic imaging, audio pronunciations, and practice quizzing. To learn more visit www.aprevealed.com

PART III

8 The Nervous System

Be careful about how quickly you eat that ice cream, or you'll get a "brain freeze"! The *trigeminal* nerve, one of twelve paired cranial nerves, is part of this phenomenon. Scientists believe that "brain freeze"—a brief, stabbing headache caused by very cold food or beverages—occurs because cold narrows blood vessels on the roof of the mouth and in the forehead. It's the job of the trigeminal nerve to react to cold, signaling brain blood vessels to widen and carry warm blood to the "frozen" area. However, the wider blood vessels stimulate pain nerves in nearby tissues. The response is the trigeminal nerve's way of telling you to slow down while you eat. You can read more about the trigeminal nerve on page 181.

Learning Outcomes

After you have studied this chapter, you should be able to:

8.1 Nervous System

1. Describe the three functions of the nervous system.
2. Describe the structure of a neuron and the functions of the three types of neurons.
3. Explain how a nerve signal is conducted along a nerve and across a synapse.

8.2 Central Nervous System

4. Describe the three layers of meninges, and state the functions of the meninges.
5. Describe the location and function of cerebrospinal fluid.
6. Describe in detail the structure of the spinal cord, and state its functions.
7. Outline the major parts of the brain and the lobes of the cerebral cortex. State functions for each structure.

8.3 Peripheral Nervous System

8. Describe the structure of a nerve, and distinguish between sensory, motor, and mixed nerves.

9. Name the twelve pairs of cranial nerves, and give a function for each.
10. Name several peripheral nerves, and describe the spinal nerves which combine to create each one. Explain the function of each of these peripheral nerves.
11. Describe the structure of a reflex arc and the function of a reflex action.
12. Define and describe the autonomic nervous system.
13. Distinguish between the sympathetic and parasympathetic divisions in four ways, and give examples of their respective effects on specific organs.

8.4 Effects of Aging

14. Describe the anatomical and physiological changes that occur in the nervous system as we age.

8.5 Homeostasis

15. Describe how the nervous system works with other systems of the body to maintain homeostasis.

Visual Focus

Synapse Structure and Function

Medical Focus

Research on Alzheimer Disease: Causes, Treatments, Prevention, and Hope for a Cure

What's New

Epidural Stimulation in Spinal Cord Injuries: Cause for Hope

Brain in a Petri Dish: A Human Model for Alzheimer Research

I.C.E. — In Case of Emergency

Traumatic Brain Injury

Visual Focus

Autonomic System Structure and Function

Human Systems Work Together

Nervous System

Medical Focus

Parkinson's Disease

8.1 Nervous System

1. Describe the three functions of the nervous system.
2. Describe the structure of a neuron and the functions of the three types of neurons.
3. Explain how a nerve signal is conducted along a nerve and across a synapse.

The nervous system has three specific functions:

1. *Sensory input.* Sensory receptors in skin and organs respond to external and internal stimuli by generating nerve signals that travel to the brain and spinal cord. For example, temperature sensors in the skin may signal to the brain that the air surrounding the body is cold.
2. *Integration.* The brain and spinal cord interpret the data received from sensory receptors all over the body, and signal the appropriate nerve responses. To continue the example of body temperature, sensory information from temperature receptors is sent to the *hypothalamus,* the brain center that controls body temperature.
3. *Motor output.* The nerve signals from the brain and spinal cord go to the *effectors,* which are muscles, glands, and organs—in other words, structures that will have an effect.

Muscle contractions, gland secretions, and changes in organ function are responses to stimuli received by sensory receptors. For example, when adjusting body temperature, the hypothalamus triggers shivering—skeletal muscles contract rhythmically, producing heat that warms the body.

It is important to stress that the nervous system maintains homeostasis by receiving sensory information, integrating that information, and making an appropriate response.

Divisions of the Nervous System

The nervous system has two major divisions: the central nervous system and the peripheral nervous system (Fig. 8.1). The **central nervous system (CNS)** includes the brain and spinal cord, which have a *central* location—they lie in the midline of the body. The **peripheral nervous system (PNS),** which is further divided into the *afferent* (sensory) and *efferent* (motor) divisions, includes all the cranial and spinal nerves. Nerves have a *peripheral* location in the body, meaning that they project out from the central nervous system. The division between the central nervous system and the peripheral nervous system is arbitrary; the two systems work together, as we'll see.

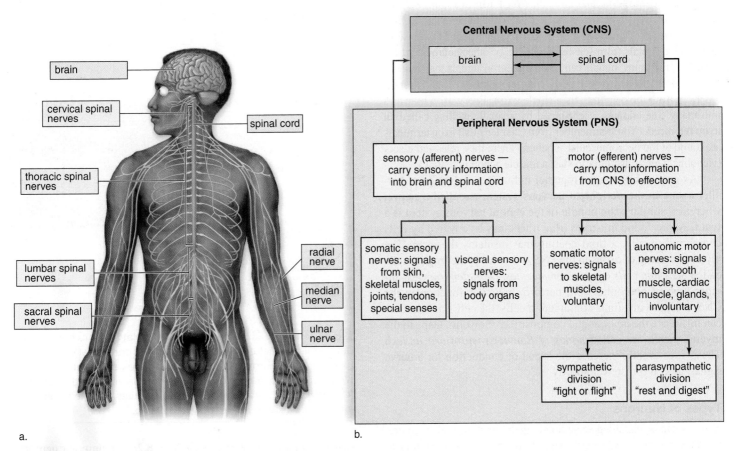

a.   b.

AP|R **Figure 8.1 Organization of the nervous system in humans. (a)** This pictorial representation shows the central nervous system (CNS; composed of brain and spinal cord) and some of the nerves of the peripheral nervous system (PNS). **(b)** The CNS and PNS communicate with each other. Somatic sensory and visceral sensory nerves carry information to the brain and spinal cord. Somatic motor nerves signal skeletal muscles and autonomic motor nerves signal smooth muscle, cardiac muscle, and glands.

Afferent nerves come from two sources in the body. Somatic sensory nerves relay signals from the body surface (skin and special sensory structures) as well as from skeletal muscles, tendons, and joints. A good example of a somatic sensation might be a gentle touch to your hand. By contrast, visceral sensations come from organs within the body. The stretch sensation transmitted to the brain from a filled stomach is a visceral sensation. Likewise, efferent nerves have two general destinations. Somatic efferents innervate skeletal muscle. Autonomic efferents signal structures under involuntary control, such as the heart and digestive organs. There are two divisions to the autonomic nervous system: sympathetic and parasympathetic. These will be detailed in section 8.3.

Nervous Tissue

Although exceedingly complex, nervous tissue is made up of just two principal types of cells: (1) **neurons,** also called nerve cells, which generate and transmit nerve signals; and (2) **neuroglia,** which nourish and support neurons (see Chapter 4, pp. 77–78).

Neuron Structure

Neurons vary in appearance, but all of them have just three parts: a cell body, dendrite(s), and an axon. As shown in Figure 8.2, the **cell body** of each type of neuron contains the nucleus, as well as other organelles. **Dendrites** are shorter, highly branched extensions that receive signals from sensory receptors or other neurons. At the dendrites, incoming signals can result in neuron signals that are conducted toward the neuron cell body. Neuron signals are then transmitted away from the cell body by a single **axon.** Every axon branches into many fine endings, each tipped by a small swelling called an **axon terminal.** (You'll remember from Chapter 7 that axon terminals are found at the neuromuscular junction. There they release the acetylcholine neurotransmitter to start a muscle action potential.)

Axons can be grouped together in bundles. A bundle of parallel axons in the peripheral nervous system is called a **nerve,** whereas a similar axon bundle in the central nervous system is a **tract.** Axons found in nerves or in tracts may be covered by cells containing myelin, a lipid coating that insulates the axon. The myelin covering of axons in the PNS is formed by neuroglial cells called *Schwann cells* or *neurolemmocytes. Oligodendrocytes,* another type of neuroglial cell, perform a similar function in the CNS. Because myelin is contained in cells, the cell's metabolism can influence the amount and composition of myelin. Gaps in the myelin sheath are called *nodes of Ranvier (neurofibril nodes).* These gaps greatly increase the speed of conduction for a nerve signal, as you'll discover shortly.

Types of Neurons

Neurons can be classified according to their function and structure. **Sensory neurons** take nerve signals from sensory receptors to the CNS. The sensory receptor, which is the distal end of the long axon of a sensory neuron, may be as simple as a naked nerve ending (a pain receptor), or it may be a part of a highly complex organ, such as the eye or ear. Almost all sensory neurons have a

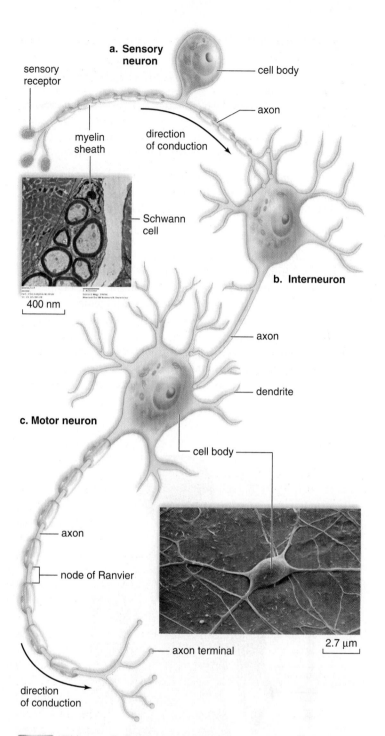

AP|R **Figure 8.2** **Neuron anatomy. (a)** Sensory neuron with dendritelike structures projecting from the peripheral end of the axon. **(b)** Interneuron (from the cortex of the cerebellum) with highly branched dendrites. Note that the myelin sheath is absent. **(c)** Motor neuron. Note the branched dendrites and the single, long axon, which branches only near its tip.

structure that is termed *unipolar* (Fig. 8.2*a*). In unipolar neurons, the extension from the cell body divides into a branch that comes to the periphery and another that goes to the CNS. Because both branches are long and transmit nerve signals, it's now generally accepted to refer to them collectively as an axon.

Interneurons, also known as association neurons, occur entirely within the CNS. These are the most common type of neuron: scientists estimate that 99% of all neurons are interneurons. Most interneurons are said to be *multipolar* because they generally have many dendrites and a single axon (Fig. 8.2b). These neurons convey nerve signals between various parts of the CNS. Some lie between sensory neurons and motor neurons, and some take messages from one side of the spinal cord to the other or from the brain to the cord, and vice versa. They also form complex pathways in the brain where processes accounting for thinking, memory, and language occur.

Motor neurons are multipolar neurons that take nerve signals from the CNS to muscles, organs, or glands (Fig. 8.2c). Motor neurons cause muscle fibers to contract, organs to modify their function, or glands to secrete, and therefore they are said to *innervate* these structures.

Nerve Signal Conduction

The function of neurons is to conduct nerve signals and activate the release of neurotransmitters, which are chemical messengers that allow neurons to communicate with other cells and transmit information. Although resting neurons aren't sending nerve signals, they are constantly prepared to do so. A nerve signal, also called an *action potential,* is conducted by the axon of the neuron.

Resting Potential

Anyone who has ever used a battery has employed an energy source that's manufactured by separating positively charged ions across a membrane from negative ions. The battery's *potential energy* can be used to perform work: lighting a flashlight, for example. When a neuron is resting, it too possesses potential energy, much like a fully charged battery. This energy, called the **resting potential,** exists because the cell membrane is *polarized:* positively charged in the space immediately outside the cell membrane, and negatively charged in a small area just inside the cell membrane. The cell's outside perimeter is positive because positively charged sodium (Na^+) ions cluster around the cell. There is a concentration gradient for sodium ions, and the ions can diffuse into the cell through protein molecules called *sodium channels.* (A cell channel is like a door that remains open all the time.) However, the cell membrane is relatively impermeable to sodium ion diffusion when the cell is at resting potential. In addition, positively charged potassium ions are concentrated just inside the perimeter of the cell membrane, and the cell membrane is permeable to potassium. Because of this concentration gradient, potassium ions are constantly diffusing out of the cell, through their own set of protein channels. There, these ions contribute to the extracellular positive charge. The inside of the cell is negatively charged because of the presence of large, negatively charged proteins and other molecules that are trapped inside the cell because of their large size.

It's important to note, though, that the charged areas on either side of the cell membrane that are responsible for creating the resting potential are very small. Most of the intracellular and extracellular fluid is electrically neutral—the positive and negative charges are perfectly balanced.

Like a battery, the neuron's resting potential energy can be measured in volts. A D-size flashlight battery has 1.5 volts; a nerve cell typically has 0.07 volts (V), or 70 millivolts (mV), of stored energy (Fig. 8.3a). By convention, the cell's voltage measurement is assigned a negative value. This is because scientists compare the inside of the cell (where negatively charged proteins and other large molecules cluster) to the outside of the cell (where positively charged sodium ions and some potassium ions are gathered). Neurons, like rechargeable batteries, must maintain their resting potential to work effectively. To do so, neurons continually transport sodium ions out of the cell and return potassium to the cytoplasm. A protein carrier in the membrane, called the **sodium-potassium pump,** actively transports sodium ions (Na^+) out of the neuron and potassium ions (K^+) into the neuron. However, the sodium-potassium pump doesn't perform an even exchange. Rather, for every three sodium ions pumped out of the cell, only two potassium ions are returned to the cell. Because potassium diffuses out of the cell faster than sodium diffuses in, and the pump returns fewer potassium ions to the cell, the pump allows the area immediately outside of the cell to keep its resting positive charge. Thus, the sodium-potassium pump constantly "recharges" the cell, keeping the area just inside the cell negatively charged and the outside perimeter positively charged. Like a freshly charged battery, the cell is always ready to perform its work (provided that there is sufficient ATP energy for the sodium-potassium pump's active transport).

Action Potential

The resting potential energy of the neuron can be used to perform the work of the neuron: conduction of nerve signals. The process of conduction is termed an **action potential,** and it occurs in the axons of neurons. An action potential begins with a **stimulus,** which activates the neuron. (For example, poking the skin with a sharp pin would be a stimulus for pain neurons in the skin.) The stimulus affects specific protein channels, called *voltage-regulated sodium gates.* These sodium gates are different from sodium channels because they are not constantly open; rather, voltage changes cause them to open and close. Once sodium gates have opened, sodium ions rush into the cell. Adding positively charged sodium ions causes the inside of the axon to become positive, compared to the outside (Fig. 8.3b).

If sufficient numbers of sodium ions pass into the cell, the cell's voltage will reach a new, higher voltage, called the cell's **threshold.** Once this voltage is reached, increasing numbers of voltage-regulated sodium gates will open and the cell potential will abruptly rise. The resulting change is called **depolarization.** Depolarization continues until the intracelluar voltage reaches a value of approximately +35 mV.

Once the change in voltage is complete, the voltage-regulated sodium gates close and a separate set of voltage-regulated potassium gates opens. Potassium rapidly leaves the cell (Fig. 8.3c). As positively charged potassium ions exit the cell, the inside of the cell becomes negative again (due to the presence of large, negatively charged ions trapped inside the cell). This change in polarity is called **repolarization** (Fig. 8.3d). Once the action potential is complete, the sodium and potassium ions are rapidly restored to their proper place through the action of the sodium-potassium pump.

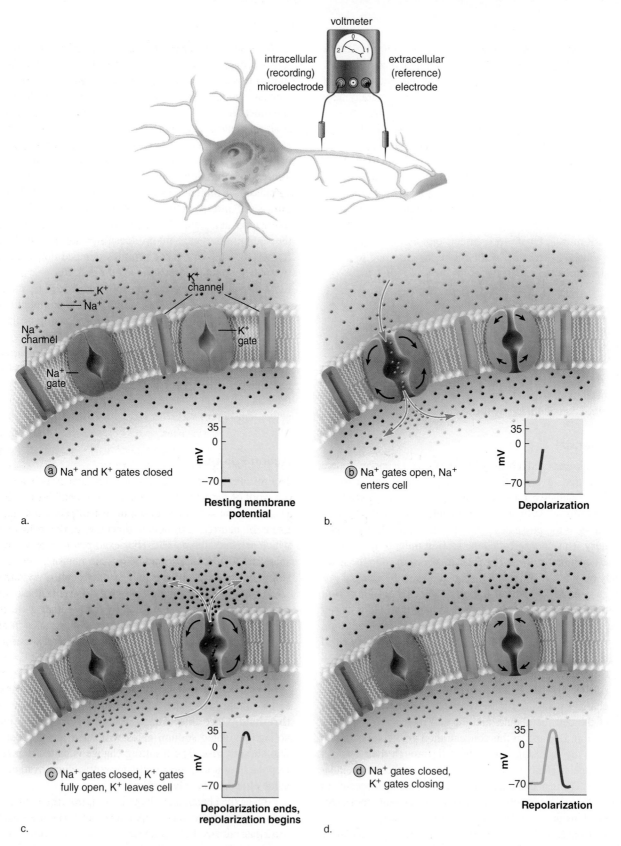

AP|R **Figure 8.3** **Resting potential and action potential in an unmyelinated axon.** (a) Resting potential. Sodium ions (red dots) are concentrated outside the cell; potassium (blue dots) and large anions are inside the cell. Potential is approximately 70 mV; the inside of the cell is negative compared to the outside. (b) Action potential: depolarization. Na⁺ gates open and sodium ions flow into cell; the inside of the cell becomes positive. (c) Action potential: repolarization. K⁺ gates open and potassium ions flow out. (d) Repolarization: Both Na⁺ and K⁺ gates are closed. When the sodium-potassium pump restores the ions to their places, repolarization is complete.

Labels within figure:

- K⁺ channel
- K⁺
- Na⁺
- Na⁺ channel
- Na⁺ gate
- K⁺ gate

voltmeter

intracellular (recording) microelectrode

extracellular (reference) electrode

(a) Na⁺ and K⁺ gates closed — **Resting membrane potential** — a.

(b) Na⁺ gates open, Na⁺ enters cell — **Depolarization** — b.

(c) Na⁺ gates closed, K⁺ gates fully open, K⁺ leaves cell — **Depolarization ends, repolarization begins** — c.

(d) Na⁺ gates closed, K⁺ gates closing — **Repolarization** — d.

Conduction of Action Potentials

For a neuron to transmit its signals, it's important that action potentials be conducted, or *propagated,* along the entire length of the axon. As depolarization proceeds, each section of the axon propagates its action potential to the neighboring area when positively charged sodium ions inside the cell diffuse into the more distal section. There, these sodium ions will raise the resting potential in that area to threshold, and another action potential will follow. Conduction of the action potential down the axon's length can be compared to pushing the first domino in a row of dominoes—push the first down, and the others follow one after another. The conduction of an action potential is an all-or-none event—that is, either an axon conducts its action potential or it doesn't. (However, if the neuron's voltage does not reach the threshold value, the axon will not generate an action potential.) Then, once they occur, action potentials don't vary in their intensity. The intensity of a message is determined by how many action potentials are generated within a given time span. A neuron can conduct a volley of action potentials very quickly because only a small number of ions is exchanged with each action potential.

As soon as the action potential has passed by each successive portion of an axon, the sodium and potassium ion concentrations for that section are reversed, and the ions must be returned to their correct places by the sodium-potassium pump. The time needed to pump sodium outside the cell and potassium inside the cell is called the **refractory period.** During this short interval, the proximal axon section is unable to conduct an action potential, so the signal can't travel in reverse. This ensures the one-way direction of a signal, from the cell body down the length of the axon to the axon terminal.

If an axon is unmyelinated, conduction along the entire axon in this fashion can be rather slow (approximately 1 meter/second in thin axons) because each section of the axon must be stimulated. In myelinated fibers, an action potential at one node of Ranvier opens a large number of Na^+ gates, allowing rapid diffusion of sodium through the myelinated portion of the axon. When sodium ions arrive at the next node, they trigger another action potential, which will trigger an action potential at the next node, and so on down the entire length of the myelinated axon (Fig. 8.4). This type of conduction, called **saltatory conduction,** is much faster: In thick, myelinated fibers, the rate is more than 100 m/s.

It's interesting to observe that all functions of the nervous system, from our deepest emotions to our highest reasoning abilities, are dependent on our ability to start and conduct nerve signals and release neurotransmitters.

Transmission Across a Synapse

As you know, every axon branches into many tiny endings called axon terminals. Each axon terminal lies very close to another cell, often another neuron. This region of close proximity is called a **synapse** (Fig. 8.5). At a synapse between two neurons, the membrane of the first neuron is called the *presynaptic* membrane, and the membrane of the next neuron is called the *postsynaptic* membrane. The small gap between is the **synaptic cleft.** The neuromuscular junction, which you studied in Chapter 7, is a synapse between motor nerves and muscle cells.

Transmission across a synapse is carried out by molecules called **neurotransmitters,** which are stored in synaptic vesicles (small membranous sacs) in the axon terminals. When nerve signals traveling along an axon reach an axon terminal, channels for calcium ions (Ca^{2+}) open, and calcium enters the terminal. This sudden rise in Ca^{2+} stimulates synaptic vesicles to merge with the presynaptic membrane, and neurotransmitter molecules are released into the synaptic cleft. This process is called *exocytosis* (see Chapter 3). Neurotransmitter molecules then diffuse across the cleft to the postsynaptic membrane. There, these molecules bind with specific receptor proteins on *ligand-regulated gates*. As you know, a gate is a form of specialized channel which doesn't remain constantly open. Instead, the neurotransmitter acts as the key—the *ligand*—which opens the gate. In the absence of the ligand, the gate remains closed.

Depending on the type of neurotransmitter ligand and the type of receptor, the response of the postsynaptic neuron can be toward excitation or toward inhibition. For example, after excitatory neurotransmitters combine with a receptor, a sodium ion gate opens, and Na^+ enters the neuron (Fig. 8.5). Other neurotransmitters have an inhibitory effect, as described in the next section.

Graded Potentials and Synaptic Integration

A single neuron can have many dendrites plus the cell body, and both can synapse with many other neurons. Typically, a neuron is on the receiving end of many excitatory and inhibitory synapses (Fig. 8.5). Each of the small signals from a synapse is called a **graded potential.** An excitatory neurotransmitter combines with receptors on the neuron's sodium gates and the gates open, producing a graded potential that drives the polarity of a neuron closer to the threshold for an action potential. An inhibitory neurotransmitter produces a graded potential that makes it harder for a neuron to have an action potential. For example, inhibitory signals can be created by neurotransmitters that attach to receptors on potassium gates. When these gates open, potassium ions can leave the cell. Opening other gates that allow negatively charged chloride ions (Cl^-) to enter the cell will also create an inhibitory signal. Either of these changes—potassium ions leaving the cell, or chloride ions entering—will cause the cell's potential to drift further from threshold and thus, the start of an action potential.

Neurons integrate these incoming signals. Integration is the summing up of excitatory and inhibitory signals. If a neuron receives many excitatory signals (either from different synapses, or at a rapid rate from one synapse), chances are the cell's axon will transmit an action potential. On the other hand, if a neuron receives both inhibitory and excitatory signals, the summing up of these signals may prohibit the axon from firing.

Sensory receptors (for example, light receptors in the eye) have special graded potentials called **receptor potentials.** These will be discussed in Chapter 9.

Neurotransmitter Molecules

At least 100 different neurotransmitters have been identified, and more will likely be identified in the future. Two very well-known ones are **acetylcholine (ACh)** and **norepinephrine (NE).**

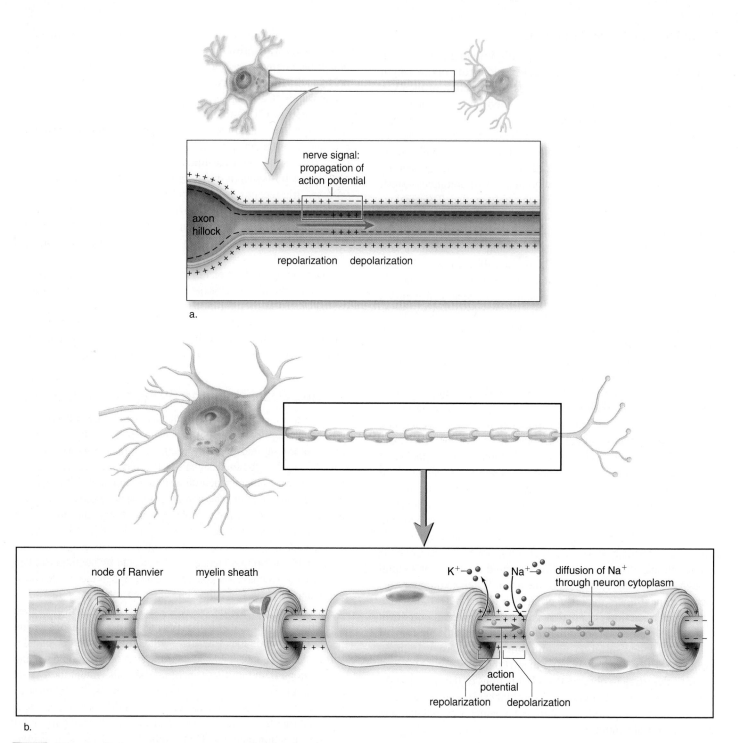

AP|R Figure 8.4 **Conduction of an action potential in unmyelinated and myelinated axons.** (**a**) In an unmyelinated axon, conduction is slow. (**b**) In a myelinated axon, the action potential quickly travels from one node of Ranvier to the next, and the speed of conduction is much more rapid. Almost all axons are myelinated in humans.

Once a neurotransmitter has been released into a synaptic cleft and has initiated a response, it is removed from the cleft. In some synapses, the postsynaptic membrane contains enzymes that rapidly inactivate the neurotransmitter. For example, the enzyme **acetylcholinesterase (AChE)** breaks down acetylcholine. In other synapses, the presynaptic membrane rapidly reabsorbs the neurotransmitters, possibly for repackaging in synaptic vesicles or for

molecular breakdown. The short existence of neurotransmitters at a synapse prevents continuous stimulation (or inhibition) of postsynaptic membranes.

The What's New feature, Research on Alzheimer Disease, discusses the disease, which is due in part to a lack of ACh in the brain. It is also of interest to note that many available drugs enhance or block the release of a neurotransmitter, mimic the

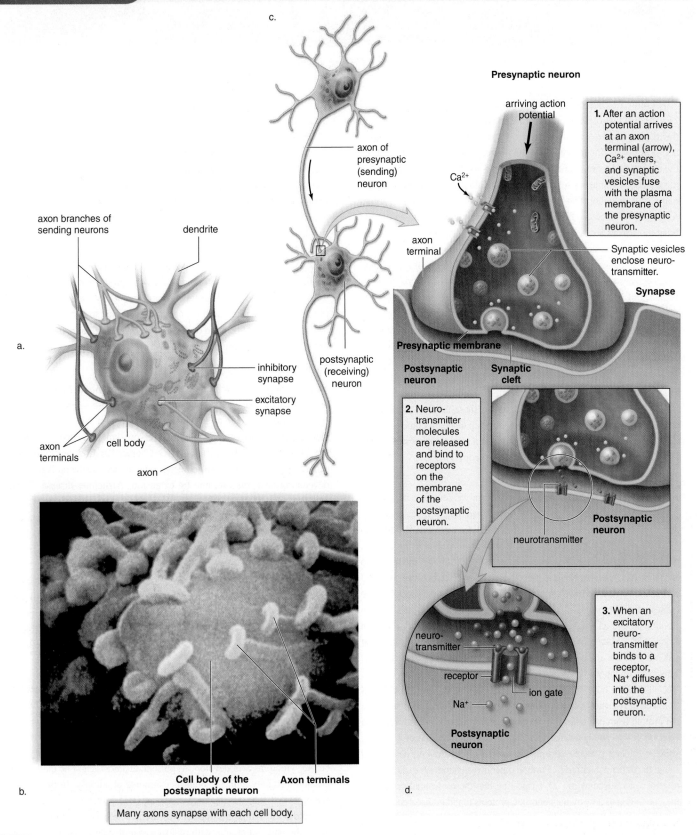

c.

Presynaptic neuron

arriving action potential

Ca²⁺

1. After an action potential arrives at an axon terminal (arrow), Ca²⁺ enters, and synaptic vesicles fuse with the plasma membrane of the presynaptic neuron.

axon of presynaptic (sending) neuron

axon branches of sending neurons

dendrite

axon terminal

Synaptic vesicles enclose neuro-transmitter.

Synapse

a.

inhibitory synapse

postsynaptic (receiving) neuron

Presynaptic membrane

Postsynaptic neuron

Synaptic cleft

excitatory synapse

axon terminals

cell body

axon

2. Neuro-transmitter molecules are released and bind to receptors on the membrane of the postsynaptic neuron.

Postsynaptic neuron

neurotransmitter

neuro-transmitter

3. When an excitatory neuro-transmitter binds to a receptor, Na⁺ diffuses into the postsynaptic neuron.

receptor

ion gate

Na⁺

Postsynaptic neuron

Cell body of the postsynaptic neuron

Axon terminals

b.

d.

Many axons synapse with each cell body.

AP|R **Figure 8.5 Synapse structure and function. (a)** A single nerve cell body may have many excitatory synapses (green) as well as inhibitory synapses (red). **(b)** Photomicrograph of a nerve cell body, showing multiple axon terminals that form synapses. **(c)** The sending neuron is called the presynaptic neuron, and the receiving neuron is the postsynaptic neuron. **(d)** Transmission across a synapse from one neuron to another occurs when a neurotransmitter is released at the presynaptic membrane, diffuses across a synaptic cleft, and binds to the receptors on the gates found in the postsynaptic membrane. Because the gates shown in the illustration are Na⁺ gates, the synapse is excitatory.

Alzheimer disease (AD) is an irreversible, fatal disorder characterized by a gradual loss of reason that begins with memory lapses and ends with the inability to perform any activities. Personality changes such agitation and hostility, and memory deficits that affect daily routines often signal the onset of AD. For example, a normal 60- to 70-year-old might forget the name of a friend not seen for years, but someone with AD forgets the name of a neighbor who visits daily. Likewise, a healthy senior might forget where he placed his car keys, while a person with AD will forget what those keys are for. People afflicted with AD become confused and tend to repeat the same question. Signs of mental disturbance eventually appear, and patients gradually become bedridden and die of a complication, such as pneumonia. At the cellular level, AD is characterized by the presence of abnormally structured neurons and a reduced amount of the neurotransmitter acetylcholine (see pp. 167–168). These defective neurons are especially seen in the portions of the brain involved in reason and memory. The AD neuron has two pathological features. The first consists of bundles of fibrous protein, called neurofibrillary tangles, which surround the nucleus in the cells. The tangles are due to an abnormal form of *tau*, a protein molecule that normally helps to stabilize microtubules that form the cell's cytoskeleton. In addition to these tangles, protein-rich accumulations, called amyloid plaques, envelop the axon branches. Over time, affected neurons will die. The cortex and hippocampus shrivel, the brain shrinks in volume, and the ventricles become enlarged.

Research Regarding Its Causes

As techniques for genetic study continue to improve, several genetic mutations specific to Alzheimer have been identified. One set, designated by the acronyms APP, PS1, and PS2, are termed *deterministic*. People who inherit one of these three mutated genes will always develop the disease, called autosomal dominant Alzheimer disease (ADAD). It's interesting to note that APP, the first of these defective genes to be discovered, is found on chromosome 21. Down syndrome results from the inheritance of three copies of chromosome 21, and people with Down syndrome tend to develop AD. (You will learn more about autosomal dominant disorders in Chapter 19.) Mutation of a fourth gene, designated APO, puts patients at risk but does not always result in disease. Scientists are now studying victims with mutations to try to discover the exact cause for the disease. Recent findings have led researchers to believe that the neuron deterioration seen in Alzheimer disease patients may be caused by the spread of the tau protein from one cell to the next, much as a virus is spread from one infected cell to another. Other studies have implicated a second protein, striatal-enriched tyrosine phosphatase, or STEP, in the cell destruction found in Alzheimer sufferers. Further, other investigators are exploring the role of cell lysosomes in AD, suspecting that these essential organelles may be failing to destroy the abnormal proteins found in diseased cells.

Research into Its Treatment

Each new finding about what causes Alzheimer disease creates new possibilities for its treatment as well. Researchers are now conducting clinical testing on antibodies that block cell-to-cell transmission of the tau protein. (You can read more about antibodies in Chapter 13.) A second treatment might involve the creation of drugs that block formation of the STEP protein. Boosting lysosomal degradation of the abnormal

Alzheimer cell protein is another possible avenue for research. At this time, only five drugs are accepted for disease treatment. One category, cholinesterase inhibitors (Aricept®, Razadyne®, Exelon®, Reminyl®), works at neuron synapses in the brain, slowing the activity of the enzyme that breaks down acetylcholine (ACh). Allowing ACh to accumulate in synapses keeps memory pathways in the brain functional for a longer period of time. The newest drug, memantine (Namenda®), blocks excitotoxicity: the tendency of diseased neurons to self-destruct. This medication is used only in moderately to severely affected patients. Using the drug allows neurons involved in memory pathways to survive longer in affected patients. However, it's important to note that neither category of medication cures AD. Both merely slow the progress of disease symptoms, allowing the patient to function independently for a longer period of time. Additional research is currently underway to test the effectiveness of anticholesterol statin drugs, as well as anti-inflammatory medications, in slowing the progress of the disease.

Research on Prevention

Much of current research on AD focuses on prevention. Early findings have shown that risk factors for cardiovascular disease—heart attacks and stroke—also contribute to an increased incidence of AD. These include elevated blood cholesterol and blood pressure, smoking, obesity, sedentary lifestyle, and diabetes mellitus (see Chapter 12). Low-level infection caused by gum disease has also been shown to increase the probability of developing heart disease, and by extension, Alzheimer disease. Thus, evidence suggests that a lifestyle tailored for good cardiovascular health may also prevent AD. Slight changes in diet may also lessen the threat of developing AD: boosting vitamins B and D, eating fatty fish such as salmon, and drinking coffee. Further, younger people must try to prevent blows to the head. It's been shown that head injuries (such as those experienced by football players) can increase the risk of developing AD in later life 19-fold. Wearing seat belts and helmets and taking steps to prevent falls are commonsense, easy ways to prevent head injury. Finally, staying mentally, physically, and socially active—as long as possible—will help to slow the course of mental impairment for AD sufferers.

Early Detection and Hope for a Cure

Currently, researchers are testing vaccines for AD that would target the patient's immune system to destroy amyloid protein. Early study results show some promising outcomes of this treatment in early-stage patients. However, scientists believe that curing AD will require an early diagnosis because it's thought that the disease may begin in the brain 15 to 20 years before symptoms ever develop. At present, diagnosis can't be made with absolute certainty until the brain is examined at autopsy. In the future, cerebrospinal fluid testing may allow amyloid proteins to be detected before disease symptoms appear. Researchers are also developing ways to tag the amyloid protein with radioactive molecules, which will allow detection of the protein using a PET scan. (You learned about PET and other imaging techniques in Chapter 1.) The Medical Focus reading in Chapter 9 describes an eye scan technique that might allow an earlier diagnosis, and the What's New reading on page 178 describes an exciting breakthrough in cell culture that will create new options for studying neurons and drug therapies in the laboratory.

neurotransmitter's action, block the ligand-gated receptor, or interfere with the removal of a neurotransmitter from a synaptic cleft.

Content CHECK-UP!

1. Which of the following types of neurons are multipolar?

   a. motor neurons

   b. sensory neurons

   c. interneurons

   d. both motor neurons and interneurons

   e. motor neurons, sensory neurons, and interneurons

2. During an action potential, depolarization occurs because:

   a. voltage-regulated gates for sodium ions in the cell membrane open, and sodium flows into the cell.

   b. voltage-regulated gates for potassium ions in the cell membrane open, and potassium leaves the cell.

   c. the sodium-potassium pump returns sodium to the outside of the cell and potassium to the inside.

3. An excitatory neurotransmitter excites the postsynaptic cell because:

   a. it opens sodium gates so sodium can enter the cell.

   b. it opens potassium gates so potassium can leave the cell.

   c. it turns on the sodium-potassium pump.

4. Imagine you've invented a new medication, and it blocks the receptor for potassium gates at a neuromuscular junction. Would the muscle be more likely or less likely to contract after this drug is administered?

Answers in Appendix A.

8.2 Central Nervous System

4. Describe the three layers of meninges, and state the functions of the meninges.
5. Describe the location and function of cerebrospinal fluid.
6. Describe in detail the structure of the spinal cord, and state its functions.
7. Outline the major parts of the brain and the lobes of the cerebral cortex. State functions for each structure.

The CNS, consisting of the brain and spinal cord, is composed of gray matter and white matter. **Gray matter** is gray because it contains cell bodies and short, nonmyelinated fibers. **White matter** is white because it contains myelinated axons that run together in bundles called tracts. The myelin covering on these axons gives them a shiny, white appearance.

Meninges and Cerebrospinal Fluid

Both the spinal cord and the brain are wrapped in protective membranes known as **meninges** (sing., *meninx*) (Fig. 8.6). The outer meninx, the **dura mater,** is tough, white, fibrous connective tissue. A small space filled with adipose tissue, called the *epidural space*, lies between the dura mater and the skull and vertebra (Fig. 8.7b). The dura mater is constructed of two separate membrane layers that are fused. However, in several areas the layers separate to form the **dural venous sinuses.** The dural venous sinuses collect venous blood and excess cerebrospinal fluid, returning both to the cardiovascular system. Deep to the dura mater is the **arachnoid mater** ("spider-like"), so called because it consists of spider-web-like connective tissue. Thin strands of the arachnoid mater

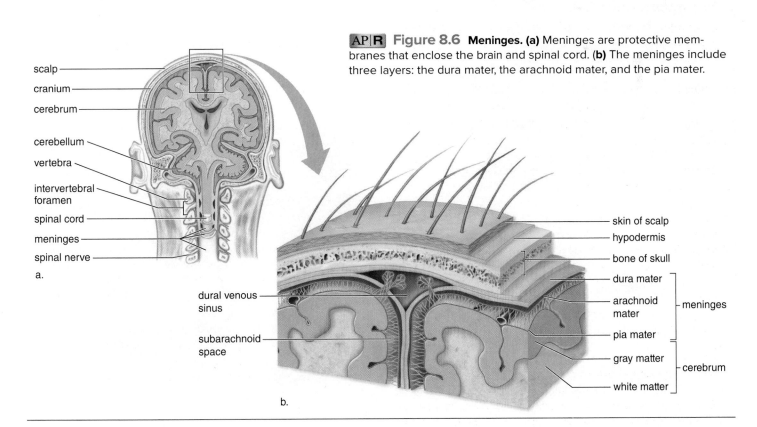

AP|R **Figure 8.6** **Meninges. (a)** Meninges are protective membranes that enclose the brain and spinal cord. **(b)** The meninges include three layers: the dura mater, the arachnoid mater, and the pia mater.

scalp
cranium
cerebrum
cerebellum
vertebra
intervertebral foramen
spinal cord
meninges
spinal nerve

a.

dural venous sinus
subarachnoid space

b.

skin of scalp
hypodermis
bone of skull
dura mater
arachnoid mater
pia mater
gray matter
white matter
meninges
cerebrum

attach it to the delicate **pia mater,** the deepest meninx. The pia mater directly attaches to the brain and spinal cord, and closely follows their contours. Between the arachnoid mater and the pia mater is the *subarachnoid space*, which is filled with **cerebrospinal fluid (CSF).** This clear tissue fluid forms a protective cushion around and within the CNS. Though formed from blood plasma, CSF differs in chemical composition. For example, it has lower concentrations of protein and glucose, as well as a lower pH.

CSF is formed by the *choroid plexus,* a tissue created by folds of pia mater, lined with a dense layer of capillaries. Choroid plexus can be found in the hollow interconnecting cavities of the brain called **ventricles.** Cerebrospinal fluid fills brain ventricles as well as the **central canal** of the spinal cord, and any excess CSF is drained into the dural venous sinuses to return to the cardiovascular system

(Fig. 8.6*b*). However, blockages can occur. In an infant, the ventricles can enlarge due to cerebrospinal fluid accumulation. As a result, the skull enlarges and the brain tissue between the ventricles and skull will be compressed, sometimes resulting in brain damage. This condition, called **hydrocephalus** ("water on the brain"), can sometimes be corrected surgically.

The Spinal Cord

The **spinal cord** (Fig. 8.7*a*) is a cylinder of nervous tissue that begins at the base of the brain and extends through a large opening in the skull called the **foramen magnum** (see Chapter 6, Fig. 6.7). The spinal cord is protected by the vertebral column, which is composed of individual vertebrae. When the vertebrae are stacked

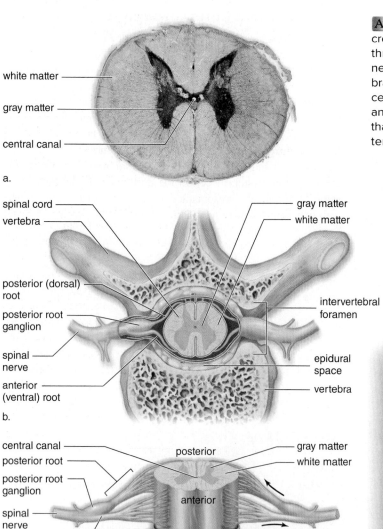

a.

b.

c.

AP|R Figure 8.7 **Spinal cord. (a)** Photomicrograph of a cross section of the spinal cord. **(b)** The spinal cord passes through the vertebral canal formed by the vertebrae. Spinal nerves branch off the spinal cord and project through intervertebral foramina. **(c)** The spinal cord has a central canal filled with cerebrospinal fluid, gray matter in an H-shaped configuration, and white matter elsewhere. The white matter contains tracts that take nerve signals to and from the brain. **(d)** Photo of posterior view of spinal cord and posterior roots of spinal nerve.

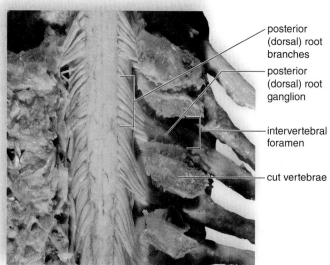

d.

on top of each other, their vertebral foramina form the vertebral canal. The spinal cord passes through the vertebral canal, ending at approximately the first lumbar vertebra (see Fig. 8.12).

Structure of the Spinal Cord

Figure 8.7*b* shows how an individual vertebra protects the spinal cord. The spinal nerves extend from the cord between the vertebrae. Intervertebral disks, which are composed of tough fibrocartilage and filled with gelatinous material, separate each vertebra. If a disk is torn open, this **herniated disk** may compress spinal nerves, causing pain and loss of function.

A cross section of the spinal cord shows a central canal, gray matter, and white matter (Fig. 8.7*a,b,c*). Both the central canal and the subarachnoid space (between the dura mater and arachnoid mater that cover the spinal cord) contain cerebrospinal fluid. The gray matter is centrally located and shaped like the letter H. Portions of sensory neurons and motor neurons are found there, as are interneurons that communicate with these two types of neurons. The posterior (dorsal) root of a spinal nerve contains sensory fibers entering the gray matter, and the cell bodies of these sensory nerves can be found in an enlarged area called the **posterior root ganglion.** The anterior (ventral) root of a spinal nerve contains motor fibers exiting the gray matter. The posterior and anterior roots join, forming a spinal nerve that leaves the vertebral canal through the intervertebral foramen. Spinal nerves are a part of the PNS.

The white matter of the spinal cord contains *ascending tracts,* which take sensory information to the spinal cord and brain, and *descending tracts,* which take motor information from the brain. Ascending tracts are generally located in the posterior white matter; descending tracts are found in the anterior white matter. Because the tracts typically cross just after they enter and exit the brain, the left brain controls the right side of the body and the right brain controls the left side of the body.

Functions of the Spinal Cord

The spinal cord provides a means of communication between the brain and the peripheral nerves that leave the cord.

When someone touches your hand, sensory receptors generate action potentials that travel by way of sensory nerve axons to the spinal cord. One of several ascending tracts next carries the information to the sensory area of the brain. When you voluntarily move your limbs, action potentials originating in the motor control area of the brain pass down one of several descending tracts to the spinal cord and out to your muscles by way of motor nerve axons. The What's New on page 178 discusses promising new therapies for patients whose spinal cord is injured.

We'll see that the spinal cord is also the center for thousands of reflex arcs (see Fig. 8.13): A stimulus causes sensory receptors to generate action potentials that travel in sensory neurons to the spinal cord. Interneurons integrate the incoming data and relay signals to motor neurons. A response to the stimulus occurs when motor axons cause skeletal muscles to contract. Each interneuron in the spinal cord has synapses with many other neurons, and

therefore they send signals to several other interneurons (as well as to the brain) in addition to motor neurons.

The Brain

The human brain consists of four major structures: the cerebrum, the diencephalon, the cerebellum, and the brain stem. The brain's four ventricles can be found within these brain structures. The paired lateral ventricles are enclosed by the cerebrum, the third ventricle lies in the center of the diencephalon, and the fourth ventricle travels within the brain stem, just anterior to the cerebellum (Fig. 8.8*a*).

The electrical activity of the brain can be recorded in the form of an **electroencephalogram (EEG).** Electrodes are taped to different parts of the scalp, and an instrument records the so-called brain waves. The EEG is a diagnostic tool; for example, an irregular pattern can signify epilepsy or a brain tumor. Absence of electrical activity on an EEG signifies brain death.

The Cerebrum

Let's begin our study of the brain with the **cerebrum,** the largest portion of the brain in humans. The cerebrum is the last center to receive sensory input and carry out integration before commanding voluntary motor responses. It communicates with and coordinates the activities of the other parts of the brain. The cerebrum carries out the higher thought processes required for learning and memory and for language and speech.

The cerebrum has two halves called the left and right **cerebral hemispheres** (Fig. 8.9). A lateral ventricle can be found inside each hemisphere. A deep groove, the longitudinal *fissure,* separates the left and right cerebral hemispheres. Still, the two cerebral hemispheres are connected by a bridge of white matter called the **corpus callosum.**

Ridges called *gyri* are separated by shallow grooves called *sulci* (sing., *sulcus*). Specific sulci divide each hemisphere into lobes (Fig. 8.9). Note that each lobe of the brain is located underneath the skull bone that shares its name. The *frontal lobe* lies under the frontal bone, anterior to the *parietal lobe* and bone. The *occipital lobe* is deep to the occipital bone, in the posterior area of the cranial vault. The *temporal lobe* is the lateral portion of the cerebral hemisphere. A fifth, very small lobe called the *insula* (not shown) lies directly deep to the lateral sulcus. Its function continues to be researched, but it's thought to be involved in speech processing, the sense of taste, and determining social emotions (such as embarrassment, resentment, empathy, and self-confidence).

The **cerebral cortex** is a thin but highly convoluted outer layer of gray matter that covers the cerebral hemispheres. The cerebral cortex contains over one billion cell bodies and is the region of the brain that accounts for sensation, voluntary movement, and all the thought processes we associate with consciousness.

Motor and Sensory Areas of the Cortex

The **primary motor area** is in the frontal lobe just anterior to the central sulcus (Fig. 8.9). Voluntary commands to skeletal muscles begin in the primary motor area, and each part of the body

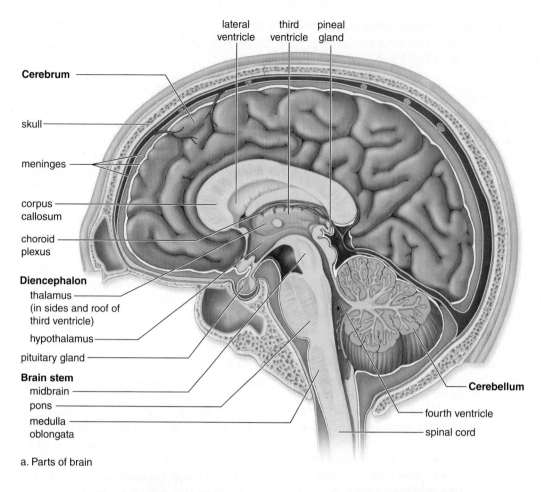

lateral ventricle   third ventricle   pineal gland

Cerebrum

skull

meninges

corpus callosum

choroid plexus

Diencephalon
thalamus (in sides and roof of third ventricle)

hypothalamus

pituitary gland

Brain stem
midbrain
pons
medulla oblongata

Cerebellum

fourth ventricle

spinal cord

a. Parts of brain

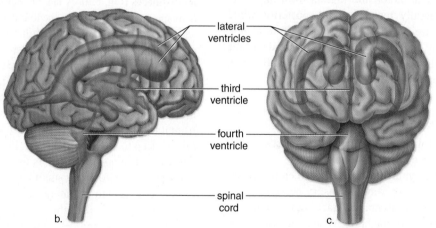

lateral ventricles

third ventricle

fourth ventricle

spinal cord

b.   c.

AP|R **Figure 8.8** **The human brain. (a)** The cerebrum, seen here in sagittal section, is the largest part of the brain in humans. **(b)** Ventricles of the brain in a lateral view. Ventricles are hollow cavities filled with cerebrospinal fluid. **(c)** Frontal view of the brain's four ventricles.

is controlled by a certain section (Fig. 8.10*a*). The right primary motor area controls the left side of the body and vice versa.

The **primary somatosensory area** is just posterior to the central sulcus in the parietal lobe. Sensory information from the skin and skeletal muscles arrives here, where each part of the body is sequentially represented (Fig. 8.10*b*). As you study Figure 8.10, notice that the areas of the body with the greatest voluntary control—the face and hands—have the largest area of

motor cortex in the brain dedicated to them. Similarly, the face and hands are among the most sensitive areas of the body, and the large area of the sensory cortex receiving information from them corresponds to that fact. As with the brain's motor areas, the left hemisphere's somatosensory area receives information from the right side of the body, and vice versa.

A primary taste area, located within adjacent areas of the parietal lobe and insula, accounts for taste sensations. A primary visual area in

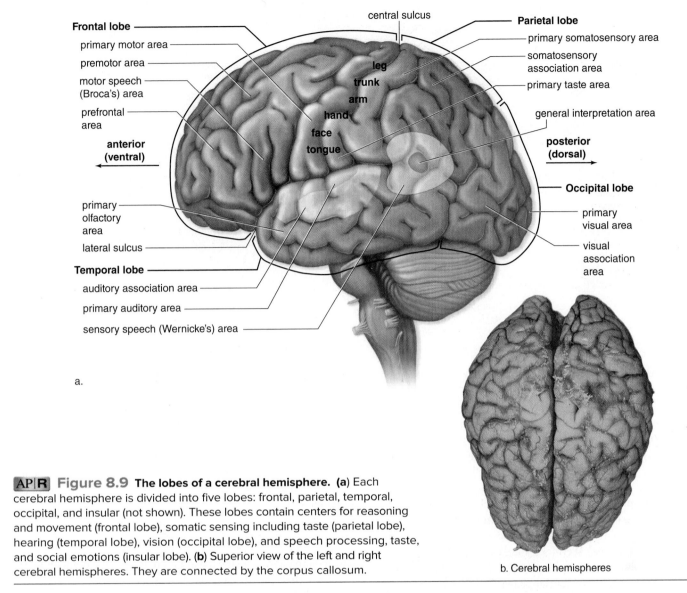

Frontal lobe
primary motor area
premotor area
motor speech (Broca's) area
prefrontal area

anterior (ventral)

central sulcus

Parietal lobe
primary somatosensory area
somatosensory association area
primary taste area
general interpretation area

posterior (dorsal)

leg
trunk
arm
hand
face
tongue

primary olfactory area
lateral sulcus

Temporal lobe
auditory association area
primary auditory area
sensory speech (Wernicke's) area

Occipital lobe
primary visual area
visual association area

a.

AP|R Figure 8.9 **The lobes of a cerebral hemisphere. (a)** Each cerebral hemisphere is divided into five lobes: frontal, parietal, temporal, occipital, and insular (not shown). These lobes contain centers for reasoning and movement (frontal lobe), somatic sensing including taste (parietal lobe), hearing (temporal lobe), vision (occipital lobe), and speech processing, taste, and social emotions (insular lobe). **(b)** Superior view of the left and right cerebral hemispheres. They are connected by the corpus callosum.

b. Cerebral hemispheres

the occipital lobe receives information from our eyes, and a primary auditory area in the temporal lobe receives information from our ears.

 Begin Thinking Clinically

Suppose that a woman slips on ice in the winter and falls backward, striking the back of her head on the pavement. The accident results in a concussion to the woman's occipital lobe. Which sense might be affected?

Answer and discussion in Appendix A.

Association Areas

Association areas are places where sensory signals are integrated and interpreted, and where memories are stored. Anterior to the primary motor area is a premotor area. The premotor area organizes motor functions for skilled motor activities, and then the primary motor area sends signals to the cerebellum and the basal nuclei, which integrate them. A

momentary lack of oxygen during birth can damage the motor areas of the cerebral cortex so that **cerebral palsy,** a condition characterized by a spastic weakness of the upper and lower limbs, develops.

The somatosensory association area, located just posterior to the primary somatosensory area, processes and analyzes sensory information from the skin and muscles. The visual association area associates new visual information with memories of previously received visual information. It might "decide," for example, whether or not we have previously seen a particular face, symbol, or other object. The auditory association area performs the same functions with regard to sounds.

Processing Centers

There are a few areas of the cortex that receive information from the other association areas and perform higher-level analytical functions. The **prefrontal area,** a processing area in the frontal lobe, receives information from the other association area and uses this information to reason and plan our actions. Integration in this area accounts for our most cherished human abilities to think critically and to formulate appropriate behaviors.

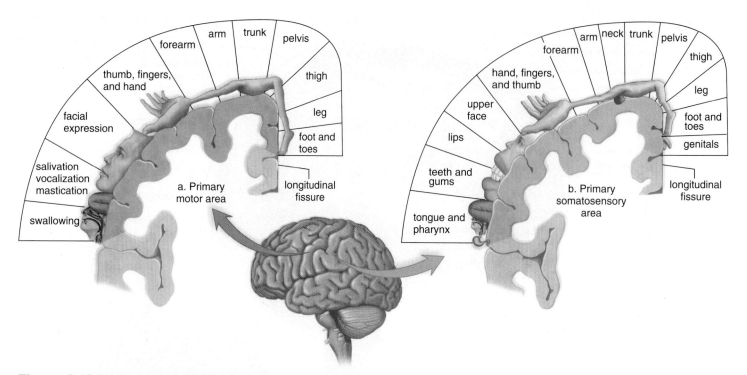

Figure 8.10 **Portions of the body controlled by the primary motor area and the primary somatosensory area of the cerebrum.** Notice that the size of the body part in the diagram reflects the amount of cerebral cortex devoted to that body part.

The unique ability of humans to speak is partially dependent upon the **motor speech area,** also called **Broca's area,** a processing area usually located in the left frontal lobe. Signals originating here pass to the premotor area before reaching the primary motor area. Damage to this area can interfere with a person's ability to coordinate the respiratory and oral movements that produce speech.

Wernicke's area, also called the **general interpretive area,** receives information from all of the other sensory association areas. Like Broca's area, Wernicke's area is usually located in the brain's left hemisphere. Its role involves recognizing and understanding spoken and written language. Damage to this area interferes with a person's ability to interpret written and spoken messages, even though the spoken words may be a part of the person's vocabulary. Wernicke's area and Broca's area cooperate to allow human communication.

Central White Matter

Much of the rest of the cerebrum beneath the cerebral cortex is composed of white matter. Tracts within the cerebrum take information between the different sensory, motor, and association areas pictured in Figure 8.9. The corpus callosum, previously mentioned, contains tracts that join the two cerebral hemispheres. Descending tracts from the primary motor area communicate with various parts of the brain and transmit instructions to the opposite side of the brain stem and spinal cord. Ascending tracts from lower brain centers send sensory information up to the primary somatosensory area (Fig. 8.10).

Basal Nuclei

While the bulk of the cerebrum is composed of tracts, there are masses of gray matter located deep within the white matter.

These so-called **basal nuclei** (formerly termed *basal ganglia*) integrate motor commands, ensuring that proper muscle groups are activated or inhibited. Huntington disease and Parkinson's disease, which are both characterized by uncontrollable movements, are believed to be due to an imbalance of neurotransmitters in the basal nuclei.

The Diencephalon

The hypothalamus and the thalamus are both in the **diencephalon,** a region that encases the third ventricle (see Fig. 8.8a). The **hypothalamus** forms the floor of the third ventricle. The hypothalamus is an integrating center that helps maintain homeostasis by regulating hunger, sleep, thirst, body temperature, and water balance. The hypothalamus produces the hormones secreted by the posterior pituitary gland and secretes hormones that control the anterior pituitary. Therefore, it is a link between the nervous and endocrine systems.

The **thalamus** consists of two masses of gray matter located in the sides and roof of the third ventricle. The thalamus is on the receiving end for all sensory input except smell; it functions as a sensory "relay center." Visual, auditory, and somatosensory information arrive at the thalamus via the cranial nerves and tracts from the spinal cord. The thalamus integrates this information and sends it on to the appropriate portions of the cerebrum. The thalamus is involved in arousal of the cerebrum, and it also participates in higher mental functions such as memory and emotions.

The pineal gland (sometimes referred to as the pineal body), which secretes the hormone melatonin and regulates our body's daily rhythms, is located in the posterior diencephalon.

Epidural Stimulation in Spinal Cord Injuries: Cause for Hope

In Memoriam: Christopher Reeve, 1952–2004

It happened in a split second: Christopher Reeve, the tall, athletic actor who achieved worldwide recognition in his role as Superman, was thrown from his horse. The accident shattered both atlas and axis, the first two cervical vertebrae, and Reeve became a ventilator-dependent quadriplegic. Although he spent the rest of his life in a wheelchair, Reeve redefined the recovery process for spinal cord injury with tremendous courage and determination. Spinal cord injuries are usually caused by compression or bruising to one side of the cord and often result from accidents or trauma. Injured cells in the center of the cord die, leaving a rim of functional cells. Depending on the nerve pathways that are interrupted, a variety of effects can result from the damage. Injury between the first thoracic vertebra (T1) and the second lumbar vertebra (L2) causes paralysis of the lower body, or **paraplegia.** If the injury is between the cervical vertebrae and the first thoracic vertebra (T1), the entire trunk is affected, resulting in **quadriplegia.** At the same time, sensation will be lost in all or parts of the body. Other serious consequences of spinal cord injury include loss of breathing ability, pain, absent or exaggerated reflexes, skeletal muscle spasms, and loss of autonomic motor nervous functions (such as blood pressure control, ability to sweat, bowel and bladder control, and sexual function). After his injury, Reeve was determined to recover, though experts called his case hopeless. He pioneered functional electrical stimulation (FES), using computer-directed muscle activation to exercise his limbs. FES therapy improves muscle tone and bone density in spinal cord–injured patients and stimulates spinal cord recovery. Reeve was the first documented case of significant improvement occurring more than two years after injury. He regained movement of his fingers and toes, as well as sensation over most of his body. He was also able to finally breathe independently for several hours at a time. Ironically, Reeve died in 2004 of complications resulting from an infected decubitus ulcer, or bedsore (see page 89), a common affliction of quadriplegics. His legacy includes a nonprofit foundation that has raised over $100 million for spinal cord research and advocacy.

Recently, the Christopher and Dana Reeve Foundation's Neural Recovery Network reported the results of a pioneering technique in spinal cord treatment conducted at the University of Louisville, Kentucky. Electrical stimulator devices were implanted into the epidural spaces above the T11-T12 regions of the spinal cords in four quadriplegic young men. The patients activated the stimulator using a handheld control. With continued electrical stimulation, all four experienced dramatic improvements in sensory and motor function in their lower limbs. In addition, each showed progress in autonomic functions, including sweating, bowel and bladder control, and sexual function. Researchers believe that this external device sparks spinal cord circuitry that was thought to be permanently damaged, enabling it to once again

respond to brain commands. Scientists are unsure of the full extent of recovery that is possible as a result of this technique. However, they are optimistic that as the technology continues to be investigated, it will become possible to restore some degree of function to those with spinal cord injuries.

In addition, the Reeve Foundation continues to fund research into improvements in emergency care of victims, as well as inventions to improve the quality of life for those with spinal injury. New drugs have been shown to help injured spinal cord nerves to survive. Hypothermia treatment, in which the victim's body temperature is lowered to 92°F (33.5°C), has been shown to slow or prevent inflammation in the damaged spinal cord. This technique has enabled some patients to fully recover from spinal cord injuries. Prosthetics for the spinal cord–injured include stair-climbing and voice-activated wheelchairs and electronic devices that allow independent breathing and restore bladder and bowel control. Studies using grafts of neural tissue, neural stem cells, and nerve growth factors—hormone-like chemicals that stimulate growth—have shown to be promising for regenerating spinal cord tissue. The goal of the Christopher and Dana Reeve Foundation (www .christopherreeve.org) is to one day produce a cure for spinal cord paralysis.

Limbic System

The **limbic system** (illustrated in purple in the following figure) is a collection of structures from both the cerebrum and the diencephalon. It lies just inferior to the cerebral cortex and contains neural pathways that connect portions of the cerebral cortex and the temporal lobes with the thalamus and the hypothalamus:

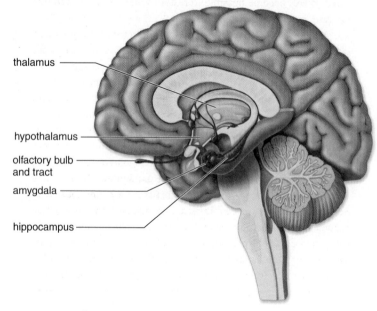

thalamus

hypothalamus

olfactory bulb and tract

amygdala

hippocampus

Stimulation of different areas of the limbic system causes the subject to experience rage, pain, pleasure, or sorrow. By causing pleasant or unpleasant feelings about experiences, the limbic system apparently guides the individual into behavior that is likely to increase the chance of survival.

The limbic system is also involved in learning and memory. In particular, the most inferior structure of the limbic system, the **hippocampus,** is vital in processing of short-term memory to become long-term memory. Learning requires memory, and memory is stored in the sensory regions of the cerebrum, but just what permits memory development is not definitely known. The involvement of the limbic system in memory explains why emotionally charged events result in our most vivid memories. The fact that the limbic system communicates with the sensory areas for touch, smell, vision, and so forth accounts for the ability of any particular sensory stimulus to awaken a complex memory.

The Cerebellum

The **cerebellum** is separated from the brain stem by the fourth ventricle (see Fig. 8.8a). The cerebellum has two hemispheres, which are joined by a narrow median portion. Each portion is primarily composed of white matter, which in longitudinal section has a treelike pattern. Overlying the white matter is a thin layer of gray matter that forms a series of complex folds.

The cerebellum receives sensory input from the eyes, ears, joints, and muscles about the present position of body parts. It also receives motor output from the cerebral cortex about where these parts should be located. After integrating this information, the cerebellum sends motor signals by way of the brain stem to the skeletal muscles. In this way, the cerebellum maintains posture and balance. It also ensures that all of the muscles work together to produce smooth, coordinated voluntary movements. In addition, the cerebellum assists the learning of new motor skills, such as playing the piano or hitting a baseball.

The Brain Stem

The brain stem is the structure that joins the upper brain regions to the spinal cord. The **brain stem** contains the midbrain, the pons, and the medulla oblongata (see Fig. 8.8*a*). Running through all of these structures is a web of gray matter called the reticular formation. The **midbrain** acts as a relay station for tracts passing between the cerebrum and the spinal cord or cerebellum. It also has reflex centers for visual,

auditory, and tactile responses. These include centers that control head movements in response to visual and auditory stimuli (for example, when you turn your head in the direction of a loud sound). The word *pons* means "bridge" in Latin, and true to its name, the **pons** contains bundles of axons traveling between the cerebellum and the rest of the CNS. In addition, the pons functions with the medulla oblongata to regulate breathing rate.

The **medulla oblongata** contains a number of reflex centers for regulating heartbeat, breathing, and vasoconstriction. It also contains the reflex centers for vomiting, coughing, sneezing, hiccupping, and swallowing. The medulla oblongata lies just superior to the spinal cord, and it contains tracts that ascend or descend between the spinal cord and higher brain centers.

The entire brain stem is critically important for control of the parasympathetic division of the autonomic motor nervous system. Cranial nerves whose neuron cell bodies are found in this area travel to the eyes, facial glands, and internal organs in both the thorax and abdominopelvic regions.

The **reticular formation** assists the cerebellum in maintaining muscle tone; it also assists the pons and medulla in regulating respiration, heart rate, and blood pressure. The sensory

I.C.E. — IN CASE OF EMERGENCY

Traumatic Brain Injury

In March 2009, Natasha Richardson, actress and wife of actor Liam Neeson, lost consciousness while she was on the beginner slope of a Montreal ski resort, after a seemingly minor fall. After regaining consciousness, she insisted that she was fine, even turning away EMS personnel. However, she complained of a severe headache hours later, and her condition rapidly deteriorated. After being declared brain dead, Richardson died in a New York hospital two days later.

Richardson's accident focused attention on the need for immediate medical attention when a traumatic brain injury (TBI) is suspected. Traumatic brain injuries cause swelling of the brain and meninges, which reduces blood supply to the brain. *Concussion* is often the first symptom of TBI. Patients who suffer a concussion become dizzy, confused or disoriented, suffer short-term memory loss, or lose consciousness. Bleeding inside the brain or skull, called *hematoma*, or bruising of the brain, called a *contusion*, may follow concussion. These are life-threatening and often fatal injuries that may not be immediately evident, but develop in the hours to days after the initial loss of consciousness. In Ms. Richardson's case, her fall resulted in an epidural hematoma: bleeding between the skull and dura mater. Had she received prompt medical treatment, the hematoma could have been surgically repaired.

Patients who have had a concussion should always be examined by an emergency room physician to rule out a critical injury. Before first responders transport the person to the hospital, they should quickly assess whether the patient is alert and able to respond to person, place, and time—in the language of the emergency room, "oriented times three." The individual should be able to identify himself (person), tell where he is (place), and correctly name the day of the week (time). Next, the victim's pupillary reflex is tested to ensure that both pupils react similarly and quickly in response to light. Emergency care providers and family members must be aware of the signs of brain damage: severe headache, nausea and vomiting, slow heartbeat and breathing rate, and decreasing consciousness. In babies and small children, the early signs of TBI include crying inconsolably and refusal to nurse or eat. In these situations, immediate medical and surgical treatment will hopefully lessen or prevent brain damage.

Athletes (and their parents and coaches) must be aware that no concussion should be considered minor; each is a traumatic brain injury. Further, repeated concussions in young people can result in permanent brain damage and predispose the victim to neurodegenerative diseases, including Alzheimer and Parkinson's disease. Under no circumstances should an athlete be returned to play in that day's game following a concussion.

component of the reticular formation processes sensory stimuli—sounds, sights, touch—and uses these signals to keep us mentally alert. Additionally, this portion of the brain helps to rouse a sleeping person. Damage to the reticular formation can result in **coma,** a state of temporary (sometimes permanent) unconsciousness.

Content CHECK-UP!

5. The deepest of the three meninges is the:
   a. dura mater.
   b. arachnoid mater.
   c. pia mater.

6. Cerebrospinal fluid is found in the _____ _____ of the spinal cord, and the _____ of the brain. It is also found in the subarachnoid space surrounding the brain and spinal cord.

7. Imagine you're standing outside when a tremendous clap of thunder occurs. Which structure routes the sound to the correct brain lobe? Which lobe contains sensory and memory areas for hearing?

8. Which structure integrates motor commands to make sure that proper muscle groups are activated or inhibited?
   a. Wernicke's area
   b. corpus callosum
   c. limbic system
   d. basal nuclei

9. Name the three main components of the diencephalon.

10. Match the brain region with its function:
    a. cerebellum   visual and auditory reflexes
    b. limbic system   motor coordination
    c. brain stem   emotional response

Answers in Appendix A.

8.3 Peripheral Nervous System

8. Describe the structure of a nerve, and distinguish between sensory, motor, and mixed nerves.
9. Name the twelve pairs of cranial nerves, and give a function for each.
10. Name several peripheral nerves, and describe the spinal nerves which combine to create each one. Explain the function of each of these peripheral nerves.
11. Describe the structure of a reflex arc and the function of a reflex action.
12. Define and describe the autonomic nervous system.
13. Distinguish between the sympathetic and parasympathetic divisions in four ways, and give examples of their respective effects on specific organs.

The peripheral nervous system (PNS) lies outside the central nervous system and is composed of nerves and ganglia. Nerves are bundles of axons, both myelinated and unmyelinated, that travel together. Ganglia (sing., **ganglion**) are swellings associated with nerves that contain collections of cell bodies. As with muscles, connective tissue separates axons at various levels of organization. Each axon is surrounded by a fine connective tissue called the endoneurium. Bundles of axons called fascicles are enclosed by a perineurium, and the entire nerve is covered by the epineurium (Fig. 8.11*a*).

The PNS is subdivided into the **afferent, or sensory, system** and the **efferent, or motor, system** (see Fig. 8.1). The **somatic sensory system** carries action potentials from the skin, skeletal muscles, joints, and tendons. The special senses (vision, hearing, taste, smell) are also part of the somatic sensory system. The **visceral sensory system** transmits action potentials from the internal organs. Nerves from both sensory systems take information from their peripheral sensory receptors to the CNS.

The **somatic motor system** carries commands away from the CNS to the skeletal muscles. The **autonomic motor system** relays action potentials from the CNS to structures that are part of the organ systems over which we have little or no control (cardiovascular, digestive, urinary, reproductive, etc.). Thus, the autonomic nervous system regulates the activities of these organ systems. thus regulating their activity.

Types of Nerves

The **cranial nerves** are attached to the brain, and the **spinal nerves** are attached to the spinal cord.

Cranial Nerves

Humans have 12 pairs of cranial nerves (Table 8.1). By convention, the pairs of cranial nerves are referred to by Roman numerals (Fig. 8.11*a*). Some cranial nerves are sensory nerves—that is, they contain only sensory fibers; some are *motor nerves,* containing only motor fibers; and others are *mixed nerves,* so called because they contain both sensory and motor fibers. Cranial nerves are largely concerned with the head, neck, and facial regions of the body. There are two exceptions: the *vagus nerve* (cranial nerve X), has sensory and motor branches to the face and most of the internal organs. This mixed nerve, whose name aptly derives from a Latin word that means "traveler," contains both somatic and visceral sensory nerves. Its efferent nerves belong to the somatic and the autonomic motor systems. The spinal accessory nerve (cranial nerve XI, sometimes simply called the accessory nerve) supplies motor signals to two skeletal muscles that move the head: the sternocleidomastoid and trapezius muscles (see Table 7.2).

Spinal Nerves

Humans have 31 pairs of spinal nerves; one of each pair is on either side of the spinal cord (Fig. 8.11*b*). The spinal nerves are grouped as shown in Table 8.2 because they are at either the

TABLE 8.1   Cranial Nerves

Nerve	Type		Brain Pathway	Transmits Nerve Impulses to (Motor) or from (Sensory)
Olfactory (I)	Sensory		I: Mucous membrane of nose to olfactory bulbs	Olfactory receptors for sense of smell
Optic (II)	Sensory		II: Retina → optic nerve → thalamus → occipital lobe	Retina for sense of sight
Oculomotor (III)	Motor		III: Midbrain → eye and eyelid	Eye muscles (including eyelids and lens); pupil (parasympathetic division)
Trochlear (IV)	Motor		IV: Midbrain → eye	Eye muscles
Trigeminal (V)	Mixed	Sensory	V: Sensory: Teeth, eye, skin, tongue → pons	Teeth, eyes, skin, and tongue
		Motor	Motor: Pons → jaw muscles	Jaw muscles (chewing)
Abducens (VI)	Motor		VI: Pons → eye	Eye muscles
Facial (VII)	Mixed	Sensory	VII: Sensory: Tongue → pons	Taste buds of anterior tongue
		Motor	Motor: Pons → facial muscles, salivary glands, tear glands	Facial muscles (facial expression) and glands (tear and salivary)
Vestibulocochlear (VIII) (also called auditory; acoustic)	Sensory		VIII: Inner ear → pons and medulla	Inner ear for sense of balance and hearing
Glossopharyngeal (IX)	Mixed	Sensory	IX: Sensory: Tongue, throat → pons	Pharynx
		Motor	Motor: Pons → salivary gland, throat muscles	Pharyngeal muscles (swallowing), salivary glands
Vagus (X)	Mixed	Sensory	X: Sensory: Eardrum, ear canal, throat, heart, lungs, abdominal organs → medulla	Internal organs, external ear canal, eardrum, back of throat
		Motor	Motor: Medulla → throat and larynx, heart, lungs, abdominal organs	Internal organs (parasympathetic division), throat muscles (somatic motor division)
Spinal accessory (XI)	Motor		XI: Medulla → muscles of throat, neck, shoulder	Neck and back muscles
Hypoglossal (XII)	Motor		XII: Medulla → tongue muscles	Tongue muscles

cervical, thoracic, lumbar, sacral, or coccygeal regions of the vertebral column. The first spinal nerve, C1, emerges between the skull and the atlas (the first cervical vertebra). The remaining spinal nerves are designated according to their location in relation to the vertebrae because each passes through an intervertebral foramen as it leaves the spinal cord. This organizational principle is illustrated in Figure 8.12.

All spinal nerves are mixed nerves because they contain both sensory fibers that conduct signals to the spinal cord from sensory receptors and motor fibers that conduct signals away from the cord to effectors. The sensory fibers enter the cord via the posterior root, and the motor fibers exit by way of the anterior root. As you'll recall (from section 8.2), the cell body of a sensory neuron is found in the posterior root ganglion (see Figs. 8.7 and 8.13). Each spinal nerve serves the particular region of the body in which it is located (Table 8.2).

Somatic Motor Nervous System and Reflexes

Most actions in the somatic motor nervous system are voluntary. These actions, such as when we decide to move a limb, always originate in the motor cortex. You will remember that the motor cortex is in the posterior part of the frontal lobe. Other actions in the somatic motor nervous system are due to **reflexes**—automatic involuntary responses to changes occurring inside or outside the body. A reflex occurs quickly; we don't even have to think about it. Reflexes are protective mechanisms that are essential to homeostasis. They keep the internal organs functioning within normal bounds and protect the body from external harm.

Some reflexes, called *cranial reflexes,* involve the brain, as when we automatically blink our eyes when an object nears the eye suddenly. Figure 8.13 illustrates the path of a reflex within the

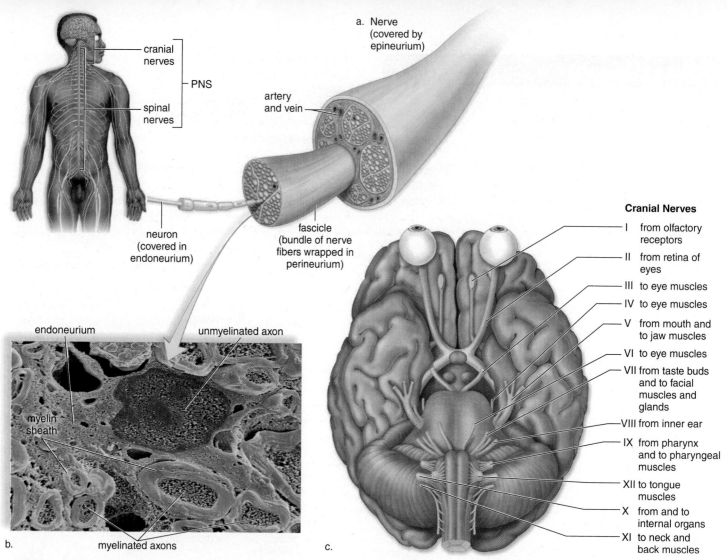

Cranial Nerves

I from olfactory receptors
II from retina of eyes
III to eye muscles
IV to eye muscles
V from mouth and to jaw muscles
VI to eye muscles
VII from taste buds and to facial muscles and glands
VIII from inner ear
IX from pharynx and to pharyngeal muscles
XII to tongue muscles
X from and to internal organs
XI to neck and back muscles

a. Nerve (covered by epineurium)

cranial nerves
PNS
spinal nerves

artery and vein

neuron (covered in endoneurium)

fascicle (bundle of nerve fibers wrapped in perineurium)

endoneurium
unmyelinated axon
myelin sheath
myelinated axons

b.

c.

AP|R **Figure 8.11 Cranial and spinal nerves. (a)** A single nerve is composed of the axons of many neuron fibers, arranged in bundles called fascicles. Axons are covered by endoneurium; fascicles are covered by perineurium, and the nerve is enclosed in epineurium. **(b)** Cross section scanning electron micrograph of a fascicle containing both myelinated axons (surrounded by gold myelin sheath) and unmyelinated axons. Endoneurium (tan tissue) surrounds each axon. **(c)** Inferior surface of the brain showing the attachment of the 12 pairs of cranial nerves.

TABLE 8.2 Spinal Nerves		
Name	**Peripheral Nerves Created from Spinal Nerves Involved[1]**	**Function**
Musculocutaneous nerves	C_5–T_1	Supply muscles of the anterior arm and carry sensation from skin of the lateral forearm
Radial nerves	C_5–T_1	Supply muscles of the posterior arm and forearm and carry sensation from skin of the posterior forearm and hand
Median nerves	C_5–T_1	Supply muscles of the anterior forearm and hand and carry sensation from skin of the palm
Ulnar nerves	C_5–T_1	Supply muscles of the forearm and hand and carry sensation from skin of the hand
Phrenic nerves	C_3–C_5	Supply the diaphragm
Intercostal nerves	T_1–T_{11}	Supply intercostal muscles, abdominal muscles, and skin of the trunk
Femoral nerves	L_2–L_4	Supply muscles and skin of the anterior thighs and legs
Sciatic nerves	L_4–S_3	Supply muscles and skin of the posterior thighs, legs, and feet

[1] C = cervical; T = thoracic; L = lumbar

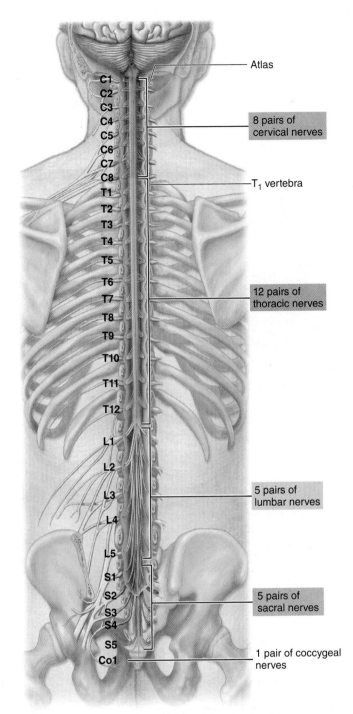

these interneurons then synapse with motor neurons whose short dendrites and cell bodies are in the spinal cord. Nerve action potentials travel along a motor fiber to an effector, which brings about a response to the stimulus. In this case, the effector is a skeletal muscle, which contracts so that you withdraw your hand from the pain stimulus.

Various other reactions are also possible—you will most likely look at the pin, wince, and cry out in pain. This whole series of responses occurs because the axons of certain interneurons carry action potentials to the brain via tracts in the spinal cord and brain. The brain makes you aware of the stimulus and directs your other reactions to it. You don't feel pain until the brain receives the information and interprets it.

Reflexes can also be used to determine if the nervous system is reacting properly. Two of these types of reflexes are:

knee-jerk reflex (patellar ligament reflex), initiated by striking the patellar ligament just below the patella. The response is contraction of the quadriceps femoris muscles, which causes the leg to extend;

ankle-jerk reflex, initiated by tapping the calcaneal (Achilles) tendon just above its insertion on the calcaneus. The response is plantar flexion due to contraction of the gastrocnemius and soleus muscles.

Many reflexes are important for avoiding injury, but the knee-jerk and ankle-jerk reflexes are also important for maintaining balance. For example, the knee-jerk reflex helps a person stand erect. If the knee begins to bend slightly when a person stands still, the quadriceps femoris is stretched, and the leg straightens.

Autonomic Motor Nervous System and Visceral Reflexes

The autonomic motor nervous system (ANS) is composed of the sympathetic and parasympathetic divisions (Fig. 8.14). These two divisions have several features in common: (1) They function automatically, with little or no voluntary control; (2) all internal organs have fibers from either or both divisions; and (3) they utilize two motor neurons and one ganglion to transmit an action potential. (By contrast, a somatic motor neuron travels directly to its effector, without synapsing at a ganglion.) The first neuron has a cell body within the CNS and a *preganglionic* axon fiber (the term *preganglionic* indicates that the axon travels to the ganglion). The second neuron has a cell body within the ganglion and a *postganglionic* axon fiber (the axon travels away from the ganglion to the effector).

Visceral reflex actions, such as those that regulate blood pressure and breathing rate, are especially important to maintenance of homeostasis. These reflexes begin when the sensory neurons in contact with internal organs send messages via spinal nerves to the CNS. They are completed when motor neurons within the autonomic system stimulate smooth muscle, cardiac muscle, or a gland. These structures are also effectors.

AP|R **Figure 8.12** **Spinal nerves.** The number and kinds of spinal nerves are shown here. Table 8.2 lists the functions of some of the major spinal nerves.

somatic motor nervous system that involves only the spinal cord (called a *spinal reflex*). If the skin on your hand is poked with a sharp pin, a pain sensory receptor in the skin generates action potentials that move along a sensory fiber through the posterior root ganglia toward the spinal cord. Sensory neurons enter the cord posteriorly and form synapses with many interneurons. Some of

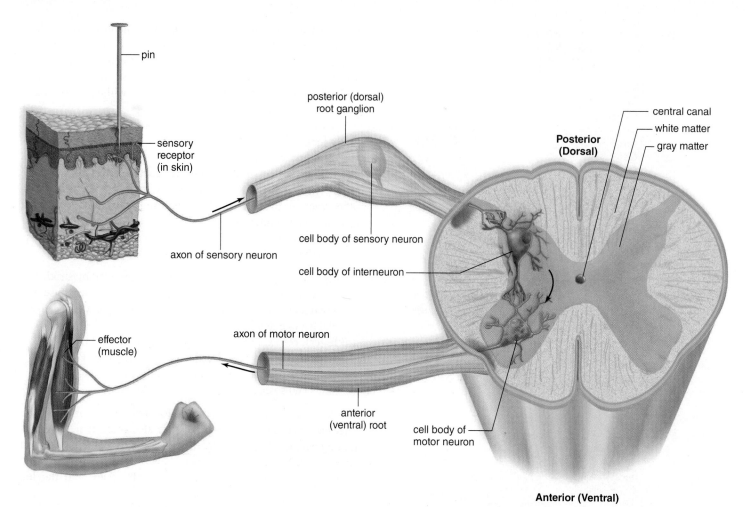

AP|R **Figure 8.13** **A reflex arc showing the path of a spinal reflex.** A stimulus (e.g., a pinprick to the skin of the hand) causes sensory receptors in the skin to generate nerve signals that travel in sensory axons to the spinal cord. Interneurons integrate data from sensory neurons and then relay signals to motor neurons. Motor axons convey nerve signals from the spinal cord to a skeletal muscle, which contracts. Movement of the hand away from the pin is the response to the stimulus.

Sympathetic Division: "Fight or Flight"

The preganglionic fibers of the **sympathetic division** arise from the middle, or thoracic-lumbar, portion of the spinal cord and almost immediately terminate in ganglia that lie near the cord. Therefore, in this division, the preganglionic fiber is short, but the postganglionic fiber that makes contact with an effector is long (Fig. 8.15a).

The sympathetic division is especially important during emergency situations when a person might be required to fight or run away. It accelerates the heartbeat and dilates the bronchi—active muscles, after all, require a ready supply of glucose and oxygen. On the other hand, the sympathetic division inhibits the digestive tract functions—digestion is not an immediate necessity if you're under attack. The neurotransmitter released by the postganglionic axon is primarily norepinephrine

(NE). The structure of NE is like that of epinephrine (adrenaline), an adrenal medulla hormone that usually increases heart rate and contractility.

Parasympathetic Division: "Rest and Digest"

The **parasympathetic division** includes several cranial nerves whose preganglionic neurons are located in the brain stem (e.g., the vagus nerve) as well as fibers that arise from the sacral (inferior) portion of the spinal cord. Therefore, this division is often referred to as the *craniosacral* portion of the autonomic system. In the parasympathetic division, the preganglionic fiber (axon) is long, and the postganglionic fiber (axon) is short because the ganglia lie near or within the effector (Fig. 8.15b.)

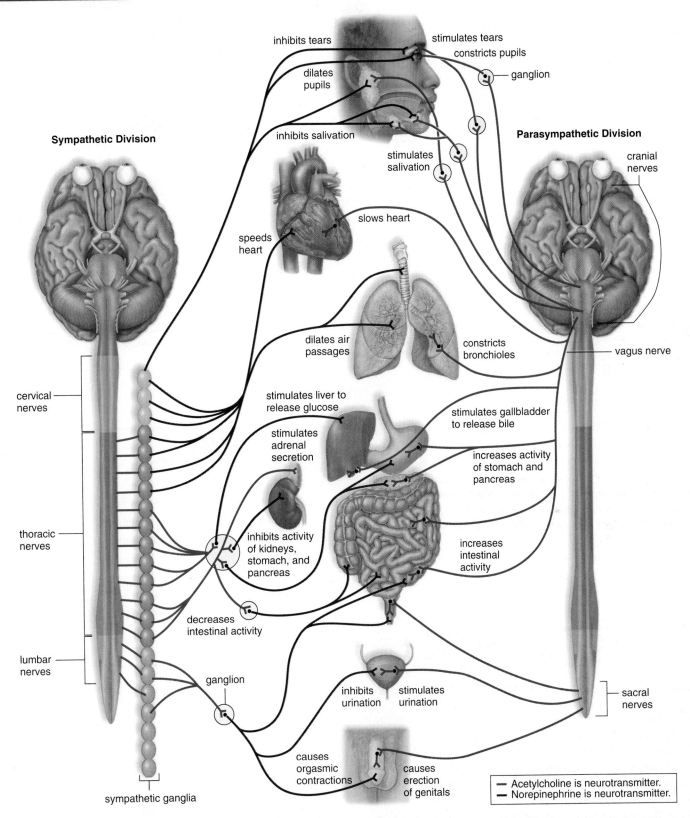

Sympathetic Division

inhibits tears

stimulates tears
constricts pupils

ganglion

dilates
pupils

inhibits salivation

Parasympathetic Division

stimulates
salivation

cranial
nerves

speeds
heart

slows heart

dilates air
passages

constricts
bronchioles

vagus nerve

cervical
nerves

stimulates liver to
release glucose

stimulates gallbladder
to release bile

stimulates
adrenal
secretion

increases activity
of stomach and
pancreas

thoracic
nerves

inhibits activity
of kidneys,
stomach, and
pancreas

increases
intestinal
activity

decreases
intestinal activity

lumbar
nerves

ganglion

inhibits   stimulates
urination   urination

sacral
nerves

sympathetic ganglia

causes
orgasmic
contractions

causes
erection
of genitals

— Acetylcholine is neurotransmitter.
— Norepinephrine is neurotransmitter.

AP|R **Figure 8.14** **Autonomic system structure and function.** Sympathetic preganglionic fibers (*left*) arise from the thoracic and lumbar portions of the spinal cord; parasympathetic preganglionic fibers (*right*) arise from the cranial and sacral portions of the spinal cord. Most of the effectors are innervated by both divisions of the ANS (as shown). Where this is the case, the two divisions usually have opposite effects.

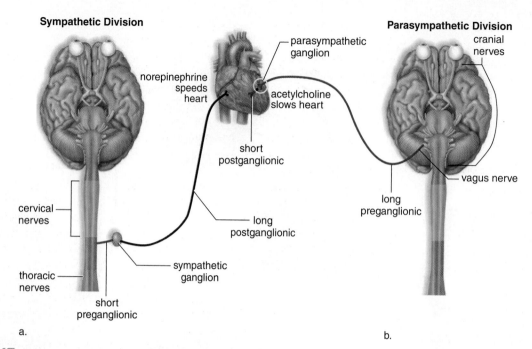

Sympathetic Division

Parasympathetic Division

parasympathetic ganglion

norepinephrine speeds heart

acetylcholine slows heart

cranial nerves

short postganglionic

cervical nerves

vagus nerve

long preganglionic

long postganglionic

thoracic nerves

sympathetic ganglion

short preganglionic

a.

b.

AP|R Figure 8.15 Sympathetic and parasympathetic pathways. (a) Because sympathetic ganglia usually lie close to the spinal cord, the preganglionic neuron is short and the postganglionic neuron is long. **(b)** Parasympathetic ganglia are close to their target organs, so the preganglionic neuron is long and the postganglionic neuron is short.

The parasympathetic division, sometimes called the "housekeeper" division, promotes all of the internal responses we associate with a relaxed state. Parasympathetic function can be referred to as "rest and digest." For example, parasympathetic nerves cause the pupil of the eye to contract, promote digestion of food, slow heart rate, and decrease the strength of cardiac contraction. The neurotransmitter utilized by the parasympathetic division is acetylcholine (ACh).

Table 8.3 contrasts the two divisions of the autonomic system.

TABLE 8.3 Autonomic Motor Pathways

	Sympathetic	Parasympathetic
Type of control	Involuntary	Involuntary
Number of neurons per message	Two (preganglionic shorter than postganglionic)	Two (preganglionic longer than postganglionic)
Location of motor fiber	Thoracolumbar spinal nerves	Cranial (e.g., vagus) and sacral spinal nerves
Neurotransmitter	Norepinephrine	Acetylcholine
Effectors	Smooth and cardiac muscle, glands	Smooth and cardiac muscle, glands

Content **CHECK-UP!**

11. Which of the following cranial nerves transmits sensory signals about hearing and balance?

   a. oculomotor   c. trochlear

   b. abducens   d. vestibulocochlear

12. Which spinal nerve supplies the diaphragm?

   a. radial   c. phrenic

   b. median   d. intercostal

13. Imagine that you accidentally touch a hot stove and automatically pull your hand away. Detail the steps in this reflex pathway.

Answers in Appendix A.

8.4 Effects of Aging

14. Describe the anatomical and physiological changes that occur in the nervous system as we age.

After age 60, brain mass begins to decrease. By age 80, the brain weighs about 10% less than when the person was a young adult. The cerebral cortex shrinks more than other areas of the brain, losing as much as 45% of its cells. Neurotransmitter production also decreases, resulting in slower synaptic transmission. As a person ages, thought processing and translating a thought into action take longer. This partly explains why younger athletes tend to outshine older athletes in sports.

However, it's very important to note that although structural changes occur, mental impairment is *not* an automatic consequence of getting older. Maintaining the health of the cardiovascular system is fundamental to retaining mental function—after all, the heart and blood vessels supply oxygen and nutrients to brain cells. In addition, older people can stay mentally alert by challenging the brain: taking courses, reading, solving puzzles, etc. Avoiding depression is important because it is a contributor to mental impairment; thus, the elderly should try to form and sustain a great social network and interact with others in meaningful ways. Exercise is also an important ingredient in good health—including mental health—at every age.

Neurological disorders, especially Alzheimer disease and Parkinson's disease, are more apt to occur in the elderly. The What's New reading on Alzheimer disease and the Medical Focus reading on Parkinson's disease on pages 170 and 189 will give you more information about these conditions.

Content CHECK-UP!

14. What can people do to slow the deterioration of the nervous system while they age?

Answer in Appendix A.

8.5 Homeostasis

15. Describe how the nervous system works with other systems of the body to maintain homeostasis.

The illustration in Human Systems Work Together on page 188 shows how the nervous system works with other systems in the body to maintain homeostasis. The nervous system detects, interprets, and responds to changes in internal and external conditions to keep the internal environment relatively constant. The hypothalamus, an important nervous system structure, is also the master controller for most of the endocrine system. Together, these two systems coordinate and regulate the functioning of the other organ systems.

The everyday regulation of internal organs that maintains the composition of blood and tissue fluid usually takes place below the level of consciousness. Subconscious control is dependent on reflex actions that involve the hypothalamus and the brain stem. The hypothalamus and the brain stem act through the autonomic motor nervous system to control such important parameters as the heart rate, the constriction of the blood vessels, and the breathing rate.

Because the nervous system stimulates skeletal muscles to contract, it controls the major movements of the body. When we are in a "fight-or-flight" mode, the nervous system stimulates the adrenal glands and voluntarily controls the skeletal muscles to keep us from danger. On a daily basis, you might think that voluntary movements don't play a role in homeostasis, but actually we usually take all necessary actions to stay in as moderate an environment as possible. Otherwise, we are testing the ability of the nervous system to maintain homeostasis despite extreme conditions.

In turn, the body's other systems support the functions of the nervous system. The integumentary, skeletal, and muscular systems protect both central and peripheral nervous system structures. These systems also supply sensory input to the CNS. The lymphatic system protects against infection. The metabolic needs of the nervous system are supplied by the digestive, cardiovascular, and respiratory systems. The kidneys dispose of metabolic waste, but also maintain the electrolyte balance that is critical for conduction of an action potential.

Content CHECK-UP!

15. A patient whose stroke caused loss of function to his upper limb developed decreased bone density in the limb. Why did this occur?

Answer in Appendix A.

Human Systems Work Together

Integumentary System

Brain controls nerves that regulate size of cutaneous blood vessels; activates sweat glands and arrector pili muscles.

Skin protects nerves, helps regulate body temperature; skin receptors send sensory input to brain.

Skeletal System

Receptors send sensory input from bones and joints to brain.

Bones protect sense organs, brain, and spinal cord; store Ca^{2+} for nerve function.

Muscular System

Brain controls nerves that innervate muscles; receptors send sensory input from muscles to brain.

Muscle contraction moves eyes, permits speech, and creates facial expressions.

Endocrine System

Hypothalamus is part of endocrine system; nerves innervate certain glands of secretion.

Sex hormones affect development of brain.

Cardiovascular System

Brain controls nerves that regulate the heart and dilation of blood vessels.

Blood vessels deliver nutrients and oxygen to neurons; carry away wastes.

How the Nervous System works with other body systems.

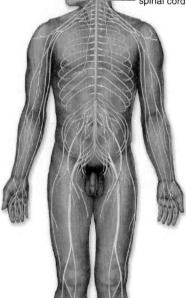

— brain

— spinal cord

Lymphatic System/Immunity

Microglial cells engulf and destroy pathogens.

Lymphatic vessels pick up excess tissue fluid; immune system protects against infections of nerves.

Respiratory System

Respiratory centers in brain regulate breathing rate.

Lungs provide oxygen for neurons and rid the body of carbon dioxide produced by neurons.

Digestive System

Brain controls nerves that innervate smooth muscle and permit digestive tract movements.

Digestive tract provides nutrients for growth, maintenance, and repair of neurons and neuroglial cells.

Urinary System

Brain controls nerves that innervate muscles that permit urination.

Kidneys maintain blood levels of Na^+, K^+, and Ca^{2+}, which are needed for nerve conduction.

Reproductive System

Brain controls onset of puberty; nerves are involved in erection of penis and clitoris, contraction of ducts that carry gametes, and contraction of uterus.

Sex hormones masculinize or feminize the brain, exert feedback control over the hypothalamus, and influence sexual behavior.

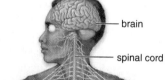

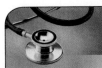

MEDICAL FOCUS

Parkinson's Disease

"My body is completely out of control. That's the hardest thing about this disease. Sometimes I can't move at all, or I move so slowly that it takes forever just to cross the room. Next thing you know, I'm jerking around like a puppet."

Your patient has just described the classic symptoms of **Parkinson's disease,** a progressive central nervous system disorder. The Parkinson's patient is usually a person age 60 or older. However, the disease is seen increasingly in younger people as well. Actor Michael J. Fox was just 38 years old when he announced publicly in 1999 that he suffers from Parkinson's disease.

If the patient's facial muscles are involved, his or her face may not be able to show emotion, resulting in a fixed, masklike appearance. Routine tasks such as dressing and bathing become very difficult. The sufferer has an increased risk of falling and injuring himself because balance and coordination are also affected. The disease takes its toll on the patient psychologically; most suffer depression as their activities and independence become more and more limited.

Parkinson's disease is caused by destruction of specific areas of the brain called the basal nuclei (see page 176). Researchers have determined that these basal nuclei nerve cells produce the neurotransmitter dopamine. The lack of this neurotransmitter seems to cause the signs and symptoms of the disorder. Treatment for the disease has, until recently, focused on ways to replace dopamine in the brain. Drug treatment produces temporary dopamine replacement and relieves the symptoms completely for a few weeks to months. However, as the disease progresses, patients need increasingly stronger medications in higher dosages to relieve the symptoms. These stronger medications produce undesirable side effects, such as dizziness, sleepiness, and memory loss.

Implants of dopamine-producing cells have also been placed into the brain. These implants have had low success rates in relieving symptoms. Research into this practice has largely been discontinued. (Because the cells are often obtained from human embryos, scientists have also raised ethical concerns about the source of the implanted cells.)

Increasingly, therapy for Parkinson's disease involves directly treating brain tissue. In **deep-brain stimulation (DBS),** a set of electrodes is implanted into precise centers in the brain. The electrodes are connected to an electrical neurostimulator implanted into the chest near the clavicle, or collarbone. The neurostimulator delivers continuous electrical signals into the patient's brain, blocking the signals that cause Parkinsonian movement. Once implanted, the stimulator can be adjusted from outside the body to achieve maximum symptom relief. **Pallidotomy** is performed on an awake patient. In this procedure, a fine probe is placed into the brain tissue. Next, electrical signals are applied to pinpoint the area for treatment, and the patient describes the signal's effects to the surgeon. Based on these descriptions, the surgeon can burn a tiny hole in the brain tissue, permanently stopping the signals that cause abnormal movements.

Selected New Terms

Basic Key Terms

acetylcholine (ăs″ē-tĭl-kō′lēn), p. 167

acetylcholinesterase (ăs″ē-tĭl-kō″lĭn-ěs′těr-ās), p. 168

action potential (ăk′shŭn pō-těn′shŭl), p. 165

afferent (sensory) system (ăf′ěr-ěnt sĭs′těm), p. 180

arachnoid mater (ŭh-răk′nŏyd māh′ tŭr), p. 171

association area (ŭh-sō″sē-ā′shŭn ā′rē-ŭh), p. 175

autonomic motor system (ăw″tō-nŏm′ĭk mō′těr sĭs′těm), p. 180

axon (ăk′sŏn), p. 164

axon terminal (ăk′sŏn těr′mĭn-ŭhl), p. 164

basal nuclei (bās′ŭl nū′klē-ī), p. 176

brain stem (brān′ stěm′), p. 179

Broca's area (brō′kŭz ā′ rē-ŭh), p. 176

cell body (sěl bŏd′ē), p. 164

central canal (sěn′ trŭhl kŭh-năl′), p. 172

central nervous system (sěn′trŭhl něr′vŭs sĭs′těm), p. 163

cerebellum (sěr″ě-běl′ŭm), p. 178

cerebral cortex (sěr′ē-brŭl kŏr′těks), p. 173

cerebral hemisphere (sěr′ē-brŭl hěm′ĭs-fēr), p. 173

cerebrospinal fluid (sěr″ē-brō-spī′nŭl flū′ĭd), p. 172

cerebrum (sěr′ē-brŭm), p. 173

corpus callosum (kŏr′pŭs kŭ-lō′sŭm), p. 173

cranial nerve (krā′nē-ŭl něrv), p. 180

dendrite (děn′drīt), p. 164

depolarization (dē-pō″lŭh-rīz-ā′shŭn), p. 165

diencephalon (dī″ěn-sěf′ŭh-lŏn), p. 176

dural venous sinuses (dūr′ŭhl vē′nŭs sī′ nŭs-ez), p. 171

dura mater (dūr′ŭh māh′ tŭr), p. 171

efferent (motor) system (ěf′ěr-ěnt sĭs′těm), p. 180

ganglion (găng′glē-ŏn), p. 180

general interpretive area (gěn′ěr-ŭhl ĭn-těr′prě-tĭv ā′ rē-ŭh), p. 176

graded potential (grād-ěd pō-těn′shŭl), p. 167

gray matter (grā măt′ěr), p. 171

hippocampus (hĭp′ō-kăm′pŭhs), p. 178

hypothalamus (hī″pō-thăl′ŭh-mŭs), p. 176

interneuron (ĭn″těr-nū′rŏn), p. 165

limbic system (lĭm′bĭk sĭs′těm), p. 178

medulla oblongata (mĭ-dūl′ŭh ŏb′lŏng-gă′tŭh), p. 179

meninges (měn-ĭn′jēz), p. 171

midbrain (mĭd′brān), p. 179

motor neurons (mō′těr nū′rŏnz), p. 165

motor speech area (mō′těr spēch ā′ rē-ŭh), p. 176

nerve (něrv), p. 164

neurotransmitter (nū″rō-trăns′mĭt-ěr), p. 167

norepinephrine (nŏr″ěp-ĭ-něf′rĭn), p. 167

parasympathetic division (pār″ŭh-sĭm″pŭh-thět′ĭk dĭ-vĭzh′ŭn), p. 184

peripheral nervous system (pě-rĭf′ěr-ăl nŭr′vŭs sĭs′těm), p. 163

pia mater (pī′ŭh māh′ tŭr), p. 172

pons (pŏnz), p. 179

posterior root ganglion (pŏs-tēr′ē-ĕr-rūt gāng′glē-ŏn), p. 173

prefrontal area (prē-frŭn′tŭl ā′ rē-ŭh), p. 175

primary motor area (prī′mā-rē mō′ tĕr ā′rē-ŭh), p. 173

primary somatosensory area (prī′mā-rē sō″mă-tō-sĕn′sō-rē ā′rē-ŭh), p. 174

receptor potential (rĭ-sĕp′tĕr pō-tĕn′shŭl), p. 167

reflex (rē′flĕks), p. 181

refractory period (rĭ-frăk′tŭh-rē pēr′ē-ŭd), p. 167

repolarization (rē-pō′lĕr-ĭ-zā′shŭn), p. 165

resting potential (rĕst-ĭng pō-tĕn′shŭl), p. 165

reticular formation (rĭ-tĭk′yŭh-lŭr fōr-mā9σhŭn), p. 179

saltatory conduction (săl′tŭh-tōr′ē kŭn-dŭk′shŭn), p. 167

sensory neurons (sĕn′sŭh-rē nū′rŏnz), p. 164

sodium-potassium pump (sō′dē-ŭm pō-tă′sē-ŭm pŭmp), p. 165

somatic motor system (sō-măt′ ĭk mō′ tĕr sĭs′ tĕm), p. 180

somatic sensory system (sō-măt′ĭk sĕn′sŭh-rē sĭs′tĕm), p. 180

spinal cord (spī′nŭl kōrd), p. 172

spinal nerve (spī′nŭl nĕrv), p. 180

stimulus (stĭm′yŭh-lŭs), p. 165

sympathetic division (sĭm″pŭh-thĕt′ĭk dĭ-vĭzh′ŭn), p. 184

synapse (sĭn′ăps), p. 167

synaptic cleft (sĭ-năp′tĭk klĕft), p. 167

thalamus (thăl′ŭh-mŭs), p. 176

threshold (thrĕsh′hōld), p. 165

tract (trăkt), p. 164

ventricle (vĕn′trĭ-kŭl), p. 172

visceral sensory system (vĭs′ĕr-ŭhl sĕn′ sŭh-rē sĭs′ tĕm), p. 180

Wernicke's area (vŭhr′nĭ-kŭhz ā′ rē-ŭh), p. 176

white matter (whīt măt′ĕr), p. 171

Clinical Key Terms

Alzheimer disease (ăltz′hī-mĕr dĭ-zēz′), p. 170

ankle-jerk reflex (ān′kŭl-jĕrk rē′flĕks), p. 183

cerebral palsy (sĕr′ē-brŭl păll′zē), p. 175

coma (kō′-mŭh), p. 180

deep-brain stimulation (DBS) (dēp brān stĭm-ū-lā′ shŭn), p. 189

electroencephalogram (ĕ-lĕk″trō-ĕn-sĕf′ŭh-lō-grăm), p. 173

herniated disk (hĕr′nē-āt-ĕd dĭsk), p. 173

hydrocephalus (hī″drō-sĕf′ ŭh-lŭs), p. 172

knee-jerk reflex (nē jĕrk rē′flĕks), p. 183

pallidotomy (păl″ĭ-dŏt′ō-mē), p. 189

paraplegia (păr″ŭh-plē′jē-ŭh), p. 177

Parkinson's disease (păr′kĭn-sŭnz dĭ-zēz′), p. 189

quadriplegia (kwăh-drŭh-plē′jē-ŭh), p. 177

Summary

8.1. Nervous System

The nervous system permits sensory input, performs integration, and stimulates motor output.

A. The nervous system is divided into the central nervous system (brain and spinal cord) and the peripheral nervous system (afferent and efferent nervous systems). The CNS lies in the midline of the body, and the PNS is located peripherally to the CNS.

B. Nervous tissue contains neurons and neuroglia. Each type of neuron (motor, sensory, and interneuron) has three parts (dendrites, cell body, and axon). Neuroglia support, protect, and nourish the neurons.

C. All axons transmit the same type of nerve signal: a change in polarity (called an action potential) that moves along the membrane of a nerve fiber. Saltatory conduction in myelinated axons is a faster type of conduction.

D. Transmission of a nerve signal across a synapse is dependent on the release of a neurotransmitter into a synaptic cleft.

E. Graded potentials vary in size and can be excitatory or inhibitory. Excitatory neurotransmitters may cause an action potential, while inhibitory neurotransmitters may prevent one.

F. There are at least 100 known types of neurotransmitters.

8.2. Central Nervous System

A. The CNS, consisting of the spinal cord and brain, is protected by the meninges and the cerebrospinal fluid.

B. The spinal cord, located in the vertebral column, is composed of white matter and gray matter.

C. White matter contains bundles of nerve fibers, called tracts, which conduct nerve signals to and from the higher centers of the brain. Gray matter is mainly made up of short fibers and cell bodies.

D. The spinal cord is a center for reflex action and allows communication between the brain and the peripheral nerves leaving the spinal cord.

E. The brain has four ventricles. The lateral ventricles are found in the left and right cerebral hemispheres. The third ventricle is found in the diencephalon. The fourth ventricle is found in the brain stem.

F. The cerebrum is divided into the left and right hemispheres. The cerebral cortex, a thin layer of gray matter, has five lobes in each hemisphere. The frontal lobe initiates motor output. The parietal lobe is the final receptor for sensory input from the skin and muscles. The other lobes receive specific sensory input.

Various association areas integrate sensory data. Processing centers integrate data from other association areas: The prefrontal area carries out higher mental processes; Broca's area and Wernicke's area are concerned with speech.

G. The diencephalon includes the hypothalamus, thalamus, and pineal gland. The hypothalamus helps control the functioning of most internal organs and controls the secretions of the pituitary gland. The thalamus receives sensory impulses from all parts of the body and relays them to the appropriate area of the cerebrum.

H. The limbic system includes portions of the cerebrum, the thalamus, and the hypothalamus. It is involved in learning and memory and in causing the emotions that guide behavior.

I. The cerebellum controls balance and complex muscular movements.

J. The brain stem contains the medulla oblongata, pons, and midbrain. The medulla oblongata contains vital centers for regulating heartbeat, breathing, and blood pressure. The pons assists the medulla oblongata in regulating the breathing rate. The midbrain contains tracts that conduct impulses to and from the higher parts of the brain. The reticular formation, a web of gray matter that travels through brain stem structures, maintains alertness and assists other brain stem centers.

8.3. Peripheral Nervous System
A nerve contains bundles of axons covered by successive layers of connective tissue.
A. Cranial nerves take signals to and/or from the brain. Spinal nerves take signals to and from the spinal cord.

B. Reflexes (automatic reactions to internal and external stimuli) depend on the reflex arc. Some reflexes are important for avoiding injury, and others are necessary for physiological functions such as maintaining balance.
C. The autonomic nervous system controls the functioning of internal organs.
  The divisions of the autonomic nervous system: (1) function automatically and usually without voluntary control; (2) innervate all internal organs; and (3) utilize two motor neurons and one ganglion for each impulse.
D. The sympathetic division brings about the responses associated with the "fight-or-flight" response.
E. The parasympathetic division "rest and digest" brings about the

responses associated with normally restful activities.

8.4. Effects of Aging
The brain loses nerve cells, and this affects learning, memory, and reasoning. Mental impairment is not an automatic consequence of aging. Maintaining cardiovascular health, pursuing lifelong learning, and sustaining a meaningful social network are essential for overall health of the brain and nervous system. Alzheimer disease is more often seen among the elderly.

8.5. Homeostasis
The nervous system, along with the endocrine system, regulates and coordinates the other systems to maintain homeostasis. Skeletal muscle contraction also plays a role because movement helps us take precautions or stay in a moderate environment.

Study Questions

1. What are the functions of the nervous system? (p. 163)
2. What are the two main divisions of the nervous system? How are these divisions subdivided? (pp. 163–164)
3. What is the general structure of a neuron, and what are the functions of three different types of neurons? (pp. 164–165)
4. Describe the resting potential. Explain the steps of an action potential. How is it propagated? Why do myelinated fibers have a faster speed of conduction? (pp. 165–168)
5. How is the nerve impulse transmitted across a synapse? Name two

well-known neurotransmitters. (pp. 167–169, 171)
6. Name the meninges, and describe their locations. Where do you find cerebrospinal fluid? (pp. 171–172)
7. Describe the structure and function of the spinal cord. (pp. 172–173)
8. What is the difference between the cerebrum and the cerebral cortex? Name the lobes of the cerebral cortex, and state their functions. Describe the primary motor area and the primary somatosensory area. (pp. 173–176)
9. What is the limbic system, and what is its function? (p. 178)

10. Name the other parts of the brain, and give a location and function for each part. (pp. 176, 178–180)
11. Describe the structure of a nerve. In general, discuss the location and function of the cranial nerves and the spinal nerves. (pp. 180–181)
12. Describe a spinal reflex, including the role played by a sensory nerve fiber, interneurons, and a motor fiber. (pp. 181–184)
13. Contrast the actions of the sympathetic and the parasympathetic divisions of the autonomic system. (pp. 184–186)
14. What role does the nervous system play in homeostasis? (pp. 187–188)

Learning Outcome Questions

Fill in the blanks.
1. Whereas the central nervous system is composed of the _____ and _____, the peripheral nervous system is composed of the _____ and _____.
2. A(n) _____ carries action potentials away from the cell body.
3. Motor nerves stimulate _____.

4. During the depolarization portion of an action potential, _____ ions are moving to the _____ of the nerve fiber.
5. The space between the axon ending of one neuron and the dendrite of another is called the _____.
6. ACh is broken down by the enzyme _____ after it has

initiated an action potential on a neighboring neuron.
7. The brain and spinal cord are covered by protective layers called _____.
8. In a simple reflex pathway that includes sensory neurons, motor neurons, and interneurons, only the _____ is found entirely in the CNS.

9. The electrical activity of the brain can be recorded in the form of a(n) _____.
10. The limbic system records emotions and also is involved in _____ and _____.
11. The _____ is the part of the brain responsible for coordinating body movements.
12. The _____ is a gray matter network traveling through the brain stem, which helps to regulate consciousness.
13. The vagus nerve is a _____ nerve that controls _____.
14. While the _____ division of the autonomic nervous system brings about organ responses that are part of the "fight-or-flight" response, the _____ division brings about responses associated with normal restful conditions.
15. Label the following diagram.

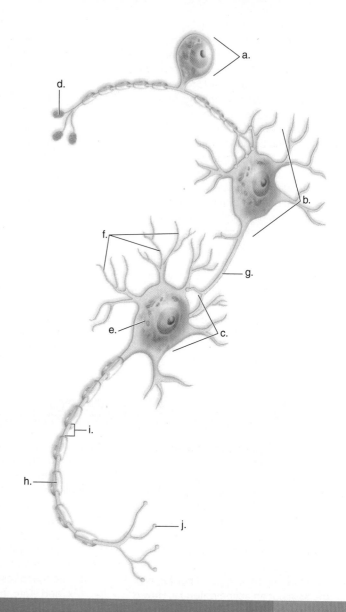

Medical Terminology Exercise

After studying this chapter, see if you can derive the definitions for the medical terms listed at right. Many of the prefixes and suffixes used to create these terms can be found throughout the chapter. For additional help, use McGraw-Hill Connect™ at www.mcgrawhillconnect.com and consult Appendix B.

1. neuropathogenesis (nū″rō-păth″ō-jen″ ĕ-sĭs)
2. anesthesia (ăn″ĕs-thē′zē-ŭh)
3. encephalomyeloneuropathy (ĕn-sĕf″ŭh-lō-mī″ĕ-lō-nŭ-rŏp′ŭh-theē)
4. hemiplegia (hĕm″ĭ-plē′jē-ŭh)
5. glioblastoma (glī″ō-blăs-tō′mŭh)
6. subdural hemorrhage (sŭb-dū′răl hĕm′ōr-ĭj)
7. cephalometer (sĕf″ŭh-lŏm′ĕ-tĕr)
8. meningoencephalocele (mĕ-nĭng″gō-ĕn-sĕf′ŭh-lō-sēl)
9. neurorrhaphy (nū-rō-răh′-fē)
10. ataxiaphasia (ā-tăk″sē-ŭh-fā′zē-ŭh)
11. cerebrovascular accident (sĕr″ē-brō-văs′kyū-lĕr ăk′sŭh-dĕnt)
12. duraplasty (dū′rŭh-plăs-tē)
13. brachycephalic (brāk″ē-sĕf-ăl′ĭk)
14. arachnoiditis (ŭh-răk″nōy-dī′tĭs)

Online Study Tools

LEARNSMART® Mc Graw Hill Education connect®

APR

Anatomy & Physiology REVEALED includes cadaver photos that allow you to peel away layers of the human body to reveal structures beneath the surface. This program also includes animations, radiologic imaging, audio pronunciations, and practice quizzing. To learn more visit www.aprevealed.com

9 The Sensory System

Once again, you hear that really annoying, repetitive song you've heard a million times before, and its lyrics replay over and over in your mind. You're affected by an *earworm*—a phenomenon that is well recognized by the public, but poorly understood by scientists. Researchers propose that earworms are auditory, or sound, memories stored in the temporal lobe, which you'll read about on page 200. Replaying the memories seems to allow the brain to "idle" (much as a car engine idles when it's standing still) while staying prepared for more important tasks. Want to get rid of a particular earworm? Activate your brain's frontal lobe by concentrating on a complex problem, or eat something very spicy to activate the taste areas. Or you can replace one earworm with another, just by thinking of a song more irritating than the first!

Learning Outcomes

After you have studied this chapter, you should be able to:

9.1 General Senses

1. Explain how sensory receptors function to deliver information to the brain, and compare and contrast receptor potentials and action potentials. Categorize sensory receptors according to five types of stimuli.
2. Discuss the function of proprioceptors.
3. Relate specific sensory receptors in the skin to particular senses of the skin.
4. Compare and contrast somatic and visceral nociceptors, and discuss the phenomenon of referred pain.

9.2 Senses of Taste and Smell

5. Name the chemoreceptors, and state their location, anatomy, and mechanism of action.

9.3 Sense of Vision

6. Describe the anatomy and function of the accessory organs of the eye.
7. Detail the anatomy of the eye, and give a function of each part.
8. Describe the sensory receptors for sight, their mechanism of action, and the mechanism for stereoscopic vision.
9. Discuss some common disorders of sight.

9.4 Sense of Hearing

10. Detail the anatomy of the ear, and give a function of each part.
11. Describe the sensory receptors for hearing and their mechanism of action.

9.5 Sense of Equilibrium

12. Describe the sensory receptors for equilibrium and their mechanism of action.

9.6 Effects of Aging

13. Explain the anatomical and physiological changes that occur in the sensory system as we age.

What's New
Detecting Alzheimer Disease with Eye Exams

Medical Focus
Corrective Lenses

Medical Focus
Eye Diseases and Disorders

Focus on Forensics
Retinal Hemorrhage in Shaken Baby Syndrome

Medical Focus
Hearing Damage and Deafness

9.1 General Senses

1. Explain how sensory receptors function to deliver information to the brain, and compare and contrast receptor potentials and action potentials. Categorize sensory receptors according to five types of stimuli.
2. Discuss the function of proprioceptors.
3. Relate specific sensory receptors in the skin to particular senses of the skin.
4. Compare and contrast somatic and visceral nociceptors, and discuss the phenomenon of referred pain.

When a sensory receptor is stimulated, it generates nerve signals that travel to your brain. Interpretation of these signals is the function of the brain, which has a special region for receiving information from each of the sense organs. Signals arriving at a particular sensory area of the brain can be interpreted in only one way; for example, those arriving at the olfactory area result in smell sensation, and those arriving at the visual area result in sight sensation.

Specialized sensory receptors throughout the body start signal transmission to the brain by using a system called a **receptor potential.** Receptor potentials begin with a stimulus (for example, a light stimulus for receptors in the eye). Unlike an action potential (which is an all-or-nothing event), receptor potentials can be weak or strong. Like the signals that occur when nerves synapse, receptor potentials can add together. Although receptors do not generate action potentials, they are parts of neurons or they synapse with other neurons that do create action potentials. In this way, sensory information is transmitted to the brain. The brain integrates data from various sensory receptors in order to perceive whatever caused the stimulation of olfactory and visual receptors—for example, a flower.

Sensory receptors may be categorized into five types based on their stimuli:

Mechanoreceptors, such as pressure receptors in the skin and proprioceptors (specialized stretch receptors) in muscle, are stimulated by changes in pressure or body movement. Other mechanoreceptors in the inner ear also detect body movements. These will be discussed in section 9.5.

Thermoreceptors in the skin and in the internal organs are stimulated by changes in the external or internal temperature.

Pain receptors, such as those in the skin, are stimulated by damage or oxygen deprivation to the tissues.

Chemoreceptors are stimulated by changes in the chemical concentration of substances. Taste buds in the tongue and olfactory (smell) receptors in the nose are chemoreceptors. Other chemoreceptors can sense the concentration of oxygen, carbon dioxide, and hydrogen ions in the blood.

Photoreceptors, which are located only in the eye, are stimulated by light energy.

Sensory receptors in the muscles, joints, tendons, some internal organs, and skin send action potentials to the spinal cord. From there, the action potentials travel up the spinal cord in tracts to the **thalamus,** the sensory relay center of the brain. The information from these receptors is then relayed to the somatosensory areas of the cerebral cortex, located in the parietal lobe of the brain. These general sensory receptors can be categorized into three types: proprioceptors, cutaneous receptors, and pain receptors.

Proprioceptors

Proprioceptors are mechanoreceptors involved in reflex actions that maintain muscle tone and thereby the body's equilibrium and posture. They help us know the position of our limbs in space by detecting the degree of muscle relaxation, the stretch of tendons, and the movement of ligaments. The two forms of proprioceptors, called **muscle spindles** and **Golgi tendon organs,** both detect the degree to which a muscle or tendon is stretched. In response to signals from muscle spindles, motor nerves will increase the degree of muscle contraction. By contrast, when Golgi tendon organs send signals, motor nerves will decrease muscle contraction, and thus, muscle tension. The result is a muscle that consistently has the proper length and tension, or **muscle tone.**

Figure 9.1 illustrates the activity of a muscle spindle. In a muscle spindle, sensory nerve endings are wrapped around thin muscle cells within a connective tissue sheath. When the muscle relaxes and undue stretching of the muscle spindle occurs, nerve action potentials are generated. The rapidity of the action potentials generated by the muscle spindle is proportional to the stretching of a muscle. A reflex action then occurs, which results in contraction of muscle fibers adjoining the muscle spindle. The knee-jerk reflex, which involves muscle spindles, offers an opportunity for physicians to test a reflex action. By contrast, a Golgi tendon organ, which is built into the tough connective tissue of a tendon, is activated by excessive muscle contraction. When Golgi tendon organs generate action potentials, inhibitory signals are sent to the muscle. The muscle is prevented from contracting too forcefully, which could possibly injure both muscle and tendon.

The information sent by muscle spindles and Golgi tendon organs to the CNS is used to maintain the body's equilibrium and posture despite the force of gravity that constantly acts on the skeleton and muscles.

Cutaneous Receptors

The skin is composed of two layers: the epidermis and the dermis. In Figure 9.2, the artist has dramatically indicated these two layers by separating the epidermis from the dermis in one location. The epidermis is stratified squamous epithelium in which cells become keratinized as they rise to the surface. Once at the surface, cells are sloughed off. The dermis is a thick connective tissue layer. The deepest layer of epidermis, and the entire dermis layer, contain **cutaneous receptors,** which make the skin sensitive to touch, pressure, pain, and temperature (warmth and cold). The dermis is a mosaic of these tiny sensory receptors, as you can determine by slowly passing a metal probe over your skin. At certain points, you'll feel touch or pressure, and at others, you'll feel heat or cold (depending on the probe's temperature).

Three types of cutaneous receptors are sensitive to fine touch (like the gentle brush of a fingertip across your cheek). *Meissner corpuscles* are concentrated in the dermal papillary layer of hairless skin: fingertips, palms, lips, tongue, nipples, the penis, and the clitoris (the female sexual organ). *Merkel disks* are found in the deepest epidermal layer, where the epidermis meets the dermis. A *root hair plexus* winds around the base of a hair follicle and signals if the hair is touched.

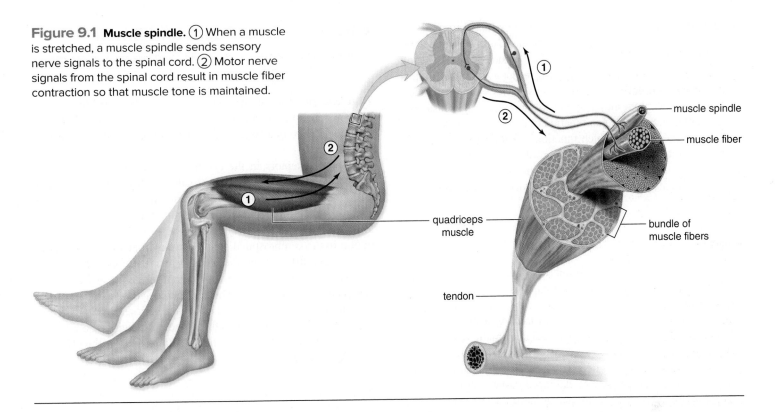

Figure 9.1 Muscle spindle. ① When a muscle is stretched, a muscle spindle sends sensory nerve signals to the spinal cord. ② Motor nerve signals from the spinal cord result in muscle fiber contraction so that muscle tone is maintained.

muscle spindle

muscle fiber

bundle of muscle fibers

quadriceps muscle

tendon

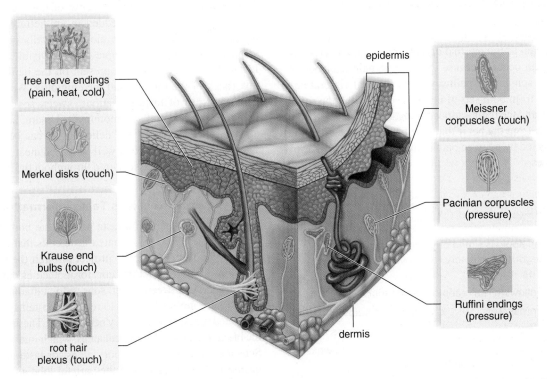

free nerve endings (pain, heat, cold)

Merkel disks (touch)

Krause end bulbs (touch)

root hair plexus (touch)

epidermis

Meissner corpuscles (touch)

Pacinian corpuscles (pressure)

Ruffini endings (pressure)

dermis

AP|R **Figure 9.2** Sensory receptors in human skin.

The three different types of cutaneous receptors that are sensitive to pressure are Pacinian corpuscles, Ruffini endings, and Krause end bulbs. *Pacinian corpuscles* are onion-shaped sensory receptors that lie deep inside the dermis. *Ruffini endings* and *Krause end bulbs* are encapsulated by sheaths of connective tissue and contain lacy networks of nerve fibers. Ruffini endings can be found in the dermis and hypodermis, and in joint capsules as well. Krause end bulbs are located in the superficial layers of the dermis.

The skin's temperature receptors are simply free nerve endings in the superficial dermis and epidermis. Some free nerve endings are responsive to cold; others are responsive to warmth. Cold receptors are far more numerous than warmth receptors, but the two types have no known structural differences.

As you've discovered, each type of receptor described in this section has a well-defined function. However, research continues to show that receptors may respond to more than one sensation. For example, microscopic examination of the skin of the ear shows only free nerve endings, which function as pain receptors. But as you know from everyday life, the skin of the ear responds to all sensations: temperature, fine touch, pressure, and so on. Therefore, it appears that the skin receptors are somewhat, but not completely, specialized.

Pain Receptors

The skin and many (but not all) internal organs have pain receptors, also called **nociceptors.** *Somatic nociceptors* from the skin and skeletal muscles respond to mechanical, thermal, electrical, or chemical damage to these tissues. Skinning a knee, burning a finger, or straining a muscle all stimulate somatic nociceptors. *Visceral nociceptors* react in response to excessive stretching of the internal organs, oxygen deprivation when blood supply is reduced, or chemicals released by damaged tissues. Visceral nociceptors create the pain sensation when the stomach is too full, as well as the crushing pain of a heart attack when blood supply to the heart is reduced.

Stimulation of internal pain receptors is sometimes felt as pain from the skin as well as the internal organs; this is called **referred pain.** Referred pain occurs because the somatic pain nociceptors travel in the same spinal cord pathway as the internal visceral nociceptors. Signals from both sets of neurons converge on the same nervous pathway, and the brain cannot distinguish between the two. For example, pain from the heart that occurs during a heart attack is often accompanied by referred pain in the left shoulder and arm, especially in males.

Content CHECK-UP!

1. Imagine that an apple is sitting on your desktop. Describe, in order, the categories of sensory receptors you'll use as you pick it up and eat it.

2. Which of the following is a receptor for fine touch?
   a. Meissner corpuscles   c. Krause end bulbs
   b. Ruffini endings   d. Pacinian corpuscles

3. Following are pairs of sensory receptors and stimuli to which they respond. Choose the correct pair(s).
   a. Golgi tendon organs—excessive muscle contraction
   b. Visceral nociceptors—oxygen deprivation to an organ
   c. Merkel disks—pressure
   d. a and b
   e. All of the above.

Answers in Appendix A.

9.2 Senses of Taste and Smell

5. Name the chemoreceptors, and state their location, anatomy, and mechanism of action.

Taste and smell are called chemical senses because their receptors are sensitive to molecules in the food we eat and the air we breathe. The body also has other chemoreceptors that govern respiratory rate.

Chemoreceptors in the carotid arteries and in the aorta are sensitive to the oxygen, carbon dioxide, and hydrogen ion concentrations of the blood. These receptors communicate with the respiratory center in the medulla oblongata. Similar chemoreceptors located directly in the medulla respond to hydrogen ion concentration in the cerebrospinal fluid. Both hydrogen ions and carbon dioxide can diffuse from the blood to the CSF; once in the CSF, carbon dioxide can combine with water to generate hydrogen ions (see sections 2.2 and 14.2). Thus, when the hydrogen ion or carbon dioxide concentration of blood increases, both sets of chemoreceptors signal an increase in breathing rate. Exhaling carbon dioxide lowers hydrogen ion concentration, and thus, returns the pH of the blood to its normal level.

Sense of Taste

The sensory receptors for the sense of taste are located in **taste buds.** Taste buds are embedded in epithelium primarily on the tongue (Fig. 9.3). Many lie along the walls of the papillae, the small elevations on the tongue that are visible to the naked eye. Isolated taste buds are also present on the hard palate, the pharynx (back of the throat), and the epiglottis. We have at least five primary types of taste sensations: sweet, sour, salty, bitter, and umami (pronounced "yoo-mommy"). Umami sensation is named after the Japanese word for "delicious"; it can best be described as the pleasant, savory taste of well-seasoned meat. The taste buds for each taste sensation are scattered throughout the tongue (Fig. 9.3*d*).

How the Brain Receives Taste Information

Each taste bud opens at a taste pore. Taste buds have supporting cells and a number of elongated taste cells that end in microvilli. The microvilli of taste cells project through the taste pore. These microvilli have receptor proteins for molecules that cause the brain to distinguish between sweet, sour, salty, umami, and bitter tastes. When these molecules bind to receptor proteins, nerve signals are generated in associated sensory nerve fibers. These nerve signals go to the brain, including the cortical areas, which interpret them as tastes. Sensory receiving and memory areas for taste are in overlapping *gustatory* (taste) regions of the parietal lobe and the insular lobe, the smaller inner lobe of the brain (see section 8.2, pp. 173–175).

Since we can respond to a range of sweet, sour, salty, umami, and bitter tastes, the brain appears to survey the overall pattern of incoming sensory signals and to take a "weighted average" of their taste messages as the perceived taste. Again, we can note that even though our senses are dependent on sensory receptors, the brain integrates the incoming information and gives us our sense perceptions.

a.

AP|R Figure 9.3 Taste buds. (a) When you taste food, activates chemoreceptors called taste buds. **(b)** Papillae on the tongue are sensitive to sweet, sour, salty, bitter, and umami tastes. **(c)** Tongue papillae, which are visible to the naked eye, have taste buds along their walls. **(d)** A single taste bud, shown in a scanning electron micrograph, has **(e)** microvilli that bear receptors. When food molecules bind to the receptors, nerve signals go to the brain where the sensation of taste occurs.

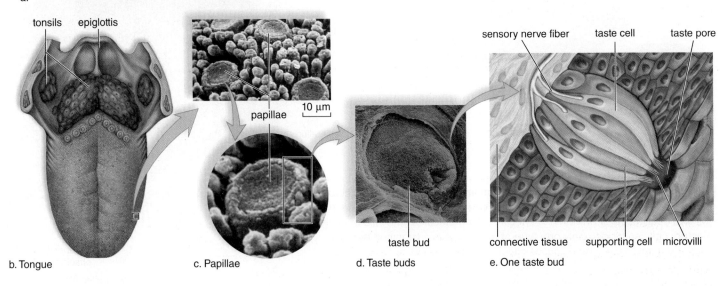

tonsils epiglottis

papillae

10 μm

sensory nerve fiber taste cell taste pore

taste bud

connective tissue supporting cell microvilli

b. Tongue c. Papillae d. Taste buds e. One taste bud

Sense of Smell

Our sense of smell is dependent on **olfactory cells** located within olfactory epithelium high in the roof of the nasal cavity (Fig. 9.4). Olfactory cells are modified neurons. Each cell ends in a tuft of about five olfactory cilia, which bear receptor proteins for odor molecules. The brain distinguishes odors after odor molecules bind to the receptor proteins.

How the Brain Receives Odor Information

Each olfactory cell has only one type out of more than 1,000 different types of receptor proteins. Nerve fibers from olfactory cells lead to neurons in the olfactory bulb, an extension of the brain. An odor contains many odor molecules, which activate a characteristic combination of receptor proteins. For example, a rose might release a certain combination of odor molecules, designated by the blue and green spheres in Figure 9.4, while a carnation might release a different combination. The odor's signature in the olfactory bulb is determined by which neurons are stimulated. When the neurons communicate this information via the olfactory tract to the olfactory areas of the cerebral cortex, we know we've smelled a rose or a carnation. As you know from Chapter 8, the olfactory nerve is cranial nerve I, and the olfactory cortex is located in the temporal lobe. Some areas of the olfactory cortex receive smell sensations; other areas contain olfactory memories.

Have you ever noticed that a certain aroma vividly brings to mind a certain person or place, and can re-create emotions you felt about that person or place? The smell of a certain cologne might depress you by reminding you of a failed relationship, or the smell of boxwood might create happier emotions because it reminds you of your grandfather's farm. The olfactory bulbs have direct connections with the limbic system and its centers for emotion and memory. For example, one investigator conducted an experiment in which his subjects smelled an orange while viewing a painting. Afterward, the descriptions of the painting's details that the subjects shared were much more accurate and vivid.

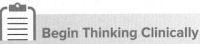

Begin Thinking Clinically

An **uncinate fit** is an olfactory hallucination in which the brain "tricks" the person into smelling odors that have no cause. The hallucinations are usually unpleasant. For example, the sufferer might smell spoiled meat even when there is none nearby. What lobe of the brain might be affected in this condition? What might cause the phenomenon?

Answer and discussion in Appendix A.

Sense of Taste and Sense of Smell

The senses of taste and smell work together to create a combined effect when interpreted by the cerebral cortex. For example,

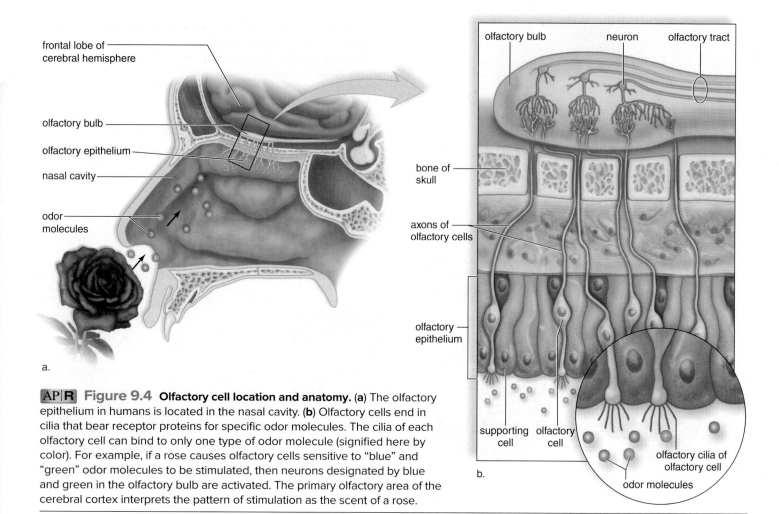

olfactory bulb   neuron   olfactory tract

frontal lobe of cerebral hemisphere

olfactory bulb

olfactory epithelium

nasal cavity

odor molecules

a.

bone of skull

axons of olfactory cells

olfactory epithelium

supporting cell   olfactory cell

olfactory cilia of olfactory cell

odor molecules

b.

AP|R **Figure 9.4** **Olfactory cell location and anatomy.** **(a)** The olfactory epithelium in humans is located in the nasal cavity. **(b)** Olfactory cells end in cilia that bear receptor proteins for specific odor molecules. The cilia of each olfactory cell can bind to only one type of odor molecule (signified here by color). For example, if a rose causes olfactory cells sensitive to "blue" and "green" odor molecules to be stimulated, then neurons designated by blue and green in the olfactory bulb are activated. The primary olfactory area of the cerebral cortex interprets the pattern of stimulation as the scent of a rose.

when you have a cold, you think food has lost its taste, but most likely you've lost the ability to sense its smell. These senses work together because each sensory pathway travels through the limbic system (see page 178 in Chapter 8). Cooperation between smell and taste signals protects us in several ways. Appealing tastes and smells stimulate appetite, help to remind you that it's time to eat, and make eating a pleasurable activity (other signal pathways for hunger are discussed in Chapter 15). By contrast, the smell and taste of spoiled or contaminated food or drink are an early warning system that helps to defend you from becoming sick—you reject the food before you actually swallow it.

Content CHECK-UP!

4. Which of the following statements about the sensation of taste is correct?

  a. Taste buds are located in the back of the throat (pharynx).

  b. Taste buds respond to five primary taste sensations.

  c. Taste buds are a type of chemoreceptor.

  d. a and b

  e. All of the preceding statements are correct.

5. As you know from Section 2.2, many common poisons are water-soluble salts. Most have an alkaline pH. Based on that information, which taste buds would have to be the most sensitive in order to enable someone to taste a poison, and then reject it?

6. Select a correct statement about the sense of smell:

  a. Taste and smell sensations travel through some of the same brain areas.

  b. Olfactory epithelium is located right at the entrance to the nasal cavity.

  c. An odor is made by a single type of odor molecule.

  d. a and c

  e. All of the above.

Answers in Appendix A.

9.3 Sense of Vision

6. Describe the anatomy and function of the accessory organs of the eye.

7. Detail the anatomy of the eye, and give a function of each part.

8. Describe the sensory receptors for sight, their mechanism of action, and the mechanism for stereoscopic vision.

9. Discuss some common disorders of sight.

The photoreceptors for sight are in the eyes. The eyes are located in orbits formed by seven of the skull's bones (frontal, lacrimal, ethmoid, zygomatic, maxilla, sphenoid, and palatine). The bony ridge superior to the orbits, called the *supraorbital ridge,* protects the eye from blows and serves as a location for the eyebrows. The eye has its own accessory organs, too.

Accessory Organs of the Eye

Accessory organs of the eye include: (1) the eyebrows, eyelids, and eyelashes; (2) the lacrimal apparatus, which produces tears; and (3) the extrinsic muscles that move the eye.

Eyebrows, Eyelids, and Eyelashes

Eyebrows have short, thick hairs positioned transversely above the eye along the supraorbital ridge (Fig. 9.5*a*). Eyebrows shade the eyes from the sun and prevent perspiration or debris from falling into the eye. (As you saw in the collection of photos in Chapter 7 introduction, eyebrows are very effective in communication as well!)

Eyelids are a continuation of the skin, and their eyelashes can trap debris and keep it from entering the eyes. Sebaceous glands associated with each eyelash produce an oily secretion that lubricates the eye. Inflammation of one of the glands is called a **sty.**

Blinking of eyelids keeps the eye lubricated and free of debris. The eyelids are operated by the *orbicularis oculi* muscle, which closes the lid, and by the *levator palpebrae superioris* muscle, which raises the lid. People with myasthenia gravis (see page 157) have weakness in these muscles due to the destruction of receptors for the neurotransmitter acetylcholine, and their eyelids often have to be taped open.

The inner surface of an eyelid is lined by a transparent mucous membrane, called the **conjunctiva.** The conjunctiva folds back to cover the anterior of the eye, except for the cornea, which is covered by a delicate epithelium.

Lacrimal Apparatus

A **lacrimal apparatus** consists of the lacrimal gland and the lacrimal sac with its ducts (Fig. 9.5*b*). The lacrimal gland, which lies in the orbit above the eye, produces tears that flow over the eye when the eyelids are blinked. The tears, collected by two small ducts, pass into the lacrimal sac before draining into the nose by way of the nasolacrimal duct.

Extrinsic Muscles

Within an orbit, the eye is anchored in place by the **extrinsic muscles,** whose contractions move the eyes. Each of these muscles originates from the bony orbit and inserts by tendons to the outer layer of the eyeball. There are three pairs of antagonistic extrinsic muscles (Fig. 9.6):

First pair:

Superior rectus	Rolls eye upward
Inferior rectus	Rolls eye downward

Second pair:

Lateral rectus	Turns eye outward, away from midline
Medial rectus	Turns eye inward, toward midline

Third pair:

Superior oblique	Rotates eye counterclockwise
Inferior oblique	Rotates eye clockwise

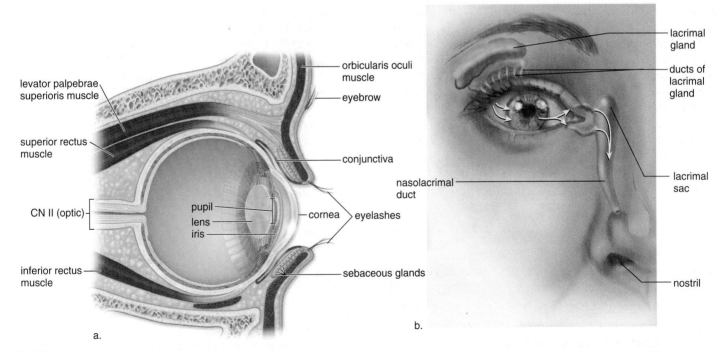

AP|R Figure 9.5 **Accessory structures of the orbit. (a)** Sagittal section of the eye and orbit. **(b)** The lacrimal apparatus. The blue arrows show the path that tears follow as they move across the eye surface.

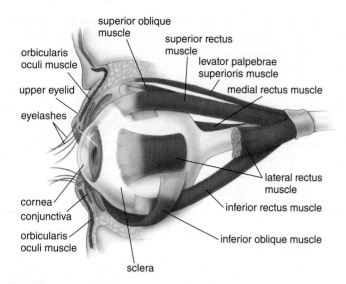

superior oblique muscle
superior rectus muscle
orbicularis oculi muscle
levator palpebrae superioris muscle
upper eyelid
medial rectus muscle
eyelashes
lateral rectus muscle
cornea
inferior rectus muscle
conjunctiva
orbicularis oculi muscle
inferior oblique muscle
sclera

AP|R **Figure 9.6** Extrinsic muscles of the eye, along with the anatomy of the eyelids and eyelashes.

Although stimulation of each muscle causes a precise movement of the eyeball, most movements of the eyeball involve the combined contraction of two or more muscles. For example, if your left eye is directed upward and medially toward your nose, which muscles are required? The answer is the superior and medial rectus muscles.

Three cranial nerves—the oculomotor, trochlear, and abducens—control these muscles. The oculomotor nerve supplies the superior, inferior, and medial rectus muscles, as well as the inferior oblique muscles. The trochlear nerve directs the superior oblique muscle. The abducens nerve innervates the lateral rectus muscle. The motor units of these muscles are the smallest in the body. A single motor axon serves only about 10 muscle fibers, allowing eyeball movements to be very precise.

Anatomy and Physiology of the Eye

The eyeball, which is an elongated sphere about 2.5 cm in diameter, has three layers or coats: the sclera, the choroid, and the retina (Fig. 9.7). Only the retina contains photoreceptors for light energy. Table 9.1 gives the functions of the parts of the eye.

The outer layer, the **sclera,** is white and fibrous except for the **cornea,** which is made of transparent collagen fibers. The cornea is the window of the eye. The middle, thin, darkly pigmented layer, the **choroid,** is vascular and absorbs stray light rays that photoreceptors have not absorbed. Anteriorly, the choroid becomes the donut-shaped **iris.** The iris is a pigmented smooth muscle structure that regulates the size of the **pupil,** a hole in the center of the iris through which light enters the eyeball. The color of the iris (and therefore the color of your eyes) is determined by its pigmentation. Heavily pigmented eyes are brown, while lightly pigmented eyes are green or blue. Behind the iris, the choroid thickens and forms the circular ciliary body. The **ciliary body** contains the **ciliary muscle,** which

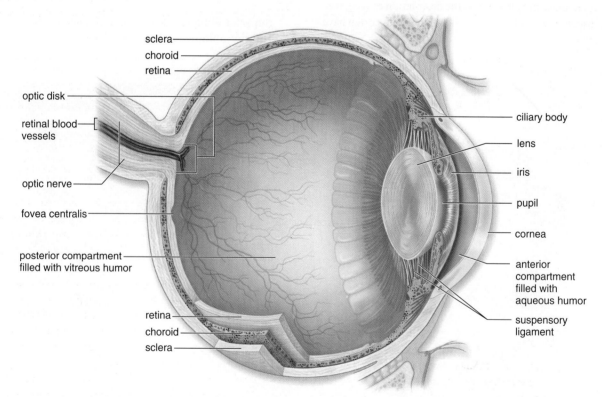

sclera
choroid
retina
optic disk
retinal blood vessels
ciliary body
lens
iris
optic nerve
pupil
fovea centralis
cornea
posterior compartment filled with vitreous humor
anterior compartment filled with aqueous humor
retina
suspensory ligament
choroid
sclera

AP|R **Figure 9.7** **Anatomy of the human eye.** Notice that the sclera, the outer layer of the eye, becomes the cornea and that the choroid, the middle layer, is continuous with the ciliary body and the iris. The retina, the inner layer, contains the photoreceptors for vision; the fovea centralis is the region where vision is most acute. The retina is supplied with a network of tiny blood vessels.

TABLE 9.1 Functions of the Parts of the Eye

Part	Function
Sclera	Tough outermost connective tissue layer; protects and supports eyeball
Cornea	Refracts (bends) light rays
Choroid	Blood vessel layer; absorbs stray light
Ciliary body	Holds the lens in place
Ciliary muscle	Accommodation: changes the shape of the lens for near or far vision
Iris	Regulates entrance of light into retina
Pupil	Opening in iris; admits light into retina
Retina	Contains sensory receptors for light
Rods	Receptors for black and white, dim-light vision; peripheral vision
Cones	Receptors for color vision; bright-light vision
Fovea centralis	Largest concentration of cone cells; makes acute vision possible
Optic nerve	Transmits visual signals to brain
Lens	Refracts (bends) and focuses light rays
Suspensory ligaments	Support lens; attach lens to ciliary body
Aqueous humor	Transmits light rays; supports anterior compartment
Vitreous humor	Transmits light rays; supports posterior compartment

controls the shape of the lens for near and far vision. Changing the shape of the lens is a process called **accommodation.**

The **lens,** which is attached to the ciliary body by the suspensory ligaments, divides the eye into two compartments—the one in front of the lens is the anterior compartment, and the one behind the lens is the posterior compartment. The anterior compartment is filled with a clear, watery fluid called the **aqueous humor.** A small amount of aqueous humor is continually produced each day. Normally, it leaves the anterior compartment by way of tiny ducts. Blockage of the ducts results in glaucoma, as described in the Medical Focus on page 206.

The third layer of the eye, the **retina,** is located in the posterior compartment. The posterior compartment contains the **vitreous body,** which is filled with a clear, gelatinous material called the **vitreous humor.** The retina contains photoreceptors called rod cells and cone cells. The rods are very sensitive to light, but they don't see color; therefore, at night or in a darkened room, we see only shades of gray. Rods also give us peripheral vision, as when you sense movement beside you. The cones, which require bright light, are sensitive to different wavelengths of light, and therefore we have the ability to distinguish colors. The retina has a very special region called the *macula lutea.* At its center is an area termed the **fovea centralis,** which contains only cone cells. Light is normally focused on the fovea when we look directly at an object. This is helpful because vision is most acute in the fovea centralis. Sensory fibers from the retina form the **optic nerve,** which takes nerve signals to the brain. The *optic disk* is a tiny area surrounding the optic nerve. Because there are no photoreceptors in this area, it is the *blind spot* of the retina.

What's New

Detecting Alzheimer Disease with Eye Exams

As mentioned in the Medical Focus in Chapter 8, Alzheimer disease is a progressive, incurable disease of the brain. Researchers now believe that the disease process begins over a decade before the first signs start to appear. Currently, Alzheimer disease is diagnosed by its symptoms—forgetfulness, confusion, mood changes—all of which worsen over time. Available treatments merely delay the patient's inevitable slow decline. Currently, a definitive diagnosis of the disease can only be made at autopsy, when the patient's brain can be physically examined.

Several new diagnostic techniques offer the promise of a much earlier diagnosis of Alzheimer disease. Neurologists have known for some time that the amyloid protein, which destroys the brains of Alzheimer victims, can also be observed in the lenses of late-stage patients. The condition is called **Alzheimer cataract.** One new test takes advantage of this fact by employing laser eye scan technology to detect the early presence of amyloid protein in the lenses of much younger, seemingly healthy patients. First, the patient's lenses are studied using exams similar to those performed in an optometrist's office. Fluorescent drops are then applied, and the lenses are re-examined. The dye binds to amyloid, thus showing any protein accumulation. Additional tests use precise measurements of blood flow in retinal vessels. These studies have shown that the amount and velocity of blood flow diminishes in Alzheimer disease. Along with blood flow studies, researchers are now employing improved retinal scanning technology to accurately view the outermost peripheral areas of the retina. (Previously, examination techniques did not allow for precise examination of the entire peripheral retina to be made.) In these outlying retinal areas, deposits of cellular debris called *drusen* accumulate. Though drusen naturally accumulates with aging, the amount of drusen in the peripheral retina is higher for individuals with Alzheimer disease.

Developers of these technologies envision yearly screening exams, much like mammograms or prostate exams. Most importantly, earlier detection may lead to a better understanding of the disease, which may then allow effective preventative treatments to be developed.

Function of the Lens

The lens, assisted by the cornea and the humors, focuses images on the retina (Fig. 9.8*a*). Focusing starts with the cornea and continues as the rays pass through the lens and the humors. The image produced is much smaller than the object because light rays are bent (refracted) when they are brought into focus. Notice that the image on the retina is inverted (upside down) and reversed from left to right.

Accommodation Imagine taking a walk in the country: as you stroll along, you focus first on a distant tree, then on a nearby flower. To maintain focus as you view distant and then near objects, the lens of the eye must change its shape—a process called visual accommodation. The shape of the lens is controlled by the ciliary muscle within the ciliary body. When we view a distant object such as a tree, the ciliary muscle is relaxed, causing the suspensory ligaments attached to the ciliary body to be taut. Therefore, the lens remains relatively flat (Fig. 9.8*b*). When we view a near object, the ciliary muscle contracts, releasing the tension on the suspensory ligaments, and the lens rounds up due to its natural elasticity (Fig. 9.8*c*). Because close work requires continuous contraction of the ciliary muscle, it very often causes muscle fatigue, known as eyestrain. As discussed in the Medical Focus on page 205, if a person's eyeball is too long or too short, he or she may need corrective lenses to focus the image on the retina.

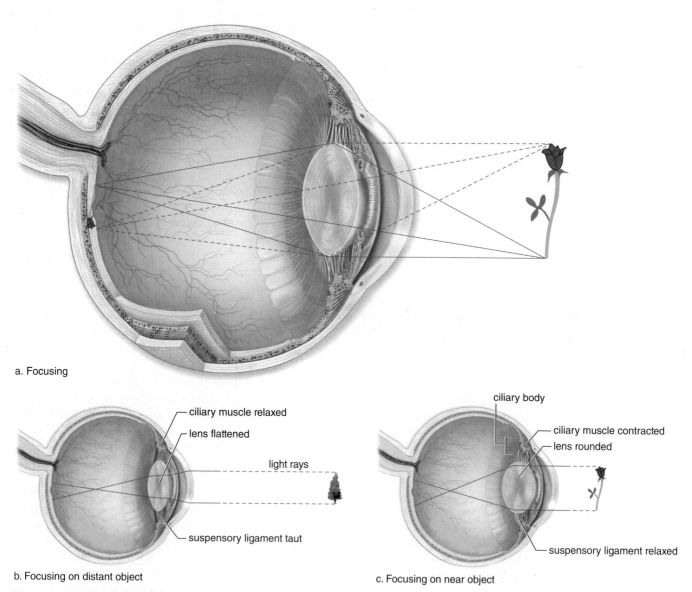

a. Focusing

b. Focusing on distant object

ciliary muscle relaxed

lens flattened

light rays

suspensory ligament taut

c. Focusing on near object

ciliary body

ciliary muscle contracted

lens rounded

suspensory ligament relaxed

AP|R **Figure 9.8** **Focusing.** **(a)** Light rays from each point on an object are bent by the cornea and the lens in such a way that an inverted and reversed image of the object forms on the retina. **(b)** When focusing on a distant object (such as the tree) the lens is flat because the ciliary muscle is relaxed and the suspensory ligament is taut. **(c)** When focusing on a near object, the lens accommodates; it becomes rounded because the ciliary muscle contracts, causing the suspensory ligament to relax.

Usually after the age of 40, the lens loses some of its elasticity and becomes less able to accommodate. Bifocal lenses may then be necessary for those who already have corrective lenses.

Vision Pathway

The pathway for vision begins once light has been focused on the photoreceptors in the retina. Some integration occurs in the retina where nerve signals begin before the optic nerve transmits them to the brain.

Function of Photoreceptors Figure 9.9 illustrates the structure of the photoreceptors called **rod cells** and **cone cells.** Both rods and cones have an outer segment joined to an inner segment by a stalk. Pigment molecules are embedded in the membrane of the many disks present in the outer segment. Synaptic vesicles are located at the synaptic endings of the inner segment.

The visual pigment in rods is a deep purple pigment called rhodopsin. **Rhodopsin** is a complex molecule made up of the protein opsin and a light-absorbing molecule called **retinal,** which is a derivative of vitamin A. When a rod absorbs light, rhodopsin splits into opsin and retinal, leading to a cascade of reactions and the closure of ion channels in the rod cell's plasma membrane. In a sequence of events unique to photoreceptors, the light stimulus *stops* the release of neurotransmitter molecules from the rod's synaptic vesicles. (By contrast, other sensory receptors react to stimuli by releasing neurotransmitter from synaptic vesicles.) Neurons that synapse with the photoreceptors react to the *absence* of neurotransmitter by creating action potentials. Thereafter, nerve signals go to the visual area of the cerebral cortex. Rods are very sensitive to light and therefore are suited to night vision. (Because carrots are rich in vitamin A, it's true that eating carrots can improve your night vision.) Rod cells are plentiful throughout the retina. However, there are no rods in the fovea centralis. Rods provide us with peripheral vision and perception of motion.

The cones, on the other hand, are located primarily in the fovea centralis and are activated by bright light. They allow us to detect the fine detail and the color of an object. Therefore, the condition called macular degeneration, which affects the fovea, is particularly devastating. The Medical Focus on page 206 discusses macular degeneration.

Color vision depends on three different kinds of cones, which contain pigments called the B (blue), G (green), and R (red) pigments. Each pigment is made up of retinal and opsin, but there is a slight difference in the opsin structure of each, which accounts for their individual absorption patterns. Various combinations of cones are believed to be stimulated by in-between shades of color. Inherited absence of the color pigments in the cones of the eye causes **color blindness.** Red-green color blindness, which occurs primarily in men, is the most common form of color blindness.

Function of the Retina The retina has three layers of neurons (Fig. 9.10). The deepest layer (closest to the choroid) contains the

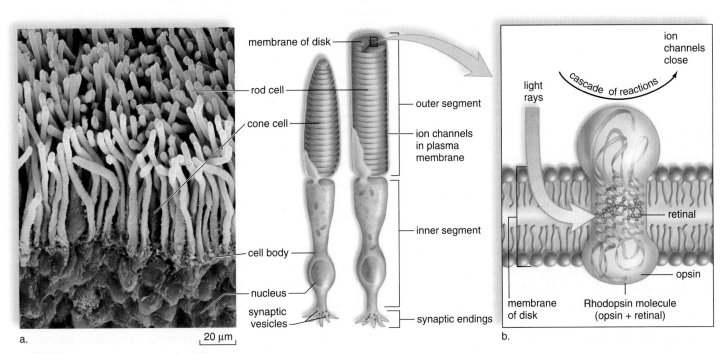

AP|R Figure 9.9 **Photoreceptors in the eye. (a)** The outer segment of rods and cones contains stacks of membranous disks, which contain visual pigments. In rods, the membrane of each disk contains rhodopsin, a complex molecule containing the protein opsin and the pigment retinal. **(b)** When retinal absorbs light energy, it splits, releasing opsin, which sets in motion a cascade of reactions that cause ion channels in the plasma membrane to close. Thereafter, nerve signals go to the brain.

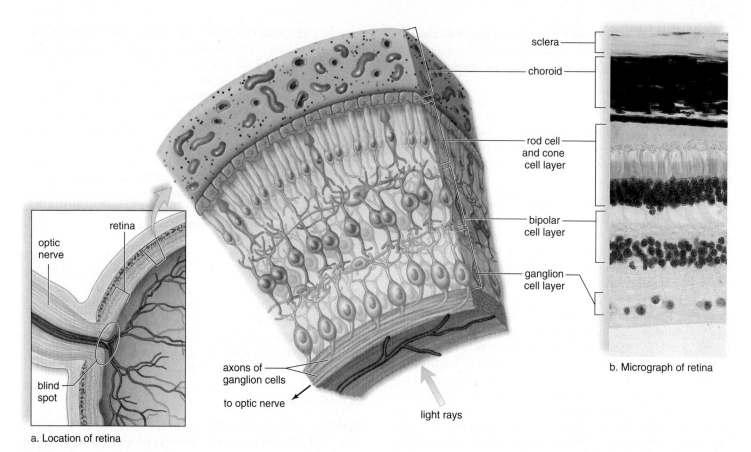

sclera

choroid

rod cell
and cone
cell layer

bipolar
cell layer

ganglion
cell layer

b. Micrograph of retina

retina

optic
nerve

blind
spot

a. Location of retina

axons of
ganglion cells

to optic nerve

light rays

AP|R **Figure 9.10** **Structure and function of the retina. (a)** The retina is the inner layer of the eyeball. Rod cells and cone cells, located at the back of the retina nearest the choroid, synapse with bipolar cells, which synapse with ganglion cells. Integration of signals occurs at these synapses; therefore, much processing occurs in bipolar and ganglion cells. Further, notice that many rod cells share one bipolar cell, but cone cells do not. Certain cone cells synapse with only one bipolar cell. Cone cells, in general, distinguish more detail than do rod cells. **(b)** This micrograph shows that the sclera and choroid are relatively thin compared to the retina, which is composed of several layers of cells.

rod cells and cone cells; the middle layer contains bipolar cells; and the innermost layer contains ganglion cells, whose sensory fibers become the *optic nerve.* Only the rod cells and the cone cells are sensitive to light, and therefore light must penetrate to the back of the retina before they are stimulated.

The rod cells and the cone cells synapse with the bipolar cells, which in turn synapse with ganglion cells that initiate nerve signals. Notice in Figure 9.10 that there are many more rod cells and cone cells than ganglion cells. In fact, the retina has as many as 150 million rod cells and 6 million cone cells but only one million ganglion cells. The sensitivity of cones versus rods is mirrored by how directly they connect to ganglion cells. As many as 150 rods may activate the same ganglion cell. It's no wonder stimulation of rods results in vision that's blurred and indistinct. In contrast, some cone cells in the fovea centralis activate only one ganglion cell. This explains why cones, especially in the fovea, provide us with a sharper, more detailed image of an object.

As signals pass to bipolar cells and ganglion cells, integration occurs. Each ganglion cell receives signals from rod cells covering about one square millimeter of retina (about the size of a thumbtack hole). This region is the ganglion cell's receptive field. Some time ago, scientists discovered that a ganglion cell is stimulated only by

nerve signals received from the center of its receptive field; otherwise, it is inhibited. If all the rod cells in the receptive field receive light, the ganglion cell responds in a neutral way—that is, it reacts only weakly or perhaps not at all. In this way, visual signals are processed and refined in the retina before nerve signals are sent to the brain. The majority of integration occurs in the visual areas of the cerebral cortex.

Blind Spot In Figure 9.10, you can see that there are no rods and cones where the optic nerve exits the retina. Therefore, no vision is possible in this area. You can prove this to yourself by putting a dot to the right of center on a piece of paper. Use your right hand to move the paper slowly toward your right eye while you look straight ahead. The dot will disappear at one point—when its image falls on the retina where receptors are absent. This is your **blind spot.**

From the Retina to the Visual Cortex As stated, sensory fibers from the ganglion cells in the retina assemble to form the optic nerves. The optic nerves carry nerve signals from the eyes to the optic chiasma. The **optic chiasma** has an X-shape formed by a crossing over of some of the optic nerve fibers. At the chiasma, fibers from the right half of each retina converge and continue on together in the *right optic tract,* and fibers from the left half of each retina converge and continue on together in the *left optic tract.*

MEDICAL FOCUS

Corrective Lenses

Normal vision is commonly designated as 20/20. This ratio is determined using a **Snellen chart,** which uses letters of different sizes to test visual acuity (alternative charts for children or illiterate individuals use symbols). The numerator of the ratio is always the distance from the chart: 20 feet. The denominator is the distance at which a normal individual can read the letter: 20 feet. The larger this number, the poorer the subject's vision. Thus, a person with 20/200 vision must stand 20 feet away to read a letter that the normal person can read at 200 feet. (In countries that use metric measures, normal vision is 6/6.) Younger people may actually have vision that is better than 20/20; 20/15 vision is not uncommon in teenagers and young adults. Persons who can see close objects but cannot see the letters from this distance have **myopia**—that is, nearsightedness. Nearsighted people can see close objects better than they can see objects at a distance. These individuals have an elongated eyeball, and when they attempt to look at a distant object, the image is brought to focus in front of the retina (Fig. 9A*a*). They can see close objects because they can adjust the lens to allow the image to focus on the retina, but to see distant objects, these people must wear concave lenses, which spread the light rays so that the image can be focused on the retina.

Rather than wear glasses or contact lenses, many nearsighted people are now choosing to have laser surgery. First, specialists determine how much the cornea needs to be flattened to achieve visual acuity. Controlled by a computer, the laser then removes this amount of the cornea. Most patients achieve at least 20/40 vision, but a few complain of glare, varying visual acuity and chronic dry eyes.

Persons who can easily see the optometrist's chart but cannot see close objects well have **hyperopia**—that is, farsightedness. These individuals can see distant objects better than they can see close objects. They have a shortened eyeball, and when they try to see close objects, the image is focused behind the retina (Fig. 9A*b*). When the object is distant, the lens can compensate for the short eyeball, but when the object is close, these persons must wear a convex lens to increase the bending of light rays so that the image can be focused on the retina.

In some individuals, the cornea assumes an oval shape along one axis (imagine a football-shaped cornea, instead of the normal round "tennis ball" shape).

In rare cases, the lens itself may take on this oblong shape. The light rays cannot be evenly focused on the retina; thus, the image is blurred. This condition, called **astigmatism,** can be corrected by an unevenly ground lens to compensate for the uneven cornea (Fig. 9A*c*).

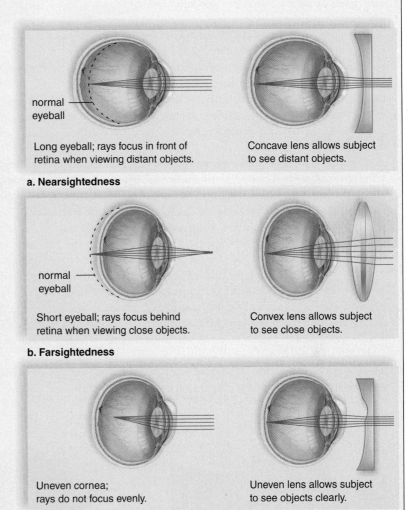

normal eyeball

Long eyeball; rays focus in front of retina when viewing distant objects.

Concave lens allows subject to see distant objects.

a. Nearsightedness

normal eyeball

Short eyeball; rays focus behind retina when viewing close objects.

Convex lens allows subject to see close objects.

b. Farsightedness

Uneven cornea; rays do not focus evenly.

Uneven lens allows subject to see objects clearly.

c. Astigmatism

Figure 9A Common abnormalities of the eye, with possible corrective lenses. (a) A concave lens in nearsighted persons focuses light rays on the retina. **(b)** A convex lens in farsighted persons focuses light rays on the retina. **(c)** An uneven lens in persons with astigmatism focuses light rays on the retina.

Eye Diseases and Disorders

Vision is a human being's strongest sense. When we study the eye, it's impressive to discover what a complex structure it is. Blindness can result from damage to structures involved in the pathway on which light travels to reach the retina, or by damage to the retina itself.

Light must first pass through the cornea to enter the eye, and scarring of the cornea can dim or distort incoming light signals. Common causes for corneal scarring include injury and/or infection. Corneal transplants, using cadaver corneas, can often restore vision. Light next passes through the lens as it continues in its path to the retina. A **cataract** occurs when the lens of the eye becomes clouded, making it difficult to read, drive, or see details distinctly. Cataract development increases with age, but smoking and excessive sunlight exposure are also known risk factors. Fortunately, the lenses of cataract patients can be replaced with an artificial lens implant. When replacement isn't possible, the lenses can be removed entirely. The patient will then see by using special glasses.

The aqueous humor in the eye's anterior chamber is continually being produced and then drained. When a person has **glaucoma,** the drainage ducts are blocked and aqueous humor builds up. Without medication and/or surgical treatment, the resulting pressure inside the eye compresses the retinal arteries. Retinal cells begin to die due to lack of nutrients, and the person becomes partially blind. Eventually, total blindness can result.

If the retinal photoreceptors are destroyed, a person will be blind, even if the rest of the visual pathway is undamaged. The most common cause of blindness resulting from retinal damage is age-related **macular degeneration (MD).** You'll remember that the *macula* (macula lutea, page 201) contains the fovea centralis, where cone cells are concentrated. Individuals with macular degeneration have a distorted visual field: Blurriness or a blind spot is present, straight lines might look wavy, objects look distorted or wrinkled,

and colors may look faded (Fig. 9B). So-called "wet" MD results from abnormal growth of new blood vessels around the macula. The blood vessels leak serum and blood, causing severe scarring and macular destruction. "Dry" MD, in which blood vessel growth doesn't occur, usually precedes the wet form. Vision loss is less severe in dry MD. A family history of MD, smoking, hypertension, lighter eye color, and excessive sunlight exposure are all risk factors for MD. Vitamin and mineral supplements, especially vitamin C and zinc, may help slow the progression of dry MD. When wet MD is diagnosed early, laser treatment can sometimes stop abnormal blood vessel growth.

Like macular degeneration, **diabetic retinopathy** is caused by damage to retinal blood vessels. Diabetic retinopathy is a common complication of poorly controlled diabetes mellitus—almost half of all diabetes patients have some degree of diabetic retinopathy. At first, the diabetic person may notice only a mild vision loss, but blurred or reduced vision (especially night vision) develop as the disease progresses. As damaged blood vessels bleed into the retina and surrounding cells, the retina may detach. Laser surgery is often necessary to stop bleeding and repair a detached retina.

The retina can also be dislodged from the choroid layer underneath by a sharp blow to the eye (for example, being struck by a ball during a baseball game). Because detachment interrupts blood supply, retinal cells will die. Permanent blindness can be prevented by laser reattachment surgery. Tragically, retinal detachment, hemorrhage, blindness, and severe brain damage occur in very young children who are victims of shaken baby syndrome (see Focus on Forensics, p. 208).

A yearly eye examination assists in the early detection of many eye diseases, including diabetic retinopathy, macular degeneration, cataracts, and glaucoma.

Scene viewed by someone with normal vision

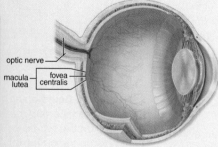

optic nerve

macula lutea — fovea centralis

Same scene viewed by someone with macular degeneration

Figure 9B When a person with macular degeneration looks at a scene, the objects may look larger or smaller than they really are. Colors will appear dimmer, and there may be areas that look gray and wrinkled.

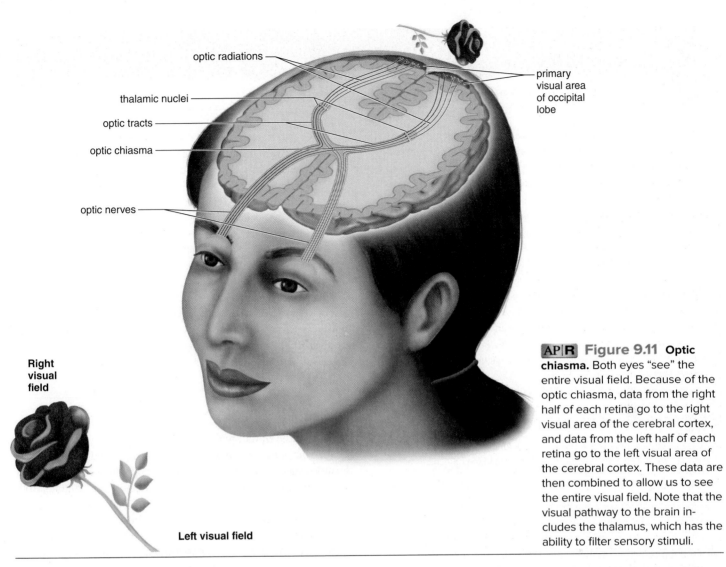

optic radiations

thalamic nuclei

optic tracts

optic chiasma

optic nerves

primary visual area of occipital lobe

Right visual field

Left visual field

AP|R Figure 9.11 Optic chiasma. Both eyes "see" the entire visual field. Because of the optic chiasma, data from the right half of each retina go to the right visual area of the cerebral cortex, and data from the left half of each retina go to the left visual area of the cerebral cortex. These data are then combined to allow us to see the entire visual field. Note that the visual pathway to the brain includes the thalamus, which has the ability to filter sensory stimuli.

The optic tracts sweep around the hypothalamus, and most fibers synapse with neurons in nuclei (masses of neuron cell bodies) in the thalamus. Axons from the thalamic nuclei form *optic radiations* that take nerve signals to the primary visual areas of the occipital lobes (Fig. 9.11). The occipital lobes are a part of the cerebral cortex (see Fig. 8.9).

The *visual cortex* consists of the primary visual area and the visual association areas of the occipital lobes. Notice that the image arriving at the thalamus, and therefore the primary visual areas, has been split because the left optic tract carries information about the right portion of the visual field and the right optic tract carries information about the left portion of the visual field. Therefore, the right visual and left visual cortex must communicate with each other for us to see the entire visual field. Because the image is inverted and reversed (see Figs. 9.8 and 9.11) it must be righted for us to correctly perceive the visual field. Additionally, inputs from each half of the visual cortex are viewed at slightly different angles. This creates *stereoscopic visions,* the impression of three-dimensional width and depth.

The most surprising finding has been that each primary visual area of the cerebral cortex acts like a post office, parceling out information regarding color, form, motion, and possibly other attributes to different portions of the adjoining visual association areas. In other words, the visual field has been taken apart even though we see a unified field. The visual association areas are believed to rebuild the field and give us an understanding of it. Visual association areas also store visual memories of objects previously seen.

Content **CHECK-UP!**

7. The posterior compartment of the eye is filled with:

   a. aqueous humor.
      c. blood.

   b. tears.
      d. vitreous humor.

8. Which statement is true?

   a. Rod cells can be found in the fovea.

   b. Light stimulus to rod cells stops the release of neurotransmitter from the rods.

   c. Cone cells provide peripheral vision.

   d. Cone cells are spread evenly throughout the retina.

9. Imagine that you're looking at your computer screen. After the visual image has reached the fovea centralis, describe the nerve path that the image must take through your brain as it travels to the visual cortex.

Answers in Appendix A.

Retinal Hemorrhage in Shaken Baby Syndrome

It's one of the fastest-growing epidemics in children in North America, and the fifteenth-leading cause of death to young children—child abuse. Approximately 1,600 American children die every year at the hands of a parent or other caregiver, and 75% of those fatalities occur in children four years old or younger. In babies up to a year old, the leading cause of child abuse death is a phenomenon called "shaken baby syndrome," or SBS. As the name implies, the affected infant has been shaken violently by a caregiver. As little as 5 seconds of violent shaking can permanently injure or kill a baby.

Shaking an infant produces the same effect as whiplash in an adult because an infant's head is very large in proportion to the rest of its body and the neck muscles are weak. However, in the infant the whiplash effect occurs over and over. Like an adult whiplash injury, a shaken baby's brain slams back and forth inside the skull. This extreme force damages nerve tissue and tears delicate blood vessels throughout the brain and in the eyes.

One key to making a correct diagnosis of SBS is a retinal exam. The retina is a highly vascular tissue with a complex system of blood vessels. A healthy retina shows distinct blood vessels in a lacy network. The retina of an infant with SBS shows irregular, blotchy areas of hemorrhaged blood. Evidence of retinal hemorrhage should always lead to suspicion of abuse—this injury does not occur in a typical accidental fall.

Studies of adult abusers have shown that child abuse is rarely premeditated; the adult simply loses control while trying to stop a particular behavior, such as excessive crying. Because adult caregivers routinely deny involvement in a child's injury, health-care workers must be vigilant and observant to detect and stop SBS. Unexplained drowsiness, unconsciousness, or seizures in an infant should always be investigated with an eye exam, using eye drops to dilate the pupil and examine the retina.

9.4 Sense of Hearing

10. Detail the anatomy of the ear, and give a function of each part.
11. Describe the sensory receptors for hearing and their mechanism of action.

The ear has two sensory functions: hearing and equilibrium (balance). The sensory receptors for both of these are located in the inner ear, and each consists of **hair cells** with *stereocilia* (long microvilli) that are sensitive to mechanical stimulation. The hair cells are mechanoreceptors, which respond to pressure or body movement.

Anatomy and Physiology of the Ear

Figure 9.12 shows that the ear has three divisions: outer, middle, and inner. The **outer ear** consists of the **pinna** (external flap) and the external **auditory canal.** The opening of the auditory canal is lined with fine hairs and sweat glands. Modified sweat glands are located in the upper wall of the canal. There, they secrete *cerumen,* or earwax, a substance that helps guard the ear against the entrance of foreign materials, such as air pollutants.

The middle ear is carved into the temporal bone of the skull. It begins at the **tympanic membrane** (eardrum). Three tiny bones, collectively called the auditory **ossicles,** are found in the middle ear. Individually, they are the **malleus** (hammer), the **incus** (anvil), and the **stapes** (stirrup) because their shapes resemble these objects. The malleus attaches to the tympanic membrane, then forms a joint with the incus. In turn, the incus then connects to the stapes. The stapes fastens to the **oval window**—a small, membrane-covered opening in the bone of the inner ear. Another, even smaller membrane, the **round window,** is also located in this bone. The round window helps to diminish pressure waves in the inner ear. An **auditory tube** (Eustachian tube), which extends from each middle ear to the nasopharynx (area at the back of the throat, which joins the nasal cavity), allows us to equalize air pressure. Chewing gum, yawning, and swallowing in elevators and on airplanes help to move air through the auditory tubes upon ascent and descent. As this occurs, we often hear our ears "pop."

The inner ear lies in the **bony labyrinth,** a delicately carved cavity within the temporal bone of the skull. Lining the bone is a tube of tissue called the **membranous labyrinth.** Two distinct fluids are found in the inner ear: The space between the bony labyrinth and the membranous labyrinth contains **perilymph** (a fluid similar to cerebrospinal fluid) and the membranous labyrinth is filled with **endolymph.** The inner ear is divided into three areas: the semicircular canals, the vestibule, and the cochlea. The **semicircular canals** and the **vestibule** are both concerned with equilibrium and the **cochlea** is involved with hearing. The cochlea resembles the shell of a snail because it spirals.

Sound Pathway

Sound waves pass through the auditory canal and middle ear to the cochlea in the inner ear, which transforms them into nerve signals conducted in the auditory nerve to the brain.

Through the Auditory Canal and Middle Ear The process of hearing begins when sound waves enter the auditory canal. Just as ripples travel across the surface of a pond, sound waves travel by the successive vibrations of air molecules. Sound waves do not usually carry much energy, but when a large number of waves strike the tympanic membrane, it moves back and forth (vibrates) ever so slightly. The malleus then vibrates from the pressure wave from the inner surface of the tympanic membrane, in turn vibrating the incus and then the stapes. The slight pressure wave is multiplied

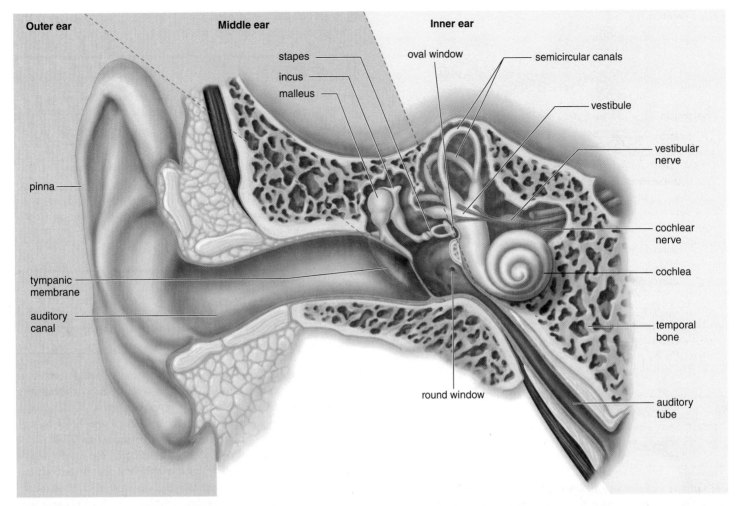

AP|R **Figure 9.12** **Anatomy of the human ear.** The outer ear consists of the pinna and the auditory canal. In the middle ear, the malleus (hammer), the incus (anvil), and the stapes (stirrup) amplify sound waves. In the inner ear, the mechanoreceptors for equilibrium are in the semicircular canals and the vestibule, and the mechanoreceptors for hearing are in the cochlea.

about 20 times as it moves from bone to bone. Finally, the stapes strikes the membrane of the oval window, causing it to vibrate. In this way, the pressure wave is passed to the perilymph within the cochlea of the inner ear.

From the Cochlea to the Auditory Cortex If the cochlea is unwound and examined in cross section (Fig. 9.13), you can see that it has three ducts: the vestibular duct, the **cochlear duct,** and the tympanic duct. Both the vestibular and tympanic ducts are filled with perilymph, while the cochlear duct is filled with endolymph. The sense organ for hearing, called the **spiral organ** (organ of Corti), is located in the cochlear duct. The spiral organ consists of tiny hair cells rooted in a *basilar membrane.* A gelatinous material called the *tectorial membrane* sits above the hair cells, and their stereocilia are embedded in this membrane.

When the stapes strikes the membrane of the oval window, pressure waves in the perilymph travel from the vestibular duct to the tympanic duct across the basilar membrane. The basilar membrane moves up and down, and the stereocilia of the hair cells embedded in the tectorial membrane bend. Hair cells communicate

with neurons of the cochlear nerve (part of cranial nerve VIII, called the *vestibulocochlear,* or *auditory, nerve*). Bending the stereocilia of a hair cell causes changes in the amount of signaling from the hair cell to the neuron, and this affects the frequency of action potentials that are generated in the neuron. Action potential signals are transmitted into the brain stem, where they are relayed through the thalamus to the auditory cortex in the temporal lobe. There, the signals are interpreted as sound. Auditory memories of the sounds we've heard in the past, such as the earworms discussed in the chapter introduction, are also housed in the temporal lobe.

Each part of the spiral organ is sensitive to different wave frequencies, or *pitches.* Near the tip, the spiral organ responds to low pitches, such as a tuba, and near the base, it responds to higher pitches, such as a bell or a whistle. The nerve fibers from each region along the length of the spiral organ lead to slightly different areas in the brain. The pitch sensation we experience depends upon which region of the basilar membrane vibrates and which area of the auditory cortex is stimulated.

Volume is a function of the amplitude (height) of sound waves. Loud noises cause the fluid within the vestibular canal to exert more

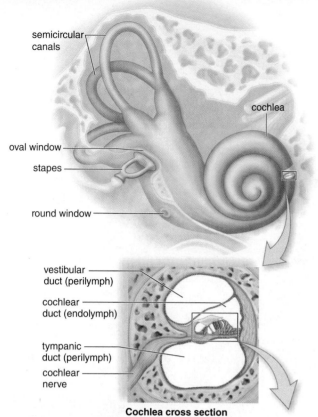

- semicircular canals
- cochlea
- oval window
- stapes
- round window

- vestibular duct (perilymph)
- cochlear duct (endolymph)
- tympanic duct (perilymph)
- cochlear nerve

Cochlea cross section

- tectorial membrane
- stereocilia
- basilar membrane
- hair cell
- cochlear nerve
- tympanic canal

Spiral organ

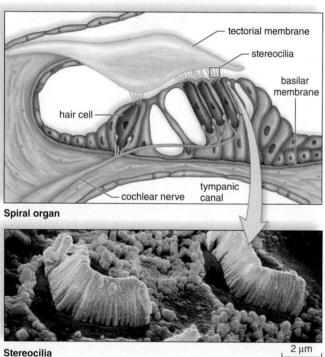

Stereocilia   2 μm

AP|R **Figure 9.13 Mechanoreceptors for hearing.** The spiral organ (organ of Corti) is located within the endolymph-filled cochlear duct. In the uncoiled cochlea, note that the spiral organ consists of hair cells (resting on the basilar membrane) and the tectorial membrane above the hair cells. The vestibular and tympanic ducts are filled with perilymph. Pressure waves in perilymph move from the vestibular duct to the tympanic duct, in turn causing the basilar membrane to vibrate. This causes the stereocilia (of at least a portion of the more than 20,000 hair cells) embedded in the tectorial membrane to bend. The nerve signals that are thus generated travel in the cochlear nerve, resulting in hearing.

pressure and the basilar membrane to vibrate farther and faster. The resulting increased stimulation is interpreted by the brain as volume.

Content **CHECK-UP!**

10. Imagine you're listening to your favorite song. Describe the pathway the sound waves must travel to reach the inner ear.

11. Which structure allows air pressure to equalize in the middle ear?

  a. auditory canal   c. cochlear canal

  b. auditory tube   d. semicircular canal

12. Nerve signals from the cochlea travel along cranial nerve number _____, called the _____ nerve. This nerve takes the signals to the thalamus, and then on to the auditory cortex in the _____ lobe of the cerebrum.

Answers in Appendix A.

9.5 Sense of Equilibrium

12. Describe the sensory receptors for equilibrium and their mechanism of action.

Mechanoreceptors in the semicircular canals detect rotational and/or angular movement of the head (**rotational,** or **dynamic, equilibrium**), while mechanoreceptors in the vestibule detect head position, as well as linear movement of the head in any direction (**gravitational,** or **static, equilibrium**) (Fig. 9.14).

Through their communication with the brain, these mechanoreceptors help us achieve equilibrium, but other structures in the body are also involved. The cerebellum must integrate information from mechanoreceptors in the inner ear, proprioceptors in muscles and joints, and photoreceptors in the eye to maintain balance and equilibrium.

Rotational Equilibrium Pathway

Rotational equilibrium involves the three semicircular canals, which are arranged so that there is one in each dimension of space. Within each canal is a tube of membranous labyrinth called a semicircular duct, and each duct is filled with endolymph. The base of each of the three ducts, called the **ampulla,** is slightly enlarged. Tiny hair cells, whose stereocilia are embedded within a gelatinous material called a *cupula,* are found within the ampullae. Because of the way the semicircular ducts are arranged, each ampulla responds to head rotation in a different plane of space. As endolymph within a semicircular duct flows over and displaces a cupula, the stereocilia of the hair cells bend, and the pattern of signals carried by the vestibular nerve to the brain changes. The cerebellum and other brain centers use information from the hair cells within the ampullae of the semicircular ducts to maintain balance. The sensory input from semicircular canals tells the brain that you are rotating; motor output to various skeletal muscles can right your present position in space as need be (so that you don't fall).

Sometimes data regarding rotational equilibrium bring about unfortunate results. For example, continuous movement of fluid in the ducts within the semicircular canals causes one form of motion sickness. **Vertigo** is dizziness and a sensation of rotation. It is possible to simulate a feeling of vertigo by spinning rapidly and

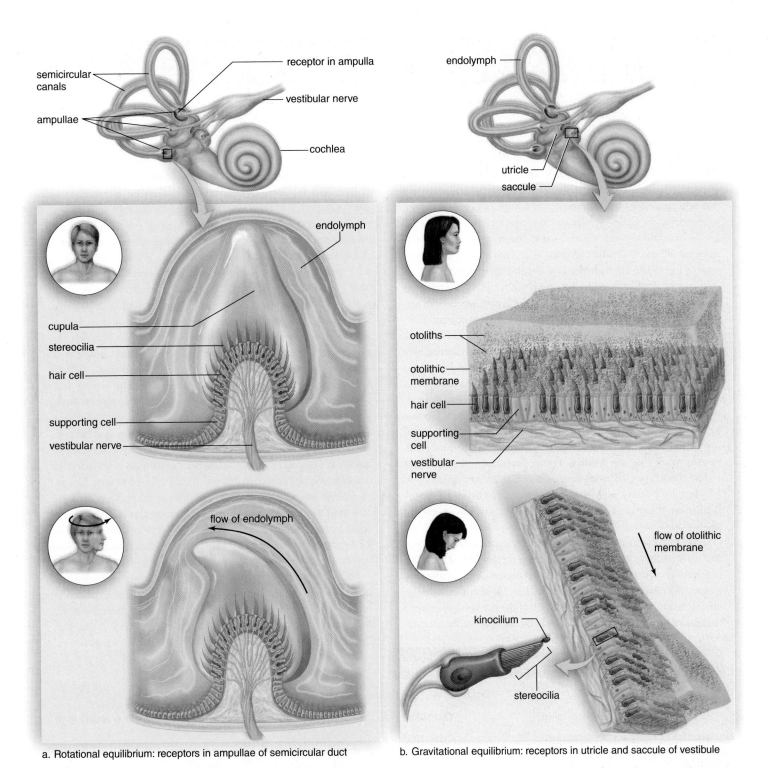

a. Rotational equilibrium: receptors in ampullae of semicircular duct

b. Gravitational equilibrium: receptors in utricle and saccule of vestibule

AP|R **Figure 9.14** **Rotational (dynamic) and gravitational (static) equilibrium. (a)** Rotational equilibrium. The ampullae of the semicircular ducts contain hair cells with stereocilia embedded in a cupula. When the head rotates, the cupula is displaced, bending the stereocilia. Thereafter, nerve signals travel in the vestibular nerve to the brain. **(b)** Gravitational equilibrium. The utricle and the saccule contain hair cells with stereocilia embedded in an otolithic membrane. When the head bends, otoliths are displaced, causing the membrane to sag and the stereocilia to bend. If the stereocilia bend toward the kinocilium, the longest of the stereocilia, nerve signals increase in the vestibular nerve. If the stereocilia bend away from the kinocilium, nerve signals decrease in the vestibular nerve. The difference tells the brain in which direction the head moved.

There are two major types of deafness: conduction deafness and sensorineural, or nerve, deafness. **Conduction deafness** occurs when a mechanical blockage keeps sound waves from reaching the oval window, the membrane at the beginning of the inner ear. Conductive deafness can result from the presence of foreign objects in the external ear canal, impacted ear wax, or cancerous tumors in the external ear canal or middle ear. Conduction deafness can also be due to repeated infections or **otosclerosis.** With otosclerosis, the normal bone of the middle ear is replaced by vascular, spongy bone.

Nerve deafness most often occurs when cilia on the receptors within the cochlea have worn away. Because this may happen with normal aging, older people are more likely to have trouble hearing. However, studies also suggest that age-associated hearing loss can be slowed, or even prevented, if ears are protected from loud noises, starting even during infancy. Hospitals are now aware of the problem and are taking steps to ensure that neonatal intensive care units and nurseries are as quiet as possible. Sensorineural deafness may also result from congenital defects, particularly when a pregnant woman contracts German measles (rubella) during the first trimester of pregnancy. For this reason, every female should be immunized against rubella before her childbearing years (see the Medical Focus: Immunization: The Great Protector in Chapter 13 for information regarding immunizations).

In today's society, exposure to the types of noises listed in Table 9A is common. Everyone should consider three aspects of noise to prevent hearing loss: (1) how loud the noise is (2) how long the noise is heard, and (3) how close the noise is to the ear. Loudness is measured in decibels, and any level above 80 decibels could damage the hair cells of the organ of Corti. Exposure to intense sounds of short duration, such as a burst of gunfire, can result in an immediate hearing loss. Hunters often have significant hearing loss. Interestingly, the butt of the gun partially protects the adjacent ear, so hearing loss occurs in the ear opposite the shoulder

the hunter uses to support the gun. Even listening to city traffic for extended periods can damage hearing, and frequent attendance at rock concerts and constant listening to loud music from an MP3 player (such as an IPod®) are obviously dangerous. The danger is compounded by using earbuds tightly fitted into the external ear canal. Noisy indoor or outdoor equipment, such as a rug-cleaning machine or a chain saw, is also troublesome. Even motorcycles and recreational vehicles, such as snowmobiles and motocross bikes, can contribute to a gradual hearing loss.

The first hint of a problem could be temporary hearing loss, a "full" feeling in the ears, muffled hearing, or **tinnitus** (ringing in the ears). If you have any of these symptoms, modify your listening habits immediately to prevent further damage. If exposure to noise is unavoidable, use specially designed noise-reduction earmuffs or purchase earplugs made from a compressible, sponge-like material at a drugstore or sporting goods store. These earplugs are not the same as those worn for swimming, and they should not be used interchangeably.

Finally, people need to be aware that some medicines are **ototoxic** (damaging to any of the elements of hearing or balance). Anticancer drugs—most notably, cisplatin—and certain antibiotics (for example, streptomycin, kanamycin, and gentamicin) make the ears especially susceptible to a hearing loss. Hearing loss has also occurred in men who take drugs for erectile dysfunction (Viagra,™ Cialis, ™ and Levitra ™). People taking such medications should protect their ears from any excessive noise, and they should discontinue taking the medication and contact a health care professional if they notice hearing loss. It's critically important for cocaine users to seek medical help for their addiction because cocaine can also cause deafness (along with a host of other critical health issues).

Cochlear implants, which directly stimulate the auditory nerve, are available for persons with sensorineural deafness. Earlier models of cochlear implants were costly, and people wearing these electronic devices reported that the speech they heard was very mechanical. Newer models produce better sound quality at a much lower cost.

TABLE 9A	Sound Intensity and Hearing Damage	
Type of Noise	**Sound Level (decibels)**	**Effect**
Rock concert, shotgun, jet engine	Over 125	Beyond threshold of pain; potential for hearing loss is high.
Nightclub, thunderclap, loud music from "ear bud" headphones	Over 120	Hearing loss is likely.
Chain saw, pneumatic drill, jackhammer, symphony orchestra, snowmobile, garbage truck, cement mixer	100–200	Regular exposure of longer than 1 minute risks permanent hearing loss.
Farm tractor, newspaper press, subway, motorcycle	90–100	Fifteen minutes of unprotected exposure is potentially harmful.
Lawnmower, food blender	85–90	Continuous daily exposure for more than 8 hours can cause hearing damage.
Diesel truck, average city traffic noise	80–85	Annoying; constant exposure may cause hearing damage.

Source: National Institute on Deafness and Other Communication Disorders, National Institutes of Health, January 1990.

TABLE 9.2 Functions of the Parts of the Ear

Part	Medium	Function	Mechanoreceptor
OUTER EAR	**AIR**		
Pinna		Collects sound waves	—
Auditory canal		Filters air; directs sound waves to tympanic membrane	—
MIDDLE EAR	**AIR**		
Tympanic membrane and ossicles		Amplify sound waves	—
Auditory tube (Eustachian tube)		Equalizes air pressure	—
INNER EAR	**FLUID**		
Cochlea (contains spiral organ)		Hearing	Stereocilia embedded in tectorial membrane
Semicircular canals		Rotational equilibrium	Stereocilia embedded in cupula
Vestibule (contains utricle and saccule)		Gravitational equilibrium	Stereocilia embedded in otolithic membrane

stopping suddenly. Inflammation of cranial nerve VIII, the vestibulocochlear (or auditory) nerve from the ear, can cause vertigo as well. Motion sickness is also possible if the sensory input from the inner ear is different from visual sensation. For example, a person standing *inside* a large ship on rough seas will sense movement with the inner ear (as the ship tosses back and forth on the waves), but the visual input will register no movement (because the ship's walls don't move relative to the person's body). The person is *seasick,* with severe nausea and vomiting. People who are seasick sometimes find relief if they stand outdoors and focus on the horizon, so that both visual and inner ear inputs to the brain signal movement.

Gravitational Equilibrium Pathway

Gravitational equilibrium depends on the **utricle** and **saccule,** two endolymph-filled membranous sacs located in the vestibule. Both of these sacs contain tiny hair cells, whose stereocilia are embedded within a gelatinous material called an otolithic membrane. Calcium carbonate ($CaCO_3$) granules, or **otoliths,** rest on this membrane. The utricle is especially sensitive to horizontal (back-forth) movements and the bending of the head, while the saccule responds best to vertical (up-down) movements.

When the body is still, the otoliths in the utricle and the saccule rest on the otolithic membrane above the hair cells. Thus, if you're standing completely still with your head upright, or if you're upside down, nerve signals in the vestibular nerve cease. When the head bends or the body moves forward-backward and/or up-down, the otoliths are displaced and the otolithic membrane sags, bending the stereocilia of the hair cells beneath. If the stereocilia move toward the largest stereocilium, called the *kinocilium,* nerve signals increase in the vestibular nerve. If the stereocilia move away from the kinocilium, nerve signals decrease in the vestibular nerve (Fig. 9.14*b*). These data tell the brain the direction of the movement of the head at the moment. The cerebellum and other brain centers use this sensory information to maintain gravitational equilibrium and balance.

Appropriate motor output from the brain to various skeletal muscles can right our present position in space.

Table 9.2 summarizes the functions of the parts of the ear.

Content CHECK-UP!

13. Which of the following receptors are found in the semicircular canals, utricle, and saccule?

a. mechanoreceptors   c. chemoreceptors

b. pressure receptors   d. nociceptors

14. The vestibule generates action potentials that determine _____ equilibrium, while the semicircular canals generate action potentials that determine _____ equilibrium.

Answers in Appendix A.

9.6 Effects of Aging

13. Explain the anatomical and physiological changes that occur in the sensory system as we age.

As people age, many will need assistance to improve their sight and hearing. The lens of the eye in older individuals does not accommodate as well—a condition called **presbyopia.** Eyeglasses, contact lenses, and/or corrective surgery can be used to improve vision.

Three serious visual disorders are seen more frequently in older persons: (1) Possibly due to exposure to the sun, the lens is subject to cataracts. (2) Age-related macular degeneration is the most frequent cause of blindness in older people. (3) Glaucoma is more likely to develop because of a tendency for eye pressure to increase in older individuals. These three disorders are described in the Medical Focus on page 206.

The need for a hearing aid also increases with age. Atrophy of the organ of Corti can lead to **presbycusis** (age-related hearing decline). First, people tend to lose the ability to detect high-frequency tones, and later the lower tones are affected. Eventually, they can hear speech but cannot understand the words being said.

Otosclerosis, an overgrowth of bone that causes the stapes to adhere to the oval window, is the most frequent cause of conduction deafness in adults (see the Medical Focus on page 212). The condition actually begins during youth but may not become evident until later in life. Dizziness and the inability to maintain balance may also occur in older people due to changes in the inner ear.

Content CHECK-UP!

15. What is the term used to describe an age-related decrease in accommodation?

Answer in Appendix A.

Selected New Terms

Basic Key Terms

accommodation (ŭh-cŏm-ŭh-dā'shŭn), p. 201

ampulla (ăm-pūl'ŭh), p. 210

aqueous humor (ā'kwē-ŭs hyū'mer), p. 201

auditory canal (ăw'dĭ-tō"rē kŭh-năl'), p. 208

auditory tube (ăw'dĭ-tō"rē tūb), p. 208

blind spot (blīnd spŏt), p. 204

bony labyrinth (bŏh'nē lăb'ŭh rĭnth), p. 208

chemoreceptor (kē'mō-rĭ-sĕp" tŭr), p. 194

choroid (kō'rŏyd), p. 200

ciliary body (sĭl'ē-ĕr"ē bŏd'ē), p. 200

ciliary muscle (sĭl'ē-ĕr"ē mŭs'ŭhl), p. 200

cochlea (kōk'lē-ŭh), p. 208

cochlear duct (kōk'lē-ĕr dŭkt), p. 209

cone cell (kōn sĕl), p. 203

conjunctiva (kŏn'jŭngk-tī'vŭh), p. 199

cornea (kŏr'nē-ŭh), p. 200

cutaneous receptor (kyū-tā' nē-ŭs rĭ-sĕp' tŭr), p. 194

dynamic equilibrium (dī năm'ĭk ē'kwĭ-lĭb" re-ŭhm) p. 210

endolymph (ĕn'dōh-lĭmf), p. 208

extrinsic muscles (ĕks-trĭn'sĭk mŭs'ĕls), p. 199

fovea centralis (fō'vē-ŭh sĕn-trā'lĭs), p. 201

Golgi tendon organs (gōl'gē tĕn'dŭhn ōr'gŭhnz), p. 194

gravitational equilibrium (grăv'ĭ-tā' shŭn-ŭhl ē'kwĭ-lĭb" re-ŭhm), p. 210

hair cells (har sĕlz), p. 208

incus (ĭng'kŭs), p. 208

iris (ī'rĭs), p. 200

lacrimal apparatus (lăk'rĭ-mŭl ăp"ŭh-ră'tŭs), p. 199

lens (lĕnz), p. 201

malleus (măl'ē-ŭs), p. 208

mechanoreceptor (mĕk"ŭ-nō-rĭ-sĕp"tŭr), p. 194

membranous labyrinth (mĕm'brŭ-nŭs lăb' ŭh-rĭnth), p. 208

muscle spindles (mŭs'ĕl spin'dŭhlz), p. 194

muscle tone (mŭs'ĕl tōn), p. 194

nociceptor (nō"sē-sĕp'tur), p. 196

olfactory cell (ōl-făk'tō-rē sĕl), p. 197

optic chiasma (ŏp'tĭk kī-ăz'mŭh), p. 204

optic nerve (ŏp'tĭk nĕrv), p. 201

ossicle (ŏs'ĭ-kl), p. 208

otolith (ō'tō-lĭth), p. 213

outer ear (ŏw'tĕr ēr), p. 208

oval window (ō'vŭl wĭn'dō), p. 208

pain receptors (pān rĭ-sĕp'tŭrz), p. 194

perilymph (pĕr'ē-lĭmf), p. 208

photoreceptor (fō'tō-rĭ-sĕp"tŭr), p. 194

pinna (pĭn'ŭh), p. 208

proprioceptor (prō'prē-ō-sĕp"tŭr), p. 194

pupil (pyū'pl), p. 200

receptor potential (rĭ-sĕp'tŭr pŭh-tĕn'shŭl), p. 194

retina (rĕt'ĭn-ŭh), p. 201

retinal (rĕt'ĭn-ŏl'), p. 203

rhodospin (rō-dŏp'sĭn), p. 203

rod cell (rŏd sĕl), p. 203

rotational equilibrium (rō-tā'shŭn-ŭhl ē'kwĭ-lĭb"re-ŭhm), p. 210

round window (rŏwnd wĭn'dō), p. 208

saccule (săk'yūl), p. 213

sclera (sklĕr'ŭh), p. 200

semicircular canal (sĕm"ē-sĕr'kyū-lĕr kŭh-nal'), p. 208

Snellen chart (snĕl'ĕn chărt), p. 205

spiral organ (spī'rŭl ŏr'gŭn), p. 209

stapes (stā'pēz), p. 208

static equilibrium (sta'tik e'kwi-lib"re-ŭhm), p. 210

taste bud (tāst bŭd), p. 196

thalamus (thălŭh-mŭs), p. 194

thermoreceptor (thĕr'mō-rĭ-sĕp"tŭr), p. 194

tympanic membrane (tĭm-păn'ĭk mĕm'brăn), p. 208

utricle (ū'trĭ-kŭl), p. 213

vestibule (vĕst'ŭh-byūl), p. 208

vitreous body (vĭt'rē-ŭs bŏd-ē), p. 201

vitreous humor (vĭt'rē-ŭs hyū'mŏr), p. 201

Clinical Key Terms

Alzheimer cataract (ăltz'hī-měr kăt'ŭh-răkt), p. 201

astigmatism (ŭh-stĭg'mŭh-tĭz'ŭm), p. 205

cataract (kăt'ŭh-răkt), p. 206

cochlear implant (kōk'lē-ĕr ĭm'plănt), p. 212

color blindness (kŭl'ŭhr blīnd'nĕs), p. 203

conduction deafness (kŏn-dŭk'shŭn dĕf'nĕs), p. 212

diabetic retinopathy (dī-ŭh-bĕt'ĭk rĕt-ĭ-nŏp'ă-thē), p. 206

glaucoma (glăw-kō'mŭh), p. 206

hyperopia (hi"pĕr-o'pē-ŭh), p. 205

macular degeneration (mă"kyū-lĕr dē"jĕn-ĕr-ā'shŭn), p. 206

myopia (mī-ō'pē-ŭh), p. 205

nerve deafness (nĕrv dĕf'nĕs), p. 212

otosclerosis (ō"tō-sklĕ-rō'sĭs), p. 212

ototoxic (ō"tō-tŏk'sĭk), p. 212

presbycusis (prĕz"bē-kū'sĭs), p. 213

presbyopia (prĕz"bē-ōh'-pē-ŭh), p. 213

referred pain (rĭ-fŭrred pān), p. 196

sty (stī), p. 199

tinnitus (tĭn'ĭ-tŭs), p. 212

uncinate fit (ŭhn'sĭn-āyt fĭt), p. 197

vertigo (vĕr'tĭ-gō), p. 210

Summary

8.1 General Senses
Each type of sensory receptor detects a particular kind of stimulus. When stimulation occurs, sensory receptors initiate receptor potentials. Action potentials are generated and transmitted to the spinal cord and/or brain. Sensation occurs when nerve signals reach the cerebral cortex. Perception is an interpretation of the meaning of sensations.

A. Proprioceptors are mechanoreceptors involved in reflexes that maintain muscle tone, equilibrium, and posture.

B. Cutaneous receptors sense touch, pressure, pain, and temperature.

C. Pain receptors are also called nociceptors, and they sense damage to tissues.

8.2 Senses of Taste and Smell

A. Taste buds contain taste cells, which can sense five primary tastes: sweet, sour, salty, bitter, and umami. Taste signals are sent to gustatory regions in the parietal and insular lobes of the cerebrum.

B. Olfactory cells are found in the olfactory epithelium in the roof of the nasal cavity. Each cell has one type of receptor protein. The senses of taste and smell work together and are directly linked to the limbic system, which explains why these senses are closely linked to emotions and memories.

8.3 Sense of Vision

A. Accessory organs of the eye include eyebrows, eyelids, eyelashes, the lacrimal apparatus, and extrinsic eye muscles. The brows, lids, and lashes protect the eyes, while the lacrimal apparatus produces tears to lubricate the eye.

B. Vision is dependent on the eye, the optic nerves, and the visual areas of the cerebral cortex. The eye has three layers. The outer layer, the sclera, can be seen as the white of the eye; it also becomes the transparent bulge in the front of the eye called the cornea. The middle pigmented layer, called the choroid, absorbs stray light rays. The choroid's anterior portion contains the pigmented iris, which controls the amount of light entering the eye, and the ciliary body, which controls lens shape. The rod cells (sensory receptors for dim light) and the cone cells (sensory receptors for bright light and color) are located in the retina, the inner layer of the eyeball. Cone cells are concentrated in the center of the macula lutea in a region called the fovea centralis. The cornea, the humors, and especially the lens bring the light rays to focus on the retina. To see a close object, accommodation occurs as the lens rounds up.

C. When light strikes rhodopsin within the membranous disks of rod cells, rhodopsin splits into opsin and retinal. A cascade of reactions leads to the closing of ion channels in a rod cell's plasma membrane. Neurotransmitter molecules are no longer released. Nerve cells, which synapse with photoreceptors, transmit action potentials, which are carried in the optic nerve to the brain.

D. Integration occurs in the retina, which is composed of three layers of cells: the rod and cone layer, the bipolar cell layer, and the ganglion cell layer. Integration also occurs in the brain. The visual field is taken apart by the optic chiasma and by the primary visual area in the cerebral cortex, which parcels out signals for color, form, and motion to the visual association area. Then the cortex rebuilds the field.

8.4 Sense of Hearing

A. Hearing is dependent on the ear, the cochlear nerve, and the auditory areas of the cerebral cortex. The ear is divided into three parts: outer, middle, and inner. The outer ear consists of the pinna and the auditory canal, which direct sound waves to the middle ear. The middle ear begins with the tympanic membrane and contains the ossicles (malleus, incus, and stapes). The malleus is attached to the tympanic membrane, and the stapes is attached to the oval window, which is covered by a membrane. The inner ear contains the cochlea and the semicircular canals, plus the utricle and the saccule.

B. Hearing begins when the outer ear receives and the middle ear amplifies the sound waves that then strike the oval window membrane. Its vibrations set up pressure waves in perilymph, which travel across the cochlear canal. The cochlear canal contains the spiral organ, consisting of hair cells whose stereocilia are embedded within the tectorial membrane. When the basilar membrane vibrates, the stereocilia of the hair cells bend. Nerve signals begin in the cochlear nerve and are carried to the brain.

8.5 Sense of Equilibrium

The ear also contains mechanoreceptors for our sense of equilibrium.

A. Rotational equilibrium is dependent on the stimulation of hair cells within the ampullae of the semicircular ducts.

B. Gravitational equilibrium relies on the stimulation of hair cells within the utricle and the saccule.

8.6 Effects of Aging

As we age, we will likely need assistance to improve our diminished senses of sight and hearing. Three more serious visual disorders—cataracts, age-related macular degeneration, and glaucoma—may occur, making medical intervention necessary.

Study Questions

1. Name the five different types of general sensory receptors. (p. 194)

2. Name two types of proprioceptors. What is their general function in the body? (pp. 194–195)

3. Name three types of cutaneous receptors that are sensitive to fine touch, and three that are sensitive to pressure. Where is each type found? (pp. 194–195)

4. How are somatic and visceral nociceptors alike? How are they different? (p. 196)

5. Name the receptors for the senses of taste and smell. Which brain areas receive their sensory information? (pp. 196–197)

6. Describe the relationship between taste and smell. (pp. 197–198)

7. Describe the protective accessory eye structures. How does each work to prevent eye damage? (p. 199)

8. How many pairs of extrinsic eye muscles are there? What movements does each pair cause? (pp. 199–200)

9. What are the eye's three layers? Which contains photoreceptors? (pp. 200–201)

10. Identify the layer where each of the following structures can be found: cornea, fovea centralis, ciliary body, iris. (pp. 200–203)
11. Explain the accommodation process. Describe the shape of the lens when a person is viewing a faraway object, then when he or she is viewing a near object. (pp. 202–203)
12. Describe sight in dim light. What chemical reaction is responsible for vision in dim light? (p. 203)
13. Compare and contrast rods and cones. Explain what role each plays in vision, what chemical each contains, where each is located on the retina, and how each reacts to light. (pp. 203–204)
14. Place the structures of the visual pathway (optic radiations, optic nerve, optic chiasma, optic tracts, thalamus, occipital lobes) in the correct order as an action potential signal travels from the eye to the brain. How does the retina integrate visual information? How does the brain process visual information? (pp. 204, 207)
15. Place the structures of the outer, middle, and inner ears in the correct order as a sound wave travels through the ear. How does the spiral organ (organ of Corti) in the cochlea translate a sound wave into an action potential? Which cranial nerve carries the action potential into the brain? (pp. 208–210)
16. Compare and contrast the utricle and saccule and the semicircular canals. How are they alike? How are they different? (pp. 210–211, 213)
17. Define conduction deafness and sensorineural deafness. Why do young people frequently suffer hearing loss? (pp. 213–214)

Learning Outcome Questions

Fill in the blanks.

1. The sensory organs for position and movement are called _____.
2. Taste buds and olfactory cells are termed _____ because they are sensitive to chemicals in the air and food.
3. The sensory receptors for sight, the _____ and _____, are located in the _____, the inner layer of the eye.
4. The cones give us _____ vision and work best in _____ light.
5. The lens _____ for viewing close objects.
6. People who are nearsighted cannot see objects that are _____. A _____ lens will restore this ability.
7. The ossicles are the _____, _____, and _____.
8. The cochlea is located in the _____ ear and contains hearing receptors called _____ cells.
9. The spiral organ is located in the _____ canal of the _____.
10. The semicircular canals are involved in the sense of _____.
11. Vision, hearing, taste, and smell do not occur unless nerve signals reach the proper portion of the _____.

Medical Terminology Exercise

After studying this chapter, see if you can derive the definitions for the medical terms listed at right. Many of the prefixes and suffixes used to create these terms can be found throughout the chapter. For additional help, use McGraw-Hill Connect™ at www.mcgrawhillconnect.com and consult Appendix B.

1. ophthalmologist (ŏf″thăl-mŏl′ō-jĭst)
2. presbyopia (prĕs″bē-ō′pē-ŭh)
3. blepharoptosis (blĕf″ŭh-rō-tō′sĭs)
4. keratoplasty (kĕr′ŭh-tō-plăs″tē)
5. optometrist (ŏp-tŏm′ĕ-trĭst)
6. lacrimator (lăk′rĭ-mā″tŏr)
7. otitis media (ō-tī′tĭs mē′dē-ŭh)
8. tympanocentesis (tĭm″pŭh-nō-sĕn-tē′sĭs)
9. microtia (mī″krō′shē-ŭh)
10. presbyosmia (prĕs′bē-oz″mē-ŭh)
11. iridomalacia (ĭr′ĭ-dō-mŭh-lā′shē-ŭh)
12. hypogeusia (hī-pō-gō′sē-ŭh)

Online Study Tools

■LEARNSMART® ■connect®

APR

Anatomy & Physiology REVEALED includes cadaver photos that allow you to peel away layers of the human body to reveal structures beneath the surface. This program also includes animations, radiologic imaging, audio pronunciations, and practice quizzing. To learn more visit www.aprevealed.com

Anatomy & Physiology REVEALED®
aprevealed.com

10 The Endocrine System

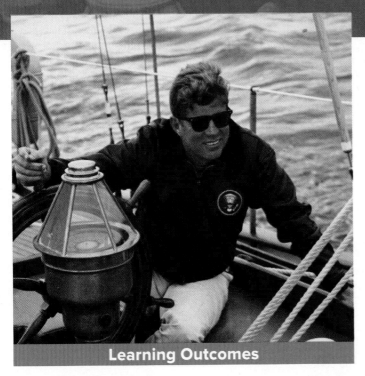

John F. Kennedy, the youngest man to be elected president, appeared healthy, vigorous, and active throughout his entire political career. Photos of the president showed a handsome, tanned sailor at the family estate in Massachusetts; others showed the Kennedy family playing football. However, throughout Kennedy's entire political life, he and his staff took great pains to hide his many ailments—in particular, the fact that Kennedy suffered from Addison's disease. This rare illness is caused by a deficiency of adrenal cortex hormones, and you can read about it on page 228. (Ironically, one symptom of Addison's disease is skin that looks like JFK's signature suntan.) Although he was hospitalized many times, Kennedy repeatedly denied having the disease or any other serious health problems. Yet, medical records tell a different story—the president was in almost constant pain, in part because his vertebrae had been destroyed by the medication for Addison's disease. At autopsy, his adrenal glands were shrunken and nonfunctional. Still, there is no evidence that Kennedy's illness affected his performance as president, and he left a legacy that includes the space program and the Peace Corps.

Learning Outcomes

After you have studied this chapter, you should be able to:

10.1 Endocrine Glands

1. Define a hormone, and state the function of hormones.
2. Discuss the difference in mode of action between peptide and steroid hormones.
3. Name the major endocrine glands, and identify their locations.
4. Discuss the control of glandular secretion by humoral, hormonal, and nervous mechanisms, and give an example of how negative feedback functions in these control mechanisms.

10.2 Hypothalamus and Pituitary Gland

5. Explain the anatomical and functional relationships between the hypothalamus and the pituitary gland.
6. Name and discuss two hormones produced by the hypothalamus that are secreted by the posterior pituitary.
7. Name the hormones produced by the anterior pituitary, and describe their function. Indicate which of these hormones control other endocrine glands.

10.3 Thyroid and Parathyroid Glands

8. Discuss the anatomy of the thyroid gland, and the chemical structure and physiological function of its hormones. Describe the effects of thyroid abnormalities.
9. Discuss the function of parathyroid hormone, and describe the effects of parathyroid hormone abnormalities.

10.4 Adrenal Glands

10. Describe the anatomy of the adrenal glands.
11. Discuss the function of the adrenal medulla and its relationship to the nervous system.
12. Name three categories of hormones produced by the adrenal cortex, give an example of each category, and discuss their actions. Describe the effects of adrenal cortex malfunction.

10.5 Pancreas

13. Describe the anatomy of the pancreas.
14. Name three hormones produced by the pancreas, and discuss their functions.
15. Discuss the two types of diabetes mellitus, and contrast hypoglycemia with hyperglycemia.

10.6 Additional Endocrine Glands

16. Name the most important male and female sex hormones. Discuss their functions.
17. State the location and function of the pineal gland and the thymus gland.

18. Discuss atrial natriuretic hormone, leptin, ghrelin, growth factors, and prostaglandins as hormones not produced by endocrine glands.

10.7 The Importance of Chemical Signals

19. Give examples to show that chemical signals can act between organs, cells, and individuals.

10.8 Effects of Aging

20. Discuss the anatomical and physiological changes that occur in the endocrine system as we age.

10.9 Homeostasis

21. Discuss how the endocrine system works with other systems of the body to maintain homeostasis.

Visual Focus

The Hypothalamus and the Pituitary

I.C.E.—In Case of Emergency

Insulin Shock and Diabetic Ketoacidosis

What's New

Options for Diabetics: The Artificial Pancreas System, Beta Cell Transplants, and the BioHub

Medical Focus

Side Effects of Anabolic Steroids

Human Systems Work Together

Endocrine System

10.1 Endocrine Glands

1. Define a hormone, and state the function of hormones.
2. Discuss the difference in mode of action between peptide and steroid hormones.
3. Name the major endocrine glands, and identify their locations.
4. Discuss the control of glandular secretion by humoral, hormonal, and nervous mechanisms, and give an example of how negative feedback functions in these control mechanisms.

The endocrine system consists of glands and tissues that secrete hormones. This chapter will give many examples of the close association between the endocrine and nervous systems. Like the nervous system, the endocrine system is intimately involved in homeostasis.

Hormones are chemical signals that affect the behavior of other glands or tissues. Hormones influence the metabolism of cells, the growth and development of body parts, and homeostasis. **Endocrine glands** are ductless; they secrete their hormones directly into tissue fluid. From there, the hormones diffuse into the bloodstream for distribution throughout the body. Endocrine glands can be contrasted with exocrine glands, which have ducts and secrete their products into these ducts. For example, the salivary glands send saliva into the mouth by way of the salivary ducts.

Each type of hormone has a unique composition. Even so, hormones can be categorized as either **peptide hormones** (which include proteins, glycoproteins, and modified amino acids) or **steroid hormones.** All steroid hormones are lipids that have the same four-carbon ring complex, but each has different side chains. The majority of hormones are peptides.

Figure 10.1 depicts the locations of the major endocrine glands in the body, and Table 10.1 lists the hormones they release. It's important for you to remember that other organs produce hormones, too. These additional hormones and their actions will be described in detail in later chapters. Further, hormones aren't the only type of chemical signal. Neurotransmitters (which you learned about in Chapter 9) are one example of signaling molecules that allow direct communication from cell to cell, and you'll learn about others here as well.

How Hormones Function

Along with fundamental differences in structure, peptide and steroid hormones also function differently. Most peptide hormones bind to a receptor protein in the plasma membrane and activate a "second messenger" system (Fig. 10.2). The "second messenger" causes the cellular changes for which the hormone is credited. As an analogy, suppose you're the person in charge of a crew assigned to redecorate a room. As such, you stand outside the room and direct the workers inside the room. The workers clean, paint, apply wallpaper, etc. Like the "boss" in this analogy, the peptide hormone stays outside the cell and directs activities within. The peptide hormone, or "first messenger," activates a "second messenger"—the crew workers inside the cell. Common second messengers found in many body cells include **cyclic AMP** (made from ATP, and abbreviated cAMP) and calcium. The second messenger sets in motion an enzyme cascade, so called because each enzyme in turn activates several others, and so on. These intracellular enzymes cause the changes in the cell that are associated with the hormone. Because of the second messenger system, the binding of a single peptide hormone can result in as much as a thousandfold response. The cellular response can be a change in cellular behavior or the formation of an end product that leaves the cell. For example, by activating a second messenger, insulin causes the facilitated diffusion of glucose into body cells, while thyroid-stimulating hormone causes thyroxine release from the thyroid gland.

Because steroid hormones are lipids, they diffuse across the plasma membrane and other cellular membranes (Fig. 10.3). Inside the cell, steroid hormones such as estrogen and progesterone bind to receptor proteins. The hormone-receptor complex then binds to DNA, activating particular genes. Activation leads to production of a cellular enzyme in varying quantities. Again, it is largely intracellular enzymes that cause the cellular changes for which the hormone receives credit. For example, estrogen directs cellular enzymes that cause the growth of axillary and pubic hair in an adolescent female.

When protein hormones such as insulin are used for medical purposes, they must be administered by injection. If these hormones were taken orally, they would be acted on by digestive enzymes. Once digested, insulin cannot carry out its functions. Steroid hormones, such as those in birth control pills, can be taken orally because they're water-insoluble lipids and poorly digested. Steroids can pass through the digestive tract largely undigested, and then diffuse through the plasma membrane into the cell.

Hormone Control

The release of hormones is usually controlled by one or more of three mechanisms: (1) the concentration of dissolved molecules or ions in the blood, referred to as *humoral* control; (2) by the actions of other hormones; and/or (3) by the nervous system.

It's important for ions or molecules in the body to be kept within a normal range to maintain homeostasis. Thus, humoral control determines the secretion of many body hormones. For example, when the blood glucose level rises following a meal, the pancreas secretes insulin. Insulin causes the liver to store glucose and the cells to take up glucose, and blood glucose is lowered back to normal. Once the blood glucose concentration is corrected, the pancreas stops producing insulin. In much the same way, a low level of calcium ions in the blood stimulates the secretion of parathyroid hormone (PTH) from the parathyroid gland. When blood calcium rises to a normal level, secretion of PTH stops. These examples of humoral control illustrate regulation by negative feedback.

Hormone release can also be controlled by specific stimulating or inhibiting hormones. Thyroid-stimulating hormone (TSH) from the pituitary gland (also called **thyrotropin**) does exactly what its name implies—it stimulates the thyroid gland to produce thyroid hormone. By contrast, the release of insulin is inhibited by the production of glucagon by the pancreas. Insulin lowers the blood glucose level, whereas glucagon raises it. In subsequent sections

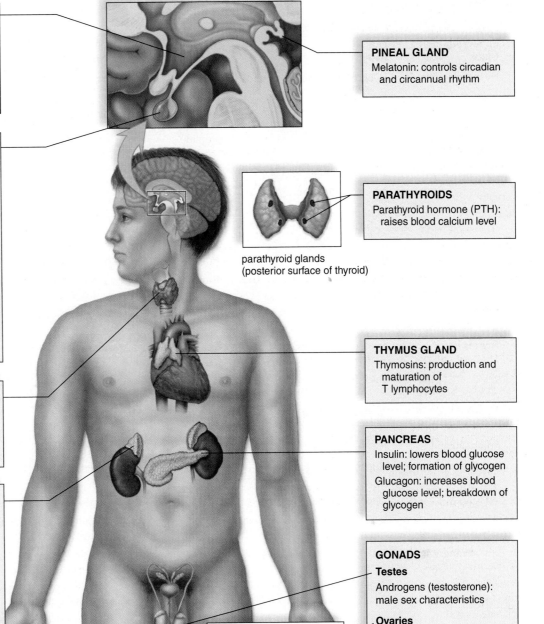

HYPOTHALAMUS

Releasing and inhibiting hormones: regulate the anterior pituitary

Antidiuretic (ADH): water reabsorption by kidneys

Oxytocin: stimulates uterine contraction and milk letdown

PITUITARY GLAND

Posterior Pituitary

Release ADH and oxytocin produced by the hypothalamus

Anterior Pituitary

Thyroid stimulating (TSH): stimulates thyroid

Adrenocorticotropic (ACTH): stimulates adrenal cortex

Gonadotropic (FSH, LH): egg and sperm production; sex hormone production

Prolactin (PL): milk production

Growth (GH): bone growth, protein synthesis, and cell division

THYROID GLAND

Thyroxine (T_4) and triiodothyronine (T_3): increase metabolic rate; regulates growth and development

Calcitonin: lowers blood calcium level

ADRENAL GLAND

Adrenal cortex

Glucocorticoids (cortisol): raises blood glucose level; stimulates breakdown of protein

Mineralocorticoids (aldosterone): reabsorption of sodium and excretion of potassium

Sex hormones: reproductive organs and bring about sex characteristics

Adrenal medulla

Epinephrine and norepinephrine: active in emergency situations; raise blood glucose level

PINEAL GLAND

Melatonin: controls circadian and circannual rhythm

PARATHYROIDS

Parathyroid hormone (PTH): raises blood calcium level

parathyroid glands
(posterior surface of thyroid)

THYMUS GLAND

Thymosins: production and maturation of T lymphocytes

PANCREAS

Insulin: lowers blood glucose level; formation of glycogen

Glucagon: increases blood glucose level; breakdown of glycogen

GONADS

Testes

Androgens (testosterone): male sex characteristics

Ovaries

Estrogens and progesterone: female sex characteristics

testis
(male)

ovary (female)

AP|R **Figure 10.1 Anatomical location of major endocrine glands in the body.** The hypothalamus, pituitary, and pineal glands are in the brain, the thyroid and parathyroids are in the neck, and the adrenal glands and pancreas are in the abdominal cavity. The gonads include the ovaries in females, located in the pelvic cavity, and the testes in males, located outside this cavity in the scrotum. The thymus gland lies within the thoracic cavity.

TABLE 10.1 Principal Endocrine Glands and Hormones

Endocrine Gland	Hormone Released	Chemical Class	Target Tissues/Organs	Chief Function(s) of Hormone
Hypothalamus	Hypothalamic-releasing and inhibiting hormones	Peptide	Anterior pituitary	Regulate anterior pituitary hormones
Produced by hypothalamus, released from posterior pituitary	Antidiuretic (ADH)	Peptide	Kidneys	Stimulates water reabsorption by kidneys and blood vessel constriction
	Oxytocin	Peptide	Uterus, mammary glands	Stimulates uterine muscle contraction, release of milk by mammary glands
Anterior pituitary	Thyroid-stimulating (TSH)	Glycoprotein	Thyroid	Stimulates thyroid gland
	Adrenocorticotropic (ACTH)	Peptide	Adrenal cortex	Stimulates adrenal cortex
	Gonadotropic Follicle-stimulating (FSH) Luteinizing (LH)	Glycoprotein	Gonads	Egg and sperm production; sex hormone production
	Prolactin (PRL)	Protein	Mammary glands	Milk production
	Growth (GH)	Protein	Soft tissues, bones	Cell division, protein synthesis, and bone growth
	Melanocyte-stimulating (MSH)	Peptide	Melanocytes in skin	Unknown function in humans; regulates skin color in lower vertebrates
Thyroid	Thyroxine (T_4) and triiodothyronine (T_3)	Iodinated amino acid	All tissues	Increases metabolic rate; regulates growth and development
	Calcitonin	Peptide	Bones, kidneys, intestine	Lowers blood calcium level
Parathyroids	Parathyroid (PTH)	Peptide	Bones, kidneys, intestine	Raises blood calcium level
Adrenal gland				
Adrenal cortex	Glucocorticoids (cortisol)	Steroid	All tissues	Raise blood glucose level; stimulate breakdown of protein
	Mineralocorticoids (aldosterone)	Steroid	Kidneys	Reabsorb sodium and excrete potassium
	Sex hormones	Steroid	Gonads, skin, muscles, bones	Stimulate reproductive organs and bring about sex characteristics
Adrenal medulla	Epinephrine and norepinephrine	Modified amino acid	Cardiac and other muscles	Released in emergency situations; raise blood glucose level
Pancreas	Insulin	Protein	Liver, muscles, adipose tissue	Lowers blood glucose level; promotes formation of glycogen
	Glucagon	Protein	Liver, muscles, adipose tissue	Raises blood glucose level
	Somatostatin	Protein	Pancreatic alpha + beta cells	Inhibits insulin and glucagon, prevents wide fluctuations in blood glucose
Gonads				
Testes	Androgens (testosterone)	Steroid	Gonads, skin, muscles, bones	Stimulate male sex characteristics
Ovaries	Estrogens and progesterone	Steroid	Gonads, skin, muscles, bones	Stimulate female sex characteristics
Thymus	Thymosins	Peptide	T lymphocytes	Stimulate production and maturation of T lymphocytes
Pineal gland	Melatonin	Modified amino acid	Brain	Controls circadian and circannual rhythms; possibly involved in maturation of sexual organs

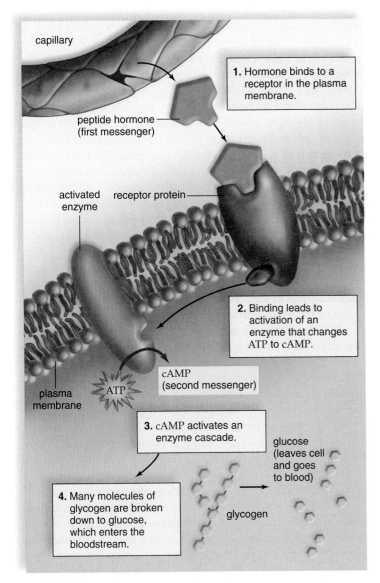

AP|R **Figure 10.2** **Action of a peptide hormone.** The binding of a peptide hormone leads to cAMP and then to activation of an enzyme cascade. In this example, the hormone causes glycogen to be broken down to glucose.

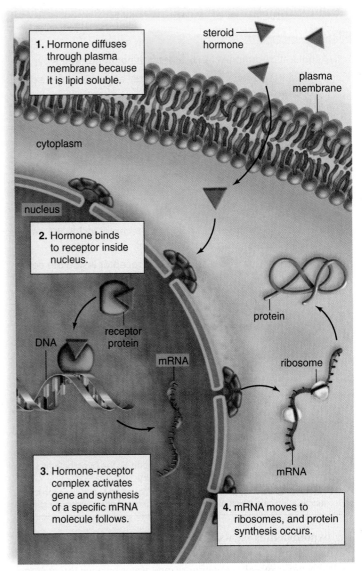

AP|R **Figure 10.3** **Action of a steroid hormone.** A steroid hormone results in a hormone-receptor complex that activates DNA and protein synthesis.

of this chapter, you'll learn about other instances in which pairs of hormones work opposite to one another, and thereby bring about the regulation of a substance in the blood.

The nervous system is an important controller of the endocrine system. Upon receiving sensory information from the body, the brain can make appropriate adjustments to hormone secretion to ensure homeostasis. For example, while you eat a meal, sensory information is relayed to the brain. In turn, the brain signals parasympathetic motor neurons to cause the release of insulin from the pancreas. (Recall that the parasympathetic neurons control "rest and digest" functions.) Insulin will allow body cells to take up glucose from digested food.

It's important to stress that many hormones are influenced by more than one control mechanism. In the previous examples, you can see that insulin release is influenced by all three controllers: humoral, hormonal, and neural control. For the majority of hormones, control is regulated by negative feedback. As you know from Chapter 1, in a negative feedback system, a stimulus causes a body response. The body response, in turn, corrects the initial stimulus. The result is that the activity of the hormone is maintained within normal limits and homeostasis is ensured. However, in a few instances, positive feedback controls the release of a hormone—for example, release of oxytocin during labor and delivery (discussed on page 222).

1. Antidiuretic hormone (ADH), a peptide hormone, works by:

   a. binding to a receptor outside the cell and activating a second messenger.

   b. diffusing into the cell, binding to a receptor inside the cell, and activating a second messenger.

   c. diffusing into the cell, binding to a receptor inside the cell, and activating genes in DNA.

2. Testosterone, a steroid hormone, works by:

   a. binding to a receptor outside the cell and activating a second messenger.

   b. diffusing into the cell, binding to a receptor inside the cell, and activating a second messenger.

   c. diffusing into the cell, binding to a receptor inside the cell, and activating genes in DNA.

3. Antidiuretic hormone stimulates the kidneys to reabsorb water and return it to the blood plasma. ADH release is controlled by a negative feedback system. Describe a situation in your daily life during which you would produce ADH.

Answers in Appendix A.

10.2 Hypothalamus and Pituitary Gland

5. Explain the anatomical and functional relationships between the hypothalamus and the pituitary gland.
6. Name and discuss two hormones produced by the hypothalamus that are secreted by the posterior pituitary.
7. Name the hormones produced by the anterior pituitary, and describe their function. Indicate which of these hormones control other endocrine glands.

The **hypothalamus** regulates the internal environment. For example, through the autonomic nervous system, the hypothalamus helps control heartbeat, body temperature, and water balance (by creating thirst). The hypothalamus also controls the glandular secretions of the **pituitary gland (hypophysis).** The pituitary, a small gland about 1 cm in diameter, is connected to the hypothalamus by a stalklike structure. The pituitary has two portions: the **posterior pituitary (neurohypophysis)** and the **anterior pituitary (adenohypophysis).**

Posterior Pituitary

Neurons in the hypothalamus called neurosecretory cells produce the hormones antidiuretic hormone (ADH) and oxytocin (Fig. 10.4, *left*). These hormones pass through axons into the posterior pituitary (neurohypophysis) where they are stored in axon endings. Thus, the hypothalamic hormones antidiuretic hormone and oxytocin are produced in the hypothalamus, but are released into the bloodstream from the posterior pituitary.

Antidiuretic Hormone and Oxytocin

Certain neurons in the hypothalamus are sensitive to the water-salt balance of the blood. When these cells determine that the sodium in blood has become too concentrated, **antidiuretic hormone (ADH,** also called *vasopressin*) is released from the posterior pituitary. In your body, blood becomes concentrated if you have just finished exercising heavily and body water has been lost as sweat. Upon reaching the kidneys, ADH causes more water to be reabsorbed into kidney capillaries. As the blood becomes diluted once again, ADH is no longer released. This is an example of control by negative feedback because the effect of the hormone (to dilute blood) acts to shut down the release of the hormone. An additional effect of ADH is to raise blood pressure, by vasoconstriction of blood vessels throughout the body (hence, the hormone's additional name of vasopressin). This mechanism also illustrates negative feedback: Blood pressure falls because body water is lost as sweat (stimulus); vasopressin is released (response); blood vessels constrict and blood pressure rises to normal (stimulus corrected). Negative feedback maintains stable conditions and homeostasis.

Inability to produce ADH causes **diabetes insipidus** (watery urine), in which a person produces copious amounts of urine with a resultant loss of ions from the blood. The condition can be corrected by the administration of ADH.

It's interesting to note that alcohol suppresses ADH production and release. When ADH is absent, the kidneys don't reabsorb as much water. The person drinking alcohol urinates more often and may become dehydrated as a result. The symptoms of the drinker's "hangover"—headache, nausea, dizziness—are largely due to dehydration.

Oxytocin, the other hormone made in the hypothalamus, causes uterine contraction during childbirth and milk letdown when a baby is nursing. Uterine contractions during labor push the baby's head against the uterine opening (called the uterine *cervix*). In turn, the cervix dilates, triggering nerve signals to the hypothalamus that cause oxytocin release. The greater the cervical dilation, the more oxytocin is released. Similarly, the more a baby suckles, the more oxytocin is released. In both instances, the release of oxytocin from the posterior pituitary is controlled by **positive feedback**—that is, the stimulus continues to bring about an effect that continues to increase in intensity. Positive feedback is not the best way to maintain stable conditions and homeostasis. However, it works during childbirth and nursing because external mechanisms interrupt the process. In childbirth, the delivery of the baby and afterbirth (the placenta and membranes surrounding the baby) eventually stops oxytocin secretion. When a baby with a full tummy stops nursing, that, too, halts oxytocin secretion. For a nursing mother, the letdown response to oxytocin becomes automatic over time. Thus, it is referred to as a *neuroendocrine reflex*. When the baby begins to suckle, the pressure and touch sensations signal the hypothalamus, oxytocin is released, and the stream of breast milk begins. Similarly, hearing a baby cry—even someone else's baby—will cause a nursing mother's breast milk to flow.

Anterior Pituitary

A portal system lies between the hypothalamus and the anterior pituitary (Fig. 10.4, *center*). In this example, the term **portal system** is used to describe the following unique pattern of circulation:

capillaries → vein → capillaries → vein

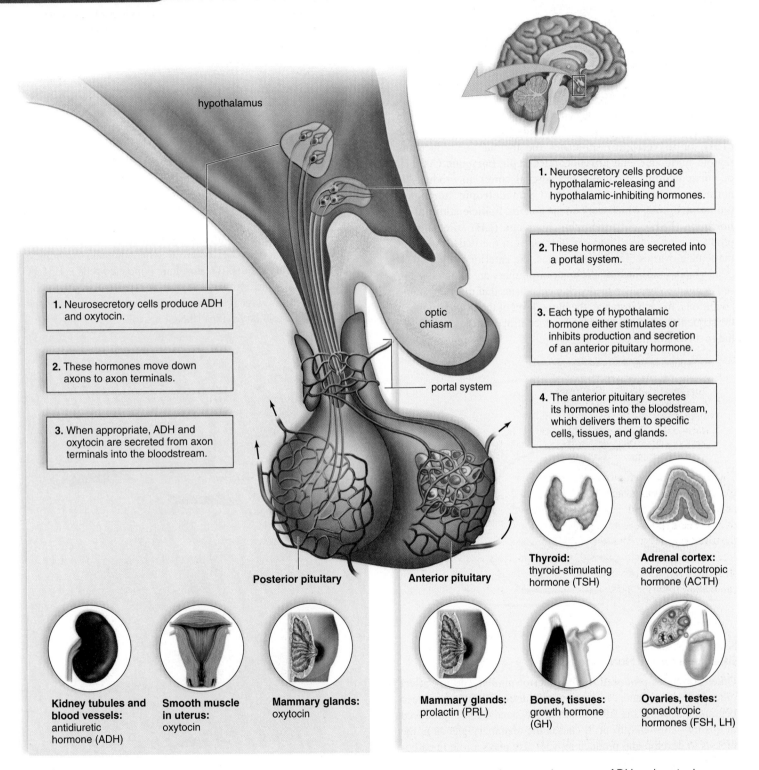

hypothalamus

1. Neurosecretory cells produce hypothalamic-releasing and hypothalamic-inhibiting hormones.

2. These hormones are secreted into a portal system.

optic chiasm

1. Neurosecretory cells produce ADH and oxytocin.

3. Each type of hypothalamic hormone either stimulates or inhibits production and secretion of an anterior pituitary hormone.

2. These hormones move down axons to axon terminals.

portal system

3. When appropriate, ADH and oxytocin are secreted from axon terminals into the bloodstream.

4. The anterior pituitary secretes its hormones into the bloodstream, which delivers them to specific cells, tissues, and glands.

Thyroid: thyroid-stimulating hormone (TSH)

Adrenal cortex: adrenocorticotropic hormone (ACTH)

Posterior pituitary

Anterior pituitary

Kidney tubules and blood vessels: antidiuretic hormone (ADH)

Smooth muscle in uterus: oxytocin

Mammary glands: oxytocin

Mammary glands: prolactin (PRL)

Bones, tissues: growth hormone (GH)

Ovaries, testes: gonadotropic hormones (FSH, LH)

AP|R **Figure 10.4** **The hypothalamus and the pituitary.** *Left:* The hypothalamus produces two hormones, ADH and oxytocin, which are stored and secreted by the posterior pituitary. *Right:* The hypothalamus controls the secretions of the anterior pituitary, and the anterior pituitary controls the secretions of the thyroid, adrenal cortex, and gonads, which are also endocrine glands. It also secretes growth hormone and prolactin.

The hypothalamus controls the anterior pituitary by producing multiple **hypothalamic-releasing hormones** and **hypothalamic-inhibiting hormones.** For example, there is a thyrotropin-releasing hormone (TRH) and a prolactin-inhibiting hormone (PIH). TRH stimulates the anterior pituitary to secrete thyroid-stimulating hormone, and PIH inhibits the pituitary from secreting prolactin.

Hormones That Affect Other Glands

Four of the hormones produced by the anterior pituitary have an effect on other glands. **Thyroid-stimulating hormone (TSH,** or **thyrotropin)** stimulates the thyroid to produce the thyroid hormones, and **adrenocorticotropic hormone (ACTH, or corticotropin)** stimulates the adrenal cortex to produce its hormones. There are actually two **gonadotropic hormones (gonadotropins),** which work on the *gonads*. **Follicle-stimulating hormone (FSH)** and **luteinizing hormone (LH)** stimulate the production of gametes and sex hormones in the male testes and the female ovaries. The hypothalamus, the anterior pituitary, and other glands controlled by the anterior pituitary are all involved in self-regulating negative feedback mechanisms that maintain stable conditions. In each instance, the blood level of the last hormone in the sequence exerts negative feedback control over the secretion of the first two hormones:

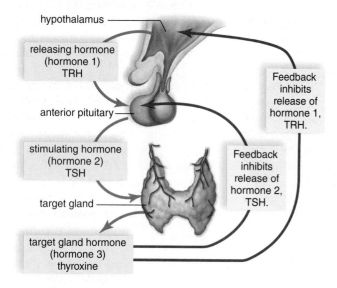

Effects of Other Hormones

Other hormones produced by the anterior pituitary do not affect other endocrine glands. In women, **prolactin (PRL)** is produced beginning at about the fifth month of pregnancy, and is produced in quantity after childbirth. It causes the mammary glands in the breasts to develop and produce milk. It also plays a role in carbohydrate and fat metabolism.

Growth hormone (GH), or somatotropic hormone, stimulates protein synthesis within cartilage, bone, and muscle. It stimulates the rate at which amino acids enter cells and protein synthesis occurs. It also promotes fat metabolism as opposed to glucose metabolism.

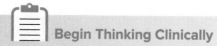

Begin Thinking Clinically

An adenoma, which is one type of pituitary gland tumor, can affect the production of one or more pituitary hormones. What would be the effect of a prolactin adenoma in a man?

Answer and discussion in Appendix A.

Effects of Growth Hormone

The amount of GH produced by the anterior pituitary affects the height of the individual. The quantity of GH produced is greatest during childhood and adolescence, when most body growth is occurring. If too little GH is produced during childhood, the individual has **pituitary dwarfism,** characterized by perfect proportions but small stature. If too much GH is secreted, a person can become a giant (Fig. 10.5). Giants usually have poor health, primarily because GH has a secondary effect on the blood sugar level, promoting an illness called diabetes mellitus (see pages 230–231).

On occasion, GH is overproduced in the adult and a condition called **acromegaly** results. Because long bone growth is no longer possible in adults, the feet, hands, and face (particularly the chin, nose, and eyebrow ridges) can respond, and these portions of the body become overly large (Fig. 10.6).

Figure 10.5 Effect of growth hormone. The amount of growth hormone produced by the anterior pituitary during childhood affects the height of an individual. Too much growth hormone can lead to giantism, while an insufficient amount results in limited stature and even pituitary dwarfism.

Figure 10.6 Acromegaly. Acromegaly is caused by overproduction of GH in the adult. It is characterized by enlargement of the bones in the face, the fingers, and the toes as a person ages.

Content CHECK-UP!

4. Name three anterior pituitary hormones that cause the release of another hormone(s).

5. Oxytocin release from the hypothalamus during labor and delivery is a mechanism that works by:

   a. positive feedback.   b. negative feedback.

6. Which of the following is an effect of growth hormone?

   a. It promotes fat metabolism.

   b. It stimulates protein synthesis in bone and cartilage.

   c. It causes a person to grow taller.

   d. All of the above

Answers in Appendix A.

10.3 Thyroid and Parathyroid Glands

8. Discuss the anatomy of the thyroid gland, and the chemical structure and physiological function of its hormones. Describe the effects of thyroid abnormalities.
9. Discuss the function of parathyroid hormone, and describe the effects of parathyroid hormone abnormalities.

The **thyroid gland** is a large gland located in the neck, where it is attached to the trachea just inferior to the larynx (see Fig. 10.1). The parathyroid glands are embedded in the posterior surface of the thyroid gland.

Thyroid Gland

The thyroid gland is composed of a large number of follicles, each a small spherical structure made of thyroid cells filled with **triiodothyronine (T_3),** which contains three iodine atoms, and **thyroxine (T_4),** which contains four iodine atoms. These are the two forms of thyroid hormone; T_3 is thought to have the greater effect on the body.

Effects of Thyroid Hormones

Thyroid hormones increase the metabolic rate. They do not have a single target organ; instead, they stimulate all cells of the body to metabolize at a faster rate. More glucose is broken down, and more energy is utilized. Thyroid hormone is essential for proper nervous system development in infants and children. It also works synergistically with growth hormone and the gonadotropins as an adolescent goes through puberty.

To produce triiodothyronine and thyroxine, the thyroid gland actively acquires iodine from the food and water we consume. The concentration of iodine in the thyroid gland can increase to as much as 25 times that of the blood. If iodine is lacking in the diet, the thyroid gland is unable to produce the thyroid hormones. In response to constant stimulation by the anterior pituitary, the thyroid enlarges, resulting in a **simple,** or **endemic, goiter** (Fig. 10.7). Some years ago, it was discovered that the use of iodized salt allows the thyroid to produce the thyroid hormones, and therefore helps prevent simple goiter.

If the thyroid fails to develop properly, a condition called **congenital hypothyroidism** results (Fig. 10.8). Individuals with this condition are short and stocky and have had extreme hypothyroidism (undersecretion of thyroid hormone) since infancy or childhood. Thyroid hormone therapy can initiate growth, but unless treatment is begun within the first two months of life, severe developmental delay results. The occurrence of hypothyroidism in adults produces the condition known as **myxedema,** which is characterized by lethargy, weight gain, hair loss, slower pulse rate, lowered body temperature, and thickness and puffiness of the skin. The administration of adequate doses of thyroid hormones restores normal function and appearance.

In the case of hyperthyroidism (oversecretion of thyroid hormone), as seen in **Graves' disease,** the thyroid gland is overactive, and a goiter forms. This type of goiter is called **exophthalmic goiter.** The eyes protrude because of edema in eye socket tissues and swelling of the muscles that move the eyes. The patient usually becomes hyperactive, anxious and irritable, and suffers from insomnia. Surgical removal or destruction of a portion of the thyroid using radioactive iodine is generally effective in curing the condition. The toxic effects of radioactive iodine are generally limited to

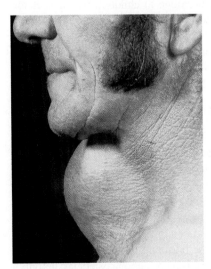

Figure 10.7 Simple goiter. An enlarged thyroid gland is often caused by a lack of iodine in the diet. Without iodine, the thyroid is unable to produce its hormones, and continued anterior pituitary stimulation causes the gland to enlarge.

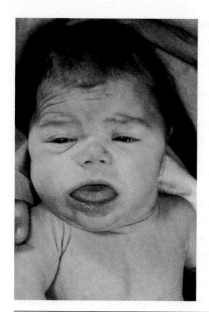

AP|R **Figure 10.8**
Congenital hypothyroidism. Individuals who have hypothyroidism since infancy or childhood do not grow and develop as others do. Unless medical treatment is begun, the body is short and stocky. Developmental delay is also likely.

the thyroid gland because other tissues incorporate so little iodine in their cells. Hyperthyroidism can also be caused by a thyroid tumor, which is usually detected as a lump during physical examination. Again, the treatment is surgery in combination with administration of radioactive iodine. The prognosis for most patients is excellent.

Calcitonin

Calcium (Ca^{2+}) plays a significant role in both nervous conduction and muscle contraction. It is also necessary for coagulation (clotting) of blood. The blood calcium level is regulated in part by **calcitonin,** a hormone secreted by the thyroid gland when the blood calcium level rises (Fig. 10.9). The primary effect of calcitonin is to bring about the deposit of calcium in the bones. It does this by temporarily reducing the activity and number of osteoclasts. Recall from Chapter 6 that these cells break down bone. When the blood calcium lowers to normal, the release of calcitonin by the thyroid is inhibited, but a low calcium level stimulates the release of parathyroid hormone (PTH) by the parathyroid glands. Calcitonin is an important hormone in children, whose skeleton is undergoing rapid growth. By contrast, calcitonin is of minor importance in adults because parathyroid hormone is the major controller of calcium homeostasis. However, calcitonin can be used therapeutically in adults to reduce the effects of osteoporosis (see the Medical Focus on page 106).

Parathyroid Glands

Parathyroid hormone (PTH), the hormone produced by the **parathyroid glands,** causes the blood phosphate (HPO_4^{2-}) level to decrease and the ionic blood calcium (Ca^{2+}) level to increase. The antagonistic actions of calcitonin, from the thyroid gland, and parathyroid hormone, from the parathyroid glands, maintain the blood calcium level within normal limits.

Note in Figure 10.9 that after a low blood calcium level stimulates the release of PTH, the hormone promotes release of calcium from the bones. (It does this by promoting osteoclast activity.) PTH promotes the reabsorption of calcium by the kidneys, where it also activates a form of vitamin D called **calcitriol.** Calcitriol, in

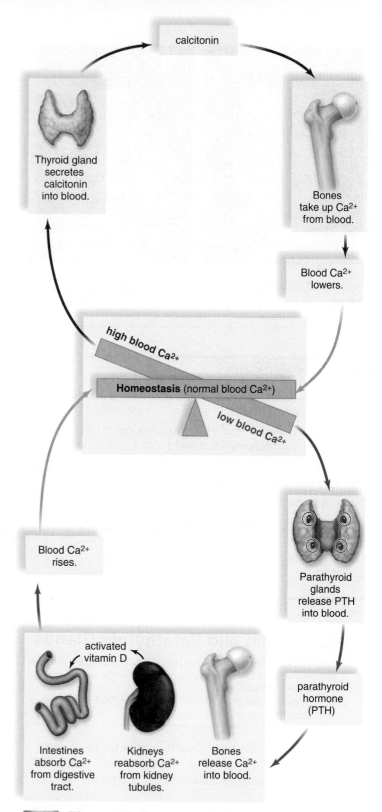

AP|R **Figure 10.9** **Regulation of blood calcium level.**
Top: When the blood calcium (Ca^{2+}) level is high, the thyroid gland secretes calcitonin. Calcitonin promotes the uptake of Ca^{2+} by the bones, and therefore the blood Ca^{2+} level returns to normal. *Bottom:* When the blood Ca^{2+} level is low, the parathyroid glands release parathyroid hormone (PTH). PTH causes the bones to release Ca^{2+}. It also causes the kidneys to reabsorb Ca^{2+} and activate vitamin D; thereafter, the intestines absorb Ca^{2+}. Therefore, the blood Ca^{2+} level returns to normal.

turn, stimulates the absorption of calcium from the small intestine. These effects bring the blood calcium level back to the normal range so that the parathyroid glands no longer secrete PTH.

Many years ago, the four parathyroid glands were sometimes mistakenly removed during thyroid surgery because of their size and location in the thyroid. When insufficient parathyroid hormone production leads to a dramatic drop in the blood calcium level, hypocalcemia results. **Hypocalcemia** can result in seizures, abnormal heart rhythms, and hypocalcemic tetany. In **tetany,** the body shakes from continuous muscle contraction. This effect is brought about by increased excitability of the nerves, which initiate nerve impulses spontaneously and without rest. In severe cases, hypocalcemic tetany is fatal due to heart failure and muscular spasms in the airways.

Excessive parathyroid hormone secretion can result from a tumor involving parathyroid tissue, or from a genetic disorder. In this case, **hypercalcemia** results. Excessive blood calcium in hypercalcemia can cause muscle weakness, abnormal heart rhythms, renal failure, and coma. Extreme hypercalcemia causes heart failure and death.

Content CHECK-UP!

7. Which of the following conditions is caused by excessive thyroid hormone?
   a. Graves' disease
   c. simple goiter
   b. cretinism (congenital hypothyroidism)
   d. myxedema

8. Explain how diet affects the production of thyroid hormone. What populations might be at risk for hypothyroidism?

9. The target organs for parathyroid hormone are:
   a. kidney, liver, stomach.
   c. kidney, bone, small intestine.
   b. kidney, bone, liver.
   d. bone, liver, small intestine.

Answers in Appendix A.

10.4 Adrenal Glands

10. Describe the anatomy of the adrenal glands.
11. Discuss the function of the adrenal medulla and its relationship to the nervous system.
12. Name three categories of hormones produced by the adrenal cortex, give an example of each category, and discuss their actions. Describe the effects of adrenal cortex malfunction.

The **adrenal glands** sit atop the kidneys (see Fig. 10.1). Each adrenal gland consists of an inner portion called the **adrenal medulla** and an outer portion called the **adrenal cortex.** These portions, like the anterior pituitary and the posterior pituitary, have no physiological connection with one another. The adrenal medulla is under nervous control, and the adrenal cortex is under the control of ACTH (also called **corticotropin**), an anterior pituitary hormone. Stress of all types, including emotional and physical trauma, prompts the hypothalamus to stimulate the adrenal glands (Fig. 10.10).

Adrenal Medulla

The hypothalamus initiates action potential signals that travel by way of the brain stem, spinal cord, and sympathetic nerve fibers to the adrenal medulla, which then secretes its hormones.

Epinephrine (adrenaline) and **norepinephrine** (noradrenaline) produced by the adrenal medulla rapidly cause all the body changes that occur when an individual reacts to an emergency situation. The release of epinephrine and norepinephrine achieves the same results as sympathetic stimulation—the "fight-or-flight" responses: increased heart rate, rapid respiration, dilation of the pupils, etc. Thus, these hormones assist sympathetic nerves in providing a short-term response to stress.

Adrenal Cortex

There are three layers in the adrenal cortex, and each produces a different set of steroid hormones. The hormones produced by the adrenal cortex provide a long-term response to stress (Fig. 10.10). The two major types of hormones produced by the adrenal cortex are the mineralocorticoids and the glucocorticoids. The **mineralocorticoids** regulate salt and water balance, leading to increases in blood volume and blood pressure. The **glucocorticoids** regulate carbohydrate, protein, and fat metabolism, leading to an increase in blood glucose level. Cortisol and other forms of glucocorticoids suppress the body's normal reaction to disease—the inflammatory reaction (see Fig. 13.3) and the immune process. Cortisone, the medication often administered for inflammation of joints, is a glucocorticoid.

The adrenal cortex also secretes small amounts of both male and female sex hormones—regardless of one's gender. Both male and female sex hormones promote skeletal growth in adolescents. The male hormones from the adrenal gland stimulate the growth of axillary and pubic hair at puberty. In addition, male hormones help to sustain the sex drive, or *libido,* in both men and women.

Glucocorticoids

Cortisol is a biologically significant glucocorticoid produced by the adrenal cortex. Cortisol raises the blood glucose level in at least two ways: (1) It promotes the breakdown of muscle proteins to amino acids, which are taken up by the liver from the bloodstream. The liver then converts these excess amino acids to glucose, which enters the blood. (2) Cortisol promotes the metabolism of fatty acids rather than carbohydrates, and this spares glucose for the brain.

Glucocorticoid Therapy

Cortisone is the glucocorticoid that is used as a medication. Because it reduces inflammation, cortisone reduces swelling and pain in joint disorders such as tendonitis and osteoarthritis. Clinicians also treat autoimmune disorders, such as rheumatoid arthritis, organ transplant rejection, allergies, and severe asthma by suppressing the immune response with cortisone therapy. However, cortisone should be used for the minimum time possible because long-term administration for therapeutic purposes causes some degree of Cushing's syndrome (see pages 229–230). Further, impaired immunity resulting from cortisone use predisposes the individual to infection and increased cancer risk. In addition, sudden withdrawal from cortisone therapy causes symptoms of diminished secretory activity by the adrenal cortex. This occurs because cortisone medication suppresses the release of adrenocorticotropic hormone (ACTH) by the anterior pituitary, leading to a decrease in natural glucocorticoid production by the adrenal cortex. Therefore, withdrawal of cortisone following long-term use must be tapered. During an alternate-day schedule, the dosage is gradually reduced and then finally discontinued as the patient's adrenal cortex resumes activity.

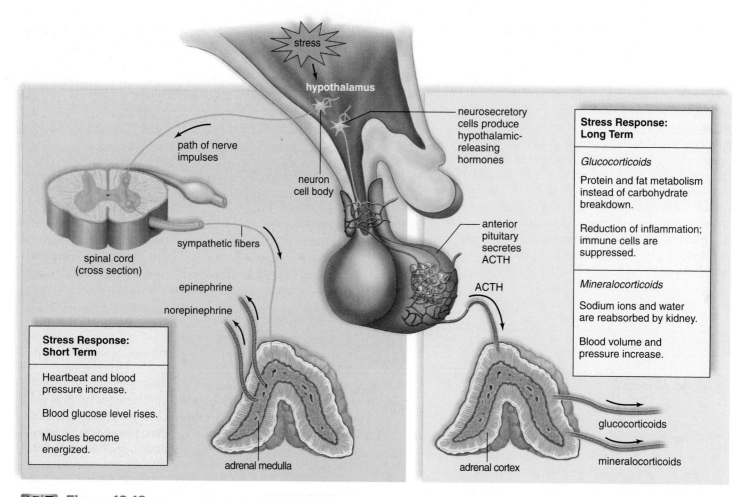

AP|R **Figure 10.10** **Adrenal glands.** Both the adrenal medulla and the adrenal cortex are under the control of the hypothalamus when they help us respond to stress. *Left:* The adrenal medulla provides a rapid, but short-term stress response. *Right:* The adrenal cortex provides a slower, but long-term stress response.

Mineralocorticoids

Aldosterone is the most important of the mineralocorticoids. Aldosterone primarily targets the kidney, where it promotes renal absorption of sodium (Na^+) and water, and renal excretion of potassium (K^+).

As one might expect, secretion of mineralocorticoids from the adrenal cortex is influenced by ACTH (adrenocorticotropic hormone or corticotropin) from the pituitary gland. However, the pituitary hormone is not the primary controller for aldosterone secretion. When the blood sodium level and therefore the blood pressure are low, the kidneys secrete **renin** (Fig. 10.11). Renin is an enzyme that converts *angiotensinogen* (a plasma protein produced by the liver) to angiotensin I. Angiotensin I is fully activated to angiotensin II by a converting enzyme found in lung capillaries. Angiotensin II stimulates the adrenal cortex to release aldosterone. The effect of this system, called the renin-angiotensin-aldosterone system, is to raise blood pressure in two ways: Angiotensin II constricts arterioles, and aldosterone causes the kidneys to reabsorb sodium. When the blood sodium level rises, water is reabsorbed, in part because the hypothalamus secretes ADH (see page 222). Reabsorption means that water enters kidney capillaries and thus returns to the blood. Then blood pressure increases to normal.

As you might have already guessed, there is an antagonistic hormone to aldosterone. When the atria of the heart are stretched due to increased blood volume, cardiac cells release a hormone called **atrial natriuretic hormone (ANH, or atriopeptide),** which inhibits the secretion of aldosterone from the adrenal cortex. The effect of ANH is the excretion of sodium in the urine—that is, *natriuresis.* When sodium is excreted, so is water, and therefore blood pressure lowers to normal.

Malfunction of the Adrenal Cortex

Malfunction of the adrenal cortex can lead to a **syndrome,** a set of symptoms that occur together. The syndromes commonly associated with the adrenal cortex are Addison's disease and Cushing's syndrome.

Addison's Disease and Cushing's Syndrome

When the level of adrenal cortex hormones is low due to hyposecretion, a person develops **Addison's disease.** The presence of excessive but ineffective ACTH causes skin bronzing because ACTH can lead to melanin buildup (see Fig. 10.12 and the Chapter Introduction). Without cortisol, glucose cannot be replenished when a stressful situation arises. Even a mild infection can lead to death. The lack of aldosterone results in a loss of sodium and water, the development of low blood pressure, and possibly severe dehydration. Left untreated, Addison's disease can be fatal.

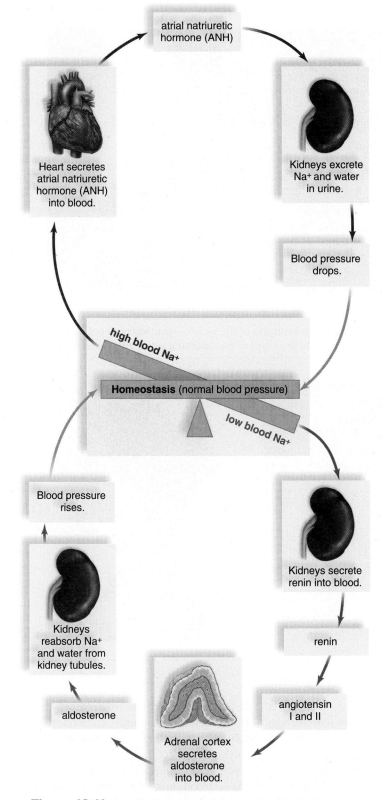

Figure 10.11 **Regulation of blood pressure and volume.**
Bottom: When the blood sodium (Na⁺) level is low, low blood pressure causes the kidneys to secrete renin. Renin leads to the secretion of aldosterone from the adrenal cortex. Aldosterone causes the kidneys to reabsorb Na⁺, and water follows, so that blood volume and pressure return to normal. *Top:* When the blood Na⁺ is high, a high blood volume causes the heart to secrete atrial natriuretic hormone (ANH). ANH causes the kidneys to excrete Na⁺, and water follows. The blood volume and pressure return to normal.

When the level of adrenal cortex hormones is high due to hypersecretion, a person develops **Cushing's syndrome** (Fig. 10.13). The excess cortisol results in a tendency toward diabetes mellitus as muscle protein is metabolized and subcutaneous fat is deposited in the midsection. The trunk is obese, while the arms and legs remain a normal size. Excessive aldosterone causes too much sodium and water to be reabsorbed by the kidneys, and simultaneous potassium loss. The person develops a basic blood pH and hypertension. Excessive tissue fluid, called *edema*, collects in the facial tissues, giving the patient's face a puffy, moon-shaped appearance. Masculinization may occur in women because of excess adrenal male sex hormones.

Content CHECK-UP!

10. From the following list of hormones of the adrenal cortex and their corresponding effects, choose the pair, or pairs, that are correct.
  a. male hormones → stimulate sex drive
  b. aldosterone → increases sodium concentration in the blood
  c. female hormones → promote long bone growth in adolescents
  d. Pairs b and c are correct.
  e. All are correct.

11. Which hormone opposes the effect of aldosterone in the body?
  a. renin
  c. atrial natriuretic hormone
  b. angiotensin I
  d. cortisol

12. Aldosterone returns blood pressure to normal by causing the kidneys to reabsorb water and sodium. As you know, it works by negative feedback. Give an example from your daily life during which your adrenal cortex produces aldosterone.

Answers in Appendix A.

10.5 Pancreas

13. Describe the anatomy of the pancreas.
14. Name three hormones produced by the pancreas, and discuss their functions.
15. Discuss the two types of diabetes mellitus, and contrast hypoglycemia with hyperglycemia.

The **pancreas** is a long organ that lies transversely in the abdomen just inferior to the stomach. Its widest portion, called the *head*, is located immediately lateral to the duodenum of the small intestine. It is composed of two types of tissue. Exocrine tissue produces and secretes digestive juices that travel through ducts to the small intestine. Pancreatic endocrine tissue includes three types of hormone-producing cells, found in clusters called the **pancreatic islets** (islets of Langerhans). Pancreatic alpha cells produce **glucagon,** beta cells produce **insulin,** and delta cells produce **somatostatin.** All three hormones are released directly into the blood.

The two antagonistic hormones, insulin and glucagon, both produced by the pancreas, help maintain the normal level of blood glucose. Insulin is secreted when the blood glucose level is high, which usually occurs just after eating. Insulin stimulates the uptake of glucose by most body cells. Insulin is not necessary for glucose transport into brain or red blood cells, but muscle cells and adipose tissue cells require insulin for glucose transport. In liver and muscle cells, insulin stimulates enzymes that promote the

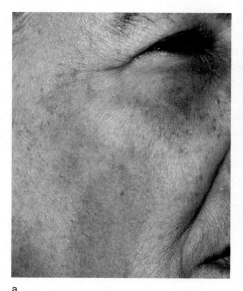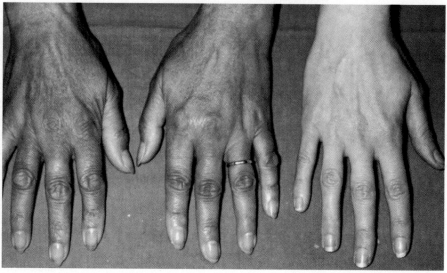

a. b.

Figure 10.12 **Addison's disease.** Addison's disease is characterized by a peculiar bronzing of the skin, particularly noticeable in these light-skinned individuals. Note the color of (**a**) the face and (**b**) the hands compared to the hand of an individual without the disease.

Figure 10.13 **Cushing's syndrome.** Cushing's syndrome results from hypersecretion of adrenal cortex hormones. (**a**) Patient at the time of surgery to remove a pituitary tumor. The tumor secreted excess ACTH, which caused excess adrenal cortex secretion and Cushing's syndrome. (**b**) Patient one year after surgery.

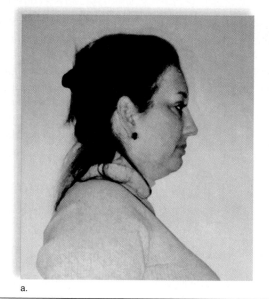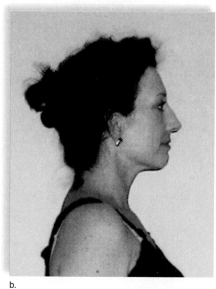

a. b.

storage of glucose as glycogen. In muscle cells, the glucose supplies energy for muscle contraction, and in fat cells, glucose enters the metabolic pool and thereby supplies glycerol for the formation of fat. In these ways, insulin lowers the blood glucose level.

Glucagon is secreted from the pancreas, usually between meals, when the blood glucose level is low. The major target tissues of glucagon are the liver and adipose tissue. Glucagon stimulates the liver to break down glycogen to glucose and to use fat and protein in preference to glucose as energy sources. Adipose tissue cells break down fat to glycerol and fatty acids. The liver takes these up and uses them as substrates for glucose formation. In these ways, glucagon raises the blood glucose level (Fig. 10.14).

Somatostatin prevents the release of the other two hormones. In this way, it prevents wide swings in blood sugar that might occur between meals.

Diabetes Mellitus

Diabetes mellitus is a fairly common hormonal disease in which insulin-sensitive body cells are unable to take up and/or metabolize glucose. Therefore, the blood glucose level is elevated—a condition called **hyperglycemia.** Because body cells cannot access glucose, starvation occurs at the cell level. The person becomes extremely hungry—a condition called **polyphagia.** As the blood glucose level rises, glucose will be lost in the urine **(glycosuria).** Glucose in the urine causes excessive water loss through urination **(polyuria).** The loss of water in this way causes the diabetic to be extremely thirsty **(polydipsia).** Because glucose is not being metabolized, the body turns to the breakdown of protein and fat for energy. Fat metabolism leads to the ketone buildup in the blood, and ketone excretion in the urine **(ketonuria).** Ketones are acidic, so their buildup in blood causes **acidosis,** or acid blood. This condition, called *ketoacidosis,* can lead to coma and death.

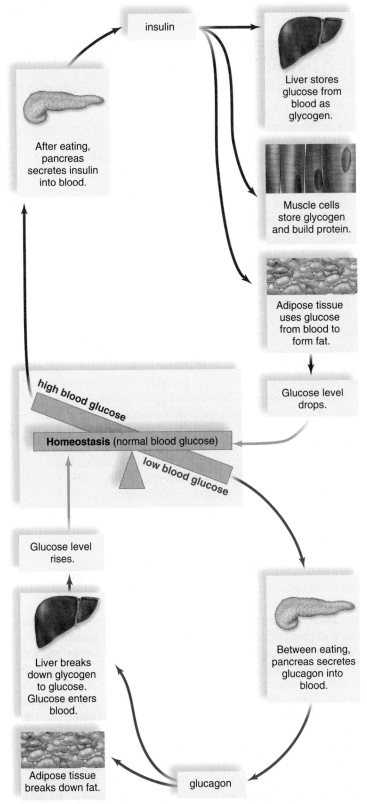

We now know that diabetes mellitus exists in two forms. In type I, often called **insulin-dependent diabetes mellitus (IDDM),** the pancreas does not produce insulin. This condition is believed to be brought on, at least in part, by exposure to an environmental agent. This agent—very likely a virus—causes an extreme immune response, and immune cells destroy the pancreatic islets. As a result, the individual must have daily insulin injections. Daily injections control the diabetic symptoms, but diabetics can still experience life-threatening problems, as described in the I.C.E. box, Insulin Shock and Diabetic Ketoacidosis.

Of the 29.1 million people who now have diabetes in the United States, most have type II, often called **noninsulin-dependent diabetes (NIDDM).** This type of diabetes mellitus usually (but not always) occurs in people of any age who tend to be obese. Researchers theorize that perhaps adipose tissue produces a substance that interferes with the transport of glucose into cells. The amount of insulin in the blood of these patients is normal or elevated, but the insulin receptors on the cells do not respond to it. It is possible to prevent, or at least control, type II diabetes by adhering to a low-fat, low-sugar diet, maintaining a healthy weight, and exercising regularly. If these attempts fail, oral drugs are available to stimulate the pancreas to secrete more insulin. Other oral medications enhance the metabolism of glucose in the liver and muscle cells. It is projected that as many as 8 million Americans may have type II diabetes without being aware of it. Yet another 86 million Americans have **prediabetes,** a condition in which blood glucose is chronically elevated. Prediabetes will often lead to full-blown diabetes. It's important to note that the effects of untreated type II diabetes are as serious as those of type I diabetes. In addition, without stringent control, the NIDDM diabetic will ultimately require insulin injections, thus becoming insulin-dependent.

Long-term complications of both types of diabetes are blindness, kidney disease, and circulatory disorders, including atherosclerosis, heart disease, stroke, and reduced circulation. The latter can lead to gangrene in the arms and legs. Pregnancy carries an increased risk of diabetic coma, and the child of a diabetic is somewhat more likely to be stillborn or to die shortly after birth. However, these complications of diabetes are not expected to appear if the mother's blood glucose level is carefully regulated and kept within normal limits during the pregnancy.

AP|R **Figure 10.14** **Regulation of blood glucose level.** *Top:* When the blood glucose level is high, the pancreas secretes insulin. Insulin promotes the storage of glucose as glycogen in the liver and muscles and the use of glucose to form fat in adipose tissue. Therefore, insulin lowers the blood glucose level. *Bottom:* When the blood glucose level is low, the pancreas secretes glucagon. Glucagon acts opposite to insulin; therefore, glucagon raises the blood glucose level to normal.

Content **CHECK-UP!**

13. Insulin-sensitive cells in the human body include:

   a. muscle cells.   d. a and b.

   b. adipose tissue cells.   e. All of the above.

   c. brain and nerve cells.

14. Which of the following is an effect of glucagon?

   a. causes the liver to break down stored glycogen

   b. causes adipose tissue to store fat

   c. lowers blood glucose level

   d. stimulates the liver to store glucose as glycogen

15. The release of insulin and glucagon is controlled by negative feedback. At what times during your average day would your pancreas secrete these hormones?

Answers in Appendix A.

I.C.E. — IN CASE OF EMERGENCY

Insulin Shock and Diabetic Ketoacidosis

If you're someone with *diabetes mellitus,* the disorder involving the hormone insulin, you already know the importance of maintaining a stable blood glucose level to ensure your long-term health. If your roommate, friend, or loved one is a diabetic, you need to be informed, too, because a diabetic's possible problems don't always take a long time to develop. Insulin shock and diabetic ketoacidosis (DKA) can develop very rapidly, and both can be fatal or result in permanent brain damage. It's important for diabetics and their friends and family to recognize the symptoms of insulin shock and DKA, and know how to use a glucometer to measure blood glucose and how to give an injection.

Insulin shock (also called an *insulin reaction*) results when blood glucose falls to critically low levels—a condition called **hypoglycemia.** It often results when the diabetic patient accidentally injects too much insulin, or takes her insulin but misses a meal. The patient is likely to feel anxious, sweat profusely, and complain of a headache. She may become hyperactive, confused, and even psychotic as the condition worsens. Eventually, she'll lose consciousness and lapse into a so-called **diabetic coma.**

It's critical for first responders to try to raise the patient's blood glucose as quickly as possible. If she is conscious and alert, she can quickly drink milk, juice, or soda, or eat something sweet. She may be able to self-inject with glucagon, the hormone that raises blood glucose.

If the blood glucose isn't too low, the insulin shock can be corrected at home. However, one should never attempt to give food or drinks to someone who is semiconscious or unconscious—she could easily choke. Instead, the inside of her cheeks can be smeared with glucose gel, honey, syrup, or frosting, which will melt and be swallowed. If the patient doesn't quickly become alert enough to eat or drink, she must be transported to an emergency room. There, an injection of glucagon or an intravenous solution will quickly raise the blood glucose level.

Diabetic ketoacidosis is caused by blood glucose that is too high. It commonly results when the diabetic eats a meal, but forgets to inject insulin. Infection, injury, or extreme stress can also lead to DKA. The symptoms of DKA are increased thirst, frequent urination, nausea, and vomiting. The patient breathes rapidly, and his breath smells like fruit-flavored gum. His pulse is very fast, but his blood pressure is low. Without an identifying bracelet or tag, he could easily be mistaken as someone who's had too much to drink: sluggish, lethargic, and increasingly sleepy. If untreated, he'll eventually fall into a diabetic coma. Fortunately, under a physician's direction, a trained paramedic can start an intravenous solution to help dilute the patient's blood. As he is being transported to the emergency room, EMS personnel can then measure the glucose and ketones in his blood to provide a complete history, in preparation for more complete treatment in the hospital.

10.6 Additional Endocrine Glands

16. Name the most important male and female sex hormones. Discuss their functions.
17. State the location and function of the pineal gland and the thymus gland.
18. Discuss atrial natriuretic hormone, leptin, ghrelin, growth factors, and prostaglandins as hormones not produced by endocrine glands.

The body has a number of other endocrine glands, including the **gonads** (testes in males and the ovaries in females). Other lesser-known glands, such as the thymus gland and the pineal gland, also produce hormones. Many other organs have their own roles as endocrine glands, and researchers continue to discover additional hormones and/or growth factors, suggesting that numerous other tissues and organs function as endocrine glands. Even individual body cells produce local messenger chemicals termed *prostaglandins.*

Testes and Ovaries

The **testes** (sing., testis) are located in the scrotum, and the **ovaries** are located in the pelvic cavity. The testes produce **androgens** (e.g., **testosterone**), which are the male sex hormones, and the ovaries produce **estrogens** and **progesterone,** the female sex hormones. The hypothalamus and the pituitary gland control the hormonal secretions of these organs in the manner previously described on pages 222–224.

Androgens

Puberty is the time of life when sexual maturation occurs. Greatly increased testosterone secretion during puberty stimulates the growth of the penis and the testes. Testosterone also brings about and maintains the male secondary sex characteristics that develop during puberty, including the growth of a beard, axillary (underarm) hair, and pubic hair. It prompts the larynx and the vocal cords to enlarge, causing a young man's voice to deepen. It is partially responsible for the muscular strength of males. This is why some athletes take supplemental amounts of **anabolic steroids,** which are either testosterone or related chemicals. The contraindications of taking anabolic steroids are discussed in the Medical Focus on pages 234–235. Testosterone also stimulates oil and sweat glands in the skin; therefore, it is largely responsible for acne and body odor. Another side effect of testosterone is baldness. Genes for baldness are probably inherited by both sexes, but baldness is seen more often in males because of the presence of testosterone.

Estrogen and Progesterone

The female sex hormones, estrogens and progesterone, have many effects on the body. In particular, estrogens secreted during puberty stimulate the growth of the uterus and the vagina. Estrogen is necessary for ovum maturation and is largely responsible

Options for Type I Diabetics: *The Artificial Pancreas System, Beta Cell Transplants, and the BioHub*

"I can remember getting sick with the flu when I was 11. I missed two or three days of school, and I just never got my strength back. I ate and drank constantly because I was thirsty and hungry all the time. I was always in the bathroom, and I started wetting the bed—can you imagine, at age 11? I fell asleep in school, and the teacher could barely get me to wake up. That's when my doctor diagnosed my diabetes for the first time."

The patient, age 25, is a typical type I insulin-dependent diabetic. Her symptoms are characteristic of this form of diabetes mellitus (IDDM) (see pages 230–231). As you've read, in insulin-dependent diabetes, the body's own immune cells destroy pancreatic beta cells. To treat their illness, type I diabetic patients inject insulin three or more times daily or use a continuous insulin pump device. Patients use blood tests to check their blood glucose levels throughout each day. Further, diabetics must carefully monitor their diets, activities, and stress levels, and regular exercise is also a must.

Fortunately, ongoing research into diabetes therapy holds the promise of treatments that will be much safer and more effective. Recently, the results were reported from clinical studies on an artificial pancreas system (APS) that effectively combines two existing technologies. The first is the insulin pump, a cell-phone-sized device that continually injects insulin into the patient's body through fine tubing which the patient positions under the abdominal skin. The second device is a continuous glucose monitor (CGM), which can constantly sample glucose levels in subcutaneous tissue fluid and give real-time information about the patient's status. The new device, called Diabetes Assistant (DiAs) is a smartphone-based application that is wirelessly connected to both the patient's CGM and insulin pump. In the study, each patient's DiAs received real-time information from both devices, then estimated the patient's metabolic rate and adjusted the insulin delivery to maintain blood glucose in a normal range. Preliminary results were promising: the blood glucose levels of study participants were maintained within normal levels, without the need for researchers to intervene. When the APS is perfected, researchers hope it will respond much as a real pancreas does: delivering just the right amount of insulin at just the right moment, day in and day out.

Whole pancreatic transplant can be a permanent cure for individual diabetics, but there is a shortage of donor organs. Further, the transplant is major surgery, and recipients must take lifelong anti-rejection medications, which have severe side effects (recurring infections, increased cancer risk, kidney damage, etc.). While less common, pancreatic islet cell transplantation also continues to show some promise as a permanent cure for type I diabetes. With this approach, cadaver beta cells are directly injected into the liver, where they form colonies and produce insulin. Regrettably, as with whole pancreas transplants, this type of transplant is limited by the scarcity of donor cells and complicated by the ever-present risk of rejection. However, in October 2014, researchers at Harvard University announced the development of a technique to produce large quantities of pancreatic beta cells from transformed embryonic stem cells. While this is encouraging, there are practical and ethical issues that remain regarding the use of embryonic stem cells. Perhaps it will be possible in the future to transform the patient's own stem cells into a new beta cell population using this process. If successful, such a technique would completely eliminate the need for anti-rejection drugs and would make islet cell transplants much more commonplace.

Meanwhile, scientists continue to investigate ways to shield transplanted cells from the recipient's immune system, and thus, reduce the risk of rejection. In previous studies of a technique called *microencapsulation*, scientists placed beta cells into tiny, porous containers consisting of minute carbon tubes called nanotubes. Early results showed that the cells were initially able to survive and produce insulin for a time. However, the cells quickly failed and died due to a lack of oxygen and nutrients. Now, new microencapsulation studies sponsored by the Diabetes Research Institute have led to techniques that enable beta cells to be individually enclosed in tight capsules, and then layered into a bioengineered scaffold called a BioHub. This technique enables oxygen and nutrients to reach the cells, and it allows the cells to respond to changes in blood sugar, as though they were housed in a person's body. If these studies are successful, human or pig beta cells could be transplanted, and patients would not need either anti-rejection medication or insulin injections (Fig. 10A).

While these long-term solutions are being researched, new ways to deliver insulin into the body are being developed. For example, researchers are actively exploring insulin delivery options that are less painful, and potentially more effective and more reliable than injections. Currently, diabetics can't swallow their insulin in a pill, because harsh stomach acid destroys the protein before it can be absorbed. Pharmaceutical companies continue in their efforts to develop gels that would reliably shield the insulin as it passes through the stomach while allowing it to be absorbed in the intestine. Researchers are optimistic that some form of insulin pill might soon be available for use.

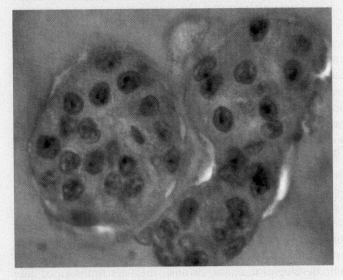

Figure 10 A Encapsulated insulin-producing pancreatic islet cells from pigs can be transplanted into patients without the need for immune system—suppressing drugs.

They're called "performance-enhancing steroids," and allegations of their use remain widespread in athletics, both amateur and professional. Whether the sport is baseball, football, track and field events, or professional cycling, no activity seems to be safe from drug abuse. In January 2013, after years of repeated denials, Lance Armstrong finally admitted to using testosterone during the years when he won an unprecedented seven Tour de France cycling races. Even the Olympic Games have been affected. Olympic history was forever changed when Marion Jones, the first female athlete to win five medals for track and field events, admitted steroid abuse during the 2000 Sydney Games. In 2008, she was stripped of all medals she had earned and disqualified from a fifth-place finish in the 2004 Athens Games. Future Olympic record books will not include her name, and the records of her relay teammates have also been tainted. Baseball records will also likely require revisions. The great 1998 home-run competition between Mark McGwire and Sammy Sosa was largely credited with reviving national interest in baseball. Both McGwire and Sosa were later accused of using anabolic steroids, and in 2010, McGwire admitted that the allegations were true. Likewise, both Jose Canseco of the Oakland Athletics and Jason Giambi of the New York Yankees have confessed to using performance-enhancing drugs. Similar charges of drug abuse may prevent baseball greats Roger Clemens and Barry Bonds from ever entering the Baseball Hall of Fame, despite each player's award-winning career.

Many athletes and officials continue to deny that anabolic steroids are widely used in professional sports. However, many people from both inside and outside the industry maintain that such abuse has been going on for many years, and that it continues despite the negative publicity. Of tremendous concern to lawmakers, educators, and parents is the increased use of steroids by teens wishing to bulk up quickly, perhaps seeking to be just like the sports figures they admire.

Anabolic steroids are synthetic forms of the male sex hormone testosterone. Taking doses 10 to 100 times the amount prescribed by doctors for various illnesses promotes larger muscles when the person also exercises. Trainers may have been the first to acquire anabolic steroids for weight lifters, bodybuilders, and other athletes, such as professional baseball players. However, being a steroid user can have serious detrimental effects. Men often experience decreased sperm counts and decreased sexual desire due to atrophy of the testicles. Some develop an enlarged prostate gland or grow breasts. On the other hand, women can develop male sexual characteristics. They grow chest and facial hair, and lose scalp hair; many experience abnormal enlargement of the clitoris. Some women cease ovulating or menstruating, sometimes permanently. Researchers have predicted that two or three months of high-dosage use of anabolic steroids as a teen can cause death by age 30 or 40. These drugs have even been linked to heart disease and sudden death in both sexes, and have been implicated in the deaths of young athletes from liver cancer and one type of kidney tumor. Steroids can cause the body to retain fluid, which results in increased blood pressure. Users then try to get rid of "steroid bloat" by taking large doses of diuretics. A young California weight lifter had a fatal heart attack after using

---

for the secondary sex characteristics in females, including female body hair and fat distribution. In general, females have a more rounded appearance than males because of a greater accumulation of fat beneath the skin. Also, the pelvic girdle is wider in females than in males, resulting in a larger pelvic cavity. Both estrogen and progesterone are required for breast development and for regulation of the uterine cycle, which includes monthly menstrual periods (discharge of blood and mucosal tissues from the uterus).

Thymus Gland

The lobular **thymus gland,** which lies just beneath the sternum (see Fig. 10.1), reaches its largest size and is most active during childhood. It then shrinks in size throughout one's adult life. **Lymphocytes** are white blood cells that originate in the bone marrow and are responsible for specific defenses against a particular invader. When lymphocytes mature and complete their development in the thymus, they are transformed into **thymus-derived lymphocytes,** or **T lymphocytes.** The lobules of the thymus are lined by epithelial cells that secrete hormones called **thymosins.** These hormones aid in the differentiation of lymphocytes packed inside the lobules. Although the hormones secreted by the thymus ordinarily work only in the thymus, researchers hope that these hormones could be injected into AIDS or cancer patients where they would enhance T-lymphocyte function.

Pineal Gland

The **pineal gland,** which is located in the brain, produces the hormone **melatonin,** primarily at night. Melatonin is involved in our daily sleep-wake cycle; normally we grow sleepy at night when melatonin levels increase and awaken once daylight returns and melatonin levels are low (Fig. 10.15). Daily 24-hour cycles such as this are called **circadian rhythms.** Circadian rhythms are controlled by an internal timing mechanism called a biological clock.

Based on animal research, it appears that melatonin also regulates sexual development. It has also been noted that children whose pineal gland has been destroyed due to a brain tumor experience early puberty.

steroids, and the postmortem showed a lack of electrolytes, salts that help regulate the heart. Finally, steroid abuse has psychological effects, including depression, hostility, aggression, eating disorders, and addiction (approximately 30% of users become dependent). Unfortunately, these drugs can also make a person feel invincible. One abuser even had his friend videotape him as he drove his car at 40 miles an hour into a tree! The many harmful effects of anabolic steroids are given in Figure 10B. The Federal Food and Drug Administration now bans most steroids, and steroid use has also been banned by the National Collegiate Athletic Association (NCAA), the National Football League (NFL), the National Basketball Association (NBA) and the International Olympic Committee (IOC).

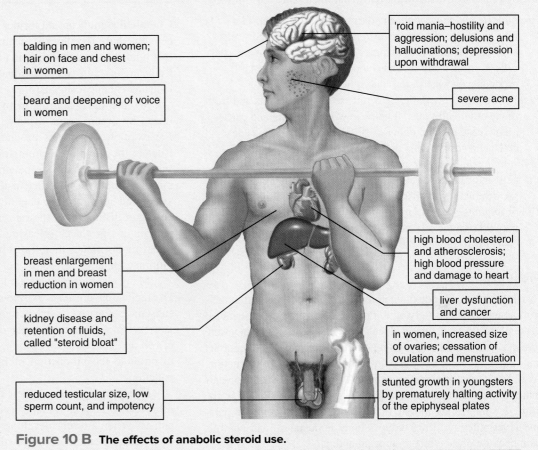

balding in men and women; hair on face and chest in women

beard and deepening of voice in women

'roid mania–hostility and aggression; delusions and hallucinations; depression upon withdrawal

severe acne

breast enlargement in men and breast reduction in women

kidney disease and retention of fluids, called "steroid bloat"

high blood cholesterol and atherosclerosis; high blood pressure and damage to heart

liver dysfunction and cancer

in women, increased size of ovaries; cessation of ovulation and menstruation

reduced testicular size, low sperm count, and impotency

stunted growth in youngsters by prematurely halting activity of the epiphyseal plates

Figure 10 B The effects of anabolic steroid use.

AP|R **Figure 10.15 Melatonin production.** Melatonin production is greatest at night when we are sleeping. Light suppresses melatonin production (**a**) so its duration is longer in the winter (**b**) than in the summer (**c**).

Hormones from Other Tissues

As you now know, the heart is an endocrine gland that produces atrial natriuretic hormone (see page 228). The kidney also influences cardiovascular system function by producing the hormone erythropoietin (EPO), which stimulates red blood cell production by the bone marrow. This hormone and its effects are detailed in Chapter 11. Further, you'll see in Chapter 15 that organs and tissues of the digestive system produce an entire set of so-called *enteric* hormones as well. For example, the stomach produces *gastrin* and the small intestine produces *secretin*. Both are hormones that promote digestion.

Leptin and Ghrelin

Leptin is a protein hormone produced by adipose tissue. Leptin acts on the hypothalamus, where it signals *satiety*—the feeling that the person has had enough to eat. Strangely, the blood of obese individuals may actually be rich in leptin. It is possible that the leptin they produce is ineffective because of a genetic mutation, or else their hypothalamic cells lack a suitable number of receptors for leptin. **Ghrelin** is an antagonist to leptin that is produced by the stomach. Where leptin signals fullness, ghrelin signals hunger.

Growth Factors

A number of different types of organs and cells produce peptide **growth factors,** which stimulate cell division and mitosis. Some, such as lymphokines, are released into the blood; others diffuse to nearby cells. Growth factors of particular interest are the following:

Granulocyte and macrophage colony-stimulating factor (GM-CSF) is secreted by many different tissues. GM-CSF causes bone marrow stem cells to form either granulocyte or macrophage cells (both are forms of white blood cells, or leukocytes), depending on whether the concentration is low or high.

Platelet-derived growth factor is released from platelets and from many other cell types. It helps in wound healing and causes an increase in the number of fibroblasts, smooth muscle cells, and certain cells of the nervous system.

Epidermal growth factor and *nerve growth factor* stimulate the cells indicated by their names, as well as many others. These growth factors are also important in wound healing.

Tumor angiogenesis factor stimulates the formation of capillary networks and is released by tumor cells. One treatment for cancer is to prevent the activity of this growth factor.

Prostaglandins

Prostaglandins are potent chemical signals produced within cells from arachidonate, a fatty acid. Prostaglandins are not distributed in the blood; instead, they act locally, quite close to where they were produced. In the uterus, prostaglandins cause muscles to contract and may be involved in the pain and discomfort of menstruation. Also, prostaglandins mediate the effects of *pyrogens,* chemicals that are believed to reset the temperature regulatory center in the brain. It's interesting to note that aspirin, the most common pain- and fever-relieving drug, reduces body temperature and controls pain because of its effect on prostaglandins.

Certain prostaglandins reduce gastric secretion and have been used to treat ulcers; others lower blood pressure and have been used to treat hypertension; and still others inhibit platelet aggregation and have been used to prevent thrombosis (the formation of stationary clots in blood vessels). However, different prostaglandins have contrary effects, and it has been very difficult to successfully standardize their use.

Content CHECK-UP!

16. Describe three functions of the female sex hormones.

17. Which of the following is a local tissue messenger that stimulates nearby cells?

    a. leptin   c. melatonin

    b. prostaglandin   d. thymosin

18. From the following list of endocrine glands and their hormones, choose the pair that is correct:

    a. ovaries → androgens   c. kidney → aldosterone

    b. thymus → insulin   d. adipose tissue → leptin

Answers in Appendix A.

10.7 The Importance of Chemical Signals

19. Give examples to show that chemical signals can act between organs, cells, and individuals.

Chemical signals are molecules that affect the behavior of those cells that have receptors to receive them. For example, a hormone that binds to a receptor protein affects the metabolism of the cell. Cells, organs, and even individuals communicate with one another by using chemical signals.

We are most familiar with chemical signals that are produced by organs some distance from one another in the body. For example, hormones produced by the anterior pituitary influence the function of numerous organs throughout the body. Insulin, produced by the pancreas, is transported in blood to muscle, adipose, and other insulin-sensitive cells. The nervous system at times utilizes chemical signals that are produced by an organ distant from the one being affected, as when the hypothalamus produces releasing hormones. As you know, these releasing hormones then travel in a portal system to the anterior pituitary gland.

Many chemical signals act locally—that is, from cell to cell. Prostaglandins are *local hormones,* and certain neurotransmitter substances released by one neuron affect a neuron nearby. Growth factors, which fall into this category, are very important regulators of cell division. Some growth factors are being used as medicines to promote the production of blood cells in AIDS and cancer patients.

Cancer cells produce their own sets of growth factors, and discovering them all remains a challenge for cancer research scientists. When a tumor develops, cell division occurs even when no detectable stimulatory growth factor has been received. Further, cancerous tumors are known to produce a growth factor called tumor angiogenesis factor, which promotes the formation of capillary networks to service its cells. Currently, several forms of cancer treatment involve shutting down the activity of cancer growth factors, and more of these treatments will likely be discovered in the future (see the What's New article entitled "Targeting the Traitor Inside" on pages 80–81 for more about this type of therapy).

Chemical Signals Between Individuals

Chemical signals that act between individuals are called *pheromones* (see the Introduction, Chapter 17). There are many examples of pheromones in other animals, but they may also be effective between people. Humans produce airborne chemicals from a variety of areas, including the scalp, oral cavity, armpits, genital areas, and feet. For example, the armpit secretions of one woman could possibly affect the menstrual cycle of another woman. Women who live in the same household often have menstrual cycles during the same times of the month. Further, when women with irregular cycles are exposed to extracts of male armpit secretions, their cycle length tends to become more normal.

10.8 Effects of Aging

20. Discuss the anatomical and physiological changes that occur in the endocrine system as we age.

Thyroid disorders and diabetes are the most significant endocrine problems affecting health and function as we age. Both hypothyroidism and hyperthyroidism are seen in the elderly. Graves' disease is an autoimmune disease that targets the thyroid, resulting in symptoms of cardiovascular disease, increased body temperature, and fatigue. In addition, a patient may experience weight loss of as much as 20 pounds, depression, and mental confusion. Hypothyroidism (myxedema) may fail to be diagnosed because the symptoms of hair loss, skin changes, and mental deterioration are attributed simply to the process of aging.

Both the thymus gland and the pineal gland decrease in size as a person ages. Thymic atrophy is thought to contribute to declining immune response observed with age, as described in Chapter 13.

The true incidence of IDDM diabetes among the elderly is unknown. Its symptoms can be confused with those of other medical conditions that are present. As in all adults, NIDDM diabetes is associated with being overweight and often can be controlled by proper diet.

The effect of age on the sex organs is discussed in Chapter 17.

10.9 Homeostasis

21. Discuss how the endocrine system works with other systems of the body to maintain homeostasis.

The endocrine system and the nervous system work together to regulate the organs of the body and thereby maintain homeostasis. It is clear from reviewing the Human Systems Work Together illustration on page 238 that the endocrine system particularly influences the digestive, cardiovascular, and urinary systems in a way that maintains homeostasis.

The endocrine system helps regulate digestion. The digestive system adds nutrients to the blood, and hormones produced by the digestive system influence the gallbladder and pancreas to send their secretions to the digestive tract. Another hormone, gastrin, promotes muscle contractions and protein digestion in the stomach. Through its influence on the digestive process, the endocrine system promotes the presence of nutrients in the blood. Leptin (from adipose tissue) and ghrelin (from the stomach) regulate satiety (the feeling of fullness) and hunger.

The endocrine system helps regulate fuel metabolism. Together, insulin and glucagon control the level of glucose in the blood. Just after eating, insulin encourages the uptake of glucose by cells and the storage of glucose as glycogen in the liver and muscles. In between meals, glucagon stimulates the liver to break down glycogen to glucose so that the blood glucose level stays constant. Somatostatin helps to prevent wide swings in blood glucose between meals by inhibiting both insulin and glucagon secretion. Adrenaline (epinephrine) from the adrenal medulla also stimulates the liver to release glucose. Glucagon (from the pancreas) and cortisol (from the adrenal cortex) promote the breakdown of protein to amino acids, which can be converted to glucose by the liver. They also promote the metabolism of fatty acids to conserve glucose, a process called **glucose sparing.** Finally, the thyroid hormones thyroxine and triiodothyronine set the body's metabolic rate, and thus are the hormones that ultimately regulate fuel metabolism.

The endocrine system helps regulate blood pressure and volume. ADH produced by the hypothalamus (but secreted by the posterior pituitary) promotes reabsorption of water by the kidneys, especially during times when we are dehydrated. Simultaneously, ADH causes blood vessels to constrict, which raises blood pressure. Aldosterone produced by the adrenal cortex causes the kidneys to reabsorb sodium, and when the level of sodium rises, water is automatically reabsorbed so that blood volume and pressure rise together. Regulation by the endocrine system often involves antagonistic hormones; in this case, ANH (atriopeptide) produced by the heart causes sodium and water excretion.

The endocrine system helps regulate calcium balance. The concentration of calcium (Ca^{2+}) in the blood is critical because this ion is important to nervous conduction, muscle contraction, and hormone action. As you know from studying Chapter 6, the bones serve as a reservoir for calcium. When the blood calcium concentration lowers, parathyroid hormone (PTH) promotes the breakdown of bone and calcium reabsorption by the kidneys. PTH also stimulates calcium absorption by the intestines by activating Vitamin D. Opposing the action of parathyroid hormone, calcitonin secreted by the thyroid brings about calcium deposition in the bones (although this function of calcitonin is more important in growing children than in adults).

The endocrine system helps regulate responses to the external environment. In "fight-or-flight" situations, the nervous

Human Systems Work Together

Integumentary System

Androgens activate sebaceous glands and help regulate hair growth.

Skin provides sensory input that results in the activation of certain endocrine glands.

Skeletal System

Growth hormone regulates bone development; parathyroid hormone and calcitonin regulate Ca^{2+} content.

Bones provide protection for glands; store Ca^{2+} used as second messenger.

Muscular System

Growth harmone and androgens promote growth of skeletal muscle; epinephrine stimulates heart and constricts blood vessels.

Muscles help protect glands.

Nervous System

Hormones affect development of brain.

Hypothalamus, pituitary, and pineal gland are part of endocrine system; nerves innervate glands of secretion.

Cardiovascular System

Epinephrine increases blood pressure; ADH, aldosterone, and atrial natriuretic hormone help regulate blood volume; growth factors control blood cell formation.

Blood vessels transport hormones from glands; blood services glands; heart produces atrial natriuretic hormones.

How the Endocrine System works with other body systems.

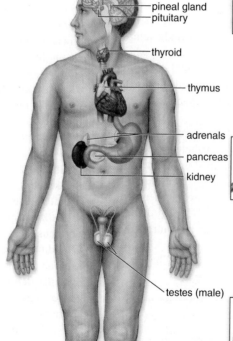

- hypothalamus
- pineal gland
- pituitary
- thyroid
- thymus
- adrenals
- pancreas
- kidney

testes (male)

 ovaries (female)

 parathyroids (posterior part of thyroid)

Lymphatic System/Immunity

Thymus is necessary for maturity of T lymphocytes.

Lymphatic vessels pick up excess tissue fluid; immune system protects against infections.

Respiratory System

Epinephrine promotes ventilation by dilating bronchioles; growth factors control production of red blood cells that carry oxygen.

Gas exchange in lungs provides oxygen and rids body of carbon dioxide.

Digestive System

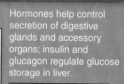

Hormones help control secretion of digestive glands and accessory organs; insulin and glucagon regulate glucose storage in liver.

Stomach and small intestine produce hormones.

Urinary System

ADH, aldosterone, and atrial natriuretic hormone regulate reabsorption of water and Na^+ by kidneys.

Kidneys keep blood values within normal limits so that transport of hormones continues.

Reproductive System

Hypothalamic, pituitary, and sex hormones control sex characteristics and regulate reproductive processes.

Gonads produce sex hormones.

system stimulates the adrenal medulla to release epinephrine (adrenaline) and norepinephrine, which have a powerful effect on various organs. This, too, is important to homeostasis because it allows us to behave in a way that keeps us alive. Any damage due to stress is then repaired by the action of other hormones, including cortisol. Glucocorticoid (e.g., cortisone) therapy is useful for its anti-inflammatory and immunosuppressive effects.

Content CHECK-UP!

21. Match the hormone with its effects:

a. glucagon   1. increases blood calcium levels

b. aldosterone   2. raises blood pressure

c. parathyroid hormone   3. raises blood glucose levels

Answer in Appendix A.

Selected New Terms

Basic Key Terms

adenohypophysis (ad″ĕn-ō-hī-pō′fĭ-sĭs), p. 222

adrenal cortex (ŭh-drē′nŭl kŏr′tĕks), p. 227

adrenal gland (ŭh-drē′nŭl glănd), p. 227

adrenal medulla (ŭh-drē′nŭl mĕ dūl′ŭh), p. 227

adrenocorticotropic hormone (ŭh-drē′nō-kŏr″tĭ-kō trōp′ĭk hŏr′mōn), p. 224

aldosterone (ăl′dŏs′tĕr-ōn), p. 228

anabolic steroid (ăn″ŭh-bŏl′ĭk stĕ′rōyd), p. 232

androgen (ăn′drō-jĕn), p. 232

anterior pituitary (ăn-tēr′ē-ŏr pĭ-tū′ĭ-tār″ē), p. 222

antidiuretic hormone (ăn″tĭ-dī′yū-rĕt′ĭk hŏr′mōn), p. 222

atrial natriuretic hormone (ā′trē-ŭhl nā″trē-yū-rĕt′ĭk hŏr′mōn), p. 228

atriopeptide (ā-trē-yō-pĕp-tīd), p. 228

calcitonin (kăl″sĭ-tō′nĭn), p. 226

calcitriol (kăl″sĭ-trĭ′ăwl), p. 226

chemical signal (kĕm′ ĭ-kŭhl sĭg′ nŭhl), p. 236

circadian rhythm (sĕr″kā′dē-ăn rĭ′thm), p. 234

corticotropin (kŏr′tĭ-kō-trōh-pĭn), p. 224

cortisol (kŏr′tĭ-sŏl), p. 227

cyclic AMP (sĭk′lĭk AMP), p. 218

endocrine gland (ĕn′dō-krĭn glănd), p. 218

epinephrine (ĕp″ĭ-nĕf′rĭn), p. 227

estrogen (ĕs′trō-jĕn), p. 232

follicle-stimulating hormone (fŏl′ĭk-kl stĭm″yoō-lā′tĭng hŏr′mōn), p. 224

ghrelin (grĕl′ŭhn), p. 235

glucagon (glū′kŭh-gŏn), p. 229

glucocorticoid (glū′kō-kŏr′tĭ-kōyd), p. 227

glucose sparing (glū′ kŏs spă′ rĭng), p. 237

gonad (gō′năd), p. 232

gonadotropic hormone (gō″năd-ō-trōp′ĭk hŏr′mōn), p. 224

growth factor (grōth făk′tŏr), p. 236

growth hormone (grōth hŏr′mōn), p. 224

hormone (hŏr′mōn), p. 218

hypothalamic-inhibiting hormone (hī″pō-thĕ-lăm′ĭk-ĭn-hĭb′ĭt-ĭng hŏr′mōn), p. 224

hypothalamic-releasing hormone (hī″pō-thĕ-lăm′ĭk-rē-lēs′ĭng hŏr′mōn), p. 224

hypothalamus (hī′pō-thăl′ă-mŭs), p. 222

insulin (in′sŭh-lĭn), p. 229

leptin (lĕp′tĭn), p. 235

luteinizing hormone (lū′tŭh-nī″zĭng hŏr′mōn), p. 224

lymphocyte (lĭmf′ ō-sīt), p. 234

melatonin (mĕl″ŭh-tō′nĭn), p. 234

mineralocorticoids (mĭn″ĕr-ăl-ō-kŏr′tĭ-kōyds), p. 227

neurohypophysis (nū′rō-hī-pŏf′ĭ-sĭs), p. 222

norepinephrine (nŏr′ĕp-ĭ-nĕf′rĭn), p. 227

ovary (ō′văr-ē), p. 232

oxytocin (ŏk″sī-tō′sĭn), p. 222

pancreas (păn′krē-ŭs), p. 229

pancreatic islets (of Langerhans), p. 229

parathyroid gland (păr″ŭh-thī′rōyd glănd), p. 226

parathyroid hormone (păr″ŭh-thī′rōyd hŏr′mōn), p. 226

peptide hormone (pĕp′tĭd hŏr′mōn), p. 218

pineal gland (pīn′ē-ul glănd), p. 234

pituitary gland (pĭ-tū′ĭ-tār″ē glănd), p. 222

portal system (pŏr′tŭl sĭs′tĕm), p. 222

posterior pituitary (pōs-tēr′ē-ŏr pĭ-tū′ĭ-tār″ē), p. 222

progesterone (prō-jĕs′tĕr-ōn), p. 232

prolactin (prō-lăk′tĭn), p. 224

prostaglandins (prŏs″tŭh-glăn′dĭnz), p. 236

renin (rē′nĭn), p. 228

somatostatin (sō′măt-ō-stăt′ĭn), p. 229

steroid hormone (stēr′ōyd hŏr′mōn), p. 218

testes (tĕs′tēz), p. 232

testosterone (tĕs-tŏs′tĕ-rōn), p. 232

thymosin (thī′mō-sĭn), p. 234

thymus gland (thī′mŭs glănd), p. 234

thyroid gland (thī′rōyd glănd), p. 225

thyroid-stimulating hormone (thī′rōyd stim′yū-lăt-ĭng hŏr′mōn), p. 224

thyrotropin (thī′rō-trō′pĭn), p. 218

thyroxine (thī″-rŏk′sĭn), p. 225

triiodothyronine (trī″ĭ-ō-dō-thī′rō-nēn), p. 225

Clinical Key Terms

acidosis (ăs′ĭ-dō′sĭs), p. 230

acromegaly (ăk″rō-mĕg′ŭh-lē), p. 224

Addison's disease (ă′dĭ-sŏns dĭ-zēz′), p. 228

congenital hypothyroidism (kŏn-gĕn′ĭ-tŭl hī″pō-thī′rōy-dĭzm), p. 225

Cushing's syndrome (kŭsh′ĭngs sĭn′drōm), p. 229

diabetes insipidus (dī″ŭh-bē′tēz ĭn-sĭp′ĭ-dus), p. 222

diabetic coma (dī″ŭh-bĕ-tĭk kō′-mŭh), p. 232

diabetic ketoacidosis (dī″ŭh-bĕ-tĭk kē′tō′ăs-ĭ-dō′sŭs), p. 232

exophthalmic goiter (ĕk″sŏf-thăl′mĭk gōy′tĕr), p. 225

glycosuria (glī″kō-sūr′ē-ŭh), p. 230

Graves' disease (grāvz dĭ-zēz′), p. 225

hypercalcemia (hī″pĕr-kăl-sē′mē-ŭh), p. 227

hyperglycemia (hī″pĕr-glī-sē′mē-ŭh), p. 230

hypocalcemia (hī″pō-kăl-sē′mē-ŭh), p. 227

hypoglycemia (hī″pō-glī-sē′mē-ŭh), p. 232

insulin-dependent diabetes mellitus (ĭn′sŭl-ĭn-dē-pĕn′dĕnt dī″ŭh-bē′tēz mĕ-lī′tŭs), p. 231

insulin shock (ĭn′sŭl-ĭn shŏk), p. 232

ketonuria (kē″tō-nū′rē-ŭh), p. 230

myxedema (mĭk″sĕ-dē′mŭh), p. 225

noninsulin-dependent diabetes (nŏn′ĭn′sŭl-ĭn-dē-pĕn′dĕnt dī″ŭh-bē′tēz), p. 231

pituitary dwarfism (pĭ-tū′ĭ-tār″e dwărf′ĭzm), p. 224

polydipsia (pŏl′ē-dĭp′sē-ŭh), p. 230

polyphagia (pŏl″ē-fā-jē-ŭh), p. 230

polyuria (pŏl″ē-yū′rē-ŭh), p. 230

prediabetes (prē′dī-ŭh-bē″tēz), p. 231

simple (endemic) goiter (sĭm′pl ĕn-dĕm′ĭk gōy′tĕr), p. 225

syndrome (sĭn′drōm), p. 228

tetany (tĕt′ŭh-nē), p. 227

10.1 Endocrine Glands

Endocrine glands secrete hormones into the bloodstream, and from there they are distributed to target organs or tissues. Hormones are either peptides or steroids.

  A. Reception of a peptide hormone at the plasma membrane activates an enzyme cascade inside the cell. Steroid hormones combine with a receptor in the cell, and the complex attaches to and activates DNA. Protein synthesis follows. The major endocrine glands and hormones are listed in Table 10.1.

  B. Neural mechanisms, hormonal mechanisms, and/or negative feedback control the levels of hormones.

10.2 Hypothalamus and Pituitary Gland

  A. Neurosecretory cells in the hypothalamus produce antidiuretic hormone (ADH) and oxytocin, which are stored in axon endings in the posterior pituitary until they are released.

  B. The hypothalamus produces hypothalamic-releasing and hypothalamic-inhibiting hormones, which pass to the anterior pituitary by way of a portal system. The anterior pituitary produces at least six types of hormones, and some of these stimulate other hormonal glands to secrete hormones.

10.3 Thyroid and Parathyroid Glands

  A. The thyroid gland requires iodine to produce triiodothyronine (T_3) and thyroxine (T_4), which increase the metabolic rate. If iodine is available in limited quantities, a simple goiter develops. If the thyroid is overactive, an exophthalmic goiter develops. The thyroid gland also produces calcitonin, which helps lower the blood calcium level.

  B. The parathyroid glands secrete parathyroid hormone, which raises the blood calcium and decreases the blood phosphate levels. Parathyroid hormone is the primary hormone responsible for calcium regulation.

10.4 Adrenal Glands

  A. The adrenal glands respond to stress: during short-term stress, the adrenal medulla secretes epinephrine and norepinephrine, which bring about responses we associate with emergency situations.

  B. During long-term stress, the adrenal cortex produces the glucocorticoids (e.g., cortisol) and the mineralocorticoids (e.g., aldosterone). Cortisol stimulates hydrolysis of proteins to amino acids, which are converted to glucose; in this way, it raises the blood glucose level. Aldosterone causes the kidneys to reabsorb sodium ions (Na^+) and to excrete potassium ions (K^+). Addison's disease develops when the adrenal cortex is underactive, and Cushing's syndrome develops when the adrenal cortex is overactive.

10.5 Pancreas

The pancreatic islets secrete insulin, which lowers the blood glucose level, and glucagon, which has the opposite effect.

  A. The most common illness caused by hormonal imbalance is diabetes mellitus, which is due to the failure of the pancreas to produce insulin and/or the failure of cellular insular receptors to properly respond to it.

10.6 Additional Endocrine Glands

  A. The gonads (ovaries and testes) produce the sex hormones.

  B. The thymus secretes thymosins, which stimulate T-lymphocyte production and maturation.

  C. The pineal gland produces melatonin, which is involved in circadian rhythms and may affect the development of the reproductive organs.

  D. Tissues also produce hormones. Adipose tissue produces leptin and the stomach produces ghrelin. Both act on the hypothalamus. Various tissues produce growth factors. Prostaglandins are produced and act locally.

10.7 The Importance of Chemical Signals

In the human body, some chemical signals, such as traditional endocrine hormones and secretions of neurosecretory cells, act at a distance. Others, such as prostaglandins, growth factors, and neurotransmitters, act locally. Whether humans have pheromones is under study.

10.8 Effects of Aging

Two concerns often seen in the elderly are thyroid malfunctioning and diabetes mellitus. The thymus gland atrophies (shrinks). As a result, the immune response is diminished in the elderly.

10.9 Homeostasis

Hormones help maintain homeostasis in several ways: they help maintain the level of nutrients (e.g., amino acids and glucose in the blood), help maintain blood volume and pressure by regulating the sodium content of the blood, help maintain the blood calcium level, help regulate fuel metabolism, and help regulate our response to the external environment.

Study Questions

1. Explain how peptide hormones and steroid hormones affect the metabolism of the cell. (pp. 218–221)

2. Contrast hormonal and neural signals, and show that there is an overlap between the mode of operation of the nervous system and that of the endocrine system. (pp. 218–221)

3. Explain the relationship of the hypothalamus to the posterior pituitary gland and to the anterior pituitary gland. List the hormones secreted by the posterior and anterior pituitary glands. (pp. 219–220, 222–224)

4. Give an example of the negative feedback relationship among the hypothalamus, the anterior pituitary, and other endocrine glands. (pp. 222–224)

5. Discuss the effect of growth hormone on the body and the result of having too much or too little growth hormone when a young person is growing. What is the result if the anterior pituitary

produces growth hormone in an adult? (pp. 224–225)

6. What types of goiters are associated with a malfunctioning thyroid? Explain each type. (pp. 225–226)

7. How do the thyroid and the parathyroid work together to control the blood calcium level? (pp. 226–227)

8. How do the adrenal glands respond to stress? What hormones are secreted by the adrenal medulla, and what effects do these hormones have? (pp. 227–228)

9. Name the most significant glucocorticoid and mineralocorticoid, and discuss their functions. Explain the symptoms of Addison's disease and Cushing's syndrome. (pp. 227–229)

10. Draw a diagram to explain how insulin and glucagon maintain the blood glucose level. Use your diagram to explain the major symptoms of type I diabetes mellitus. (pp. 229–231)

11. Name the additional endocrine glands discussed in this chapter, and discuss

the functions of the hormones they secrete. (pp. 219–220, 232, 234)

12. What are leptin, ghrelin, growth factors, and prostaglandins? How do these substances act? (pp. 235–236)

13. Discuss five ways the endocrine system helps maintain homeostasis. (pp. 237–239)

Learning Outcome Questions

Fill in the blanks.

1. Whereas _____ hormones are lipid soluble and bind to receptor proteins within the cytoplasm of target cells, _____ hormones bind to membrane-bound receptors, thereby activating second messengers.

2. Generally, hormone production is self-regulated by a _____ mechanism.

3. The hypothalamus produces the hormones _____ and _____, released by the posterior pituitary.

4. The _____ secreted by the hypothalamus control the anterior pituitary.

5. Growth hormone is produced by the _____ pituitary.

6. Simple goiter occurs when the thyroid is producing (too much or too little) _____ because it has (too much or too little) _____.

7. Parathyroid hormone increases the level of _____ in the blood.

8. Whereas the adrenal _____ is under the control of the autonomic nervous system, the adrenal _____ secretes its hormones

in response to _____ from the anterior pituitary gland.

9. Adrenocorticotropic hormone (ACTH), produced by the anterior pituitary, stimulates the _____ of the adrenal glands.

10. An overproductive adrenal cortex results in the condition called _____.

11. Type I diabetes mellitus is due to a malfunctioning _____, but type II diabetes is due to malfunctioning _____.

12. Prostaglandins are not carried in the _____ as are hormones secreted by the endocrine glands.

Medical Terminology Exercise

After studying this chapter, see if you can derive the definitions for the medical terms listed at right. Many of the prefixes and suffixes used to create these terms can be found throughout the chapter. For additional help, use McGraw-Hill Connect™ at www.mcgrawhillconnect.com and consult Appendix B.

1. hypophysectomy (hī-pŏf″ĭ-sĕk′tō-mē)
2. hypokalemia (hī″pō-kāl″ē′mē-ŭh)
3. lactogenic (lăk″tō-jĕn′ĭk)
4. adrenopathy (ăd″rĕn-ŏp′ŭh-thē)
5. adenomalacia (ăd″ĕ-nō-mŭh-lā′shē-ŭh)

6. parathyroidectomy (păr″ŭh-thī′rōy-dĕk′tō-mē)
7. polydipsia (pŏl″ē-dĭp′sē-ŭh)
8. dyspituitarism (dĭs-pĭ-tu′ĭ-tĕr′ĭzm)
9. thyroiditis (thī-rōy-dī′tĭs)
10. microsomia (mī′krō-sō′mē-ŭh)

Online Study Tools

LEARNSMART® connect

APR

Anatomy & Physiology REVEALED includes cadaver photos that allow you to peel away layers of the human body to reveal structures beneath the surface. This program also includes animations, radiologic imaging, audio pronunciations, and practice quizzing. To learn more visit www.aprevealed.com

Anatomy & Physiology REVEALED
aprevealed.com

11 Blood

If you know about ABO blood types, and know your own, maybe you've wondered which of the four possible types is the most common worldwide. The answer depends on the genetic inheritance of a particular population. Worldwide, the most prevalent ABO type is type O. However, type A blood is most common in several European countries, including the Scandinavian countries. In certain geographic areas, virtually the entire population has the same blood type. For example, certain native Central and South American tribes are nearly all type O. The least common ABO type is type AB. If you also know your Rh type (positive or negative), you'll be interested to know that approximately 85% of Caucasians, 92% of African Americans, and 97% of Asians are Rh positive. In addition to the A, B, and Rh proteins, there are several thousand known blood proteins that have been identified. The International Society of Blood Transfusion has organized 32 separate major group systems to classify them (ABO and Rh are two examples of group systems). Doubtless, more proteins remain to be discovered, and more classification systems will be developed for them. You can read more about blood types and blood compatibility on pages 252–254.

Learning Outcomes

After you have studied this chapter, you should be able to:

11.1 The Composition and Functions of Blood

1. Describe, in general, the composition of blood and define the term *hematocrit*.
2. Divide the functions of blood into three categories, and discuss each category.

11.2 Components of Blood

3. Describe the composition of plasma, and explain the functions of the plasma proteins.
4. Explain the hematopoietic role of stem cells in the red bone marrow.
5. Describe the structure, function, and life cycle of red blood cells and white blood cells. Briefly outline the effects of abnormal red and white blood cell counts.

11.3 Platelets and Hemostasis

6. Describe the structure, function, and life cycle of platelets.
7. Describe the three events of hemostasis and the reactions necessary for coagulation to occur.
8. Discuss disorders of hemostasis.

11.4 Blood Typing and Transfusions

9. Explain the ABO and Rh systems of blood typing.
10. Explain agglutination and its relationship to transfusions.

11.5 Effects of Aging

11. Name the blood disorders that are commonly seen as we age.

Visual Focus

Hematopoiesis

Medical Focus

Abnormal Red and White Blood Cell Counts

What's New

Improvements in Transfusion Technology

I.C.E.—In Case of Emergency

Hemorrhage

Focus on Forensics

Blood at the Crime Scene

11.1 The Composition and Functions of Blood

1. Describe, in general, the composition of blood and define the term hematocrit.
2. Divide the functions of blood into three categories, and discuss each category.

When a blood sample is first prevented from clotting and then spun in a centrifuge tube, it separates into three layers (Fig. 11.1). The lower layers consist of the **formed elements:** red blood cells (erythrocytes), white blood cells (leukocytes), and platelets (thrombocytes). Formed elements make up about 45% of the total volume of whole blood. Red blood cells are the heaviest cells due to their iron-containing hemoglobin, and can be found on the bottom of the tube. The percentage of the volume of a blood sample consisting of red blood cells is that sample's **hematocrit.** The white blood cells and the blood platelets settle on top of the red blood cells in a thin, shiny layer called the **buffy coat.** The upper layer is **plasma,**

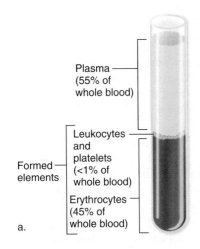

Plasma
(55% of
whole blood)

Leukocytes
and
platelets
(<1% of
whole blood)

Formed
elements

Erythrocytes
(45% of
whole blood)

a.

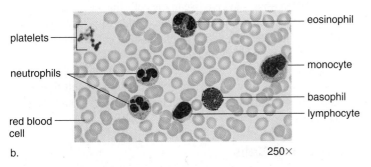

platelets

neutrophils

red blood
cell

b.   250×

eosinophil

monocyte

basophil

lymphocyte

Figure 11.1 Composition of blood. (a) When blood is transferred to a test tube containing an anticoagulant (to prevent clotting) and then centrifuged, it consists of three layers. The transparent straw-colored or yellow top layer is the plasma, the liquid portion of blood. The thin middle buffy coat layer consists of leukocytes and platelets. The bottom layer contains the erythrocytes. **(b)** Micrograph of the formed elements in blood. Neutrophils, basophils, eosinophils, monocytes, and lymphocytes are all different forms of white blood cells.

a light-yellow fluid that contains a variety of inorganic and organic molecules dissolved or suspended in water. Plasma accounts for about 55% of the total volume of whole blood.

It's interesting to note that, on average, males have a slightly higher hematocrit than females. The difference is primarily due to the effects of the sex hormone testosterone. Testosterone (and similar precursor molecules found in high levels in males) stimulates red blood cell production. In addition, adult females of menstrual age also experience blood loss during menstruation, and this lowers a woman's hematocrit.

Functions of Blood

The functions of blood fall into three categories: transport, defense, and regulation.

Transport

Blood moves from the heart to all the various organs, where exchange with tissues takes place across thin capillary walls. (Recall from Chapter 4 that capillaries are the smallest blood vessels and are formed from a single layer of flattened epithelial cells. Capillaries are so narrow that erythrocytes must travel through them in single file, as shown in Fig. 11B.) Blood picks up oxygen from the lungs and nutrients from the digestive tract and transports these to the tissues. It also picks up and transports cellular wastes, including carbon dioxide, away from the tissues to organs with exchange surfaces, such as the lungs and kidneys. In the lungs, carbon dioxide is excreted every time we exhale. In the kidneys, metabolic wastes can be excreted in the urine. We'll see that capillary exchanges keep the composition of tissue fluid within normal limits.

Various organs and tissues secrete hormones and other messenger chemicals into the blood, and blood transports these to other organs and tissues, where they serve as signals that influence cellular metabolism.

Defense

Because blood contains the white blood cells, it defends the body against invasion by **pathogens** (microscopic infectious agents, such as disease-causing bacteria and viruses). The white blood cells also remove dead and dying cells, clearing room for the growth of healthy cells. Mutated cells—which could potentially develop into cancer—are destroyed by white blood cells as well. Certain white blood cells are capable of engulfing and destroying pathogens or cancer cells; others produce and secrete antibodies into the blood. Antibodies are proteins made by a specific class of white blood cells. They incapacitate pathogens, making them subject to destruction by other white blood cells. This function of the blood will be discussed completely in Chapter 13.

When an injury occurs, blood forms a clot, and this prevents blood loss. Blood clotting involves platelets and the plasma proteins prothrombin and fibrinogen. Without blood clotting, we could bleed to death even from a small cut.

Regulation

Blood helps regulate body temperature by picking up heat, mostly from active muscles, and transporting it about the body. If the blood is too warm, the heat dissipates to the environment from dilated blood vessels in the skin.

The salts and plasma proteins in blood act to keep the osmotic pressure in the normal range. Recall that osmotic pressure causes diffusion of water (review section 3.2). Thus, the osmotic pressure of blood causes diffusion of water into the blood. In this way, blood plays a role in maintaining its own water-salt balance.

Because blood contains buffers, it also helps regulate body pH and keeps it relatively constant (in the range of 7.35 to 7.45).

Content CHECK-UP!

1. What are the three types of formed elements found in blood? Describe where each would be found in a centrifuge tube containing a blood sample.

2. Which of the following are defense mechanisms of the blood?

   a. Blood cells can engulf and destroy bacteria and viruses.

   b. Antibodies target pathogens for destruction.

   c. The blood clotting mechanism prevents blood loss.

   d. All of the above are defense mechanisms of the blood.

3. How does blood regulate body temperature? osmotic pressure? body pH?

Answers in Appendix A.

11.2 Components of Blood

3. Describe the composition of plasma, and explain the functions of the plasma proteins.
4. Explain the hematopoietic role of stem cells in the red bone marrow.
5. Describe the structure, function, and life cycle of red blood cells and white blood cells. Briefly outline the effects of abnormal red and white blood cell counts.

Plasma

Plasma is the liquid portion of blood, and about 92% of plasma is water. The remaining 8% of plasma is composed of various ions (from salts that dissolve in water) and organic molecules (Fig. 11.2a.) The salts, which are simply dissolved in plasma, help maintain the osmotic pressure and pH of the blood. Small organic molecules such as glucose, amino acids, and urea are also dissolved in plasma. Glucose and amino acids are nutrients for cells; urea is a nitrogenous waste product on its way to the kidneys for excretion. The large organic molecules in plasma include hormones and the plasma proteins.

Three major types of plasma proteins are the albumins, the globulins, and fibrinogen. Most plasma proteins are made in the liver. One exception is the antibodies produced by B lymphocytes, which function in immunity. Certain hormones are also plasma proteins made by various glands.

The plasma proteins have many functions that help maintain homeostasis. They are able to take up and release hydrogen ions; therefore, the plasma proteins help buffer the blood and keep its pH around 7.40. **Osmotic pressure** is the force caused by a difference in solute concentration on either side of a membrane. As mentioned previously, plasma proteins, particularly the **albumins,** contribute to the osmotic pressure. Their role will be described completely in Chapter 12.

There are three types of **globulins,** designated alpha, beta, and gamma globulins. The alpha and beta globulins, produced by the liver, bind to metal ions and to fat-soluble vitamins. They also bind to lipids, forming the **lipoproteins.** Gamma globulins are one of several kinds of *antibodies,* which are infection-fighting proteins made by B-lymphocyte white blood cells.

Both albumins and globulins combine with and transport large organic molecules. For example, albumin transports the molecule *bilirubin,* a breakdown product of hemoglobin. Lipoproteins, whose protein portion is a globulin, transport cholesterol.

The other plasma proteins, fibrinogen and prothrombin, are necessary to coagulation (blood clotting), which is discussed on pages 250–251.

Formed Elements

The formed elements consist of red and white blood cells and platelets (Fig. 11.2b). (Platelets are discussed in detail on pages 250–251.) In the adult, the formed elements are produced continuously in the red bone marrow of the skull, ribs, and vertebrae, the iliac crests, and the ends of long bones.

The process by which formed elements are made is called **hematopoiesis** (Fig. 11.3). **Multipotent stem cells** are the red bone marrow cells that mature into all the various types of blood cells. At the top of Figure 11.3 is a multipotent stem cell that first replicates itself by mitosis. Each cell then *differentiates,* producing two other types of stem cells: myeloid and lymphoid (lymphatic) stem cells. (When a stem cell differentiates, it commits itself to follow a single pathway in development.) The myeloid stem cell further differentiates to give rise to the cells that become red blood cells, granular leukocytes, monocytes, and megakaryocytes, the parent cell for platelets. The lymphoid stem cell differentiates to produce the lymphocytes and natural killer cells.

Many scientists are very interested in developing ways to use blood stem cells, as well as stem cells from other adult tissues, to regenerate the body's tissues in the laboratory (see What's New in Chapter 4).

Red Blood Cells

Red blood cells (RBCs, or **erythrocytes)** are small, biconcave disks that lack a nucleus when mature. They occur in great quantity; there are 4 to 6 million red blood cells per mm^3 of whole blood.

Red blood cells transport oxygen, and each contains about 200 million molecules of **hemoglobin (Hb),** the respiratory pigment. If this much hemoglobin were suspended within the plasma rather than enclosed within the cells, blood would be so thick and viscous that the heart would have difficulty pumping it.

Plasma		
Type	**Function**	**Source**
Water (90–92% of plasma)	Maintains blood volume; transports molecules	Absorbed from digestive tract
Plasma proteins (7–8% of plasma)	Maintain blood osmotic pressure and pH	
Albumin	Maintains blood volume and pressure; transport	Liver
Globulins	Transport; fight infection	
Fibrinogen	Clotting	
Inorganic ions (salts) (less than 1% of plasma)	Maintain blood osmotic pressure and pH; aid metabolism	Absorbed from intestine
Gases		
Oxygen	Cellular respiration	Lungs
Carbon dioxide	End product of metabolism	Tissues
Nutrients		
Lipids	Food for cells	Absorbed from intestine
Glucose		
Amino acids		
Phospholipids		
Vitamins, etc.		
Nitrogenous wastes		
Urea	Excretion by kidneys	Liver
Uric acid		
Ammonia		
Regulatory substances		
Enzymes	Aid metabolism	Varied
Hormones		
Other chemical signals		

a.

Formed Elements			
	Type	**Function and Description**	**Source**
	Red blood cells (erythrocytes)	Transport O_2 and help transport CO_2	Red bone marrow
	 4 million–6 million per mm³ blood	7–8 µm in diameter Bright-red to dark-purple biconcave disks without nuclei	
	White blood cells (leukocytes) 5,000–11,000 per mm³ blood	Fight infection	Red bone marrow
Granular leukocytes	Neutrophils  ★ 40–70%	10–14 µm in diameter Spherical cells with multilobed nuclei; fine, lilac granules in cytoplasm; phagocytize pathogens	
Granular leukocytes	Eosinophils 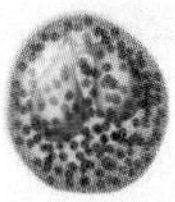 ★ 1–4%	10–14 µm in diameter Spherical cells with bilobed nuclei; coarse, deep-red, uniformly sized granules in cytoplasm; phagocytize antigen-antibody complexes and allergens	
Granular leukocytes	Basophils  ★ 0–1%	10–12 µm in diameter Spherical cells with lobed nuclei; large, irregularly shaped, deep-blue granules in cytoplasm; release histamine and heparin, which promotes blood flow to injured tissues	
Agranular leukocytes	Lymphocytes  ★ 20–45%	5–17 µm in diameter (average 9–10 µm) Spherical cells with large, round nuclei; responsible for specific immunity	
Agranular leukocytes	Monocytes 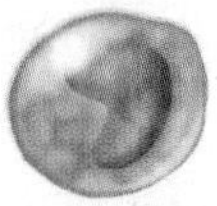 ★ 4–8%	10–24 µm in diameter Large spherical cells with kidney-shaped, round, or lobed nuclei; become macrophages that phagocytize pathogens and cellular debris; antigen-presenting cells in immune system responses	
Granular lymphocyte	Natural killer (NK) cell 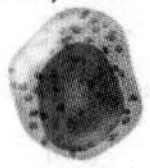 ★ 1–5%	5–17 µm in diameter Granular leukocytes; similar in structure to lymphocytes; destroy virus infected cells and cancer cells; involved in transplant rejection	
	Platelets (thrombocytes)	Aid hemostasis	Red bone marrow
	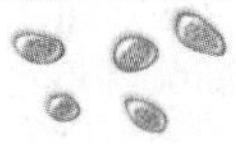 150,000–300,000 per mm³ blood	2–4 µm in diameter Disk-shaped cell fragments with no nuclei; purple granules in cytoplasm	

b.   ★ Appearance with Wright's stain.

AP|R **Figure 11.2** **Plasma and formed elements.** (a) Ingredients of plasma. Plasma is approximately 92% water and 8% dissolved or suspended solutes. (b) Types, functions, and descriptions of the formed elements.

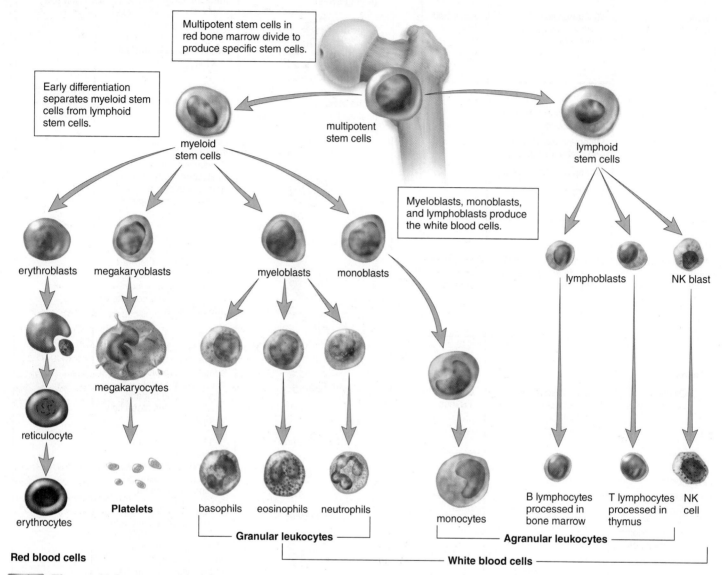

Multipotent stem cells in red bone marrow divide to produce specific stem cells.

Early differentiation separates myeloid stem cells from lymphoid stem cells.

multipotent stem cells

myeloid stem cells

lymphoid stem cells

Myeloblasts, monoblasts, and lymphoblasts produce the white blood cells.

erythroblasts   megakaryoblasts   myeloblasts   monoblasts   lymphoblasts   NK blast

megakaryocytes

reticulocyte

Platelets   basophils   eosinophils   neutrophils

erythrocytes   monocytes

B lymphocytes   T lymphocytes   NK
processed in   processed in   cell
bone marrow   thymus

Red blood cells

Granular leukocytes   **Agranular leukocytes**

White blood cells

AP|R **Figure 11.3** **Hematopoiesis.** Multipotent stem cells give rise to two specialized stem cells. The myeloid stem cell gives rise to still other cells, which become red blood cells, platelets, and all the whole blood cells except lymphocytes. The lymphoid stem cell gives rise to lymphoblasts, which become lymphocytes. Though they have granules like the granular leukocytes, natural killer cells also arise from the lymphoid stem cells.

Hemoglobin

In a molecule of hemoglobin, each of four polypeptide chains making up globin has an iron-containing heme group in the center (Fig. 11.4). Oxygen combines loosely with iron when hemoglobin is oxygenated:

$$\text{HHb} + \text{O}_2 \underset{\text{tissues}}{\overset{\text{lungs}}{\rightleftharpoons}} \text{HbO}_2$$

In this equation, the hemoglobin on the left, which is not combined with oxygen, is called deoxyhemoglobin (its symbol is HHb). Deoxyhemoglobin forms in tissue capillaries and has a dark maroon color. The hemoglobin on the right, which is combined with oxygen, is called oxyhemoglobin (its symbol is HbO_2). Oxyhemoglobin forms in lung capillaries and has a bright red color.

Hemoglobin is remarkably adapted to its function of picking up oxygen in lung capillaries and releasing it in the tissues. The higher concentration of oxygen, plus the slightly cooler temperature and

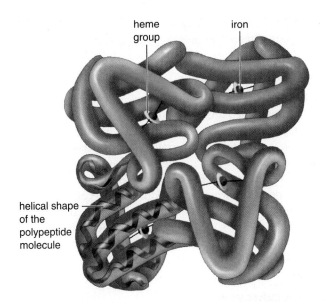

Figure 11.4 Hemoglobin contains four polypeptide chains.
There is an iron-containing heme group in the center of each
chain. Oxygen combines loosely with iron when hemoglobin is
oxygenated. Oxyhemoglobin is bright red, and deoxyhemoglobin
is a dark maroon color.

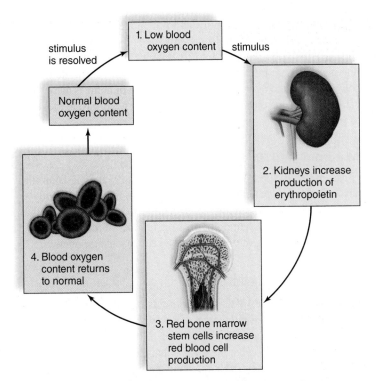

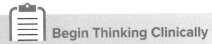

Figure 11.5 Regulation of red blood cell production. The
kidneys release increased amounts of erythropoietin whenever
the oxygen capacity of the blood is reduced. Erythropoietin
stimulates the red bone marrow to speed up its production of
red blood cells, which carry oxygen. Once the oxygen-carrying
capacity of the blood is sufficient to support normal cellular activity, the kidneys cut back on their production of erythropoietin.

slightly higher pH within lung capillaries, causes hemoglobin to
take up oxygen. By contrast, in tissues the lower concentration of
oxygen, plus the slightly warmer temperature and slightly lower pH
within tissue capillaries, causes hemoglobin to give up its oxygen.

Production of Red Blood Cells

Erythrocytes are formed from red bone marrow stem cells (see
Fig. 11.3): A multipotent stem cell descendant, called a myeloid
stem cell, gives rise to erythroblasts, which divide many times. As
they mature, erythroblasts fill with millions of molecules of hemoglobin and lose their nucleus and most of their organelles. An
immature red blood cell lacking a nucleus is called a **reticulocyte**
because it contains ribosomes in a network called a reticulum.
In the bloodstream, reticulocytes can be detected using special
stains. Reticulocytes leave the bone marrow through capillaries,
and they continue to produce additional hemoglobin for a day or
two, at which time they are fully mature. Possibly because mature
red blood cells lack a nucleus, they live only about 120 days. It is
estimated that about 2 million red blood cells are destroyed per
second; therefore, an equal number must be produced to keep the
red blood cell count in balance.

Whenever blood carries a reduced amount of oxygen, more red
blood cells must be produced to maintain homeostasis. For example,
lowered blood oxygen occurs when an individual first takes up residence at a high altitude, or after a hemorrhage when a person loses
red blood cells, or in a patient with impaired lung function. These
situations are stimuli for the kidneys to accelerate their release of
erythropoietin (EPO), a hormone first mentioned in section 10.6
(Fig. 11.5). This hormone stimulates mitosis of stem cells and speeds
up the maturation of red blood cells. The liver and other tissues also
produce erythropoietin. Erythropoietin, now mass-produced through

biotechnology, is often prescribed for cancer patients who are anemic (low red blood cell count) due to chemotherapy or radiation
therapy. Unfortunately, erythropoietin is sometimes abused by athletes in order to raise their red blood cell counts and thereby increase
the oxygen-carrying capacity of their blood.

Begin Thinking Clinically

Polycythemia is a condition caused by excessive numbers
of red blood cells in the blood. In this disorder, blood is
excessively thick and unable to flow properly, and it often
forms unnecessary, dangerous clots. What do you think
might cause the condition, and what effect could it have on
the patient's heart? How might polycythemia be treated?
Answer and discussion in Appendix A.

Destruction of Red Blood Cells

With age, red blood cells are destroyed in the liver and spleen,
where they are engulfed by macrophages. When red blood cells are
broken down, hemoglobin is released and separates into its three
components. The globin portion of the hemoglobin is broken down
into its component amino acids, which are recycled by the body.

The iron is recovered and returned to the bone marrow for reuse. The heme portion of the molecule undergoes chemical degradation and is excreted by the liver into the bile as **bile pigments.** These bile pigments are bilirubin and biliverdin, which contribute to the color of urine and feces. Chemical breakdown of heme is also what causes a bruise on the skin to change color from red/purple to green to yellow.

Abnormal Red Blood Cell Counts

As discussed in the Medical Focus on page 249, anemia is an illness in which the patient has a tired, run-down feeling. The body's cells are not getting enough oxygen due to a reduction in the amount of functional hemoglobin or the number of red blood cells. **Hemolysis** (bursting of red blood cells) can also cause anemia.

White Blood Cells

White blood cells (**WBCs,** or **leukocytes**) differ from red blood cells in that they are usually larger, have a nucleus, lack hemoglobin, and are translucent unless stained. White blood cells are not as numerous as red blood cells; there are only 5,000–11,000 per mm^3 of blood. White blood cells fight infection, destroy dead or dying body cells, and recognize and kill cancerous cells. Thus, they are important contributors to homeostasis. This function of white blood cells is discussed at greater length in Chapter 13, which concerns immunity.

White blood cells are derived from stem cells in the red bone marrow, and they, too, undergo several maturation stages (see Fig. 11.3). Each type of white blood cell is apparently capable of producing a specific growth factor that circulates back to the bone marrow to stimulate its own production.

Red blood cells are confined to the blood, but white blood cells are able to squeeze through pores in the capillary wall (Fig. 11.6). Therefore, they can form colonies in tissues throughout out the body, including the lungs and liver. They can also be found in tissue fluid and lymph (the fluid within lymphatic vessels), and in lymphatic organs. When an infection is present, white blood cells greatly increase in number. Many white

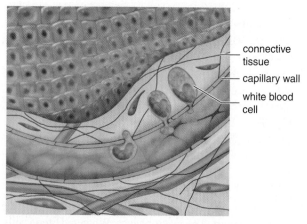

connective tissue

capillary wall

white blood cell

Figure 11.6 Mobility of white blood cells. When injury or infection occurs, white blood cells can squeeze between the cells of a capillary wall and enter the tissues of the body.

blood cells live only a few days—they probably die while fighting pathogens. Others live months or even years.

Types of White Blood Cells

White blood cells are classified into the **granular leukocytes** and the **agranular leukocytes.** Both types of cells have granules in the cytoplasm surrounding the nucleus, but the granules are more visible upon staining in granular leukocytes. (The white cells shown in Figures 11.1 through 11.3 have been stained with a dye called Wright's stain.) The granules contain various enzymes and proteins, which help white blood cells defend the body. There are four types of granular leukocytes and two types of agranular leukocytes. They differ somewhat by the size of the cell and the shape of the nucleus (see Figs. 11.2 and 11.3), and they also differ in their functions.

Granular Leukocytes

Neutrophils (see Fig. 11.3 and Fig. 13.2a) are the most abundant of the white blood cells. They have a multilobed nucleus joined by nuclear threads; therefore, they are also called polymorphonuclear ("many-shaped nuclei"). Some of their granules take up acid stain, and some take up basic stain (creating an overall lilac color). Neutrophils are the first type of white blood cell to respond to an infection, and they engulf pathogens during phagocytosis.

Eosinophils (see Fig. 11.3 and Fig. 13.2b) have a bilobed nucleus, and their large, abundant granules take up the dye called eosin and become a red color. (This accounts for their name, *eosinophil.*) Among several functions, they increase in number in the event of a parasitic worm infection. Eosinophils also lessen an allergic reaction by phagocytizing antigen-antibody complexes involved in an allergic attack.

Basophils (see Fig. 11.3 and Fig. 13.2c) have a U-shaped or lobed nucleus. Their granules take up the basic stain and become dark blue in color. (This accounts for their name, *basophil.*) In the connective tissues, basophils, as well as a similar type of cells called mast cells, release histamine and heparin. Histamine, which is associated with allergic reactions, dilates blood vessels and causes contraction of smooth muscle. Heparin prevents clotting and promotes blood flow.

Natural killer cells (see Fig. 11.3) are large, granular leukocytes that are similar in structure to lymphocytes. Like lymphocytes, they develop from the lymphoid stem cells. They are particularly important for their role in destruction of virus-infected and mutated (and thus, potentially cancerous) cells. Their function will be detailed in Chapter 13.

Agranular Leukocytes The agranular leukocytes include lymphocytes, which have a spherical nucleus, and monocytes, which have a kidney-shaped nucleus. **Lymphocytes** are responsible for specific immunity to particular pathogens and their toxins (poisonous substances), as well as recognizing and destroying cancer cells. Lymphocytes (see Fig. 11.3 and Fig. 13.2d) are of two types, B lymphocytes and T lymphocytes. Pathogens have **antigens,** surface molecules that the immune system can recognize as foreign. When an antigen is recognized as foreign, B lymphocytes will

Abnormal Red and White Blood Cell Counts

Normal erythrocyte count should be in the range of 4–6 million cells per cubic millimeter of blood (about the size of one large drop), and normal blood hemoglobin level is 12–17 grams per 100 milliliters. However, it's important to recognize gender variation: due to the effects of androgen hormones such as testosterone, men have a higher red blood cell count (and thus, a higher hemoglobin level) than women. Various disease processes can affect erythrocyte count and hemoglobin levels. As you know from the Begin Thinking Clinically question on page 247, polycythemia is a disorder in which an excessive number of red blood cells makes the blood so thick that it is unable to flow properly. An increased risk of clot formation is also associated with this condition.

In **anemia,** the number of red cells is insufficient and/or the cells do not have enough normal hemoglobin. Anemia can be classified into one of several categories. The first category, **hemolytic anemia,** occurs because the rate of red blood cell destruction increases (hemolysis is the rupturing of red blood cells). **Hemolytic disease of the newborn,** discussed at the end of this chapter (see page 253), is one form of this type of anemia.

Sickle-cell disease is an inherited hemolytic anemia caused by an abnormal form of hemoglobin. Affected individuals have fragile, sickle-shaped red blood cells (a sickle is a shape resembling the letter **C;** Fig. 11A) that easily tear open as they pass through the narrow capillaries. As a result, the person has far fewer red blood cells than normal, and anemia symptoms result. Both parents must carry the gene for sickle-cell disease for a child to be afflicted. Someone with a single gene is said to have **sickle-cell trait,** and doesn't have disease symptoms.

Both sickle-cell disease and sickle-cell trait are most prevalent among people who live in areas bordering the equator (for example, in Africa, Asia, India, and Central and South America) where malaria is common. Malaria is a disease caused by a parasite and spread by mosquitoes that infect humans when they bite. Having sickle-cell trait gives the person a better chance of surviving malaria, though it doesn't protect against the actual infection or subsequent illness. Scientists aren't sure why the trait enables some people to survive, but when

survivors have children, the sickle-cell gene passes on to the next generation. In the United States, sickle-cell disease is most common in African Americans whose ancestors are from equatorial Africa, but it occurs in people of any race.

Dietary anemias occur because the patient's diet lacks substances needed for red blood cell development. In **iron deficiency anemia,** a common type of anemia, the person's diet does not contain enough iron, or there is excessive iron loss from the body. The hemoglobin count is low, red blood cells are small and pale in color, and the individual feels tired and run-down. Including iron supplements in the diet can help prevent this type of anemia. **Pernicious anemia** is another form of dietary anemia. The digestive tract is unable to absorb enough vitamin B_{12}, which is essential to red cell development. Without vitamin B_{12}, immature red cells tend to accumulate in the bone marrow. A special diet, vitamin supplements, and/or injections of vitamin B_{12} are effective treatments.

In **aplastic anemia,** the red bone marrow has been damaged due to radiation or chemicals, and red blood cells, white blood cells, and platelets are all deficient. Bone marrow transplant is one option to treat this condition.

Hemorrhagic anemia is decreased red blood cell count following a hemorrhage. Transfusions may be administered to increase red blood cell count. Research continues to develop improvements in transfusion technology, as detailed in What's New on page 255.

Certain viral illnesses, such as influenza, measles, and mumps, cause the white blood cell count to decrease. **Leukopenia** is a total white blood cell count below 5,000 per cubic millimeter. Other illnesses, including appendicitis and bacterial infections, cause the white blood cell count to increase dramatically. **Leukocytosis** is a white blood cell count above 10,000 per cubic millimeter.

Illness often causes an increase in a particular type of white blood cell. For this reason, a **differential white blood cell count** may help with diagnosis. In this microscopic procedure, 100 white blood cells are counted, and numbers of each type of cell are recorded. For example, the characteristic finding in the viral disease **mononucleosis** is a great number of large, dark-staining abnormal lymphocytes. This condition takes its name from the fact that the abnormal lymphocytes resemble monocytes (although monocytes are much larger than lymphocytes).

Leukemia is a form of cancer characterized by uncontrolled production of abnormal white blood cells. These cells accumulate in the bone marrow, lymph nodes, spleen, and liver so that these organs are unable to function properly. **Acute lymphoblastic leukemia (ALL),** which represents over 80% of the acute leukemias in children, also occurs in adults. Chemotherapy is used to destroy abnormal cells and restore normal blood cell production. Intraspinal injection of drugs and craniospinal irradiation are measures that prevent leukemic cells from invading the central nervous system. In general, the prognosis is more favorable for children between the ages of 2 and 10 years, and females fare better because leukemia recurs in the testes of 8–16% of males. Remission occurs in 70–90% of adult patients after chemotherapy, and the median period of remission is 20 months. With chemotherapy, 85% of children survive past five years, and of those among this group who do not have a relapse, 85% are considered cured.

Figure 11A **Sickle-shaped red blood cells,** as seen by a scanning electron microscope.

form antibodies against it. **Antibodies** are proteins that neutralize antigens. T lymphocytes, on the other hand, directly attack and destroy any cell, such as a pathogen that has foreign antigens. B lymphocytes and T lymphocytes are discussed more fully in Chapter 13.

Monocytes (see Fig. 11.4 and Fig. 13.2e) are the largest of the white blood cells, and after taking up residence in the tissues, they differentiate into even larger macrophages. **Macrophages** phagocytize pathogens, old cells, and cellular debris. They also stimulate other white blood cells, including lymphocytes, to defend the body.

Abnormal White Cell Counts

Abnormal white blood cell counts are discussed in the Medical Focus on page 249. Because specific white blood cells increase with particular infections, a differential white cell count, also discussed in the Medical Focus, can be quite helpful in diagnosing the cause of a particular illness.

> *Content* **CHECK-UP!**

4. Imagine you have just donated a pint of blood at a blood drive in your community. Describe how your body would replace the donated blood.

5. Which of the following is/are bile pigments?
   a. hemoglobin
   b. biliverdin
   c. bilirubin
   d. b and c
   e. All of the above.

6. Granular white blood cells include all of the following except:
   a. eosinophils.   c. neutrophils.
   b. macrophages.   d. basophils.

Answers in Appendix A.

11.3 Platelets and Hemostasis

6. Describe the structure, function, and life cycle of platelets.
7. Describe the three events of hemostasis and the reactions necessary for coagulation to occur.
8. Discuss disorders of hemostasis.

Platelets (thrombocytes) are formed elements necessary to the process of **hemostasis,** the cessation of bleeding.

Platelets

Platelets result from fragmentation of certain large cells, called **megakaryocytes,** that develop in red bone marrow. It is important to note that although platelets are called thrombocytes, they are *not* cellular; rather, they are cell *fragments*. Platelets are produced at a rate of 200 billion per day, and the blood contains 150,000–300,000 per mm^3. Because platelets have no nucleus, they last at most ten days, assuming they are not used sooner than that in hemostasis.

Hemostasis

Hemostasis is divided into three events: vascular spasm, platelet plug formation, and coagulation (Fig. 11.7).

Vascular spasm, the constriction of the smooth muscle layer in a broken blood vessel, is the immediate response to blood vessel injury. Platelets release serotonin, a chemical that prolongs smooth muscle contraction.

Platelet plug formation is the next event in hemostasis. Platelets don't normally adhere to undamaged blood vessel walls, but when the lining of a blood vessel breaks, connective tissue, including collagen fibers, is exposed. Platelets adhere to collagen fibers and release a number of substances, including one that promotes platelet aggregation so that a so-called *platelet plug* forms. As a part of normal activities, small blood vessels often break, and a platelet plug is usually sufficient to stop the bleeding.

Coagulation, also called blood clotting, is the last event to bring about hemostasis. As you'll see, two plasma proteins, called **fibrinogen** and **prothrombin,** participate in blood clotting.

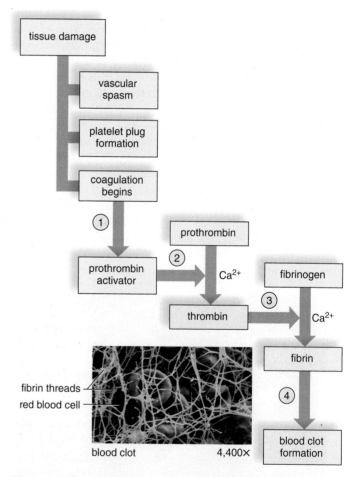

Figure 11.7 Process of hemostasis. Hemostasis requires three events: vascular spasm, platelet plug formation, and coagulation. Coagulation is further broken down into four steps. When coagulation is complete, threads of insoluble fibrin form a framework that effectively seals the injured area.

Vitamin K, found in green vegetables and also formed by intestinal bacteria, is necessary for the production of prothrombin. If, by chance, vitamin K is missing from the diet, **hemorrhagic bleeding** disorders develop.

Coagulation

Coagulation requires many protein clotting factors which are constantly present in the blood in an inactive state. Most of these clotting factors are produced by the liver, although several are released "on demand" by platelets if tissues or blood vessels are damaged. There are two mechanisms for activation of clotting. The **intrinsic mechanism,** so named because all clotting factors are *intrinsic* to the blood, is the slower pathway. The intrinsic mechanism explains why blood will coagulate if placed in a test tube that has not previously been treated with an anticoagulant. In the body, the intrinsic mechanism is initiated by contact with the exposed collagen in a torn blood vessel. The **extrinsic mechanism** is activated when damaged tissues release a substance called **tissue thromboplastin** (thus, its activation is *extrinsic* to blood itself). Tissue thromboplastin will interact with platelets, other clotting factors, and calcium ions (Ca^{2+}).

In the body, both clotting pathways are activated simultaneously; their common end point is production of prothrombin activator. As you know (from Chapter 1), blood clotting is a positive feedback mechanism. Thus, as each step occurs, it stimulates the next step in the process. Figure 11.7 breaks down the subsequent clotting process into four steps: ① **prothrombin activator** is formed. ② Prothrombin activator then converts prothrombin to **thrombin.** ③ Thrombin, in turn, severs two short amino acid chains from each fibrinogen molecule, and these activated fragments join end-to-end, forming long threads of **fibrin.** ④ Fibrin threads wind around the platelet plug in the damaged area of the blood vessel and provide the framework for the blood clot. Red blood cells also are trapped within the fibrin threads. These cells make a clot appear red. Though blood clotting is a positive feedback process, once clotting is initiated, the clotting process will be self-limiting. Several mechanisms ensure that clotting is confined to just the area of injury. In particular, thrombin is removed from the area as it is absorbed by fibrin, trapped in the clot itself, or digested by enzymes in the blood. These important safeguards prevent the clot from becoming too large, possibly blocking blood flow to other body areas. Limiting the clotting process to only the injured area also prevents excessive consumption of the available clotting factors in the blood.

Clot retraction follows, and the clot gets smaller as platelets contract. A fluid called **serum** (plasma minus fibrinogen and prothrombin) is squeezed from the clot. A fibrin clot is present only temporarily. As soon as blood vessel repair is initiated, an enzyme called *plasmin* begins to destroy the fibrin network and restore the fluidity of the plasma.

Naturally, it is vital to activate clotting when tissues have been injured. However, it is just as important that clotting *not* be activated if there is no injury. Normally, the smooth endothelial lining of an intact blood vessel prevents clots from forming in the blood vessel. Anticoagulants, such as heparin produced by basophils and mast cells, also help to prevent accidental clotting.

Disorders of Hemostasis

Both impaired clotting and excessive clotting can be life-threatening or fatal. **Thrombocytopenia,** a low platelet count, is one cause of impaired clotting. It occurs when one's own antibodies attack platelets, or if megakaryocytes in red bone marrow (parent cells that fragment to form platelets) are destroyed. The **hemophilias** are inherited clotting disorders caused by deficiencies of clotting factors. (The most severe form, hemophilia A, is due to the lack of clotting Factor VIII, a step in the intrinsic clotting pathway. You can read more about how the disease is inherited in Chapter 19.) In these clotting impairments, the slightest bump can cause bleeding into the tissues. Bleeding into a joint causes cartilage degeneration, and resorption of underlying bone can follow. Bleeding into muscles can lead to nerve damage and muscular atrophy. The most frequent cause of death is bleeding into the brain with accompanying neurological damage. Treatment involves giving the patient regular transfusions containing Factor VIII.

Undesirable, excessive clotting may block blood flow to tissues. Despite the presence of anticoagulants in the blood, sometimes a clot forms in an unbroken blood vessel. Such a clot is called a **thrombus** if it remains stationary. If the clot dislodges and travels in the blood, it is called an **embolus.** If this embolus blocks a blood vessel, it is a **thromboembolism.** Pulmonary thromboembolism (blockage of the pulmonary arteries by a clot) will seriously inhibit the oxygenation of the blood; cerebral thromboembolism will cause a **cerebrovascular accident** or **stroke.** Coronary thrombosis or thromboembolism causes a heart attack, as discussed in Chapter 12.

Content **CHECK-UP!**

7. The parent cell that fragments to form thrombocytes is called a:
   a. stem cell.
   b. monocyte.
   c. megalocyte.
   d. megakaryocyte.

8. Choose the correct sequence for hemostasis (first step to last):
   a. coagulation → vascular spasm → platelet plug formation
   b. vascular spasm → coagulation → platelet plug formation
   c. vascular spasm → platelet plug formation → coagulation
   d. coagulation → platelet plug formation → vascular spasm

9. Which vitamin is necessary to form prothrombin?
   a. B_{12} c. C
   b. folic acid   d. K

Answers in Appendix A.

I.C.E. — IN CASE OF EMERGENCY

Hemorrhage

Anyone who drives a car might be involved in a minor fender-bender type of accident. Thankfully, in those situations only the car gets "hurt." However, as you know, a car can be a deadly piece of equipment. Car accidents are the leading cause of death for people ages 16–26, and the fifth-leading cause of death overall, according to the Centers for Disease Control (CDC). Tragically, many of these deaths are caused by severe bleeding, or **hemorrhage,** and might have been prevented.

When you're the first person at an accident scene, your help could make the difference between life or death for a hemorrhaging patient. It's important to react appropriately—first get yourself and the victim out of danger, and then call for assistance. Next, immediately check the patient's ABCs: Airway—ensure that the airway is intact and open; Breathing—observe regular respiration; and Circulation—check for heartbeat and pulse. Once you've taken these steps, act quickly to stop or slow any serious bleeding. Excessive blood loss can cause blood pressure to fall, a condition called **hypovolemic shock.** Shock causes organ failure and death if not treated.

Experts agree that the best way to stop a hemorrhage is to quickly apply direct pressure to the injury site, using gauze, towels, clothing, or even your bare hand if you must. Not only does pressing on a bleeding wound stop or slow the bleeding, it will also start the blood clotting process. If the cloth becomes soaked, don't remove it; add more layers. A clot may have begun to form in the fabric, and peeling it away might cause bleeding to start again. In addition, elevate the victim's legs, and keep him or her warm. Taking these steps will help prevent shock.

At the accident scene, EMS providers can begin an IV while transporting the patient to the emergency room. In addition, paramedic crews, hospital emergency rooms, and military medics now carry prepared dressings that are designed to stop bleeding, even in deep wounds. All work by quickly drawing water out of blood plasma, concentrating the clotting factors at the site of the wound. This promotes rapid clot formation, sometimes stopping even a severe hemorrhage within minutes.

11.4 Blood Typing and Transfusions

9. Explain the ABO and Rh systems of blood typing.
10. Explain agglutination and its relationship to transfusions.

A **blood transfusion** is the transfer of blood from one individual into the blood of another. In order for transfusions to be done safely, it is necessary for blood to be typed so that **agglutination** (clumping of red blood cells) does not occur. Blood typing usually involves determining the ABO blood group and whether the individual is Rh$^-$ (negative) or Rh$^+$ (positive).

ABO Blood Groups

ABO blood typing is based on the presence or absence of two possible antigens, called type A antigen and type B antigen, on the surface of red blood cells. Whether these antigens are present or not depends on the particular inheritance of the individual.

A person with type A antigen on the surface of the red blood cells has type A blood; one with type B blood has type B antigen on the surface of the red blood cells. What antigens would be present on the surface of red blood cells if the person has type AB blood or type O blood? Notice in Figure 11.8 that a person with type AB blood has both antigens. A person with type O blood has no AB antigens on the surface of the red blood cell, though other protein antigens will be present.

It so happens that an individual with type A blood has anti-B antibodies in the plasma; a person with type B blood has anti-A antibodies in the plasma; and a person with type O blood has both antibodies in the plasma (Fig. 11.8). These antibodies are not present at birth, but they appear over the course of several months after birth. Unlike those

with the other ABO blood types, a person with type AB blood has neither anti-A nor anti-B antibodies in the plasma.

Blood compatibility is very important when transfusions are done. The antibodies in the plasma must not combine with the antigens on the surface of the red blood cells, or else agglutination occurs. With agglutination, anti-A antibodies have combined with type A antigens, or anti-B antibodies have combined with type B antigens, or both types of binding have occurred. Therefore, agglutination is expected if the donor has type A blood and the recipient has type B blood (Fig. 11.9). What about other combinations of blood types? Table 11.1 shows when a blood transfusion is most likely safe. Type O blood is sometimes called the *universal donor* because it has no antigens on the red blood cells, and type AB blood is sometimes called the *universal recipient* because this blood type has no antibodies in the plasma. In practice, however, there are other possible blood groups, aside from ABO blood groups, so it is necessary to physically put the donor's blood on a slide with the recipient's blood and observe whether the blood types match (no

TABLE 11.1 Transfusion	
If your blood type is:	**You can safely receive blood of this type***
A	A or O
B	B or O
AB	A, B, AB, or O
O	O

*Regardless of ABO type, a person who is Rh$^-$ can never be transfused with Rh$^+$ blood.

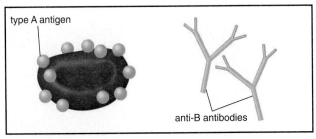

Type A blood. Red blood cells have type A surface antigens. Plasma has anti-B antibodies.

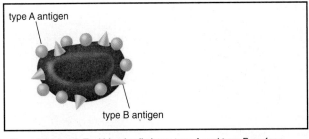

Type AB blood. Red blood cells have type A and type B surface antigens. Plasma has neither anti-A nor anti-B antibodies.

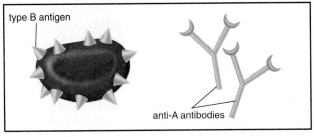

Type B blood. Red blood cells have type B surface antigens. Plasma has anti-A antibodies.

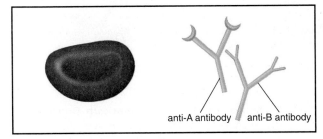

Type O blood. Red blood cells have neither type A nor type B surface antigens. Plasma has both anti-A and anti-B antibodies.

Figure 11.8 Types of blood. In the ABO system, blood type depends on the presence or absence of antigens A and B on the surface of red blood cells. In these drawings, A and B antigens are represented by different shapes on the red blood cells. The possible anti-A and anti-B antibodies in the plasma are shown for each blood type. Notice that an anti-B antibody cannot bind to an A antigen, and vice versa.

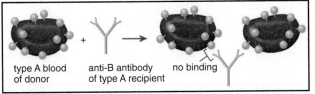

a. No agglutination

No clumping seen.
Successful blood type match.

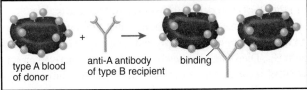

b. Agglutination

Clumping seen.
Hemolysis occurs.
Unsuccessful blood type match.

Figure 11.9 Cross-matching before a blood transfusion. No agglutination (**a**) versus agglutination (**b**) is determined by whether antibodies are present that can combine with antigens.

agglutination occurs) before blood can be safely given from one person to another. This is called cross-matching blood.

As explained in the What's New reading on page 255, new technology may help to eliminate the problems of matching blood types.

Rh Blood Groups

The designation of blood type usually also includes whether the person has or does not have the Rh factor on the red blood cell. Rh⁻ individuals normally do not have antibodies to the Rh factor, but they make them when exposed to the Rh factor.

If a mother is Rh⁻ and the father is Rh⁺, the fetus conceived can be Rh⁺. The Rh⁺ red blood cells may begin leaking across the placenta into the mother's cardiovascular system, as placental tissues normally break down before and at birth. The presence of these Rh⁺ antigens causes the mother to produce anti-Rh antibodies. During this pregnancy or a subsequent pregnancy with another Rh⁺ baby, the anti-Rh antibodies may cross the placenta and destroy the child's red blood cells. This is called hemolytic disease of the newborn (HDN) because hemolysis continues after the baby is born (Fig. 11.10). Due to red blood cell destruction the baby can be severely anemic at birth. Excess bilirubin in the blood can lead to brain damage and developmental delay, or even death.

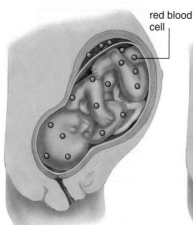
red blood
cell

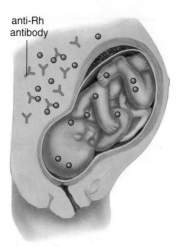
anti-Rh
antibody

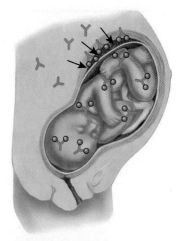

Child is Rh positive;
mother is Rh negative.

Red blood cells leak
across placenta.

Mother makes anti-Rh
antibodies.

Antibodies attack Rh-positive
red blood cells in child.

Figure 11.10 Hemolytic disease of the newborn. Due to a pregnancy in which the child is Rh positive, an Rh-negative mother can begin to produce antibodies against Rh-positive red blood cells. In a subsequent pregnancy, these antibodies can cross the placenta and cause hemolysis of an Rh-positive child's red blood cells.

The Rh problem is prevented by giving Rh⁻ women an Rh immunoglobulin injection at 28 weeks of pregnancy, followed by a second injection no later than 72 hours after giving birth to an Rh⁺ child. This injection, called Rho-Gam, contains anti-Rh antibodies that attack any of the baby's red blood cells in the mother's blood before these cells can stimulate her immune system to produce her own antibodies. This injection is not beneficial if the woman has already begun to produce antibodies; therefore, the timing of the injection is extremely important.

Content **CHECK-UP!**

10. Which of the following correctly describes type AB blood?

 a. A antigen and B antigen on the red cell membrane, anti-A and anti-B antibodies in plasma

 b. A antigen and B antigen on the red cell membrane, no antibodies in plasma

 c. A antigen on the red cell membrane, anti-B antibodies in plasma

 d. B antigen on the red cell membrane, anti-A antibodies in plasma

11. Which of the following transfusions would most likely be safe to administer?

 a. type A blood to a type B blood recipient

 b. type AB blood to a type B blood recipient

 c. type B blood to a type AB recipient

 d. type AB blood to a type O recipient

12. Hemolytic disease of the newborn occurs in which situation?

 a. Rh⁺ mother, Rh⁻ father, Rh⁺ fetus

 b. Rh⁻ mother, Rh⁻ father, Rh⁺ fetus

 c. Rh⁻ mother, Rh⁺ father, Rh⁺ fetus

 d. Rh⁺ mother, Rh⁻ father, Rh⁻ fetus

Answers in Appendix A.

11.5 Effects of Aging

11. Name the blood disorders that are commonly seen as we age.

Anemias, leukemias, and clotting disorders increase in frequency with age. As with other disorders, good health habits can help prevent these conditions from appearing.

Iron deficiency anemia most frequently results from an iron-deficient diet, but it can also result from a gradual and sometimes undetectable loss of blood. It is vital that the bleeding source be discovered as soon as possible. For example, iron deficiency anemia is often a first symptom of colon cancer in its early stage. The blood is often not visible in a person's stool because it has oxidized and is concealed in the stool. An *occult blood test* (test for oxidized blood present in the stool) can detect this source of blood loss.

Pernicious anemia most often signals that the digestive tract is unable to absorb enough vitamin B_{12}, but it can also be due to a diet deficient in B_{12}. Diets that completely exclude animal-derived protein sources (called *vegan* diets) are often B_{12} deficient. A vitamin supplement containing B_{12} will prevent pernicious anemia in vegan dieters.

Leukemia is a form of cancer that generally increases in frequency with age because of both intrinsic (genetic) and extrinsic (environmental) reasons.

Thromboembolism, a clotting disorder, may be associated with the progressive development of atherosclerosis in an elderly person. When arteries develop plaque (see Fig. 12A), thromboembolism often follows. For many people, atherosclerosis can be controlled by diet and exercise, as discussed in the Chapter 12 Medical Focus, "Preventing Cardiovascular Disease."

Content **CHECK-UP!**

13. Elderly people often have poor appetites and don't eat well. How can this cause anemia?

Answer in Appendix A.

Improvements in Transfusion Technology

If you work in an emergency room (ER) setting, having a supply of donor blood available for transfusion is a problem that you and your co-workers will face every day. Your patient may be a victim of a car accident, a shooting, or a stabbing. Perhaps she's a pregnant woman who is hemorrhaging, or a leukemia sufferer whose *hematocrit* (see page 243) has dropped dangerously low because of chemotherapy. Without an immediate blood transfusion, the patient's tissue cells will die from lack of oxygen. If the ER is a major trauma center in a large city hospital, donor blood for transfusion is usually available, and the ready supply of O-negative blood can theoretically be donated to anyone.

But what if the transfusion is needed in a remote area where whole blood isn't available, such as a wartime field hospital or an accident scene on an isolated stretch of highway? Likewise, small regional hospitals face regular shortages of donor blood. In an emergency (for example, a hemorrhaging patient), the patient can be transfused with plasma if whole blood is not available. However, plasma shortages are possible as well, especially in a mass casualty situation. Further, although the majority of transfusions are safe, there is always a small chance of a transfusion reaction, in which the recipient's immune system rejects the donor blood. Moreover, even though blood is carefully tested for viruses and other infectious agents, infection from donated blood is still a very slight possibility.

Recent developments show promise in combating the problems of blood availability and safety by providing a constant supply of O-negative universal donor cells. Scientists at the University of Edinburgh have successfully produced red blood cells from stem cells. These stem cells, called fibroblasts, were obtained from an adult donor. (Fibroblasts are described in Chapter 4.) The fibroblasts were then "de-differentiated"—coaxed into becoming more primitive versions of themselves—to create *induced pluripotent stem cells* (iPSCs).

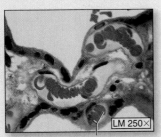

LM 250×

Erythrocytes

Figure 11B Erythrocytes passing through a capillary. The cells line up in a single file to travel through these tiny vessels.

Unlike fibroblasts, iPSCs are capable of converting into hematopoietic cells. After a month of growth in an environment similar to that of bone marrow, the iPSCs produced red blood cells that were then harvested from the cell culture. Because the fibroblasts had been obtained from an O-negative donor, all of the red blood cells were universal donor cells and could therefore be transfused into anyone, regardless of A-B-O or Rh blood type. The researchers are optimistic that this technique can be applied to large-scale manufacturing. If that's possible, it would lead to a steady, reliable, infection-free supply of universal donor red blood cells.

In the meantime, research continues to explore ways of *chemically* altering red blood cell membranes. Once again, the goal is to find a way to convert all donated blood to O-negative universal donor blood, making it safe for anyone needing a transfusion. One strategy involves *antigen camouflage,* in which the cell membrane proteins are covered with a plastic-like "coat." Once coated, the proteins can't trigger a transfusion reaction in the recipient. A second procedure uses bacterial-derived enzymes to "scrub" the red blood cells of A, B, and Rh antigens, essentially converting every donated unit into O-negative blood. Treated cells are called Enzyme Converted group O (ECO) cells, and blood containing these cells has already been safely transfused. Similarly, the problem of blood plasma shortage is being addressed by a company developing techniques to spray-dry blood plasma. Once dry, the plasma can be safely stored without refrigeration for long periods of time, then dissolved in sterile solution to create plasma for transfusion.

Whenever possible, surgeons avoid human-to-human blood transfusion entirely by using the patient's own blood during an operation. This procedure is called *autologous transfusion* or **autotransfusion.** For example, patients can have plasma or whole blood stored before an elective surgery such as a hip replacement. Then the patient receives a transfusion using his or her own blood if needed. *Intraoperative blood salvage,* using an apparatus called a "Cell Saver," allows an emergency room physician or surgeon to suction blood from the patient's body. The suctioned blood is immediately mixed with an anticoagulant, washed with a saline solution to remove any contaminants, mixed with more saline, and then reintroduced into a vein just like a traditional transfusion. The procedure is quite fast: To prepare an entire unit of blood for reinfusion into the patient takes only three minutes. In addition, autotransfusion completely eliminates the possibility of transfusion reaction.

Until blood replacement technologies are perfected, there will always be a critical need for blood donation. To find out how you can help, visit the Red Cross blood donor website at http://www.redcross .org/donate/give.

FOCUS on FORENSICS

Blood at the Crime Scene

In cases of violent crime such as homicide, assault, or sexual assault, the presence of blood at the crime scene is often the most important evidence. Blood evidence can tell an entire story about the crime—how a victim was attacked, whether the victim fought back or tried to escape, the type of weapon used, and perhaps even the identity of the attacker. Extensive research on the chemistry and properties of blood evidence, along with detailed examination of the crime scene, can enable a *forensic serologist* to help solve violent crime cases. (Forensic serologists are scientists trained in the examination of blood and body fluid at a crime scene.)

Upon arrival at the scene, investigators must first carry out a careful visual inspection for the presence of blood. Because blood hemoglobin is a protein molecule (globin) containing an iron pigment (heme), its stains are very difficult to remove completely from contaminated areas. As a result, a criminal who tries to clean up blood will almost always leave traces behind. Investigators sample between floorboards, under baseboards, in carpet padding, and on all surface areas. Tests can be done to confirm that a stain is due to human blood and not some other reddish-brown chemical. One test involves spraying the stain with luminol, a chemical that binds with blood and then glows in the dark. The Takayama and Teichmann tests cause crystals to form from hemoglobin, confirming the presence of blood. Furthermore, by testing with anti-human antibodies (protein molecules from rabbits), researchers can prove that a sample is in fact human in origin.

Careful documentation and analysis of the location and shape of any drops, spatters, or smears follows next. This type of evidence is carefully recorded and can also help investigators to piece together the events of the crime—the position of the suspect and the victim, whether a struggle occurred, if the victim moved or was dragged, etc. For example, research has shown that the shape of blood drops can yield useful information about a crime. If sprayed from a short distance, a blood drop will form a perfect circle on a wall or floor; however, blood that travels a greater distance will form a "sunburst" pattern. Droplet shape also yields information about the speed of travel from a person's body. When a droplet travels slowly from the body to an object (for example, a wall), it will form a circular drop. Blood that travels at a fast speed from the body to the object will form an elongated, "teardrop" shape. Blood flying in a trail of droplets from a knife or a club will leave a trail of spatters called a *castoff pattern*. Smeared blood may indicate that the body has been dragged, or that the victim attempted to escape.

If the suspect was wounded in a struggle, the suspect's blood may be present at the crime scene along with that of the victim. Blood chemistry is as individual as a fingerprint, and can positively confirm the identity of the suspect. ABO and Rh blood typing, studies of unique blood proteins, and DNA analysis (see page 61) are carried out on all blood samples. Other body fluids, if present, may also need to be tested because approximately 80% of the human population is comprised of *secretors*—individuals whose blood proteins and enzymes are present in saliva, tears, urine, semen, and skin, as well as in their blood. The data from these tests identify the source of blood samples, which can be statistically narrowed down to one person in several billion. Thus, for all practical purposes, a positive match against a suspect's blood can be made. Evidence such as this can prove that a suspect was at the crime scene and result in a conviction. Just as important, blood analysis may serve to acquit an innocent person.

Selected New Terms

Basic Key Terms

agglutination (ŭh-glū″tĭ-nā′shŭn), p. 252

agranular leukocyte (ā-grăn′yū-lĕr lū′kō-sīt), p. 248

albumin (ăl-byū′mĭn), p. 244

antibody (ăn′tĭ-bŏd″ē), p. 250

antigen (ăn′tĭ-jĕn), p. 248

basophil (bā′sō-fĭl), p. 248

bile pigments (bīl pĭg′mĕnts), p. 248

buffy coat (bŭ′fē kōt), p. 243

coagulation (kō-ăg″yū-lā′shŭn), p. 250

eosinophil (ē″ō-sĭn′ō-fĭl), p. 248

erythrocyte (ĕ-rĭth″rō-sīt′), p. 244

erythropoietin (ĕ-rĭth″rō-pōy′ĕ-tĭn), p. 247

extrinsic mechanism (ĕks-trĭn′sĭk mĕ′kŭn-ĭzm), p. 251

fibrin (fĭ′brĭn), p. 251

fibrinogen (fĭ-brĭn′ō-jĕn), p. 250

formed element (fŏrmd ĕl′ĕ-mĕnt), p. 243

globulin (glŏb′ū-lĭn), p. 244

granular leukocyte (grăn′ū-lĕr lū′kō-sīt), p. 248

hematocrit (hē-măt′ō-krĭt), p. 243

hematopoiesis (hē′mŭh-tō-poĭ-ē′sĭs), p. 244

hemoglobin (hē′mō-glō″bĭn), p. 244

hemolysis (hē-mŏl′ĭ-sĭs), p. 248

hemostasis (hē′mŭh-stā′sĭs), p. 250

intrinsic mechanism (ĭn-trĭn′sĭk mĕ′kŭn-ĭzm), p. 251

leukocytes (lū′ kō-sītz), p. 248

lipoproteins (lĭp″ō-prō′tēnz), p. 244

lymphocyte (lĭm′fō-sīt), p. 248

macrophage (măk′rō-fāj), p. 250

megakaryocyte (mĕg″ŭh-kār′ē-ō-sīt), p. 250

monocyte (mŏn′ō-sīt), p. 250

multipotent stem cell (mŭl′tĭ-po′ tent stĕm sĕl), p. 244

natural killer cell (nătch′ŭh-rŭl kĭl′ŭr sĕl), p. 248

neutrophil (nū′trō-fĭl), p. 248

osmotic pressure (ŏz-mŏt′ĭk prĕsh′ur), p. 244

pathogen (păth′ō-jĕn), p. 243

plasma (plăz′mŭh), p. 243

platelet (plāt′lĕt), p. 250

platelet plug (plāt′lĕt plŭg), p. 250

prothrombin (prō-thrŏm′bĭn), p. 250

prothrombin activator (prō-thrŏm′bĭn ăk′tĭ-vā″tŏr), p. 251

red blood cell (rĕd blŭd sĕl), p. 244

reticulocyte (rŭh-tĭk′yū-lō-sīt), p. 247

serum (sē′rŭm), p. 251

thrombin (thrŏm′bĭn), p. 251

thrombocyte (thrŏm′bō-sīt), p. 250

tissue thromboplastin (tĭ′shū thrŏm′bō-plăs″tĭn), p. 251

vascular spasm (văs′kyū-lĕr spăzm), p. 250

white blood cell (whīt blŭd sĕl), p. 248

Clinical Key Terms

acute lymphoblastic leukemia (ŭh-kyūt′ lĭm-fō-blăs′tĭk lū-kē′mē-ŭh), p. 249

anemia (ŭh-nē′mē-ŭh), p. 249

aplastic anemia (ā-plăs′tĭk ŭh-nē′mē-ŭh), p. 249

autotransfusion (ăh′tō-trănz-fyū″zhŭn), p. 255

blood transfusion (blŭd trăns-fyū′zhŭn), p. 252

cerebrovascular accident (sŭh-rē′ brō-văs″kyŭh-lĕr ăk′sĭ-dĕnt), p. 251

dietary anemia (dī′ ŭh-tĕr″ē ŭh-nē′mē-ŭh), p. 249

differential white blood cell count (dĭf″ĕr-ĕn′shŭl whīt blŭd sĕl kŏwnt), p. 249

embolus (ĕm′bō-lŭs), p. 251

hemolytic anemia (hē-mō-lĭt′ĭk ŭh-nē′mē-ŭh), p. 249

hemolytic disease of the newborn (hē-mō-lĭt′ĭk dĭ-zēz′ ŏv thăh nū′bŏrn), p. 249

hemophilia (hē-mō-fĭl′ē-ŭh), p. 251

hemorrhage (hĕm′ō-rĭj), p. 252

hemorrhagic anemia (hĕm′ō-răj-ĭk ŭh-nē′mē-ŭh), p. 249

hemorrhagic bleeding (hĕm′ō-răj-ĭk blē′dĭng), p. 251

hypovolemic shock (hī′pō-vō-lē′mĭk shŏk), p. 252

iron deficiency anemia (ī′ĕrn dĭ-fĭ′shŭn-sē ŭh-nē′mē-ŭh), p. 249

leukemia (lū-kē′mē-ŭh), p. 249

leukocytosis (lū″kō-sī-tō′sĭs), p. 249

leukopenia (lū″kō-pē′nē-ŭh), p. 249

mononucleosis (mŏn″ō-nū″klē-ō′sĭs), p. 249

pernicious anemia (pĕr-nĭ′shŭs ŭh-nē′mē-ŭh), p. 249

polycythemia (pŏl″ē-sī-thē′mē-ŭh), p. 247

sickle-cell disease (sĭ′kl sĕl dĭ-zēz′), p. 249

sickle-cell trait (sĭ′kl sĕl trāt), p. 249

stroke (strōk), p. 251

thrombocytopenia (thrŏm″bō-sī-tō-pē′nē-ŭh), p. 251

thromboembolism (thrŏm″bō-ĕm′bō-lĭzm), p. 251

thrombus (thrŏm′bŭs), p. 251

Summary

11.1 The Composition and Functions of Blood

A. Blood, which is composed of formed elements and plasma, has several functions. It transports hormones, oxygen, and nutrients to the cells and carbon dioxide and other wastes away from the cells. It fights infections. It regulates body temperature, and keeps the pH of body fluids within normal limits. All of these functions help maintain homeostasis.

11.2 Components of Blood

A. Plasma is mostly water (92%) and solutes (8%). Small organic molecules such as glucose and amino acids are dissolved in plasma and serve as nutrients for cells; urea is a waste product. Large organic molecules include the plasma proteins. The plasma proteins, most of which are produced by the liver, occur in three categories: albumins, globulins, and fibrinogen. The plasma proteins maintain osmotic pressure, help regulate pH, and transport molecules. Certain plasma proteins have specific functions: The gamma globulins, which are antibodies produced by B lymphocytes, function in immunity; fibrinogen and prothrombin are necessary to blood clotting.

B. All blood cells, including red blood cells, are produced within red bone marrow from stem cells, which are constantly capable of dividing and producing new cells.

C. Red blood cells are small, biconcave disks that lack a nucleus. Men have more RBCs than women because of the effects of androgens such as testosterone. Red blood cells contain hemoglobin, the respiratory pigment, which combines with oxygen and transports it to the tissues. Red blood cells live about 120 days and are destroyed in the liver and spleen when they are old or abnormal. The production of red blood cells is controlled by the oxygen concentration of the blood. When the oxygen concentration decreases, the kidneys increase their production of erythropoietin, and more red blood cells are produced.

D. White blood cells are larger than red blood cells, have a nucleus, and are translucent unless stained. Like red blood cells, they are produced in the red bone marrow. White blood cells are divided into the granular leukocytes and the agranular leukocytes. The granular leukocytes have conspicuous granules; in eosinophils, granules are red when stained with eosin, and in basophils, granules are blue when stained with a basic dye. In neutrophils, some of the granules take up eosin, and others take up the basic dye, giving them a lilac color. Neutrophils are the most plentiful of the white blood cells, and they are able to phagocytize pathogens. Many

neutrophils die within a few days when they are fighting an infection. Natural killer (NK) cells are granulocytes structurally similar to lymphocytes which help to destroy damaged body cells. The agranulocytes include the lymphocytes and the monocytes, which function in specific immunity. On occasion, the monocytes become large phagocytic cells of great significance. They engulf worn-out red blood cells and pathogens at a ferocious rate.

11.3 Platelets and Hemostasis

A. The extremely plentiful platelets result from fragmentation of megakaryocytes.

B. The three events of hemostasis are vascular spasm, platelet plug formation, and coagulation. Extrinsic and intrinsic pathways that activate the clotting mechanism both cause formation of prothrombin activator, which breaks down prothrombin to thrombin. Thrombin changes fibrinogen to fibrin threads, entrapping cells. The fluid that escapes from a clot is called serum and consists of plasma minus fibrinogen and prothrombin.

11.4 Blood Typing and Transfusions

A. ABO typing is the most common blood typing system used. Type A, type B, both type A and B, or no antigens can be on the surface of red blood cells. In the plasma, there are two possible antibodies: anti-A or anti-B. If the corresponding antigen and antibody are put together during a transfusion, ag-glutination occurs. Therefore, it is necessary to determine an individual's blood type before a transfusion is given.

B. Another important antigen is the Rh antigen. This particular anti-gen must also be considered in transfusing blood, and it is important during pregnancy because an Rh⁻ mother may form antibodies to the Rh antigen when giving birth to an Rh⁺ child. These antibodies can cross the placenta and destroy the red blood cells of any subsequent Rh⁺ child.

11.5 Effects of Aging

As we age, anemias, leukemias, and clotting disorders increase in frequency.

Study Questions

1. Name the two main components of blood, and describe the functions of blood. (pp. 243–244)
2. List and discuss the major components of plasma. Name several plasma proteins, and give a function for each. (pp. 244–245)
3. What is hemoglobin, and how does it function? (pp. 246–247)
4. Describe the life cycle of red blood cells, and tell how the production of red blood cells is regulated. (pp. 247–248)
5. Name the six types of white blood cells. Describe the structure and give a function for each type. (pp. 248, 250)
6. Name the steps that take place when blood clots. Which substances are present in blood at all times, and which appear during the clotting process? (pp. 250–251)
7. What are the four ABO blood types? For each, state the antigen(s) on the red blood cells and the antibody(ies) in the plasma. (pp. 252–253)
8. Explain why a person with type O blood cannot receive a transfusion of type A blood. (pp. 252–253)
9. Problems can arise if the mother is which Rh type and the father is which Rh type? Explain why this is so. (pp. 253-254)

Learning Outcome Questions

I. Fill in the blanks.

1. The liquid part of blood is called _____.
2. Red blood cells carry _____, and white blood cells _____.
3. Hemoglobin that is carrying oxygen is called _____.
4. Human red blood cells lack a _____ and only live about _____ days.
5. The most common granular leukocyte is the _____, a phagocytic white blood cell.

6. B lymphocytes produce _____, and T lymphocytes attack and _____ pathogens.
7. When a blood clot occurs, fibrinogen has been converted to _____ threads.
8. AB blood has the antigens _____ and _____ on the red blood cells and _____ of these antibodies in the plasma.
9. Hemolytic disease of the newborn can occur when the mother is _____ and the father is _____.

II. Match the terms in the key to the descriptions in questions 10–13.

Key:
   a. hematocrit
   b. red blood cell count
   c. white blood cell count
   d. hemoglobin

10. 5,000 to 11,000 per cubic millimeter
11. 4 to 6 million per cubic millimeter in males
12. Just under 45% of blood volume
13. 200 million molecules in one red blood cell

Medical Terminology Exercise

After studying this chapter, see if you can derive the definitions for the medical terms listed at right. Many of the prefixes and suffixes used to create these terms can be found throughout the chapter. For additional help, use McGraw-Hill Connect™ at www.mcgrawhillconnect.com and consult Appendix B.

1. hematemesis (hĕm″ŭh-tĕm′ĕ-sĭs)
2. erythrocytometry (ĕ-rĭth″rō-sī-tŏm′ĕ-trē)
3. leukocytogenesis (lū″kō-sī″tō-jĕn′ĕ-sĭs)
4. hemophobia (hē″mō-fō′bē-ŭh)
5. afibrinogenemia (ŭh-fī″brĭn-ō-jĕ-nē′mē-ŭh)
6. lymphosarcoma (lĭm″fō-sär-kō′mŭh)
7. phlebotomy (flĕ-bŏt′ō-mē)
8. hemocytoblast (hē′mō-sī′tō-blăst)

9. megaloblastic anemia (mĕg′ŭh-lō-blăs′tĭk ŭh-nē′mē-ŭh)
10. microcytic hypochromic anemia (mī′krō-sĭt′ĭk hī″pō-krō′mĭk ŭh-nē′mē-ŭh)
11. hematology (hē′mŭh-tŏl′ō-jē)
12. lymphedema (lĭmf′ŭh-dē′mŭh)
13. antithrombin (ăn″tē-thrŏm′bĭn)

Online Study Tools

LEARNSMART® McGraw Hill Education **connect**

APR

Anatomy & Physiology REVEALED includes cadaver photos that allow you to peel away layers of the human body to reveal structures beneath the surface. This program also includes animations, radiologic imaging, audio pronunciations, and practice quizzing. To learn more visit www.aprevealed.com

Anatomy & Physiology | REVEALED
aprevealed.com

12 The Cardiovascular System

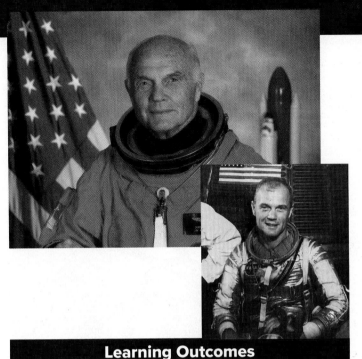

As you'll recall from Chapter 5, tattoos have been part of human cultural practice for centuries. However, it may surprise you to hear that tattoos once had an interesting medical application as well: ensuring the health of the original Mercury 7 astronauts before their space voyages in the early 1960s. John Glenn, pictured here, was one of the original pioneer astronauts, and the first man to orbit the Earth. He and his fellow astronauts underwent hundreds of electrocardiograms throughout the years of the program. An electrocardiogram is a record of the heart's electrical activity, and is an important indicator of heart health. Cardiologists (physicians who specialize in heart care) tattooed Glenn's chest so that when each repeated electrocardiogram was taken, the recording pads (called electrodes) would always be placed in the same position. By creating these permanent markings, the doctors hoped to ensure the accuracy of the recordings. Glenn was a pioneer once again, as the oldest space traveler—at age 77, he flew on the space shuttle *Discovery*, launched on October 29, 1998. This time, however, improved ECG technology eliminated the need for his tattoos. Instead, computers provided constant, real-time information about every aspect of his heart's performance: electrocardiogram, heart rate, strength of contraction, blood pressure, etc. You can find out more about how electrocardiograms are recorded and used in the Medical Focus on page 269.

Learning Outcomes

After you have studied this chapter, you should be able to:

12.1 Anatomy of the Heart

1. Describe the location of the heart and its functions.
2. Detail the wall and coverings of the heart.
3. Trace the path of blood through the heart, naming its chambers and valves.
4. Explain the operation of the heart valves.
5. Outline the coronary circulation, and discuss several coronary circulation disorders and possible treatments.

12.2 Physiology of the Heart

6. Describe the conduction system of the heart.
7. Label and explain a normal electrocardiogram.
8. Describe the cardiac cycle and the heart sounds.
9. Describe the cardiac output and regulation of the heartbeat.

12.3 Anatomy of Blood Vessels

10. Name the three types of blood vessels, and describe their structure and function.

12.4 Physiology of Circulation

11. Explain how blood pressure changes throughout the vascular system, and describe the factors that determine blood pressure.
12. Describe how blood pressure is regulated.
13. Define pulse, and tell where the pulse may be taken.
14. Describe shock due to hypotension and various medical consequences of hypertension.

12.5 Circulatory Routes

15. Name the two circuits of the cardiovascular system, and trace the path of blood from the heart to any organ in the body and back to the heart.
16. Describe the major systemic arteries and the major systemic veins.
17. Describe several special circulations: blood supply to the liver, blood supply to the brain, and fetal circulation.

12.6 Effects of Aging

18. Describe the anatomical and physiological changes that occur in the cardiovascular system as we age.

12.7 Homeostasis

19. Describe how the cardiovascular system works with other systems of the body to maintain homeostasis.

I.C.E.—In Case of Emergency

Cardiopulmonary Resuscitation and Automated External Defibrillation

Medical Focus

Arteriosclerosis, Atherosclerosis, and Coronary Artery Disease
The Electrocardiogram
Preventing Cardiovascular Disease

What's New

Novel Stent for the Severest Strokes

Human Systems Work Together

Cardiovascular System

Chapter 11 described the components of blood and detailed the role that blood has in transport, defense, and regulation. In this chapter, we'll study the body's transportation system: the cardiovascular system. We'll begin with the anatomy and physiology of the heart and of the blood vessels. Then, we'll take a look at various branches of the circulatory system. A crucial function of circulation is to connect the body's trillions of cells so that each organ system can carry out its homeostatic function. In the lungs, oxygen enters and carbon dioxide exits the blood, and this exchange helps to maintain acid-base balance in a normal range. (You'll read more about acid-base balance in Chapters 14 and 16.). The small intestine absorbs nutrient molecules into the blood. The kidneys allow metabolic wastes to exit the blood, while reabsorbing critical ions, molecules, and nutrients and helping to maintain a normal pH. If the body becomes too cool, blood can be directed away from the surface to be warmed by the muscles. Conversely, blood that is too warm can be diverted to the skin so that heat can be lost to the surroundings. None of these homeostatic functions—maintaining O_2/CO_2, nutrient, pH, and temperature balance—would be possible without transportation from the circulatory system.

The cardiovascular system consists of three components: (1) the heart, which pumps blood so that it flows to tissue capillaries and lung capillaries, (2) the blood vessels through which the blood flows, and (3) the blood itself, which as you learned in the previous chapter, is a tissue. As you can see in Figure 12.1, the cardiovascular system is divided into two functional systems. The right side of the heart and its blood vessels form the **pulmonary circuit,** which pumps blood to the lungs. There, oxygen diffuses into the bloodstream, and CO_2 diffuses into the *alveoli* (the lung's air sacs) to be exhaled. The left side of the heart and its vessels form the **systemic circuit,** which supplies blood containing oxygen and nutrients to the entire body.

12.1 Anatomy of the Heart

1. Describe the location of the heart and its functions.
2. Detail the wall and coverings of the heart.
3. Trace the path of blood through the heart, naming its chambers and valves.
4. Explain the operation of the heart valves.
5. Outline the coronary circulation, and discuss several coronary circulation disorders and possible treatments.

The heart is located in the thoracic cavity within the mediastinum, a serous membrane sac between the lungs. It is a hollow, cone-shaped, muscular organ. To approximate the size of your heart, make a fist and then clasp the fist with your opposite hand. Figure 12.1 shows that the base (the widest part) of the heart is superior to its apex (the pointed tip), which rests on the diaphragm. Also, the heart lies on a slant; the base is directed toward the right shoulder, and the apex points to the left hip. The base is deep to the second rib, and the apex is at the level of the fifth intercostal space (though these positions can vary depending on a person's size).

As the heart pumps the blood through the pulmonary and systemic vessels, it performs these functions:

1. keeps oxygenated blood separate from partially deoxygenated blood;

2. keeps the blood flowing in one direction—blood flows away from and then back to the heart in each circuit;
3. creates blood pressure, which moves the blood through the circuits;
4. regulates the blood supply based on the current needs of the body;
5. serves as an endocrine gland, producing atrial natriuretic hormone (ANH, or atriopeptide) for regulation of blood pressure.

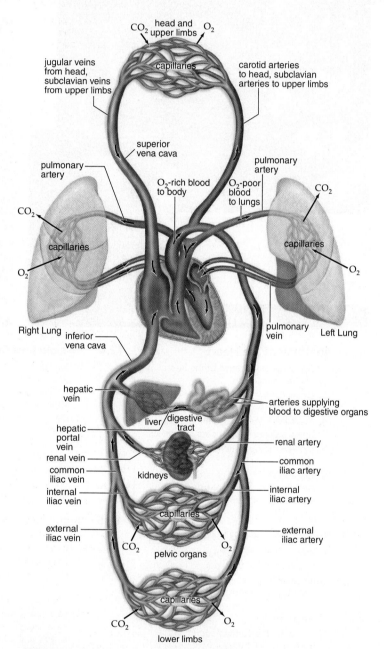

AP|R **Figure 12.1 Cardiovascular system.** The right side of the heart pumps blood through vessels of the pulmonary circuit. The left side of the heart pumps blood through vessels of the systemic circuit. Gas exchange and nutrient-for-waste exchange occur as blood passes through lung (pulmonary) capillaries and tissue (systemic) capillaries. In this illustration, red vessels carry O_2-rich blood, and blue vessels carry O_2-poor blood.

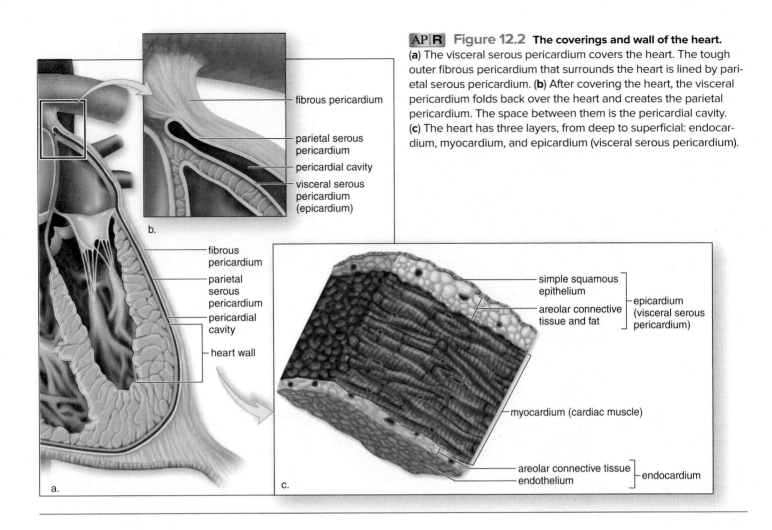

AP|R **Figure 12.2** **The coverings and wall of the heart.**
(a) The visceral serous pericardium covers the heart. The tough outer fibrous pericardium that surrounds the heart is lined by parietal serous pericardium. **(b)** After covering the heart, the visceral pericardium folds back over the heart and creates the parietal pericardium. The space between them is the pericardial cavity. **(c)** The heart has three layers, from deep to superficial: endocardium, myocardium, and epicardium (visceral serous pericardium).

The Wall and Coverings of the Heart

The heart is composed of three layers, as shown in Figure 12.2. The innermost layer, the **endocardium,** is a single layer of simple squamous epithelium, called endothelium. **Endothelium** not only lines the heart but it also continues into and lines the blood vessels. Its smooth nature helps prevent blood from clotting unnecessarily. The central **myocardium** is the thickest part of the heart wall and is made up of cardiac muscle (see Fig. 4.15). When cardiac muscle fibers contract, the heart beats. The outermost layer is the **epicardium,** which is also called the **visceral serous pericardium** (the term *visceral* means organ, and refers to the fact that this layer covers the heart). After covering the heart, the visceral pericardium folds back over the heart, creating the parietal serous pericardium. The two serous membranes (epicardium and parietal pericardium) secrete pericardial fluid (a fluid similar to plasma). The pericardial fluid reduces friction as the heart beats. The parietal pericardium is fused to the outermost fibrous pericardium. The fibrous pericardium is a thick layer of fibrous connective tissue that adheres to the great blood vessels at the heart's base and anchors the heart to the diaphragm and the mediastinal wall. The coverings of the heart protect the heart, confine it to its location, and prevent it from overfilling, while still allowing the heart to contract and carry out its function of pumping the blood.

A layer of the heart can become inflamed due to infection, cancer, injury, or a complication of surgery. The suffix "itis" added to the name of a heart condition tells which layer is affected. For example, pericarditis refers to inflammation of the pericardium, and endocarditis refers to inflammation of the endocardium.

Chambers of the Heart

The heart has four hollow chambers: two superior **atria** (sing., **atrium**) and two inferior **ventricles** (Fig. 12.3). Each atrium has an anterior pocket-like flap called an **auricle.** The auricles expand fully when the atrium fills with blood. Auricles also contain cells that produce atrial natriuretic hormone (see p. 228), as well as cardiac stem cells. Internally, the atria are separated by the **interatrial septum,** and the ventricles are separated by the **interventricular septum** (plural, *septa*). Therefore, the heart's pulmonary circuit (its right side) is completely separated from its systemic circuit (the left side) by the septa. However, it's important to note that though they are physically separated, the pulmonary and systemic circuits perform their work together. Thus, the two atria contract simultaneously, and then the two ventricles contract simultaneously.

The thickness of each chamber's myocardium is suited to its function. The atria have thin walls, and each pumps blood into the

left subclavian artery

left common carotid artery

brachiocephalic artery

superior vena cava

aorta

left pulmonary artery

pulmonary trunk

left pulmonary veins

right pulmonary artery

right pulmonary veins

pulmonary
semilunar valve

left atrium

right atrium

atrioventricular
(bicuspid) valve

atrioventricular
(tricuspid) valve

chordae tendineae

papillary muscles

right ventricle

interventricular septum

left ventricle

inferior vena cava

a.

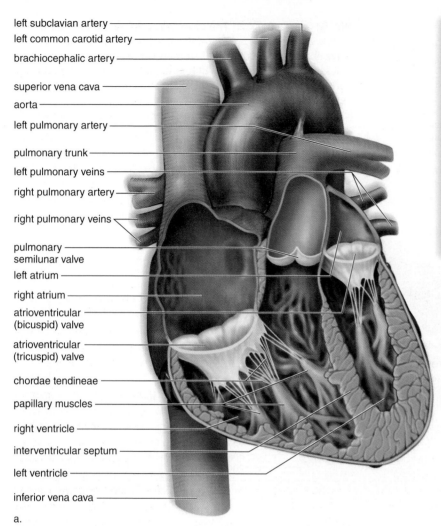

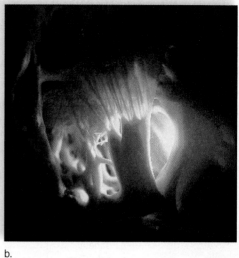

b.

AP|R **Figure 12.3 Internal heart anatomy.**
(a) The heart has four valves. The two atrioventricu-lar valves allow blood to pass from the atria to the ventricles, and the two semilunar valves allow blood to pass out of the heart (the aortic semilunar valve lies deep to the pulmonary semilunar valve in this view). (b) Photomicrograph of a heart valve, show-ing the fibrous chordae tendineae that attach the valve to the heart wall.

ventricle below. The ventricles are thicker, and they pump blood into blood vessels that travel to other parts of the body. The thinner myocardium of the right ventricle is suited for pumping blood to the lungs, which are nearby in the thoracic cavity. The left ventricle has a thicker wall than the right ventricle. Thicker myocardium enables the left ventricle to pump its blood to all other parts of the body.

Right Atrium

At its posterior wall, the **right atrium** receives O_2-poor blood from three veins: the *superior vena cava, the coronary sinus,* and the *inferior vena cava.* Venous blood passes from the right atrium into the right ventricle through an **atrioventricular (AV) valve.** This valve, like the other heart valves, directs the flow of blood and prevents any backflow. The AV valve on the right side of the heart is specifi-cally called the **tricuspid valve** because it has three *cusps,* or flaps.

Right Ventricle

In the **right ventricle,** the cusps of the tricuspid valve are con-nected to fibrous cords, called the **chordae tendineae** (meaning "heart strings"). The chordae tendineae in turn are connected to the **papillary muscles,** which are conical extensions of the myocardium.

Blood from the right ventricle passes through a **semilunar valve** into the *pulmonary trunk.* Semilunar valves are so called because their cusps are thought to resemble half-moons. This par-ticular semilunar valve, called the **pulmonary semilunar valve,** prevents blood from flowing back into the right ventricle.

Note in Figure 12.3 that the pulmonary trunk divides into the left and right pulmonary arteries. For help in remembering how blood flows through the heart, trace the path of O_2-poor blood from the vena cava to the pulmonary arteries that take blood to the lungs (see Fig. 12.1).

In the Lungs

Within the right and left lungs, the pulmonary arteries divide to form smaller and smaller arterioles. The smallest arterioles supply *pulmonary capillaries*: tiny blood vessels which cover the alveoli, or air sacs of the lungs. As you know from Chapter 4, both capillaries and alveoli are composed of simple squamous epithelium, an exceed-ingly thin tissue. The respiratory gases oxygen and carbon dioxide freely diffuse between the alveoli and pulmonary capillaries. The capillaries then empty into *pulmonary venules,* which join to form larger veins. The largest of these veins are the four *pulmonary veins.*

Left Atrium

At its posterior wall, the **left atrium** receives O_2-rich blood from the pulmonary veins. Two veins come from each lung. Blood passes from the left atrium into the left ventricle through an AV valve. The AV valve on the left side is specifically called the **bicuspid valve** because it has two cusps. (In the United States, the bicuspid valve is more commonly referred to as the **mitral valve,** so called because the valve is similar in shape to a bishop's hat, or mitre.)

Left Ventricle

The **left ventricle** forms the apex of the heart. The papillary muscles in the left ventricle are quite large, and the chordae tendineae attached to the AV valve are thicker and stronger than those in the right ventricle.

Blood passes from the left ventricle through a semilunar valve into the **aorta.** This semilunar valve is appropriately called the **aortic semilunar valve.** The semilunar cusps of this valve are larger and thicker than those of the pulmonary semilunar valve.

Just beyond the aortic semilunar valve lie the first branches from the aorta. These are the **coronary arteries**—blood vessels that lie on and nourish the heart itself. The rest of the blood stays in the aorta, which continues as the arch of the aorta and then the descending aorta.

Operation of the Heart Valves and Heart Sounds

Let's take a look at how the valves of the heart operate to create a one-way flow of blood from the atria to the ventricles to the arteries. The AV valves (tricuspid and mitral valve) are open when the ventricles are filling with blood. When a ventricle contracts, however, the increasing pressure of the blood inside the ventricle forces the cusps of the AV valve to slam shut. The force of the blood can be compared to a strong wind that can blow a door (the valve cusps) shut. However, when the ventricle contracts, the papillary muscles also contract, causing the chordae tendineae to tighten and pull on the valve. Thus, AV valves in the normal heart are prevented from inverting back up into the atrium.

The semilunar valves (pulmonary and aortic) are normally closed while the ventricles are filling with blood. However, the contraction of the ventricles pushes blood at high pressure against the valve cusps, forcing the valves open. Then, when the ventricle relaxes, the blood in the artery falls backward toward the valve, closing the valve once again.

A heartbeat produces the familiar "LUB-DUP" sounds as the chambers contract and the valves close. The first heart sound, "lub," is heard when the ventricles begin to contract and the atrioventricular (tricuspid and mitral) valves close. This sound lasts the longest and has a lower pitch. The second heart sound, "dup," is heard when the relaxation of the ventricles allows the semilunar (pulmonary and aortic) valves to close.

Like mechanical valves, the heart valves sometimes leak—in the language of medicine, the valves are *incompetent.* When valves don't close properly there is a backflow of blood. Blood leaks from the ventricle to the atrium (if an atrioventricular valve is incompetent) or from the pulmonary artery or aorta back into the ventricle (if the semilunar valve fails). Most often, the valves of the systemic circulation—bicuspid (or mitral) valves and the aortic semilunar valve—are affected. The backflow causes a heart murmur, which is typically a clicking or swishing sound heard after the first heart sound ("lub"). A trained physician or health professional can diagnose heart murmurs from their sound and timing. A person can be born with a valve deformity (called a *congenital valve defect*), or the valve can be damaged by bacterial infection. For example, rheumatic fever begins in the throat, and then spreads throughout the body. The bacteria attack connective tissue in the heart valves as well as other organs. Further, antibodies produced by the infected individual can also damage the valves. Fortunately, defective heart valves can be repaired surgically, or replaced with a synthetic valve or one taken from a pig's heart.

Coronary Circulation

Cardiac muscle fibers and the other types of cells in the wall of the heart are not nourished by the blood in the chambers; diffusion of oxygen and nutrients from this blood to all the cells that make up the heart would be too slow. Instead, these cells receive nutrients and rid themselves of wastes at capillaries embedded in the heart wall.

As mentioned previously, the right and left coronary arteries branch from the aorta just superior to the aortic semilunar valve (Fig. 12.4). Each of these arteries branches and then rebranches, until the heart is encircled by small arterial blood vessels. Some of these join so that there are several routes to reach any particular capillary bed in the heart. Alternate routes are helpful if an obstruction should occur along the path of blood reaching cardiac muscle cells. The Medical Focus on pp. 267–268 describes the cause, and some treatment options, for obstruction of the coronary arteries.

After blood has passed through cardiac capillaries, it is taken up by vessels that join to form veins. The coronary veins are specifically called **cardiac veins.** The cardiac veins enter a coronary sinus, which is essentially a thin-walled vein. The coronary sinus enters the right atrium, where deoxygenated blood mixes with the blood from the superior vena cava and inferior vena cava. This blood will then be sent to the lungs for oxygenation.

Content **CHECK-UP!**

1. Describe the coverings of the heart in order from superficial to deep. Why is the fibrous pericardium important for survival?

2. Which valve is located between the right atrium and the right ventricle?

   a. pulmonary semilunar   c. aortic semilunar

   b. bicuspid (mitral)   d. tricuspid

3. The first heart sound is caused by the closure of the:

   a. AV valves (tricuspid and bicuspid).

   b. semilunar valves (aortic and pulmonary).

Answers in Appendix A.

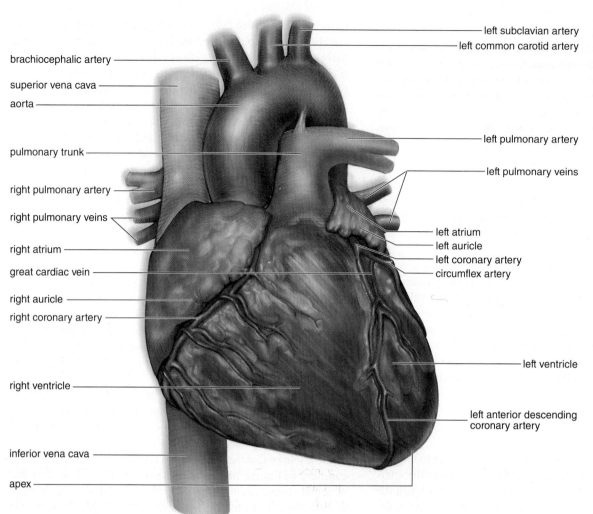

brachiocephalic artery

superior vena cava

aorta

pulmonary trunk

right pulmonary artery

right pulmonary veins

right atrium

great cardiac vein

right auricle

right coronary artery

right ventricle

inferior vena cava

apex

left subclavian artery

left common carotid artery

left pulmonary artery

left pulmonary veins

left atrium

left auricle

left coronary artery

circumflex artery

left ventricle

left anterior descending coronary artery

AP|R Figure 12.4
Anterior view of exterior heart anatomy.
The great vessels (venae cavae, pulmonary trunk, pulmonary arteries, and aorta) are attached to the base of the heart. The right ventricle forms most of the anterior surface of the heart, and the left ventricle forms most of the posterior surface. The coronary arteries bring oxygen and nutrients to cardiac cells, which derive little from the blood coursing through the heart.

12.2 Physiology of the Heart

6. Describe the conduction system of the heart.
7. Label and explain a normal electrocardiogram.
8. Describe the cardiac cycle and the heart sounds.
9. Describe the cardiac output and regulation of the heartbeat.

The physiology of the heart pertains to its pumping action—that is, the heartbeat. It is estimated that the heart beats almost three billion times in a lifetime, continuously recycling some 5 to 6 liters (L) of blood to keep us alive. In this section, we will consider what causes the heartbeat, what it consists of, and its consequences.

Conduction System of the Heart

The **conduction system** of the heart is a route of specialized cardiac muscle fibers that initiate and stimulate contraction of the atria and ventricles. The conduction system is said to be *intrinsic,* meaning that the heart beats automatically without the need for external nervous stimulation. The conduction system coordinates the atria and ventricles so they work as a "team," that is, the atria contract simultaneously, and the ventricles then contract simultaneously. Thus, the heart is an effective pump.

Without this conduction system, the atria and ventricles would contract at different rates.

Nodal Tissue

The heartbeat is controlled by nodal tissue, which has both muscular and nervous characteristics. This unique type of cardiac muscle is located in two regions of the heart: The **SA (sinoatrial) node** is located in the upper posterior wall of the right atrium. The **AV (atrioventricular) node** is located in the base of the right atrium very near the interatrial septum (Fig. 12.5).

The SA node initiates the heartbeat and automatically sends out an excitation signal every 0.85 second. The SA node normally functions as the pacemaker for the entire heart because its intrinsic rate is the fastest in the system. From the SA node, signals spread out over the atria, causing them to contract.

When the signals reach the AV node, there is a slight delay that allows the atria to finish their contraction before the ventricles begin their contraction. The signal for the ventricles to contract travels from the AV node through the two branches of the **atrioventricular bundle (AV bundle)** before reaching the numerous and smaller **Purkinje fibers.** The AV bundle, its branches, and the Purkinje fibers consist of specialized cardiac muscle fibers that

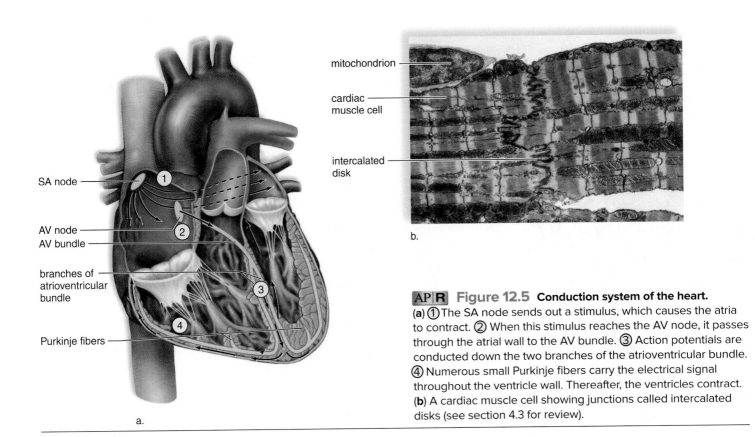

mitochondrion

cardiac muscle cell

intercalated disk

b.

SA node

AV node

AV bundle

branches of atrioventricular bundle

Purkinje fibers

a.

AP|R **Figure 12.5** **Conduction system of the heart.**
(a) ① The SA node sends out a stimulus, which causes the atria to contract. ② When this stimulus reaches the AV node, it passes through the atrial wall to the AV bundle. ③ Action potentials are conducted down the two branches of the atrioventricular bundle. ④ Numerous small Purkinje fibers carry the electrical signal throughout the ventricle wall. Thereafter, the ventricles contract. **(b)** A cardiac muscle cell showing junctions called intercalated disks (see section 4.3 for review).

efficiently spread an electrical signal throughout the ventricles. Recall from Chapter 4 (p. 73) that cardiac muscle cells are bound end-to-end at intercalated disks, and within the disks are adhesion junctions and gap junctions. **Gap junctions** are specialized intercellular connections that allow electrical current to flow directly from cell to cell. Once stimulated electrically, the ventricular muscle contracts purposefully to pump blood.

The SA node **pacemaker** usually keeps the heartbeat regular. If the SA node fails to work properly, the ventricles still beat due to impulses generated by the AV node. But the beat is slower (40 to 60 beats per minute). This condition is referred to as a **heart block.**

An area other than the SA node can become the pacemaker when it develops a rate of contraction that is faster than the SA node. This site, called an **ectopic pacemaker,** may cause an extra beat, if it operates only occasionally, or it can even pace the heart for a while. Caffeine and nicotine are two substances that can stimulate an ectopic pacemaker.

Occasionally, the cardiac conduction system fails to operate properly. As a result, the heartbeat will be abnormal: failing to beat at regular intervals, or perhaps beating too slowly or too quickly. To correct this, an artificial electronic pacemaker can be surgically implanted to restore a normal heart rate.

Electrocardiogram

With the contraction of any muscle, including the myocardium, electrolyte changes occur that can be detected by electrical recording devices. These changes occur as a muscle action potential sweeps over the cardiac muscle fibers. The resulting record, called an electrocardiogram, helps a physician detect and possibly diagnose the cause of an irregular heartbeat. There are many types of irregular heartbeats, called **arrhythmias.** The Medical Focus on page 269 discusses the electrocardiogram and some types of arrhythmias.

Cardiac Cycle

A **cardiac cycle** includes all the events that occur during one heartbeat. On average, the heart beats about 70 times a minute, although a normal adult heart rate can vary from 60 to 100 beats per minute. After tracing the path of blood through the heart, it might seem that the right and left sides of the heart beat independently of one another, but as you know, they actually contract together. First the two atria contract simultaneously; then the two ventricles contract together. The term **systole** refers to contraction of heart muscle, and the term **diastole** refers to relaxation of heart muscle. During the cardiac cycle, *atrial systole* is followed by *ventricular systole.*

As shown in Figure 12.6, the three phases of the cardiac cycle are:

Phase 1: Atrial Systole. Time = 0.15 sec. During this phase, both atria are in systole (contracted), while the ventricles are in diastole (relaxed). Rising blood pressure in the atria forces the blood to enter the two ventricles through the AV valves. At this time, both atrioventricular valves are open, and the semilunar valves are closed. Atrial systole ends when the atrioventricular valves (tricuspid and bicuspid/mitral) slam shut. Closure of the AV valves is caused by the rising pressure of blood filling the ventricle. Remember that closure of the AV valves causes the first heart sound, "lub" (page 263).

Time	Atria	Ventricles	Atrioventricular Valves (tricuspid, bicuspid)	Semilunar Valves (pulmonary, aortic)
0.15 sec	Systole	Diastole	open	closed
0.30 sec	Diastole	Systole	closed	open
0.40 sec	Diastole	Diastole	open	closed

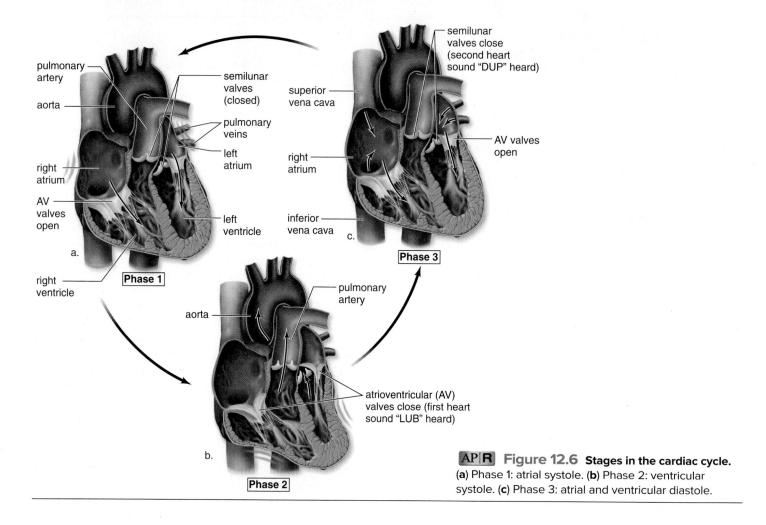

AP|R Figure 12.6 **Stages in the cardiac cycle.**
(a) Phase 1: atrial systole. (b) Phase 2: ventricular systole. (c) Phase 3: atrial and ventricular diastole.

Phase 2: Ventricular Systole. Time = 0.30 sec. During this phase, both ventricles are in systole (contracted), while the atria are in diastole (relaxed). Rising blood pressure in the ventricles forces the semilunar valves (aortic and pulmonary) to open. Blood in the right ventricle exits through the pulmonary artery trunk to the right and left pulmonary arteries. Simultaneously, blood in the left ventricle exits into the aorta. During ventricular systole, both semilunar valves are open, and the atrioventricular valves are closed. Ventricular systole ends as the ventricles complete their pumping job; recall that backflow of blood in the pulmonary artery and aorta forces the semilunar valves to slam shut once more (page 263). Closure of the semilunar valves causes the second heart sound "dup."

Phase 3: Atrial and Ventricular Diastole. Time = 0.40 sec. During this period, both atria and both ventricles are in diastole (relaxed). At this point, pressure in all the heart chambers is low. Blood returning to the heart from the superior and inferior venae cavae, the coronary sinus, and the pulmonary veins fills the right and left atria and flows passively into the ventricles. At this time, both atrioventricular valves are open and the semilunar valves are closed.

Cardiac Output

Cardiac output (CO) is the volume of blood pumped out of a ventricle in one minute. (The same amount of blood is pumped out of each ventricle in one minute.) Cardiac output is dependent on two factors:

- *heart rate* (*HR*) = beats per minute
- *stroke volume* (*SV*) = amount of blood pumped by a ventricle each time it contracts

Thus, cardiac output = HR × SV.

MEDICAL FOCUS

Arteriosclerosis, Atherosclerosis, and Coronary Artery Disease

The leading cause of heart attack and stroke, and the number-one killer in North America and Western Europe, is arteriosclerosis. **Arteriosclerosis** is the older, more generalized term for abnormal thickening and hardening of the arterial wall over time. (However, this term is now seldom used in clinical literature.) The most common form of arteriosclerosis is **atherosclerosis.** Scientists agree that atherosclerosis begins with injury to the arterial wall. Research has suggested several possible causes for injury: smoking, high blood pressure (called **hypertension**), low levels of HDL cholesterol (high-density lipoprotein, often referred to as "good cholesterol") and elevated levels of blood lipids, LDL cholesterol (low-density lipoprotein, or "bad cholesterol"), and homocysteine (a by-product of protein metabolism). Diabetics (especially those with type II or non-insulin-dependent diabetes mellitus) are at increased risk for atherosclerosis, probably because their disease causes high levels of blood lipids and LDL cholesterol.

Research also indicates that low-level bacterial or viral infection that spreads to the blood may injure arterial walls and start the atherosclerosis process. This infection may originate with gum disease, or it can be caused by a bacterium called *Helicobacter pylori* (the microbe that also causes ulcers in the stomach). Antibodies specific to these microbes are found in people with atherosclerosis. In addition, a blood protein called C-reactive protein, or CRP, is an important piece of evidence suggesting an infection. For example, blood CRP levels rise if you suffer from a cold or are recovering from a wound. High blood levels of CRP in an otherwise healthy person imply that the arteries are damaged. Indeed, recent studies show that people with the highest blood levels of CRP have double the risk of heart attack. Further, new research has shown that excesses of additional blood components may also be linked to an increased risk of atherosclerosis.

These components include fibrinogen (the inactive clotting protein you learned about in Chapter 11) and a specific lipid called lipoprotein (a).

Once the arterial wall is injured, the body's defense mechanisms respond. White blood cells called **macrophages** invade the injured area and stick to the arterial wall. These macrophages ingest LDL and are then called foam cells (because mixing fat with the cell's watery cytoplasm creates a foamy appearance). A collection of foam cells creates a **fatty streak.** Sadly, post-mortem studies on the arteries of young people have shown that these fatty streaks begin to develop during the early teenage years.

Over time, the artery's smooth muscle cells migrate to cover the fatty streak. Finally, fibroblasts and scar tissue will cover the smooth muscle cells. Calcium ions will invade the tissue, causing it to harden into an **atherosclerotic plaque.** Atherosclerotic plaques can grow so large that they completely block arterial blood flow, causing the tissue supplied by the artery to die. Because the plaque's surface is very rough compared to the smooth endothelium, the plaque may also trigger the clotting mechanism and cause a stationary blood clot, or thrombus, to form inside a blood vessel. Thrombi may also form if the surface of the plaque ulcerates (cracks open and bleeds). As mentioned in Chapter 11, **thromboembolism** occurs when a blood clot breaks away from its place of origin and is carried to a new location. Further, the plaque can prevent the arterial wall cells from receiving oxygen and nutrients. The cells die, causing the wall itself to weaken, which might result in an **aneurysm.** Aneurysms are weakened areas in the arterial wall, which balloon outward and may even rupture.

Coronary artery disease is the term for atherosclerosis of the coronary arteries (Fig. 12A). If the coronary artery is partially

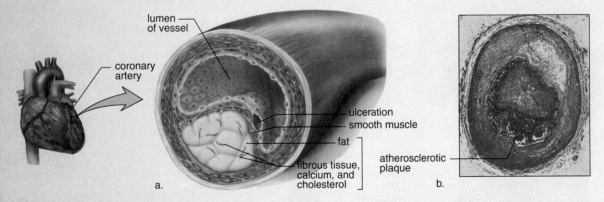

lumen of vessel

coronary artery

ulceration

smooth muscle

fat

fibrous tissue, calcium, and cholesterol

atherosclerotic plaque

a.

b.

Figure 12A **Coronary artery disease.** **(a)** Atherosclerosis begins with injury to an artery; a fatty streak develops at the site and smooth muscle grows over the lesion. Fibrous tissue and ionic calcium enlarge and stiffen the atherosclerotic plaque. **(b)** When plaque is present in a coronary artery, restricted blood flow may result in a heart attack.

—Continued

Continued—

occluded (blocked) by atherosclerosis, the individual may suffer from **ischemic heart disease.** Although enough oxygen normally reaches the resting heart, the person's heart is oxygen-deprived during periods of exercise or stress. Some patients experience *silent ischemia* during the earliest stage of the disease; that is, they don't detect any symptoms of a problem. However, most patients suffer from **angina pectoris,** chest pain that is often accompanied by a radiating pain in the left arm. Angina pain is a warning sign of reduced blood flow to the heart and must not be ignored. Should the coronary artery become completely blocked by atherosclerotic plaque or thrombus, a portion

of the heart will die from lack of oxygen. Dead tissue is called an **infarct** and, therefore, a heart muscle attack is termed a **myocardial infarction.**

Two surgical procedures can reopen occluded coronary arteries. In *balloon angioplasty*, a plastic tube is threaded into an artery of an arm or thigh, then guided through a major blood vessel toward the heart. Once the tube reaches a blockage, a balloon attached to the end of the tube can be inflated to break up a blood clot or open a vessel clogged with plaque (Fig. 12B). Often, a small metal-mesh cylinder called a vascular **stent** is inserted into a blood vessel during balloon angioplasty. The stent holds the vessels open and decreases the risk of future occlusion. Stent devices currently in use have built-in medication, which is slowly released in the artery. These medications prevent blood clots and additional scar tissue from forming, thus helping to keep the stent open and blood flowing. In a similar fashion, a stent can be placed into an artery weakened by an aneurysm (Fig. 12C). This type of stent supports the arterial wall and can be monitored externally, allowing a physician to ensure that it is working properly.

In a **coronary bypass operation,** a portion of a blood vessel from another body part (usually one of the mammary arteries from the chest) is sutured from the aorta to the coronary artery, past the obstruction point. Blood can then flow normally again from the aorta to the heart muscle. Figure 12D shows a triple bypass in which three blood vessels connect the aorta to the coronary artery, restoring blood flow to the myocardium.

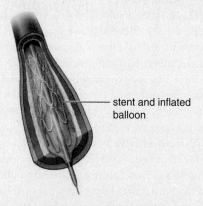

stent and inflated balloon

Figure 12B **Balloon angioplasty.** A balloon inserted in an artery can be inflated to open up a clogged coronary blood vessel.

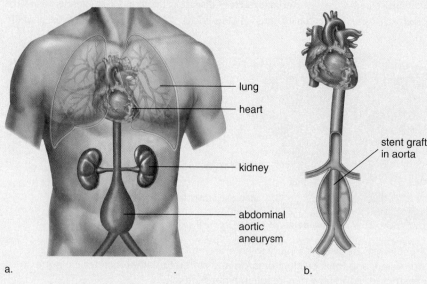

lung

heart

kidney

abdominal aortic aneurysm

stent graft in aorta

a.

b.

Figure 12C **Abdominal aortic aneurysm. (a)** Before and **(b)** after stent placement.

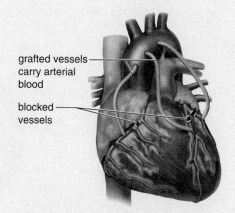

grafted vessels carry arterial blood

blocked vessels

Figure 12D **Coronary bypass surgery.** During this operation, the surgeon grafts segments of another vessel between the aorta and the coronary vessels, bypassing areas of blockage.

MEDICAL FOCUS

The Electrocardiogram

A graph that records the electrical activity of the myocardium during a cardiac cycle is called an **electrocardiogram,** or **ECG.**[*]An ECG is obtained by placing several electrodes on the patient's skin, then wiring the electrodes to a voltmeter (an instrument for measuring voltage). As the heart's chambers contract and then relax, the change in polarity is measured in millivolts.

An ECG consists of a set of waves: the P wave, a QRS complex, and a T wave (Fig. 12E). The P wave represents depolarization of the atria as a signal started by the SA node travels throughout the atria. The P wave signals that the atria are going to be in systole and that the atrial myocardium is about to contract. The QRS complex represents depolarization of the ventricles following excitation of the Purkinje fibers. It signals that the ventricles are going to be in systole and that the ventricular myocardium is about to contract. The QRS complex shows greater voltage changes than the P wave because the ventricles have more muscle mass than the atria. The T wave represents repolarization of the ventricles. It signals that the ventricles are going to be in diastole and that the ventricular myocardium is about to relax. Atrial diastole (repolarization) does not show up on an ECG as an independent event because the voltage changes are masked by the QRS complex.

An ECG records the duration of electrical activity and therefore can be used to detect arrhythmia, an irregular or abnormal heartbeat. A rate of fewer than 60 heartbeats per minute is called **bradycardia,** and more than 100 heartbeats per minute is called **tachycardia.** Another type of arrhythmia is **fibrillation,** in which the heart beats rapidly but the contractions are uncoordinated. Fibrillation can be very dangerous and potentially deadly, because if the heart muscle is not contracting properly, the heart will not pump blood. Cells and tissues will subsequently die of oxygen starvation. The heart can sometimes be defibrillated by briefly applying a strong electrical current to the chest.

It is important to understand that an ECG only supplies information about the heart's electrical activity. To be used in diagnosis, an ECG must be coupled with other information, including X rays, studies of blood flow, and a detailed history from the patient.

[*]An ECG is sometimes also referred to as an EKG (from the German, ElektroKardioGramm), because its Dutch inventor spoke German.

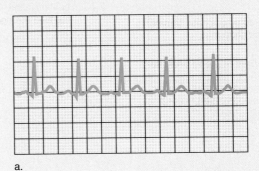

Figure 12E **Electrocardiogram. (a)** A portion of an electrocardiogram. **(b)** An enlarged normal cycle.

The CO of an average human is 5,250 ml (or 5.25 L) per ventricle, per minute, which equates to about the total volume of blood in the human body. Each minute, the right ventricle pumps about 5.25 L through the pulmonary circuit, while the left ventricle pumps about 5.25 L through the systemic circuit. And this is only the resting cardiac output! During exercise, cardiac output can increase tremendously to meet the body's need for more oxygen.

Cardiac output can vary because stroke volume and heart rate can vary, as discussed next. In this way, the heart regulates the blood supply, dependent on the body's needs. For example, increases in heart rate and stroke volume during exercise can increase cardiac output as much as seven to eight times the normal resting amount.

Heart Rate

A **cardioregulatory center** in the medulla oblongata of the brain can alter the heart rate by way of the autonomic nervous system (Fig. 12.7). Parasympathetic motor signals conducted by the vagus nerve cause the heart rate to slow, and sympathetic motor signals conducted by sympathetic motor fibers cause the heart rate to increase.

The cardioregulatory center receives sensory input from receptors within the cardiovascular system. For example, stretch

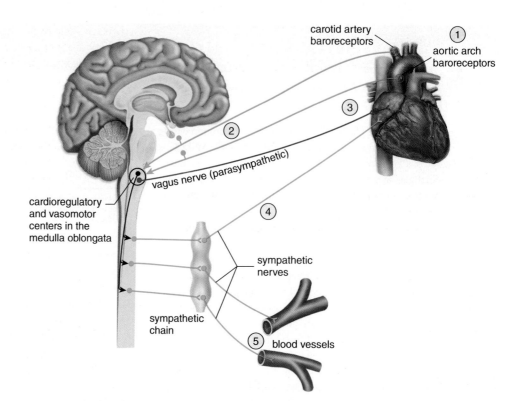

Regulation of heart rate:

① Baroreceptors in the aortic arch and carotid arteries monitor blood pressure.

② Nerve signals from the baroreceptors signal the cardioregulatory center.

③ Increased parasympathetic signals decrease heart rate.

④ Increased sympathetic signals increase heart rate.

Regulation of blood pressure:

⑤ Increased sympathetic signals cause blood vessels to constrict.

carotid artery baroreceptors

aortic arch baroreceptors

vagus nerve (parasympathetic)

cardioregulatory and vasomotor centers in the medulla oblongata

sympathetic nerves

sympathetic chain

blood vessels

AP|R **Figure 12.7** **Control of heart activity.** The cardioregulatory center regulates the heart rate, and the vasomotor center regulates constriction of blood vessels, according to input received from baroreceptors in the carotid artery and aortic arch.

receptors called *baroreceptors* are present in the aorta just after it leaves the heart, and also in the carotid arteries, which take blood from the aorta to the brain. If blood pressure falls, as it sometimes does when we stand up quickly, the baroreceptors signal the cardioregulatory center. In response, sympathetic motor signals to the heart cause the heart rate to increase. Once blood pressure begins to rise above normal, nerve signals from the cardioregulatory center cause the heart rate to decrease. Such reflexes help control cardiac output and, therefore, blood pressure, as discussed in section 12.4.

The cardioregulatory center is under the influence of the cerebrum and the hypothalamus. Therefore, when we feel anxious, the sympathetic motor nerves are activated. In addition, the adrenal medulla releases the hormones norepinephrine and epinephrine. The result is an increase in heart rate. On the other hand, activities such as yoga and meditation lead to activation of the vagus nerve, which slows the heart rate.

Other factors affect the heart rate as well. For example, a low body temperature slows the rate. Also, the proper electrolyte concentrations are needed to keep the heart rate regular.

Stroke Volume

Stroke volume, which is the amount of blood that leaves a ventricle, depends on the strength of heart contraction. As with heart rate, the autonomic nervous system helps to determine contraction force and stroke volume. Sympathetic activity strengthens each contraction and increases stroke volume. By contrast, parasympathetic activity decreases both heart rate and contractile force, which in turn decreases

stroke volume. Further, the degree of contraction also depends on the correct blood electrolyte concentration. Recall from Chapter 8 that the proper concentration of ions, or electrolytes, is essential to create cardiac muscle action potentials. Without these electrolytes, cardiac conduction and contraction are impaired and stroke volume decreases. Two additional factors, venous return and difference in blood pressure, also influence the strength of contraction.

Venous Return Venous return is the amount of blood entering the heart by way of the venae cavae (right side of heart) or pulmonary veins (left side of heart). The heart adjusts the strength of its own contraction beat by beat, based upon venous return. This principle is called the Frank-Starling law. The more blood returned to the heart before a given beat, the more the ventricles stretch. As the ventricles are stretched, they contract more and more forcefully. Thus, any event that increases the volume of blood entering the heart will increase the stroke volume leaving the heart. For example, exercise increases the strength of cardiac contraction because skeletal muscle contraction squeezes the veins within muscles and increases venous return. The opposite is also true: If venous return decreases, stroke volume decreases for the next beat. A low venous return, as might happen if there is blood loss, decreases the strength of cardiac contraction.

Difference in Blood Pressure The strength of ventricular contraction has to be strong enough to overcome the blood pressure within the attached arteries. If a person has hypertension or atherosclerosis, the opposing arterial pressure may reduce the effectiveness of contraction and the stroke volume.

4. The SA node is the pacemaker of the heart and controls heart rate because:

    a. it sits at the top of the heart.

    b. it has the fastest signaling rate.

    c. it has the largest myocardial cells.

    d. it has the slowest signaling rate.

5. During atrial systole,

    a. the atrioventricular valves are open, but the semilunar valves are closed.

    b. the atrioventricular valves are closed, but the semilunar valves are open.

    c. both sets of valves are closed.

6. Consider two athletes running a race. The first has trained for months, and the second has had little to no training for the event. How might training affect each athlete's cardiac output?

Answers in Appendix A.

12.3 Anatomy of Blood Vessels

10. Name the three types of blood vessels, and describe their structure and function.

There are three types of blood vessels: arteries, capillaries, and veins (Fig. 12.8). These vessels function to:

1. transport blood and its contents (see page 260);
2. carry out exchange of gases in the pulmonary capillaries and exchange of gases plus nutrients for waste at the systemic capillaries;
3. regulate blood pressure; and
4. direct blood flow to those systemic tissues that most require it at the moment.

Arteries and Arterioles

Arteries (Fig. 12.8) transport blood away from the heart. They have thick, strong walls composed of three layers: (1) The **tunica intima** (sometimes also referred to as tunica interna) is an endothelium layer with a basement membrane. (2) The **tunica media** is a thick middle layer of smooth muscle and elastic fibers. (3) The **tunica externa** is an outer connective tissue layer composed principally of elastic and collagen fibers. Arterial walls are sometimes so thick that they are supplied with their own blood vessels. The radius of an artery allows the blood to flow rapidly, and the elasticity of an artery allows it to expand when the heart contracts and recoil when the heart rests. This means that blood continues to flow in an artery even when the heart is in diastole.

Arterioles are small arteries just visible to the naked eye. The middle layer of these vessels has some elastic tissue but is composed mostly of smooth muscle whose fibers encircle the arteriole. If the muscle fibers contract, the lumen (cavity) of the arteriole decreases; if the fibers relax, the lumen of the arteriole enlarges. Whether arterioles are constricted or dilated affects the distribution of blood flow and blood pressure throughout the body. When a muscle is actively contracting, for example, the arterioles in the vicinity dilate to allow additional blood flow to the muscle. In this way, the muscle's need for oxygen and glucose is met.

Constriction and dilation of arterioles throughout the body is controlled by the autonomic nervous system and helps to determine blood pressure. The greater the number of arterioles that are constricted, the higher the resistance to blood flow. Higher resistance causes higher blood pressure. By contrast, the greater the number of dilated arterioles, the lower the resistance to blood flow. Blood pressure falls as a result.

Capillaries

Arterioles branch into **capillaries** (Fig. 12.8), which are extremely narrow, microscopic blood vessels. Their walls are composed of only one layer of endothelial cells connected by tight junctions. These thin walls easily allow gases, nutrients, and wastes to diffuse between the capillary and the surrounding cells. Capillaries are extremely numerous: The body probably contains a billion, and their combined surface area is estimated at 6,300 square meters. Capillary beds (networks of many capillaries) are present in all regions of the body, and each supplies the needs of neighboring cells. Therefore, most of the body's cells are near a capillary, and the heart and other vessels of the cardiovascular system can be considered a means for conducting blood to and from the capillaries.

Not all capillary beds are open or in use at the same time. For instance, after a meal, the capillary beds of the digestive tract are usually open, while during muscular exercise, the capillary beds of the skeletal muscles are open. Most capillary beds have a shunt that allows blood to move directly from an arteriole to a venule (a small vessel leading to a vein) when the capillary bed is closed. Sphincter muscles, called *precapillary sphincters,* encircle the entrance to each capillary (Fig. 12.8). When the precapillary sphincters are constricted, the capillary bed is closed, preventing blood from entering the capillaries. Conversely, when the precapillary sphincters are relaxed, the capillary bed is open. As would be expected, the larger the number of open capillary beds, the lower the blood pressure in the body.

Capillary Exchange

In the tissues of the body, metabolically active cells require oxygen and nutrients and give off wastes, including carbon dioxide. During capillary exchange between tissue capillaries and body cells, oxygen and nutrients leave a capillary. Cellular wastes, including carbon dioxide, enter a capillary. For this reason, systemic arterial blood contains more oxygen and nutrients than does systemic venous blood, and venous blood contains more wastes than does arterial blood. In pulmonary capillaries—the capillaries supplying the lung—the exchange is reversed: Oxygen enters the blood and carbon dioxide leaves.

The internal environment of the body consists of blood and tissue fluid. **Tissue fluid** is simply the fluid that surrounds the cells of the body. In other words, substances that leave a capillary pass through tissue fluid before entering the body's cells, and substances that leave the body's cells pass through tissue fluid before entering a capillary. Water and other small molecules can cross through the cells of a capillary wall or through tiny clefts that occur

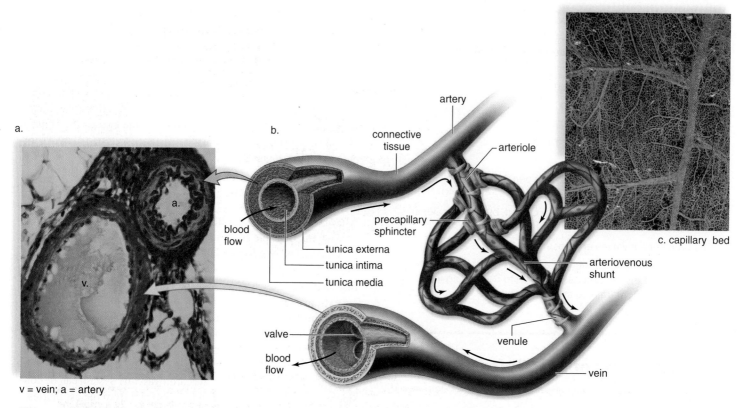

a.

b.

artery

connective
tissue

arteriole

blood
flow

precapillary
sphincter

tunica externa

tunica intima

tunica media

arteriovenous
shunt

valve

venule

blood
flow

vein

c. capillary bed

v = vein; a = artery

Figure 12.8 **Blood vessels.** (a) The walls of arteries and veins have three layers: tunica intima (interna), tunica media, and tunica externa. Arteries have a thicker muscle layer and a thick wall, while veins have valves and the larger diameter. (b) A capillary bed lies between an arteriole and a venule, and flow through it is controlled by a precapillary sphincter. When the sphincter relaxes, blood passes through the capillary bed and gives up its oxygen to the tissues. Thus, blood goes from being oxygen-rich in the arteriole (red color) to oxygen-poor in the venule (purple color). When the sphincter contracts, blood can bypass the capillary bed through an arteriovenous shunt. (c) Photomicrograph of a capillary bed.

between the cells. Large molecules in plasma, such as the plasma proteins, are too large to pass through capillary walls. Yet, the composition of tissue fluid stays relatively constant because of capillary exchange. Tissue fluid is a water-based solution that contains sodium chloride, other electrolytes, and scant protein. Any excess tissue fluid is collected by lymphatic capillaries, which are always found near blood capillaries.

Three processes influence capillary exchange—blood pressure, diffusion, and osmotic pressure:

Blood pressure, which is created by the pumping of the heart, pushes the blood through the capillary. Blood pressure also pushes blood against a vessel's (e.g., capillary) walls.

Diffusion, as you know, is simply the movement of substances from the area of higher concentration to the area of lower concentration.

Osmotic pressure is a force caused by a difference in solute concentration on either side of a membrane.

To understand osmotic pressure, consider that water will diffuse across a membrane toward the side that has the greater concentration of solutes, and the accumulation of this water results in a pressure. The presence of the plasma proteins, and also salts to some degree, means that blood has a greater osmotic pressure than

does tissue fluid. Therefore, the osmotic pressure of blood pulls water into and retains water inside a capillary.

Notice in Figure 12.9 that a capillary has an arterial end (contains arterial blood) and a venous end (contains venous blood). In between, a capillary has a midsection. We will now consider the exchange of molecules across capillary walls at each of these locations.

Arterial End of Capillary

When arterial blood enters tissue capillaries, it is bright red because the hemoglobin in red blood cells is carrying oxygen. Blood is also rich in nutrients, which are dissolved in plasma.

At the arterial end of a capillary, blood pressure, an outward force, is higher than osmotic pressure, an inward force. Pressure is measured in terms of mm Hg (mercury). In this case, blood pressure is 30 mm Hg, and osmotic pressure is 21 mm Hg. Because blood pressure is higher than osmotic pressure at the arterial end of a capillary, water and other small molecules (e.g., glucose and amino acids) filter out of a capillary at its arterial end.

Red blood cells and a large proportion of the plasma proteins generally remain in a capillary because they are too large to pass through its wall. The exit of water and other small molecules from a capillary creates tissue fluid. Therefore, tissue fluid consists of all the components of plasma, except that it contains fewer plasma proteins.

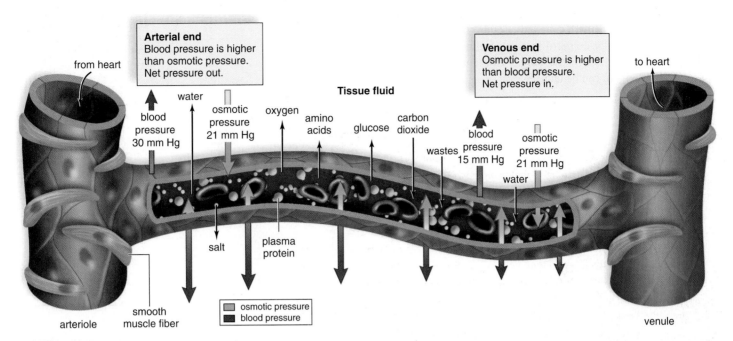

AP|R Figure 12.9 **Capillary exchange.** The capillary shows the exchanges that take place and the forces that aid the process. At the arterial end of a capillary, the blood pressure is higher than the osmotic pressure. Tissue fluid tends to leave the bloodstream. In the midsection, solutes, including oxygen (O_2) and carbon dioxide (CO_2), diffuse from high to low concentration. Carbon dioxide and wastes diffuse into the capillary while nutrients and oxygen enter the tissues. At the venous end of a capillary, the osmotic pressure is higher than the blood pressure. Tissue fluid tends to re-enter the bloodstream. The red blood cells and the plasma proteins are too large to exit a capillary.

Midsection of Capillary

Diffusion takes place along the length of the capillary, as small molecules follow their concentration gradient by moving from the area of higher to the area of lower concentration. In the tissues, the area of higher concentration of oxygen and nutrients is always blood, because after these molecules have passed into tissue fluid, they are constantly being taken up and metabolized by cells. The cells use oxygen and glucose in the process of cellular respiration, and they use amino acids for protein synthesis.

As a result of metabolism, tissue cells constantly give off carbon dioxide and other wastes. Because tissue fluid is always the area of greater concentration for waste materials, they diffuse from tissue fluid into a capillary.

Venous End of Capillary

At the venous end of the capillary, blood pressure is much reduced to only about 15 mm Hg, as shown in Figure 12.9. Blood pressure is reduced at the venous end because capillaries have a greater cross-sectional area at their venous end than their arterial end. However, there is no reduction in osmotic pressure, which remains at 21 mm Hg and is now higher than blood pressure. Therefore, water tends to diffuse into a capillary at the venous end. As water enters a capillary, it brings with it additional waste molecules. Blood that leaves the capillaries to drain into veins is deep maroon in color because red blood cells now contain reduced hemoglobin—hemoglobin that has given up its oxygen and taken on hydrogen ions.

In the end, about 85% of the water that left a capillary at the arterial end returns to it at the venous end. Therefore, retrieving fluid by means of osmotic pressure is not completely effective. The body has an auxiliary means of collecting tissue fluid; any excess usually enters lymphatic capillaries.

Veins and Venules

Veins and smaller vessels called **venules** return blood from the capillary beds to the heart. The venules first drain the blood from the capillaries and then join together to form a vein. The wall of a vein is much thinner than that of an artery because the middle layer of muscle and elastic fibers is thinner (see Fig. 12.8). Within some veins, especially the major veins of the arms and legs, **valves** allow blood to flow only toward the heart when they are open and prevent the backward flow of blood when they are closed.

At any given time, more than half of the total blood volume is found in the veins and venules. If blood is lost due to, for example, hemorrhaging, sympathetic nervous stimulation causes the veins to constrict, providing more blood to the rest of the body. In this way, the veins act as a blood reservoir.

Varicose Veins and Phlebitis

Varicose veins are abnormal and irregular dilations in superficial (near the surface) veins, particularly those in the lower legs. Varicose veins in the rectum, however, are commonly called piles, or more properly, **hemorrhoids.** Varicose veins develop when the valves of the veins become weak and ineffective due to backward pressure of the blood.

Phlebitis, or inflammation of a vein, is a more serious condition because thromboembolism can occur. In this instance, the embolus may eventually come to rest in a pulmonary arteriole, blocking circulation through the lungs. This condition, termed **pulmonary embolism,** can be fatal.

Content CHECK-UP!

7. The order for the three layers of a blood vessel, from superficial to deep, is:

   a. tunica externa—tunica intima—tunica media

   b. tunica media—tunica externa—tunica intima

   c. tunica externa—tunica media—tunica intima

8. If you described the differences between arteries and veins, you could say:

   a. Arteries have a thicker wall than veins.

   b. Veins have valves but arteries do not.

   c. Veins return blood to the heart; arteries carry blood away from the heart.

   d. All of these statements are correct.

9. Blood pressure is greatest at:

   a. the venous end of a capillary bed.

   b. the arterial end of a capillary bed.

   c. the center of a capillary bed.

   d. no single area; it is equal throughout the capillary bed.

Answers in Appendix A.

12.4 Physiology of Circulation

11. Explain how blood pressure changes throughout the vascular system, and describe the factors that determine blood pressure.
12. Describe how blood pressure is regulated.
13. Define pulse, and tell where the pulse may be taken.
14. Describe shock due to hypotension and various medical consequences of hypertension.

Circulation is the movement of blood through blood vessels, from the heart to the body and then back to the heart. In this section, we discuss various factors affecting circulation.

Velocity of Blood Flow

The velocity of blood flow is slowest in the capillaries. What might account for this? Consider that the aorta branches into the other arteries, and these in turn branch into the arterioles, and so forth until blood finally flows into the capillaries. As you know, the diameter of these vessels decreases, getting smaller and smaller with each branching. Capillary diameter is so small that blood cells must travel through in single file. In addition, each time an artery branches, the total cross-sectional area of the blood vessels increases, reaching the maximum cross-sectional area in the capillaries (Fig. 12.10). The slow rate of blood flow in the capillaries is beneficial because it allows time for the exchange of gases in pulmonary capillaries and for the exchange of gases and nutrients for wastes in systemic capillaries (see Fig. 12.9).

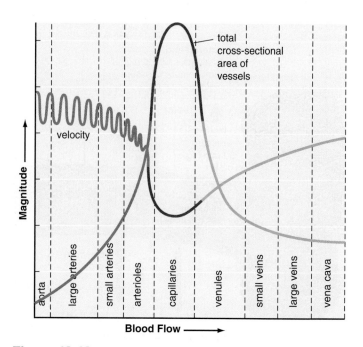

Figure 12.10 Velocity of blood flow changes throughout the systemic circuit. Velocity changes according to the total cross-sectional area of vessels.

Once blood has left the capillaries, blood velocity increases as venules combine to form larger and larger veins. Thus, velocity in the venous system is greatest in the venae cavae, which are the largest veins. However, the velocity of venous blood flow returning to the heart is always lower than that of arterial blood leaving the heart. Contractions of the powerful left ventricle generate a greater velocity for arterial blood. The velocity of the arterial and venous systems working together is very high—in a resting individual, it takes only about a minute for a drop of blood to go from the heart to the foot and back again to the heart!

Blood Pressure

Blood pressure is the force of blood against the walls of blood vessels. You would expect arterial blood pressure to be highest in the aorta. Why? Because the pumping action of the powerful, thick-muscled left ventricle forces blood into the aorta. Further, Figure 12.11 shows that systemic blood pressure decreases progressively with distance from the left ventricle. Blood pressure is lowest in the venae cavae because they are farthest from the left ventricle.

Note also in Figure 12.11 that blood pressure fluctuates in the arterial system between systolic blood pressure and diastolic blood pressure. Certainly, we can correlate this with the action of the heart. During systole, the left ventricle is pumping blood out of the heart, and during diastole the left ventricle is resting.

More important than the systolic and diastolic pressure is the *mean arterial blood pressure (MABP)*, the pressure in the arterial system averaged over time. It is important to note that MABP is not determined by taking the average of systolic and diastolic pressures. Rather, MABP is the product of *cardiac output (CO)* times *peripheral resistance.* (Recall that cardiac output equals heart rate

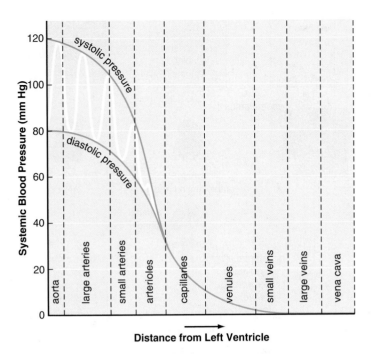

Figure 12.11 Blood pressure changes throughout the systemic circuit. Blood pressure decreases with distance from the left ventricle.

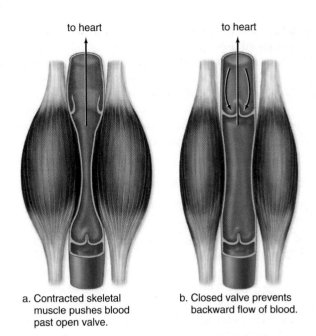

a. Contracted skeletal muscle pushes blood past open valve.

b. Closed valve prevents backward flow of blood.

Figure 12.12 Skeletal muscle pump. (a) When skeletal muscles contract and compress a vein, blood is squeezed past a valve. **(b)** When muscles relax, the valve closes.

---

times stroke volume; see page 266). To put it as a simple math equation, MABP = CO × PR. According to this equation, increasing CO will also increase MABP. In other words, the greater the amount of blood leaving the left ventricle, the greater the pressure of blood against the wall of an artery.

Another factor that determines blood pressure is peripheral resistance, which is the resistance to flow between blood and the walls of a blood vessel. All things being equal, the smaller the blood vessel diameter, the greater the resistance and the higher the blood pressure. As an analogy, imagine a skinny, 1-inch-diameter garden hose (high resistance) compared to a firefighter's 12-inch canvas hose (low resistance). Similarly, total blood vessel length increases blood pressure because a longer vessel offers greater resistance. For this reason, an obese person is apt to have high blood pressure because about 200 miles of additional blood vessels develop for each extra pound of adipose (fat) tissue.

Let's summarize our discussion so far. The two factors that affect blood pressure are:

Cardiac output
  Heart rate
  Stroke volume

Peripheral resistance
  Arterial diameter and length

Blood Pressure and Cardiac Output

Our previous discussion on pages 266 and 269 emphasized that the heart rate and the stroke volume determine cardiac output. We learned that the heart rate is intrinsic but is under extrinsic (nervous and endocrine) control. Therefore, heart rate can speed up. The faster the heart rate, the greater the cardiac output. As cardiac output increases, blood pressure increases as well (assuming constant peripheral resistance). Similarly, the larger the stroke volume, the greater the blood pressure. However, stroke volume and heart rate increase blood pressure *only* if the venous return is adequate.

Venous Return Venous return depends on three factors:

1. a blood pressure difference—blood pressure is higher in systemic veins than in the right atrium, which enables venous blood to empty into the heart
2. the skeletal muscle pump and the respiratory pump, both of which are effective because of the presence of valves in veins;
3. total blood volume in the cardiovascular system.

Here's how the **skeletal muscle pump** works: When skeletal muscles contract, they compress the weak walls of the veins. This causes the blood to move past a valve (Fig. 12.12). Once past the valve, backward pressure of blood closes the valve and prevents its return. Blood in veins will always return to the heart.

As you might suspect, gravity can assist the return of venous blood from the head to the heart but not the return of blood from the extremities and trunk to the heart. The importance of the skeletal muscle pump in maintaining CO and blood pressure can be demonstrated by forcing a person to stand rigidly still for a number of hours. Frequently, the person faints because blood collects in the limbs, limiting venous return to the heart. Cardiac output decreases and **hypotension** (low blood pressure) develops, robbing the brain of oxygen. In this case, fainting helps: the horizontal body position caused by the faint aids in getting blood to the brain.

The **respiratory pump** works like this: When inhalation occurs, thoracic pressure falls and abdominal pressure rises as the chest expands. This aids in the flow of venous blood back to the heart because blood flows from areas of higher pressure (in the abdominal cavity) to areas of lower pressure (in the thoracic cavity).

During exhalation, the pressure reverses, but the valves in the veins prevent backward flow.

As stated, the amount of venous return also depends on the total blood volume in the cardiovascular system. As you know, this volume in the pulmonary circuit and the systemic circuit is 5 L. If this amount of blood decreases (for example, due to a hemorrhage), blood pressure falls. On the other hand, if blood volume increases (due to water retention, for example), blood pressure rises.

Blood Pressure and Peripheral Resistance

The nervous system and the endocrine system both affect peripheral resistance.

Neural Regulation of Peripheral Resistance A **vasomotor center** in the medulla oblongata controls vasoconstriction. This center is under the control of the cardioregulatory center. As mentioned on pages 269–270, if blood pressure falls, baroreceptors in the blood vessels signal the cardioregulatory center. The cardioregulatory center will activate its vasomotor center. The vasomotor center then stimulates sympathetic nerve fibers, which cause the heart rate to increase *and* the arterioles to constrict. Increasing heart rate increases cardiac output; constricting the arterioles increases peripheral resistance. The result is a rise in blood pressure. What factors lead to a reduction in blood pressure? If blood pressure rises above normal, the baroreceptors signal the cardioregulatory center in the medulla oblongata. Subsequently, the heart rate decreases *and* the arterioles dilate.

Nervous control of blood vessels also causes blood to be shunted from one area of the body to another. During exercise, arteries in the viscera and skin are more constricted than those in the exercising muscles. Therefore, blood flow to the muscles increases. Also, dilation of the precapillary sphincters in muscles means that blood will flow to the muscles and not to the viscera.

Hormonal Regulation of Peripheral Resistance Certain hormones cause blood pressure to rise. Epinephrine and norepinephrine, the hormones from the adrenal medulla, increase the heart rate. As heart rate increases, so too does cardiac output and blood pressure. Epinephrine and norepinephrine also constrict arterioles in the capillary beds supplying the skin, abdominal viscera, and kidneys. Arteriolar vasoconstriction increases blood pressure by increasing peripheral resistance in these large vascular beds.

When the blood volume and blood sodium level are low, the kidneys secrete the enzyme **renin.** Renin converts the plasma protein angiotensinogen to angiotensin I, which is changed to angiotensin II by a converting enzyme found in the lungs. Angiotensin II stimulates the adrenal cortex to release aldosterone. The effect of this system, called the renin-angiotensin-aldosterone system, is to raise the blood volume and pressure in two ways. First, angiotensin II constricts the arterioles directly. Second, aldosterone causes the kidneys to reabsorb sodium. When the blood sodium level rises, water is reabsorbed, and blood volume and pressure are maintained.

Two other hormones play a role in the homeostatic maintenance of blood volume and blood pressure. As discussed in Chapter 10, antidiuretic hormone (ADH) helps increase blood volume by causing the kidneys to reabsorb water. ADH also causes vasoconstriction of smooth muscle in arteries and veins throughout the body.

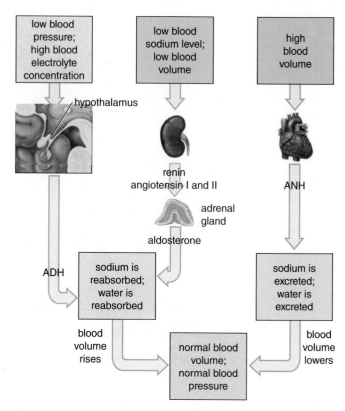

Figure 12.13 Blood volume maintenance. Normal blood volume is maintained by ADH (antidiuretic hormone) and aldosterone, whose actions raise blood volume, and by ANH (atrial natriuretic hormone), whose actions lower blood volume.

Thus, ADH boosts blood pressure by simultaneously increasing blood volume (which increases cardiac output) *and* by causing vasoconstriction (which increases peripheral resistance).

The hormonal mechanism for decreasing blood pressure involves an endocrine hormone secreted by the heart. You'll remember (from Chapter 10) that when the atria of the heart are stretched due to increased blood volume, cardiac cells release a hormone called atrial natriuretic hormone (ANH, or atriopeptide), which inhibits renin secretion by the kidneys and aldosterone secretion by the adrenal cortex. The effect of ANH, therefore, is to cause sodium excretion—that is, *natriuresis* (a term that means "excretion of sodium in the urine"). When sodium is excreted, so is water, and therefore blood volume and blood pressure decrease (Fig. 12.13).

Evaluating Circulation

Taking a patient's pulse and arterial blood pressure are two ways to quickly acquire important information needed to assess the status of the cardiovascular system.

Pulse

The surge of blood entering the arteries causes their elastic walls to stretch, but then they almost immediately recoil. This alternating expansion and recoil of an arterial wall can be felt as a **pulse** in any artery that runs close to the body's surface. These

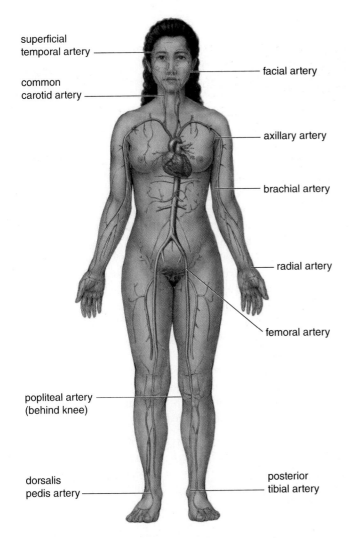

superficial
temporal artery

facial artery

common
carotid artery

axillary artery

brachial artery

radial artery

femoral artery

popliteal artery
(behind knee)

dorsalis
pedis artery

posterior
tibial artery

Figure 12.14 **Pulse points.** Pulse points are the locations where the pulse can be taken. Each pulse point is named after the appropriate artery.

superficial arteries are called pulse points (Fig. 12.14). It is customary to feel the pulse by placing several fingers on the radial artery, which lies near the outer border of the palm side of a wrist. The common carotid artery, on either side of the trachea in the neck, is another accessible location for feeling the pulse. Normally, the pulse rate indicates the rate of the heartbeat because the arterial walls pulse whenever the left ventricle contracts. The resting pulse is usually 70 times per minute, but can vary between 60 and 80 times per minute.

Blood Pressure

Blood pressure is usually measured in the brachial artery with a sphygmomanometer, an instrument that records changes in terms of millimeters (mm) of mercury (Fig. 12.15). A blood pressure cuff connected to the sphygmomanometer is wrapped around the patient's arm, and a stethoscope is placed over the brachial artery. The blood pressure cuff is inflated until no blood flows through it; therefore, no sounds can be heard through the stethoscope. The cuff pressure is then gradually lowered. As soon as the cuff pressure declines below systolic pressure, blood flows through the brachial artery each time the left ventricle contracts. The blood flow is turbulent below the cuff. This turbulence produces vibrations in the blood and surrounding tissues that can be heard through the stethoscope. These sounds are called *Korotkoff sounds,* and the cuff pressure at which the Korotkoff sounds are heard the first time is the systolic pressure. As the pressure in the cuff is lowered still more, the Korotkoff sounds change tone and loudness. When the cuff pressure no longer constricts the brachial artery, no sound is heard. The cuff pressure at which the Korotkoff sounds disappear is the diastolic pressure.

Normal resting blood pressure for a young adult is 120/80. The higher number is the *systolic pressure,* the pressure recorded in an artery when the left ventricle contracts. The lower number is the

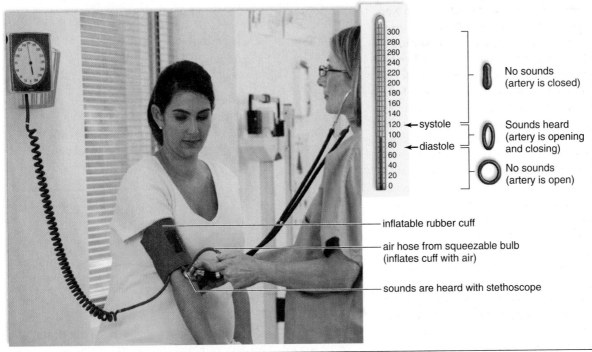

300
280
260
240
220
200
180
160
140
120 ← systole
100
80 ← diastole
60
40
20
0

No sounds
(artery is closed)

Sounds heard
(artery is opening
and closing)

No sounds
(artery is open)

inflatable rubber cuff

air hose from squeezable bulb
(inflates cuff with air)

sounds are heard with stethoscope

Figure 12.15 **Use of a sphygmomanometer.** The technician inflates the cuff with air, gradually reduces the pressure, and listens with a stethoscope for the sounds that indicate blood is moving past the cuff in an artery. This is systolic blood pressure. The pressure in the cuff is further reduced until no sound is heard, indicating that blood is flowing freely through the artery. This is diastolic pressure.

Cardiopulmonary Resuscitation and Automated External Defibrillation

How many people's lives have been saved during cardiac arrest by cardiopulmonary resuscitation (CPR)? The exact number would probably be impossible to track, according to the American Heart Association (AHA). Regrettably, it's easier to track the tragedies that occur without it: Fewer than 30% of people who suffer cardiac arrest ever receive bystander help. Onlookers simply don't know what to do, or perhaps are afraid that they might do something wrong. For this reason, the AHA has published the **Chain of Survival,** a new set of four guidelines that make it easier to help a person in cardiac arrest:

- Recognize that you're in an emergency, **and immediately call for help.** Call 9-1-1, or use the EMS service in your area.
- Start CPR. The traditional method alternates chest compressions with mouth-to-mouth breathing, also known as rescue breathing. If you're not sure how to do rescue breathing, simply pump *fast* (100 or more times a minute) and *hard* (at a depth of least 2 inches) on the body of the sternum. Let the patient's chest rise in between each pump, but don't stop—keep any interruptions to 10 seconds or fewer. Recent research from the AHA has shown that simple manual chest compression can be as effective as traditional CPR (chest compression accompanied by rescue breathing). If you know how to do rescue breathing, use 2 quick breaths

for every 30 chest compressions. Continue CPR until you're relieved by emergency responders.
- If you have access to an automated external defibrillator (AED), use it. This computerized device is available in many public places, such as airports and shopping malls, and is prominently labeled. An AED will explain, step by step, how to check for the person's breathing and a pulse first. Next, it will describe how to apply pads to the victim's chest so that the computer can analyze the heart activity. If a shock is needed to restart the victim's heart, the AED delivers a burst of intense electrical current to the chest. The rescuer simply moves back and pushes a button when prompted. Voice instructions will also detail how to do CPR until paramedics arrive.
- Transfer to advanced care. Clinicians refer to the first hour after admission to an emergency room as the "golden hour." It's not surprising that research has shown that patients who receive the best care during this critical time are the most likely to survive.

The AHA recommends that everyone learn both CPR and how to use an AED. The Red Cross and many hospitals regularly offer introductory and refresher classes in traditional classrooms and online. With training, you just might be able to save a person's life!

---

diastolic pressure, the pressure recorded in an artery when the left ventricle relaxes.

Hypotension, or low blood pressure, may be caused by a number of factors. Simply standing up very quickly causes *orthostatic hypotension,* and a person may temporarily feel light-headed. Normally, the brain's blood pressure control mechanisms rapidly compensate. However, factors such as hemorrhage or excessive fluid loss (for example, following a burn, or as a consequence of vomiting and/or diarrhea) can cause severe hypotension and shock (see page 252). If uncontrolled, shock is fatal.

Currently, an estimated 33% of all Americans suffer from hypertension, or high blood pressure. Most have *essential hypertension,* for which the cause is not precisely known. Hypertension is present when systolic blood pressure is 140 or greater, or diastolic blood pressure is 90 or greater. Hypertension is sometimes a silent killer because it may not be detected until a stroke or heart attack occurs. Among other factors, one's genetic makeup affects the development of hypertension. Two genes are involved: the first codes for the plasma protein angiotensinogen (see page 276), while the second gene's product helps activate angiotensinogen into a powerful vasoconstrictor. Gene therapy might one day cure individuals with this gene combination. Meanwhile, regular blood pressure checks and a lifestyle that lowers hypertension risk are our best safeguards against hypertension (see the Medical Focus on pages 285–286). Medication might also be necessary—often for an entire lifetime.

Content CHECK-UP!

10. Blood flow is slowest through which blood vessels?

   a. arterioles   c. capillaries

   b. veins   d. venules

11. Mean arterial blood pressure is equal to:

   a. cardiac output times peripheral resistance.

   b. cardiac output plus peripheral resistance.

   c. cardiac output/peripheral resistance.

   d. cardiac output minus peripheral resistance.

12. Why does cardiac output increase when you exercise? What effect does this increase have on your blood pressure?

Answers in Appendix A.

12.5 Circulatory Routes

15. Name the two circuits of the cardiovascular system, and trace the path of blood from the heart to any organ in the body and back to the heart.
16. Describe the major systemic arteries and the major systemic veins.
17. Describe several special circulations: blood supply to the liver, blood supply to the brain, and fetal circulation.

Blood vessels belong to either the pulmonary circuit or the systemic circuit. The path of blood through the pulmonary circuit can be traced as follows: Blood from all regions of the body first collects in the right atrium and then passes into the right ventricle, which pumps it into the pulmonary trunk.

The pulmonary trunk divides into the **pulmonary arteries,** which in turn divide into the many arterioles of the lungs. These pulmonary arterioles then take blood to the pulmonary capillaries where carbon dioxide and oxygen are exchanged. The blood then enters the pulmonary venules and flows through the **pulmonary veins** back to the left atrium. Because the blood in the pulmonary arteries is O_2-poor but the blood in the pulmonary veins is O_2-rich, it is not correct to say that all arteries carry blood that is high in oxygen and that all veins carry blood that is low in oxygen. In fact, just the opposite is true in the pulmonary circuit.

The systemic circuit includes all of the other arteries and veins of the body. The largest artery in the systemic circuit is the aorta, and the largest veins are the **superior vena cava** and **inferior vena cava.** The superior vena cava collects blood from the head, chest, shoulders, and upper limbs, and the inferior vena cava collects blood from the lower body regions. Both venae cavae enter the right atrium. The aorta and venae cavae are the major pathways for blood in the systemic system.

The path of systemic blood to any organ in the body begins in the left atrium, which pumps blood into the left ventricle. In turn, the left ventricle pumps blood into the aorta. Branches from the aorta go to the major body regions and organs. Tracing the path of blood to any organ in the body requires mentioning only the aorta, the proper branch of the aorta, the organ, and the returning vein to the vena cava. In many instances, the artery and vein that serve the same organ have the same name. For example, the path of blood to and from the kidneys is: left ventricle; aorta; renal artery; arterioles, capillaries, venules; renal vein; inferior vena cava; right atrium. In the systemic circuit, unlike the pulmonary circuit, arteries contain O_2-rich blood and appear bright red, while veins contain O_2-poor blood and appear dark maroon.

The Major Systemic Arteries

After the aorta leaves the heart, it divides into the *ascending aorta,* the *aortic arch,* and the *descending aorta* (Fig. 12.16). The left and right coronary arteries, which supply blood to the heart, branch off the ascending aorta (Table. 12.1).

Three major arteries branch off the aortic arch: the **brachiocephalic artery** (also called the *brachiocephalic trunk*), the **left common carotid artery,** and the **left subclavian artery.** The brachiocephalic artery is also correctly referred to as the brachiocephalic trunk because it divides into two branches: the **right common carotid** and the **right subclavian arteries.** These blood vessels serve the head (right and left common carotids) and the shoulders and upper limbs (right and left subclavians).

The descending aorta is divided into the *thoracic aorta,* which branches off to the organs within the thoracic cavity, and the *abdominal aorta,* which branches off to the organs in the abdominal cavity.

The descending aorta ends when it divides into the **common iliac arteries** that branch into the **internal iliac arteries** and the **external iliac arteries.** Each internal iliac artery serves the pelvic organs, and the external iliac artery serves the lower limbs. These and other arteries are shown in Figure 12.16.

The Major Systemic Veins

Figure 12.17 shows the major veins of the body. The **external** and **internal jugular veins** drain blood from the brain, head, and neck.

TABLE 12.1 The Aorta and Its Principal Branches

Portion of Aorta	Major Branch	Regions Supplied
Ascending aorta	Left and right coronary arteries	Heart
Aortic arch	Brachiocephalic artery	
	Right common carotid	Right side of head
	Right subclavian	Right shoulder and upper limb
	Left common carotid artery	Left side of head
	Left subclavian artery	Left shoulder and upper limb
Descending aorta		
Thoracic aorta	Intercostal arteries	Thoracic wall
Abdominal aorta	Celiac artery	Stomach, spleen, and liver
	Superior mesenteric artery	Small and large intestines (ascending and transverse colons)
	Left and right renal arteries	Kidneys
	Left and right gonadal arteries	Ovaries or testes
	Inferior mesenteric artery	Lower digestive system (transverse, descending and sigmoid colons, and rectum)
	Common iliac arteries	Pelvic organs and lower limbs

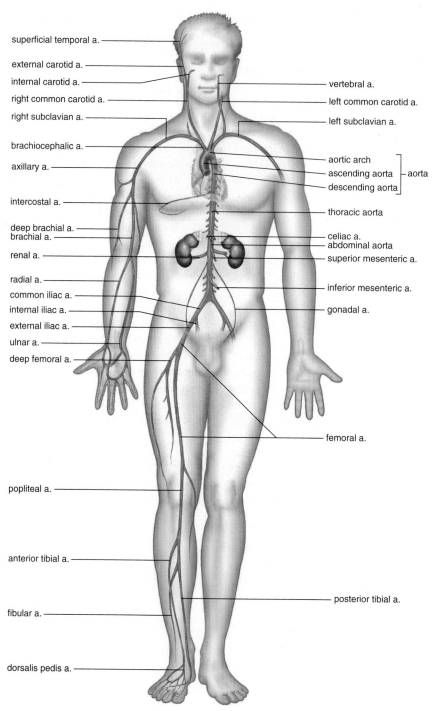

Figure 12.16 Major arteries (a.) of the body.

superficial temporal a.

external carotid a.

internal carotid a.

right common carotid a.

right subclavian a.

brachiocephalic a.

axillary a.

intercostal a.

deep brachial a.

brachial a.

renal a.

radial a.

common iliac a.

internal iliac a.

external iliac a.

ulnar a.

deep femoral a.

popliteal a.

anterior tibial a.

fibular a.

dorsalis pedis a.

vertebral a.

left common carotid a.

left subclavian a.

aortic arch

ascending aorta — aorta

descending aorta

thoracic aorta

celiac a.

abdominal aorta

superior mesenteric a.

inferior mesenteric a.

gonadal a.

femoral a.

posterior tibial a.

Each external jugular vein enters a **subclavian vein.** In turn, the subclavian veins join with the internal jugular veins to form the brachiocephalic veins. Right and left brachiocephalic veins merge, giving rise to the superior vena cava.

In the abdominal cavity, paired veins return blood from bilateral structures such as the kidneys and gonads. In addition, as discussed in more detail later, the **hepatic portal vein** receives blood from the stomach, spleen, intestines, and other abdominal organs,

and then enters the liver. Emerging from the liver, the **hepatic veins** enter the inferior vena cava.

In the pelvic region, veins from the various organs enter the **internal iliac** veins, while the veins from the lower limbs enter the **external iliac veins.** The internal and external iliac veins become the **common iliac veins** that merge, forming the inferior vena cava. Table 12.2 lists the principal veins that enter the venae cavae.

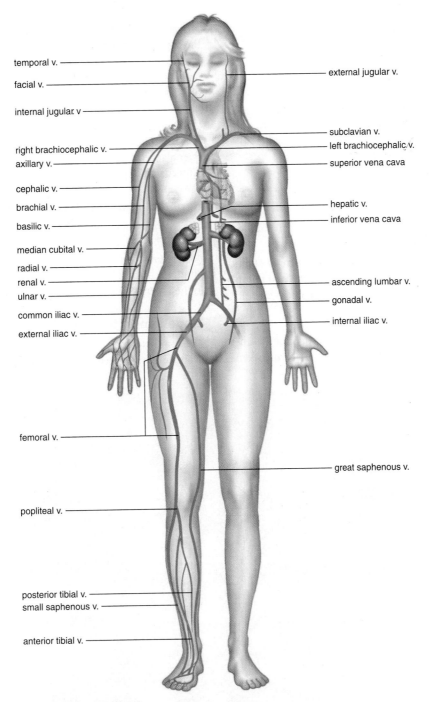

temporal v.
facial v.
internal jugular v
right brachiocephalic v.
axillary v.
cephalic v.
brachial v.
basilic v.
median cubital v.
radial v.
renal v.
ulnar v.
common iliac v.
external iliac v.
femoral v.
popliteal v.
posterior tibial v.
small saphenous v.
anterior tibial v.

external jugular v.
subclavian v.
left brachiocephalic v.
superior vena cava
hepatic v.
inferior vena cava
ascending lumbar v.
gonadal v.
internal iliac v.
great saphenous v.

AP|R **Figure 12.17** **Major veins (v.) of the body.**

TABLE 12.2 Principal Veins That Join the Venae Cavae

Vein	Region Drained	Vena Cava
Right and left brachiocephalic veins	Head, neck, and upper limbs	Form superior vena cava
Right and left common iliac veins	Lower limbs and pelvic organs	Form inferior vena cava
Right and left renal veins	Kidneys	Enters inferior vena cava
Right gonadal vein	Right ovary or testis	Enters inferior vena cava
Left gonadal vein	Left ovary or testis	Left renal vein to inferior vena cava
Right and left hepatic veins	Liver, digestive tract, and spleen	Enters inferior vena cava

Special Systemic Circulations

Hepatic Portal System

The **hepatic portal system** (Fig. 12.18) carries blood from the stomach, intestines, spleen, and other organs to the liver. The term *portal system* is used to describe the following unique pattern of circulation:

$$\text{capillaries} \rightarrow \text{vein} \rightarrow \text{capillaries} \rightarrow \text{vein}$$

Capillaries of the digestive tract drain into the superior mesenteric and inferior mesenteric veins and the splenic vein, which join to form the hepatic portal vein. The gastric veins empty directly into the hepatic portal vein. The hepatic portal vein carries blood to capillaries in the liver. The hepatic capillaries allow nutrients and wastes to diffuse into liver cells for further processing. In addition, colonies of white blood cells destroy any pathogens that may be present in the blood. Then, hepatic capillaries join to form venules that enter a hepatic vein. The hepatic veins enter the inferior vena cava.

In addition to receiving venous blood from the intestine, the liver also receives arterial blood via the hepatic artery. The hepatic artery is not a part of the hepatic portal system.

Hypothalamus-Pituitary Portal System

The hypothalamus-pituitary portal system (illustrated in the Visual Focus, Fig. 10.4) is an important endocrine portal system. This portal venous arrangement links the hypothalamus to the anterior pituitary. Through it, the hypothalamus sends releasing hormones to the anterior pituitary.

Blood Supply to the Brain

The brain is supplied with O_2-rich blood by the anterior and posterior cerebral arteries and the carotid arteries. These arteries give

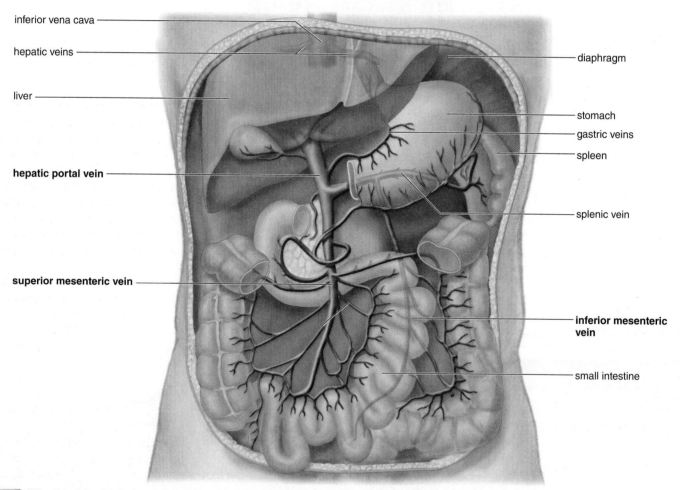

AP|R **Figure 12.18 Hepatic portal system.** This system provides venous drainage of the digestive organs and takes venous blood to the liver.

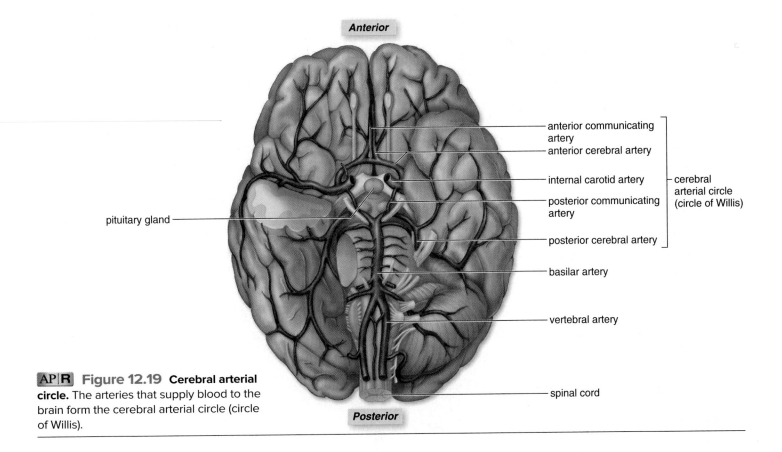

Anterior

anterior communicating artery

anterior cerebral artery

internal carotid artery

posterior communicating artery

pituitary gland

posterior cerebral artery

basilar artery

vertebral artery

spinal cord

Posterior

cerebral arterial circle (circle of Willis)

AP|R **Figure 12.19** **Cerebral arterial circle.** The arteries that supply blood to the brain form the cerebral arterial circle (circle of Willis).

off branches that join to form the **cerebral arterial circle** (circle of Willis), a vascular route in the region of the pituitary gland (Fig. 12.19). Because the blood vessels form a circle, alternate routes are available for bringing arterial blood to the brain and thus supplying the brain with oxygen. The presence of the cerebral arterial circle also equalizes blood pressure in the brain's blood supply.

Fetal Circulation

As Figure 12.20 shows, the fetus has four circulatory features that are not present in adult circulation:

1. The **foramen ovale,** or *oval window,* is an opening between the two atria. This window is covered by a flap of tissue that acts as a valve.
2. The **ductus arteriosus,** or *arterial duct,* is a connection between the pulmonary artery and the aorta. It is found on the superior pulmonary trunk near the origin of the left pulmonary artery.
3. The **umbilical arteries** and **vein** are vessels that travel to and from the placenta, leaving waste and receiving nutrients.
4. The **ductus venosus,** or *venous duct,* is a connection between the umbilical vein and the inferior vena cava.

All of these features can be related to the fact that the fetus does not use its lungs for gas exchange—it receives oxygen and nutrients from the mother's blood at the placenta. During development, the lungs receive only enough blood to supply their developmental need for oxygen and nutrients.

The path of blood in the fetus can be traced, beginning from the right atrium (Fig. 12.20). Most of the blood that enters the right atrium passes directly into the left atrium by way of the foramen ovale because the blood pressure in the right atrium is somewhat greater than that in the left atrium. The rest of the fetal blood entering the right atrium passes into the right ventricle and out through the pulmonary trunk. However, because of the ductus arteriosus, most pulmonary trunk blood passes directly into the aortic arch. Notice that whatever route blood takes, most of it reaches the aortic arch instead of the pulmonary circuit vessels.

Blood within the aorta travels to the various branches, including the iliac arteries, which connect to the umbilical arteries leading to the placenta. Diffusion of oxygen, nutrients, carbon dioxide, and metabolic wastes between maternal and fetal blood takes place at the placenta. There, oxygen and nutrients diffuse from mother to fetus, while carbon dioxide and waste simultaneously travel from fetus to mother. It's important to note that maternal and fetal blood do not mix at the placenta during a normal pregnancy. Oxygenated blood from the placenta then returns to the fetus's body through the umbilical vein.

Thus, blood in the umbilical arteries is O_2-poor, but blood in the umbilical vein, which travels from the placenta, is O_2-rich. The umbilical vein enters the ductus venosus, which passes directly through the liver. The ductus venosus then joins with the inferior vena cava, a vessel that contains O_2-poor blood. The vena cava returns this mixture to the right atrium.

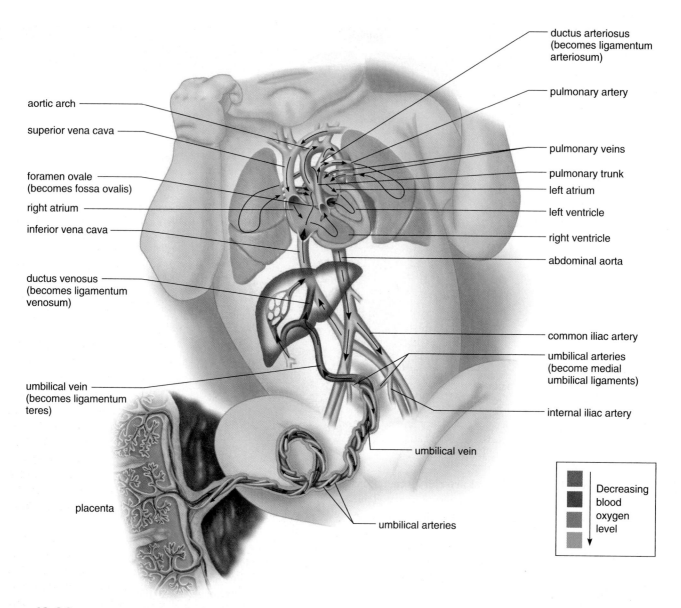

Figure 12.20 Fetal circulation. Arrows indicate the direction of blood flow. The lungs are not functional in the fetus, but they are developing. The blood passes directly from the right atrium to the left atrium via the foramen ovale or from the right ventricle to the aorta via the pulmonary trunk and ductus arteriosus. The umbilical arteries take fetal blood to the placenta where exchange of molecules between fetal and maternal blood takes place. Oxygen and nutrient molecules diffuse into the fetal blood, and carbon dioxide and urea diffuse from the fetal blood. The umbilical vein returns blood from the placenta to the fetus.

Changes at Birth Sectioning and tying the umbilical cord permanently separates the newborn from the placenta. The newborn's first breath inflates the lungs, and oxygen enters the blood at the lungs instead of the placenta. O₂-rich blood returning from the lungs to the left side of the heart usually causes a flap on the left side of the interatrial septum to close the foramen ovale. What remains is a depression called the *fossa ovalis*. Incomplete closure occurs in nearly one out of four individuals, but even so, blood rarely passes from the right atrium to the left atrium because either the opening is small or it closes when the atria contract. In a small number of cases, the passage of O₂-poor blood from the right side to the left side of the heart is sufficient to cause **cyanosis,** a bluish cast to the skin. This condition can now be corrected by open-heart surgery.

The fetal blood vessels and shunts constrict and become fibrous connective tissue called ligamentums in all cases except the distal portions of the umbilical arteries, which become the medial umbilical ligaments. Regardless, these structures run between internal organs. For example, the ligamentum teres (which is the remnant of the umbilical vein) attaches the umbilicus to the liver.

All of us can take steps to prevent cardiovascular disease, the most frequent cause of death in the United States. Genetic factors that predispose an individual to cardiovascular disease include family history of heart attack under age 55, male gender, and ethnicity (African Americans are at greater risk). However, people with one or more of these risk factors don't have to give up. It only means that they need to pay particular attention to the following guidelines for a heart-healthy lifestyle.

The Don'ts

Smoking

Hypertension is recognized as a major contributor to cardiovascular disease. When a person smokes, the drug nicotine, present in cigarette smoke, enters the bloodstream. Nicotine causes the arterioles to constrict and the blood pressure to rise. Restricted blood flow and cold hands are associated with smoking in most people. Cigarette smoke also contains carbon monoxide, and hemoglobin combines preferentially and nonreversibly with carbon monoxide. Therefore, the presence of carbon monoxide lowers the oxygen-carrying capacity of the blood, and the heart must pump harder to propel the blood through the lungs. Smoking also damages the arterial wall and accelerates the formation of atherosclerosis and plaque.

Drug Abuse

Stimulants, such as cocaine and amphetamines, can cause an irregular heartbeat and lead to heart attacks even in people who are using drugs for the first time. Intravenous drug use may also result in a cerebral blood clot and stroke.

Too much alcohol can destroy just about every organ in the body, the heart included. But investigators have discovered that people who take an occasional drink have a 20% lower risk of heart disease than do those who completely abstain from alcohol. Two to four drinks a week is the recommended limit for men; one to three drinks is the recommendation for women.

Research has shown that wines (and especially red wine) contain antioxidants which can further help to reduce the risk of cardiovascular damage.

Weight Gain

Hypertension also occurs more often in persons who are more than 20% above the recommended weight for their height. Because more tissue requires servicing, the heart must send extra blood out under greater pressure in those who are overweight. It may be very difficult to lose weight once it is gained, and therefore weight control should be a lifelong endeavor. Even a slight decrease in weight can bring a reduction in hypertension. A 4.5-kilogram weight loss doubles the chance that blood pressure can be normalized without drugs.

The Do's

Healthy Diet

Of all the possible dietary changes one could make to prevent cardiovascular disease, the most important will be to switch to a diet low in saturated fats, trans fats, and cholesterol. These three are found most often as the "solid" forms of fat: butter, margarine, shortening, lard, and the marbling found on fatty meat. Trans fats can be found in baked goods as well. Instead, replace these fats with monounsaturated fats (found in nuts, olives, and olive oil) polyunsaturated fats (found in vegetable oils) and omega-3 fatty acids (found in fish, especially fatty fish such as salmon, mackerel, and trout). Healthy fats like these can help to lower total cholesterol and low-density lipoprotein (LDL), while boosting high-density lipoprotein (HDL). Those changes are very important: In our bodies, cholesterol is ferried in the blood by LDL and HDL. LDL (often referred to as "bad cholesterol" by physicians) takes cholesterol from the liver to the tissues. Recall (from the Medical Focus on pages 267–268) that in blood vessels, LDL is oxidized by macrophages to form foam cells, and later, a fatty streak that begins an atherosclerotic plaque. HDL ("good" cholesterol) transports cholesterol out of the tissues to the liver, where cholesterol is metabolized. When the LDL level in the blood is abnormally high or the HDL level is abnormally low, cholesterol accumulates in artery walls. Substituting heart-healthy fats in the diet has been shown to decrease the risk of cardiovascular disease—but it's important to recognize that all fats are very calorie-dense, and must be used in moderation to avoid weight gain.

Limiting sodium intake is another important step to take in modifying diet. In 2009, data from the Centers for Disease Control showed that approximately 70% of Americans are salt-sensitive, and their blood pressure tends to rise after consumption of excess sodium. In response, the American Heart Association lowered its recommended daily sodium intake to less than 1500 milligrams (mg). Almost all processed food contains sodium, so watch those nutrition labels!

Here's another recommendation from the American Heart Association, and it may surprise you how few Americans actually do it. Only about 27% of us eat those five servings of fruits and vegetables daily. Evidence is mounting to suggest a role for antioxidant vitamins (A, E, and C) in the prevention of cardiovascular disease, and the best sources are in plant-based food. Antioxidants protect the body from free radicals that may damage HDL cholesterol through oxidation or damage the lining of an artery, leading to a blood clot that can block the vessel. It's not all about vitamins either: By creating a feeling of fullness, all that fiber found in fruits and vegetables can help with weight loss.

Regular Health Screenings

It is recommended that everyone know his or her blood cholesterol level, as well as levels of HDL, LDL, and triglyceride. Individuals with a high

—Continued

Continued—

total blood cholesterol level (240 mg/100 ml) should be further tested to determine their LDL cholesterol level. Blood tests can also determine high levels of homocysteine and C-reactive protein, two important indicators that atherosclerosis is occurring. Levels of these two markers, along with LDL cholesterol level, must be considered together with other cardiovascular risk factors such as age, family history, general health, and whether the patient smokes. Further, for those with a strong family history of cardiovascular disease, testing may be recommended for elevated levels of fibrinogen and lipoprotein (a), two blood components that may indicate atherosclerosis. A person with a moderate risk of cardiovascular disease may have success with dietary therapy to lower LDL. Cholesterol-lowering drugs are reserved for high-risk patients.

Proper Dental Care

Periodontal disease, the inflammation of the gums caused by poor dental hygiene, has been suggested as a cause for atherosclerosis. Scientists suspect that people with tooth decay and gum disease may have a low-level bacterial infection in the blood—not severe enough to cause illness, but enough to injure the endothelial lining and start the formation of atherosclerotic plaques. Proper care of the teeth and gums, along with regular visits to the dentist, just might prevent cardiovascular disease.

Exercise

People who exercise are less apt to have cardiovascular disease. Exercise helps keep weight under control, may help minimize stress, and reduces hypertension. The heart beats faster when exercising, but exercise slowly increases the heart's capacity. This means that the heart can beat more slowly when we're at rest and still do the same amount of work. The American Heart Association recommends at least 150 minutes per week of moderate exercise: 30 minutes a day, for 5 days per week. Those 30 minutes can even be broken into three 10-minute sessions, making regular daily exercise an attainable goal for practically everyone. In addition, practice meditation and yoga-like stretching and breathing exercises to reduce stress.

For more information about cardiovascular disease prevention, visit the American Heart Association website: http://www.heart.org

---

Content CHECK-UP!

13. The superior and inferior venae cavae return their blood to the:

  a. right atrium.   c. left atrium.

  b. pulmonary artery.   d. pulmonary veins.

14. In an adult heart, the oxygen content is highest in blood found in the:

  a. pulmonary arteries.   c. pulmonary veins.

  b. pulmonary trunk.   d. right ventricle.

15. Occasionally, the ductus arteriosus fails to close in a newborn baby. What happens to the pulmonary circuit when this occurs?

Answers in Appendix A.

12.6 Effects of Aging

18. Describe the anatomical and physiological changes that occur in the cardiovascular system as we age.

The heart generally grows larger with age, primarily because of fat deposition in the epicardium and myocardium. In many middle-aged people, the heart is covered by a layer of fat, and the number of collagenous fibers in the endocardium increases. With age, the valves, particularly the aortic semilunar valve, become thicker and more rigid.

As a person ages, the myocardium loses some of its contractile power and some of its ability to relax. The resting heart rate decreases throughout life, and the maximum possible rate during exercise also decreases. With age, the contractions become less forceful; the heart loses about 1% of its reserve pumping capacity each year after age 30.

In the elderly, arterial walls tend to thicken with plaque and become inelastic, signaling that atherosclerosis and arteriosclerosis are present. Increased blood pressure was once believed to be inevitable with age, but now hypertension is known to result from other conditions, such as kidney disease and atherosclerosis. The Medical Focus on pages 285–286 describes how diet and exercise in particular can help prevent atherosclerosis.

Myocardial infarction (described in the Medical Focus box on pages 267–268) and other diseases related to atherosclerosis increase in frequency as a person ages. Congestive heart failure can result from myocardial infarction.

In **congestive heart failure,** a damaged left side of the heart fails to pump adequate blood, and blood backs up in the pulmonary veins. Therefore, pulmonary blood vessels have become *congested.* The congested vessels leak fluid into tissue spaces, causing **pulmonary edema.** The result is shortness of breath, fatigue, and a constant cough with pink, frothy sputum. Treatment consists of the three Ds: diuretics (to increase urinary output), digoxin (to increase the heart's contractile force), and dilators (to relax the blood vessels). If necessary, a surgically implanted mechanical pump called a *left ventricular assist device* (LVAD) can help to maintain the pumping ability of the damaged left ventricle until it can recover. In some cases, heart transplant is also an option.

The occurrence of varicose veins increases with age, particularly in people who are required to stand for long periods. Thromboembolism as a result of varicose veins can lead to death if a blood clot settles in a major branch of a pulmonary artery. (This disorder is called pulmonary embolism.)

What's New

Novel Stent for the Severest Strokes

Imagine you are a physician, nurse, or other health-care provider in a busy family practice center, when a 68-year-old patient comes into your office complaining that the right side of his face is numb and his right arm is tingling. His wife reports that he has had trouble answering her questions all morning. On physical examination, you can see that his right-side facial muscles are functioning weakly, and his face is drooping. When asked to smile, he can only raise the corner of his mouth on the left side, and when asked to stick his tongue straight out, you can see that it bends to the left. His previous history includes all of the risk factors for atherosclerosis: he's an overweight, sedentary type II diabetic, with a 20-year history of cigarette smoking and poorly controlled hypertension. During his most recent physical exam several months ago, you lectured him about his elevated blood LDL, triglyceride, homocysteine, and CRP levels. You explained that his lifestyle might lead to hardening and narrowing of his inflamed arteries, and warned him that atherosclerosis causes heart attacks and strokes. You know that he is currently experiencing a **transient ischemic attack (TIA),** and his symptoms are a warning that he is at tremendous risk for a **cerebrovascular accident (CVA),** or stroke. Without hesitating, you send him to the nearby hospital known for its excellent stroke care.

Acting fast in this scenario is critical for the patient's survival and continued quality of life. Just as a myocardial infarction causes death to heart muscle, a cerebrovascular accident results in the death of nerve cells. Scientists estimate that for each minute that delicate brain tissue is deprived of oxygenated blood, up to 1.9 million neurons are lost. Stroke is a leading cause of death and disability in the United States, affecting more than 800,000 people every year. A small percentage of strokes occur when an intracranial blood vessel bursts, causing a **hemorrhagic stroke.** However, the vast majority of strokes are caused by clots that form in cerebral arteries. As you know from the Medical Focus on pages 267–268, atherosclerotic plaques often trigger the clotting cascade. **Thrombotic stroke** results from a stationary clot in a cerebral artery. If the clot forms elsewhere in the body, then breaks off and blocks a cerebral vessel, it causes **embolic stroke.** Patients with smaller clots can often be successfully treated with tPA, a drug that

Figure 12F Clot-Removing Stent

dissolves clots. However, because large clots often don't respond to tPA treatment, the stroke victim may die or be seriously impaired.

A new catheter stent device shows promise for removing these large clots (Fig. 12F). The catheter is inserted into the femoral artery and threaded up to the blocked cerebral artery until it reaches the clot. Next, a tiny metal cage at the end of the stent is opened and pushed into the clot. Once the clot has been captured, the catheter is withdrawn, pulling the clot along with it. In a recent clinical study of stroke patients, 20% of the patients who received tPA alone recovered enough to return to independent living. However, of those patients treated with both tPA and the clot-removing stent, fully 33% were able to resume their normal daily activities. Furthermore, subsequent CT scans showed less brain damage in patients who received the stent treatment. Neurologists and neurosurgeons are optimistic that the device will greatly improve available care for those patients at greatest risk of brain damage.

It's important to remember that stroke and heart attack risk can be lessened by minimizing or avoiding their known risk factors. See the Medical Focus on pages 285–286 for a complete discussion on preventing cardiovascular disease.

Content **CHECK-UP!**

16. In the elderly, the aortic valve can be come thicker and more rigid. Which chamber of the heart pumps blood through this valve? What happens to the chamber?

Answer in Appendix A.

12.7 Homeostasis

19. Describe how the cardiovascular system works with other systems of the body to maintain homeostasis.

Homeostasis is possible only if the cardiovascular system delivers oxygen and nutrients to and takes metabolic wastes from the tissue fluid surrounding the cells. Human Systems Work Together on page 288 summarizes how the cardiovascular system works with other systems of the body to maintain homeostasis.

Maintaining Blood Composition, pH, and Temperature

As you study Human Systems Work Together, it's important to understand that the composition of the blood is maintained by the other systems of the body. The red bone marrow of the skeletal system contributes red blood cells for O_2 and CO_2 transport. The respiratory

Integumentary System

Blood vessels deliver nutrients and oxygen to skin; carry away wastes; blood clots if skin is broken.

Skin prevents water loss; helps regulate body temperature; protects blood vessels.

Skeletal System

Blood vessels deliver nutrients and oxygen to bones; carry away wastes.

Rib cage protects heart; red bone marrow produces blood cells; bones store Ca^{2+} for blood clotting.

Muscular System

Blood vessels deliver nutrients and oxygen to muscles; carry away wastes.

Muscle contraction keeps blood moving in heart and blood vessels.

Nervous System

Blood vessels deliver nutrients and oxygen to neurons; carry away wastes.

Brain controls nerves that regulate the heart and dilation of blood vessels.

Endocrine System

Blood vessels transport hormones from glands; blood services glands; heart produces atrial natriuretic hormone.

Epinephrine increases blood pressure; ADH, aldosterone, and atrial natriuretic hormone factors help regulate blood volume; growth factors control blood cell formation.

How the Cardiovascular System works with other body systems.

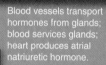
Capillaries

Heart

Vein

Artery

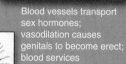

Reproductive System

Blood vessels transport sex hormones; vasodilation causes genitals to become erect; blood services reproductive organs.

Sex hormones influence cardiovascular health; sexual activities stimulate cardiovascular system.

Lymphatic System/Immunity

Blood vessels transport leukocytes and antibodies; blood services lymphatic organs and is source of tissue fluid that becomes lymph.

Lymphatic vessels collect excess tissue fluid and return it to blood vessels; lymphatic organs store lymphocytes; lymph nodes filter lymph, and the spleen filters blood.

Respiratory System

Blood vessels transport gases to and from lungs; blood services respiratory organs.

Gas exchange in lungs rids body of carbon dioxide, helping to regulate the pH of blood; breathing aids venous return.

Digestive System

Blood vessels transport nutrients from digestive tract to body; blood services digestive organs.

Digestive tract provides nutrients for plasma protein formation and blood cell formation; liver detoxifies blood, makes plasma proteins, destroys old red blood cells.

Urinary System

Blood vessels deliver wastes to be excreted; blood pressure aids kidney function; blood services urinary organs.

Kidneys filter blood and excrete wastes; maintain blood volume, pressure, and pH; produce renin and erythropoietin.

system is responsible for gas exchange. Both the respiratory and urinary systems cooperate to maintain proper pH. The digestive system supplies nutrients, and the liver detoxifies waste. Skeletal muscle contraction supplies heat to keep us warm, and the integumentary system allows us to eliminate any excess heat. The cardiovascular system is the body's transportation mechanism: O_2 and nutrients to body tissues, CO_2 and wastes to lungs and kidneys, cooler blood to the skeletal muscles to be warmed, warmed blood to the skin to be cooled, and so on. Thus, the body's O_2-CO_2 level, nutrient distribution, pH, and temperature remain relatively constant.

Maintaining Blood Pressure

The pumping of the heart is critical to creating the blood pressure that moves blood throughout the body. The importance of the heart to survival can be seen in the speed with which it develops during prenatal life. Long before other major organs, the heart and its vessels have taken shape and are ready to function.

The body has multiple ways to maintain blood pressure. Blood vessel sensory receptors signal the brain when blood pressure falls. The brain's regulatory center subsequently increases heartbeat and constricts blood vessels, and blood pressure is restored. The lymphatic system collects excess tissue fluid, returning it to the cardiovascular system. In this way, the lymphatic system makes an important contribution to regulating blood volume and pressure.

The endocrine system assists the nervous system in regulating blood pressure. Epinephrine and norepinephrine act on blood vessels in variable ways, depending on the region of the body in which the blood vessels are found. Norepinephrine causes vasoconstriction, while epinephrine can cause either vasoconstriction or vasodilation. Aldosterone, ADH, and ANH (atriopeptide) regulate urine excretion. Retention of water helps to raise blood pressure, while water excretion helps to lower blood pressure.

Venous return from the capillaries to the heart is assisted by the muscular and respiratory systems. Skeletal muscle contraction pushes blood past the vein valves, and breathing movements encourage the flow of blood toward the heart. Smooth muscle in the walls of arterioles constricts and helps raise blood pressure. Platelets are necessary to blood clotting, preventing excessive bleeding and loss of pressure.

An individual who loses more than 10% of his or her blood will suffer a sudden drop in blood pressure and go into shock. The decreased pressure triggers the body's last defense: A powerful wave of sympathetic impulses constricts the veins and arterioles throughout the body to slow the drop in blood pressure. Heart rate soars as high as 200 beats a minute to maintain blood flow, especially to the brain and the heart itself. Because of this reflex, you might lose as much as 40% of your total blood volume and still live.

Content **CHECK-UP!**

17. During exercise, the muscles generate heat. What system assists the cardiovascular system in ridding the body of excess heat? How does this work?

Answer in Appendix A.

Selected New Terms

Basic Key Terms

aorta (ā-ōr′tŭh), p. 263

aortic semilunar valve (ā-ōr′tĭk sĕm′ ī-lū″ nĕr vălv), p. 263

arteriole (ăr-tē′rē-ōl), p. 271

artery (ăr′tĕr-ē), p. 271

atria (ā′ trē-ŭh), p. 261

atrioventricular (AV) node (ā″trē-ō-vĕn-trĭk′ū-lĕr nōd), p. 264

atrioventricular bundle (ā″trē-ō-vĕn-trĭk′ū-lĕr bŭn′dl), p. 264

atrioventricular valve (ā″trē-ō-vĕn-trĭk′ū-lĕr vălv), p. 262

atrium (ā′trē-ŭm), p. 261

auricle (ōr′ŭh-kŭl), p. 261

bicuspid valve (bī-kŭs′pĭd vălv), p. 263

blood pressure (blŭhd prĕ′ shĕr), p. 274

capillary (kăp′ĭ-lār″ē), p. 271

cardiac cycle (kăr′dē-ăk sī′kŭl), p. 265

cardioregulatory center (kăr″dē-ō-rĕg′ū-lŭh-tōr-ē sĕn′tĕr), p. 269

cerebral arterial circle (sĕr′ē-brăl ăr-tē′rē-ăl sĕr′kl), p. 283

chordae tendineae (kŏr′dē tĕn′dĭ-nē), p. 262

conduction system (kŭhn′ dŭk-shŭn sĭs′ tĕm), p. 264

coronary artery (kŏr′ŭh-nĕr-ē″ ăr′tĕr-ē), p. 263

diastole (dī-ăs′tō-lē), p. 265

ductus arteriosus (dŭk′tŭs ăr-tēr-ē-ō′sŭs), p. 283

ductus venosus (dŭk′tŭs vē-no′sŭs), p. 283

endocardium (ĕn″dō-kăr′dē-ŭm), p. 261

endothelium (ĕn″dō-thē′ lē-ŭhm), p. 261

epicardium (ĕp″ĭ-kăr′dē-ŭm), p. 261

foramen ovale (fō-rā′mĕn ō-vā′lē), p. 283

gap junction (găp jŭngk′ shŭn), p. 265

hepatic portal system (hĕ-păt′ĭk pŏr′tăl sĭs′tĕm), p. 282

inferior vena cava (ĭn-fēr′ē-ŏr vē′nŭh kă′vŭh), p. 279

interatrial septum (ĭn″tĕr-ā′trē-ăl sĕp′tŭm), p. 261

interventricular septum (ĭn″tĕr-vĕn-trĭk′ū-lĕr sĕp′tŭm), p. 261

macrophage (măk′rō-făj), p. 267

mitral valve (mī′trl vălv), p. 263

myocardium (mī″ō-kăr′dē-ŭm), p. 261

pacemaker (pās′mā′kŭhr), p. 265

papillary muscles (păp′ŭh-lĕr″ē mŭs′ĕlz), p. 262

pulmonary artery (pŭhl′ mŭh-nār″ ē ăr′tĕr-ē), p. 279

pulmonary circuit (pŭhl′ mŭh-nār″ ē sĕr′kŭt), p. 260

pulmonary semilunar valve (pŭhl′ mŭh-nār″ ē sĕm′ ī-lū″ nĕr vălv), p. 262

pulmonary vein (pŭhl′ mŭh-nār″ ē vān), p. 279

pulse (pŭls), p. 276

Purkinje fiber (pĕr-kĭn′jē fī′bĕr), p. 264

renin (rē′nĭn), p. 276

respiratory pump (rĕs′pŭr-ŭh-tōr-ē pŭmp), p. 275

semilunar valve (sĕm′ ī-lū″ nĕr vălv), p. 262

sinoatrial (SA) node (sī″nō-ā′trē-ŭl nōd), p. 264

skeletal muscle pump (skĕ′ lē-tŭl mŭh′ sŭl pŭmp), p. 275

superior vena cava (sū-pēr′ē-ŏr vē′nŭh kă′vŭh), p. 279

systemic circuit (sĭs-tĕm′ĭk sĕr′kŭt), p. 260

systole (sĭs′tō-le), p. 265

tissue fluid (ti′ shōō flōo′ ĭd), p. 271

Clinical Key Terms

Summary

12.1 Anatomy of the Heart

The heart keeps oxygen-poor blood separate from oxygen-rich blood and keeps blood flowing in one direction. It creates blood pressure and regulates the supply of blood to meet current needs. The right side of the heart pumps blood to the lungs via the pulmonary circuit; the left side pumps blood to the tissues via the systemic circuit.

A. The heart is covered by the pericardium. The visceral pericardium also functions as the epicardium of the heart wall. The myocardium is made of cardiac muscle, and the endocardium is the heart's inner lining.

B. The heart has right and left sides and four chambers: two atria and two ventricles. There are two atrioventricular valves separating the atria from the ventricles: the right tricuspid valve and the left bicuspid, or mitral, valve. On the right side, the pulmonary semilunar valve controls blood flow from the right ventricle into the pulmonary artery. On the left side, the aortic semilunar valve controls blood flow from the left ventricle into the aorta.

C. The first heart sound, "lub," is caused by AV valve closure. The second heart sound, "dup," occurs when the semilunar valves close. When valves don't work properly, they can be replaced with synthetic or pig valves.

D. The myocardium is supplied with blood from the coronary circulation. Coronary artery disease is caused by atherosclerosis.

12.2 Physiology of the Heart

A. The conduction system of the heart includes the SA node, the AV node, the AV bundle, the bundle branches, and the Purkinje fibers. The SA node causes the atria to contract. The AV node and the rest of the conduction system cause the ventricles to contract.

B. The heartbeat (cardiac cycle) is divided into three phases: (1) In atrial systole, the atria contract; (2) in ventricular systole, the ventricles contract; and (3) in atrial and ventricular diastole, both the atria and the ventricles rest.

C. The cardiac output (amount of blood discharged by the heart in one minute) is the product of stroke volume and heart rate. The heart rate is regulated largely by the cardioregulatory center and the autonomic nervous system.

12.3 Anatomy of Blood Vessels

Blood vessels transport blood, carry out exchange in pulmonary and systemic capillaries, regulate blood pressure, and direct blood flow.

A. Arteries and arterioles carry blood away from the heart.

B. Capillaries are made of endothelial cells and have precapillary sphincters.

C. Capillary exchange occurs via a balance of hydrostatic and osmotic pressures. At the arterial end of a capillary, outward hydrostatic pressure is greater and tissue fluid accumulates. At the venous end, osmotic pressure is greater and tissue fluid is reabsorbed.

12.4 Physiology of Circulation

A. Velocity of blood flow varies according to total cross-sectional area; therefore, blood flow is slowest in the capillaries.

B. Blood pressure decreases with distance from the left ventricle. Cardiac output (CO) and resistance to flow determine blood pressure. Venous return affects CO. The skeletal muscle pump and the respiratory pump assist venous return. A vasomotor center regulates peripheral resistance. Neural regulation of peripheral resistance is via a vasomotor center in the medulla that is under the control of the cardioregulatory center. Several different hormones regulate blood pressure through their influence over the kidney's reabsorption of water.

C. To evaluate a person's circulation, it is customary to take the pulse and blood pressure.

12.5 Circulatory Routes
The pulmonary arteries transport oxygen-poor blood to the pulmonary capillaries, and the pulmonary veins return oxygen-rich blood to the heart. In the systemic circuit, blood travels from the left ventricle to the aorta, systemic arteries, arterioles, and capillaries, and then from the capillaries to the venules and veins to the right atrium of the heart. The systemic circuit serves the body proper.

A. Blood leaves the left ventricle and enters the ascending aorta, which sends smaller branches off to supply various parts of the body, including the heart itself.

B. All systemic veins drain into the inferior and superior venae cavae as well as the coronary sinus.

C. The hepatic portal system carries blood from the stomach and intestines to the liver. Circulation to the brain includes the cerebral arterial circle, which protects all regions of the brain from reduced blood supply.

D. Fetal circulation includes four unique features: (1) the foramen ovale, (2) the ductus arteriosus, (3) the umbilical arteries and vein, and (4) the ductus venosus. These features are necessary because the fetus does not use its lungs for gas exchange.

12.6 Effects of Aging
Aging increases the risk of atherosclerosis and arteriosclerosis. Congestive heart failure is common in the elderly and causes blood to back up in the pulmonary vessels.

12.7 Homeostasis
The cardiovascular system is the body's transportation mechanism. It assures exchange at the pulmonary capillaries and the systemic capillaries. There are many examples of the interaction of the cardiovascular system with other systems. For example, the endocrine system is dependent on the cardiovascular system to transport its hormones, and hormones help maintain blood pressure. Blood vessels deliver wastes to the kidneys, and the kidneys help maintain blood pressure. The respiratory system is dependent on the cardiovascular system to transport gases to and from cells, and the respiratory system assists venous return.

Study Questions

1. State the location and functions of the heart. (p. 260)
2. Describe the wall and coverings of the heart. (p. 261)
3. Name the chambers and valves of the heart. Trace the path of blood through the heart. (pp. 261–263)
4. Describe the coronary circuit, and discuss several coronary circuit disorders. (pp. 263, 267–268)
5. Describe the conduction system of the heart and how conduction can be recorded using an electrocardiogram. (pp. 264–265, 269)
6. Describe the cardiac cycle (using the terms *systole* and *diastole*), and explain the heart sounds. (pp. 265–266)
7. What is cardiac output (CO)? What two factors determine CO? How are these factors regulated? (pp. 266, 269–271)

8. What types of blood vessels are in the body? Discuss their structure and function. (pp. 271–274)
9. What factors determine velocity of blood flow? Blood pressure? In what vessel is blood pressure highest? Lowest? (pp. 274–275)
10. What mechanisms assist venous return to the heart? Discuss nervous and hormonal control of blood pressure. (pp. 275–276)
11. What is pulse? How do you take a person's pulse? How do you take a person's blood pressure? What does a blood pressure of 120/80 mean? (pp. 276–278)
12. What are hypertension, stroke, aneurysm, and congestive heart failure? (pp. 267–268, 278, 286–287)

13. Trace the path of blood from the aorta to the femoral artery to the dorsalis pedis artery in the foot. Then, trace a return pathway from the foot through the great saphenous vein and back to the aorta. Indicate which of the vessels are in the systemic circuit and which are in the pulmonary circuit. (pp. 279–281)
14. Compare the fetal circulation with that of an adult. How are they alike? How are they different? (pp. 283–284)
15. Give examples to show that the cardiovascular system functions to maintain homeostasis and that interactions with other systems help it and the other systems to maintain homeostasis. (pp. 287–289)

Learning Outcome Questions

Fill in the blanks.
1. The right side of the heart pumps blood to the _____.
2. The valve between the left atrium and left ventricle is the _____, or mitral, valve.
3. When the left ventricle contracts, blood enters the _____.

4. The _____ node is known as the pacemaker.
5. Arteries are blood vessels that take blood _____ the heart.
6. The force of blood against the walls of an artery is termed _____.
7. The blood pressure recorded when the left ventricle contracts is called

the _____ pressure, and the pressure recorded when the left ventricle relaxes is called the _____ pressure.
8. The two factors that affect blood pressure are _____ and _____.

9. Blood moves in arteries due to _____ and in veins movement is assisted by _____.

10. The major blood vessels taking blood to and from the shoulders and upper limbs are the _____ arteries and veins. Those taking blood to and from the legs are the _____ arteries and veins.

11. The blood vessels that serve the heart are the _____ arteries and veins.

12. The human body contains a hepatic portal system that takes blood from the _____ to the _____.

13. Blood vessels to the brain end in a circular path known as the _____.

14. In the fetus, the opening between the two atria is called the _____ and the connection between the pulmonary artery and the aorta is called the _____.

Medical Terminology Exercise

After studying this chapter, see if you can derive the definitions for the medical terms listed at right. Many of the prefixes and suffixes used to create these terms can be found throughout the chapter. For additional help, use McGraw-Hill Connect™ at www.mcgrawhillconnect.com and consult Appendix B.

1. echocardiography (ĕk″ō-kăr″dē-ŏg′rŭh-fē)
2. coronary angioplasty (kōr′ŭh-nĕr-ē″ ăn′jē-ō-plăs″tē)
3. vasoconstriction (văs″ō-kŏn-strĭk′shŭn)
4. valvuloplasty (văl′vū-lō-plăs″tē)
5. antihypertensive (ăn″tĭ-hī″pĕr-tĕn′sĭv)

6. thromboendarterectomy (thrŏm″bō-ĕnd″ăr-tĕr-ĕk′tō-mē)
7. cardiovalvulitis (kăr′dē-ō-vălv-yū-lī′tĭs)
8. vasospasm (va′-sō-spăzm)
9. pericardiocentesis (pĕr-ĭ-kăr′dē-ō-sĕn-tē′sĭs)
10. ventriculotomy (vĕn-trĭk-yū-lŏt′ō-mē)

Online Study Tools

APR

Anatomy & Physiology REVEALED includes cadaver photos that allow you to peel away layers of the human body to reveal structures beneath the surface. This program also includes animations, radiologic imaging, audio pronunciations, and practice quizzing. To learn more visit www.aprevealed.com

13 The Lymphatic System and Body Defenses

As you'll discover in this chapter about the immune system, an *allergy* is an exaggerated immune response to ordinary, everyday substances in the environment. Mold, pollen, animal hair and dander, dust, bee stings, antibiotics, food—the list of things that can cause an allergy is a long one (and based on your own experience, you might be able to add to the list). Peanut allergy is one of the most prevalent and dangerous food allergies, and the leading cause of food-related death. For a person with this allergy, the world is a minefield of peanut products, in one form or another—nuts, oil, flour, empty peanut shells. Even peanut dust can contaminate other food: For example, baked goods produced in a factory that also makes peanut butter cookies are unsafe. Public places such as schools, restaurants, hotels, stores, and even public transportation are dangerous, too, because all might have some peanut contamination. Enter man's best friend: a peanut-sniffing dog. Like a seeing eye dog is for a blind person, so a peanut-sniffer is for a person with this deadly allergy. The dogs are trained to detect even minute amounts of peanut substances and then lead their owners away. Equipped with a sniffing dog, the allergic person can navigate a peanut-filled world in relative safety. You can read more about allergies on page 312.

Learning Outcomes

After you have studied this chapter, you should be able to:

13.1 Lymphatic System

1. Describe the functions of the lymphatic system.
2. Explain the structure of lymphatic vessels and the path of lymph from the tissues to the cardiovascular veins.

13.2 Organs, Tissues, and Cells of the Immune System

3. Describe the structure and function of the primary lymphatic organs: red bone marrow and the thymus gland.
4. Describe the structures and functions of the secondary lymphatic organs: the spleen and the lymph nodes.

13.3 Nonspecific and Specific Defenses

5. Outline the body's nonspecific defense mechanisms: barriers to entry, inflammatory reaction, natural killer cells, and protective proteins.

6. Describe the body's specific defense mechanisms: antibody-mediated immunity and cell-mediated immunity.
7. Give examples of immunotherapeutic drugs.

13.4 Creating an Immune Response

8. Explain natural active and passive immunity, and describe how artificial active and passive immunity are created.
9. Give examples of disorders caused by excessive and inadequate immune system responses.

13.5 Effects of Aging

10. Describe the anatomical and physiological changes that occur in the immune system as we age.

13.6 Homeostasis

11. Describe how the lymphatic system works with other systems of the body to maintain homeostasis.

Medical Focus
The Lymphatic System and Illness

Visual Focus
Steps of the Inflammatory Reaction

Medical Focus
AIDS Epidemic
Immunization: The Great Protector
Influenza: A Constant Threat of Pandemic

What's New
Parasite Prescription for Autoimmune Disease

Human Systems Work Together
Lymphatic System

13.1 Lymphatic System

1. Describe the functions of the lymphatic system.
2. Explain the structure of lymphatic vessels and the path of lymph from the tissues to the cardiovascular veins.

The **lymphatic system** consists of lymphatic vessels and the lymphatic organs. This system, which is closely associated with the cardiovascular system, has three main functions that contribute to homeostasis:

1. **Fluid balance.** The lymphatic system takes up excess tissue fluid and returns it to the bloodstream. Recall (from Chapter 12) that lymphatic capillaries lie very near blood capillaries, and they serve as an auxiliary way to take up fluid that has exited the blood capillaries.
2. **Fat absorption.** The lymphatic system absorbs fats from the digestive tract and transports them to the bloodstream. Special lymphatic capillaries called **lacteals** are located in the intestinal villi (see Fig. 15.7). This function ensures the absorption of dietary lipids as well as lipid-soluble vitamins.
3. **Defense.** The lymphatic system helps defend the body against disease. This function is carried out by the white blood cells present in lymphatic vessels and lymphatic organs, as well as those present in the blood and scattered between tissue cells. In addition to destroying foreign invaders, white blood cells are important in the destruction of dead and dying tissue cells. White blood cells also defend against body cells that have mutated to become cancerous.

Lymphatic Vessels

Lymphatic vessels form a one-way system that begins with lymphatic capillaries. Most regions of the body are richly supplied with lymphatic capillaries. These are tiny, closed-ended vessels consisting of a single layer of simple squamous epithelium, whose walls have large pores to allow particles to enter (Fig. 13.1). Lymphatic capillaries take up excess tissue fluid. Tissue fluid is mostly water, but it also contains solutes (e.g., nutrients, electrolytes, and oxygen) derived from plasma and cellular products (i.e., hormones, enzymes, and wastes) secreted by cells. Once inside lymphatic vessels, this tissue fluid is called **lymph.**

The lymphatic capillaries join to form lymphatic vessels that merge before entering one of two ducts: the **thoracic duct** or the **right lymphatic duct.** The larger thoracic duct returns lymph collected from the body below the thorax and the left upper limb and left side of the head and neck into the left subclavian vein. The right lymphatic duct returns lymph from the right upper limb and right side of the head and neck into the right subclavian vein.

The construction of the larger lymphatic vessels is similar to that of cardiovascular veins, including the presence of valves. The movement of lymph within lymphatic capillaries is largely dependent upon skeletal muscle contraction. Lymph forced through lymphatic vessels as a result of muscular compression is prevented from flowing backward by one-way valves (in the same way that vein valves prevent backflow). Failure of the lymphatic system to properly collect and return excessive tissue fluid to the bloodstream results in **edema,** as described in the Medical Focus on page 297.

Content CHECK-UP!

1. If you wanted to describe the lymphatic system, which statements would you use?
   a. It takes up excess tissue fluid and returns it to the circulation.
   b. It houses white blood cells.
   c. Lacteals in the intestines absorb dietary lipids.
   d. All of these statements are correct.
2. Imagine that lymph is draining from your right arm to return to the heart. Starting with lymphatic capillaries, describe the correct pathway for lymph flow.
3. Large lymphatic vessels are similar in structure to:
   a. capillaries in the cardiovascular system.
   b. arteries in the cardiovascular system.
   c. veins in the cardiovascular system.

Answers in Appendix A.

13.2 Organs, Tissues, and Cells of the Immune System

3. Describe the structure and function of the primary lymphatic organs: red bone marrow and the thymus gland.
4. Describe the structures and functions of the secondary lymphatic organs: the spleen and the lymph nodes.

The **immune system,** which plays an important role in keeping us healthy, consists of a network of lymphatic organs, tissues, and cells as well as products of these cells, including antibodies and regulatory agents. **Immunity** is the ability to react to antigens so that the body remains free of disease. Disease results from a failure of homeostasis. It can be due to infection by foreign microbes, and/or to the failure of the immune system to function properly.

Primary Lymphatic Organs

Lymphatic (lymphoid) organs contain large numbers of lymphocytes, the type of white blood cell that plays a pivotal role in immunity. The *primary lymphatic organs* are the red bone marrow and the thymus gland (Fig. 13.1, *bottom*). Lymphocytes originate and/or mature in these organs.

Red Bone Marrow

Red bone marrow is the site of stem cells that are capable of constantly dividing and producing blood cells. Some of

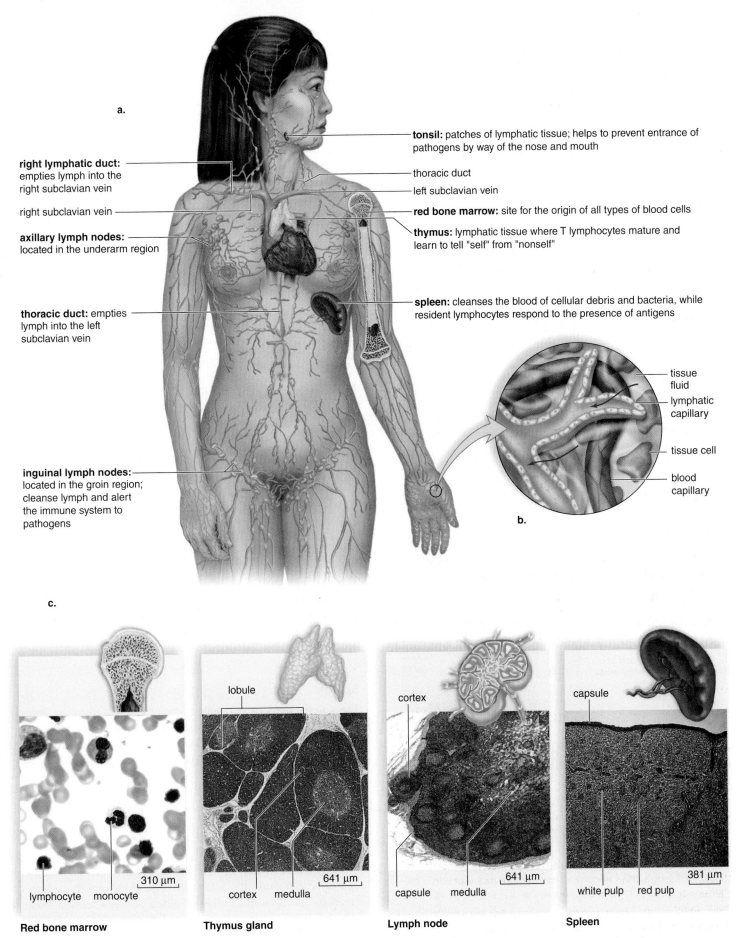

a.

right lymphatic duct: empties lymph into the right subclavian vein

right subclavian vein

axillary lymph nodes: located in the underarm region

thoracic duct: empties lymph into the left subclavian vein

inguinal lymph nodes: located in the groin region; cleanse lymph and alert the immune system to pathogens

tonsil: patches of lymphatic tissue; helps to prevent entrance of pathogens by way of the nose and mouth

thoracic duct

left subclavian vein

red bone marrow: site for the origin of all types of blood cells

thymus: lymphatic tissue where T lymphocytes mature and learn to tell "self" from "nonself"

spleen: cleanses the blood of cellular debris and bacteria, while resident lymphocytes respond to the presence of antigens

tissue fluid

lymphatic capillary

tissue cell

blood capillary

b.

c.

lymphocyte monocyte

310 µm

Red bone marrow

lobule

cortex medulla

641 µm

Thymus gland

cortex

capsule medulla

641 µm

Lymph node

capsule

white pulp red pulp

381 µm

Spleen

AP|R **Figure 13.1 The lymphatic system. (a)** Lymphatic vessels drain excess fluid from the tissues and return it to the cardiovascular system. **(b)** The enlargement shows a lymphatic vessel. **(c)** The red bone marrow, thymus gland, lymph nodes, and spleen are among those lymphatic organs that assist immunity.

these cells become the various types of white blood cells: neutrophils, eosinophils, basophils, lymphocytes, and monocytes (Fig. 13.2).

In a child, most bones have red bone marrow, but in an adult it is limited to the sternum, vertebrae, ribs, the skull, part of the pelvic girdle, and the proximal heads of the humerus and femur.

The red bone marrow consists of a network of reticular tissue fibers, which support the stem cells and their progeny. They are packed around thin-walled sinuses filled with venous blood. Differentiated blood cells enter the bloodstream at these sinuses.

Lymphocytes differentiate into the B lymphocytes, T lymphocytes, and natural killer (NK) cells, the specialized granular white blood cells similar in structure to a lymphocyte. Bone marrow is not only the source of B lymphocytes, but also the place where B lymphocytes mature. T lymphocytes mature in the thymus.

Thymus Gland

The soft, bilobed **thymus gland** is located in the thoracic cavity between the trachea and the sternum superior to the heart. The thymus varies in size, but it is largest in children and shrinks as we get older. Connective tissue divides the thymus into lobules, which are filled with lymphocytes. The thymus gland produces a group of thymic hormones, collectively called *thymosins,* that are thought to aid in the maturation of T lymphocytes. Thymosins also have other functions in immunity.

Immature T lymphocytes migrate from the bone marrow through the bloodstream to the thymus, where they mature. Only about 5% of these cells ever leave the thymus. These T lymphocytes have survived a critical test: If any show the ability to react with "self" cells (i.e., one's own tissue cells), they die. If they have potential to attack a foreign cell, they leave the thymus.

The thymus is absolutely critical to immunity; without a thymus, the body does not reject foreign tissues, blood lymphocyte levels are drastically reduced, and the body's response to most antigens is poor or absent.

Secondary Lymphatic Organs

The secondary lymphatic organs are the spleen, the lymph nodes, and other organs, such as the tonsils, Peyer patches, and the appendix. All the secondary organs are places where lymphocytes encounter and bind with antigens, after which they proliferate and become actively engaged cells.

Spleen

The **spleen,** the largest lymphatic organ, is located in the upper left region of the abdominal cavity posterior to the stomach. Connective tissue divides the spleen into partial compartments, each of which contains tissue known as white pulp and red pulp (Fig. 13.1, *bottom*). The white pulp contains a concentration of lymphocytes. The red pulp, which surrounds venous sinuses, is involved in filtering the blood. Blood entering the spleen must pass through the sinuses before exiting. Lymphocytes and macrophages react to pathogens, and macrophages engulf debris and also remove any old, worn-out red blood cells.

The spleen's outer capsule is relatively thin, and an infection or a blow can cause the spleen to burst. Although the spleen's functions are replaced by other organs, a person without a spleen is often slightly more susceptible to infections and may have to receive antibiotic therapy indefinitely.

a. Neutrophil
40–70%
Phagocytizes
primarily bacteria

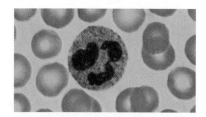

b. Eosinophil
1–4%
Phagocytizes and
destroys antigen-antibody
complexes and parasitic
organisms

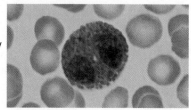

c. Basophil
0–1%
Releases histamine and
other chemicals when
stimulated

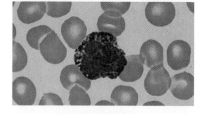

d. Lymphocyte
20–45%
B type produces
antibodies in blood and
lymph; T type kills virus-
containing cells and
cancer cells

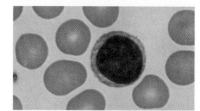

e. Monocyte
4–8%
Becomes macrophage—
phagocytizes bacteria
and viruses

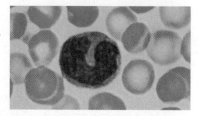

AP|R **Figure 13.2** **The five types of white blood cells.** These cell types differ according to structure and function. The frequency of each type of cell is given as a percentage of the total found in circulating blood. (See Chapter 11 for a detailed description of these cells.)

MEDICAL FOCUS

The Lymphatic System and Illness

As the "next-door neighbor" to the cardiovascular system, the parallel lymphatic system defends against infectious disease and cancer, and regulates fluid balance in the tissues. Failure of the lymphatic system to maintain homeostatic balance in these critical areas results in disease.

The internal structure of a lymph node is designed to filter out from lymph and then destroy any foreign material, including microbes and cancer cells. An infection that causes swelling and tenderness of nearby lymph nodes is called **lymphadenitis.** If the infection is not contained, **lymphangitis,** an infection of the lymphatic vessels, may result. Red streaks can be seen through the skin, indicating that the infection may have spread to the bloodstream.

Like microbes, cancer cells sometimes enter lymphatic vessels and move undetected to other regions of the body, where they produce secondary tumors. In this way, the lymphatic system sometimes assists *metastasis,* the spread of cancer far from its place of origin. Because of this potential for metastasis, regional lymph nodes are usually removed for examination whenever surgery is used to diagnose or treat cancer. The presence or absence of tumor cells in the nodes can be used to determine how far the cancer has spread. Lymph node *biopsy* (the microscopic study of the tissue) also aids in making decisions concerning additional treatment, such as radiation or chemotherapy.

Failure of the lymphatic vessels to properly remove tissue fluid results in a condition called edema. Edema forms because too much tissue fluid is made and/or not enough tissue fluid is drained away. Excessive tissue fluid can compress the blood vessels within a tissue, preventing the flow of blood, oxygen, and nutrients to tissue cells. Thus, edema can lead to tissue damage and eventual death, illustrating the importance of tissue fluid collection by the lymphatic system.

Blockage of the lymphatic vessels prevents proper tissue fluid drainage, resulting in edema. As mentioned previously, surgical removal of lymph nodes is a common treatment for cancer. However, node removal often blocks lymphatic drainage, producing a painful and debilitating **lymphedema** for the patient. A dramatic example of edema caused by lymphatic obstruction occurs when a parasitic roundworm clogs the lymphatic vessels. Tremendous swelling of the arm, leg, or external genitals results, causing a condition called **elephantiasis.**

Edema can also be due to a low osmotic pressure of the blood, as when plasma proteins are excreted by the kidneys instead of being retained in the blood. Extra tissue fluid forms, and lymphatic vessels may not be able to absorb it all. **Pulmonary edema** is a life-threatening condition associated with congestive heart failure. When the heart muscle is damaged and unable to effectively empty the heart, blood backs up in the pulmonary circulation. Pulmonary capillary blood pressure increases, which leads to excess fluid in lung tissue. Congested lung tissue cannot properly exchange oxygen and carbon dioxide with the blood, and the patient may suffocate.

Cancer of lymphatic tissue is called **lymphoma.** In **Hodgkin disease,** billions of lymphoma cells create swollen lymph nodes throughout the body. The lymphoma cells can migrate and grow in the spleen, liver, and bone marrow. The prognosis is good, however, if Hodgkin disease is diagnosed early.

Lymph Nodes

Lymph nodes, which are small, ovoid structures, occur along lymphatic vessels. Connective tissue forms the capsule of a lymph node and also divides the organ into compartments (see Fig. 13.1). Each compartment contains a nodule packed with B lymphocytes and a sinus that increases in size toward the center of the node. As lymph courses through the sinuses, it is filtered by macrophages (phagocytic white blood cells), which engulf pathogens and debris. T lymphocytes, also present in sinuses, fight infections and attack cancer cells.

Each portion of the anterior body cavity (see Fig. 1.5) contains superficial and deep lymph nodes, named for their location. For example, inguinal nodes are in the groin, and axillary nodes are in the armpits. Physicians often examine for the presence of swollen, tender lymph nodes as evidence that the body is fighting an infection. This is a noninvasive, preliminary way to help make such a diagnosis.

Begin Thinking Clinically

Why do surgeons remove lymph nodes at the same time that a cancer is taken out? What is a sentinel node?

Answer and discussion in Appendix A.

Lymphatic Nodules

Lymphatic nodules are concentrations of lymphatic tissue not surrounded by a capsule. The **tonsils** are lymphatic nodules located in the posterior pharynx (see Fig. 14.2). A single **pharyngeal tonsil** (also referred to as adenoids) is in the nasopharynx, whereas **lingual tonsils** sit at the base of the tongue. **Palatine tonsils** are visible in the posterior oral cavity. Like lymph nodes, the tonsils contain both T and B lymphocytes. Because of their location, they are the first to encounter pathogens and antigens that enter the body by way of the nose and mouth.

Peyer patches are located in the intestinal wall, and similar lymphatic nodules are found in the **appendix.** These structures encounter and help to defend against pathogens that enter the body by way of the intestinal tract.

Content CHECK-UP!

4. List and describe two functions of the thymus gland.

5. Which lymphatic organ functions as a blood reservoir and removes dead and dying red blood cells?

   a. thymus   c. appendix

   b. Peyer patches   d. spleen

6. Which of the following structures has a connective tissue capsule?

   a. axillary lymph nodes   c. Peyer patches

   b. palatine tonsils   d. adenoids (pharyngeal tonsil)

Answers in Appendix A.

13.3 Nonspecific and Specific Defenses

5. Outline the body's nonspecific defense mechanisms: barriers to entry, inflammatory reaction, nonspecific phagocytic white blood cells, natural killer cells, and protective proteins.

6. Describe the body's specific defense mechanisms: antibody-mediated immunity and cell-mediated immunity.

7. Give examples of immunotherapeutic drugs.

Immunity includes nonspecific defenses and specific defenses. The five types of nonspecific defenses—barriers to entry, the inflammatory reaction, nonspecific phagocytic white blood cells, natural killer cells, and protective proteins—are effective against many types of infectious agents. Specific defenses are effective against a particular infectious agent.

Nonspecific Defenses

Barriers to Entry

The body has built-in barriers, both physical and chemical, that help to prevent infection by microbes. These barriers are the first line of defense against infection. The intact skin is generally a very effective physical roadblock that prevents infection. Mucous membranes lining the respiratory, digestive, reproductive, and urinary tracts are also physical barriers to entry by pathogens. (However, it must be noted that their effectiveness is limited, especially in the oral and nasal cavities, because most cold viruses enter the body by crossing these membranes.) The ciliated cells that line the upper respiratory tract sweep mucus and trapped particles up into the throat, where they can be swallowed or expectorated (spit out).

Chemical barriers to infection include the secretions of sebaceous (oil) glands in the skin, which contain chemicals that weaken or kill certain bacteria on the skin. As perspiration evaporates, the concentrated salt on the skin helps prevent infection. Further, perspiration, saliva, and tears contain an antibacterial enzyme called *lysozyme.* Saliva also helps to wash microbes off the teeth and tongue. Similarly, as urine is voided from the body, it flushes bacteria from the urinary tract. The acidic pH of the stomach inhibits bacterial growth or kills many types

of bacteria. Finally, a significant chemical barrier to infection is created by the *normal flora*—microbes that normally reside in the mouth, large intestine, and other areas. By using up available nutrients and releasing their own waste, these normal germs prevent potential pathogens from taking up residence. For this reason, abusing antibiotics can make a person susceptible to pathogenic infection by killing off the normal flora.

Inflammatory Reaction

In a situation where the first line of defense is inadequate, the body's second line of defenses will activate. These mechanisms involve the inflammatory response, accompanied by the actions of phagocytic cells, natural killer cells, and protective proteins. Like the first line of defense, this second set of responses is nonspecific.

Whenever tissue is damaged by physical or chemical agents or by pathogens, a series of events occurs that is known as the **inflammatory reaction.** The inflammatory reaction has four outward signs: redness, swelling, heat, and pain. All of these signs are due to capillary changes in the damaged area. Figure 13.3 illustrates the five basic steps of an inflammatory response:

1. Chemical mediators, such as **histamine,** cause capillaries to dilate and become more permeable. These tissue chemicals are released by damaged tissue cells and by **mast cells** found in the tissues. Mast cells resemble basophils, one of the types of white blood cells (see section 11.2 to review the types of white blood cells). Excess blood flow to the inflamed area causes the skin to redden and become warm. Elevated temperature tends to inhibit the growth of some pathogens, and heightened blood flow brings more defensive white blood cells to the area.

2. Increased capillary permeability allows fluids and proteins, including blood clotting factors, to escape into the tissues. Escaped fluid causes swelling in the affected areas, putting pressure on nearby pain neurons and causing pain. Further, the inflammatory chemicals themselves can also stimulate the pain sensation. Although we can all agree that pain is a big nuisance, it's also a warning that tissue has been injured, and we need to try to correct the situation.

3. Edema and clots that form in the injured area help to "wall off" the area from the rest of the body. Thus, if pathogens have entered the body during injury, they will not be able to cause a systemic infection.

4. Migration of phagocytes, primarily neutrophils and monocytes, also occurs during the inflammatory reaction. Neutrophils and monocytes are amoeboid (see Figure 11.6) and can change shape to squeeze through capillary walls and enter tissue fluid. There, they phagocytize pathogens.

5. After monocytes appear on the scene, they differentiate into **macrophages,** large phagocytic cells that are able to devour as many as 100 pathogens and still survive. Some tissues, particularly connective tissue, have resident macrophages that routinely act as scavengers—devouring old blood cells, bits of dead tissue, and other debris. Macrophages also release colony-stimulating factors, chemicals which pass by way of the blood to the red bone marrow. There, the factors stimulate the production and the release of white blood cells, primarily neutrophils.

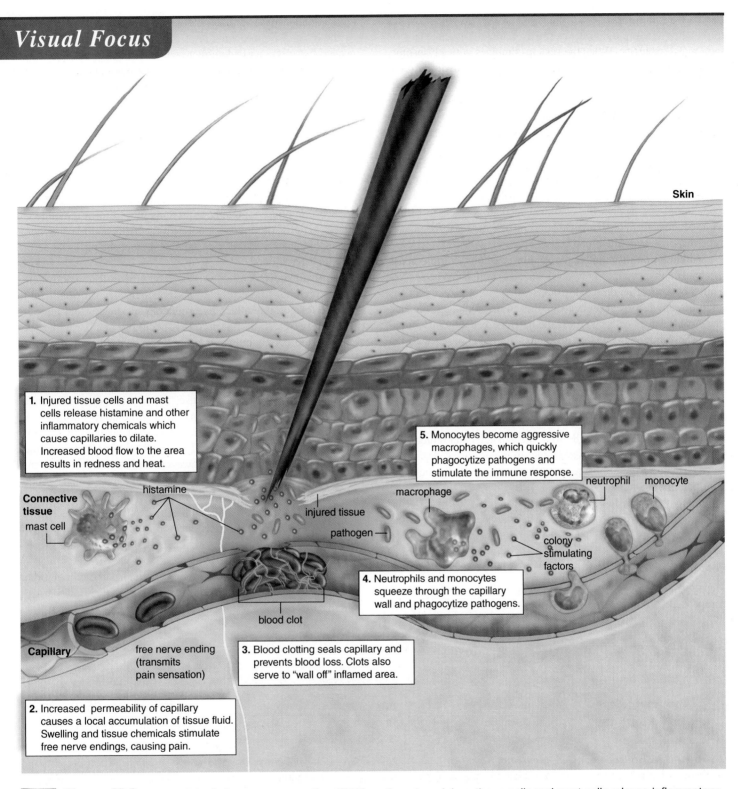

Skin

1. Injured tissue cells and mast cells release histamine and other inflammatory chemicals which cause capillaries to dilate. Increased blood flow to the area results in redness and heat.

5. Monocytes become aggressive macrophages, which quickly phagocytize pathogens and stimulate the immune response.

neutrophil   monocyte

histamine

Connective tissue

mast cell

macrophage

injured tissue

pathogen

colony stimulating factors

4. Neutrophils and monocytes squeeze through the capillary wall and phagocytize pathogens.

blood clot

Capillary

free nerve ending (transmits pain sensation)

3. Blood clotting seals capillary and prevents blood loss. Clots also serve to "wall off" inflamed area.

2. Increased permeability of capillary causes a local accumulation of tissue fluid. Swelling and tissue chemicals stimulate free nerve endings, causing pain.

AP|R **Figure 13.3** **Steps of the inflammatory reaction. (1)** When there is an injury, tissue cells and mast cells release inflammatory chemicals (such as histamine) that dilate blood vessels and cause them to leak. More blood flows to the area, and the skin appears red and feels warm. **(2)** Because capillaries are more permeable, edema forms. Swelling and inflammatory chemicals stimulate pain receptors (free nerve endings). **(3)** Blood clotting seals off the capillary, preventing blood loss and walling off the inflamed area. **(4)** Monocytes (which become macrophages) and neutrophils squeeze through the capillary wall, and phagocytosis begins. **(5)** Monocytes transform to macrophages, which stimulate specific immune responses in addition to destroying pathogens.

During the process of phagocytosis, endocytic vesicles form when neutrophils and macrophages engulf pathogens (see section 3.2, page 55). When the vesicle combines with a lysosome, a cellular organelle, the pathogen is destroyed by hydrolytic enzymes. As the infection is being overcome, some phagocytes die. These—along with dead tissue cells, dead bacteria, living white blood cells, and possibly living bacteria—form **pus**, a whitish material. The presence of pus indicates that the body is destroying dead tissue cells, and/or trying to overcome an infection.

Sometimes an inflammation persists, and the result is chronic inflammation that is often treated by administering anti-inflammatory agents such as aspirin, ibuprofen, or cortisone. These medications act against the chemical mediators released by the white blood cells in the damaged area and also decrease the phagocytic response by white blood cells.

Although an inflammatory reaction is a nonspecific defense mechanism that occurs whenever tissue is injured, it can be accompanied by a specific, directed response to the injury. As infectious microbes are destroyed by neutrophils and macrophages, bits and pieces of the dead cells can cross the barrier created by inflammation. These fragments, called **antigens**, can travel along with the released chemical mediators through the tissue fluid and lymph to the lymph nodes. Now lymphocytes mount a specific defense to the infection, as described in the next section.

Natural Killer Cells

Natural killer (NK) cells kill virus-infected cells and tumor cells by cell-to-cell contact. They are large, granular leukocytes, similar in structure to lymphocytes, and they originate from the same multipotential stem cells as lymphocytes. However, unlike B and T lymphocytes, they have no specificity for individual pathogens, and they do not form memory cells. Their number is not increased by prior exposure to any kind of cell.

Protective Proteins

The **complement system**, often simply called complement, is composed of a number of blood plasma proteins designated by the letter C and a subscript. Complement proteins are primarily made by the liver. A limited amount of activated complement protein is needed because a cascade effect occurs: Each activated protein in a series is capable of activating many other proteins.

The complement proteins are activated when pathogens enter the body. The proteins "complement" certain immune responses, which accounts for their name. For example, they are involved in and amplify the inflammatory response because complement proteins attract phagocytes to the scene. Some complement proteins bind to the surface of pathogens already coated with antibodies (specific antibacterial protein molecules), which ensures that the pathogens will be phagocytized by a neutrophil or macrophage.

Certain other complement proteins join to form a membrane attack complex that punches holes in the walls and plasma membranes of bacteria. Fluids and salts then enter the bacterial cell to the point that it bursts.

Interferon is a protein produced by virus-infected body cells. Interferon binds to receptors of noninfected cells, causing them to prepare for possible attack by producing substances that interfere with viral replication. Interferon is specific to one particular species; therefore, only human interferon can be used in humans.

Specific Defenses

When creating a specific defense, the immune system lymphocytes respond to foreign antigens. This is the body's third line of defense against infection. It's important to understand that normal cells also have antigens, primarily in the form of protein molecules found on the cell membrane. However, the immune system is able to distinguish "self" antigens from foreign antigens. Thus, lymphocytes don't ordinarily attack normal body cells. Lymphocytes are capable of recognizing alien antigens because they have *antigen receptors*—plasma membrane receptor proteins that combine with a specific "nonself" antigen. Antigens are typically large molecules like proteins, though fragments of bacteria, viruses, molds, or parasites can also be antigenic. Further, abnormal cell membrane proteins produced by cancer cells may also be antigens. Table 13.1 summarizes the nonspecific and specific defenses.

	Types	How They Work
TABLE 13.1 Three Lines of Defense		
First Line: Nonspecific	**Mechanical barriers**	Intact skin, mucous membranes, respiratory cilia prevent entry
	Chemical barriers	Stomach acid, saliva, tears, etc., destroy and/or wash away pathogens
	Normal flora of microbes	Create a hostile environment for pathogens
Second Line: Nonspecific	**Inflammatory reaction**	Walls off infection site, draws white cells to area, signals a pain response
	Nonspecific phagocytic cells	Neutrophils, eosinophils, monocytes/macrophages engulf and destroy invaders; basophils/mast cells produce chemicals to stimulate immune response
	Natural killer cells	Kill pathogens by cell-to-cell contact
	Protective proteins	Complement participates, enhances defensive response. Interferon interferes with viral replication
Third Line: Specific	**B lymphocytes**	Antibody response
	T lymphocytes	Produce messenger cytokines, stimulate B cells, direct cell destruction

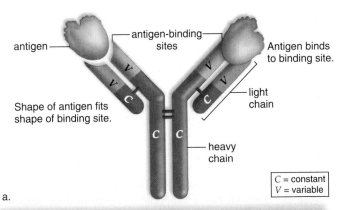

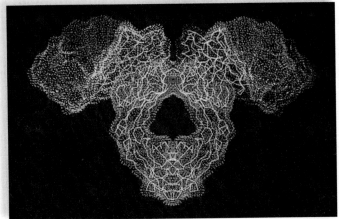

AP|R **Figure 13.4** **Structure of a single antibody molecule (monomer).** **(a)** Monomers contain two heavy (long) polypeptide chains and two light (short) chains arranged so that there are two variable regions, where a particular antigen is capable of binding with an antibody (*V* = variable region, *C* = constant region). **(b)** Computer model of an antibody molecule. The antigen combines with the two side branches.

Immunity usually lasts for some time. For example, once a child has been properly immunized against measles virus, the immune system confers a long-term protection (see the Medical Focus on pages 309–310). Booster immunizations help to maintain immunity throughout the teen years and into adulthood.

Immunity is primarily the result of the action of the **B lymphocytes** and the **T lymphocytes.** Recall that B lymphocytes mature in the bone marrow,[1] and T lymphocytes mature in the thymus gland. B lymphocytes, also called B cells, give rise to **plasma cells,** which produce antibodies. **Antibodies** combine with and neutralize foreign antigens. Each antibody protein is shaped like a specific antigen and combines with that antigen in a precise, "lock-and-key" fit (Fig. 13.4). These antibodies are secreted into the blood, lymph, and other body fluids. In contrast, T lymphocytes, also called T cells, do not produce antibodies. Instead, certain T cells directly attack cells that bear nonself proteins. Other T cells regulate the immune response by producing **cytokines,** which are proteins that regulate the immune response.

[1]Historically, the B stands for bursa of Fabricius, an organ in the chicken where these cells were first identified.

B Cells and Antibody-Mediated Immunity

There are millions of individual groups, or **clones,** of immature B lymphocytes in an adult. Each clone is pre-programmed to produce a particular type of antibody if exposed to a specific antigen. The **clonal selection theory** states that one particular lymphocyte in a clone is *selected* to respond because its antigen receptors combine with the antigen in the precise "lock-and-key" fit. Next, the selected lymphocyte goes on to reproduce multiple copies of itself. Each of the identical cells in the resulting group is a clone of the original cell. The majority of these B lymphocytes will mature to form plasma cells, all bearing the same type of antigen receptor.

Notice in Figure 13.5 that different types of antigen receptors are represented by their shaped ends: circular, triangular, square, and so on. The B cell with curved, circular receptors undergoes clonal expansion because a specific antigen (green dots) is present and binds to its receptors. B cells are stimulated to divide and become plasma cells by helper T-cell secretions called cytokines, as discussed later in this section. Some members of the clone become memory cells, which are the means by which long-term immunity is possible. If the same antigen enters the system again, **memory B cells** quickly divide and give rise to more lymphocytes capable of quickly producing antibodies. A second exposure to the same antigen produces a stronger, faster immune response.

Once the threat of an infection has passed, the development of new plasma cells ceases, and those present undergo apoptosis. **Apoptosis** is a process of programmed cell death involving a cascade of specific cellular events leading to the death and destruction of the cell. It is important that these cells die once they are no longer needed. Otherwise they could mistakenly destroy body cells in an **autoimmune response.**

Defense by B cells is called **antibody-mediated immunity** because the various types of B cells produce antibodies. It is also called **humoral immunity** because these antibodies are present in blood and lymph. A *humor* is any fluid normally occurring in the body.

Structure and Function of Antibodies

The basic unit that composes antibody molecules is a Y-shaped protein molecule with two arms. Each arm has a "heavy" (long) polypeptide chain and a "light" (short) polypeptide chain (see Fig. 13.4). These chains have *constant regions,* located at the trunk of the Y, where the sequence of amino acids is set. The class of antibody for each individual molecule is determined by the structure of the antibody's constant region. The *variable regions* form an antigen-binding site, and their shape is specific to a particular antigen. The antigen combines with the antibody at the antigen-binding site in a lock-and-key manner. Antibodies may consist of single Y-shaped molecules called *monomers,* or they can be paired together in a molecule called a *dimer.* Very large antibodies belonging to the M class are *pentamers*—clusters of five Y-shaped molecules linked together.

The antigen-antibody reaction can take several forms. Antibodies react with viruses and toxins (the poisons made by bacteria) by coating them completely—a process called *neutralization.* Other antibodies cause *agglutination,* a reaction that produces an immune complex. An *immune complex* is a clump of antigens combined with antibodies. Clustering antigens into an immune complex marks the

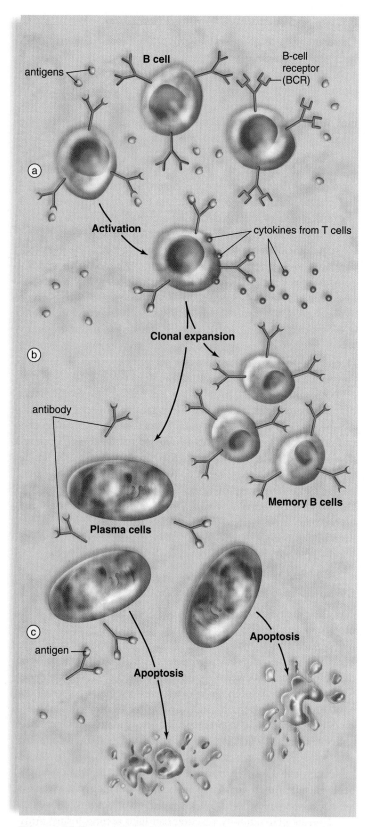

Figure 13.5 Clonal selection theory as it applies to B cells.
(a) Each B cell has different antigen receptors. (b) An antigen combines with only one B-cell receptor. Cytokines cause the B cell to multiply.
(c) The chosen B cell forms plasma cells, which undergo apoptosis when the immune response is complete. Memory B cells remain in the body.

antigens for destruction. For example, an antigen-antibody complex may be engulfed by neutrophils or macrophages, or it may activate complement. Complement makes pathogens more susceptible to phagocytosis, as discussed previously.

Classes of Antibodies

There are five different classes of circulating antibody proteins, or immunoglobulins (Igs) (Table 13.2). IgG antibodies are the major type in blood, and lesser amounts are also found in lymph and tissue fluid. IgG antibodies bind to pathogens and their toxins. IgG antibodies can cross the placenta from a mother to her fetus, so that the newborn has a temporary, partial immune response. As mentioned previously, IgM antibodies are pentamers. These antibodies are "first responders": the first antibodies produced by a newborn's body. Subsequently, IgM antibodies are the first to appear in blood soon after an infection begins and the first to disappear before the infection is over. They are good activators of the complement system. IgA antibodies are monomers or dimers containing two Y-shaped structures. They are the main type of antibody found in body secretions: saliva, tears, mucus, and breast milk. IgA molecules bind to pathogens and prevent them from reaching the bloodstream. The main function of IgD molecules seems to be to serve as antigen receptors on immature B cells. IgE antibodies are responsible for prevention of parasitic infections, but they can also cause immediate allergic responses.

TABLE 13.2 Functional Classes of Antibodies		
Classes	**Presence**	**Function**
IgG	Main antibody type in circulation; crosses the placenta from mother to fetus	Binds to pathogens, activates complement proteins, and enhances phagocytosis
IgM	Antibody type found in circulation; largest antibody; first antibody formed by a newborn; first antibody formed with any new infection	Activates complement proteins; clumps cells
IgA	Main antibody type in secretions such as saliva and milk	Prevents pathogens from attaching to epithelial cells in digestive and respiratory tract
IgD	Antibody type found on surface of immature B cells	Presence signifies readiness of B cell
IgE	Antibody type found as antigen receptors on basophils in blood and on mast cells in tissues	Responsible for immediate allergic response and protection against certain parasitic infections

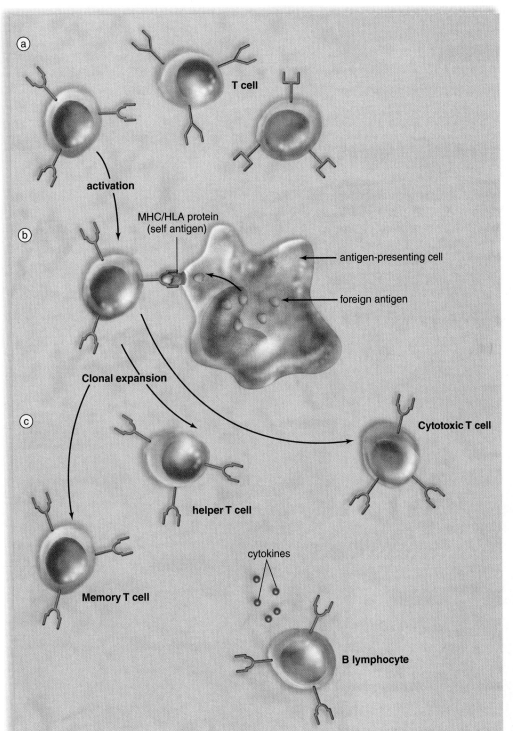

(a)

T cell

activation

(b)

MHC/HLA protein
(self antigen)

antigen-presenting cell

foreign antigen

Clonal expansion

(c)

Cytotoxic T cell

helper T cell

cytokines

Memory T cell

B lymphocyte

AP|R **Figure 13.6 Clonal selection theory as it applies to T cells.** (a) Each T cell has different antigen receptors that combine with only one antigen. (b) The macrophage presents this antigen in the groove of an HLA molecule, and the T cell that can combine with the antigen-HLA pair is stimulated to divide. (c) Clonal expansion gives rise to three types of T cells. Helper T cells produce cytokines that stimulate immune responses, including B-cell response. Cytotoxic T cells attack cells having the same antigen. Memory cells remain after the immune response is complete.

T Cells and Cell-Mediated Immunity

When T cells leave the thymus, they have unique antigen receptors just as B cells do (Fig. 13.6*a*). Unlike B cells, however, T cells are unable to recognize an antigen present in lymph, blood, or the tissues without help. The antigen must be presented to them by an **antigen-presenting cell (APC).** APCs can be specialized macrophage cells or B lymphocytes. When an APC presents a viral or cancer cell antigen to the T cell, the antigen is first linked to a **major histocompatibility complex (MHC)** protein in the APC cell's plasma membrane (Fig 13.6*b*).

Human MHC proteins are called **HLA (human leukocyte antigens).** These proteins are found on all of the body's cells. Thousands of different genes that code for these proteins have been identified. Each individual human being has a unique combination of genes, and thus, a unique set of HLAs. No two sets are exactly alike. An exception is the set belonging to monozygotic, or identical twins.

Because identical twins arise from the division of a single zygote, their HLA proteins are identical. Because HLA antigens mark the cell as belonging to a particular individual, they are *self* proteins. The importance of self proteins in plasma membranes was first recognized when it was discovered that they contribute to the specificity of tissues and make it difficult to transplant tissue from one human to another. Comparison studies of HLA antigens must always be carried out before a transplant is attempted. The more of the 50-plus proteins that the donor and recipient share in common, the better the tissue match. In other words, when the donor and the recipient are histo (tissue)-compatible, a transplant is more likely to be successful.

When an antigen-presenting cell links a foreign antigen to the self protein on its plasma membrane, it carries out an important safeguard for the rest of the body. Now the T cell to be activated can compare the antigen and self protein side by side. The activated T cell and all of the daughter cells it will form can distinguish "foreign" from "self," and go on to destroy cells carrying foreign antigens while leaving normal body cells unharmed (Fig. 13.6*b*).

Presentation of the antigen leads to activation of the T cell. An activated T cell undergoes clonal expansion (Fig. 13.6*c*). Many copies of the activated T cell are produced during clonal expansion. They destroy any cell, such as a virus-infected cell or a cancer cell, that displays the antigen presented earlier.

Types of T Cells

The two main types of T cells are cytotoxic T cells and helper T cells. **Cytotoxic T (T_c) cells,** also called CD_8 leukocytes, can bring about the destruction of antigen-bearing cells, such as virus-infected or cancer cells (Fig. 13.7*a*). Cancer cells also have nonself proteins.

Cytotoxic T cells have storage vacuoles containing perforin molecules. **Perforin** molecules punch holes into a plasma membrane, forming a pore that allows destructive enzymes to enter. The cell then undergoes apoptosis. Cytotoxic T cells are responsible for so-called **cell-mediated immunity** (Fig. 13.7*b*).

Helper T (T_h) cells, also called CD_4 leukocytes, regulate immunity by secreting cytokines, the chemicals that enhance the response of other immune cells. Cytokines stimulate macrophages, B cells, and other T cells to perform their functions. Because HIV, the virus that causes AIDS, infects helper T cells and certain other cells of the immune system, it inactivates the immune response. AIDS is described in the Medical Focus, pages 307–308.

Notice in Figure 13.6*c* that a few of the clonally expanded T cells are **memory T cells.** They remain in the body and can jumpstart an immune reaction to an antigen previously present in the body.

When a person's immune response has successfully overcome the illness, ultimately that response must be shut down. Now, the activated T cells become susceptible to apoptosis. As mentioned previously,

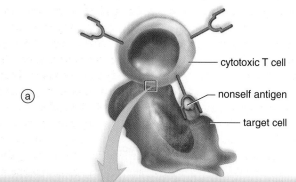

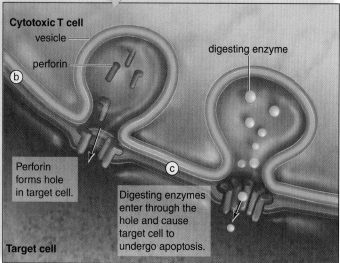

cytotoxic T cell

nonself antigen

target cell

Cytotoxic T cell
vesicle

perforin

digesting enzyme

Perforin forms hole in target cell.

Digesting enzymes enter through the hole and cause target cell to undergo apoptosis.

Target cell

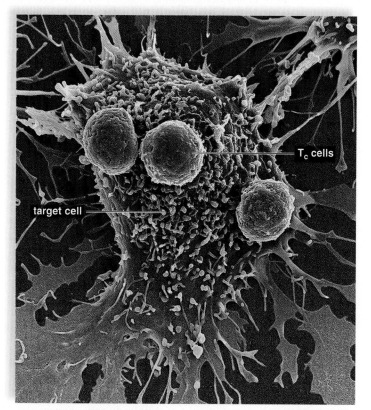

T_c cells

target cell

d. Scanning electron micrograph   SEM 1,250×

AP|R **Figure 13.7** **Cell-mediated immunity: How a cytotoxic T cell destroys a virus-infected or cancer cell. (a)** The activated T cell binds with a nonself antigen presented by the foreign cell. **(b)** The T cell releases perforin molecules, which form pores in the target cell membrane. **(c)** Digesting enzymes enter through the hole, triggering apoptosis in the target cell. **(d)** This scanning electron micrograph shows cytotoxic cells destroying a cancer cell.

apoptosis is programmed cell death that contributes to homeostasis by regulating the number of cells present in an organ, or in this case, in the immune system. When apoptosis does not occur as it should, the potential exists for self-destruction (called an autoimmune response). Occasionally, T-cell cancers (i.e., lymphomas and leukemias) can result.

Apoptosis also occurs in the thymus as T cells are maturing. Any T cell that has the potential to destroy the body's own cells undergoes suicide.

Cytokines in Cancer Chemotherapy

Whenever cancer develops, it is possible that cytotoxic T cells have not been activated. With this possibility in mind, cytokines have been used as immunotherapeutic drugs to enhance the ability of T cells to fight cancer. Interferon, discussed on page 300, and also **interleukins,** which are cytokines produced by various white blood cells, are also being administered for this purpose.

Content **CHECK-UP!**

7. Which type of cell produces the chemical mediators that cause the four outward signs of an inflammatory response?

   a. macrophage   c. neutrophil

   b. mast cell   d. natural killer cells

8. Which sign of inflammation occurs when tissue capillaries become more permeable and leak protein and tissue fluid into the tissue spaces?

   a. swelling   c. heat

   b. redness   d. formation of pus

9. Which type of immune response involves B-lymphocytes?

   a. cell-mediated

   b. antibody-mediated

10. Virus-infected cells produce a specific protein that helps to slow virus infection. What is it, and how does it work?

Answers in Appendix A.

13.4   Creating an Immune Response

8. Explain natural active and passive immunity, and describe how artificial active and passive immunity are created.

9. Give examples of disorders caused by excessive and inadequate immune system responses.

Immunity occurs naturally or is brought about artificially (induced) by medical intervention. The two types of immune responses are active and passive. In **active immunity,** the individual alone produces antibodies against an antigen; in **passive immunity,** the individual receives prepared antibodies from another person.

Active Immunity

Active immunity often develops naturally after a person is infected with a pathogen. Once recovery from illness is complete, the person will have a *natural* active immunity thanks to the B and T memory cells that remain after the infection. However, active

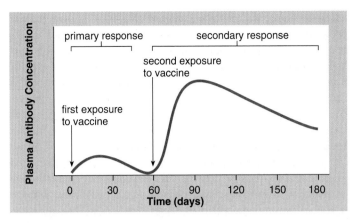

Figure 13.8 Vaccine responses. During immunization, the primary response, after the first exposure to a vaccine, is minimal, but the secondary response, which may occur after the second exposure, shows a dramatic rise in the amount of antibody present in plasma.

---

immunity can also be induced while a person is well, so that future infection will not take place. To prevent infections, people are immunized against them. Successful immunization provides *artificial* active immunity, because the individual produces his or her own antibodies against the antigen given. The United States is committed to immunizing all children against the common types of childhood disease, as discussed in the Medical Focus on pages 309–310.

Immunization involves the use of **vaccines,** substances that contain an antigen to which the immune system responds. Traditionally, vaccines are the pathogens themselves, or their products, that have been treated so they are no longer virulent (able to cause disease). Today, it's possible to genetically engineer bacteria to mass-produce a protein from pathogens, and this protein can be used as a vaccine. This method has now produced a vaccine against hepatitis B, a viral-induced disease, and is being used to prepare a vaccine against malaria, a protozoan-induced disease.

After a vaccine is given, it's possible to follow an immune response by determining the amount of antibody present in a sample of plasma. This is called the **antibody titer.** After the first exposure to a vaccine, a primary response occurs. For a period of several days, no antibodies are present; then the titer rises slowly, levels off, and gradually declines as the antibodies bind to the antigen or simply break down (Fig. 13.8). After a second exposure to the vaccine, a secondary response is expected. The titer rises rapidly to a level much greater than before; then it slowly declines. The second exposure is called a "booster" because it boosts the antibody titer to a high level. The high antibody titer now is expected to help prevent disease symptoms even if the individual is exposed to the disease-causing antigen.

As mentioned previously, active immunity is dependent upon the presence of memory B cells and memory T cells that are capable of responding to lower doses of antigen. Active immunity is usually long-lasting, although a booster may be required every few years.

Passive Immunity

Passive immunity occurs when an individual is given prepared antibodies (immunoglobulins) to combat a disease. Since these antibodies are not produced by the individual's own plasma cells, passive immunity is temporary. For example, newborn infants are passively immune to some diseases because antibodies have crossed the placenta from the mother's blood. This type of immunity is called natural passive immunity. These antibodies soon disappear, however, so that within a few months, infants become more susceptible to infections. Breast-feeding prolongs the natural passive immunity an infant receives from the mother because antibodies are present in the mother's milk (Fig. 13.9).

Even though passive immunity does not last, it is sometimes used to prevent illness in a patient who has been unexpectedly exposed to an infectious disease. Usually, the patient receives a **gamma globulin** injection (serum that contains IgG class antibodies), perhaps taken from individuals who have recovered from the illness. For example, a health-care worker who suffers an accidental needle stick may come into contact with the blood from a patient infected with hepatitis B virus. Immediate treatment with an antiviral gamma globulin injection (along with simultaneous vaccination against the virus) can typically prevent the virus from causing infection in the health-care worker. The gamma globulin injection imparts an artificial *passive* immunity. The vaccination will cause the worker to develop her own antibodies, thus creating artificial *active* immunity as well.

Monoclonal Antibodies

Every plasma cell derived from the same B cell secretes antibodies against one specific antigen. These are **monoclonal antibodies** because all are the same type and because they are produced by plasma cells derived from the same B cell. One method of producing monoclonal antibodies *in vitro* (outside the body in a laboratory) is depicted in Figure 13.10. B lymphocytes are removed from an animal (today, mice are usually used) and exposed to a particular antigen. The resulting plasma cells are fused with myeloma cells (malignant plasma cells that live and divide indefinitely). The fused cells are called hybridomas—*hybrid*- because they result from the fusion of two different cells, and *oma* because one of the cells is a cancer cell. It is important to note that the antibodies from these cells *cannot* cause cancer, however.

At present, monoclonal antibodies are being used for quick and certain diagnosis of various conditions. For example, a particular

Figure 13.9 Passive immunity. Breast-feeding is believed to prolong the passive immunity an infant receives from the mother because antibodies are present in the mother's milk.

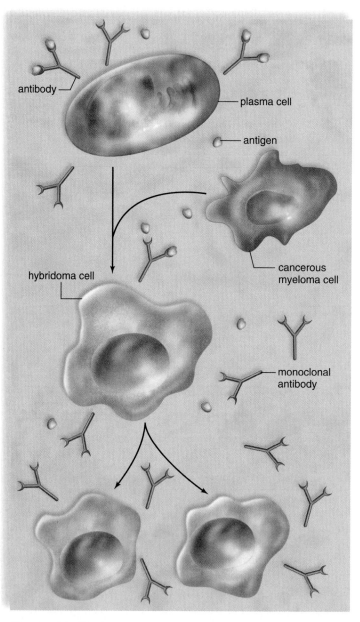

Figure 13.10 Production of monoclonal antibodies. Plasma cells of the same type (derived from immunized mice) are fused with myeloma (cancerous) cells, producing hybridoma cells that are "immortal." Hybridoma cells divide and continue to produce the same type of antibody, called monoclonal antibodies.

AIDS Epidemic

Acquired immunodeficiency syndrome (AIDS) is caused by a group of related retroviruses known as HIV (human immunodeficiency viruses). There are two forms of the virus, termed HIV-1 and HIV-2. In the United States, AIDS is usually caused by HIV-1. Both virus forms enter a host by attaching to a cell membrane receptor called a CD_4 receptor. HIV infects helper T cells, the type of lymphocyte that stimulates B cells to produce antibodies and cytotoxic T cells to destroy virus-infected cells. Macrophages, which present antigens to helper T cells and thereby stimulate them, are also under attack. Other cells that have CD_4 receptors can also be infected and later destroyed.

HIV is a **retrovirus,** meaning that its genetic material consists of RNA instead of DNA. Once inside the host cell, HIV uses a special enzyme called *reverse transcriptase* to make a DNA copy (called cDNA) of its genetic material. Next, cDNA integrates into a host chromosome, where it directs the production of more viral RNA. Each strand of viral RNA causes the synthesis of an outer protein coat called a **capsid.** The viral enzyme **protease** is necessary to the formation of capsids. Capsids assemble with RNA strands to form viruses, which bud from the host cell and spread to other cells.

Transmission of AIDS

HIV is transmitted by contact with infected body fluids: semen, blood, saliva, breast milk, vaginal fluid, and/or tissue fluid from an open wound.

The most common method of transmission involves sexual contact with an infected person, including vaginal or rectal intercourse and oral/genital contact. Also, needle-sharing among intravenous drug users is high-risk behavior. Babies born to HIV-infected women may become infected before or during birth, or through breast-feeding after birth. The rise in the incidence of AIDS among women of reproductive age is paralleled by a rise in the incidence of AIDS in children younger than age 13.

To date, an estimated 100 million people worldwide have contracted HIV, and almost 39 million have died. A new infection is believed to occur every 15 seconds, the majority in heterosexuals. HIV infections are not distributed equally throughout the world. Most infected people live in Africa (70%) where the infection first began, but new infections are now occurring at the fastest rate in Southeast Asia and the Indian subcontinent.

Phases of an HIV Infection

The Centers for Disease Control and Prevention recognize three stages of an HIV-1 infection, called categories A, B, and C. During the category A stage, the helper T-lymphocyte count is 500 per mm^3

of blood or greater (Fig. 13A). For a period of time after the initial infection with HIV, people don't usually have any symptoms at all. A few (1–2%) do have mononucleosis-like symptoms that may include fever, chills, aches, swollen lymph nodes, and an itchy rash. These symptoms disappear, however, and no other symptoms appear for quite some time. Although there are no symptoms, the person is highly infectious. Despite the presence of a large number of viruses in the plasma, the HIV blood test is not yet positive because it tests for the presence of antibodies and not for the presence of HIV itself. This means that HIV can still be transmitted before the HIV blood test is positive.

Several months to several years after a nontreated infection, the individual will probably progress to category B, in which the helper T-lymphocyte count is 200 to 499 per mm^3. During this stage, the patient may experience swollen lymph nodes in the neck, armpits, or groin that persist for three months or more. Other symptoms that indicate category B are severe fatigue not related to exercise or drug use; unexplained persistent or recurrent fevers, often with night sweats; persistent cough not associated with smoking, a cold, or the flu; and persistent diarrhea.

The development of non-life-threatening but recurrent infections is a signal that the disease is progressing. One possible infection is thrush, a fungal infection that is identified by the presence of white spots and ulcers on the tongue and inside the mouth. The fungus may also spread to the vagina, resulting in a chronic infection there. Another frequent infection is herpes simplex, with painful and persistent sores on the skin surrounding the anus, the genital area, and/or the mouth.

Without treatment, the majority of infected persons proceed to category C, in which the helper T-lymphocyte count is below 200 per mm^3 and the lymph nodes degenerate. The patient is now suffering from AIDS, characterized by severe weight loss and weakness due to persistent diarrhea and coughing, and will most likely contract an opportunistic infection. An **opportunistic infection** is one that only has the opportunity to occur because the immune system is severely weakened. Such infections are extremely rare in healthy individuals. Persons with AIDS die from one or more opportunistic diseases, such as *Pneumocystis jirovecii* pneumonia, tuberculosis, toxoplasmic encephalitis, Kaposi's sarcoma, or invasive cervical cancer. This last condition has been added to the list because the incidence of AIDS has now increased in women.

Treatment for AIDS

Treatment for AIDS is termed highly active antiretroviral therapy, or HAART. This therapy consists of combining three or more categories

—Continued

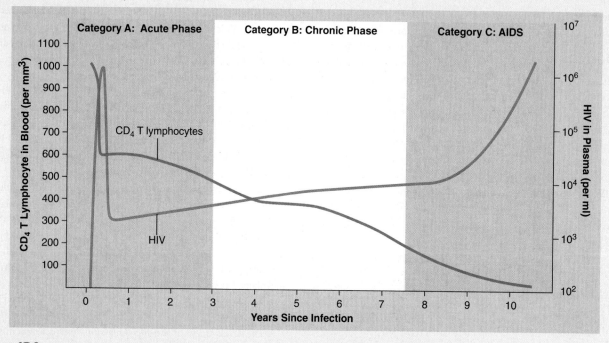

Figure 13A Stages of an HIV infection. In category A individuals, the number of HIV in plasma rises upon infection and then falls. The number of CD_4 T lymphocytes falls, but stays above 400 per mm^3. In category B individuals, the number of HIV in plasma is slowly rising, and the number of T lymphocytes is decreasing. In category C individuals, the number of HIV in plasma rises dramatically as the number of T lymphocytes falls below 200 per mm^3.

of drugs. Currently, four categories exist: (1) drugs that inhibit viral entry into the cell; (2) drugs that inhibit the reverse transcriptase enzyme; (3) drugs to inhibit protease, an enzyme needed for formation of a viral capsid; (4) drugs that prevent integration of the viral RNA into human DNA. This multidrug therapy, when taken according to the manner prescribed, can often effectively reduce detectable blood HIV and restore concentrations of CD_4 cells to normal or near-normal levels. The sooner drug therapy begins after infection, the better the chances that the immune system will not be destroyed by HIV. Also, HAART medication must be continued indefinitely. However, it's important to note that HAART does not cure AIDS, and all of the medications potentially have very serious side effects. These include bone and joint destruction, heart attack, liver damage, and anemia. Unfortunately, a patient's strain of HIV may become resistant to HAART over time.

At present, the best strategy to slow the AIDS epidemic focuses on infection prevention. Education regarding safe sex practices, especially proper condom use and abstinence, has already reduced the rate of infection worldwide (see Chapter 17 Medical Focus: Prevention of STIs for details). Counseling of pregnant women is essential to prevent mother-to-child HIV transmission.

The likelihood of transmission can be lessened if the pregnant mother takes a modified form of HAART, and if the child is delivered by cesarean section. The mother also must not breast-feed her child because the virus is transmissible in breast milk. Medical and dental professionals should carefully follow universal precautions against infection (gloves, surgical masks, proper disinfection techniques, etc.).

Many investigators are working on a vaccine for AIDS. Some are trying to develop a vaccine in the traditional way. Others are working on subunit vaccines that utilize just a single HIV protein as the vaccine. So far, no method has resulted in creating sufficient antibodies to keep an infection at bay. After many clinical trials, none too successful, most investigators now agree that a combination of various vaccines may be the best strategy to bring about a response in both B lymphocytes and cytotoxic T cells.

MEDICAL FOCUS

Immunization: The Great Protector

Immunization protects children and adults from diseases. The success of immunization is witnessed by the fact that the smallpox vaccination is no longer required because the disease has been eradicated. However, parents today often fail to get their children immunized because they don't realize the importance of immunizations, fear potential side effects (which occur rarely), or can't bear the expense. As a result, preventable diseases have recently come roaring back: a major outbreak of measles in 2014 spread throughout 14 states and sickened over 600 people. This was the greatest number of cases since 2000, when it was thought that measles had been eliminated in the

	Birth	1 month	2 mos.	4 mos.	6 mos.	12 mos.	15 mos.	18 mos.	24 mos.	4-6 yrs.	11-12 yrs.	13-18 yrs.	Adult
Hepatitis B	1st Dose	Booster 1			Booster 2			Booster 3	Catch Up*				3 Dose booster ‡
Rotavirus			1st Dose	2nd Dose	Catch Up*								
Diphtheria, Pertussis, Tetanus			1st Dose	2nd Dose	3rd Dose		Booster 1			Booster 2	Tetanus, Diphtheria booster		Tetanus, Diphtheria booster every 10 years
Haemophilus influenzae			1st Dose	2nd Dose	3rd Dose**	Final Dose							
Pneumococcal			1st Dose	2nd Dose	3rd Dose	Final Dose			Catch Up*				Booster after age 60
Polio			1st Dose	2nd Dose	Final Dose	Catch Up*				Booster			
Measles, Mumps, Rubella						1st Dose				2nd Dose	Catch Up*		1-2 Doses
Varicella (chickenpox)						1st Dose	Catch Up*	Catch Up*	2nd Dose				2 Doses△
Hepatitis A						1st Dose		2nd Dose	Catch Up*				2 Dose booster ‡
Human Papillomavirus											3 doses for females ages 9-18 / 3 doses for males ages 9-21		
Meningococcal											1st Dose	Booster at age 16.	
Influenza					Annually	Annually	Annually	Annually	Annually	Annually	Annually	Annually	Annually
Zoster													1 Dose after age 60

* for children or adolescents not previously immunized as infants
**optional, not necessary if doses 1 and 2 have been administered previously
***optional, not necessary if three doses administered previously
‡ recommended if some risk factor is present (e.g., medical condition, occupation, lifestyle).
△ for persons with no prior immunity

Figure 13B Suggested immunization schedule for infants, young children, and adolescents. Children who are immune-compromised may need additional immunizations. Always consult with your health-care provider for up-to-date information on recommended immunizations for your area.

Source: Centers for Disease Control and Prevention. Recommended immunization schedules for persons aged 0–18 years—United States, 2015.

—Continued

Continued—

United States. Pertussis (whooping cough), mumps, and chickenpox have also made a comeback in children and adults.

Figure 13B shows a recommended immunization schedule for children and adults. The United States is now committed to the goal of immunizing all children against the common types of childhood diseases listed. Diphtheria, whooping cough, *Haemophilus influenzae,* and *Pneumococcus* infection are all life-threatening respiratory diseases. Tetanus is characterized by muscular rigidity, including a locked jaw. These extremely serious infections are all caused by bacteria; the rest of the diseases listed are caused by viruses. *Varicella* virus causes chickenpox. Polio is a type of paralysis; measles and rubella, sometimes called German measles, are characterized by high fever and skin rashes; and mumps is characterized by fever and enlarged parotid and other salivary glands. Rotavirus causes a severe diarrheal illness, which can be fatal in infants and small children. Further, each one of these illnesses can cause fatal complications, though the incidence is rare.

Currently, immunizations for two forms of hepatitis viruses is recommended. Hepatitis A virus (HAV) is transmitted by contaminated water and food, especially raw shellfish. Hepatitis B virus (HBV) is a blood-borne pathogen that is spread in the United States mainly by sexual contact and intravenous drug use. Health-care workers who are exposed to blood or blood products are also at risk, and maternal-neonatal transmission is a possibility as well. Both of these viruses can cause a serious acute infection. However, HBV infection can also lead to chronic hepatitis and then cancer of the liver. It is relatively common for children to come into contact with HAV, but rare for a child to come into contact with hepatitis B virus. However, physicians feel that the potential for liver damage from

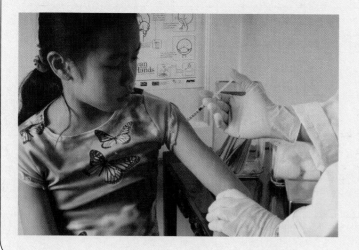

exposure to these viruses warrants vaccination early in life. Infants now receive their first dose of HBV vaccine before leaving the hospital, with booster immunizations at approximately 3 months and 12 months of age.

Cervical cancer has recently been linked to the occurrence of genital warts, a sexually transmitted disease caused by human papillomavirus (HPV). Therefore, a new vaccine for papillomavirus type 16, the most frequent cause of genital warts, has been developed. Currently, the U.S. Centers for Disease Control recommends three doses of human papillomavirus vaccine for young women between the ages of 9 and 18, and also for young men 9 to 21 years old. Perhaps more vaccines for sexually transmitted diseases will one day become available.

It's important to note that the need for immunizations and boosters doesn't stop with adolescence. Regular boosters for tetanus and diptheria are recommended every 10 years for adults. Further, everyone should get a yearly flu shot, beginning at age 6 months.

Even though most bacterial infections (e.g., tetanus) can be cured by antibiotic therapy, it's far better to be immunized. Some patients develop allergic reactions to antibiotics, which can be fatal. In addition, antibiotics kill beneficial bacteria as well as those that cause disease. These beneficial bacteria may have checked the spread of pathogens that now are free to multiply and to invade the body. This is why antibiotic therapy is often followed by a secondary infection, such as a vaginal yeast infection in women. Likewise, a serious diarrheal illness caused by a bacterium called *Clostridium difficile* (or "C-diff," as it is called by heath-care workers) often follows antibiotic use, especially in the elderly.

Antibiotic abuse also leads to resistant bacterial strains that are difficult to cure. For example, a once-harmless skin bacterium called *Staphylococcus aureus* now causes thousands of life-threatening infections and deaths every year. Post-surgical patients are particularly at risk, but anyone who is hospitalized can become infected. After decades of antibiotic exposure, this microbe (abbreviated MRSA, for *m*ethicillin-*r*esistant *S*taphylococcus *a*ureus) and others have become resistant to all but the strongest antibiotics.

Therefore, everyone should take advantage of appropriate vaccinations. Preventing a disease by becoming actively immune to it is certainly preferable to becoming ill and needing antibiotic therapy to be cured. Vaccination also benefits all of society by preventing the spread of disease to the most vulnerable among us: newborns and infants, the elderly, and those whose immune systems are weakened.

Influenza: A Constant Threat of Pandemic

"Over there, over there, Send the word, send the word over there—
That the Yanks are coming, The Yanks are coming . . . And we won't
come back till it's over, Over there!"
—"Over There," George M. Cohan

In April 1918, American soldiers who proudly marched off to World War I battlefields "over there" (referring to Europe, in the words of the popular patriotic ballad of the time), took an unwanted hitchhiker with them—a particularly virulent strain of influenza A, nicknamed the "Spanish Flu" because of the huge numbers of victims in that country. It quickly became a worldwide epidemic, or **pandemic,** causing an estimated 50 million deaths. Among the soldiers, Spanish flu caused twice as many deaths as those that resulted from combat. This particularly savage virus is once again a concern because of characteristics it shares with a modern-day virus: influenza A (H1N1), the so-called "swine flu" virus, which surfaced and rapidly became pandemic in 2009.

Influenza viruses, like all viruses, have two basic parts. The core of the virus is its genetic material, which can be either DNA or RNA. Influenza A viruses are RNA viruses (Fig. 13C). Like the HIV virus (see Medical Focus, pp. 307–308), their genetic material is covered by a protective protein coat called a capsid. An additional third piece, called the *lipid envelope,* contains protein spikes that allow the virus to attach to its own specific kind of host cell and not another nonmatching kind. The two types of protein spikes, abbreviated H (for hemagglutinin) and N (for neuraminidase), allow each type of influenza virus to be categorized. There are 16 known types of H proteins, and 9 known types of N proteins. Thus, many different combinations are possible, meaning there are many different ways for viruses to attach to different kinds of host cells. Each different combination of H and N represents a new form of flu virus. Avian influenza A, the "bird flu," which emerged in 1997 and again in 2006, is designated H5N1. The original Spanish flu virus is denoted H1N1, and the swine flu virus is its molecular "grandchild."

How do these new viruses emerge? It's well known that the genetic material of influenza virus has an extremely high mutation rate. A phenomenon called *antigenic drift* causes small changes in flu viruses, and explains why each year a new and different flu shot is needed. By contrast, *antigenic shift* is a major change that occurs very infrequently (between 10 and 40 years during the last century). Some of these mutations affect the structure of the protein spikes, so that a virus that previously could only infect a particular animal can "jump species." As the virus moves from one animal to the next, it can also swap genetic material and create new combinations. Scientists studying the ancestor H1N1 from 1918 have noted that it probably originated as a "bird flu" virus, which then mutated to infect human beings. Modern H1N1 flu virus contains bird, human, and swine DNA—and the swine DNA is from both Europe and Asia. Scientists call this four-way mutant a "quadruple reassortant" virus. The radical changes of an antigenic shift are what cause pandemics, because human populations have little or no immunity to a dramatically changed virus.

In 2009, public health officials were particularly concerned because the illness caused by H1N1 virus was eerily reminiscent of the 1918 Spanish flu epidemic. As in 2009, the original pandemic first began in the spring, not in the winter like most new viral illnesses. Both forms killed healthy young adults and children, not just the elderly and infirm. In June 2009, the World Health Organization declared the H1N1 swine flu outbreak to be a global pandemic. Fortunately, the pandemic persisted for only about a year.

In recent studies, scientists reconstructed the 1918 ancestor virus using genetic engineering, then compared its disease-causing ability to that of modern-day swine flu H1N1. The results were staggering: the Spanish flu virus is roughly *39,000* times more virulent than the newer H1N1 strains that now exist. That's good news for us as a species, because it seems to indicate that the viral mutants have grown weaker over time. However, the possibility of deadly antigenic shift will always remain. As we humans continue to travel the world more and more easily, we'll take our viruses (and those of our native animals) with us. If rapid genetic changes occur, new viral strains could spread efficiently from person to person, and another pandemic could happen. However, thanks to modern technology, we'll have a few more weapons in our antiviral arsenal than our great-grandparents had back in 1918. During the 2009 pandemic, communication tools like radio, television, and the Internet enabled public health officials all over the world to organize their efforts to fight this new influenza. Preventive vaccinations were quickly prepared, and drugs that prevent viral replication, and antibiotics for secondary infections (such as pneumonia) were available for treatment. Researchers continue to work on ways to better anticipate and recognize viral antigenic shift, so we can stay one step ahead of our influenza viruses.

In the meantime, your best approach to avoid influenza is to use common sense. Get your flu shot every fall, because each new vaccination is tailored to that season's viruses. Wash your hands often and well, don't touch your nose or eyes (because virus can enter your body when you do), and stay away from sick people. To keep from infecting others if you do get sick, don't go out in public. Cough or sneeze into tissues or your sleeve, and wash your hands before you touch anything. Further, if you do develop severe flu symptoms such as difficulty breathing, extremely high fever, or chest pain, seek prompt medical attention. Antiviral drugs have the best chance to successfully treat you if you seek help quickly.

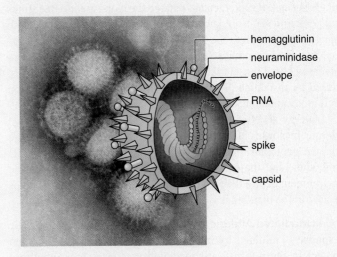

hemagglutinin
neuraminidase
envelope
RNA
spike
capsid

Figure 13C Influenza A virus is an RNA virus. It has protein spikes on its outer envelope that make it specific to one kind of host cell.

hormone is present in the urine of a pregnant woman. A monoclonal antibody can be used to detect this hormone; if it is present, the woman knows she is pregnant. Monoclonal antibodies are also used to identify infections. And because they can distinguish between cancerous and normal tissue cells, they are used to carry radioactive isotopes or toxic drugs to tumors, which can then be selectively destroyed.

Hypersensitivity Reactions

The immune system usually protects us from disease because it can distinguish self from nonself. Sometimes, however, it responds in a manner that harms the body, as when individuals develop allergies, suffer tissue rejection, or have an autoimmune disease. These are all forms of hypersensitivity reactions.

Allergies

Allergies are hypersensitivities to substances such as pollen or animal hair that ordinarily would do no harm to the body. The response to these antigens, called **allergens,** usually includes some degree of tissue damage. There are four types of allergic responses, but we will consider only two of them: IgE-mediated allergic response and T-cell mediated allergic response.

IgE-Mediated Allergic Response An **IgE-mediated allergic response** is often referred to as an **immediate allergic response** because it can occur within seconds of contact with an antigen. The response is caused by antibodies known as IgE (see Table 13.2). IgE antibodies are attached to the plasma membrane of mast cells in the tissues and also to basophils in the blood. When an allergen attaches to the IgE antibodies on these cells, mast cells release histamine and other substances that bring about the allergic symptoms. When pollen is an allergen, histamine stimulates the mucosal membranes of the nose and eyes to release fluid, causing the runny nose and watery eyes typical of **hay fever.** If a person has **asthma,** the airways leading to the lungs constrict, resulting in difficult breathing accompanied by wheezing. When food contains an allergen, nausea, vomiting, and diarrhea result.

Anaphylactic shock is an IgE-mediated allergic response that occurs because the allergen has entered the bloodstream. Bee stings and penicillin shots are known to cause this reaction because both inject the allergen into the blood. Anaphylactic shock is characterized by a sudden and life-threatening drop in blood pressure due to increased blood vessel dilation and permeability of the capillaries, caused by histamine. Taking epinephrine can delay this reaction until medical help is available.

Allergy shots sometimes prevent the onset of IgE-mediated allergic responses. It has been suggested that injections of the allergen may cause the body to build up high quantities of IgG antibodies, and these combine with allergens received from the environment before they have a chance to reach the IgE antibodies located in the membrane of mast cells and basophils.

T-Cell Mediated Allergic Response A **T-cell mediated allergic response** is initiated by memory T cells at the site of allergen contact in the body. T-cell mediated responses are sometimes referred to as *delayed* allergic responses, because they develop more slowly than IgE-mediated (*immediate*) responses. The allergic response is regulated by the cytokines secreted by both T cells and macrophages.

A classic example of a T-cell mediated allergic response is the skin test for tuberculosis (TB). When the test result is positive, the tissue where the antigen was injected becomes red and hardened. This shows that there was prior exposure to tubercle bacilli, the cause of TB. Contact dermatitis, which occurs when a person is allergic to poison ivy, jewelry, cosmetics, and many other substances that touch the skin, is also an example of a T-cell mediated allergic response.

Tissue Rejection

Certain organs, such as skin, the heart, and the kidneys, could be transplanted easily from one person to another if the body did not attempt to reject them. Rejection of transplanted tissue results because the recipient's immune system recognizes that the transplanted tissue is not "self." Cytotoxic T cells respond by causing disintegration of the transplanted tissue.

Organ rejection can be controlled by carefully selecting the organ to be transplanted and administering **immunosuppressive drugs.** It is best if the transplanted organ has the same type of HLA antigens as those of the recipient, because T_c cells recognize foreign HLA antigens. Two well-known immunosuppressive drugs, cyclosporine and tacrolimus, both act by inhibiting the response of T cells to cytokines.

Researchers hope that tissue engineering, including the production of organs that lack antigens or that can be protected in some way from the immune system, will one day do away with the problem of rejection.

Autoimmune Disease

When a person has an **autoimmune disease,** cytotoxic T cells or antibodies mistakenly attack the body's own cells as if they bear foreign antigens. Exactly what causes autoimmune diseases is not known. However, sometimes they occur after an individual has recovered from an infection. Recent research has also shown that autoimmune diseases are genetically inherited.

Virtually any tissue of the body can be a target for autoimmune disease. In **myasthenia gravis,** the synaptic junctions between motor nerves and skeletal muscles are destroyed. Neuromuscular junctions do not work properly, and muscular weakness results. In **multiple sclerosis (MS),** the myelin sheath of nerve fibers is attacked. This can cause a wide variety of neurological and neuromuscular symptoms: muscle weakness, paralysis, blurred vision, dizziness, and deafness. A person with **systemic lupus erythematosus (SLE)** has an extremely serious disease characterized by antibodies directed against multiple self antigens, including her own DNA molecules. Symptoms can include facial rashes, arthritis, anemia, and kidney damage. In **rheumatoid arthritis,** the joints are affected. Researchers suggest that heart damage following rheumatic fever, and type I diabetes are also autoimmune illnesses. **Crohn's disease** and **ulcerative colitis** are two diseases affecting the digestive system. Both cause severe diarrhea, which results in fluid and nutrient loss. As of yet, there are no cures for autoimmune diseases, but they can be managed with medication.

It's a fact that few of us appreciate: Autoimmune disease is approaching epidemic levels. It's the second-leading cause of chronic illness in the United States, and over 50 million Americans suffer from at least one autoimmune disease. Further, the incidence of autoimmune diseases has risen dramatically in the entire Western world over the past several decades, and research continues to show that many illnesses have an autoimmune link. Scientists have long suspected that autoimmune disease results from excessive cleanliness—the so-called "hygiene hypothesis." These researchers argue that because sanitation has eliminated exposure to microbes, the immune system is not properly regulated. As a result, lymphocytes produce T1 cell cytokines, a form that attacks body cells. In effect, an immune system that is poorly challenged by pathogens attacks and destroys body cells and tissues instead. Therapy for any one of these many disorders can be complicated and tricky, because each is a separate disease. Traditional treatments that suppress or eliminate the immune response can create fatal complications, including infections and cancer.

Enter swine whipworm, *Trichuris suis,* a lowly parasitic worm that normally infects pigs. Clinical researchers have chosen the eggs, or ova, as the agent for *helminthic therapy.* This unusual (and some would say, downright weird) treatment involves deliberately infecting people with parasites to create a counteractive immune response that appears to calm the inflammation that causes autoimmune disease. There are many human parasitic organisms, but *Trichuris suis* ova (TSO) are the parasite of choice because they can't reproduce in the human intestinal tract, so they don't normally infect humans. Subjects drink a solution containing TSO, and the eggs hatch in the intestine and latch onto its wall. There, they prompt production of T2 cell cytokines, which create a separate immune response, and in turn, diminish the inflammatory responses of autoimmune disease. After approximately two weeks, TSO die and are excreted. Clinical trials continue to show success in the treatment of Crohn's disease and ulcerative colitis using TSO, and more studies are planned.

Immune Deficiency

When a person has an immune deficiency, the immune system is unable to protect the body against disease. AIDS (see the Medical Focus on pages 307–308) is an example of an acquired immune deficiency. As a result of a weakened immune system, AIDS patients show a greater susceptibility to a variety of diseases, and they also have a higher risk of cancer. Immune deficiency may also be genetic (that is, inherited from one's parents) or congenital (due to failure of lymphatic tissue to develop). For example, in some infants the thymus gland fails to develop, producing a child with severe immune deficiency. Infrequently, a child may be born with an impaired B- or T-cell system caused by a defect in lymphocyte development. In **severe combined immunodeficiency disease (SCID),** a genetic disorder, both antibody and cell-mediated immunity are lacking or inadequate. Without treatment, even common infections can be fatal. Replacing defective stem cells with healthy ones through bone marrow transplantation can often produce a cure.

Content CHECK-UP!

11. Which of the following is a form of active immunity?

   a. transfer of antibodies from mother to fetus across the placenta

   b. transfer of antibodies from mother to baby in breast milk

   c. immunization for hepatitis B virus

   d. gamma globulin injection for hepatitis B virus

12. Describe the two forms of passive immunity: natural and artificial. How is each one obtained? Is this form of immune response as effective or long-lasting as active immunity? Why or why not?

13. Which term describes the amount of antibody present in a blood plasma sample?

   a. booster   c. hematocrit

   b. vaccine   d. titer

Answers in Appendix A.

13.5 Effects of Aging

10. Describe the anatomical and physiological changes that occur in the immune system as we age.

With advancing age, people become more susceptible to all types of infections and disorders because the immune system exhibits lower levels of function. One reason is that the thymus gland degenerates. Having reached its maximum size in early childhood, it begins to shrink after puberty and has virtually disappeared by old age. As the gland decreases in size, so does the number of T cells. The T cells remaining do not respond to foreign antigens; therefore, the chance of having cancer also increases with age.

Among the elderly, the B cells sometimes fail to form clones. Other B cells may form clones, but the antibodies released may not function well. Therefore, infections are more common among the elderly. In addition, the antibodies of older people are more likely to attack the body's own tissues, increasing the incidence of autoimmune diseases.

The response of elderly individuals to vaccines is decreased. However, considering that their overall level of immune response is low, it is better that these people be vaccinated than not. For this reason, elderly individuals are encouraged to get an influenza (flu) vaccination each year.

Human Systems Work Together

Integumentary System

Lymphatic vessels pick up excess tissue fluid; immune system protects against skin infections.

Skin serves as a barrier to pathogen invasion; Langerhans cells phagocytize pathogens; protects lymphatic vessels.

Skeletal System

Lymphatic vessels pick up excess tissue fluid; immune system protects against infections.

Red bone marrow produces leukocytes involved in immunity.

Muscular System

Lymphatic vessels pick up excess tissue fluid; immune system protects against infections.

Skeletal muscle contraction moves lymph; physical exercise enhances immunity.

Nervous System

Lymphatic vessels pick up excess tissue fluid; immune system protects against infections of nerves.

Microglia engulf and destroy pathogens.

Endocrine System

Lymphatic vessels pick up excess tissue fluid; immune system protects against infections.

Thymus is necessary to maturity of T lymphocytes.

How the Lymphatic System works with other body systems.

tonsils

thymus

thoracic duct

cervical lymph nodes

axillary lymph nodes

spleen

inguinal lymph nodes

popliteal lymph node

lymphatic vessel

Cardiovascular System

Lymphatic organs produce and store formed elements; lymphatic vessels transport leukocytes and return tissue fluid to blood vessels; spleen serves as blood reservoir, filters blood.

Blood vessels transport leukocytes and antibodies; blood services lymphatic organs and is source of tissue fluid that becomes lymph.

Respiratory System

Lymphatic vessels pick up excess tissue fluid; immune system protects against respiratory tract and lung infections.

Tonsils and adenoids occur along respiratory tract; breathing aids lymph flow.

Digestive System

Lacteals absorb fats; Peyer patches prevent invasion of pathogens; appendix contains lymphatic tissue.

Digestive tract provides nutrients for lymphatic organs; stomach acidity prevents pathogen invasion of body.

Urinary System

Lymphatic system picks up excess tissue fluid, helping to maintain blood pressure for kidneys to function; immune system protects against infections.

Kidneys control volume of body fluids, including lymph.

Reproductive System

Female immune system does not attack sperm or fetus, even though they are foreign to the body.

Sex hormones influence immune functioning; acidity of vagina helps prevent pathogen invasion of body; milk passes antibodies to newborn.

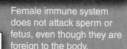

13.6 Homeostasis

11. Describe how the lymphatic system works with other systems of the body to maintain homeostasis.

The three functions of the lymphatic system listed on page 294 assist homeostasis. The lymphatic system helps the digestive system by absorbing fats. In the process of absorbing dietary fats, lacteals also absorb fat-soluble vitamins. The lymphatic system assists the cardiovascular system by absorbing excess tissue fluid. The lymphatic vessels return excess tissue fluid as lymph to cardiovascular veins in the thorax, thus helping to maintain the blood volume and pressure needed for capillary exchange.

The lymphatic organs and the immune system protect us from infectious diseases. Nonspecific defenses precede specific immunity. The skin and the mucous membranes of the respiratory, digestive, reproductive, and urinary systems all resist infection. If a pathogen should enter the body, the infection is localized as much as possible by an inflammatory reaction. Phagocytic white blood cells engulf as many pathogens as possible. If the pathogens do enter the blood, complement proteins work in diverse ways to keep the blood free of disease-causing organisms and their toxins.

Specific defenses depend on blood cells, and the lymphocytes and macrophages play central roles. B and T cells have antigen receptors and can distinguish self from nonself. Antigen binding selects which B or T cells will undergo clonal expansion. B cells recognize an antigen directly, but T cells must have the antigen displayed by an APC in the groove of an HLA antigen. Plasma cells (mature B cells) produce antibodies, but T cells kill virus-infected and cancer cells outright.

The lymphatic organs play a central role in immunity, and the skeletal system as well. White blood cells are made in the red bone marrow where B cells also mature. T cells mature in the thymus. The spleen filters the blood directly. Clonal expansion of lymphocytes occurs in the lymph nodes, which also filter the lymph.

A strong connection exists between the immune, nervous, and endocrine systems. Lymphocytes have receptors for a wide variety of hormones, and the thymus produces hormones that influence the immune response. Cytokines produced by T lymphocytes affect the brain's temperature control center. The high body temperature of a fever is thought to create an unfavorable environment for the foreign invaders. Further, cytokines promote behaviors that tend to make us take care of ourselves by causing sleepiness and appetite loss. A close connection between the immune and endocrine systems is illustrated by the ability of cortisone to decrease the inflammatory reaction in the joints.

Selected New Terms

Basic Key Terms

active immunity (ăk′tĭv ĭ-myū′nĭ-tē), p. 305

antibodies (ăn′tĭ-bŏd′ēz), p. 301

antibody-mediated immunity (ăn′tĭ-bŏd″ē mē′dē-āt″ĕd ĭ-myū′nĭ-tē), p. 301

antigen-presenting cell (ăn′tĭ-jĕn prē-zĕnt-ĭng sĕl′), p. 303

antigens (ăn′tĭ-jĕn), p. 300

apoptosis (ăp′ō-tō′sĭs), p. 301

appendix (ă-pĕn′dĭks), p. 298

autoimmune response (ŏt-ō-ĭ-myūn′ rē-spŏns′), p. 301

B lymphocytes (B lĭm′fō-sītz), p. 301

capsid (kăp′sĭd), p. 307

cell-mediated immunity (sĕl mē′dē-āt″ĕd ĭ-myū′nĭ-tē), p. 304

clonal selection theory (klōn′ŭl sĕ-lĕk′shŭn thē′ō-rē), p. 301

clone (klōn), p. 301

complement system (kŏm′plĕ-mĕnt sĭs′tĕm), p. 300

cytokines (sī′tō-kīnz), p. 301

cytotoxic T cells (sī′tō-tŏk′sĭk T sĕlz), p. 304

helper T cells (hĕlp′ĕr T sĕlz), p. 304

histamine (hĭs′tŭh-mēn), p. 298

HLA (human leukocyte antigens) (hyū′mŭn lū′kō-sīt ăn′tĭ-jĕnz), p. 303

humoral immunity (hyū′mō-răl ĭ-myū′nĭ-tē), p. 301

immune system (ĭ-myūn′ sĭs′tĕm) p. 294

immunity (ĭ-myū′nĭ-tē), p. 294

inflammatory reaction (ĭn-flăm′ŭh-tō″rē rē-ăk′shŭn), p. 298

lacteal (lăk′tē-ăl), p. 294

lingual tonsil (lĭng′gwăl tŏn′sĭl), p. 297

lymph (lĭmf), p. 294

lymphatic nodule (lĭm-făt′ĭk nŏd′yūl), p. 297

lymphatic organ (lĭm-făt′ĭk ŏr′gŭn), p. 294

lymphatic system (lĭm-făt′ĭk sĭs′tĕm), p. 294

lymphatic vessels (lĭm-făt′ĭk vĕs′lz), p. 294

lymph node (lĭmf nōd), p. 297

macrophages (măk′rā-fāj″ēz), p. 298

major histocompatibility complex (mā′jōr hĭs′tō-kŏm-păt-ĭ-bĭl′ĭ-tē kŏm′plĕks), p. 303

mast cells (măst sĕlz), p. 298

memory B cells (mĕm′ō-rē B sĕlz), p. 301

memory T cells (mĕm′ō-rē T sĕlz), p. 304

monoclonal antibodies (mŏn″ō-klōn′ăl ăn′tĭ-bŏd″ēz), p. 306

natural killer cells (năt′ū-răl kĭl′ĕr sĕlz), p. 300

palatine tonsil (păl′ă-tīn tŏn′sĭl), p. 297

passive immunity (păs′ĭv ĭ-myū′ nĭ-tē), p. 305

perforin (pĕr′fŏr-ŭn), p. 304

Peyer patches (pī′ĕr păch′ēz), p. 298

pharyngeal tonsil (fū-rĭn′gē-ăl tŏn′sĭl), p. 297

plasma cells (plăz′mŭh sĕlz), p. 301

protease (prō′tē-ās), p. 307

pus (pŭs), p. 300

red bone marrow (rĕd bōn măr′ō), p. 294

retrovirus (rĕt′rō-vī′rŭs), p. 307

right lymphatic duct (rīt lĭm-făt′ĭk dŭkt), p. 294

spleen (splēn), p. 296

thoracic duct (thō-răs′ĭk dŭkt), p. 294

thymus gland (thī′mŭs glănd), p. 296

T lymphocytes (T lĭm′fō-sītz), p. 301

tonsils (tŏn′sĭlz), p. 297

Clinical Key Terms

AIDS (acquired immunodeficiency syndrome) (ŭh-kwīrd′ ĭm′ yŭh-nō-dĭ-fĭsh″ ŭn-sē sĭn′drōm), p. 307

allergens (ăl′ĕr-jĕnz), p. 312

allergies (ăl′ĕr-jēz), p. 312

anaphylactic shock (ăn″ŭh-fĭ-lăk′tĭk shŏk), p. 312

antibody titer (ăn′tĭ-bŏd″ē tī′tĕr), p. 305

asthma (ăz′mŭh), p. 312

autoimmune disease (ăw′tō-ĭ-myūn′ dĭ-zēz′), p. 312

Crohn's disease (krōnz dĭ-zēz′), p. 312

edema (ĕ-dē′mŭh), p. 294

elephantiasis (ĕ″lŭh-fŭn-tī′ŭh-sĭs), p. 297

gamma globulin (găm′ŭh glŏb′yū-lĭn), p. 306

hay fever (hā fē′vĕr), p. 312

Hodgkin disease (hŏj′kĭn dĭ′zĕz′), p. 297

IgE-mediated allergic response (ī gē ē mē′dē-āt″ĕd ă-lĕr′jĭk rē-spŏns′), p. 312

immediate allergic response (ĭm-mē′dē-ĭt ŭh-lŭr′ jĭk rē-spŏns′), p. 312

immunization (ĭ″myū-nĭ-zā′shŭn), p. 305

immunosuppressive drugs (ĭ-myū″nō-sŭ-prĕs″ĭv drŭgz), p. 312

interferon (ĭn″tĕr-fēr′ŏn), p. 300

interleukin (ĭn-tĕr-lūk′ĭn), p. 305

lymphadenitis (lĭm-făd″ē-nī′tĭs), p. 297

lymphangitis (lĭm″făn-jī′tĭs), p. 297

lymphedema (lĭm″fē-dē′măh), p. 297

lymphoma (lĭm-fō′mŭh), p. 297

multiple sclerosis (mŭl′tĭ-pŭl sklĕr-ō′sĭs), p. 312

myasthenia gravis (mī′-ŭhs-thē″ nē-ŭh grăʹ vis) p. 312

opportunistic infection (ŏp″ĕr-tū-nĭs′tĭk ĭn-fĕk′shŭn), p. 307

pandemic (păn-dĕm′ĭk), p. 311

pulmonary edema (pŭl′mō-nĕr″ē ĕ-dē′mŭh), p. 297

rheumatoid arthritis (rū′mŭh-tōyd ăr-thrī′tĭs), p. 312

severe combined immunodeficiency disease (sĕ-vēr′ kŭm-bīnd′ ĭm′ yŭh-nō-dĭ-fĭsh″ ŭn-sē dĭ-zēz′), p. 313

systemic lupus erythematosus (sĭs-tĕm′ĭk lū′pŭs ĕr-ĭ-thē-mŭh-tō′sŭs), p. 312

T-cell mediated allergic response (tē sĕl mē′dē-ā-tĕd ŭh-lŭr′ jĭk rē-spŏns′), p. 312

ulcerative colitis (ŭl′sĕr-ă-tĭv cō-lī′tĭs), p. 312

vaccines (văk-sēnz′), p. 305

Summary

13.1 Lymphatic System

The lymphatic system consists of lymphatic vessels and lymphatic organs.

  A. The lymphatic vessels return excess tissue fluid to the bloodstream, absorb fats at intestinal villi, and help the immune system defend the body against disease.

  B. Lymphatic capillaries have thin walls. Larger vessels are structured the same as cardiovascular veins, with valves that prevent backward flow. The largest lymphatic vessels, called the right lymphatic duct and thoracic duct, return blood to the right and left subclavian veins.

13.2 Organs, Tissues, and Cells of the Immune System

Lymphocytes are produced and accumulate in the lymphatic organs.

  A. The primary lymphatic organs are the red bone marrow, where blood cells are produced, and the thymus, which houses and matures T lymphocytes.

  B. The secondary lymphatic organs include the spleen, lymph nodes, tonsils, Peyer patches, and appendix. Lymph is cleansed of pathogens and/or their toxins in both lymph nodes and spleen, and blood pathogens are destroyed in the spleen.

  C. T lymphocytes mature in the thymus, while B lymphocytes mature in the red bone marrow, where all blood cells are produced. White blood cells are necessary for both nonspecific and specific defenses.

13.3 Nonspecific and Specific Defenses

Immunity involves nonspecific and specific defenses.

  A. Nonspecific defenses include barriers to entry, the inflammatory reaction, nonspecific white blood cells, natural killer cells, and protective proteins.

  B. Specific defenses require B lymphocytes and T lymphocytes, also called B cells and T cells. B cells undergo clonal selection with production of plasma cells and memory B cells after their antigen receptors combine with a specific antigen. Plasma cells secrete antibodies and eventually undergo apoptosis. Plasma cells are responsible for antibody-mediated immunity. Antibodies are Y-shaped molecules with two binding sites for a specific antigen. Memory B cells remain in the body and produce antibodies if the same antigen enters the body at a later date.

  C. T cells are responsible for cell-mediated immunity. The two main types of T cells are cytotoxic T cells and helper T cells. Cytotoxic T cells kill virus-infected or cancer cells on contact because they bear a nonself antigen. Helper T cells produce cytokines and stimulate other immune cells. Like B cells, each T cell bears antigen receptors. However, for a T cell to recognize an antigen, the antigen must be presented by an antigen-presenting cell (APC), usually a macrophage, in the groove of an HLA (human leukocyte antigen). Thereafter, the activated T cell undergoes clonal expansion until the illness has been stemmed. Then most of the activated T cells undergo apoptosis. A few cells remain, however, as memory T cells. Cytokines, including interferon and interleukins, are used in an attempt to promote the body's ability to recover from cancer and to treat AIDS.

13.4 Creating an Immune Response

  A. Immunity can be induced in various ways. Vaccines are available to induce long-lasting, active immunity.

  B. Antibodies sometimes are available to provide an individual with temporary, passive immunity.

  C. Monoclonal antibodies are produced in the laboratory and used for diagnosis and treatment purposes.

  D. Allergic responses occur when the immune system reacts vigorously to substances not normally recognized as foreign. IgE-mediated or immediate allergic responses, usually consisting of coldlike symptoms, are due to the activity of antibodies. T-cell mediated allergic responses, such as contact dermatitis, are due to the activity of T cells.

  E. Autoimmune responses occur when immune system cells mistakenly attack body tissues and organs. Myasthenia gravis, multiple sclerosis, and rheumatoid arthritis are a few of the many autoimmune diseases.

  F. Immune system deficiencies can be acquired, such as AIDS, or genetic, such as SCID.

13.5 Effects of Aging

The thymus gets smaller as we age, and fewer antibodies are produced. The elderly are at great risk of infections, cancer, and autoimmune diseases.

13.6 Homeostasis
The lymphatic system assists the cardiovascular system by returning excess tissue fluid to the bloodstream. It assists the digestive system by absorbing fats from the intestinal tract, and it assists the immune system through the functioning of its lymphatic organs.

Study Questions

1. What is the lymphatic system, and what are its three functions? (p. 294)
2. Describe the structure and the function of red bone marrow, the thymus, the spleen, lymph nodes, and the tonsils. (pp. 294–298)
3. What are the body's nonspecific defense mechanisms? (pp. 298–300)
4. Describe the inflammatory reaction and give a role for each type of cell and molecule that participates in the reaction. (pp. 298–300)
5. What is the clonal selection theory as it applies to B cells? B cells are responsible for which type of immunity? (p. 301)
6. Describe the structure of an antibody, and define the terms *variable regions* and *constant regions*. (pp. 301–302)
7. Describe the clonal selection theory as it applies to T cells. (pp. 303–304)
8. Name the two main types of T cells and state their functions. (pp. 304–305)
9. What are cytokines and how are they used in immunotherapy? (p. 305)
10. How is active immunity artificially achieved? How is passive immunity achieved? (pp. 305–306)
11. How are monoclonal antibodies produced, and what are their applications? (pp. 306, 312)
12. Discuss allergies, tissue rejection, and autoimmune diseases as they relate to the immune system. (p. 312)
13. Explain the different causes for immune deficiency diseases, and describe their effects. (pp. 307–308, 313)
14. How do the lymphatic and immune systems help maintain homeostasis? (pp. 314–315)
15. How does the skeletal system assist the immune system in maintaining homeostasis? (pp. 314–315)

Learning Outcome Questions

Fill in the blanks.

1. Lymphatic vessels contain _____, which close, preventing lymph from flowing backward.
2. _____ and _____ are two types of white blood cells produced and stored in the lymphatic organs.
3. T lymphocytes have matured in the _____.
4. Lymph nodes cleanse the _____ while the spleen cleanses the _____.
5. Barriers to entry, protective proteins, and the inflammatory reaction are all examples of _____ defenses.
6. _____ and _____ are phagocytic white blood cells.
7. Proteins that function to form holes in bacterial cell walls comprise the _____ system.
8. A stimulated B cell divides and differentiates into antibody-secreting _____ cells and also into _____ cells that are ready to produce the same type of antibody at a later time.
9. B cells are responsible for _____ mediated immunity.
10. Cytotoxic T cells are responsible for _____ mediated immunity.
11. Immunization with _____ brings about active immunity.
12. Whereas _____ immunity occurs when an individual is given antibodies to combat a disease, _____ immunity occurs when an individual develops the ability to produce antibodies against a specific antigen.
13. Allergic reactions are associated with the release of _____ from mast cells.

Medical Terminology Exercise

After studying this chapter, see if you can derive the definitions for the medical terms listed at the right. Many of the prefixes and suffixes used to create these terms can be found throughout the chapter. For additional help, use McGraw-Hill Connect™ at www.mcgrawhillconnect.com and consult Appendix B.

1. metastasis (mĕ-tăs′tŭh-sĭs)
2. allergist (ăl′ĕr-jĭst)
3. immunotherapy (ĭ-myū″nō-thĕr′ŭh-pē)
4. splenorrhagia (splē″nō-rā′jē-ŭh)
5. lymphadenopathy (lĭm-făd″ĕ-nŏp′ŭh-thē)
6. lymphangiography (lĭm-făn″jē-ŏg′rŭh-fē)
7. eosinophilia (ē′ōh-sĭn′ō-fĭl′ē-ŭh)
8. thymectomy (thī-mĕk′tō-mē)
9. lymphopenia (lĭmf′ō-pē′nē-ŭh)
10. agammaglobulinemia (ā-găm′ŭh-glŏb′yū-lĭ-nē′mē-ŭh)
11. pyemia (pī-ē′mē-ŭh)
12. tonsillectomy (tŏn′sĭ-lĕk′tĭ-mē)
13. hypersensitivity (hī′pĕr-sĕn-sĭ-tiv′ĭ-tē)

Online Study Tools

▣ LEARNSMART® ▣ connect®

APR

Anatomy & Physiology REVEALED includes cadaver photos that allow you to peel away layers of the human body to reveal structures beneath the surface. This program also includes animations, radiologic imaging, audio pronunciations, and practice quizzing. To learn more visit www.aprevealed.com

14 The Respiratory System

Hold your breath, drink from a glass (sideways), eat a spoonful of sugar, have someone startle or scare you, pull hard on your tongue—all of these are "cures" for **singultus,** also known as the hiccups (and any of them might work). Technically, hiccups are spasms of the respiratory diaphragm muscle, caused by irritation of the phrenic nerve which controls it. (You learned about the phrenic nerve in Chapter 8.) Everyone gets hiccups, even babies in the womb. Common causes include excessive alcohol and nicotine, taking a bite of food that's too big, and excessive stress (for example, if you've been crying). Normally, hiccups last only a few minutes at most, and they're harmless. However, they can continue for a long time—the longest-recorded bout lasted six months—and might be a sign of much more serious conditions such as a heart attack or a tumor in the chest. If your hiccups last more than three hours, and you're in pain, too, it's time to see a doctor. You can read more about the diaphragm and phrenic nerve on pages 327–328. And if you want to give up smoking to stop your hiccups, see the Medical Focus on pages 333–334.

Learning Outcomes

After you have studied this chapter, you should be able to:

14.1 The Respiratory System

1. Describe the events that occur during respiration.
2. Describe the structure and function of the respiratory system organs.
3. Explain the structure and importance of the respiratory membrane.

14.2 Mechanism of Breathing

4. Outline the process of ventilation, including inspiration and expiration.
5. Describe vital capacity and its relationship to other measurements of breathing capacity.
6. Tell where the respiratory center is located, and explain how it controls the normal breathing rate.

14.3 Gas Exchange and Transport

7. Describe the process of gas exchange in the lungs and the tissues.
8. Explain how oxygen and carbon dioxide are transported in the blood.

14.4 Respiration and Health

9. Name and describe the various infections of the respiratory tract.
10. Describe the effects of smoking on the respiratory tract and on overall health.

14.5 Effects of Aging

11. Describe the anatomical and physiological changes that occur in the respiratory system as we age.

14.6 Homeostasis

12. Describe how the respiratory system works with other systems of the body to maintain homeostasis.

I.C.E.—In Case of Emergency
Lung Collapse

Medical Focus
The Most-Often-Asked Questions About Tobacco and Health

What's New
Bronchial Thermoplasty: A Surgical Treatment for Asthma

Human Systems Work Together
Respiratory System

14.1 The Respiratory System

1. Describe the events that occur during respiration.
2. Describe the structure and function of the respiratory system organs.
3. Explain the structure and importance of the respiratory membrane.

The primary function of the respiratory system is gas exchange—allowing oxygen from the air to enter the blood and carbon dioxide from the blood to exit into the air. During **inspiration,** or inhalation (breathing in), and **expiration,** or exhalation (breathing out), air is conducted toward or away from the lungs by a series of cavities, tubes, and openings, illustrated in Figure 14.1.

The respiratory system also works with the cardiovascular system to accomplish these four respiratory actions:

1. pulmonary ventilation (breathing): the entrance and exit of air into and out of lungs
2. external respiration: the exchange of gases (oxygen and carbon dioxide) between air and blood
3. internal respiration: the exchange of gases between blood and tissue fluid
4. transport of gases to and from the lungs and the tissues

Cellular respiration, which produces ATP, uses the oxygen and produces the carbon dioxide that makes gas exchange with the environment necessary. Without a continuous supply of ATP, the cells cease to function. The four actions listed here allow cellular respiration to continue.

The Respiratory Tract

Table 14.1 traces the path of air from the nose to the lungs. As air moves in along the airways, it is cleansed, warmed, and moistened by the mucous membrane lining of the nasal cavity. Cleansing is accomplished by coarse hairs just inside the nostrils and by cilia and mucus in the nasal cavities and the other airways of the respiratory tract. Nasal hairs, cilia, and mucus act as screening devices for dust, dirt, pollen, fungal spores, etc., in inhaled air. Cilia inside the nasal cavity beat backwards, sweeping the mucous sheet into the throat where it can be swallowed. Lysozyme in the mucus helps to kill bacteria, while lymphocytes populating underlying tissue create an immune response whenever pathogens are accidentally inhaled. In the trachea and other airways, the cilia beat upward, carrying mucus, dust, and other trapped potential contaminants

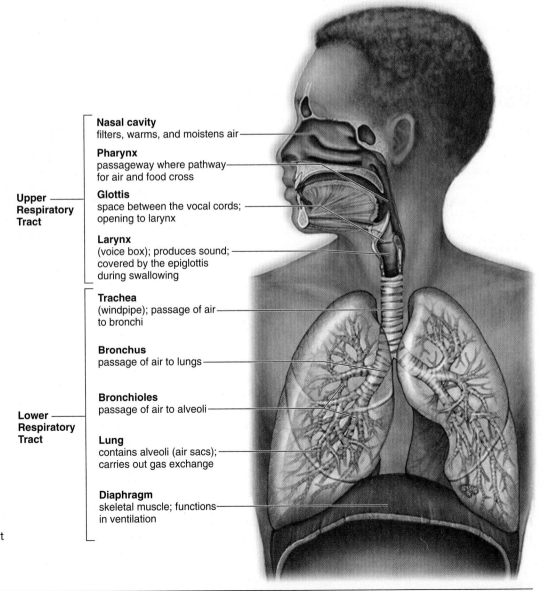

Upper Respiratory Tract

Nasal cavity
filters, warms, and moistens air

Pharynx
passageway where pathway for air and food cross

Glottis
space between the vocal cords; opening to larynx

Larynx
(voice box); produces sound; covered by the epiglottis during swallowing

Lower Respiratory Tract

Trachea
(windpipe); passage of air to bronchi

Bronchus
passage of air to lungs

Bronchioles
passage of air to alveoli

Lung
contains alveoli (air sacs); carries out gas exchange

Diaphragm
skeletal muscle; functions in ventilation

APǀR **Figure 14.1 The respiratory tract.** Note the structures of the upper respiratory tract and the lower respiratory tract. The tract extends from the nasal cavities to the lungs.

TABLE 14.1 Path of Air

Structure	Description	Function
UPPER RESPIRATORY TRACT		
Nasal cavities	Hollow spaces in nose	Filter, warm, and moisten air
Pharynx	Chamber posterior to oral cavity; lies between nasal cavity and larynx	Connection to surrounding regions
Glottis	Opening into larynx	Passage of air into larynx
Larynx	Cartilaginous organ that houses the vocal cords; voice box	Sound production
LOWER RESPIRATORY TRACT		
Trachea	Flexible tube that connects larynx with bronchi	Passage of air to bronchi
Bronchi	Paired tubes inferior to the trachea that enter the lungs	Passage of air to lungs
Bronchioles	Branched tubes that lead from bronchi to alveoli	Passage of air to each alveolus
Lungs	Soft, cone-shaped organs that occupy lateral portions of thoracic cavity	Gas exchange
		Acid-base balance; conversion of angiotensin I to angiotensin II
Alveoli	Air sacs that branch from bronchioles	Passage of gases into and out of pulmonary capillaries
Pulmonary capillaries	Capillaries that cover surface of alveoli	Receive oxygen from alveoli; oxygenation of hemoglobin; deliver CO_2 from tissues to alveoli

into the pharynx. This important protective mechanism is called the **mucociliary escalator.** Once in the pharynx, the mucus accumulation can be swallowed (thus, destroyed by stomach acid and digestive enzymes) or spit out. Inhaled air is also warmed by heat given off by blood vessels lying close to the surface of the lining of the airways, and it is moistened by the wet surface of the mucous membrane.

Conversely, as air moves out during expiration, it cools and loses its moisture. As the air cools, it deposits its moisture on the lining of the trachea and the nose, and the nose may even drip as a result of this condensation. However, the exhaled air still retains so much moisture that when we exhale on a cold day, it condenses and forms a small cloud.

The Nose

The **nose,** a prominent feature of the face, is the only external portion of the respiratory system. Air enters the nose through external openings called **nostrils.** The nose contains two **nasal cavities,** which are narrow canals separated from one another by a septum composed of bone and cartilage (Fig. 14.2). Mucous membrane lines the nasal cavities. The nasal **conchae** (sing., concha) are bony ridges that project laterally into the nasal cavity. They increase the surface area for moistening and warming air during inhalation and for trapping water droplets during exhalation. Odor receptors are on the cilia of cells in the olfactory epithelium, located high in the recesses of the nasal cavities.

The tear (lacrimal) glands drain into the nasal cavities by way of the nasolacrimal canals (see Fig. 9.5). For this reason, crying produces a runny nose. The nasal cavities also communicate with the **paranasal sinuses,** air-filled spaces that reduce the weight of the skull and act as resonating chambers for the voice. Sinuses can be

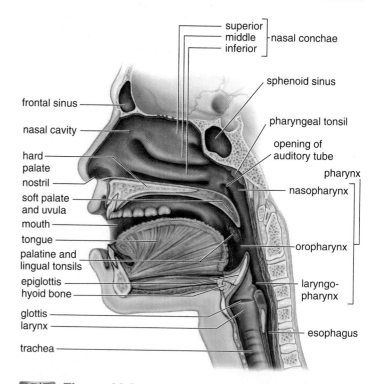

AP|R **Figure 14.2 The path of air.** Air flows through the nose and mouth, which are upper respiratory tract structures, to enter the pharynx. The pharynx opens into the trachea. The trachea is part of the lower respiratory tract.

found in the maxillae and in the frontal, ethmoid, and sphenoid bones. If the ducts leading from the sinuses become inflamed, fluid may accumulate, causing a sinus headache. The nasal cavities are separated from the oral cavity by the palate, a partition that has two portions. Anteriorly, the hard palate is supported by the maxilla and palatine bones. Posteriorly, the soft palate is composed of muscle and glandular tissue.

The Pharynx

The **pharynx** is a funnel-shaped passageway that connects the nasal and oral cavities to the larynx. Consequently, the pharynx, commonly referred to as the "throat," has three parts: the naso-pharynx, where the nasal cavities open posterior to the soft palate; the oropharynx, where the oral cavity joins the pharynx; and the laryngopharynx, which opens into the larynx. The soft palate has a soft extension called the *uvula* that can be seen hanging down into the oropharynx.

The single pharyngeal tonsil (or *adenoids*) in the posterior nasal cavity is the primary lymphatic tissue defense for breathing. Its action is supported by the paired palatine tonsils at the rear of the oropharynx, as well as the lingual tonsils at the base of the tongue. The tonsils contain lymphocytes that protect against invasion of inhaled pathogens. Here, both B cells and T cells can respond to antigens that may subsequently invade internal tissues and fluids. Thus, the respiratory tract assists the immune system in maintaining homeostasis.

In the pharynx, the air and food passages cross because the larynx, which receives air, is anterior to the esophagus, which receives food. The larynx lies at the top of the trachea. The larynx and trachea are normally open, allowing air to pass, but the esophagus is normally collapsed and opens only when a person swallows.

Begin Thinking Clinically

If your upper airways narrow or collapse when you inhale during sleep, you may have **obstructive sleep apnea (OSA). Apnea** is defined as absent respiration. The affected person has long periods of apnea during sleep, during which he or she actually stops breathing. Loud snoring and gasping are symptoms of OSA. The condition causes a very poor night's sleep, making the sufferer much more likely to have a daytime accident. Long-term, OSA causes hypertension (see Chapter 12 for a review of hypertension). As you study the upper airway structures, which do you think might be involved in causing OSA? How might OSA be treated?

Answer and discussion in Appendix A.

The Larynx

The cartilaginous **larynx** lies posterior to the thyroid cartilage, and serves as a passageway for air between the pharynx and the trachea (Fig.14.3a). The larynx can be pictured as a triangular box whose apex, the Adam's apple, is located at the anterior neck. The larynx is called the "voice box" because it houses the vocal cords (Fig. 14.3b). The **vocal cords** are mucosal folds supported by elastic ligaments, and the slit between the vocal cords is called the **glottis** (Fig. 14.3c).

AP|R **Figure 14.3 Anatomy of the larynx. (a)** Anterior view of the larynx. The larynx lies posterior to the thyroid cartilage of the neck (commonly called the Adam's apple). **(b)** Sagittal view of the larynx, showing location of the vocal cords. **(c)** Viewed from above, the vocal cords can be seen to stretch from anterior to posterior across the larynx. When air is forced past the vocal cords, they vibrate, producing sound.

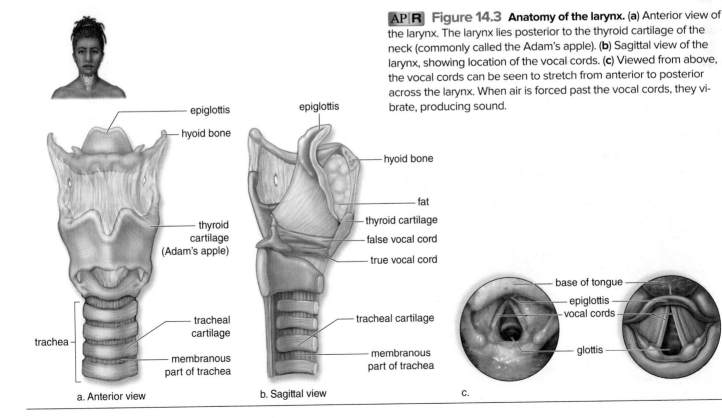

a. Anterior view

b. Sagittal view

c.

When air is expelled past the vocal cords through the glottis, the vocal cords vibrate, producing sound. At puberty, the growth of the larynx and the vocal cords is much faster and accentuated in males than in females, causing males to have a more prominent Adam's apple and deeper voices. The voice "breaks" in the young male because he cannot control the longer vocal cords. These changes cause the lower pitch of the voice in males.

Voice pitch is regulated when we speak and sing by changing the tension on the vocal cords. The greater the tension, as when the glottis becomes more narrow, the higher the pitch. When the vocal cords relax, the glottis is wider, and the pitch is lower. The loudness, or intensity, of the voice depends upon the amplitude of the vibrations—that is, how much the vocal cords vibrate.

Coughing, sneezing, and hiccuping also involve the glottis. When we cough, the glottis closes, then suddenly opens as a blast of air is forced upward from the lower respiratory tract. Sneezing is much like coughing, except that the air blast is directed through the nasal passages. As you read in the chapter introduction, a hiccup is caused by spasm of the diaphragm muscle while the glottis is closed.

When food is swallowed, the larynx moves upward against the **epiglottis,** a flap of elastic cartilage that prevents food from passing through the glottis into the larynx. You can detect this movement by placing your hand gently on your larynx and swallowing.

The Trachea

The **trachea,** or windpipe, is a tube connecting the larynx to the primary bronchi. The trachea lies anterior to the esophagus and is held open by C-shaped cartilaginous rings. The open part of the C-shaped rings forms the anterior wall of the esophagus, and this allows the esophagus to expand when swallowing. The mucosa that lines the trachea has a layer of pseudostratified ciliated columnar epithelium (Fig. 14.4a, b). Pseudostratified—"false layers"—means that while the epithelium appears to be layered, actually each cell touches the basement membrane. As mentioned previously, the epithelial cilia sweep mucus produced by goblet cells, along with trapped debris, toward the pharynx. You'll recall that this mechanism is called the *mucociliary escalator* (Fig. 14.4c). Smoking is known to destroy these cilia, and consequently the soot in cigarette smoke collects in the lungs. The effects of smoking are discussed more fully in the Medical Focus on pages 333–334.

If the trachea is blocked because of illness or the accidental swallowing of a foreign object, it is possible to insert a breathing

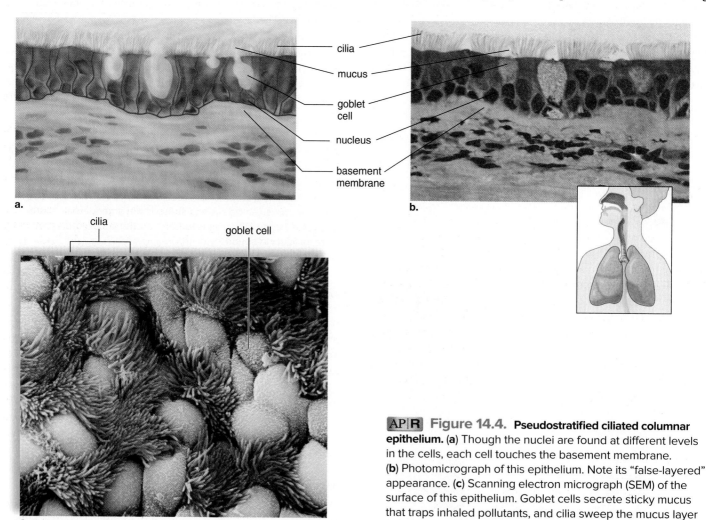

a.

cilia
mucus
goblet cell
nucleus
basement membrane

b.

cilia

goblet cell

c.   2,865×

AP|R **Figure 14.4. Pseudostratified ciliated columnar epithelium. (a)** Though the nuclei are found at different levels in the cells, each cell touches the basement membrane. **(b)** Photomicrograph of this epithelium. Note its "false-layered" appearance. **(c)** Scanning electron micrograph (SEM) of the surface of this epithelium. Goblet cells secrete sticky mucus that traps inhaled pollutants, and cilia sweep the mucus layer to the pharynx to be swallowed or expectorated (spit out).

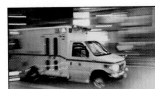
Imagine that you're a military medic who's called upon to respond when troops have been injured due to a bomb blast. As you arrive at the scene, two fallen soldiers need your attention. One has an open chest wound, caused by shrapnel cutting his chest. The second was nearby when the blast occurred, but has no obvious wounds. Yet, both have the same symptoms: sharp pain when they inhale, difficulty speaking, and a feeling of breathlessness. Both soldiers' blood pressure is low and pulse is rapid, indicating that they might slip into shock. You take a quick history from both victims and from their buddies.

Right away, you suspect each soldier has *atelectasis*—the technical term for a collapsed lung. As you'll recall, the lungs are held up against the chest wall by the attraction force of surface tension. If air enters the thorax, surface tension will fail, and the lungs will collapse. The first soldier's chest wound is allowing air from the atmosphere to enter the thorax. A section of the second soldier's lung

burst as a result of the high pressure from the bomb blast, and air has filled his thorax from the hole in his lung. When the lungs collapse, the air filling the chest compresses the heart and prevents it from filling with blood. This is termed *tension pneumothorax*, or air in the thorax. You'll need to act fast, or both victims will slip into shock.

With the help of the first soldier's buddies, you put a special airtight pressure bandage over his open chest wound, which will prevent additional air from entering the wound and help stop bleeding. Next, you'll start his intravenous solution (IV). By listening to the second soldier's chest with your stethoscope, you'll be able to tell where the lung has collapsed because it will sound hollow. When you trained as a medic, you learned to do a *thoracocentesis*, and you'll rapidly insert a catheter between the soldier's ribs to let the trapped air out into the atmosphere. Now your patients are ready for their helicopter trip to a field hospital for more advanced care.

tube by way of an incision made in the trachea. This tube acts as an artificial air intake and exhaust duct. The operation is called a **tracheostomy.**

The Bronchial Tree

The trachea divides into right and left primary **bronchi** (sing., **bronchus**), which lead into the right and left lungs (see Fig. 14.1). The primary bronchi then branch into secondary bronchi: one for each lobe of the lung. Thus, there are three secondary bronchi for the right lung, which has three lobes. Two secondary bronchi supply the left lung, which has only two lobes in order to allow room for the heart. Each secondary bronchus then divides into smaller tertiary bronchi. These smaller bronchi are supported by smaller plates of cartilage, in place of the cartilage rings of the trachea. **Bronchioles** are the smallest conducting airways. They lack cartilage support, but possess a ciliated epithelium and a well-developed smooth muscle layer. During an asthma attack, the smooth muscle of the bronchioles contracts, causing bronchiolar constriction and characteristic wheezing. Each bronchiole leads to an elongated space enclosed by a multitude of air pockets, or sacs, called **alveoli** (sing., **alveolus**). The components of the bronchial tree beyond the primary bronchi, including the alveoli, compose the lungs.

The Lungs

The **lungs** are paired, cone-shaped organs. Each fills its own pleural cavity inside the thoracic cavity, separated by the mediastinum. Recall that the mediastinum is the central compartment that separates the thoracic cavity. It contains the heart and its major vessels, primary bronchi, thymus gland, trachea, and esophagus (see Chapter 1, page 7). The apex is the superior narrow portion

of a lung, and the base is the inferior broad portion that curves to fit the dome-shaped diaphragm, the muscle of respiration that separates the thoracic cavity from the abdominal cavity. The lateral surfaces of the lungs follow the contours of the ribs in the thoracic cavity.

Each lobe of the lung is further divided into lobules, and each lobule has a bronchiole supplying many alveoli. Pulmonary arteries travel alongside the bronchi; likewise, pulmonary arterioles parallel the bronchioles. Each pulmonary arteriole then further branches to form **pulmonary capillaries.** Pulmonary capillaries surround and cover each alveolus of the lung. Elastic connective tissue binds the air passages to the blood vessels within each lung; this elastic tissue helps the lungs return to their resting position, or *recoil,* when a person exhales.

Each lung is enclosed by a double layer of serous membrane called the **pleurae** (sing., **pleura**). The visceral pleura adheres to the surface of the lung; the parietal pleura lines the inside of the thoracic cavity. The pleurae produce a lubricating serous fluid that reduces friction and allows the two layers to slide across one another. Serous fluid, a water-based solution, also creates **surface tension:** the tendency for water molecules to cling to each other (due to hydrogen bonding between the molecules) and to form a droplet (see section 2.2). Surface tension holds the two pleural layers together, thus holding the lungs open against the chest wall. If this surface tension force is disrupted, the lung will collapse, as described in the In Case of Emergency reading above.

The Alveoli

With each inhalation, air passes through the bronchial tree to the alveoli. An alveolar sac is made up of simple squamous epithelium surrounded by pulmonary capillaries. Gas exchange occurs

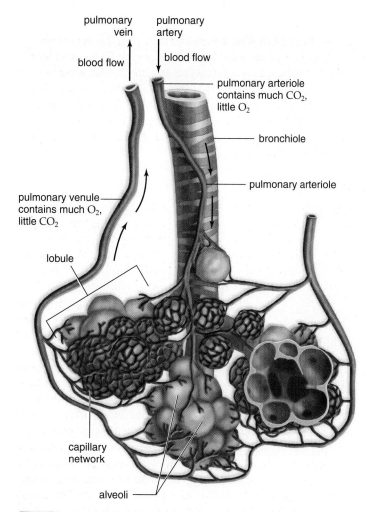

AP|R **Figure 14.5** **Gas exchange in the lungs.** The lungs consist of portions of the bronchial tree leading to the alveoli, each of which is surrounded by an extensive capillary network. Notice that the pulmonary artery and arteriole carry O_2-poor blood (colored blue), and the pulmonary vein and venule carry O_2-rich blood (colored red).

between the air in the alveoli and the blood in the capillaries (Fig. 14.5). Oxygen diffuses across the alveolar and capillary walls to enter the bloodstream, while carbon dioxide diffuses from the blood across these walls to enter the alveoli.

Each alveolus is lined with an extremely thin layer of water-based tissue fluid. Gas exchange takes place across moist cellular membranes, and the attractive force created by the fluid's surface tension helps the distended lung tissue to return to its resting position when a person exhales. However, the alveoli must stay open to receive the inhaled air if gas exchange is to occur, and the surface tension of the fluid lining the alveoli is capable of causing them to close up. (As an analogy, rinse a balloon several times with water, then empty the balloon completely. Note how the balloon's sides stick together and how hard it is to blow it up.) Normal alveoli are lined with **surfactant**—a film of lipoprotein that lowers surface tension to an acceptable level. Thus, in healthy lungs surface tension is high enough to help the lungs recoil (return to a resting position) yet low enough to prevent the alveoli from entirely collapsing. The lungs collapse in some newborn babies, especially premature infants, who lack this surfactant film. The condition, called **infant respiratory distress syndrome,** is now treatable using steroid drugs and by surfactant replacement therapy.

Alveoli are also equipped with specialized white blood cells called **dust cells** that help to defend us against any debris or pathogens we might inhale. Dust cells are alveolar macrophages that can be found in the walls between adjacent alveoli and on the surface of the alveolus as well (Fig. 14.6). When they are filled with dust and/or microbes, they are transported to the pharynx and swallowed.

Respiratory Membrane Gas exchange occurs very rapidly because of the characteristics of the so-called respiratory membrane (Fig. 14.6). The **respiratory membrane** consists of the alveolar epithelium and the capillary endothelium, which are layered together. Throughout most of the lungs, their basement membranes are

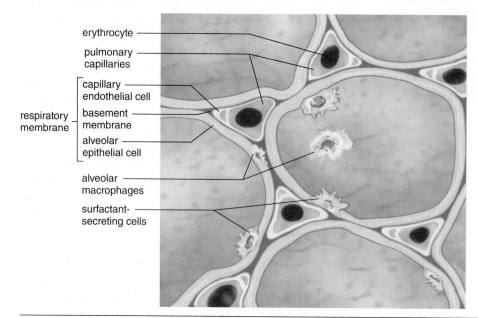

AP|R **Figure 14.6** **Structure of the alveolus and respiratory membrane.** The alveolar wall and the capillary wall allow gases to easily diffuse between the alveolus and the capillary.

fused, meaning that very little tissue fluid separates the two portions[1] of the respiratory membrane, and they are indeed one membrane. This membrane is extremely thin—only 0.2–0.6 μm thick.

The total surface area of the respiratory membrane is the same as the area of the alveoli, namely 50–70 m². The blood that enters the many pulmonary capillaries spreads thin. The red blood cells within the capillaries are pressed up against the narrow capillary walls, and little plasma is present. This too facilitates the speed of gas exchange during external respiration.

Content **CHECK-UP!**

1. Which structure covers the larynx as swallowing occurs?
   a. epiglottis
   c. uvula
   b. soft palate
   d. hard palate

2. Put the following structures in order, from external to internal: trachea-alveoli-tertiary bronchi-primary bronchi-secondary bronchi-bronchioles.

3. Which membrane structure divides the thoracic cavity into two separate cavities?
   a. parietal pleura
   c. parietal pericardium
   b. mediastinum
   d. visceral pericardium

Answers in Appendix A.

14.2 Mechanism of Breathing

4. Outline the process of ventilation, including inspiration and expiration.
5. Describe vital capacity and its relationship to other measurements of breathing capacity.
6. Tell where the respiratory center is located, and explain how it controls the normal breathing rate.

Ventilation

To understand **ventilation**—the manner in which air enters and exits the lungs—it helps to remember these facts:

1. *The lungs lie within the sealed-off thoracic cavity.* The rib cage, consisting of the ribs joined to the vertebral column posteriorly and to the sternum anteriorly, forms the top and sides of the thoracic cavity. The intercostal muscles lie between the ribs. The diaphragm and connective tissue form the floor of the thoracic cavity.

2. *The lungs adhere to the thoracic wall by way of the pleurae.* The visceral pleura covering the lungs attaches to the parietal pleura covering the chest wall, by utilizing the force of surface tension created by the fluid between them. Any space between the two layers of the pleura is minimal. (As an analogy, wet a clean glass microscope slide with a drop of water, then cover it with a second glass slide. Note the two pieces of glass will slide back and forth, but can't be easily pulled apart.) This surface tension force creates **intrapleural pressure**—the pressure

between the pleurae. Intrapleural pressure is less than atmospheric pressure, and this also helps to keep the lungs inflated. The importance of a reduced intrapleural pressure is demonstrated when atmospheric air accidentally enters the intrapleural space and the lung then collapses. (The In Case of Emergency reading on page 323 describes just such a scenario.)

3. *A continuous column of air extends from the pharynx to the alveoli of the lungs.*

Inspiration

Inspiration is the active phase of ventilation because this is the phase in which the diaphragm and the external intercostal muscles contract (Fig. 14.7a). In its relaxed and resting state, the diaphragm is dome-shaped; during inspiration, it contracts and becomes a flattened sheet of muscle. Simultaneously, the external intercostal muscles contract, and the rib cage moves upward and outward.

As the thoracic volume increases, the lungs increase in volume as well because the lung adheres to the wall of the thoracic cavity. As lung volume increases, the air pressure within the alveoli (called *intrapulmonary pressure*) decreases, creating a partial vacuum. In other words, alveolar pressure is now less than atmospheric pressure (air pressure outside the lungs). Because a continuous column of air fills the lungs, air will naturally flow from outside the body into the respiratory passages and into the alveoli. Airflow continues, until intrapulmonary pressure equals atmospheric pressure.

It's important to realize that air comes into the lungs because they have already opened up; air does not force the lungs open. This is why it's sometimes said that *humans breathe by subatmospheric pressure*. The creation of a partial vacuum in the alveoli causes air to enter the lungs. While inspiration is the active phase of breathing, the actual flow of air into the alveoli is passive.

Expiration

Usually, expiration is the passive phase of ventilation, and no muscular effort is required to bring it about. During expiration, the diaphragm and the external intercostal muscles relax. Therefore, the diaphragm resumes its dome shape and the rib cage moves down and in (Fig. 14.7b). As the volume of the thoracic cavity decreases, the lungs are free to recoil. Lung recoil occurs due to the elastic tissue built into the lungs' walls, and also due to the slight alveolar surface tension. As thoracic cavity volume and, thus, *lung* volume, decrease, the air pressure within the alveoli (intrapulmonary pressure) increases above atmospheric pressure. Because the airways are filled with a continuous column of air, some of that air will naturally flow out of the body. Airflow continues until intrapulmonary pressure equals atmospheric pressure.

What keeps the alveoli from collapsing as they decrease in size during expiration? Recall that the presence of surfactant lowers the surface tension within the alveoli. Also, as the lungs recoil, pressure between the two layers of pleura decreases, and this also helps the alveoli stay open.

Maximum Inspiratory Effort and Forced Expiration

If you recall the last time you exercised vigorously—perhaps running in a race or even just climbing all those stairs to your classroom—you'll probably remember that you were breathing a

---

[1] The respiratory membrane has two portions but four layers: alveolar epithelium plus its basement membrane and capillary endothelium plus its basement membrane.

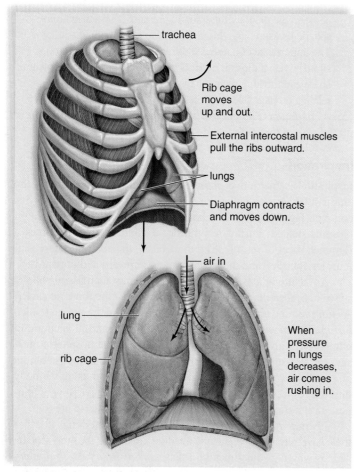

- trachea
- Rib cage moves up and out.
- External intercostal muscles pull the ribs outward.
- lungs
- Diaphragm contracts and moves down.
- air in
- lung
- rib cage
- When pressure in lungs decreases, air comes rushing in.

a. Inspiration

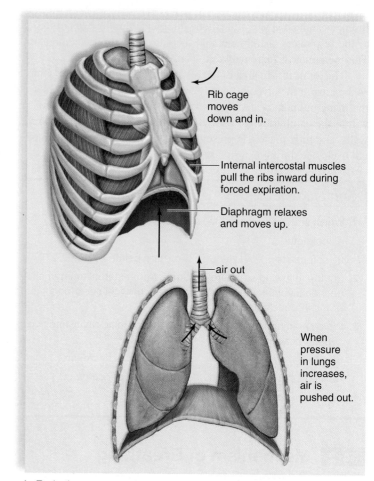

- Rib cage moves down and in.
- Internal intercostal muscles pull the ribs inward during forced expiration.
- Diaphragm relaxes and moves up.
- air out
- When pressure in lungs increases, air is pushed out.

b. Expiration

AP|R **Figure 14.7** **Inspiration versus expiration. (a)** During inspiration, the thoracic cavity and lungs expand so that intrapulmonary pressure decreases. Now air flows into the lungs. **(b)** During expiration, the thoracic wall and lungs recoil, assuming their original positions and pressures. Now air is forced out. The internal intercostal muscles only contract during forceful expiration.

lot harder than normal during and immediately after that heavy exercise. Maximum inspiratory effort involves the *accessory muscles* of respiration: muscles of the back, pectoralis minor (chest), sternocleidomastoid muscles of the anterior neck, etc. Their combined efforts can help to make the thoracic cavity larger than normal, thus allowing maximum expansion of the lungs.

While inspiration is always the active phase of breathing, expiration is usually passive—that is, the diaphragm and external intercostal muscles are simply allowed to relax, the lungs recoil, and expiration occurs. However, expiration can also be forced. Forced expiration accompanies the maximum inspiratory efforts of heavy exercise. Forced expiration is also necessary to sing, blow air into a trumpet, or blow out birthday candles. Contraction of the internal intercostal muscles can force the rib cage to move downward and inward. In addition, when the abdominal wall muscles (rectus abdominis, external and internal obliques, and transversus abdominis) contract, they push on the abdominal organs, which are compressed upward against the diaphragm. This creates increased pressure in the thoracic cavity, which helps to forcefully expel air.

During breathing, air moves into the lungs from the nose or mouth (called inspiration, or inhalation), and then moves out of the lungs during expiration, or exhalation. A free flow of air from the nose or mouth to the lungs and from the lungs to the nose or mouth is vitally important. Therefore, a technique has been developed that allows physicians to determine if there is a medical problem that prevents the lungs from filling with air upon inspiration and releasing air from the body upon expiration. An instrument called a **spirometer** records the volume of air exchanged during normal breathing and during deep breathing. A spirogram (recording from a spirometer) shows the measurements recorded by a spirometer when a person breathes as directed by a technician (Fig. 14.8).

Respiratory Volumes and Capacities

Normally when we're relaxed, only a small amount of air moves in and out with each breath. This amount of air, called the **tidal volume,** is only about 500 ml.

It is possible to increase the amount of air inhaled, and therefore the amount exhaled, by deep breathing. The maximum volume of air that can be moved in plus the maximum volume that can be moved out during a single breath is the **vital capacity.** It is called vital capacity because your life depends on breathing, and the more air you can move, the better off you are. A number of different illnesses, discussed in section 14.4, can decrease vital capacity.

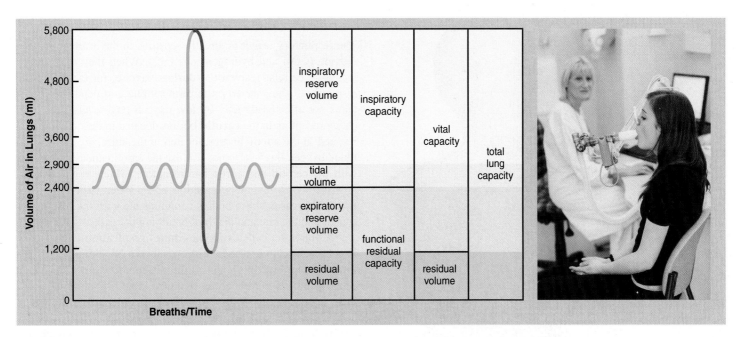

AP|R **Figure 14.8 Vital capacity.** A spirometer measures the amount of air inhaled and exhaled with each breath. During inspiration, the tracing moves up, and during expiration, the tracing moves down. The volume of one normal breath (tidal volume) multiplied by the number of breaths per minute is called the minute ventilation. A lower-than-normal minute ventilation can be a sign of pulmonary malfunction. Vital capacity (red) is the maximum amount of air a person can exhale after taking the deepest inhalation possible.

Vital capacity varies by how much we can increase inspiration and expiration over the tidal volume amount. We can increase inspiration by using the accessory muscles of respiration (sternocleidomastoid, pectoralis minor, etc.) to expand the chest, and also by lowering the diaphragm to the maximum extent possible. Forced inspiration usually increases the volume of air beyond the tidal volume by 2,900 ml, and that amount is called the **inspiratory reserve volume.** We can increase the amount of air expired by contracting the abdominal and internal intercostal muscles. This so-called **expiratory reserve volume** is usually about 1,400 ml of air. You can see from Figure 14.8 that vital capacity is the sum of the tidal, inspiratory reserve, and expiratory reserve volumes.

It's a curious fact that some of the inhaled air never reaches the alveoli; instead, it fills the nasal cavities, trachea, bronchi, and bronchioles (see Fig. 14.1). In an average adult, some 70% of the tidal volume does reach the alveoli, but 30% remains in the airways. These passages are not used for gas exchange, and therefore they are said to contain **dead-space** air. To ensure that a large portion of inhaled air reaches the lungs, it is better to breathe slowly and deeply. Also, note in Figure 14.8 that even after a very deep exhalation, some air (about 1,000 ml) remains in the alveoli; this is called the **residual volume.** This air is not as useful for gas exchange because it contains a great deal of CO_2 and has been depleted of oxygen. In some lung diseases, such as emphysema, the residual volume increases because the individual has difficulty emptying the lungs. This means that the vital capacity is reduced because the lungs have more residual volume.

With further study of Figure 14.8, you can see that the four respiratory capacities are sums of the four respiratory volumes. *Inspiratory capacity* is the sum of inspiratory reserve volume and tidal volume; likewise, *expiratory capacity* is the sum of expiratory reserve volume and tidal volume. *Functional residual capacity* can

be determined by adding expiratory reserve volume and residual volume. The maximum amount of air that can be held in a person's lungs is called the *total lung capacity*. This measure is the sum of all four volumes: tidal volume, expiratory reserve volume, inspiratory reserve volume, and residual volume.

Control of Ventilation

Normally, adults have a breathing rate of 12 to 20 ventilations per minute. The basic rhythm of ventilation is controlled by a **primary respiratory center** located in the medulla oblongata of the brain.

The primary respiratory center automatically sends out motor nerve signals by way of the phrenic nerve to the diaphragm. Simultaneously, the intercostal nerves stimulate the external intercostal muscles of the rib cage (Fig. 14.9). When these muscles contract, the thoracic volume and lung volume increase, and the person inhales. When the respiratory center stops sending neuronal signals to the diaphragm and the rib cage, the diaphragm relaxes, resuming its dome shape. The rib cage moves down and in. Decreasing thoracic and lung volumes allow the person to exhale.

The primary respiratory center allows the *basic* pattern of inhalation and exhalation. However, if the medulla functions alone (as might occur in a person with a serious head injury, if nerves from higher centers are damaged), respiration is short and gasping. Breathing rhythmically at a normal rate and volume requires nervous input from the **pons,** the brain stem center immediately superior to the medulla. Functioning together, these two brain centers allow normal, quiet breathing, a pattern referred to as **eupnea:** smooth, sustained inspiration, followed by a smooth, sustained expiration.

Although the respiratory center controls the rate and depth of breathing, its activity can be influenced by nervous input and chemical input.

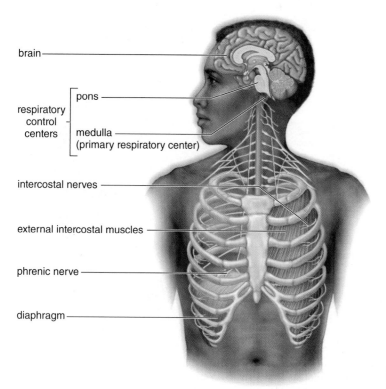

AP|R Figure 14.9 **Nervous control of breathing.** During inspiration, respiratory control areas of the pons and medulla stimulate the intercostal and phrenic nerves. Intercostal nerves cause the external intercostal muscles to contract. The phrenic nerve stimulates the diaphragm. Expiration occurs rhythmically when respiratory control areas stop nervous stimulation and allow the muscles to relax.

Labels in figure:
- brain
- respiratory control centers
  - pons
  - medulla (primary respiratory center)
- intercostal nerves
- external intercostal muscles
- phrenic nerve
- diaphragm

Nervous Input

By observing our breathing during day-to-day living, it's evident that higher brain centers influence the rate and depth of respiration. Input from the cerebral cortex, limbic system, hypothalamus, and other brain centers causes increased respiration rate and depth if you're angry, frightened, or otherwise upset. Faster-than-normal respiration is referred to as **hyperpnea.** Conversely, respiration rate and depth decrease in the soundest stages of sleep. Conscious control of respiration allows a person to hold his breath for a time, or to voluntarily hyperventilate. Neural influence over the respiratory centers also helps increase respiratory rate and depth during exercise.

Nervous control over respiration also helps to protect delicate lung tissue, as illustrated by the *Hering-Breuer reflex.* During exercise, inspiratory depth increases, due to recruitment of muscle fibers in the diaphragm, intercostal muscles, and accessory muscles. Then, stretch receptors in the bronchi, bronchioles, and the walls between adjacent alveoli are stimulated; in turn, they produce inhibitory nerve signals that travel from the inflated lungs to the respiratory center. This causes the respiratory center to stop sending out nerve signals. This reflex helps support rhythmic respiratory movements by limiting inspiration, thus preventing overexpansion of lung tissue.

Chemical Input

The respiratory center is directly sensitive to the levels of carbon dioxide (CO_2) and hydrogen ions (H^+). When they rise, due to increased cellular respiration (during exercise, for example), the respiratory center increases respiratory rate and depth. The center is not affected directly by low oxygen (O_2) levels. However, chemoreceptors in the **carotid bodies,** located in the carotid arteries, and in the **aortic bodies,** located in the aorta, are sensitive to blood oxygen levels. (Do not confuse the carotid and aortic bodies with the carotid and aortic baroreceptors, which monitor blood pressure. For a review, see section 12.2). When oxygen concentration decreases, these bodies communicate with the respiratory center, and the rate and depth of breathing increase.

A condition called **Cheyne-Stokes respiration** is characterized by alternating periods of deep, labored breathing, then shallow or absent breathing (apnea). In this condition, the respiratory center is largely controlled by chemical input. Breathing rate first increases when blood CO_2 and H^+ are high and O_2 is low, then decreases when blood CO_2 and H^+ are low and O_2 is high. Congestive heart failure and other disorders may cause this condition.

Content **CHECK-UP!**

4. Name the muscles that contract when a person inhales.
5. Which of the following has to be true for a person to inhale?
   a. Intrapulmonary pressure is greater than atmospheric pressure.
   b. Atmospheric pressure is greater than intrapulmonary pressure.
   c. The abdominal muscles must contract.
   d. a and c
6. Suppose you're a respiratory therapist and you want to measure your patient's tidal volume, inspiratory reserve volume, and expiratory reserve volume. What would you tell him or her to do during the test? Describe the steps you'd use.

Answers in Appendix A.

14.3 Gas Exchange and Transport

7. Describe the process of gas exchange in the lungs and the tissues.
8. Explain how oxygen and carbon dioxide are transported in the blood.

Respiration includes exchange of gases both in the lungs *and* in the tissues. Both are critical to homeostasis. Recall that diffusion is the movement of molecules from a higher concentration to a lower concentration. The principles of diffusion alone govern whether oxygen (O_2) or carbon dioxide (CO_2) enters or leaves the blood in the lungs and in the tissues.

External Respiration

External respiration is the exchange of gases in the lungs. Specifically, during external respiration, gases are exchanged between the alveolar air and the pulmonary capillary blood. Blood that enters the pulmonary capillaries is dark maroon because it is relatively O_2-poor.

Once inspiration has occurred, the alveoli have a higher concentration of O_2 than does blood entering the lungs. Therefore, O_2 diffuses from the alveoli into the pulmonary capillary blood. The reverse is true of CO_2. The alveoli have a lower concentration of CO_2 than does blood entering the lungs. As a result, CO_2 diffuses out of the pulmonary blood into the alveoli. This CO_2 exits the body during expiration.

Another way to explain gas exchange in the lungs is to consider the *partial pressure* of the gases involved. As the molecules of gases move randomly in all directions, they exert pressure. As an analogy, imagine blowing up a balloon, then holding its neck closed. When you open the balloon, air escapes—it rushes out from high pressure (inside) to low pressure (outside). In the alveoli, the amount of pressure each gas exerts is its partial pressure, symbolized as P_{O_2} and P_{CO_2}. Alveolar air has a much higher P_{O_2} than does blood; thus, oxygen rushes into the blood. The pressure pattern is the reverse for CO_2: blood in the pulmonary capillaries has a higher P_{CO_2} than the air in the alveoli. Carbon dioxide escapes pulmonary blood to enter the alveoli.

Internal Respiration

Internal respiration refers to the exchange of gases in the tissues. Specifically, during internal respiration, gases are exchanged between the blood in systemic capillaries and the tissue fluid. Blood that enters the systemic capillaries is a bright red color because the blood is O_2-rich. Tissue fluid, on the other hand, has a low concentration of O_2. Why? Because the cells are continually consuming O_2 during cellular respiration. Therefore, O_2 diffuses from the blood into the tissue fluid. Tissue fluid has a higher concentration of CO_2 than does the blood entering the tissues. Why? Because CO_2 is an end product of cellular respiration. Therefore, CO_2 diffuses from the tissue fluid into the blood. Figure 14.10 summarizes our discussion of gas exchange in the lungs and tissues and shows the differences in O_2 and CO_2 that lead to diffusion of these gases.

Again, we can explain exchange in the tissues by considering the partial pressure of the gases involved. In this case, oxygen diffuses out of the blood into the tissues because the P_{O_2} in tissue fluid is lower than that of the blood. Carbon dioxide diffuses into the blood from the tissues because the P_{CO_2} in tissue fluid is higher than that of the blood.

Gas Transport

The mode of transport of oxygen and carbon dioxide in the blood differs, although red blood cells are involved in transporting both of these gases.

Oxygen Transport

After O_2 enters the blood contained in pulmonary capillaries of the lungs, it enters red blood cells and combines with the iron portion of **hemoglobin,** the pigment in red blood cells. In addition, a small amount of oxygen is transported as a dissolved gas in the watery blood plasma. This dissolved oxygen amounts to only 2–3% of

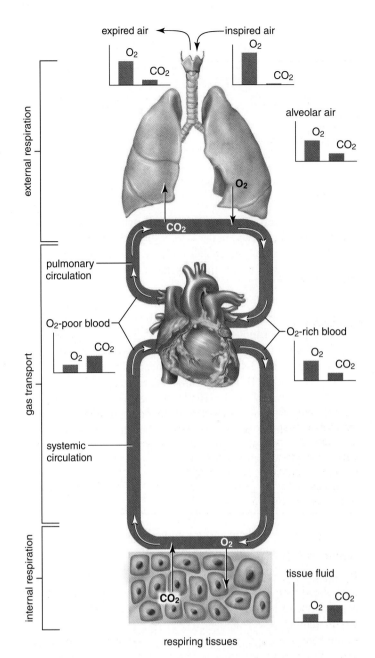

Figure 14.10 External and internal respiration. During external respiration in the lungs, CO_2 leaves the blood and O_2 enters the blood passively by diffusion. During internal respiration in the tissues, O_2 leaves the blood and CO_2 enters the blood passively by diffusion.

the body's oxygen at any given time because oxygen is not very soluble in water.

Hemoglobin is remarkably suited to the task of transporting oxygen because it can either combine with or release oxygen (depending upon its surroundings). The higher concentration of oxygen in the alveoli, plus the slightly higher pH and slightly cooler temperature, causes hemoglobin to take up oxygen and become **oxyhemoglobin** (HbO_2). The lower concentration of oxygen in the tissues, plus the slightly lower pH and slightly warmer temperature in the tissues, causes hemoglobin to release oxygen and become deoxyhemoglobin (HHb). This equation summarizes the action of hemoglobin in oxygen transport:

$$\text{HHb} + \text{O}_2 \underset{\text{tissues}}{\overset{\text{lungs}}{\rightleftharpoons}} \text{HbO}_2$$

Carbon Dioxide Transport

Transport of CO_2 to the lungs involves a number of steps. After CO_2 diffuses into the blood at the tissues, it can be transported in one of three ways:

1. Approximately 10% of CO_2 is transported as a dissolved gas in blood plasma and in the cytoplasm of red blood cells. Carbon dioxide is much more soluble in water than is oxygen and, thus, three to five times as much CO_2 can be transported as a dissolved gas.

2. Roughly 30% of CO_2 molecules formed are taken up by the protein (globin) portion of hemoglobin, forming a compound called **carbaminohemoglobin.**

3. Most of the CO_2—60%—combines with water, forming carbonic acid (H_2CO_3). The carbonic acid dissociates to hydrogen ions (H^+) and bicarbonate ions (HCO_3^-). This reaction is *catalyzed* (assisted by) an enzyme, **carbonic anhydrase,** which is present in all body cells but is especially abundant in red blood cells. The release of hydrogen ions from carbonic acid explains why the blood in systemic capillaries has a lower pH than the blood in pulmonary capillaries. However, the difference in pH is slight because the globin portion of hemoglobin combines with excess hydrogen ions and becomes **reduced hemoglobin (HHb).**

Bicarbonate ions are carried in the plasma because they diffuse out of red blood cells and go into the plasma. Most of the carbon dioxide in blood is carried as HCO_3^-, the **bicarbonate ion.** As bicarbonate ions diffuse out of red blood cells, chloride ions ($Cl-$) diffuse into them. This so-called **chloride shift** maintains the electrical balance between the plasma and the red blood cells.

In pulmonary capillaries, a reverse reaction occurs. Bicarbonate combines with hydrogen ions to form carbonic acid, which this time splits into CO_2 and H_2O, and the CO_2 diffuses out of the blood into the alveoli. The following equation summarizes the reaction between carbon dioxide and water:

$$\text{CO}_2 + \text{H}_2\text{O} \underset{\text{lungs}}{\overset{\text{tissues}}{\rightleftharpoons}} \text{H}_2\text{CO}_3 \underset{\text{lungs}}{\overset{\text{tissues}}{\rightleftharpoons}} \text{H}^+ + \text{HCO}_3^-$$

By changing the breathing rate, the respiratory system uses this chemical relationship to help regulate blood pH.[2] As you know, tissue cells are constantly producing carbon dioxide as a waste product

of metabolism. Thus, if breathing rate slows, the amount of CO_2 in the blood increases, and more carbonic acid is formed. As carbonic acid dissociates, free H+ (hydrogen ions) are released. As mentioned previously, hydrogen ions will combine with hemoglobin. However, if enough H+ are released, ultimately the blood pH decreases. By contrast, if respiratory rate increases, the blood carbon dioxide decreases as more is exhaled. Carbonic acid will re-form, and fewer free H+ will be found in the plasma. Now blood pH will increase. Breathing rate is constantly being "fine-tuned" in this way, so that blood pH remains at normal levels and homeostasis is maintained.

Many respiratory diseases interfere with normal excretion of carbon dioxide, causing excess carbonic acid to be formed. If blood pH falls below the normal range as a result, a condition termed **acidosis** develops. Other conditions may cause a person to hyperventilate, or breathe too rapidly. If too much CO_2 is eliminated in this fashion, blood pH will rise above normal level, causing **alkalosis.** Stress and anxiety can cause a person to hyperventilate. Disorders involving the respiratory control center (such as a brain tumor or brain damage near the brain stem) may cause a person to hyperventilate as well. Both acidosis and alkalosis can be fatal because they interfere with cell enzyme function.

> ## *Content* CHECK-UP!
>
> 7. If you describe internal respiration, you can use this statement(s):
>   a. Oxygen diffuses from the blood into the tissue fluid.
>   b. Carbon dioxide diffuses from tissue fluid into the blood.
>   c. Internal respiration takes place in systemic capillaries.
>   d. All of these statements are correct.
>
> 8. Hemoglobin with carbon dioxide attached is termed:
>   a. reduced hemoglobin.   c. oxyhemoglobin.
>   b. carbaminohemoglobin.   d. dioxyhemoglobin.
>
> 9. List three mechanisms used to transport CO_2 in the blood.
>
> *Answers in Appendix A.*

14.4 Respiration and Health

9. Name and describe the various infections of the respiratory tract.
10. Describe the effects of smoking on the respiratory tract and on overall health.

The respiratory tract is constantly exposed to environmental air. The quality of this air can affect our health. The presence of a disease means that homeostasis is threatened and if the condition is not brought under control, death is inevitable.

Upper Respiratory Tract Infections

The upper respiratory tract consists of the nasal cavities, the pharynx, and the larynx. Upper respiratory infections (URI) can spread from the nasal cavities to the sinuses, middle ears, and larynx. Viral infections sometimes lead to secondary bacterial infections.

---

[2] pH means the negative logarithm of free hydrogen ion concentration, so the smaller the pH number, the greater the number of free hydrogen ions. The pH of arterial blood should range from 7.35 to 7.45. Blood with pH 7.3 (acidosis) has *more* free hydrogen ions than blood with pH 7.4 (normal). See page 25 to review the concept of pH.

What we call "strep throat" is a primary bacterial infection caused by *Streptococcus pyogenes* that can lead to a generalized upper respiratory infection and even a systemic (affecting the body as a whole) infection. Although antibiotics have no effect on viral infections, they are successfully used to treat most bacterial infections, including strep throat. The symptoms of strep throat are severe sore throat, high fever, and white patches on a dark red throat.

Sinusitis

Sinusitis is an infection of the cranial sinuses, the cavities within the facial skeleton that drain into the nasal cavities (see Fig. 6.5). Only about 1–3% of upper respiratory infections are accompanied by sinusitis. Sinusitis develops when nasal congestion blocks the tiny openings leading to the sinuses. Symptoms include postnasal discharge as well as facial pain that worsens when the patient bends forward. Pain and tenderness usually occur over the lower forehead or over the cheeks. If the latter, toothache is also a complaint. Successful treatment depends on restoring proper drainage of the sinuses. Antibiotic therapy is necessary if the condition results from bacterial infection. Even a hot shower and sleeping upright can be helpful. Otherwise, spray decongestants are preferred over oral antihistamines, which thicken rather than liquefy the material trapped in the sinuses.

Otitis Media

Otitis media is a bacterial infection of the middle ear. The middle ear is not a part of the respiratory tract, but this infection is considered here because it is a complication often seen in children who have a nasal infection. Infection can spread by way of the **auditory (eustachian) tube** that leads from the nasopharynx to the middle ear (see Fig. 9.12). Pain is the primary symptom of a middle ear infection. A sense of fullness, hearing loss, vertigo (dizziness), and fever may also be present. Antibiotics almost always bring about a full recovery, and recurrence is probably due to a new infection. Tubes (called *tympanostomy tubes*) are sometimes placed in the tympanic membranes (eardrums) of children with multiple recurrences to help promote fluid drainage, as well as to prevent the buildup of pressure in the middle ear and the possibility of hearing loss. Normally, the tubes fall out with time, and the eardrum heals over completely.

Tonsillitis

Tonsillitis occurs when the **tonsils,** masses of lymphatic tissue in the pharynx, become inflamed and enlarged. If tonsillitis occurs frequently and enlargement makes breathing difficult, the tonsils can be removed surgically in a **tonsillectomy.** Fewer tonsillectomies are performed today than in the past because we now know that the tonsils remove many of the pathogens that enter the pharynx; therefore, they are a first line of defense against pathogenic invasion of the body.

Laryngitis

Laryngitis is an inflammation of the larynx, with accompanying hoarseness leading to the inability to talk in an audible voice. It can result from an upper respiratory infection, or simply from overuse of the larynx (if, for example, you've screamed for several hours at a sports event or concert). So long as the immune system is functioning properly, an upper respiratory infection usually resolves over time, with rest and adequate fluid intake. Likewise, laryngitis disappears over time as long as the larynx is rested. Persistent hoarseness without the presence of an upper respiratory infection is one of the warning signs of cancer, and therefore should be checked out by a physician.

Lower Respiratory Tract Disorders

Lower respiratory tract disorders include infections, restrictive pulmonary disorders, obstructive pulmonary disorders, and lung cancer.

Lower Respiratory Infections

Acute bronchitis, pneumonia, and tuberculosis are infections of the lower respiratory tract. **Acute bronchitis** is an infection of the primary and secondary bronchi. Usually, it is preceded by a viral URI that has led to a secondary bacterial infection. Most likely, a nonproductive cough has become a deep cough that expectorates mucus and perhaps pus.

Pneumonia is a viral or bacterial infection of the lungs in which the bronchi and alveoli fill with thick fluid (Fig. 14.11). Risk factors for pneumonia include advanced age (it is most common and most often fatal in the elderly), weakened immune system, smoking, and being immobilized. A common scenario for development of pneumonia is an elderly person who is immobilized due to a fractured hip. High fever, productive cough, difficulty in breathing, headache, and chest pain are symptoms of pneumonia. Rather than being a generalized lung infection, pneumonia may be localized in specific lobules of the lungs. Obviously, the more lobules involved, the more serious is the infection. Pneumonia can be caused by a bacterium that is usually held in check but has gained the upper hand due to stress and/or reduced immunity. AIDS patients are subject to a particularly rare form of pneumonia caused by the fungus *Pneumocystis jiroveci*. Pneumonia of this type is almost never seen in individuals with a healthy immune system.

Pulmonary tuberculosis is caused by the tubercle bacillus, a type of bacterium. When tubercle bacilli invade the lung tissue, the lung cells build a protective capsule about the foreigners, isolating them from the rest of the body. This tiny capsule is called a tubercle. If the resistance of the body is high, the imprisoned organisms die, but if the resistance is low, the organisms eventually can be liberated. If a chest X ray detects active tubercles, the individual is put on appropriate antibiotic therapy to ensure the localization of the disease and the eventual destruction of any live bacteria. It is possible to tell if a person has ever been exposed to tuberculosis with a test in which a highly diluted extract of the bacillus is injected into the skin of the patient. A person who has never been in contact with the tubercle bacillus shows no reaction, but one who has had or is fighting an infection shows an area of inflammation that peaks in about 48 hours.

More than two billion people worldwide—about 33% of the world's population—are estimated to be infected with tuberculosis (TB).

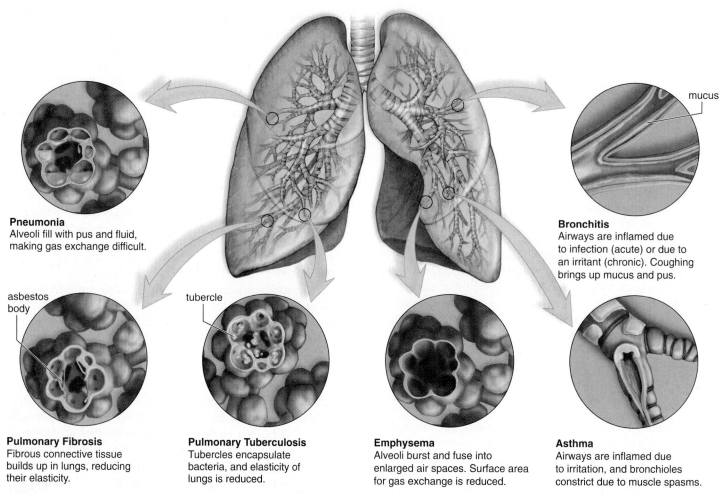

Pneumonia
Alveoli fill with pus and fluid, making gas exchange difficult.

mucus

Bronchitis
Airways are inflamed due to infection (acute) or due to an irritant (chronic). Coughing brings up mucus and pus.

asbestos body

Pulmonary Fibrosis
Fibrous connective tissue builds up in lungs, reducing their elasticity.

tubercle

Pulmonary Tuberculosis
Tubercles encapsulate bacteria, and elasticity of lungs is reduced.

Emphysema
Alveoli burst and fuse into enlarged air spaces. Surface area for gas exchange is reduced.

Asthma
Airways are inflamed due to irritation, and bronchioles constrict due to muscle spasms.

AP|R **Figure 14.11 Common bronchial and pulmonary diseases.** Exposure to infectious pathogens and/or polluted air, including tobacco smoke, causes the diseases and disorders shown here.

Poverty and HIV are the risk factors most closely associated with the disease, and 95% of TB victims live in developing countries. Many of these people are infected with multiple-drug-resistant (MDR) tuberculosis, a form that is not easily controlled by the standard set of antibiotics used to fight TB. However, there is good news for the United States, where the incidence of both common and MDR tuberculosis infection is in decline and hit a record low in 2013. Public health officials credit several interventions for this improvement: a very vigorous response to new infection, including physician and patient education, more effective infection control, and close supervision of patients in active treatment. The World Health Organization is developing similar strategies for use in other countries.

Restrictive Pulmonary Disorders

In restrictive pulmonary disorders, vital capacity is reduced because the lungs have lost their elasticity. An entire class of restrictive pulmonary disorders, called *pneumoconioses,* is caused by inhaling particles such as silica (sand), asbestos, clay, cement, flour, and fiberglass. Pneumoconioses can lead to **pulmonary fibrosis,** a condition in which fibrous connective tissue builds up in the lungs. The lungs cannot inflate properly and are always tending

toward deflation. Breathing asbestos is also associated with the development of cancer. Because asbestos was formerly used widely as a fireproofing and insulating agent, unwarranted exposure has occurred. It has been projected that more than 300,000 deaths caused by asbestos exposure—mostly in the workplace—will occur in the United States between 2012 and 2020.

Coal-dust pneumoconiosus is also known as **black lung,** and it is found in miners who are not adequately protected against inhaling coal dust. The dust collects in the bronchi, forming small deposits called nodules. These nodules can harden and destroy the lung tissue that surrounds them, gradually decreasing the patient's ability to inflate his lungs. Victims of black lung are also more susceptible to tuberculosis and pneumonia. The symptoms of black lung can be treated with various medications, but the only cure for the disorder is a lung transplant.

Obstructive Pulmonary Disorders

In obstructive pulmonary disorders, air does not flow freely in the airways, and the time it takes to inhale or exhale maximally is greatly increased. Chronic bronchitis and emphysema are referred to as **chronic obstructive pulmonary disorders (COPD)** because

The Most-Often-Asked Questions About Tobacco and Health

Is there a safe way to smoke?

No. All forms of tobacco can cause damage, and smoking even a small amount is dangerous. Tobacco is perhaps the only legal product whose advertised and intended use—that is, smoking it or chewing it—will inevitably hurt the body. The nicotine in tobacco is one of the most addictive chemicals known; it causes feelings of well-being and euphoria similar to those caused by heroin and cocaine. It will both stimulate and also relax the user throughout the day, depending on the person's mood. And as anyone who's tried to quit can tell you—withdrawing from nicotine addiction is stressful and very, very difficult.

Does smoking cause cancer?

Yes, and not only lung cancer. Besides lung cancer, smoking a pipe, cigarettes, or cigars is also a major cause of cancers of the mouth, larynx (voice box), and esophagus. In addition, smoking increases the risk of cancer of the bladder, kidney, pancreas, stomach, and uterine cervix.

What are the chances of being cured of lung cancer?

Very low; the five-year survival rate is only 13%. However, it is important to remember that lung cancer, or **bronchogenic carcinoma** in technical terms, is very rare (although not unheard of) in nonsmokers. Thus, lung cancer is largely preventable: Don't smoke, and avoid exposure to smoke.

Does smoking cause other lung diseases?

Yes. Smoking leads to chronic bronchitis, a disease in which the airways produce excess mucus, forcing the smoker to cough frequently. Smoking is also the major cause of emphysema, a disease that slowly destroys a person's ability to breathe. Chronic bronchitis and pulmonary emphysema rates are higher in smokers than in nonsmokers.

Why do smokers have "smoker's cough"?

Remember the mucociliary escalator (page 322)? This is the protective mechanism for the lungs and the airways: a sheet of mucus present in the trachea, bronchi, and bronchioles. Normally, constantly beating cilia on the cells lining the airways move the sheet upward. Once at the pharynx, mucus can be swallowed or spit out. Smoking is a double assault on the mucociliary escalator: First, inhaled, nasty toxins and particles cling to the sticky mucus. Then, a big dose of nicotine paralyzes the cilia for 30 minutes or more. Long-term smokers will eliminate the delicate, ciliated epithelial cells entirely, replacing them with a sturdier epithelium better able to stand the heat of cigarette smoke. The result: There is no way to clear accumulated mucus (and the trapped contaminants in it) except to cough it out. The longer a person smokes, the more pronounced and productive the cough becomes.

If you smoke but don't inhale, is there any danger?

Yes. Wherever smoke touches living cells, it does harm. So, even if smokers of pipes, cigarettes, and cigars don't inhale, they are at an increased risk for lip, mouth, and tongue cancer.

Does smoking affect the heart?

Yes. Smoking increases the risk of heart disease, which is the number one killer in the United States. Smoking, high blood pressure, high cholesterol, and lack of exercise are all risk factors for heart disease. Smoking alone doubles the risk of heart disease.

Is there any risk for pregnant women and their babies?

Pregnant women who smoke endanger the health and life of their unborn babies. When a pregnant woman smokes, she really is smoking for two because the nicotine, carbon monoxide, and other dangerous chemicals in smoke enter her bloodstream and then pass into the baby's body. Smoking mothers have more stillbirths and babies of low birthweight than do nonsmoking mothers. Tragically, the death rate due to sudden infant death syndrome (SIDS) is also higher for babies born to mothers who smoke.

Does smoking cause any special health problems for women?

Yes. Women who smoke and use the birth control pill have an increased risk of stroke and blood clots in the legs. In addition, women who smoke increase their chances of getting cancer of the uterine cervix.

What are some of the short-term effects of smoking cigarettes?

Almost immediately, smoking can make it hard to breathe. Within a short time, it can also worsen asthma and allergies. Only seven seconds after a smoker takes a puff, nicotine reaches the brain, where it produces a morphinelike effect.

Are there any other risks to the smoker?

Yes, there are many more risks. Smoking is a cause of stroke, which is the third leading cause of death in the United States. Smokers are more likely to have and die from stomach ulcers than are nonsmokers. Smokers have a higher incidence of cancer in general. If a person smokes and is exposed to radon or asbestos, the risk for lung cancer increases dramatically. Research continues to show that smoking is also a major risk factor for autoimmune diseases as well. These include Crohn's disease (which affects the digestive tract), psoriasis (which affects skin and connective tissue), and Buerger's disease (which causes inflammation and destruction of blood vessels, especially in the limbs).

What are the dangers of passive smoking?

Passive smoking causes lung cancer in healthy nonsmokers. Children whose parents smoke are more likely to suffer from pneumonia or bronchitis in the first two years of life than are children who come from smoke-free households, and children of smokers are more likely to develop asthma during childhood. Passive smokers have a 30% greater risk of developing lung cancer than do nonsmokers who live in a smoke-free house.

—Continued

Continued—

Are chewing tobacco and snuff safe alternatives to cigarette smoking?

No, they are *not*. Many people who use chewing tobacco or snuff believe it can't harm them because there is no smoke. Wrong. Smokeless tobacco still contains nicotine, the same addicting drug found in cigarettes and cigars. Although not inhaled through the lungs, the juice from smokeless tobacco is absorbed through the lining of the mouth. There it can cause sores and white patches, which often lead to cancer of the mouth. Snuff dippers actually take in an average of over ten times more cancer-causing substances than cigarette smokers.

But what about those e-cigarettes? Those seem like they ought to be pretty safe to use.

Electronic cigarettes, or *e-cigarettes,* vary in their style, but all basically consist of a cartridge containing a nicotine solution, a heat source that heats the solution into a vapor, and a battery to power the heat source. The user inhales the addictive nicotine vapor, along with whatever other additives are in the liquid. Over 700 different flavors—including soda, candy, fruit, and alcohol flavors—can be added to make the solution more attractive to the smoker. In addition, propylene glycol and glycerol are two ingredients that are added to dissolve the nicotine and keep it soluble. However, when exposed to the high heat necessary to vaporize e-cigarette liquid, these compounds break down to form carcinogens. Propylene glycol and glycerol form formaldehyde and acetaldehyde, both of which are carcinogenic, and flavorings can also create cancer-causing agents when burned. Furthermore, e-cigarettes contain many of the same contaminants as their traditional counterparts: heavy metals (including toxic lead), organic carbon–containing compounds, small particles, and ash. The bottom line is that they are clearly toxic, and are likely no safer than traditional cigarettes.

One thing is certain: public health officials are increasingly viewing e-cigarettes as a societal menace, especially for children and youth. Babies and children have been accidently poisoned from drinking the nicotine-laced solution, and burned by the hot cartridge. The vapor flavorings are tailored to appeal to young people, with flavors such as bubble gum, cherry, and grape. Studies show that adolescents and teens who use e-cigarettes are more than twice as likely as their peers to transition to traditional cigarettes. While over 40 states prohibit direct sales to minors, e-cigarettes are easily purchased over the Internet, and there is no minimum age for those types of purchases. Moreover, if you're trying to quit smoking, you need to be aware that e-cigarettes have not been approved by the Food and Drug Administration for smoking cessation. In fact, physicians are concerned that smokers may try to use them to quit, instead of traditional, FDA-approved medications. So think again if you have been considering e-cigarettes to be a safe, appealing, and fun alternative to cigarettes, cigars, pipes, and chewing tobacco.

Okay, okay, enough already! You've convinced me that I should quit. But how? I've tried before. Is there a better way to quit smoking?

The U.S. Office of the Surgeon General has done extensive research on the most effective ways to quit smoking. This research has determined that to be successful, a smoker should use the START plan:

1. **Set a date, and get ready to quit:** Think of all of the reasons *you* want to quit, and write them down. (For starters, you'll save money and you'll feel better!) Once you've decided on a date, plan ahead. Review what worked, or didn't, in the past.
2. **Tell others:** Inform your family, friends, and co-workers of your intention to quit, and ask them not to smoke around you. Once you quit, don't take another puff from a cigarette.
3. **Anticipate the problems you might encounter.** Learn new behaviors that aren't associated with smoking: When you first quit, change your routine completely so that subconscious cues to smoke don't keep appearing. Take a different route to work, eat breakfast in a different place, drink tea instead of coffee. When the urge to smoke recurs, distract yourself with physical or mental activity. Exercise or talk to someone to reduce stress. Each day, plan something fun and pleasurable for yourself.
4. **Remove anything related to smoking from your home and workplace.** Throw out all smoking materials: cigarettes, pipes, ashtrays, lighters, etc. Avoid spending time with people who might undermine your efforts, especially other smokers!
5. **Talk to your health-care provider to get support, encouragement, and help to quit.** Never be embarrassed to ask for, and use, antismoking medication. Nicotine comes in gum and patches (both sold over the counter) and in prescription inhaler and nasal spray forms. Using these products can help you to overcome the cravings for nicotine that occur immediately after quitting. In addition, use of antidepressants (available by prescription) has been shown to significantly increase a patient's chances of quitting for good. Chantix®, an antidepressant, also reduces the effects of nicotine addiction by blocking nicotine receptors. Private and group smoking cessation counseling often helps, and is available for free or at a low cost. Your local health department will have information about programs in your area. There are even apps for your mobile device to help support your efforts to quit, and you can sign up for a text service that will send you regular text messages with tips, advice, and encouragement while you quit.

And remember—if at first you don't succeed, try, try again! The first three to six months after quitting will be the toughest, and that's when many people go back to smoking. Don't be discouraged if this happens to you. On average, most people try to quit at least twice before finally succeeding. To avoid a relapse, don't drink alcohol and/or hang around with other smokers. Studies have shown that both can lead to failure in the attempt to quit. Weight gain and increased irritability or depression can also be expected. Exercise and eating a healthy diet will minimize both of these side effects.

For more information, visit the Surgeon General's stop-smoking website: http://smokefree.gov.

they develop slowly, over a long period of time, and are recurrent. Asthma is an obstructive disorder as well. However, asthma is generally an acute illness—that is, one that occurs in intermittent episodes that flare up and disappear quickly.

In **chronic bronchitis,** the airways are inflamed and filled with mucus. A cough that brings up mucus is common. The bronchi have undergone degenerative changes, including the loss of cilia and their normal cleansing action. Under these conditions, an infection is more likely to occur. Smoking cigarettes and cigars is the most frequent cause of chronic bronchitis. Long-term exposure to other pollutants can also cause chronic bronchitis.

Emphysema is a chronic and incurable disorder that results from smoking (though there is also a rare form caused by a genetic mutation). Emphysema is often preceded by chronic bronchitis. With emphysema, the alveoli are distended and their walls damaged so that the surface area available for gas exchange is reduced (Fig. 14.12*b*). Air trapped in the lungs leads to alveolar damage and a noticeable ballooning of the chest. On an X ray, the diaphragm (which is normally curved when at rest; see Figure 14.1) appears flattened as the lungs increase in size. Breakdown in the elastic tissue of the lung diminishes the elastic recoil. Not only are the airways narrowed, but the driving force behind expiration is also reduced. The patient has a very difficult time exhaling, and residual volume increases. The victim often feels a breathless sensation and may have a cough. Because the surface area for gas exchange in the lungs is reduced, less oxygen reaches the heart and the brain. Even so, the heart works furiously to force more blood through the lungs, and an increased workload on the heart can result. Lack of oxygen to the brain can make the person feel depressed, sluggish, and irritable. Before therapy can be effective, the patient must stop smoking. Medical treatment for emphysema is limited to drugs that dilate the bronchi and bronchioles by relaxing the smooth muscle, along with inhaled oxygen to improve oxygenation of the blood. Surgical options include lung volume reduction surgery (LVRS) and lung transplantation. In LVRS, the diseased upper portions of the lung are surgically removed, and the remaining lung tissue is sealed with a flap of pericardium. Removing the diseased portion of the lung allows the remaining healthy tissue to expand and fill the thoracic cavity. Lung transplantation is the other surgical option for emphysema, but it's rarely used because of the scarcity of donor organs. As with all transplant recipients, the patient must remain on lifelong anti-rejection therapy after a lung transplant.

Asthma is a disease of the bronchi and bronchioles that is marked by wheezing, breathlessness, and sometimes a cough and expectoration of mucus. The airways are unusually sensitive to specific irritants, which can include a wide range of allergens such as pollen, animal dander, dust, cigarette smoke, and industrial fumes. Even cold air can be an irritant. When exposed to the irritant, the smooth muscle in the bronchioles undergoes spasms. It now appears that chemical mediators given off by immune cells in the bronchioles cause the spasms. Most asthma patients have some degree of bronchial inflammation that reduces the diameter of the airways and contributes to the seriousness of an attack. Asthma is not curable, but it is treatable. Special inhalers can control the inflammation and hopefully prevent an attack, while other types of inhalers can stop the muscle spasms should an attack occur.

Lung Cancer

Lung cancer used to be more prevalent in men than in women, but recently it has surpassed breast cancer as a cause of death in women. The recent increase in the incidence of lung cancer in women is directly correlated to increased numbers of women who smoke. Autopsies on smokers have revealed the progressive steps by which the most common form of lung cancer develops. The first event appears to be thickening and callusing of the cells lining the primary bronchi. (Callusing occurs whenever cells are exposed to irritants.) Then cilia are lost, making it impossible to prevent dust and dirt from settling in the lungs. Following this, cells with atypical nuclei appear in the callused lining. A tumor consisting of disordered cells with atypical nuclei is considered cancer in situ (at one location) (Fig. 14.12*c*). A final step occurs when some of these cells break loose and penetrate other tissues, a process called **metastasis.** Now the cancer has spread. The original tumor may grow until a bronchus is blocked,

  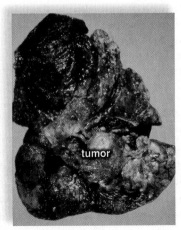

a. Normal   b. Emphysema   c. Lung cancer

Figure 14.12 Normal lung versus diseased lungs. (a) Normal lung. Note the healthy red color. **(b)** Lung of an emphysema victim. Small black deposits from smoking can be seen throughout. The apex (upper portion) of the lung shows lung tissue that has broken down. **(c)** Lung from a heavy smoker. Notice how black the lungs are except where the large red cancerous tumor has formed.

cutting off the supply of air to that lung. The entire lung then collapses, the secretions trapped in the lung spaces become infected, and pneumonia or a lung abscess (localized area of pus) results. The only treatment that offers a possibility of cure is to remove a lobe or the entire lung before metastasis has had time to occur. This operation is called **pneumonectomy.** If the cancer has spread, chemotherapy and radiation are also required.

The Medical Focus on pages 333–334 lists the various illnesses, including cancer, that are apt to occur in persons who smoke. Current research indicates that passive smoking—exposure to smoke created by others who are smoking—can also cause lung cancer and other illnesses associated with smoking. If a person stops voluntary smoking and avoids passive smoking, and if the body tissues are not already cancerous, they may return to normal over time.

Content **CHECK-UP!**

10. What type of lower respiratory infection should be treated with antibiotics? What kind should never be treated using antibiotics, and why?

11. Obstructive pulmonary diseases include:

   a. emphysema.   c. asthma.

   b. chronic bronchitis.   d. All of the above.

12. Which of the following increases in a person with emphysema?

   a. amount of surface area in the lung for gas exchange

   b. residual volume

   c. amount of elastic tissue in the lungs

   d. amount of oxygen delivered to the brain and heart

Answers in Appendix A.

14.5 Effects of Aging

11. Describe the anatomical and physiological changes that occur in the respiratory system as we age.

Respiratory fitness decreases with age. Maximum breathing capacities decline, while the likelihood of fatigue increases. Inspiration and expiration are not as effective in older persons. With age, weakened intercostal muscles and decreased elasticity of the rib cage combine to reduce the inspiratory reserve volume, while the lungs' inability to recoil reduces the expiratory reserve volume. As a result, more residual air is found in the lungs of older people, and the residual volume increases.

With age, gas exchange in the lungs becomes less efficient, not only due to changes in the lungs but also due to changes in the blood capillaries. The respiratory membrane thickens, and the gases cannot diffuse as rapidly as they once did.

Respiratory diseases, such as those discussed in section 14.4, are more prevalent in older people than in the general public. In the elderly, the ciliated cells of the trachea are reduced in number, and those remaining are not as effective as they once were. As you know from Chapter 13, other aspects of the immune response also diminish in the elderly, including the numbers and effectiveness of the white blood cells. Thus, pneumonia and other respiratory infections are among the leading causes of death in older persons. For this reason, the Centers for Disease Control (CDC) recommends that everyone who is older than 65 receive the pneumococcal vaccine to prevent pneumonia.

14.6   Homeostasis

12. Describe how the respiratory system works with other systems of the body to maintain homeostasis.

The illustration in Human Systems Work Together on page 338 describes how the respiratory system assists and depends on other body systems. The respiratory system contributes to homeostasis in many ways—in particular, by carrying on gas exchange and regulating blood pH.

Gas Exchange

First and foremost, the respiratory system performs gas exchange. Carbon dioxide, a waste molecule given off by cellular respiration, exits the body, and oxygen, a molecule needed for cellular respiration, enters the body at the lungs. Cellular respiration produces ATP, a molecule that allows the body to perform all sorts of work, including muscle contraction and nerve conduction. It is estimated that the brain uses 15–20% of the oxygen taken into the blood. Not surprisingly, a lack of oxygen affects the functioning of the brain before it affects other organs.

Regulation of Blood pH

The respiratory system can alter blood pH by changing blood carbon dioxide levels. In the tissues, carbon dioxide enters the blood and red blood cells where this reaction occurs. The bicarbonate ion (HCO_3^-) diffuses out of the red blood cells to be carried in the plasma:

$$CO_2 \; + \; H_2O \; \underset{\text{lungs}}{\overset{\text{tissues}}{\rightleftharpoons}} \; H_2CO_3 \; \underset{\text{lungs}}{\overset{\text{tissues}}{\rightleftharpoons}} \; H^+ \; + \; HCO_3^-$$

This reaction lowers blood pH because free H^+ (hydrogen ions) are formed during the reaction as carbonic acid dissociates. The greater the amount of CO_2 in the blood, the greater the carbonic acid formed, and the more free hydrogen ions are formed. Conversely, when carbon dioxide starts to diffuse out of the blood in the lungs, the reaction occurs in the reverse direction. Carbon dioxide concentration in blood decreases (because it is being exhaled), and carbonic acid concentration decreases, too. Now, the blood pH rises.

What happens to your blood pH if you hypoventilate—that is, breathe at a lower-than-normal rate? Acidosis results because carbon dioxide is not being exhaled at the normal rate. Too much carbonic acid is formed from the excess carbon dioxide. Any respiratory condition, such as emphysema, that hinders the passage of carbon dioxide out of the blood also results in acidosis. By contrast, what happens to your blood pH if you hyperventilate—that is, breathe at a higher than normal rate? Alkalosis (high blood pH) will develop because carbon dioxide is leaving the body at a high rate. This decreases the concentration of carbonic acid and the concentration of free H^+. Severe anxiety can cause a person to hyperventilate.

Control of Blood Pressure

The lungs play a role in the control of blood pressure by assisting in the renin-angiotensin-aldosterone pathway (see p. 219). The kidneys produce the enzyme renin whenever blood pressure falls. Renin activates **angiotensin** I (an inactive plasma protein produced by the liver). As blood that contains angiotensin I flows through the lungs, angiotensin I is further activated to form angiotensin II by angiotensin-converting enzyme (ACE), which is found in endothelial cells that compose the pulmonary capillaries. Once formed, angiotensin II causes arterioles to constrict, increasing blood vessel resistance. Angiotensin II also causes the release of aldosterone from the adrenal gland. In turn, aldosterone causes the kidneys to reabsorb salt and water. Through this mechanism, blood pressure returns to normal and homeostasis is maintained.

Defense

The respiratory tract assists in the defense against pathogens by preventing their entry into the body and by removing them from respiratory surfaces. For example, the cilia that line the trachea sweep impurities toward the throat. The respiratory tract also assists immunity. We now know that the tonsils serve as a location where T cells are presented with antigens before they enter the body as a whole. This action helps the body prepare to respond to an antigen before it enters the bloodstream. Within the alveoli, dust cell macrophages phagocytize not only dust, dirt, and pollen, but pathogens as well.

Integumentary System

Gas exchange in lungs provides oxygen to skin and rids body of carbon dioxide from skin.

Skin helps protect respiratory organs and helps regulate body temperature.

Skeletal System

Gas exchange in lungs provides oxygen and rids body of carbon dioxide.

Rib cage protects lungs and assists breathing; bones provide attachment sites for muscles involved in breathing.

Muscular System

Lungs provide oxygen for contracting muscles and rid the body of carbon dioxide from contracting muscles.

Muscle contraction assists breathing; physical exercise increases respiratory capacity.

Nervous System

Lungs provide oxygen for neurons and rid the body of carbon dioxide produced by neurons.

Respiratory centers in brain regulate breathing rate.

Endocrine System

Gas exchange in lungs provides oxygen and rids body of carbon dioxide.

Epinephrine promotes ventilation by dilating bronchioles; growth factors control production of red blood cells that carry oxygen.

How the Respiratory System works with other body systems

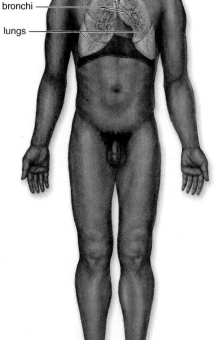

nasal cavity

nose

pharynx (throat)

larynx

trachea

bronchi

lungs

Cardiovascular System

Gas exchange in lungs rids body of carbon dioxide, helping regulate pH of blood; breathing aids venous return; lungs activate angiotensin, which helps maintain blood pressure.

Blood vessels transport gases to and from lungs; blood services respiratory organs.

Lymphatic System/Immunity

Tonsils and adenoids occur along respiratory tract; breathing aids lymph flow; tracheal cilia sweep impurities toward throat.

Lymphatic vessels pick up excess tissue fluid; immune system protects against respiratory tract and lung infections.

Digestive System

Gas exchange in lungs provides oxygen to the digestive tract and excretes carbon dioxide from the digestive tract.

Breathing is possible through the mouth because digestive tract and respiratory tract share the pharynx.

Urinary System

Lungs excrete carbon dioxide, provide oxygen, and convert angiotensin I to angiotensin II, leading to kidney regulation.

Kidneys compensate for water lost through respiratory tract; work with lungs to maintain blood pH.

Reproductive System

Gas exchange increases during sexual activity.

Sexual activity increases breathing; pregnancy causes breathing rate and vital capacity to increase.

Selected New Terms

Basic Key Terms

alveolus (ăl-vē′ō-lŭs), p. 323
alveoli (ăl-vē′ō-lī), p. 323
angiotensin (ăn-jē-ō-tĕn′sĭn), p. 337
aortic body (ā-ŏr′tĭk bŏd′ē), p. 328
auditory (eustachian) tube (ŏ′dĭ-tŏr′ē [yū-stā′shŭn] tūb), p. 331
bicarbonate ion (bī-kăr′bō-nāt ī′ŏn), p. 330
bronchiole (brŏng′kē-ōl), p. 323
bronchus (brŏng′kŭs), p. 323
bronchi (brŏng′kī), p. 323
carbaminohemoglobin (kăr-bŭh-mē″nō-hē′mō-glō″bĭn), p. 330
carbonic anhydrase (kăr-bŏn′ĭk ăn-hī′drās), p. 330
carotid body (kăr-ăh′tĭd bŏd′ē), p. 328
chloride shift (klōr′ĭd shĭft), p. 330
conchae (kŏng′kā), p. 320
dead-space (dĕd spās), p. 327
epiglottis (ĕ″pĭ-glŏt′ĭs), p. 322
expiration (ĕks″pĭ-rā′shŭn), p. 319
expiratory reserve volume (ĕk-spī′rŭh-tō-rē rē-zĕrv′ vŏl′yūm), p. 327
external respiration (ĕks-tĕr′năl rĕs″pĭ-rā′shŭn), p. 328
glottis (glŏt′ĭs), p. 321
hemoglobin (hē′mŭh-glō′bĭn), p. 329
inspiration (ĭn″spĭ-rā′shŭn), p. 319
inspiratory reserve volume (ĭn-spī′rŭh-tō-rē rē-zĕrv′ vŏl′yūm), p. 327
internal respiration (ĭn-tĕr′năl rĕs″pĭ-rā′shŭn), p. 329
intrapleural pressure (ĭn′trăh-plūr′ăl prĕsh′ŭr), p. 325
larynx (lār′ĭnks), p. 321

lungs (lŭngz), p. 323
mucociliary escalator (mū-kō-sĭl′ē-ă-rē ĕs′ kŭh-lā-tōr), p. 320
nasal cavities (nā′zŭl kăv′ĭ-tēz), p. 320
nose (nōz), p. 320
nostrils (nŏs′trĭls), p. 320
oxyhemoglobin (ŏk″sē-hē′ mŭh-glō-bĭn), p. 329
paranasal sinus (păr-ŭh-nā′zŭl sī′nŭs), p. 320
pharynx (făr′ĭnks), p. 321
pleurae (pleura) (plūr′-ā), p. 323
pons (pŏnz), p. 327
primary respiratory center (prī′mĕr-ē rĕs″ pī-rŭh-tōr′ ē sĕn′tĕr), p. 327
pulmonary capillaries (pŭl′mōh-nār-ē kăp′ĭl-lār-ē), p. 323
reduced hemoglobin (rē-dūs′d hē′mō-glō-bĭn), p. 330
residual volume (rē-zĭd′yū-ŭl vŏl′yūm), p. 327
respiratory membrane (rĕs″ pī-rŭh-tōr′ ē mĕm′brăn), p. 324
spirometer (spī-rŏm′-ĕ-tĕr), p. 326
surface tension (sĕr′fŭs tĕn′shŭn), p. 323
surfactant (sŭr-făk′tănt), p. 324
tidal volume (tī′dŭl vŏl′yūm), p. 326
tonsils (tŏn′sūlz), p. 331
trachea (trā′kē-ŭh), p. 322
ventilation (vĕn″tĭ-lā′shŭn), p. 325
vital capacity (vī′tŭhl kŭh-păs′ĭ-tē), p. 326
vocal cord (vō′kŭhl kŏrd), p. 321

Clinical Key Terms

acidosis (ăs-i-dō′sĭs), p. 330
acute bronchitis (ŭh-kyūt brŏng-kī′tĭs), p. 331
alkalosis (ăl-kă-lō′sĭs), p. 330

apnea (ăp′nē-ŭh), p. 321
asthma (ăz′mŭh), p. 335
black lung, (blăk lŭng) p. 332
bronchogenic carcinoma (brŏng-kō-jĕn′ĭk kăr′sūh-nō′mŭh), p. 333
Cheyne-Stokes respiration (shān-stōks rĕs″pī-rā′shŭn), p. 328
chronic bronchitis (krŏn′ĭk brŏng-kī′tĭs), p. 335
chronic obstructive pulmonary disorders (krŏn′ĭk ŭb-strŭkt′ĭv pŭl′mōh-nār-ē dĭs-ŏr′dĕrs), p. 332
dust cell (dŭhst sĕl), p. 324
emphysema (em″fĭ-sē′mŭh), p. 335
eupnea (yūp-nē′ŭh), p. 327
hyperpnea (hī″pĕr-nē′ŭh), p. 328
infant respiratory distress syndrome (ĭn′fŭnt rĕs″ pī-rŭh-tōr′ ē dĭs-trĕs′ sĭn′drōm), p. 324
laryngitis (lār-in-jī′tĭs), p. 331
lung cancer (lŭng kăn′sĕr), p. 335
metastasis (mĕ-tăs′-tă-sĭs), p. 335
obstructive sleep apnea (ŏb-strŭk′-tĭv slēp ăp′nē-ŭh), p. 321
otitis media (ō-tī′tĭs mē′dē-ŭh), p. 331
pneumonectomy (nū-mŭh-nĕk′tō-mē), p. 336
pneumonia (nū-mō′nē-ŭh), p. 331
pulmonary fibrosis (pŭl′mōh-nār-ē fĭ-brō′sĭs), p. 332
pulmonary tuberculosis (pŭl′mōh-nār-ē tū-bĕr″kyū-lō′sĭs), p. 331
singultus (sing-gŭl′tŭs), p. 318
sinusitis (sī-nū-sī′tĭs), p. 331
tonsillectomy (tŏn′sĭ-lĕk′tō-mē), p. 331
tonsillitis (tŏn-sĭl-ī′tĭs), p. 331
tracheostomy (trā″kē-ăhs′tō-mē), p. 323

Summary

14.1 The Respiratory System
The four main actions of the respiratory system are (1) ventilation, (2) external respiration (gas exchange in the lungs), (3) internal respiration (gas exchange in the tissues), and (4) the transport of gases in the blood.
   A. The nasal cavities, which filter, warm, and humidify incoming air, open into the pharynx.
   B. The food and air passages cross in the pharynx, which conducts air to the larynx and food to the esophagus.

   C. The larynx is the voice box. It houses the vocal cords. The glottis, a slit between the vocal cords, is covered by the epiglottis when food is being swallowed.
   D. The trachea and the primary bronchi are held open by cartilaginous rings. The trachea divides to form two primary bronchi, one to supply each lung. The cartilage rings gradually disappear as the primary bronchi branch into smaller bronchi and

then into bronchioles. The lungs are composed of bronchi, bronchioles, alveoli, pulmonary capillaries, and connective tissue.
   E. The respiratory membrane is formed by the alveolar wall and capillary wall, joined by basement membrane. The large surface area and thinness of the respiratory membrane allow rapid gas exchange.
14.2 Mechanism of Breathing
   A. Ventilation is the movement of gases into and out of the lungs.

Surface tension allows the visceral pleura, which covers the lungs, to adhere to the parietal pleura, which lines the inner chest wall. During inspiration, the thoracic cavity increases in size as the diaphragm lowers and the rib cage moves upward and outward. Therefore, the lungs expand, creating a partial vacuum, which causes air to rush in. During expiration, the diaphragm relaxes and resumes its dome shape. As the rib cage and lungs recoil, air is pushed out of the lungs. Expiration can be forceful when the internal intercostal muscles contract, causing the rib cage to move farther downward and inward. Also, contraction of the abdominal wall muscles pushes the abdominal organs up against the diaphragm, and the increased pressure expels air.

B. Tidal volume is the amount of air inhaled and exhaled with each breath. Vital capacity is the total volume of air that can be moved in and out of the lungs during a single breath. Some air remains in the lungs after expiration. This is called the residual volume. Passages within the airways are called dead-space because no gas exchange takes place in the airways. The four respiratory capacities are summed from the respiratory volumes.

C. The primary respiratory center in the medulla oblongata is assisted by the pons. It rhythmically controls the ventilation rate. The respiratory center increases the rate when CO_2 and H^+ levels increase, as detected by chemoreceptors in the respiratory center or the carotid and aortic bodies. The latter also detects a low O_2 level and stimulates the respiratory center, which then increases the ventilation rate.

14.3 Gas Exchange and Transport

A. Diffusion accounts for the movement of gases during external and internal respiration. External respiration occurs when carbon dioxide moves from the area of higher concentration in the pulmonary capillary blood to lower concentration in the alveoli. Oxygen moves from a higher concentration in the alveoli to a lower concentration in the pulmonary capillary blood.

B. Internal respiration occurs when oxygen moves from a higher concentration in the systemic bloodstream to a lower concentration in the tissue fluid, and carbon dioxide moves from a higher concentration in the tissue fluid to a lower concentration in the systemic bloodstream. These movements can be explained in terms of gas partial pressures as well. Continuous cellular respiration creates the concentration and partial pressure gradients.

C. Oxygen is transported to the tissues in combination with hemoglobin as oxyhemoglobin (HbO_2). Carbon dioxide is mainly carried within the plasma to the lungs as bicarbonate ion (HCO_3^-). Hemoglobin combines with hydrogen ions and becomes reduced (HHb), and this helps maintain the pH level of the blood within normal limits. By adjusting respiratory rate, the respiratory system helps to maintain the pH balance. Breathing slowly decreases pH, and breathing rapidly increases it.

14.4 Respiration and Health

A. A number of illnesses are associated with the respiratory tract. These disorders can be divided into those that affect the upper respiratory tract and those that affect the lower respiratory tract. Infections of the nasal cavities, sinuses, throat, tonsils, and larynx are all well known. In addition, infections can spread from the nasopharynx to the ears.

B. The lower respiratory tract is subject to infections such as acute bronchitis, pneumonia, and pulmonary tuberculosis. In restrictive pulmonary disorders, exemplified by pulmonary fibro-sis, the lungs lose their elasticity. In obstructive pulmonary disorders, exemplified by chronic bronchitis, emphysema, and asthma, the bronchi (and bronchioles) do not effectively conduct air to and from the lungs. Smoking, which is associated with chronic bronchitis and emphysema, can eventually lead to lung cancer.

14.5 Effects of Aging

All aspects of respiration decline with age. The elderly often die from pulmonary infections. Immunization with the pneumococcal vaccine is recommended for anyone over 65.

14.6 Homeostasis

A. Oxygen is brought into the lungs, where it diffuses into pulmonary capillary blood. Carbon dioxide is constantly produced by cells and tissues. It diffuses into systemic capillaries, and is returned to the lungs for excretion.

B. The respiratory system regulates blood pH by changing blood carbon dioxide levels. Hypoventilation creates acidosis, while hyperventilation creates alkalosis.

C. The respiratory system helps maintain blood pressure by converting angiotensin I to angiotensin II.

D. The respiratory tract assists in the defense against pathogens by keeping the tract clean of debris. Also, the tonsils are lymphatic tissue where antigens are presented to T-cells. Dust cell macrophages remove debris and pathogens from the alveoli.

E. The respiratory system works with other systems of the body. The cardiovascular system transports gases, and breathing helps systemic venous blood return to the heart. The nervous system maintains rhythmic ventilation, and the sensory organs for olfaction are located in the nasal cavities. The respiratory center responds to the increased gas exchange needs of the muscular system when we exercise.

Study Questions

1. Name and explain the four events that comprise respiration. (p. 319)
2. What is the path of air from the nose to the lungs? Describe the structure and state the function of all the organs mentioned. (pp. 319–324)
3. What is the respiratory membrane, and how does its structure promote rapid gas exchange? (pp. 324–325)
4. What three conditions should one have in mind in order to understand ventilation? (p. 325)
5. Explain the volume and pressure changes necessary to inspiration and expiration. How is ventilation controlled? (pp. 325–326)
6. What is the difference between tidal volume and vital capacity? Of the air we inhale, some is not used for gas exchange. Why not? (pp. 326–327)
7. How is respiration controlled? Contrast external and internal respiration. Why is there gas flow during each of these? (pp. 327–329)
8. How are oxygen and carbon dioxide transported in the blood? What role does hemoglobin play in the transport of CO_2? (pp. 329–330)
9. Name and describe several upper and several lower respiratory tract disorders (other than cancer). If appropriate, explain why breathing is difficult with these conditions. (pp. 330–332, 335)
10. List the steps by which lung cancer develops. (pp. 335–336)

Learning Outcome Questions

Fill in the blanks.

1. In tracing the path of air, the _____ immediately follows the pharynx.
2. The _____ closes the opening into the larynx during swallowing.
3. The lungs contain air sacs called _____.
4. The respiratory membrane consists of the walls of the _____ and _____.
5. Air enters the lungs after they have _____.
6. The amount of air moved into and out of the respiratory system with each normal breath is called the _____.
7. The total amount of air that can be moved into and out of the lungs during a single breath is called the _____.
8. The breathing rate is primarily regulated by the amount of _____ in the blood.
9. Gas exchange is dependent on the physical process of _____.
10. During external respiration, oxygen _____ the blood.
11. During internal respiration, carbon dioxide _____ the blood.
12. Carbon dioxide is carried in the blood as _____ ions.
13. The most likely cause of emphysema and chronic bronchitis is _____.
14. Most cases of lung cancer actually begin in the _____.
15. Label structures a. through i. in the diagram of the respiratory tract.

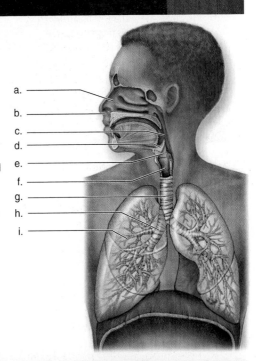

a. _____
b. _____
c. _____
d. _____
e. _____
f. _____
g. _____
h. _____
i. _____

Medical Terminology Exercise

After studying this chapter, see if you can derive the definitions for the medical terms listed at right. Many of the prefixes and suffixes used to create these terms can be found throughout the chapter. For additional help, use McGraw-Hill Connect™ at www.mcgrawhillconnect.com and consult Appendix B.

1. nasopharyngitis (nā″zō-făr″ĭn-jī′tĭs)
2. pleuropericarditis (plĕr″ō-pĕr″ĭ-kăr″dī′tĭs)
3. bronchoscopy (brŏng-kŏs′kŭh-pē)
4. dyspnea (dĭsp′nē-ŭh)
5. laryngospasm (lŭh-rĭng′gō-spăzm)
6. hemothorax (hē″mō-thō′răks)
7. otorhinolaryngology (ō″tō-rī″nō-lār″ĭn-gŏl′ō-jē)
8. hypoxemia (hī″pŏk-sē′mē-ŭh)
9. pulmonectomy (pŭl-mō-nĕk′tō-mē)
10. hypercapnea (hī-pĕr-kăp′nē-ŭh)
11. spirometry (spy-rŏm′ŭh-trē)
12. thoracentesis (thŏr′ŭh-sĕn-tē′sĭs)

Online Study Tools

🔲LEARNSMART® Mc Graw Hill Education **connect**

APR

Anatomy & Physiology REVEALED includes cadaver photos that allow you to peel away layers of the human body to reveal structures beneath the surface. This program also includes animations, radiologic imaging, audio pronunciations, and practice quizzing. To learn more visit www.aprevealed.com

15

The Digestive System

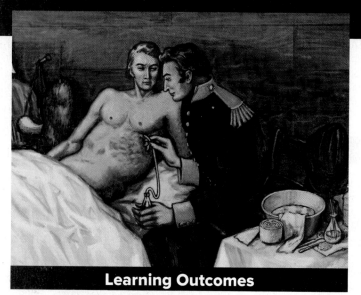

In a time before modern, well-equipped laboratories existed, one man's misfortune turned into an unparalleled opportunity to study digestion. In 1822, Alexis St. Martin was injured by a gunshot wound. The bullet tore through his stomach, leaving a *fistula* (a hole in the body wall) that refused to heal. Dr. William Beaumont, one of St. Martin's physicians, took advantage of his patient's situation to study the stomach's digestive processes. After tying various forms of food to a string, Beaumont first pushed the food through the fistula into St. Martin's stomach. Next, Beaumont retrieved the food at specified intervals by pulling on the string. He then measured the remaining food morsel's mass and volume to track its digestion rate, and sampled the liquid in the stomach to analyze its composition. Much of the early knowledge regarding the stomach resulted from this unusual research. You can find out more about the stomach's role in digestion on pages 347–349.

Learning Outcomes

After you have studied this chapter, you should be able to:

15.1 Anatomy of the Digestive System

1. Trace the path of food through the alimentary canal, and describe the general structure and function of each organ mentioned.
2. Describe the wall of the small intestine, and relate its anatomy to nutrient absorption.
3. Describe peristalsis, and state its function.
4. Name the hormones produced by the alimentary canal that help control digestive secretions.

15.2 Accessory Organs of Digestion

5. Name five accessory organs of digestion.
6. Outline the location, anatomy, and functions of the pancreas, the liver, and the gallbladder.

15.3 Chemical Digestion

7. Name and state the functions of the digestive enzymes for carbohydrates, proteins, and fats.

15.4 Effects of Aging

8. Describe the anatomical and physiological changes that occur in the digestive system as we age.

15.5 Homeostasis

9. Explain how the digestive system works with other systems of the body to maintain homeostasis.

15.6 Nutrition

10. State the functions of glucose, fats, and amino acids in the body.
11. Define the terms *essential fatty acid*, *essential amino acid*, and *vitamin*.
12. Describe the functions of the major vitamins and minerals in the body.

Focus on Forensics
The Stories That Teeth Can Tell

Medical Focus
Disorders of the Digestive Tract

Human Systems Work Together
Digestive System

Medical Focus
Tips for Effectively Using Nutrition Labels

Medical Focus
Bariatric Surgery for Obesity

15.1 Anatomy of the Digestive System

1. Trace the path of food through the alimentary canal, and describe the general structure and function of each organ mentioned.
2. Describe the wall of the small intestine, and relate its anatomy to nutrient absorption.
3. Describe peristalsis, and state its function.
4. Name the hormones produced by the alimentary canal that help control digestive secretions.

The organs of the digestive system are located within a tube called the **alimentary canal,** or **gastrointestinal tract.** The tube begins with the mouth and ends with the **anus** (Fig. 15.1).

Although the term *digestion*, strictly speaking, means the breakdown of food by enzymatic action, we'll expand the definition to include both the physical and chemical processes that reduce food to small, soluble molecules that can be absorbed into the bloodstream.

The functions of the digestive system are to:

1. ingest the food;
2. break food down into small molecules that can cross plasma membranes;
3. absorb these nutrient molecules;
4. eliminate nondigestible wastes.

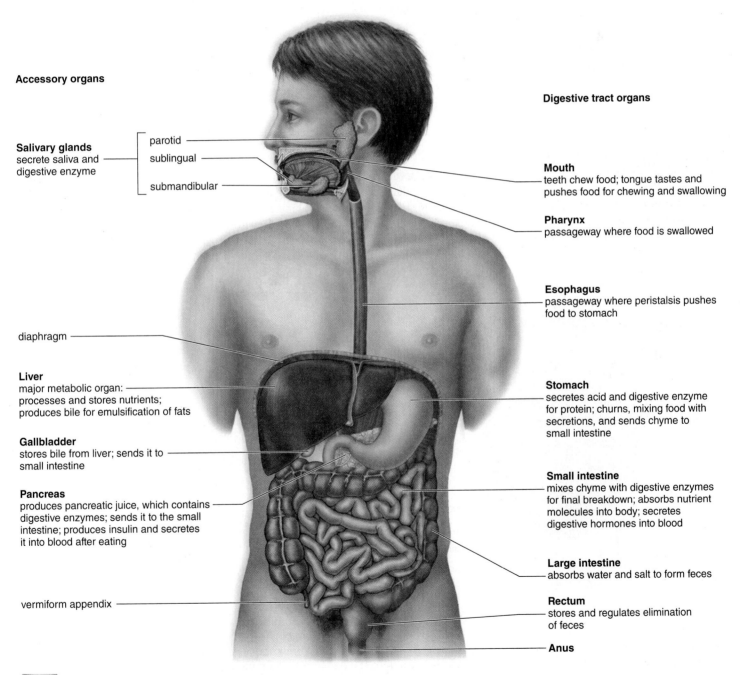

Accessory organs

Salivary glands
secrete saliva and digestive enzyme

parotid
sublingual
submandibular

diaphragm

Liver
major metabolic organ: processes and stores nutrients; produces bile for emulsification of fats

Gallbladder
stores bile from liver; sends it to small intestine

Pancreas
produces pancreatic juice, which contains digestive enzymes; sends it to the small intestine; produces insulin and secretes it into blood after eating

vermiform appendix

Digestive tract organs

Mouth
teeth chew food; tongue tastes and pushes food for chewing and swallowing

Pharynx
passageway where food is swallowed

Esophagus
passageway where peristalsis pushes food to stomach

Stomach
secretes acid and digestive enzyme for protein; churns, mixing food with secretions, and sends chyme to small intestine

Small intestine
mixes chyme with digestive enzymes for final breakdown; absorbs nutrient molecules into body; secretes digestive hormones into blood

Large intestine
absorbs water and salt to form feces

Rectum
stores and regulates elimination of feces

Anus

AP|R Figure 15.1 **Structures of the digestive system.**

The Mouth

The **mouth,** also termed the *oral cavity,* receives food and begins physical and chemical digestion. The oral cavity is bounded externally by the lips and cheeks. The space between the teeth, the lips, and cheeks is the **vestibule.**

The tongue is composed of skeletal muscle whose contraction changes the tongue's shape. Muscles exterior to the tongue cause it to move about. Rough projections on the tongue, called *papillae,* help it handle food and also contain the sensory receptors called taste buds. A fold of mucous membrane, called the lingual frenulum, on the underside of the tongue attaches it to the floor of the mouth. If the frenulum is too short, the individual cannot speak clearly and is said to be tongue-tied. Posteriorly, the tongue is anchored to the hyoid bone.

The mouth has a roof that separates it from the nasal cavities. The roof has two parts: an anterior (toward the front) **hard palate** and a posterior (toward the back) **soft palate** (see Fig. 15.3). The hard palate is formed by the maxilla and palatine bones; the soft palate is formed by muscle and glandular tissue. The soft palate ends in a finger-shaped projection called the **uvula.**

Human Teeth and Salivary Glands

During the first two years of life, the 20 deciduous, or baby, teeth appear. Eventually, the deciduous teeth are replaced by the adult teeth. Normally, adults have 32 teeth (Fig. 15.2a). The maxilla (upper jaw bone) and the mandible (lower jaw bone) each contain teeth of four different types. Anteriorly, four chisel-shaped **incisors** function for biting. Flanking them on either side are two pointed canine teeth, or **cuspids,** which help tear food. Laterally, four flat premolar teeth, or **bicuspids,** are used for grinding food. The most posterior are the six **molar** teeth, designed for crushing and grinding food. The last molars, called the *wisdom teeth,* usually erupt between ages 17 and 25. However, they often fail to come in, or if they do, they may grow in crooked and be useless. Frequently, wisdom teeth are extracted.

Each tooth (Fig. 15.2b) has a crown and a root. The crown has a layer of enamel, an extremely hard outer covering of calcium compounds; dentin, a thick layer of bonelike material; and an inner pulp, which contains the nerves and blood vessels. Dentin and pulp are also in the root. **Caries** (tooth decay; commonly called cavities) occur when bacteria within the mouth break down sugar and give off acids that corrode the teeth. Once these acids dissolve the enamel and dentin, the pulp is damaged and inflamed, triggering a toothache. Fluoride treatments, particularly in children, can make the enamel stronger and more resistant to decay.

Gum disease is more likely as we age. One example is inflammation of the gums, called **gingivitis,** that may spread to the periodontal membrane lining the tooth socket (Fig. 15.2b). When this occurs, the individual develops **periodontitis,** characterized by loss of bone and loosening of the teeth. Brushing and flossing your teeth after every meal cleans the teeth and stimulates the gums, preventing these conditions. Make sure you brush gently and away from the gumline, to keep delicate gum tissue from being destroyed. Taking good care of your teeth and gums is also important in avoiding atherosclerosis (see Chapter 12).

Three pairs of **salivary glands** send saliva by way of ducts to the mouth. The parotid glands lie anterior and somewhat inferior

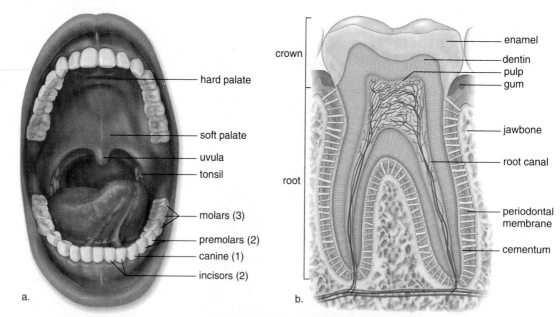

AP|R **Figure 15.2 Human teeth. (a)** The chisel-shaped incisors bite; the pointed canines tear; the fairly flat premolars grind; and the flattened molars crush food. The last molar, called a wisdom tooth, may fail to erupt, or if it does, it is sometimes crooked and useless. Often dentists recommend the extraction of the wisdom teeth. **(b)** Longitudinal section of a tooth. The crown is the portion that projects above the gumline and can be replaced by a dentist if damaged. When the periodontal membrane is inflamed, the tooth can loosen.

to the ears between the cheek and the masseter muscle. They have ducts that open on the inner surface of the cheek at the location of the second upper molar. The parotid glands swell when a person has the mumps, a disease caused by a viral infection. The sublingual glands are located beneath the tongue, and the submandibular glands are in the floor of the mouth on the inside surface of the lower jaw. The ducts from the sublingual and submandibular glands open under the tongue. You can locate the openings for the salivary glands if you use your tongue to feel for small flaps on the inside of your cheek and under your tongue. Saliva is a solution of mucus and water. It also contains bicarbonate and an enzyme called **salivary amylase,** which begins the process of digesting

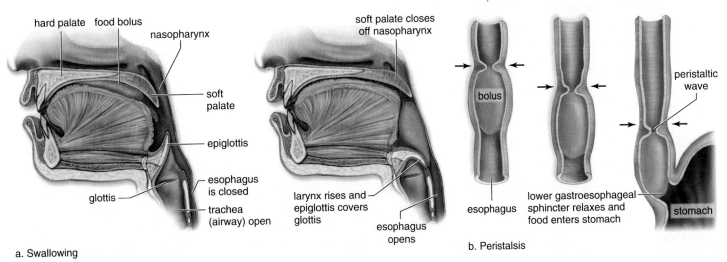

a. Swallowing

b. Peristalsis

Figure 15.3 Swallowing. (a) When food is swallowed, the soft palate closes off the nasopharynx, and the epiglottis covers the glottis, forcing the bolus to pass down the esophagus. Therefore, a person does not breathe while swallowing. **(b)** Peristalsis moves food through the gastroesophageal sphincter into the stomach.

carbohydrate. Saliva moistens food and prepares it for swallowing. In addition, saliva contains an antibacterial enzyme called *lysozyme,* as well as secretory antibodies, which help to protect the body. By constantly bathing the teeth, tongue, and oral mucous membrane, saliva removes microbes. Swallowed microbes can be destroyed by stomach acid and enzymes.

The Pharynx

Table 15.1 traces the path of food. From the mouth, food passes through the pharynx and esophagus to the stomach, small intestine, and large intestine. The food passage and the air passage cross in the **pharynx** because the trachea is anterior to the esophagus, a long, muscular tube that takes food to the stomach (Fig. 15.3).

The pharynx has three parts: (1) The **nasopharynx,** posterior to the nasal cavity, serves as a passageway for air; (2) the **oropharynx,** posterior to the soft palate, is a passageway for both air and food; and (3) the **laryngopharynx,** just superior to the esophagus, is a passageway for food entering the esophagus.

The tonsils are embedded in the mucous membrane of the tongue and pharynx. As you know (from Chapter 13), the *palatine tonsils* are on either side of the tongue close to the soft palate, and the *lingual tonsils* sit at the base of the tongue. Together, they help to protect the body from infection caused by ingested and/or inhaled microbes. The single *pharyngeal tonsil* (sometimes called the *adenoids*) sits in the posterior nasopharynx. All of this tonsillar tissue defends against microbes in inhaled air. A person who has inflamed tonsils has *tonsillitis*. If the tonsillitis keeps recurring, the tonsils may be surgically removed (called a tonsillectomy).

Swallowing

During swallowing, food normally enters the esophagus because other possible avenues are blocked. Swallowing has a voluntary phase—from day-to-day living, you know that you can swallow voluntarily (try it). However, once food or drink is pushed back to the oropharynx, swallowing then becomes a reflex action performed automatically (without our willing it). When we swallow, the soft palate moves back to close off the nasopharynx, and the larynx moves up under the epiglottis so that food is less likely to enter it. (We don't breathe when we swallow.) The tongue presses against the soft palate, sealing off the oral cavity, and the esophagus opens to receive a food **bolus** (Fig. 15.3). *Bolus* is the technical term for a morsel of chewed and swallowed food or drink.

Unfortunately, we've all had the unpleasant experience of having food "go the wrong way." The wrong way may be either into the nasal cavities or into the trachea. If it is the latter, coughing will most likely force the food up out of the trachea and into the pharynx again.

The Wall of the Digestive Tract

The entire digestive tract, from the esophagus to the rectum, is collectively referred to by the technical term *gut*. The gut is a continuous tube composed of four layers (Fig. 15.4). From deepest to most superficial, the layers are:

Mucosa (mucous membrane layer) A layer of epithelium supported by connective tissue and smooth muscle lines the **lumen** (central cavity). This layer contains glandular epithelial cells that secrete digestive enzymes and goblet cells that secrete mucus.

Submucosa (submucosal layer) A broad band of loose connective tissue that contains blood vessels, lymphatic vessels, and nerves lies beneath the mucosa. The submucosa joins the mucosa to the muscularis layer. Lymph nodules, called Peyer patches, are scattered throughout the submucosa of the small intestine (especially in the ileum; see Chapter 13). Like the tonsils, they help protect us from disease.

Muscularis (smooth muscle layer) Two layers of smooth muscle make up this section. The inner, circular layer encircles the gut; the outer, longitudinal layer lies in the same direction as the gut. The stomach also has oblique muscles.

TABLE 15.1 Path of Food in the Digestive Process

Organ	Function of Organ	Anatomic Features	Function of Anatomic Features
Oral cavity	Mechanical breakdown of food; swallowing; starch digestion	Teeth Tongue	Biting, tearing, chewing food Forms bolus; swallowing
Pharynx	Transport of food to esophagus	Soft palate and uvula Epiglottis	Prevents food from entering nasal cavity Prevents food from entering glottis
Esophagus	Transport of food to stomach	Four-layer construction	Muscularis layer carries out peristalsis
Stomach	Food storage; antibacterial; begins protein digestion; intrinsic factor production; slow release of chyme to small intestine	Rugae Parietal cells of gastric glands Chief cells of gastric glands Parietal cells of gastric glands Pyloric sphincter	Expand as stomach fills Produce hydrochloric acid Produce pepsin Produce intrinsic factor Controls release of chyme
Small intestine	Completes digestion of food; absorption of nutrients	Brush border enzymes Intestinal villi; microvilli	Complete digestive process Absorb nutrients
Large intestine	Absorption of water; storage of indigestible remains; elimination of fecal material	Epithelial cells Haustra Teniae coli and muscularis	Water absorption by osmosis Pouch-like structures store material Move feces toward anus

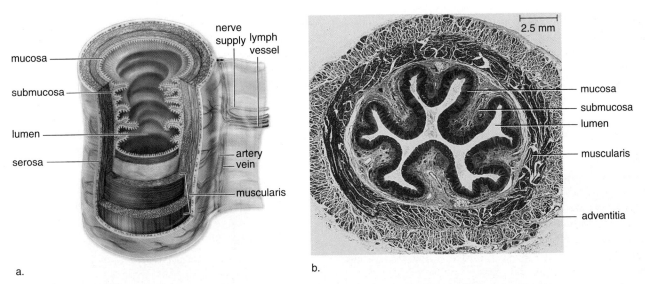

a.

b.

AP|R **Figure 15.4** **Wall of the alimentary canal.** **(a)** Several different types of tissues are found in the wall of the alimentary canal. Note the placement of circular muscle inside longitudinal muscle. **(b)** Micrograph of the wall of the esophagus.

Serosa (serous membrane layer) Most of the alimentary canal has a serosa, a very thin, outermost layer of squamous epithelium supported by connective tissue. The serosa secretes a serous fluid that keeps the outer surface of the intestines moist so that the organs of the abdominal cavity slide against one another. The esophagus has an outer layer composed only of loose connective tissue called the *adventitia.*

Although the gut is a continuous tube, its different regions have individual characteristics uniquely suited to each region's particular function. For example, the stomach's circular, longitudinal and oblique muscle layers enable it to mix food. You'll read about other specializations as you continue through this chapter.

The Esophagus

The **esophagus** is a muscular tube that lies posterior to the trachea. It passes from the pharynx through the thoracic cavity and diaphragm into the abdominal cavity, where it joins the stomach. The esophagus is ordinarily collapsed, but it opens and receives the bolus when swallowing occurs.

A rhythmic contraction called **peristalsis** pushes the food along the alimentary canal. In peristalsis, short segments of smooth muscle built into the wall of the digestive tract alternately contract and then relax. Peristalsis constantly moves food forward through the digestive tract. (As an analogy, think of squeezing a toothpaste tube, starting at the bottom, then continuing upward along the entire length of the tube. If you do it correctly, the toothpaste moves forward until the tube is empty.) Peristalsis begins in the esophagus and continues in all the organs of the alimentary canal. Occasionally, peristalsis begins even though there is no food in the esophagus. This produces the sensation of a lump in the throat.

The esophagus plays no role in the chemical digestion of food. Its sole purpose is to transport the food bolus from the mouth to the stomach. The entrance of the esophagus to the stomach is marked by a constriction, often called the esophageal **sphincter.** It's important to note that the esophageal sphincter isn't a true sphincter muscle. A true sphincter muscle (such as the pyloric sphincter, discussed with the stomach) is an actual ring of muscle that closes a tube when it contracts and opens the tube when it relaxes. Although not an actual ring of muscle, the esophageal sphincter functions as one because it allows the bolus to pass into the stomach when relaxed (see Fig. 15.3). Likewise, the sphincter prevents the stomach's acidic contents from backing up into the esophagus when contracted. When vomiting occurs, contraction of the diaphragm and abdominal muscles propels the stomach's contents past the esophageal sphincter upward through the esophagus.

The Stomach

The **stomach** (Fig. 15.5) is a thick-walled, J-shaped organ that lies on the left side of the abdominal cavity inferior to the diaphragm and posterior to the liver. The stomach is continuous with the esophagus superiorly and the duodenum of the small intestine inferiorly.

The length of the stomach remains at about 25 cm (10 in.) regardless of the amount of food it holds, but the diameter varies, depending on how full it is. As the stomach expands, deep folds in its wall, called **rugae,** stretch out and gradually disappear. When full, the stomach can hold about 4 liters (1 gallon). The stomach receives food from the esophagus, stores food, liquifies food by mixing food with its juices (thereby starting the digestion of proteins), and moves food into the small intestine.

Regions of the Stomach

The stomach has four regions. The **cardiac region,** which is near the heart, surrounds the lower esophageal sphincter where food

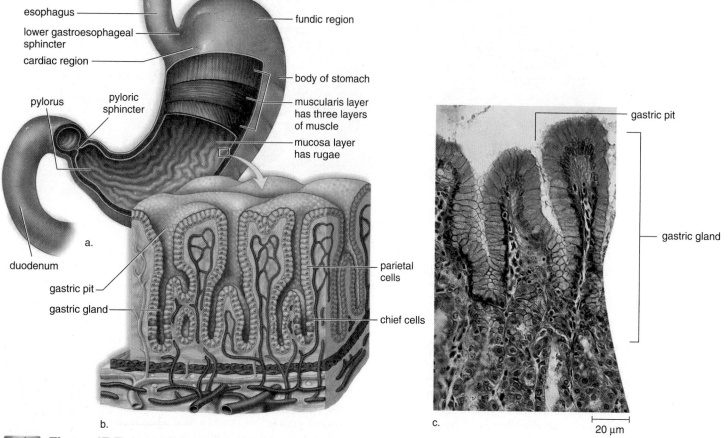

esophagus

lower gastroesophageal sphincter

cardiac region

pylorus

pyloric sphincter

a.

duodenum

gastric pit

gastric gland

b.

fundic region

body of stomach

muscularis layer has three layers of muscle

mucosa layer has rugae

parietal cells

chief cells

c.

gastric pit

gastric gland

20 μm

AP|R **Figure 15.5** **Anatomy and histology of the stomach. (a)** The stomach has a thick wall with deep folds, rugae, that allow it to expand and fill with food. **(b)** The stomach lining has gastric glands, which secrete mucus and protein-digesting gastric juice. **(c)** Photomicrograph of the stomach lining.

enters the stomach. The **fundic region,** which holds food temporarily, is an expanded portion superior to the cardiac region. The **body region,** which comes next, is the main part. The **pyloric region** narrows to become the pyloric canal leading to the pyloric sphincter. Food passes through this sphincter and enters the duodenum, the first part of the small intestine.

Digestive Functions of the Stomach

The stomach acts on food both chemically and physically. Its wall contains three muscle layers: One layer is longitudinal, another is circular, and the third is obliquely arranged. This muscular wall not only moves the food along, but it also churns, mixing the food with gastric juice and breaking it down to small pieces.

The term *gastric* always refers to the stomach. The columnar epithelial lining of the stomach has millions of **gastric pits,** which lead into **gastric glands** (Fig. 15.5a, b). Gastric glands contain four types of secretory cells: *chief cells, parietal cells, enteroendocrine cells,* and *mucous cells.* Other secretory cells found scattered throughout the glands are *ECL cells* and *D cells.* The gastric glands produce **gastric juice,** a watery solution that contains pepsinogen, hydrochloric acid (HCl), intrinsic factor, and mucus. Chief cells secrete pepsinogen, which becomes the protein-digesting enzyme

pepsin when exposed to hydrochloric acid. Parietal cells produce the hydrochloric acid. The HCl causes the gastric juice in the stomach to have a high acidity with a pH of about 2; this is beneficial because it kills most of the bacteria present in food. Although HCl does not digest food, it does break down the connective tissue of meat and activates pepsin.

Parietal cells also produce **intrinsic factor,** a glycoprotein that binds to vitamin B_{12}, preventing this vitamin from being destroyed in the harsh, acidic environment of the stomach. If the stomach fails to produce intrinsic factor, or if the diet is deficient in vitamin B_{12}, a serious disorder called *pernicious anemia* will result. Lacking the vitamin, red blood cells will fail to develop.

Enteroendocrine cells of the gastric glands produce the hormone **gastrin.** This hormone enters stomach blood vessels and is circulated throughout the stomach. Gastrin regulates muscular contraction and secretion by the stomach. Histamine from ECL cells prolongs the effect of gastrin, causing additional acid to be secreted. However, it's important to be able to inhibit acid secretion when appropriate, and that is the role of the somatostatin secreted by the D cells.

The wall of the stomach is protected by the thick layer of mucus secreted by the *mucous cells.* If, by chance, HCl penetrates this mucus, the wall can begin to break down, and an ulcer results as described in the Medical Focus on page 356.

Alcohol and water are absorbed in the stomach, but food substances are not. Normally, the stomach empties in about 2–6 hours. When food leaves the stomach, it is a thick, soupy liquid called **chyme.** Chyme enters the small intestine in squirts by way of the pyloric sphincter, dilating and opening, then contracting and closing repeatedly.

The Small Intestine

The **small intestine** extends from the pyloric valve of the stomach to the *ileocecal valve* where it joins the large intestine. It is named for its small diameter (compared to that of the large intestine), but it might make more sense to call it the long intestine. The small intestine takes up a large portion of the abdominal cavity, averaging about 6 m (18 ft) in length in a cadaver. Its actual length in a living person cannot be accurately determined, because living intestinal smooth muscle has muscle tone and is much shorter. (See page 143 to review *muscle tone.*)

All the contents of food—fats, proteins, and carbohydrates—are digested in the small intestine to molecules that can be absorbed. To help complete the task, the small intestine receives secretions from the pancreas and liver and produces intestinal juices. Absorption of nutrients for the body's cells, such as amino acids and sugars, occurs in the small intestine. It also transports nondigestible remains to the large intestine.

Regions of the Small Intestine

The small intestine has the following regions (Fig. 15.6):

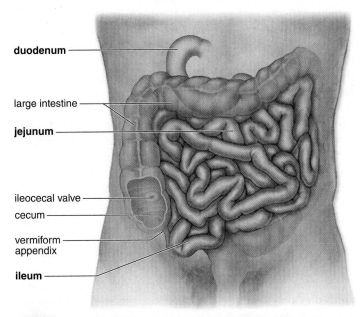

duodenum

large intestine

jejunum

ileocecal valve

cecum

vermiform
appendix

ileum

AP|R **Figure 15.6 Regions of the small intestine.** The duodenum is attached to the stomach. The jejunum leads to the ileum, which is attached to the large intestine.

Duodenum The first 25 cm (10 in.) contain distinctive glands that secrete mucus. Like that found in the stomach, this mucus protects the duodenal lining from harsh stomach acid. Pancreatic secretions and bile from the liver empty into the duodenum through a common duct. Folds and villi (Fig. 15.7) are more numerous at the end than at the beginning.

Jejunum The next 2.5 m (7.5 ft in a cadaver) contains folds and villi. These features are more common at the proximal end than at the distal end of the jejunum.

Ileum The last 3.6 m (10.8 ft in a cadaver) contain fewer folds and villi than the jejunum. The ileum wall contains Peyer patches, clusters of lymph nodules mentioned in Chapter 13.

Wall of the Small Intestine

The approximate surface area of the small intestine has been compared to that of a tennis court. Three features contribute to increasing its surface area: circular folds, villi, and microvilli (Fig. 15.7). The **circular folds** are permanent transverse folds involving the mucosa and submucosa of the small intestine. The **villi** (sing., *villus*) are fingerlike projections of the mucosa into the lumen of the small intestine (Fig. 15.7a and b). The villi are so numerous and closely packed that they give the wall a velvetlike appearance. A villus has an outer layer of columnar epithelial cells, and each of these cells has thousands of microscopic extensions called **microvilli.** Collectively, in electron micrographs, microvilli give the villi a fuzzy border known as a "brush border" (Fig. 15.7c). Because the microvilli bear the intestinal enzymes, these enzymes are called *brush-border enzymes.*

Functions of the Small Intestine

The digestive process is completely finished in the small intestine. Ducts from the liver, gallbladder, and pancreas join to form one duct that enters the duodenum (Fig. 15.8 and Fig. 15.11). The small intestine receives bile from the gallbladder and pancreatic juice from the pancreas via this duct. **Bile,** which is produced by the liver but stored in the gallbladder, *emulsifies* fat—emulsification is a process that allows fat droplets to disperse in water. (However, it's important to note that bile doesn't digest fat; that job is left to the lipase enzyme produced by the pancreas.) The intestine has a slightly basic pH because pancreatic juice contains sodium bicarbonate ($NaHCO_3$), which neutralizes the acidic chyme from the stomach. The enzymes in pancreatic juice and the enzymes produced by the intestinal wall complete the process of food digestion. The functions of the liver, gallbladder, and pancreas will be detailed in section 15.2.

The other primary function of the small intestine is *absorption of nutrients.* The tremendous increase in surface area created by the circular folds, villi, and microvilli makes this an efficient process—the greater the surface area, the greater the volume of intake in a given unit of time. Also, a villus contains a generous supply of blood capillaries and a small lymphatic capillary, called a **lacteal** (see Fig. 15.7). As you know, the lymphatic system partners with the cardiovascular system; its vessels carry a fluid called *lymph* to the cardiovascular veins. Sugars (digested in part from

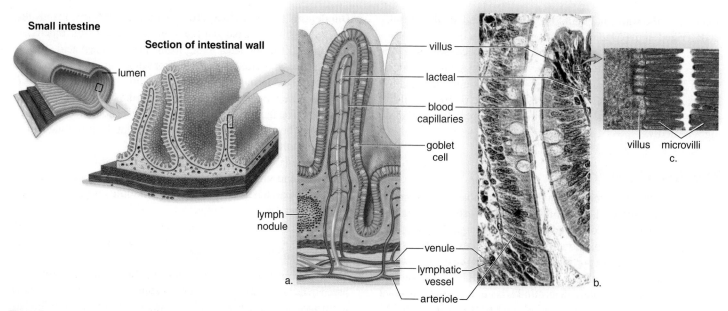

Small intestine

lumen

Section of intestinal wall

villus

lacteal

blood
capillaries

goblet
cell

lymph
nodule

venule

lymphatic
vessel

arteriole

a.

b.

villus   microvilli
c.

AP|R **Figure 15.7** **Anatomy of the small intestine. (a)** The wall of the small intestine has folds that bear fingerlike projections called *villi* (sing., *villus*). **(b)** Photomicrograph of intestinal villi. Villi in turn have projections called *microvilli*. **(c)** Photomicrograph of a villus with its microvilli. The products of digestion are absorbed by microvilli and enter the blood capillaries and the lacteals of the villi.

carbohydrates) and amino acids (digested from protein) enter the blood capillaries of a villus. Glycerol and fatty acids (digested from fats) enter the epithelial cells of the villi, and within these cells are joined and packaged as lipoprotein droplets, which enter a lacteal. Thus, lipids first enter the lymphatic capillaries and are ultimately transported in the lymph to the lymphatic ducts. Recall that the lymphatic ducts return lymph to the bloodstream by way of the subclavian veins (p. 294). After nutrients are absorbed into the blood, they are eventually carried to all the cells of the body by the bloodstream.

As we noted previously, a third function of the small intestine is movement of nondigested remains to the large intestine. The wall of the small intestine has two types of movements: segmentation and peristalsis. Segmentation refers to localized contractions and constrictions that mix the chyme thoroughly with digestive juices. Segmentation also encourages absorption of nutrients into the bloodstream or lymph. Peristalsis then moves nondigested remains toward the large intestine.

Regulation of Contraction and Secretion in the Digestive Tract

A combination of three control mechanisms cooperates to regulate the muscular contractions needed to move food throughout the gut, as well as the secretion of digestive juices needed for breaking down the food as it travels along. The central nervous system allows us to respond to hunger signals (and for many of us, stimulating eating behavior when we're stressed). The **enteric nervous system** is a nerve network built directly into the wall of the gut, allowing it to control its own movements and secretion.

Enteroendocrine hormones produced by cells in the stomach and small intestine participate as well.

Three phases occur as the control mechanisms become active. The first, the **cephalic phase** (remember from Chapter 1 that *cephalic* means "head"), begins when we look at, think about, smell, or taste food. Sensory signals are relayed to the hypothalamus, the brain center that stimulates hunger. In turn, the hypothalamus triggers autonomic nerves belonging to the parasympathetic nervous system to automatically stimulate secretion and muscle contraction. Filling the stomach helps to trigger the next regulatory mechanism: the **gastric phase** of control. The presence of food (particularly food high in amino acids) that stretches the stomach causes stimulation of the enteric nerve network. At the same time, the stomach hormone gastrin is also released. Gastrin is transported in the bloodstream to nearby smooth muscle and gastric glands. Gastric contraction and secretion by gastric glands are controlled internally by cooperation of these two mechanisms. Both muscle contraction and gastric gland secretion increase in response.

A similar nerve network and hormone set cooperate with parasympathetic nerve activity to trigger the third, or **intestinal phase** of secretion. When the duodenum is stretched by digested food that is rich in fatty acids and carbohydrate, the enteric nervous system responds to stimulate intestinal secretion and contraction. Simultaneously, cells of the duodenal wall produce three hormones that are of particular interest: **secretin, gastric inhibitory polypeptide** (GIP; sometimes referred to as glucose-dependent insulinotropic polypeptide), and **cholecystokinin** (CCK). Acid, especially hydrochloric acid (HCl) present in chyme, is a potent stimulator for secretin release, whereas partially digested protein and fat stimulate the release of CCK. Secretin stimulates secretion of bicarbonate solution by the pancreas, and cholecystokinin stimulates pancreatic enzyme secretion. Thus, when these hormones enter the bloodstream, the pancreas increases its output of pancreatic juice. Pancreatic juice buffers the stomach's acid chyme and helps to digest food. The liver and gallbladder also respond: The liver increases its production of bile in response to secretin and the gallbladder contracts to release stored bile in response to CCK. The third hormone produced by the duodenal wall, GIP, works opposite to gastrin. It inhibits gastric gland secretion and slows the contractions of the stomach. In this way, GIP prevents the stomach from emptying too quickly. Figure 15.8 summarizes the actions of gastrin, secretin, and CCK. As the figure illustrates, these digestive hormones must first enter the bloodstream to be carried to their target cells (just as all other endocrine hormones are transported in the blood).

It's important to note that these three regulatory methods—cephalic, gastric, and intestinal phases—occur in sequence only as your meal begins. As you continue eating, digesting, and absorbing your food, reflexes from each phase may be occurring at the same time.

Two additional hormones contribute to hunger (the desire for food) and appetite (the desire for specific foods that one enjoys). These are **leptin**, produced by fat cells, and **ghrelin**, produced by the stomach. Both hormones act on the hypothalamus. Leptin stimulates a feeling of fullness and contentment after eating and causes us to stop eating. In contrast, ghrelin triggers a hunger sensation and causes us to seek food. Scientists are actively studying both hormones to learn about their roles in the development of obesity.

The Large Intestine

The **large intestine,** which includes the cecum, the colon, the rectum, and the anal canal (Fig. 15.9), is larger in diameter than the small intestine (6.5 cm compared to 2.5 cm), but it is shorter in length (see Fig. 15.1). The large intestine absorbs water, salts, and some vitamins. It also stores indigestible material that is eventually eliminated at the anus.

The **cecum,** which lies below the junction with the small intestine, is the blind end of the large intestine. The cecum has a small projection called the **vermiform appendix** (*vermiform* means wormlike). In humans, the appendix also may play a role in fighting infections.

The **colon** has four portions: the **ascending colon,** which goes up the right side of the body to the level of the liver; the **transverse colon,** which crosses the abdominal cavity just below the liver and the stomach; the **descending colon,** which passes down the left side of the body; and the **sigmoid colon,** which enters the **rectum,** the last 20 cm of the large intestine. The colon is characterized by two distinct anatomic features: pouches called **haustra** (sing., **haustrum**) along its length, and a band of muscle called **taenia coli** on its surface (see Fig. 15.9). Haustra expand to store material in the colon. The taenia coli and the muscularis layer of the colon cause peristaltic movement called a *mass movement* several times daily.

Feces are 75% water and 25% solids. A breakdown product of bilirubin (see page 248) and the presence of oxidized iron cause the brown color of feces. Bacteria, shed intestinal cells, undigested cellulose fiber, and other indigestible materials are in the solid portion. Colon bacteria use undigested cellulose as an energy source, producing fatty acids, B-complex vitamins, and most of the vitamin K needed by our bodies. Bacteria also release hydrogen gas and sulfur-containing compounds that contribute to human flatulence (gas).

Although most intestinal bacteria are harmless, many cause disease. To ensure water quality, public health departments constantly monitor water supplies for coliform (nonpathogenic intestinal) bacterial count. A high count indicates that a significant amount of fecal material has entered the water, and serves as a warning that disease-causing bacteria might also be present.

Water is removed from the nondigestible intestinal contents entering the ascending colon from the small intestine. Then, fecal material begins to form in the transverse colon. From there, feces are propelled down the descending colon toward the rectum by periodic, firm peristaltic contractions (see page 347 to review peristalsis). When sufficient feces are in the rectum, a reflex defecatory

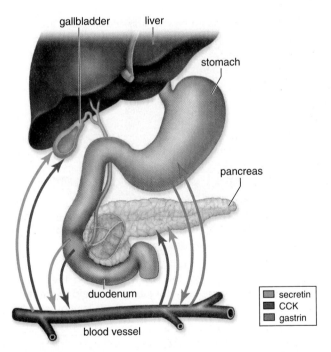

Figure 15.8 Hormonal control of digestive gland secretions. Gastrin (blue), produced by the lower part of the stomach, enters the bloodstream and thereafter stimulates the upper part of the stomach to produce more gastric juice. Secretin (green) and CCK (purple), produced by the duodenal wall, stimulate the pancreas to secrete its juice, the liver to produce bile, and the gallbladder to release stored bile.

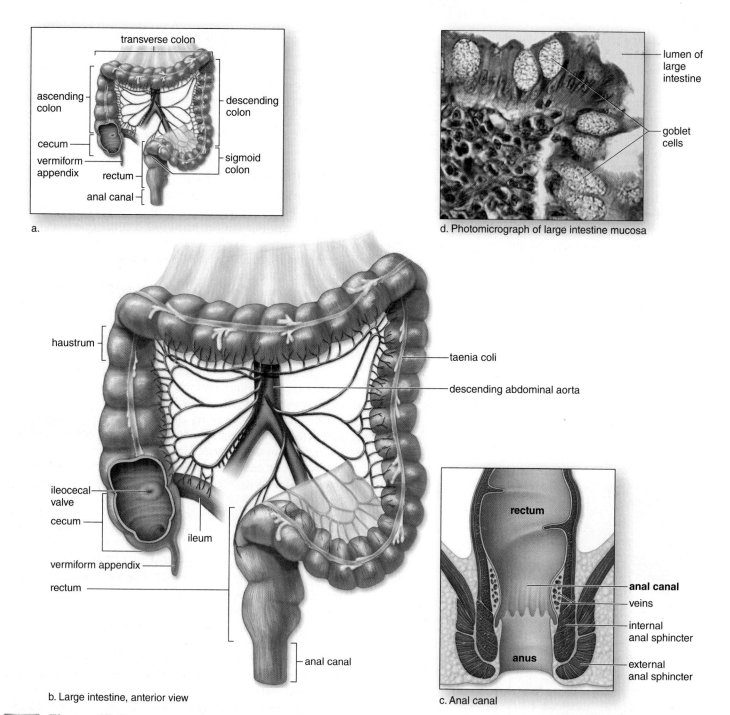

a.

d. Photomicrograph of large intestine mucosa

b. Large intestine, anterior view

c. Anal canal

AP|R **Figure 15.9** **The large intestine. (a)** The colon has four regions: the ascending colon, the transverse colon, the descending colon, and the sigmoid colon. **(b)** Haustra (sing., haustrum) expand the colon's volume, and taenia coli help to create peristalsis so that defecation occurs. **(c)** The rectum and anal canal are at the distal end of the alimentary canal. **(d)** The intestinal mucosa has many goblet cells.

urge is felt. The involuntary defecation reflex contracts the rectal muscle and relaxes the internal anal sphincter, a ring of muscle that closes off the rectum. Then, feces move toward the anus. A pushing motion, along with the relaxation of the external anal sphincter, propels feces from the body. Because these activities are under voluntary control, it is possible to control defecation. Defecation normally occurs from three times a week to three times daily, and some variation is normal.

Peritoneum

The abdominal wall and the organs of the abdomen are covered by *peritoneum,* a serous membrane (Fig. 15.10). The portion of the peritoneum that lines the wall is called the *parietal peritoneum.* The portion that covers the organs is called the *visceral peritoneum.* In between the organs, the visceral peritoneum is a double-layered **mesentery** that supports the visceral organs, including the blood vessels, nerves, and lymphatic vessels.

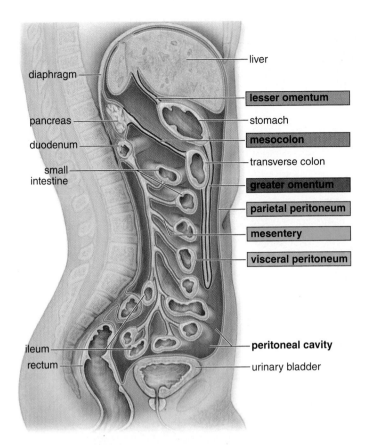

	liver
diaphragm	
	lesser omentum
pancreas	stomach
duodenum	**mesocolon**
small intestine	transverse colon
	greater omentum
	parietal peritoneum
	mesentery
	visceral peritoneum
	peritoneal cavity
ileum	
rectum	urinary bladder

Figure 15.10 Function of mesentery. Mesentery formed by two layers of the peritoneal membrane supports the abdominal viscera. Deep folds of the peritoneal membrane, called the greater omentum, cover these organs anteriorly.

Some portions of the peritoneum have specific names. The **lesser omentum** is mesentery that runs between the stomach and the liver. The **greater omentum** is indeed "greater." It hangs down in front of the intestines like a large, double-layered apron. The greater omentum has several functions: It contains fat that cushions and insulates the abdominal cavity; it contains macrophages that can take up and rid the body of pathogens; and it can wall off portions of the alimentary wall that may be infected, keeping the infection from spreading to other parts of the peritoneal cavity.

Begin Thinking Clinically

As you know, the large intestine, or colon, is filled with bacteria. This population is referred to as the *normal flora* of the gut. Like all bacteria, they're sensitive to antibiotics. If the normal flora organisms were destroyed by antibiotics, what would be the effect on the body?

Answer and discussion in Appendix A.

15.2 Accessory Organs of Digestion

5. Name five accessory organs of digestion.
6. Outline the location, anatomy, and functions of the pancreas, the liver, and the gallbladder.

The salivary glands and even the teeth are accessory organs of digestion that were discussed earlier (see pages 344–346). The pancreas, liver, and gallbladder are also accessory digestive organs. Figure 15.11 shows how the pancreatic duct from the pancreas and the common bile duct from the liver and gallbladder join before entering the duodenum.

The Pancreas

The **pancreas** lies deep in the abdominal cavity, behind the peritoneum, resting on the posterior abdominal wall. Its broad end, called the head, more than fills the loop formed by the duodenum, and its tail extends in the opposite direction (Fig. 15.11). The pancreas has both an endocrine and an exocrine function. As you know from Chapter 10, pancreatic islets (formerly called islets of Langerhans) secrete insulin, glucagon, and somatostatin; all are hormones that help keep the blood glucose level within normal limits.

In this chapter, however, we're interested in the exocrine function of the pancreas. Most pancreatic cells, called *pancreatic acinar cells*, produce pancreatic juice, which is secreted into tiny tubes that unite to form larger and larger tubes. Finally, all drain into one of two pancreatic ducts: the longer pancreatic duct, which extends the entire length of the pancreas to the duodenum, or the shorter accessory pancreatic duct (Fig. 15.11). Both ducts empty into the duodenum.

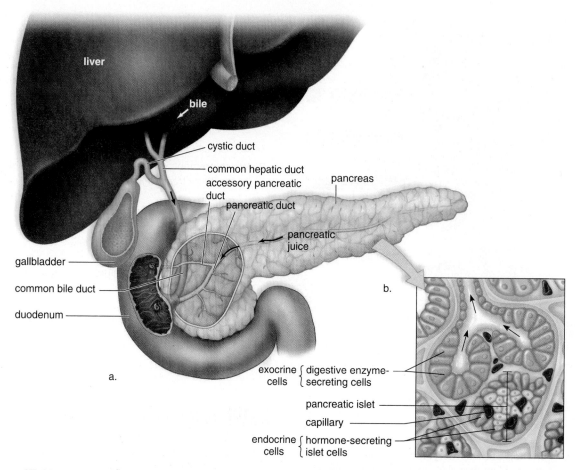

AP|R **Figure 15.11** **The pancreas.** **(a)** The pancreas is an exocrine gland when it secretes digestive enzymes into tubes that join to become the pancreatic duct. The pancreatic duct and the common bile duct empty into the duodenum of the small intestine. **(b)** Pancreatic exocrine and endocrine cells. Pancreatic juice produced by the exocrine cells contains enzymes that digest all types of food: carbohydrates, fats, proteins, and nucleic acids. Pancreatic endocrine cells secrete hormones.

Pancreatic Juice

Pancreatic juice contains sodium bicarbonate ($NaHCO_3$) and digestive enzymes for all types of food. Sodium bicarbonate neutralizes the acidic chyme from the stomach. This is important because pancreatic enzymes require a slightly basic pH to function optimally. **Pancreatic amylase** digests starch, and **lipase** digests fat. There are three protein-digesting enzymes in pancreatic juice: trypsin, chymotrypsin, and carboxypeptidase. Pancreatic juice also contains two nucleases, which are enzymes that break down nucleic acid molecules into nucleotides. Whereas the majority of pancreatic enzymes are already functional when they reach the small intestine, the three protein-digesting pancreatic enzymes are secreted in an inactive form. Once inside the small intestine, trypsin is activated by an intestinal enzyme, **enterokinase.** In turn, trypsin goes on to activate chymotrypsin and carboxypeptidase so protein digestion in the small intestine can start. It's important for the pancreas to secrete nonfunctional protein enzymes—if the enzymes were active, they could digest the pancreas itself!

In the genetic disorder *cystic fibrosis*, the production of abnormally thick mucus blocks the pancreatic duct. The patient must take supplemental pancreatic enzymes by mouth for proper digestion to occur.

The Liver

The **liver,** the largest organ inside the body, lies mainly in the upper right section of the abdominal cavity, just inferior to the diaphragm (see Fig. 15.1).

Liver Structure

The liver has two main lobes, the right lobe and the smaller left lobe, separated by a ligament. Each lobe is divided into many **hepatic lobules** that serve as its structural and functional units (Fig. 15.12). A lobule consists of many hepatic cells arranged in long rows that radiate out from a central vein. Large blood-filled capillaries called hepatic sinusoids separate the groups of cells from each other. Large, fixed phagocytic macrophages called *Kupffer cells* are attached to the lining of the hepatic sinusoids. They remove pathogens and debris that may have entered the hepatic portal vein at the small intestine.

Portal triads consisting of the following three structures are located between the lobules: a bile duct that takes bile away from the liver; a branch of the hepatic artery that brings O_2-rich blood to the liver; and a branch of the hepatic portal vein that transports

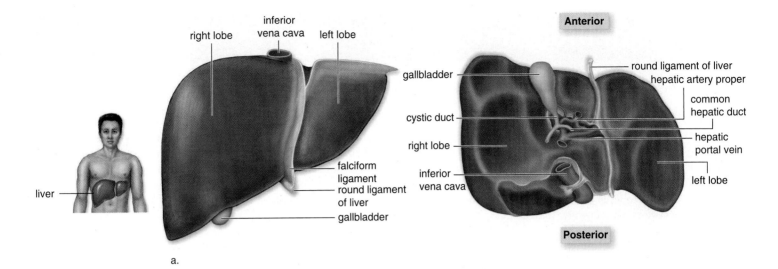

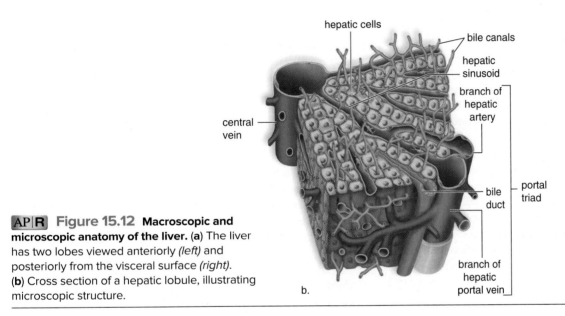

AP|R **Figure 15.12** **Macroscopic and microscopic anatomy of the liver. (a)** The liver has two lobes viewed anteriorly *(left)* and posteriorly from the visceral surface *(right)*. **(b)** Cross section of a hepatic lobule, illustrating microscopic structure.

blood from the stomach, pancreas, spleen, and intestines. The bile ducts merge to form the common hepatic duct.

The central veins from each of the lobules enter a hepatic vein. With the help of Figure 12.18, trace the path of blood from the intestines to the liver via the hepatic portal vein and from the liver to the inferior vena cava via the hepatic veins.

Liver Functions

As the blood from the hepatic portal vein passes through the liver, hepatic cells remove poisonous substances and detoxify them. The liver also removes and stores nutrients and works to keep the contents of the blood constant. It removes and stores iron and the fat-soluble vitamins A, D, E, and K; makes the plasma proteins from amino acids; and helps regulate the quantity of cholesterol in the blood.

The liver maintains the blood glucose level at about 100 mg/ 100 ml (0.1%), even though a person eats intermittently. When insulin is present, any excess glucose in the blood is removed and stored by the liver as glycogen. Between meals, glycogen is broken down to glucose, which enters the hepatic veins. In this way, the blood glucose level remains within a homeostatic range.

If the supply of glycogen is depleted, the liver converts glycerol (from fats) and amino acids to glucose molecules. This process is called *gluconeogenesis.* To convert amino acids to glycerol for gluconeogenesis, the liver must first carry out *deamination,* a complex metabolic pathway. Through this process, amino groups are removed, converted to ammonia, and combined with carbon dioxide to form urea:

$$2\ NH_3 \quad + \quad CO_2 \quad \longrightarrow \quad \overset{\displaystyle O}{\overset{\displaystyle \|}{H_2N - C - NH_2}}$$

ammonia   carbon dioxide   urea

Urea is the usual nitrogenous waste product from amino acid breakdown by the liver. Urea is normally excreted by the kidneys. However, if it accumulates excessively, it can be toxic to the body.

MEDICAL FOCUS

Disorders of the Digestive Tract

Human beings are true omnivores when it comes to the foods we consume. When you consider the myriad of things that people eat—habanero hot peppers, raw fish, insect larvae, seaweed, bacon-topped greasy cheeseburgers, and all the rest—it's remarkable to think that the digestive tract and its accessory organs are durable enough to handle it all. However, even this sturdy system has its limits. Here are a few of the more common digestive tract disorders.

Disorders of the Esophagus

The cardiac sphincter, the structure that separates the esophagus from the stomach, is a relatively weak barrier. As a result, acidic stomach contents can escape to splash into the esophagus, causing the pain of **heartburn.** Most people experience heartburn occasionally, but it often occurs in overweight people, pregnant women, and smokers (because smokers produce less saliva to wash stomach acid out of the esophagus). Occasional heartburn can be controlled by over-the-counter medications such as antacids, drugs called proton pump inhibitors (Prilosec®, Prevacid®) and histamine receptor blockers (Zantac®, Tagamet®, Pepcid®). However, chronic reflux causes a disorder called **gastroesophageal reflux disease,** or GERD. Patients with GERD will need stronger medications and perhaps even surgery to strengthen the cardiac sphincter. Without effective treatment, persons with GERD will often develop a condition called *Barrett's esophagus*. In Barrett's esophagus, the chronic acid backwashing into the esophagus causes the delicate simple squamous epithelium of the esophagus to be replaced by abnormal columnar epithelium. The constant inflammation that leads to this tissue change also predisposes the individual to esophageal cancer.

Disorders of the Stomach

The hydrochloric acid produced by the parietal cells of the stomach is extremely concentrated, and the stomach shields itself with thick protective mucus produced by the mucous cells of the gastric glands. However, if these cells fail to generate adequate mucus, gradual disintegration of the stomach wall will create an open sore called a **gastric ulcer.** Decreased mucus production can also cause esophageal ulcers in the lower esophagus and duodenal ulcers in the proximal small intestine. It now appears that most ulcers are due to an infection caused by a bacterium called *Helicobacter pylori,* which impairs the ability of mucous cells to make protective mucus. (It's interesting to note that Dr. Barry Marshall, the Nobel Prize–winning physician who first described the link

between this bacterium and ulcers, proved his theory to his skeptical peers by drinking a culture of *H. pylori*—and giving himself an inflamed stomach!) For this reason, ulcers are now treated with antibiotics as well as antacids. Ulcers can also result from taking certain medications, including anti-inflammatory drugs such as aspirin and ibuprofen. Discontinuing the medication usually allows the ulcer to heal.

Disorders of the Large Intestine

As you have read, the vermiform appendix is an organ suspended from the cecum, the proximal portion of the large intestine. This organ is subject to inflammation, a condition called *appendicitis*. If inflamed, the appendix should be removed before the fluid content rises to the point that the appendix bursts. A ruptured (burst) appendix will cause **peritonitis,** a generalized infection of the abdominal cavity lining. Peritonitis can be fatal.

Constipation occurs when the large intestine fails to reflexively empty itself and the feces become dry and hard. Temporary constipation due to traveling, pregnancy, or medication can be relieved by increasing fiber intake, drinking plenty of water, and getting moderate exercise. Though oral laxatives (agents that aid in emptying the colon) can provide temporary relief, their use must be limited because the person can become dependent on them to defecate. Very rarely, medical attention is required for constipation. Chronic constipation is associated with the development of **hemorrhoids,** which are enlarged and inflamed blood vessels at the anus.

Diarrhea is a potentially serious condition characterized by frequent, watery stools. Diarrhea often results from food poisoning caused by consuming contaminated food or water. The infection irritates the intestinal wall, peristalsis increases and water is not absorbed, and diarrhea rids the body of the infectious organisms. In nervous diarrhea, the nervous system excessively stimulates the intestinal wall. Osmotic diarrhea results from consuming excess dietary fat or fiber, which prevents the large intestine from absorbing enough water to solidify the stool. Regardless of the cause, prolonged diarrhea can lead to dehydration, and in severe cases, disturbances in the heart's contraction may occur due to electrolyte imbalance in the blood. Untreated diarrhea can be fatal, especially in infants and small children.

Intestinal **polyps** are small growths that may be benign or cancerous. Polyps arise from the colon's epithelial lining and can be removed surgically, along with a portion of the colon if necessary. Colon cancer is completely curable if detected while still confined to a polyp. Once diagnosed, colon cancer is treated with surgical removal of the affected portion of the colon, followed by chemotherapy. Increasingly, evidence shows that dietary fat increases colon cancer risk by causing an increase in bile secretion. Researchers believe that intestinal bacteria convert bile salts to carcinogens, substances that promote the development of cancer. On the other hand, dietary fiber in the diet decreases colon cancer risk by absorbing water, diluting bile salts, and promoting defecation. Regular elimination reduces the time that the colon wall is exposed to any fecal carcinogens.

Diverticulosis is characterized by the presence of *diverticula*, or saclike pouches, in the colon. Diverticula commonly form when a person strains to defecate. Ordinarily, these pouches cause no problems, but approximately 15% of people with diverticulosis develop an inflammation known as *diverticulitis*. The symptoms of diverticulitis are similar to those of appendicitis—cramps or steady pain with local tenderness. Fever, loss of appetite, nausea, and vomiting may also occur. High-fiber diets are recommended to prevent the development of diverticula.

Disorders of the Liver, Gallbladder, and Pancreas

Hepatitis is inflammation of the liver, and it has many causes. Viral hepatitis can be caused by one of five hepatitis viruses, but the hepatitis A and B viruses are considered the most dangerous in the United States. Hepatitis A is usually acquired from sewage-contaminated drinking water. Hepatitis B is typically spread by sexual contact, but it can also be spread by blood transfusions or contaminated needles. The hepatitis B virus is more contagious than the AIDS virus, which is spread in the same way. The United States Centers for Disease Control (CDC) recommends immunization for these viruses beginning in infancy (see Chapter 13, Medical Focus: "Immunization: The Great Protector"). Both hepatitis C and D viruses are usually acquired by contact with infected blood. Hepatitis E has been found in sewage-contaminated water in several developing countries, including China, India, and Pakistan. Acute viral hepatitis can cause liver failure, and chronic viral hepatitis B, C, and D can result in liver cancer. **Cirrhosis** is another chronic liver disease that often results from alcoholism. In alcoholic cirrhosis, the organ becomes fatty, and liver tissue is replaced by inactive fibrous scar tissue. Hepatitis and cirrhosis affect the entire liver, hinder its ability to repair itself, and are often fatal. Both cause **jaundice,** a yellowish tint to the skin and the whites of the eyes. Jaundice develops because the liver fails to properly excrete bile pigments, the breakdown products of hemoglobin. The liver has amazing regenerative powers and can recover if the rate of regeneration exceeds the rate of damage. During liver failure, however, there may not be enough time for the liver to heal itself. Liver transplantation is usually the preferred treatment for liver failure.

The gallbladder is a small, muscular sac with one primary function: to store bile, a watery solution containing bile salts, bile pigments, and dissolved mineral salts. These salts can settle out of solution and form crystals called *gallstones*. If the gallstones become large enough, they can block the common bile duct through which bile leaves the liver, causing obstructive jaundice. The gallbladder will then need to be surgically removed.

Pancreatitis, or pancreatic inflammation, can be caused by drugs, excess alcohol consumption, and viral infections (including the mumps). When the pancreas is inflamed, its digestive enzymes become activated within the gland and begin to destroy the gland itself. If the destruction spreads to other areas of the abdomen, the condition can be fatal.

Autoimmune Disorders of the Digestive Tract

As you know from Chapter 14, autoimmune diseases happen when the body's defense mechanisms launch an attack on cells, tissues, and organs. Several autoimmune disorders affect the digestive tract. *Gluten-sensitive enteropathy*, or *celiac disease*, is an allergic hypersensitivity to the protein gluten, found in wheat, rye, and barley. The allergy causes the small intestine lining to die off. If exposed to gluten in food or drink, the celiac disease sufferer experiences painful cramping, diarrhea, and weight loss. He or she may also develop vitamin and mineral deficiencies. A strict gluten-free diet and nutritional supplements alleviate the symptoms of celiac disease.

Inflammatory bowel diseases such as *Crohn's disease* and *ulcerative colitis* are autoimmune disorders characterized by severe diarrhea, cramping, fatigue, and weight loss. Ulcerative colitis is confined to the colon, but Crohn's disease can be found anywhere along the entire length of the digestive tract, from the mouth to the anus. Both disorders can be effectively treated with a combination of steroids, chemotherapy drugs, and sometimes surgery to remove damaged sections of the small intestine or colon. Though there is no cure for these diseases, their symptoms may resolve over time. However, sufferers must learn to anticipate and control flare-ups.

It's important to drink plenty of water following a high-protein meal so that urea can be quickly and effectively eliminated.

The liver produces bile, which is stored in the gallbladder. Bile has a yellowish-green color because it contains the bile pigment **bilirubin,** which is derived from the breakdown of hemoglobin from red blood cells. Bilirubin from bile is excreted by the kidneys and accounts for urine's yellow color. Bile also contains **bile salts,** which are derived from cholesterol. Bile salts emulsify (but don't digest) fat in the small intestine. When fat is emulsified, it disperses into droplets. Emulsification of fats provides a much larger surface area that can be more easily attacked by pancreatic digestive enzymes.

Because the liver is the most important of the metabolic organs, it works closely with the cardiovascular system. A large amount of the body's blood constantly flows to the liver. Between meals, more than three-quarters of this supply reaches the liver by way of the hepatic portal vein, which drains the intestine. The remainder comes from the abdominal aorta, via the hepatic artery. Immediately after a meal, additional blood is diverted to the intestine to cope with the tasks of digestion and absorption, and the hepatic portal vein drains even more blood to the liver.

Scientists estimate that the liver carries out more than 500 separate functions to help maintain homeostasis. Here are just a few:

1. Detoxifies blood by removing and metabolizing poisonous substances.
2. Stores iron (Fe^{2+}) and the fat-soluble vitamins A, D, E, and K.
3. Makes plasma proteins, such as albumins and fibrinogen, from amino acids.
4. Stores glucose as glycogen after a meal. Between meals, it breaks down glycogen to glucose to maintain the glucose concentration of blood. When blood glucose is depleted, the liver converts glycerol to glucose in gluconeogenesis.
5. Produces urea after breaking down amino acids.
6. Forms and secretes bile. Bile can be secreted directly into the duodenum or into the gallbladder for storage. Bile salts, formed from cholesterol, emulsify fats. Bilirubin, the bile pigment, is a product of the breakdown of hemoglobin from red blood cells.
7. Helps regulate the blood cholesterol level, converting some to bile salts.

The Gallbladder

The **gallbladder** is a pear-shaped, muscular sac located in a depression on the inferior surface of the liver (Fig. 15.12*a*). About 1,000 ml of bile are produced by the liver each day, and any excess is stored in the gallbladder. Water is reabsorbed by the gallbladder so that bile becomes a thick, mucus-like material. When needed, bile leaves the gallbladder by way of the **cystic duct.** The cystic duct and the **common hepatic duct** join to form the common bile duct, which enters the duodenum.

Function of Bile Salts

Bile salts carry out emulsification; they break up masses of fat into droplets that can be acted on by enzymes that digest fat. Through

their ability to make fats interact with water, they also enhance the absorption of fatty acids, cholesterol, and the fat-soluble vitamins A, D, E, and K.

15.3 Chemical Digestion

7. Name and state the functions of the digestive enzymes for carbohydrates, proteins, and fats.

The digestive enzymes are **hydrolytic enzymes,** which break down substances by the introduction of water at specific bonds. Digestive enzymes have an optimum pH at which they work best.

In the mouth, saliva from the salivary glands has a neutral pH and contains salivary amylase, the first enzyme to act on carbohydrate:

$$\text{starch} \ + \ H_2O \ \xrightarrow{\text{salivary amylase}} \ \text{maltose}$$

Notice that the name of the enzyme is written above the arrow to indicate that it is not used up. In this case, the enzyme speeds the breakdown of starch to maltose, a disaccharide. Maltose molecules are too large to be absorbed by the gastrointestinal tract; therefore, more digestion is required.

In the stomach, gastric juice secreted by gastric glands has a very low pH—about 2—because it contains hydrochloric acid (HCl). The inactive enzyme precursor, pepsinogen, is converted to the active enzyme pepsin when exposed to HCl. Pepsin acts on protein to produce peptides:

$$\text{protein} \ + \ H_2O \ \xrightarrow{\text{pepsin}} \ \text{peptides}$$

Peptides vary in length, but they are usually too large to be absorbed and must be broken down further.

In the small intestine, starch, proteins, nucleic acids, and fats are all enzymatically broken down. Pancreatic juice, which enters the duodenum, has a basic pH because it contains sodium

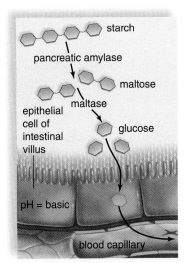

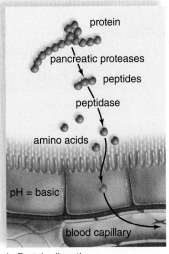

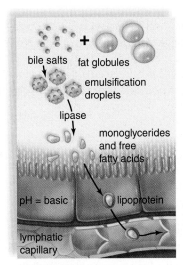

a. Carbohydrate digestion b. Protein digestion c. Fat digestion

Figure 15.13 Completing digestion and absorption in the small intestine. (a) Starch is digested to maltose by pancreatic amylase and to glucose by the brush border enzyme maltase. Glucose is actively transported to intestinal villus cells and enters intestinal capillaries. Disaccharides are similarly digested and absorbed. **(b)** Proteins are digested to amino acids by proteases and peptidases; amino acids are actively absorbed by intestinal villus cells and enter the bloodstream. **(c)** Fats are emulsified by bile and digested to monoglycerides and free fatty acids before diffusing into villus cells. In villus cells, they are reassembled into lipoproteins before diffusing into lacteals.

bicarbonate ($NaHCO_3$). One pancreatic enzyme, pancreatic amylase, digests starch (Fig. 15.13a):

$$\text{starch} + H_2O \xrightarrow{\text{pancreatic amylase}} \text{maltose}$$

The three **disaccharases** (maltase, sucrase, and lactase) are brush border enzymes produced by cells lining the small intestine. Intestinal disaccharases digest compound sugar molecules, called disaccharides, into single sugar molecules, or monosaccharides. Maltose, a disaccharide that results from the first step in starch digestion, is digested to glucose by maltase:

$$\text{maltose} + H_2O \xrightarrow{\text{maltase}} \text{glucose} + \text{glucose}$$

Similarly, the enzyme sucrase digests table sugar (the disaccharide sucrose) and lactase digests milk sugar (the disaccharide lactose). The monosaccharides that result are absorbed by intestinal cells and enter intestinal capillaries.

Recall that there are three pancreatic proteases, or enzymes that digest protein: trypsin, chymotrypsin, and carboxypeptidase. The following chemical equation summarizes their action (Fig. 15.13b):

$$\text{protein} + H_2O \xrightarrow{\text{pancreatic proteases}} \text{peptides and free amino acids}$$

Free amino acids are ready to be absorbed by the intestinal villus cells. Peptides are too large and must undergo further digestion in the small intestine. It is important to note that all three pancreatic proteases are secreted in an inactive form and cannot digest protein until they are inside the small intestine. This prevents them from accidentally digesting the pancreas itself.

The **peptidases** are brush border enzymes that complete the digestion of protein to amino acids. Peptides, which result from the first step in protein digestion, are digested to amino acids by these peptidases:

$$\text{peptides} + H_2O \xrightarrow{\text{peptidases}} \text{amino acids}$$

Like monosaccharides, amino acids are absorbed by intestinal cells and enter intestinal capillaries.

Lipase, a third pancreatic enzyme, digests fat molecules in the fat droplets after they have been emulsified by bile salts (Fig. 15.13c):

$$\text{fat} \xrightarrow{\text{bile salts}} \text{fat droplets}$$

$$\text{fat droplets} + H_2O \xrightarrow{\text{pancreatic lipase}} \text{glycerol and fatty acids}$$

As mentioned previously, glycerol and fatty acids enter the cells of the intestinal villi. Within these cells, they are rejoined and packaged as lipoprotein droplets before entering the lacteals.

Digestion of the nucleic acids DNA and RNA into free nucleotides is accomplished by two pancreatic nuclease enzymes. The free nucleotides can be further broken down into their components by a phosphatase and a nucleosidase enzyme built into the brush border cells. The phosphatase cleaves a phosphate from the nucleotide, leaving a molecule called a nucleoside. The nucleosidase completes digestion into the nucleic acid's three ingredient molecules: sugar, phosphate, and organic base.

Table 15.2 lists some of the major digestive enzymes produced by the gastrointestinal tract, salivary glands, or the pancreas. Each type of food is broken down by specific enzymes.

TABLE 15.2 Major Digestive Enzymes

Enzyme	Produced By	Site of Action	Optimum pH	Digestion Process
CARBOHYDRATE-DIGESTING ENZYMES				
Salivary amylase	Salivary glands	Mouth	Neutral	Starch + H_2O → maltose
Pancreatic amylase	Pancreas	Small intestine	Basic	Starch + H_2O → maltose
Maltase	Small intestine	Small intestine	Basic	Maltose → glucose + glucose
Sucrase	Small intestine	Small intestine	Basic	Sucrose → glucose + fructose
Lactase	Small intestine	Small intestine	Basic	Lactose → glucose + galactose
PROTEIN-DIGESTING ENZYMES				
Pepsin	Stomach	Stomach	Acidic	Protein + H_2O → peptides
Trypsin	Pancreas	Small intestine	Basic	Peptides + H_2O → smaller peptides + amino acids
Chymotrypsin	Pancreas	Small intestine	Basic	Peptides + H_2O → smaller peptides + amino acids
Carboxypeptidase	Pancreas	Small intestine	Basic	Peptides + H_2O → smaller peptides + amino acids
Peptidases	Small intestine	Small intestine	Basic	Small peptides → amino acids
LIPID-DIGESTING ENZYMES				
Pancreatic lipase	Pancreas	Small intestine	Basic	Fat droplet + H_2O → glycerol + fatty acids
NUCLEIC ACID-DIGESTING ENZYMES				
Nuclease	Pancreas	Small intestine	Basic	RNA or DNA + H_2O → nucleotides
Phosphatase	Small intestine	Small intestine	Basic	Nucleotide + H_2O → nucleoside + phosphate
Nucleosidase	Small intestine	Small intestine	Basic	Nucleoside + H_2O → base + sugar

Content CHECK-UP!

7. Choose the correct enzyme to cause this reaction:

   starch + water → maltose

   a. trypsin   c. lipase

   b. pancreatic amylase   d. pepsin

8. Which of the following enzymes is *not* produced by the pancreas?

   a. trypsin   c. nuclease

   b. lipase   d. sucrase

9. Imagine you've just eaten your favorite hamburger (or veggie burger). Describe the steps your body will take to break that protein down and absorb it into your bloodstream.

Answers in Appendix A.

15.4 Effects of Aging

8. Describe the anatomical and physiological changes that occur in the digestive system as we age.

Disorders of the digestive tract are described in the Medical Focus on pages 356–357, and the incidence of all of these gastrointestinal disorders increases with age. Periodontitis, which is common in elderly people, leads to the loss of teeth and the need for dentures.

The esophagus is more prone to disorders in the elderly, though it rarely causes difficulties in younger people. The portion of the esophagus normally found inferior to the diaphragm can protrude into the thoracic cavity, causing an esophageal *hiatal hernia*. In some older persons, chest pain may occur when the gastroesophageal sphincter fails to open and a bolus cannot enter the stomach. Eventually, the esophagus may develop a diverticulum that allows food to collect abnormally.

Peristalsis generally slows within the gastrointestinal tract as the muscular wall loses tone. When older people fail to consume enough dietary fiber, the result is often constipation that leads to diverticulosis and hemorrhoids.

The liver shrinks with age and receives a smaller blood supply. Notably, it needs more time to metabolize drugs and alcohol. As a result, medication dosages may need to be adjusted as a person ages. The older person may need to decrease alcohol consumption as well. Tragically, malnutrition is common among the elderly. Hunger and thirst sensations commonly decrease with age, and older people with memory issues may simply forget to eat. In addition, because one's senses of smell and taste diminish as well, food may not be appealing. It's important for caregivers to monitor their older patients to ensure that they receive a balanced diet.

10. Decreased senses of hunger, thirst, taste, and smell often cause _____ in the elderly.

Answer in Appendix A.

15.5 Homeostasis

9. Explain how the digestive system works with other systems of the body to maintain homeostasis.

Human Systems Work Together on page 368 tells how the digestive system works with other systems in the body to maintain homeostasis.

Within the gastrointestinal tract, the food we eat is broken down to nutrients small enough to be absorbed by the villi of the small intestine. Digestive enzymes are produced by the salivary glands, gastric glands, and intestinal glands. Three accessory organs of digestion (the pancreas, the liver, and the gallbladder) also contribute secretions that help break down food. The liver produces bile (stored by the gallbladder), which emulsifies fat. The pancreas produces enzymes for the digestion of carbohydrates, proteins, and fat. Secretions from these glands, which are sent by ducts into the small intestine, are regulated by hormones such as secretin produced by the alimentary canal. Therefore, the gastrointestinal tract is also a part of the endocrine system. Other accessory organs, such as the salivary glands and teeth, are also essential to digestion.

The nutrients absorbed by the gastrointestinal tract are converted by the body into energy and used for physical activities and for the growth and repair of body tissue. As we'll see in section 15.6, carbohydrates and fats are used to fuel all the body's processes and functions, while protein is mainly used as a building material. Besides these three basic components, the body must also have vitamins and minerals. Vitamins are essential for normal growth and development, and because they cannot be manufactured in the body, they must be supplied as part of a well-balanced diet or as supplements. Minerals assist in many body processes, such as normal nerve and muscle function, but they are needed only in small quantities. By providing all other tissues with required nutrients, the digestive tract helps to ensure homeostasis. Its interactions with other body systems are summarized in the Human Systems Work Together figure.

11. Because the alimentary canal secretes gastrin and secretin, it is considered a part of the _____ organ system.

Answer in Appendix A.

15.6 Nutrition

10. State the functions of glucose, fats, and amino acids in the body.
11. Define the terms *essential fatty acid*, *essential amino acid*, and *vitamin*.
12. Describe the functions of the major vitamins and minerals in the body.

Nutrition involves an interaction between food and the living organism. A nutrient is a substance that the body uses to maintain health. A balanced diet contains all the essential nutrients and includes a variety of foods, proportioned as shown in Figure 15.14.

Following digestion, nutrients enter the blood in the cardiovascular system, which distributes them to the tissues, where they are utilized by the body's cells. Mitochondria use glucose to produce a constant supply of ATP for the cell. In other words, glucose is the body's immediate energy source. Because the brain's preferred source of energy is glucose, it needs a constant supply.

The liver can chemically alter ingested fats to suit the body's needs, with the exception of linoleic acid and linolenic acid. Both fatty acids are required for the construction of plasma membranes and for the synthesis of messenger chemicals. Thus, they are **essential fatty acids.** Essential molecules must be present in food because the body cannot manufacture them. Other fats, especially saturated fats, should be restricted, as discussed in the Medical Focus on page 364.

If glucose isn't available, fats can be metabolized into smaller molecules, which are then used as an alternate energy source. Therefore, fats are said to be a long-term energy source. When adipose tissue cells store fats, the body increases in weight. Cells have the capability of converting excess sugar molecules into fats for storage, which accounts for the fact that carbohydrates can also contribute to weight gain.

Amino acids from protein digestion are used by the cells to construct their own proteins, including the enzymes that carry out metabolism. Protein formation requires 20 different types of amino acids. Of these, nine are required in the diet because the body is unable to produce them. These are termed the **essential amino acids.** The body produces the other 11 amino acids by simply

Figure 15.14 **Newest dietary guidelines, published by the U.S. Department of Agriculture, and available at www.choosemyplate.gov.** The website can be customized for an individual's age, gender, and activity level.

transforming one type into another type. Some protein sources, such as meat, are complete in the sense that they provide all the different types of amino acids. Vegetables and grains supply the body with amino acids, but they are typically incomplete sources because at least one of the essential amino acids is absent. A combination of certain vegetables, however, can provide all of the essential amino acids. In addition, soy protein is a complete protein.

Vitamins

Vitamins are *vital* to life because they play essential roles in cellular metabolism. Because the body is unable to produce them, most vitamins must be present in the diet. Vitamins are organic molecules, but they differ radically from carbohydrates, fats, and proteins. They are much smaller in size and are not broken down to be used as building blocks or as a source of energy. Instead, the body protects them and provides many of them with protein carriers that transport them in the blood to the cells. In the cells, vitamins become helpers in metabolic processes that break down or synthesize other organic molecules. Because vitamins can be used over and over again, they are required in very small amounts.

Vitamins fall into two groups: fat-soluble vitamins (vitamins A, D, E, and K) and water-soluble vitamins (the B complex vitamins and vitamin C; Table 15.3). Most of the water-soluble

TABLE 15.3 Vitamins: Their Role in the Body and Food Sources

Vitamins	Role in Body	Good Food Sources
FAT-SOLUBLE VITAMINS		
Vitamin A	Assists in the formation and maintenance of healthy skin, hair, and mucous membranes; aids in the ability to see in dim light (night vision); is essential for proper bone growth, tooth development, and reproduction	Deep yellow/orange and dark green vegetables and fruits (carrots, broccoli, spinach, cantaloupe, sweet potatoes); cheese, milk, and fortified margarine
Vitamin D	Aids in the formation and maintenance of bones and teeth; assists in the absorption and use of calcium and phosphorus	Milk fortified with vitamin D; tuna, salmon, or cod liver oil; also made in the skin when exposed to sunlight
Vitamin E	Protects vitamin A and essential fatty acids from oxidation; prevents plasma membrane damage	Vegetable oils and margarine; nuts; wheat germ and whole-grain breads and cereals; green, leafy vegetables
Vitamin K	Aids in synthesis of substances needed for clotting of blood; helps maintain normal bone metabolism	Green, leafy vegetables, cabbage, and cauliflower; also made by bacteria in intestines of humans, except for newborns
WATER-SOLUBLE VITAMINS		
Vitamin C	Is important in forming collagen, a protein that gives structure to bones, cartilage, muscle, and vascular tissue; helps maintain capillaries, bones, and teeth; aids in absorption of iron; helps protect other vitamins from oxidation	Citrus fruits, berries, melons, dark green vegetables, tomatoes, green peppers, cabbage, potatoes
B-COMPLEX VITAMINS		
Thiamin	Helps in release of energy from carbohydrates; promotes normal functioning of nervous system	Whole-grain products, dried beans and peas, sunflower seeds, nuts
Riboflavin	Helps body transform carbohydrates, proteins, and fats into energy	Nuts, yogurt, milk, whole-grain products, cheese, poultry, leafy green vegetables
Niacin	Helps body transform carbohydrates, proteins, and fats into energy	Nuts, poultry, fish, whole-grain products, dried fruit, leafy greens, beans; can be formed in the body from tryptophan, an essential amino acid found in protein
Vitamin B_6	Aids in the use of fats and amino acids; aids in the formation of protein	Sunflower seeds, beans, poultry, nuts, bananas, dried fruit, leafy green vegetables
Folic acid	Aids in the formation of hemoglobin in red blood cells; aids in the formation of genetic material	Nuts, beans, whole-grain products, fruit juices, dark green leafy vegetables
Pantothenic acid	Aids in the formation of hormones and certain nerve-regulating substances; helps in the metabolism of carbohydrates, proteins, and fats	Nuts, beans, seeds, poultry, dried fruit, milk, dark green leafy vegetables
Biotin	Aids in the formation of fatty acids; helps in the release of energy from carbohydrates	Occurs widely in foods, especially eggs; made by bacteria in the human intestine
Vitamin B_{12}	Aids in the formation of red blood cells and genetic material; helps in the functioning of the nervous system; requires intrinsic factor from the stomach to be absorbed	Milk, yogurt, cheese, fish, poultry, eggs; not found in plant foods unless fortified (as in some breakfast cereals)

Source: From David C. Nieman, et al., *Nutrition,* Revised 1st ed. Copyright © 1992 Wm. C. Brown Communications, Inc., Dubuque, Iowa.

vitamins are coenzymes, or enzyme helpers, that help speed up specific reactions. The functions of the fat-soluble vitamins, some of which have been previously discussed, are more specialized. Vitamin A, as noted in Chapter 9, is used to synthesize the visual pigments. Vitamin D is needed to produce a hormone that regulates calcium and phosphorus metabolism (see Chapter 5), and vitamin E is an antioxidant. Vitamin K is required to form *prothrombin*, a substance necessary for normal blood clotting (see Chapter 11).

Minerals

In contrast to vitamins, **minerals** are inorganic elements (Table 15.4). An element, you'll recall, is one of the basic

TABLE 15.4 Minerals: Their Role in the Body and Food Sources

Minerals	Role in Body	Good Food Sources
MACRONUTRIENTS		
Calcium	Is used for building bones and teeth and for maintaining bone strength; also involved in muscle contraction, blood clotting, and maintenance of plasma membranes	All dairy products; dark green, leafy vegetables; beans, nuts, sunflower seeds, dried fruit, molasses, canned fish
Phosphorus	Is used to build bones and teeth; to release energy from carbohydrates, proteins, and fats; and to form genetic material, plasma membranes, and many enzymes	Beans, sunflower seeds, milk, cheese, nuts, poultry, fish, lean meats
Magnesium	Is used to build bones, to produce proteins, to release energy from muscle carbohydrate stores (glycogen), and to regulate body temperature	Sunflower and pumpkin seeds, nuts, whole-grain products, beans, dark green vegetables, dried fruit, lean meats
Sodium	Regulates body-fluid volume and blood acidity; aids in transmission of nerve impulses	Most of the sodium in the U.S. diet is added to food as salt (sodium chloride) in cooking, at the table, or in commercial processing; animal products contain some natural sodium
Chloride	Is a component of gastric juice and aids in acid-base balance	Table salt, seafood, milk, eggs, meats
Potassium	Assists in muscle contraction, the maintenance of fluid and electrolyte balance in the cells, and the transmission of nerve impulses; also aids in the release of energy from carbohydrates, proteins, and fats	Widely distributed in foods, especially fruits and vegetables, beans, nuts, seeds, and lean meats
MICRONUTRIENTS (TRACE ELEMENTS)		
Iron	Is involved in the formation of hemoglobin in the red blood cells of the blood and myoglobin in muscles; also is a part of several enzymes and proteins	Molasses, seeds, whole-grain products, fortified breakfast cereals, nuts, dried fruits, beans, poultry, fish, lean meats
Zinc	Is involved in the formation of protein (growth of all tissues), in wound healing, and in prevention of anemia; is a component of many enzymes	Whole-grain products, seeds, nuts, poultry, fish, beans, lean meats
Iodine	Is an integral component of thyroid hormones	Table salt (fortified), dairy products, shellfish, and fish
Fluoride	Is involved in maintenance of bone and tooth structure	Fluoridated drinking water is the best source; also found in tea, fish, wheat germ, kale, cottage cheese, soybeans, almonds, onions, milk
Copper	Is vital to enzyme systems and in manufacturing red blood cells; is needed for utilization of iron	Nuts, oysters, seeds, crab, wheat germ, dried fruit, whole grains, legumes
Selenium	Functions in association with vitamin E; may assist in protecting tissues and plasma membranes from oxidative damage; may also aid in preventing cancer	Nuts, whole grains, lean pork, cottage cheese, milk, molasses, squash
Chromium	Is required for maintaining normal glucose metabolism; may assist insulin function	Nuts, prunes, vegetable oils, green peas, corn, whole grains, orange juice, dark green vegetables, legumes
Manganese	Is needed for normal bone structure, reproduction, and the normal functioning of the central nervous system; is a component of many enzyme systems	Whole grains, nuts, seeds, pineapple, berries, legumes, dark green vegetables, tea
Molybdenum	Is a component of enzymes; may help prevent dental caries	Tomatoes, wheat germ, lean pork, legumes, whole grains, strawberries, winter squash, milk, dark green vegetables, carrots

Source: From David C. Nieman, et al., *Nutrition,* Revised 1st ed. Copyright © 1992 Wm. C. Brown Communications, Inc., Dubuque, Iowa.

Tips for Effectively Using Nutrition Labels

No, it isn't quite as interesting as your morning newspaper, but there's important information tucked into the corner of your Aqua Puffs (and all other packaged food). Reading nutrition labels, like the one shown in Fig. 15A, should become a habit for anyone interested in health and wellness. You'll find the nutritional breakdown for a single serving (in this example, a serving is 1¼ cup, or 57 grams) on these labels. Labels also tell us what share of daily nutrients and Calories is found in that serving, based on a diet of 2,000 Calories. Calories are a measurement of food energy. As you can see, one serving of the cereal provides 220 Calories.

Carbohydrates

When studying nutrition labels, check carbohydrate content first. Carbohydrates are the most readily available energy source, and the best source for brain and nerve tissue. Sugar molecules are simple carbohydrates. Complex carbohydrates found in breads, cereals, vegetables, and fruits have the highest dietary fiber content. Fiber is found in two forms, and each has distinct health benefits: Soluble fiber combines with cholesterol, preventing it from being absorbed, while insoluble fiber has a laxative effect. Current recommendations include 25 to 30 grams of fiber daily.

Fats

Next, take a look at the food's fat content as you review the nutrition label. Dietary fat makes food more appealing and tasty, but it's also essential for health. Fat is needed for energy, vitamin absorption, and for manufacturing cell lipids and steroids. However, any excess nutrient energy is stored as fat, which could ultimately lead to obesity. Further, nutrition research continues to prove the danger of excess saturated fat, and even worse are the artificially created trans-fats. High saturated fat levels have been implicated in colon, pancreas, ovary, prostate, and breast cancers. Saturated fat and cholesterol contribute to atherosclerosis, the major cause of hypertension (high blood pressure), heart attack, and stroke in the Western world. (Atherosclerosis is described on pages 267–268.) Better choices for fat intake would be monosaturated and polyunsaturated plant fats such as canola and olive oils. Regardless of the fat type, a 2,000 daily Calorie diet should have no more than 65 grams (585 Calories) of fat.

Proteins

In a normal, healthy adult, a daily serving of protein is 45 grams (about the same size as a deck of playing cards, or a 3 × 5–inch rectangle). Low-fat plant source protein (contained in whole grain cereals like Aqua Puffs) is often healthier than animal source protein, which may be high in saturated fat. In adults, protein is typically needed only for tissue growth and repair. Children and women who are pregnant or nursing should increase daily protein intake to support their body's rapid growth.

Vitamins and Minerals

Vitamins are organic molecules required in small amounts for good health. Antioxidant vitamins (A, C, E, and beta-carotene) can neutralize *free radicals*, unstable molecules generated during cellular metabolism. Free radicals damage cell components and are linked to cancer and atherosclerosis. Calcium, iron, and sodium are important minerals to consider. Calcium is required for bone manufacture, blood clotting, and proper nerve function, and iron is needed for hemoglobin. However, sodium consumption should be limited: Excess sodium can worsen hypertension.

Bottom line: How nutritious is that bowl of Aqua Puffs? It's a low-fat, high-fiber cereal with no cholesterol or saturated fat. It provides 80% of daily dietary iron, and 10% of vitamin C, along with 5 grams of protein. Add a cup of low-fat milk to supply calcium and additional protein, and the cereal will be a decent quick breakfast. But all that sugar—11 grams! It's bound to be tasty for someone with a sweet tooth, but simple sugars add Calories. Further, you'd probably still need a vitamin supplement. A breakfast of fresh fruit, along with a no-fat dairy (or perhaps soy) product for protein, would be a better way to match the cereal's nutrients without the sugar. Moreover, if you eat the recommended 2 cups of fruit and 3 cups of vegetables daily, you'll be getting your vitamins, along with other valuable compounds you can't get from a supplement.

Figure 15A Nutrition label on the side panel of a cereal box.

substances of matter that cannot be broken down further into simpler substances. Minerals sometimes occur as a single atom, in contrast to vitamins, which contain many atoms. Minerals cannot lose their identity, no matter how they are handled. Because they are indestructible, no special precautions are needed to preserve them when cooking.

Minerals are divided into macronutrients, which are needed in gram amounts per day, and micronutrients (trace elements), which are needed in only microgram amounts per day. The macronutrients sodium, magnesium, phosphorus, chlorine, potassium, and calcium serve as constituents of cells and body fluids, and as structural components of tissues. The micronutrients have very specific functions, as noted in Table 15.4. As research continues, more elements will likely be added to the list of those considered essential.

Eating Disorders

Authorities recognize three primary eating disorders: obesity, bulimia nervosa, and anorexia nervosa. Although they exist in a continuum as far as body weight is concerned, all represent an inability to maintain normal body weight because of eating habits.

Obesity

It's an alarming statistic: In the United States, 69% of us are overweight, and approximately half of those who are overweight are obese. In children and young people, the numbers are equally grim: one-third of American children are overweight or obese. **Obesity** is most often defined as a body weight 20% or more above the ideal weight for a person's height. By this standard, 36% of women and 32% of men in the United States are obese. Moderate obesity is 41–100% above ideal weight, and severe obesity is 100% or more above ideal weight.

Obesity is most likely caused by a combination of hormonal, metabolic, genetic, and social factors. It is known that obese individuals have more fat cells than normal. When an obese person loses weight, the fat cells simply get smaller; they don't disappear. The social factors that cause obesity include the eating habits of other family members. Consistently eating fatty foods, for example, may cause you to gain weight. Sedentary activities, such as watching television instead of exercising, also determine how much body fat you have. The risk of heart disease is higher in obese individuals, and this alone tells us that excess body fat is not consistent with optimal health.

Treatment depends on the degree of obesity. Currently, there are several medications that may be prescribed for weight control. Orlistat (sold by prescription as Xenical and over the counter as Alli), lorcaserin (sold by prescription as Belviq), phenteramine-topiramate (sold by prescription as Qsymia) and buproprion-naltrexone (sold by prescription as Contrave) are all FDA approved. However, all have serious potential side effects, and anyone who takes them must be carefully monitored by a qualified health care provider. In addition, each must be accompanied by proper dietary modification and exercise in order for weight loss to be maintained.

Surgery to restrict stomach volume may be required for those who are moderately or greatly overweight (see the Medical Focus on page 367). But for most people, a knowledge of good eating habits along with behavior modification may be enough, particularly if a balanced diet is accompanied by a sensible exercise program. A lifelong commitment to a properly planned program is the best way to prevent a cycle of weight gain followed by weight loss. Cycling in this way can damage one's health over the long term (and it's certainly discouraging to the person trying to lose weight!).

Bulimia Nervosa

Bulimia nervosa can coexist with either obesity or anorexia nervosa, which is discussed next. People with this condition have the habit of eating to excess (called binge eating) and then purging themselves by some artificial means, such as self-induced vomiting or the use of a laxative (Fig. 15.15). Bulimic individuals are overly concerned about their body shape and weight, and therefore they may be on a very restrictive diet. A restrictive diet may bring on the desire to binge, and typically the person chooses to consume sweets, such as cakes, cookies, and ice cream. The amount of food consumed is far beyond the normal number of calories for one meal, and the person keeps on eating until every bit is gone. Then, a feeling of guilt most likely brings on the next phase, which is a purging of all the calories that have been taken in.

Bulimia is extremely dangerous to health. Blood composition is altered, leading to an abnormal heart rhythm, and damage to the kidneys can even be fatal. At the very least, vomiting can lead to inflammation of the pharynx and esophagus, and stomach acids can cause the teeth to erode. The esophagus and stomach may even rupture and tear due to strong contractions during vomiting.

The most important aspect of treatment is to get the patient on a sensible and consistent diet. Again, behavioral modification is helpful, and so perhaps is psychotherapy to help the patient understand the emotional causes of the behavior. Medications, including antidepressants, have sometimes helped to reduce the bulimic cycle and restore normal appetite.

Anorexia Nervosa

In **anorexia nervosa,** a morbid fear of gaining weight causes the person to be on a very restrictive diet (see Fig. 15.15). Athletes such as distance runners, wrestlers, and dancers are at risk of anorexia nervosa because they believe that being thin gives them a competitive edge. In addition to eating only low-calorie foods, the person may induce vomiting and use laxatives to bring about further weight loss. No matter how thin they have become, people with anorexia nervosa think they are overweight. Such a distorted self-image may prevent recognition of the need for medical help.

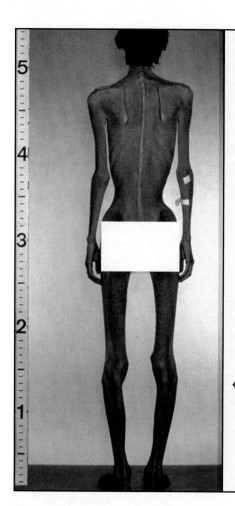

Persons with bulimia nervosa have:

- recurrent episodes of binge eating: Consuming an amount of food much higher than normal for one sitting and a sense of lack of control over eating during the episode.
- an obsession about body shape and weight.
- increase in fine body hair, halitosis (bad breath), and gingivitis (gum disease).

Body weight is regulated by

- a restrictive diet, excessive exercise.
- purging (self-induced vomiting or laxative misuse).

Persons with anorexia nervosa have:

- a morbid fear of gaining weight; body weight no more than 85% of normal.
- a distorted body image; person feels fat even when emaciated.
- in females, absence of a menstrual cycle for at least three months.

Body weight is kept too low by either/or

- a restrictive diet, often with excessive exercise.
- binge eating/purging (self-induced vomiting or laxative misuse).

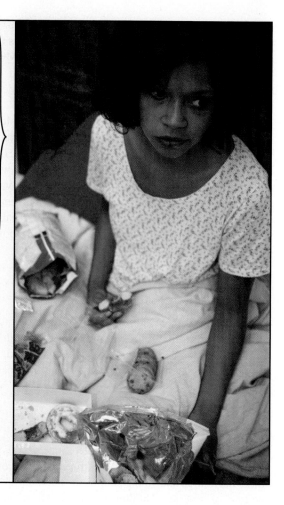

Figure 15.15 Recognizing anorexia nervosa and bulimia nervosa.

Actually, the person is starving and has all the symptoms of starvation, including low blood pressure, irregular heartbeat, constipation, and constant chilliness. Bone density decreases, and stress fractures occur. The body begins to shut down; menstruation ceases in females; the internal organs, including the brain, don't function well; and the skin dries up. Impairment of the pancreas and alimentary canal means that any food consumed does not provide nourishment. Death may be imminent. If so, the only recourse may be hospitalization and force-feeding. Eventually, it is necessary to use behavior therapy and psychotherapy to enlist the cooperation of the person to eat properly. Family therapy may be necessary, because anorexia nervosa in children and teens is believed to be a way for them to gain some control over their lives.

Content **CHECK-UP!**

12. Fatty acids and amino acids that must be taken in through the diet, since the body can't produce them, are called _____.

13. An obese person is one who is at least ____ percent above the normal weight.

a. 10

b. 20

c. 50

d. 100

Answer in Appendix A.

MEDICAL FOCUS

Bariatric Surgery for Obesity

It's a final measure sought by increasing numbers of people: **bariatric surgery,** or surgical intervention with the specific goal of causing drastic weight loss. For many overweight people, years of dieting haven't worked. Name a diet plan, and they have tried it—sure, they lose weight, only to regain it. Obesity is approaching epidemic levels in the United States, with more than 78 million adults classified as obese. In addition, pediatricians are especially concerned with the dramatic rise in childhood and teenage obesity.

Obesity is defined as a body mass index (BMI) greater than 30,[1] or a weight that is 41% or more higher than the ideal weight for one's height. It's a physical and emotional challenge for the overweight patient. Obesity is a primary risk factor for hypertension, type II diabetes mellitus, atherosclerosis, stroke, coronary artery disease, and early heart attack. It has been linked to increased risk of breast, ovarian, uterine, and prostate cancer. Obese individuals suffer disrespect and ridicule from society and discrimination on the job. After years of struggling with the problem, many are willing to undergo surgery as a last-chance option.

However, reputable programs offering bariatric surgery have strict requirements for patients. To be admitted as a surgical candidate, the patient must be morbidly obese (BMI greater than 40; or greater than 100 pounds over ideal weight) or have a BMI in the range of 35 to 39, along with a serious obesity-related health issue such as diabetes or high blood pressure. Patients cannot have respiratory or cardiac problems that might complicate surgery. Most important, the patient must understand the risks of surgery as well as its benefits. Patients must also understand that lifestyle changes will be necessary even after successful surgery and recovery. Psychological and nutritional counseling is usually required before surgery to prepare the patient for new eating habits and ways of thinking about food. Follow-up counseling tracks the patient's progress in adapting.

The two most commonly used interventions are laparoscopic banding and the Roux-en-Y gastric bypass (Fig. 15B). The more conservative approach, laparoscopic banding, requires making a series of small incisions around the stomach and using an instrument called a laparoscope to illuminate structures in the abdomen. A band is placed around the stomach. Once tightened like a belt, the stomach is divided into a smaller upper chamber, which receives food, and a lower chamber that remains connected to the duodenum. The belt can later be tightened further, or removed if necessary. In Roux-en-Y gastric bypass, the top section of the stomach is cut free and stapled shut to make a pouch about the size of an egg. The small intestine is cut free between the duodenum and jejunum, and the jejunum is sewn to the end of the stomach pouch. Finally, the duodenal segment is sewn back to the jejunum (forming the Y-shaped branch for which the procedure is named). The lower stomach and duodenum remain healthy and continue to secrete digestive enzymes, but never receive food. Regardless of the approach used, the person will only be able to eat small amounts of food, but should feel full due to the small size of the stomach after surgery. In addition, secretion of the gastric hormone ghrelin decreases after surgery, and that often diminishes the person's appetite.

It's important to note that bariatric surgery comes with extremely serious potential complications. Postoperative bleeding and infection are risks of any surgical procedure. Blood clots in the legs can form during hospitalization, causing pulmonary embolism or stroke. After gastric bypass, staple lines in the stomach can leak. In rare cases, the connection between the stomach pouch and jejunum narrows, requiring additional corrective surgery. Worst of all, a small percentage of patients die during the surgery itself. After surgery, vitamin and mineral deficiencies are possible.

Further, bariatric surgery offers no guarantees of permanent weight loss—patients can in fact regain any weight that is lost, even with a drastically smaller stomach. However, with proper nutrition and behavioral changes, bariatric surgery can result in dramatic weight loss and improvement to health.

---

[1] To calculate BMI, use the following formula: $\dfrac{\text{weight in pounds}}{(\text{height in inches}) \times (\text{height in inches})} \times 703$.
BMI is not always accurate in determining obesity. Other factors such as percent body fat may also need to be used.

After laparoscopic banding   **Before surgery**

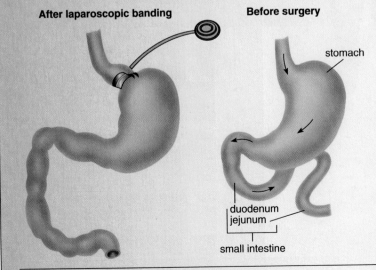

stomach

duodenum
jejunum

small intestine

After Roux-en-Y gastric bypass

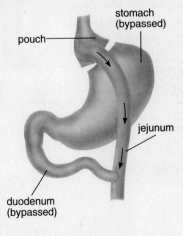

stomach
(bypassed)

pouch

jejunum

duodenum
(bypassed)

Figure 15B (*far left figure*) A laparascopic band is placed around the stomach and then tightened to decrease stomach size. (*center*) The normal movement of food prior to surgery. (*right*) The flow of food after Roux-en-Y gastric bypass surgery.

Human Systems Work Together

Integumentary System

Digestive tract provides nutrients needed by skin.

Skin helps to protect digestive organs; helps to provide vitamin D for Ca^{2+} absorption.

Skeletal System

Digestive tract provides Ca^{2+} and other nutrients for bone growth and repair.

Bones provide support and protection; hyoid bone assists swallowing.

Muscular System

Digestive tract provides glucose for muscle activity; liver metabolizes lactic acid following anaerobic muscle activity.

Smooth muscle contraction accounts for peristalsis; skeletal muscles support and help protect abdominal organs.

Nervous System

Digestive tract provides nutrients for growth, maintenance, and repair of neurons and neuroglia.

Brain controls nerves, which innervate smooth muscle and permit tract movements.

Endocrine System

Stomach and small intestine produce hormones.

Hormones help control secretion of digestive glands and accessory organs; insulin and glucagon regulate glucose storage in liver.

How the Digestive System works with other body systems.

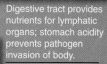

oral cavity (mouth)
salivary glands
pharynx (throat)
esophagus
liver
stomach
large intestine
small intestine

Cardiovascular System

Digestive tract provides nutrients for plasma protein formation and blood cell formation; liver detoxifies blood, makes plasma proteins, destroys old red blood cells.

Blood vessels transport nutrients from digestive tract to body; blood services digestive organs.

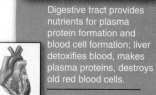

Lymphatic System/Immunity

Digestive tract provides nutrients for lymphatic organs; stomach acidity prevents pathogen invasion of body.

Lacteals absorb fats; Peyer patches prevent invasion by pathogens; appendix contains lymphatic tissue.

Respiratory System

Breathing is possible through the mouth because digestive tract and respiratory tract share the pharynx.

Gas exchange in lungs provides oxygen to digestive tract and excretes carbon dioxide from digestive tract.

Urinary System

Liver synthesizes urea; digestive tract excretes bile pigments from liver and provides nutrients.

Kidneys convert vitamin D to active form needed for Ca^{2+} absorption; compensate for any water loss by digestive tract.

Reproductive System

Digestive tract provides nutrients for growth and repair of organs and for development of fetus.

Pregnancy crowds digestive organs and promotes heartburn and constipation.

Selected New Terms

Basic Key Terms

alimentary canal (ăl-ĭ-měn′tĕr-ē kŭh-năl′), p. 343

anus (ā′nŭs), p. 343

ascending colon (ā-sĕnd′ĭng kō′lŏn), p. 351

bicuspid (bī-cŭs′pĭd), p. 344

bile (bīl), p. 349

bile salts (bīl săwltz), p. 358

bilirubin (bĭl″ē-rū′bĭn), p. 358

body region (bŏd′ē rē′jĕn), p. 348

bolus (bō′lŭs), p. 346

cardiac region (kăr′dē-ăk rē′jĕn), p. 347

cecum (sē′kŭm), p. 351

cephalic phase (sŭh-făl′ĭk fāz), p. 350

cholecystokinin (kō″lē-sĭs-tō-kī′nĭn), p. 350

chyme (kīm), p. 349

circular folds (sĕr′kyū-ler fōldz), p. 349

colon (kō′lŭn), p. 351

common hepatic duct (cŏm′mŭn hŭh-păt-ĭk dŭkt), p. 358

cuspid (cŭs′pĭd), p. 344

cystic duct (sĭs′tĭk dūkt), p. 358

descending colon (dē-sĕnd′ĭng kō′lŏn), p. 351

disaccharases (dī-săk′ŏh-rā-sĕs), p. 359

duodenum (du″ŏ-dē′nŭm), p. 349

enteric nervous system (ĕn-tĕr′ĭk nĕr′vŭs sĭs′tĕm), p. 350

enterokinase (ĕn″tĕr-ō-kī′nās), p. 354

esophagus (ĕ-sŏf′ŭh-gŭs), p. 347

essential amino acid (ĕ-sĕn′shŭl ŭh-mē′nō ăs′ĭd), p. 361

essential fatty acid (ĕ-sĕn′shŭl făt′ē ăs′ĭd), p. 361

fundic region (fŭn′dĭk rē′jĕn), p. 348

gallbladder (găwl′blăd-ĕr), p. 358

gastric gland (găs′trĭk glănd), p. 348

gastric juice (găs′trĭk jūs), p. 348

gastric inhibitory polypeptide (găs′trĭk ĭn-hĭb′ĭ-tŏr-ē pŏl-ē-pĕp′tīd), p. 350

gastric phase (găs′trĭk fāz), p. 350

gastric pits (găs′trĭk pĭts), p. 348

gastrin (găs′trĭn), p. 348

gastrointestinal tract (găs′trō-ĭn″tĕs-tĭn-ăl trăkt), p. 343

ghrelin (grē′lŭn), p. 351

greater omentum (grā′tĕr ō-měn′tŭm), p. 353

hard palate (hărd păl′ŭt), p. 344

haustra (hăw′strŭh), p. 351

haustrum (hăw′ strŭhm), p. 351

hepatic lobule (hŭh-păt′ĭk lŏb′yŭl), p. 354

hydrolytic enzymes (hī′ drō-lĭ″tĭk ĕn′ zīmz), p. 358

ileum (ĭl′ē-ŭm), p. 349

incisor (ĭn-sī′zōr), p. 344

intestinal phase (ĭn-tĕs′tĭn-ŭl fāz), p. 350

intrinsic factor (ĭn-trĭn′zĭk făk′tŭr), p. 348

jejunum (jĕ-jū′nŭm), p. 349

lacteal (lăk′tē-l), p. 349

large intestine (lărj ĭn-tĕs′tĭn), p. 351

laryngopharynx (lă-rĭng′ gō-făr″ ĭnks), p. 346

leptin (lĕp′tĭn), p. 351

lesser omentum (lĕs′ĕr ō-měn′tŭm), p. 353

lipase (līp′ās), p. 354

liver (lĭv′ĕr), p. 354

lumen (lū′měn), p. 346

mesentery (měs′ĕn-tĕ-rē), p. 352

microvilli (mī-krō-vĭl′ī), p. 349

mineral (mĭn′ĕr-ăl), p. 363

molar (mō′lăr), p. 344

mouth (mŏwth), p. 344

mucosa (myū-kō′sŭh), p. 346

muscularis (mŭs-ku-lă′rĭs), p. 346

nasopharynx (nā′ zō-făr″ ĭnks), p. 346

oropharynx (ōr′ ō-făr″ ĭnks), p. 346

pancreas (păng′krē-ŭs), p. 353

pancreatic amylase (păng′krē-ăt′ĭk ăm′ŭh-lās), p. 354

pepsin (pĕp′sĭn), p. 348

peptidases (pĕp′tĭ-dās′), p. 359

peristalsis (pĕr″ĭ-stăl′sĭs), p. 347

pharynx (făr′ĭnks), p. 346

portal triads (pōr′tŭl trī′ădz), p. 354

pyloric region (pī-lōr′ĭk rē′jĕn), p. 348

rectum (rĕk′tŭm), p. 351

rugae (rū′jē), p. 347

salivary amylase (săl′ŭh-vĕr-ē ăm′ŭh-lās), p. 345

salivary gland (săl′ŭh-vĕr-ē glănd), p. 344

secretin (sē-krēt′ĭn), p. 350

serosa (sē-rō′să), p. 347

sigmoid colon (sĭg′mōyd kō′lŏn), p. 351

small intestine (smăwl ĭn-tĕs′tĭn), p. 349

soft palate (săwft păl′ŭt), p. 344

sphincter (sfĭngk′tĕr), p. 347

stomach (stŭm′ăk), p. 347

submucosa (sŭb-myū-kō′să), p. 346

taenia coli (tē′nē-ă kō′lī), p. 351

transverse colon (trăns-vĕrs′ kō′lŏn), p. 351

urea (yū-rē′ŭh), p. 355

uvula (yū′vyū-lŭh), p. 344

vermiform appendix (vĕr′mĭ-fŏrm ŭh-pĕn′dĭks), p. 351

vestibule (vĕs′tĭ-byūl), p. 344

villi (vĭl′ī), p. 349

vitamin (vī′tŭh-mĭn), p. 362

Clinical Key Terms

anorexia nervosa (ăn′ŭh-rĕk′sē-ŭh nŭr-vō′sŭh), p. 365

bariatric surgery (băr′ē-ăt′rĭk sŭr′jŭh-rē), p. 367

bulimia nervosa (bū-lē′mē-ŭh nŭr-vō′sŭh), p. 365

caries (kār′ēz), p. 344

cirrhosis (sŭh-rō′sĭs), p. 357

constipation (kŏn-stĭ-pā′shŭn), p. 356

diarrhea (dī-ŭh-rē′ŭh), p. 356

diverticulosis (dī″vĕr-tĭ-kyū-lō′sĭs), p. 357

gastric ulcer (găs′ trĭk ŭlsĕr), p. 356

gastroesophageal reflux disease (găs′trō-ĕs-ŏf′ŭh-gē-ŭl rē′flŭcks dĭ-zēz′), p. 356

gingivitis (jĭn-jŭh-vī′tĭs), p. 344

heartburn (hărt′bĕrn), p. 356

hemorrhoids (hěm′ŭh-rōydz), p. 356

hepatitis (hě″pŭh-tī′tĭs), p. 357

jaundice (jŏn′dĭs), p. 357

obesity (ō-bē′sĭ-tē), p. 365

periodontitis (pĕr″ē-ō-dŏn-tī′tĭs), p. 344

peritonitis (pĕr″ĭ-tō-nī′tĭs), p. 356

polyp (păh′lĭp), p. 356

Summary

15.1 **Anatomy of the Digestive System**
The gastrointestinal tract consists of the mouth, pharynx, esophagus, stomach, small intestine, and large intestine. The functions of the digestive system include (1) ingestion, (2) the breakdown of food into small particles called nutrients, (3) nutrient absorption, and (4) the elimination of indigestible wastes.
A. The oral cavity begins physical and chemical digestion.
B. The teeth chew the food and the tongue forms a bolus for swallowing. The salivary glands send saliva into the mouth. Saliva contains salivary amylase, an enzyme that begins the chemical digestion of starch.
C. Both air and food pass through the pharynx.

D. When a person swallows, the nasopharynx, larynx, and trachea are normally blocked off, and food must enter the esophagus.

E. The wall of the gastrointestinal tract has four layers: mucosa, submucosa, muscularis, and serosa.

F. The esophagus performs peristalsis to propel food through the lower esophageal sphincter into the stomach.

G. The stomach expands and stores food. While food is in the stomach, the stomach churns, mixing it with acidic gastric juices. Gastric juices contain hydrochloric acid and the enzyme pepsin. Both initiate the digestion of protein.

H. The three regions of the small intestine are the duodenum, jejunum, and ileum. The duodenum receives bile from the liver and pancreatic juice from the pancreas. The walls of the small intestine have fingerlike projections called villi. The cells of the villi have microvilli, where small nutrient molecules are absorbed. Amino acids and glucose enter the blood vessels of a villus, while glycerol and fatty acids are joined and packaged as lipoproteins before entering lymphatic vessels called lacteals.

I. The large intestine consists of the cecum, the colon (including the ascending, transverse, descending, and sigmoid colon), and the rectum, which ends at the anus. The large intestine does not produce digestive enzymes; it does absorb water, salts, and some vitamins.

15.2 Accessory Organs of Digestion
Three accessory organs of digestion—the pancreas, liver, and gallbladder—send secretions to the duodenum via ducts.

A. The pancreas produces pancreatic juice, which contains digestive enzymes for carbohydrates, protein, fats, and nucleic acids. Pancreatic amylase digests starch. Three pancreatic enzymes (trypsin, chymotrypsin, and carboxypeptidase) digest protein.

B. The liver receives blood from the small intestine by way of the hepatic portal vein. It has many important functions, including detoxifying blood, forming plasma proteins, and storing iron and glycogen. The liver produces bile, which emulsifies fat and readies it for digestion by lipase, an enzyme produced by the pancreas.

C. The gallbladder stores excess bile. Bile emulsifies fats, and increases dietary absorption of fatty acids, cholesterol, and fat-soluble vitamins.

15.3 Chemical Digestion
Digestive enzymes are present in digestive juices and break down food into absorbable nutrient molecules: monosaccharides, amino acids, fatty acids, and glycerol (see Table 15.2). Salivary amylase and pancreatic amylase begin the chemical digestion of starch. Pepsin and pancreatic peptidases digest protein to peptides. Lipase digests fat to glycerol and fatty acids. Intestinal enzymes finish the digestion of starch and protein. Digestive enzymes are specific to their substrate and speed up specific reactions at optimum body temperature and pH.

15.4 Effects of Aging
The structure and function of the digestive system generally decline with age. The various illnesses associated with the digestive system are more likely to be seen among the elderly.

15.5 Homeostasis
The digestive system works with the other systems of the body in the ways described in Human Systems Work Together on page 368.

15.6 Nutrition
The nutrients released by the digestive process should provide us with an adequate amount of energy, essential amino acids and fatty acids, and all necessary vitamins and minerals.

A. Vitamins are small organic molecules that have various functions in the body. They are categorized as water- or lipid-soluble. The vitamins C, E, and A are antioxidants that protect cell contents from damage due to free radicals.

B. Minerals are inorganic, and are classified as macronutrients or micronutrients, depending on the quantity needed by the body. One mineral, calcium, is needed for strong bones.

C. The diet should be balanced and low in saturated fatty acids and cholesterol molecules, whose intake is linked to cardiovascular disease.

D. The causes of eating disorders, including obesity, bulimia nervosa, and anorexia, are being explored in order to help people maintain a normal weight for their height.

Study Questions

1. List the organs of the alimentary canal, and state the contribution of each to the digestive process. (pp. 343–350)

2. Discuss the absorption of the products of digestion into the lymphatic and cardiovascular systems. (pp. 349–350, 358–359)

3. Name and state the functions of the hormones that assist the nervous system in regulating digestive secretions. (pp. 350–351)

4. Name the accessory digestive organs, and describe the part they play in the digestion of food. (pp. 353–355, 358)

5. Name and discuss any three functions of the liver. (pp. 355, 358)

6. Discuss the digestion of starch, protein, and fat, listing all the steps that occur with each. (pp. 358–360)

7. How does the digestive system help maintain homeostasis? (pp. 361, 368)

8. What is the chief contribution of each of these constituents of the diet: (a) carbohydrates; (b) proteins; (c) fats; (d) fruits and vegetables? (pp. 361–362)

9. What role do water-soluble vitamins usually play in the body? (pp. 362–364)

10. Name and discuss three eating disorders. (pp. 365–366)

Learning Outcome Questions

Fill in the blanks.

1. In the mouth, salivary _____ digests starch.

2. The stomach begins chemical digestion of protein with the enzyme _____.

3. The _____ takes food to the stomach, where primarily _____ is digested.

4. The gastric juices are _____, and therefore, they usually destroy any bacteria in the food.

5. The proximal large intestine is called the _____, and it continues distally into these four colon segments: _____, _____, _____, and _____.

6. The pancreas transports digestive juices to the _____, the first part of the small intestine.

7. After a meal, the liver stores glucose as _____.

8. The gallbladder stores _____, a substance that _____ fat.

9. Pancreatic juice contains _____, _____, and _____ for digesting protein, _____ for digesting fat, and _____ for digesting starch.

10. The products of digestion are absorbed into the cells of the _____, fingerlike projections of the intestinal wall.

11. Small organic molecules that assist in metabolic reactions and are essential for life are called _____ nutrients.
    a. vitamins
    b. minerals
    c. hormones
    d. lipids

Medical Terminology Exercise

After studying this chapter, see if you can derive the definitions for the medical terms listed at right. Many of the prefixes and suffixes used to create these terms can be found throughout the chapter. For additional help, use McGraw-Hill Connect™ at www.mcgrawhillconnect.com and consult Appendix B.

1. stomatoglossitis (stō″mŭh-tō-glŏs-sī′tĭs)
2. glossopharyngeal (glŏs″ō-făh-rĭn′jē-ăl)
3. gastroenteritis (găs″trō-ĕn-tĕr-ī′tĭs)
4. sublingual (sŭb-lĭng′gwăl)
5. gingivoperiodontitis (jĭn″jĭ-vō-pĕr″ē-ō-dŏn-tī′tĭs)
6. dentalgia (dĕn-tăl′jē-ŭh)
7. pyloromyotomy (pī-lō″rō-mī-ŏt′ō-mē)
8. cholecystolithotripsy (kō″lē-sĭs″tō-lĭth′ō-trĭp″sē)
9. proctosigmoidoscopy (prŏk″tō-sĭg″mŏy-dŏs′kŭh-pē)
10. colocentesis (kŏ″lō-sĕn-tē′sĭs)
11. trichophagia (trī-kō-fāj′ē-ŭh)
12. ileocecal (ĭl″ē-ō-sē′kŭl)

Online Study Tools

∎ILEARNSMART® **Connect®**

APR

Anatomy & Physiology REVEALED includes cadaver photos that allow you to peel away layers of the human body to  reveal structures beneath the surface. This program also includes animations, radiologic imaging, audio pronunciations, and practice quizzing. To learn more visit www.aprevealed.com

16 The Urinary System and Excretion

The year was 1943, and the Nazis were in control of the Netherlands. A young Dutch doctor named Willem Kolff had fled from a Nazi-controlled large city hospital to a smaller regional one, and it was there that he completed the first of many revolutionary inventions: a kidney dialysis machine. It was cobbled together using 10 meters (almost 33 feet!) of sausage casing, wooden barrels, washing machine parts, a rotating drum, and a tank of electrolyte solution. Kolff had never forgotten one of his first patients: a 22-year-old man who had died of acute renal failure. He reasoned that if he could create a system to remove urea from blood, he could save patients like the young man. He tested his device on a patient by allowing his patient's blood to flow into the sausage casing, which was wrapped around the rotating drum. As the drum slowly spun in the tank of electrolyte solution, urea diffused from high concentration in the blood to low concentration in the solution. (You learned about diffusion in Chapter 3.) The patient's blood, cleansed of the poisonous urea, was then pumped back into the body. Two years and 16 failures later, the procedure finally worked. Ironically, his first success was on a Dutch woman who had been a known Nazi collaborator, whom he treated in spite of his countrymen's objections. Kolff went on to design the heart-lung machine and the first artificial heart, and is recognized as the father of artificial organs. You can read more about renal failure and dialysis on pages 385–386.

Learning Outcomes

After you have studied this chapter, you should be able to:

16.1 Urinary System

1. List and discuss the functions of the urinary system.
2. Name and describe the structure and function of each organ in the urinary system.
3. Describe how urination is controlled.

16.2 Anatomy of the Kidney and Excretion

4. Describe the macroscopic and microscopic anatomy of the kidney.
5. State the parts of a kidney nephron, and relate them to the gross anatomy of the kidney.
6. Outline the three steps in urine formation, and relate them to the parts of a nephron.

16.3 Regulatory Functions of the Kidneys

7. Describe how the kidneys help maintain the fluid and electrolyte balance of blood.

8. Name and explain how three hormones—aldosterone, antidiuretic hormone, and atrial natriuretic hormone—work together to maintain blood volume and pressure.
9. Describe three mechanisms, including how the kidneys function, to maintain the acid–base balance of blood.

16.4 Problems with Kidney Function

10. State, in general, the normal composition of urine and the benefits of doing a urinalysis.
11. Discuss the need for hemodialysis and how hemodialysis functions to bring about the normal composition of urine.

16.5 Effects of Aging

12. Describe the anatomical and physiological changes that occur in the urinary system as we age.

16.6 Homeostasis

13. Describe how the urinary system works with other systems of the body to maintain homeostasis.

Visual Focus
Steps in Urine Formation

Focus on Forensics
Urinalysis

Medical Focus
Prostate Enlargement and Prostate Cancer

Human Systems Work Together
Urinary System

16.1 Urinary System

1. List and discuss the functions of the urinary system.
2. Name and describe the structure and function of each organ in the urinary system.
3. Describe how urination is controlled.

The kidneys are the primary organs of excretion. **Excretion** is the removal of metabolic wastes from the body. People sometimes confuse the terms excretion and defecation, but they do not refer to the same process. Defecation, the elimination of feces from the body, is a function of the digestive system. Excretion, on the other hand, is the elimination of metabolic wastes, which are the products of metabolism. For example, the undigested food and bacteria that make up feces have never been a part of the functioning of the body, while the substances excreted in urine were once metabolites in the body.

Functions of the Urinary System

The urinary system filters the blood to produce urine, then transports the urine out of the body. As the kidneys produce urine, they carry out four functions that are essential to homeostasis: excretion of metabolic wastes, preservation of water–salt balance in blood and body fluids, maintenance of blood pressure, and maintenance of acid–base balance. As you know from studying endocrine hormones in chapter 10, the kidneys also produce erythropoietin (EPO), their own endocrine hormone.

Excretion of Metabolic Wastes

The kidneys excrete metabolic wastes, notably nitrogenous wastes. **Urea** is the primary nitrogenous end product of metabolism in human beings, but humans also excrete some ammonium, creatinine, and uric acid.

Urea is a by-product of amino acid metabolism. The breakdown of amino acids in the liver releases ammonia, which the liver combines with carbon dioxide to produce urea. Ammonia is very toxic to cells, but urea is much less toxic. Because urea is less toxic than ammonia, it can be safely excreted in the relatively small amounts of water passed daily as urine. (If our bodies had to excrete ammonia instead of urea, the ammonia would have to be diluted in extremely large volumes of urine to avoid being toxic to body cells.)

Creatine phosphate is a high-energy phosphate reserve molecule in muscles. The metabolic breakdown of creatine phosphate results in **creatinine.**

The breakdown of nucleotides, such as those containing adenine and thymine, produces **uric acid.** Uric acid is rather insoluble. If too much uric acid is present in blood, crystals form and *precipitate,* or become solid. Crystals of uric acid sometimes collect in the joints, producing a painful ailment called **gout.**

Urochrome is a waste product formed as the liver breaks down hemoglobin. As you know from reading Chapter 15, hemoglobin is first formed into bilirubin, which is excreted in the bile. Once in the intestine, bilirubin is further broken down into smaller compounds, including urochrome. It is urochrome that gives urine its yellow color.

Preservation of Water–Salt Balance

A principal function of the kidneys is to maintain the appropriate water–salt balance of the blood. As we will see, blood volume is intimately associated with the salt balance of the body. Salts, such as NaCl, have the ability to cause osmosis, the diffusion of water—in this case, into the blood. The more salts there are in the blood, the greater the blood volume. The kidneys also maintain the appropriate level of other ions (electrolytes), such as potassium ions (K^+), bicarbonate ions (HCO_3^-), and calcium ions (Ca^{2+}), in the blood.

Maintenance of Blood Pressure

Recall that blood pressure is the product of cardiac output and peripheral resistance, and the kidneys influence both variables. By preserving salt and water balance, the kidneys maintain normal blood volume. In turn, blood volume determines the heart's stroke volume and, thus, cardiac output. As you'll recall from Chapter 12, the kidneys also influence peripheral resistance by producing the enzyme renin, which (along with a lung enzyme) activates the plasma protein angiotensin. Angiotensin constricts the blood vessel smooth muscle, increasing blood vessel resistance and raising blood pressure.

Maintenance of Acid–Base Balance

The kidneys regulate the acid–base balance of the blood. In order for a person to remain healthy, the blood pH should be maintained in a narrow range around 7.4 (the normal blood pH range is 7.35–7.45). The kidneys monitor and control blood pH, mainly by excreting hydrogen ions (H^+) from the blood into the urine. At the same time, the kidneys reabsorb bicarbonate ions (HCO_3^-) and return them to the blood as needed to keep blood pH at about 7.4. Urine usually has a pH of 6 or lower because our diet often contains acidic foods whose hydrogen ions must be excreted.

Secretion of Hormones

The kidneys assist the endocrine system in hormone secretion. By releasing the enzyme renin whenever their own blood supply decreases, the kidneys activate angiotensin, which leads to the secretion of the hormone aldosterone. Aldosterone is produced by the adrenal cortex, the outer portion of the adrenal glands, which lie atop the kidneys. As described in section 16.3, aldosterone promotes the reabsorption of sodium ions (Na^+) and water by the kidneys.

Whenever the oxygen-carrying capacity of the blood is reduced, the kidneys secrete the hormone **erythropoietin (EPO),** which stimulates red blood cell production.

The kidneys also help activate vitamin D from the skin. Vitamin D is the precursor of the hormone calcitriol, which promotes calcium (Ca^{2+}) absorption from the digestive tract.

The Work of the Kidneys: By the Numbers

It's probably safe to say that most healthy people rarely, if ever, think about their kidney function. Yet when kidneys are analyzed by the numbers, their work becomes extremely impressive. Consider this: a healthy, average man's kidneys filter about 125 mL of blood every minute—that's 7.5 L per hour, 180 L per day, and 65,700 L of blood per year. Now assume he lives for 76 years (the average life span for a man in the United States) and for 60 of those years he's filtering at the normal adult rate. That's almost 5 million L, or roughly 1.3 million gallons, of blood filtered by the kidneys during his adult lifetime (the figures are slightly less for women). However, of that volume, only about 1% ever leaves the body as urine, and the remainder is quickly reabsorbed and returned to the bloodstream. Still, those 13,000 gallons of urine eliminate enormous quantities of waste, including almost half a ton of potentially toxic urea.

Organs of the Urinary System

The urinary system consists of the kidneys, ureters, urinary bladder, and urethra. Figure 16.1 shows these organs and also traces the path of urine.

Kidneys

The **kidneys** are paired organs located in the lumbar region on either side of the vertebral column. The kidneys are *retroperitoneal,* which means that they are covered by the parietal peritoneum (the serous membrane that lines the abdominopelvic cavity). Thus, the kidneys are located posterior to the abdominopelvic cavity, not within it. The kidneys lie in depressions against the deep muscles of the back, where they receive some protection from the lower rib cage. Each kidney is usually held in place by connective tissue, called renal fascia. Masses of adipose tissue adhere to each kidney.

The kidneys are fist-sized, bean-shaped organs that are reddish-brown in color. They are covered by a tough capsule of fibrous connective tissue, called a renal capsule. The concave side

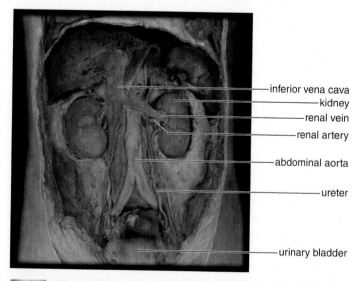

inferior vena cava
kidney
renal vein
renal artery
abdominal aorta
ureter
urinary bladder

AP|R **Figure 16.1 The urinary system.** The kidneys produce urine, and the ureters transport it to the urinary bladder for storage. The urethra (not shown in this photograph) transports urine from the bladder to the outside. Urine is found only in the kidneys, ureters, urinary bladder, and urethra.

of a kidney has a depression called the **hilum** where a **renal artery** enters and a **renal vein** and a ureter exit the kidney.

Ureters

The **ureters,** which extend from the kidneys to the bladder, are small, muscular tubes about 25 cm long and 5 mm in diameter. Each descends behind the parietal peritoneum, from the hilum of a kidney, to enter the bladder posteriorly at its inferior surface.

The wall of a ureter has three layers. The inner layer is a mucosa (mucous membrane), the middle layer consists of smooth muscle, and the outer layer is a fibrous coat of connective tissue. Peristaltic contractions of the ureters cause urine to enter the bladder even if a person is lying down. Urine enters the bladder in spurts that occur at the rate of one to five per minute.

Urinary Bladder

The **urinary bladder** is located in the pelvic cavity, below the parietal peritoneum and just posterior to the pubic symphysis. In males, it is directly anterior to the rectum; in females, it is anterior to the vagina and inferior to the uterus. Its function is to store urine until it is expelled from the body. The bladder has three openings—two for the ureters and one for the urethra, which drains the bladder. The *trigone* is a smooth triangular area at the base of the bladder outlined by these three openings (Fig. 16.2).

Collectively, the muscle layers of the bladder wall are called the *detrusor muscle.* The wall contains a middle layer of circular fiber and two layers of longitudinal muscle, and it can expand. The transitional epithelium of the mucosa becomes thinner, and folds in the mucosa called *rugae* disappear as the bladder fills and enlarges.

The bladder has other features that allow it to retain urine. After urine enters the bladder from a ureter, small folds of bladder mucosa act like a valve to prevent backward flow.

Two sphincters in close proximity are found where the urethra exits the bladder. The internal sphincter occurs around the opening to the urethra. Inferior to the internal sphincter, the external sphincter is composed of skeletal muscle that can be voluntarily controlled.

Urethra

The **urethra** is a tube that extends from the urinary bladder to an external opening (see Fig. 17.3 and Fig. 17.9). The urethra is a different length in females than in males. In females, the urethra is only about 4 cm long. The short length of the female urethra makes bacterial invasion easier and helps explain why females are more prone to urinary tract infections than males. In males, the urethra averages 20 cm when the penis is flaccid (limp, nonerect). As the urethra leaves the male urinary bladder, it is surrounded by the **prostate gland.** In older men, enlargement of the prostate gland can restrict urination. Medical and surgical treatments for prostate enlargement are detailed in the Medical Focus on page 387.

In females, the reproductive and urinary systems are not connected. In males, the urethra carries urine during urination and sperm during ejaculation. This double function of the urethra in males does not alter the path of urine.

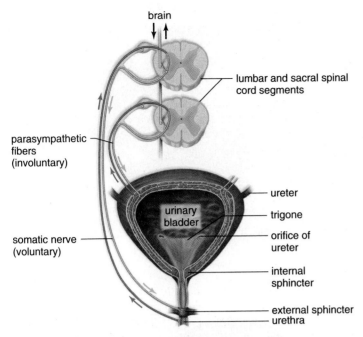

brain

lumbar and sacral spinal
cord segments

parasympathetic
fibers
(involuntary)

ureter

trigone

urinary
bladder

orifice of
ureter

somatic nerve
(voluntary)

internal
sphincter

external sphincter
urethra

AP|R **Figure 16.2** **Urination.** As the bladder fills with urine, sensory signals (blue arrows) go to the spinal cord and then to the brain. The brain can override the urge to urinate. When urination occurs, motor nerve signals (green arrows) cause the bladder to contract and an internal sphincter to open. Nerve signals also cause an external sphincter to open.

Urination

When the adult urinary bladder fills to about 250 ml with urine, stretch receptors send sensory nerve signals to the spinal cord. Subsequently, parasympathetic motor nerve signals from the lumbar and sacral regions of the spinal cord cause the urinary bladder to contract and the sphincters to relax so that urination, also called **micturition,** is possible (Fig. 16.2). In older children and adults, the brain controls this reflex, delaying urination until a suitable time.

Content CHECK-UP!

1. Which is an enzyme produced by the kidney?

   a. renin   c. erythropoietin

   b. angiotensin   d. aldosterone

2. Describe the way in which the kidneys and lungs interact to control blood pressure. What chemicals are involved?

3. If you were to describe the urinary system, you could say:

   a. The kidneys are in the lumbar region.

   b. The urinary bladder sits in the pelvic cavity.

   c. The urethra transports urine from the kidney to the urinary bladder.

   d. Only a and b are correct.

   e. All three statements are correct.

Answers in Appendix A.

4. Describe the macroscopic and microscopic anatomy of the kidney.

5. State the parts of a kidney nephron, and relate them to the gross anatomy of the kidney.

6. Outline the three steps in urine formation, and relate them to the parts of a nephron.

A sagittal section of a kidney shows that many branches of the renal artery and renal vein reach inside a kidney (Fig. 16.3*a*). Removing the blood vessels shows that a kidney has three regions (Fig. 16.3*b*). The **renal cortex** is an outer, granulated layer that dips down in between a radially striated inner layer called the renal medulla. The **renal medulla** consists of cone-shaped tissue masses called renal pyramids. The **renal pelvis** is a central space, or cavity, that is continuous with the ureter.

Anatomy of a Nephron

Each kidney is composed of over one million **nephrons,** microscopic structures that are the functional units of the kidney (Fig. 16.3*d*). There are two nephron types: cortical nephrons, which are shorter because they are located primarily in the cortex, and juxtamedullary nephrons, which extend deep into the medulla. A nephron consists of two combined structures: a **renal corpuscle** and a **renal tubule.** The renal corpuscle is formed by the **glomerulus,** a tightly coiled capillary network, and the **glomerular capsule** (also called *Bowman's capsule*). The glomerular capsule is formed by the closed end of the renal tubule, which widens to form a cuplike structure that surrounds the glomerulus. The blood supply of a nephron is unusual because it contains two distinct capillary regions (Fig. 16.4). The nephron receives oxygenated arterial blood from an afferent arteriole, which originates from a smaller branch of the renal artery. The afferent arteriole leads to the glomerulus, the first capillary bed. Blood leaves the glomerulus through the efferent arteriole. The efferent arteriole takes blood to the second capillary bed, called the **peritubular capillary network,** which surrounds the remaining renal tubule. A specialized section of the peritubular capillary network, the vasa recta, helps to maintain solute concentration in the renal medulla, and will be detailed later. From the peritubular capillary network, the blood goes into a venule. Venules join larger veins that ultimately empty into the renal vein.

Parts of a Renal Tubule

Each renal tubule is made up of several parts, beginning with the glomerular capsule. This wide proximal end of the tubule surrounds the glomerulus and receives the fluid it filters. The inner layer of the glomerular capsule is composed of *podocytes* that have long, cytoplasmic extensions. The podocytes cling to the capillary walls of the glomerulus and leave pores that allow easy passage of small molecules from the glomerulus to the inside of the glomerular capsule.

Next, there is a **proximal convoluted tubule (PCT),** called "proximal" because it is near the glomerular capsule. The cuboidal epithelial cells lining this part of the nephron have numerous microvilli about 1 μm in length. These microvilli are tightly packed and form a brush border, increasing the surface area for reabsorption.

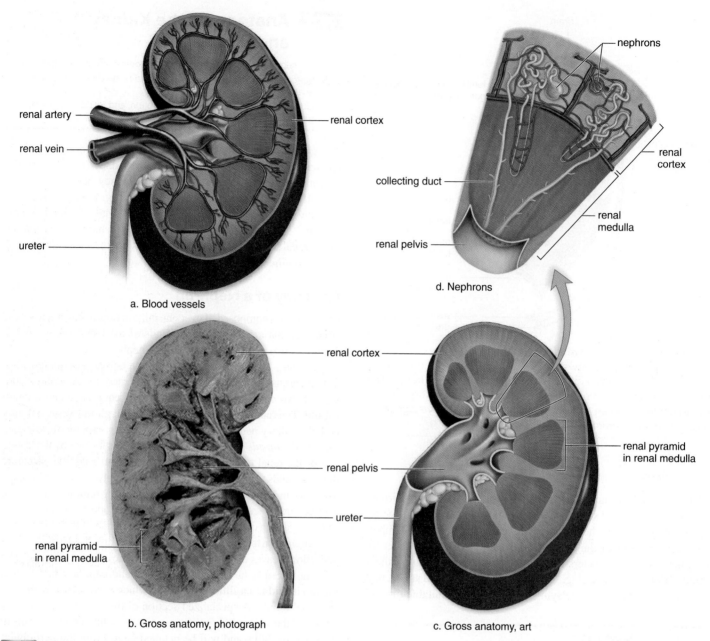

renal artery

renal vein

ureter

renal cortex

a. Blood vessels

nephrons

collecting duct

renal pelvis

renal cortex

renal medulla

d. Nephrons

renal cortex

renal pelvis

ureter

renal pyramid in renal medulla

renal pyramid in renal medulla

b. Gross anatomy, photograph

c. Gross anatomy, art

AP|R **Figure 16.3** **Gross anatomy of the kidney.** (a) A sagittal section of the kidney showing the blood supply. Note that the renal artery divides into smaller arteries, and these divide into arterioles. Venules join to form small veins, which join to form the renal vein. (b) Sagittal section photograph of the kidney. (c) The same section without the blood supply. Now it is easier to distinguish the renal cortex, the renal medulla, and the renal pelvis, which connects with the ureter. The renal medulla consists of the renal pyramids. (d) An enlargement showing the placement of nephrons.

Simple squamous epithelium appears as the tube narrows and makes a U-turn called the **loop of Henle.** Each loop consists of a descending limb and an ascending limb.

The cuboidal epithelial cells of the **distal convoluted tubule (DCT)** have numerous mitochondria, but they lack microvilli. This is consistent with the active role they play in moving molecules from the blood into the tubule, a process called tubular secretion. Within the nephron, the distal convoluted tubule is generally adjacent to the afferent arteriole. The region where

they touch is called the juxtaglomerular apparatus. (It is illustrated in Fig. 16.8, and its function will be described in section 16.3.) The distal convoluted tubules of several nephrons enter a single collecting tubule. In turn, collecting tubules empty into larger collecting ducts. Many **collecting ducts** carry urine to the renal pelvis.

As shown in Figure 16.4, the glomerular capsule and the convoluted tubules always lie within the renal cortex. The loop of Henle dips down into the renal medulla; juxtamedullary

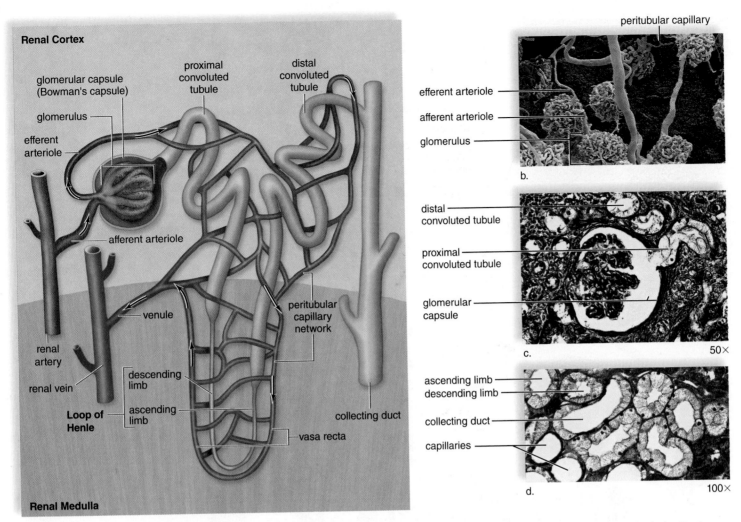

AP|R **Figure 16.4** **Nephron anatomy. (a)** A nephron is made up of a glomerular capsule, the proximal convoluted tubule, the loop of Henle, the distal convoluted tubule, and the collecting duct. Trace the path of blood around the nephron by following the arrows. **(b)** Scanning electron micrograph shows a surface view of the glomerulus and its blood supply: the arterioles and peritubular capillary. **(c)** Photomicrograph of cross section of the glomerulus. Notice that the convoluted tubules are just adjacent to it. **(d)** Photomicrograph of cross sections of ascending and descending limbs of the loop of Henle, and peritubular capillaries.

nephrons have a very long loop of Henle, which penetrates deep into the renal medulla. Collecting ducts are also located in the renal medulla, and they give the renal pyramids their lined appearance.

Urine Formation

Figure 16.5 gives an overview of urine formation, which is divided into these steps: glomerular filtration, tubular reabsorption, and tubular secretion.

Glomerular Filtration

Glomerular filtration occurs when whole blood enters the afferent arteriole and the glomerulus. Due to glomerular blood pressure, water and small molecules are pushed from the glomerulus to the

inside of the glomerular capsule. This is a filtration process because large molecules and formed elements are unable to pass through the capillary wall. In effect, then, blood in the glomerulus has two portions: the filterable components and the nonfilterable components:

Filterable Blood Components	Nonfilterable Blood Components
Water	Formed elements (blood cells and platelets)
Nitrogenous wastes	Large plasma proteins
Small molecules of nutrients	
Salts (ions)	
Creatinine and urochrome	

The **glomerular filtrate** contains small, dissolved molecules found in approximately the same concentrations as their plasma concentrations. Small molecules that escape being filtered as well as the

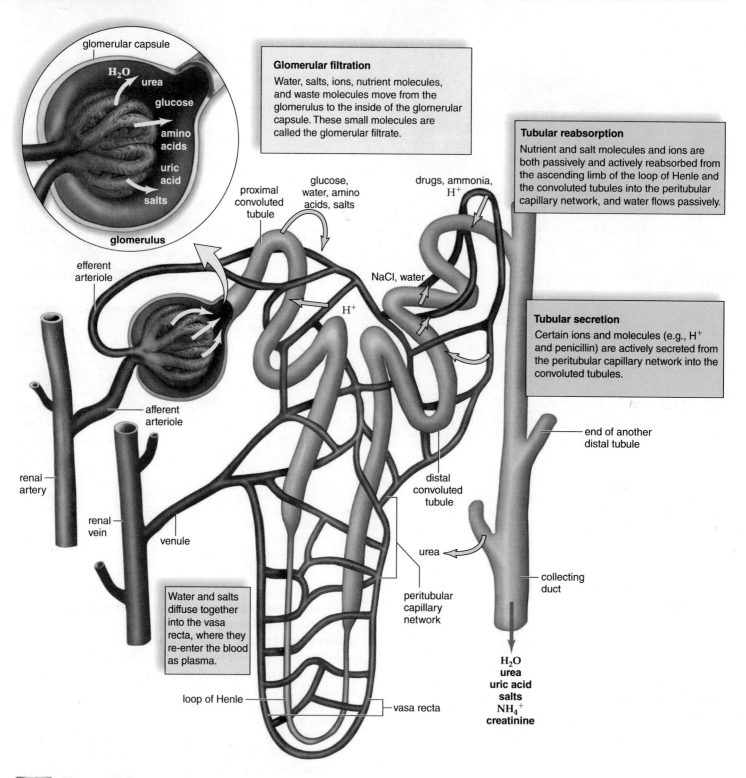

glomerular capsule

H_2O

urea

glucose

amino acids

uric acid

salts

glomerulus

efferent arteriole

afferent arteriole

renal artery

renal vein

venule

loop of Henle

Glomerular filtration

Water, salts, ions, nutrient molecules, and waste molecules move from the glomerulus to the inside of the glomerular capsule. These small molecules are called the glomerular filtrate.

proximal convoluted tubule

glucose, water, amino acids, salts

H^+

drugs, ammonia, H^+

NaCl, water

Tubular reabsorption

Nutrient and salt molecules and ions are both passively and actively reabsorbed from the ascending limb of the loop of Henle and the convoluted tubules into the peritubular capillary network, and water flows passively.

Tubular secretion

Certain ions and molecules (e.g., H^+ and penicillin) are actively secreted from the peritubular capillary network into the convoluted tubules.

end of another distal tubule

distal convoluted tubule

urea

Water and salts diffuse together into the vasa recta, where they re-enter the blood as plasma.

peritubular capillary network

vasa recta

collecting duct

H_2O
urea
uric acid
salts
NH_4^+
creatinine

AP|R **Figure 16.5** **Steps in urine formation.** *Top:* The three main steps in urine formation are described in boxes that are color-coded to arrows showing the movement of molecules into or out of the nephron at specific locations. In the end, urine is composed of the substances within the collecting duct (see brown arrow, lower right).

TABLE 16.1	Reabsorption from Nephrons		
Substance	Amount Filtered (per day)	Amount Excreted (per day)	Reabsorption (%)
Water (L)	180	1.8	99.0
Sodium (g)	630	3.2	99.5
Glucose (g)	180	0.0	100.0
Urea (g)	54	30.0	44.0

L = Liter; g = grams

nonfilterable components leave the glomerulus by way of the efferent arteriole.

As indicated in Table 16.1, nephrons in the kidneys filter 180 L of water per day, along with a considerable amount of small molecules (such as glucose) and ions (such as sodium). If the composition of the urine were the same as that of the glomerular filtrate, the body would constantly lose water, salts, and nutrients in the urine. This does not happen in a healthy person, because the composition of the original filtrate is altered significantly as this fluid passes through the remainder of the tubule.

It is important to stress that many small chemicals are freely filtered from the blood. Once in the filtrate, they cannot be reabsorbed and are passed in urine. Thus, urinalysis provides a history of chemicals recently present in the blood, and can be used for detection of illegal drug use, as described in Focus on Forensics (page 386).

Tubular Reabsorption

Tubular reabsorption occurs as molecules and ions are both passively and actively reabsorbed from the nephron into the blood of the peritubular capillary network. The osmolarity of the blood is maintained by the presence of both plasma proteins and salt. When sodium ions (Na^+) are actively reabsorbed, chloride ions (Cl^-) follow passively. The reabsorption of salt (NaCl) increases the osmolarity of the blood compared to the filtrate, and therefore water moves by osmosis from the tubule into the blood. About 67% of Na^+ is reabsorbed from the proximal convoluted tubule.

Nutrients such as glucose and amino acids also return to the blood from the proximal convoluted tubule. This is a selective process. Only molecules recognized by carrier proteins built into tubular cell membranes will be actively reabsorbed. Glucose is an example of a molecule that ordinarily is completely reabsorbed because there is a plentiful supply of carrier molecules for it. However, every substance has a maximum rate of transport. After all its carriers are in use, any excess in the filtrate will appear in the urine. For example, as reabsorbed levels of glucose approach 1.8–2 mg/ml plasma, the rest appears in the urine. Excess glucose occurs in the blood in untreated diabetes mellitus because of lack of insulin (type I diabetes), or because of failure of insulin receptors to respond normally (type II diabetes). Without insulin function, glucose cannot be transported into cells, and liver and muscle cannot store glucose as glycogen.

Excess blood glucose is filtered into the tubular fluid and subsequently passes into the urine because the kidneys cannot reabsorb it all. The resulting condition is called *glycosuria* (see Chapter 10). The presence of glucose in the filtrate increases its osmolarity compared to that of the blood, and therefore less water is reabsorbed into the peritubular capillary network. The frequent urination and increased thirst experienced by untreated diabetics are due to the fact that water isn't being reabsorbed.

We've seen that the filtrate that enters the proximal convoluted tubule is divided into two portions: components that are reabsorbed from the tubule into the blood and components that are not reabsorbed and continue to pass through the nephron to be further processed into urine:

Reabsorbed Filtrate Components	Nonreabsorbed Filtrate Components
Most water	Some water
Nutrients	Much nitrogenous waste
Required salts (ions)	Excess salts (ions)
Some urea and uric acid	Creatinine and urochrome
	Molecules not recognized by carriers

The substances that are not reabsorbed become the tubular fluid, which enters the loop of Henle.

Tubular Secretion

Tubular secretion is a second way by which substances are removed from blood and actively transported into the tubular fluid. Hydrogen ions, potassium ions, creatinine, and drugs such as penicillin are some of the substances that are moved by active transport from the blood into the distal convoluted tubule. In the end, urine contains (1) substances that have undergone glomerular filtration but have not been reabsorbed, and (2) substances that have undergone tubular secretion.

> ## Content CHECK-UP!
>
> 4. Select the correct order of the blood supply for a nephron:
>   a. afferent arteriole → efferent arteriole → glomerulus → peritubular capillary
>   b. efferent arteriole → glomerulus → afferent arteriole → peritubular capillary
>   c. afferent arteriole → glomerulus → efferent arteriole → peritubular capillary
>   d. glomerulus → afferent arteriole → efferent arteriole → peritubular capillary
>
> 5. Imagine a sodium ion that has been filtered out of the glomerulus. Describe the path it will follow to be excreted in the urine. At which points in the pathway could it leave the tubule?
>
> 6. Name two substances you would expect to find in the blood, but not the tubular fluid.
>
> *Answers in Appendix A.*

16.3 Regulatory Functions of the Kidneys

7. Describe how the kidneys help maintain the fluid and electrolyte balance of blood.

8. Name and explain how three hormones—aldosterone, antidiuretic hormone, and atrial natriuretic hormone—work together to maintain blood volume and pressure.

9. Describe three mechanisms, including how the kidneys function, to maintain the acid–base balance of blood.

The kidneys are involved in maintaining the blood's fluid and electrolyte balance, and also the acid–base balance. If the kidneys fail to carry out these vital functions, either hemodialysis or a kidney transplant is needed.

Fluid and Electrolyte Balance

The average adult male body is about 60% water by weight. The average adult female body is only about 50% water by weight because females generally have more subcutaneous adipose tissue, which contains less water. About two-thirds of this water is inside the cells (called *intracellular fluid*), and the rest is largely distributed in the plasma, tissue fluid, and lymph (called *extracellular fluid*). Water is also present in such fluids as cerebrospinal fluid and synovial fluid; in Figure 16.6, these fluids are referred to as "other" fluids.

For body fluids to be normal, it is necessary for the body to be in fluid balance. The total water intake should equal the total water loss. Table 16.2 shows how water enters the body—namely, in liquids we drink, in foods we eat, and as a by-product of metabolism. We drink water when the hypothalamus detects an increase in blood osmolarity. Table 16.2 also shows how water exits the body—namely, in urine, sweat, exhaled air, and feces. Similar to the gain and loss of water, the body also gains and loses electrolytes. Despite these changes, the kidneys keep the fluid and electrolyte balance of the blood within normal limits. In this way, they also maintain the blood volume and blood pressure.

Reabsorption of Water

Because of the process of osmosis (see Chapter 3, page 52), the reabsorption of salt (NaCl) automatically leads to the reabsorption of water until the osmolarity is the same on both sides of a plasma membrane. Most of the salt, and therefore water, present in the

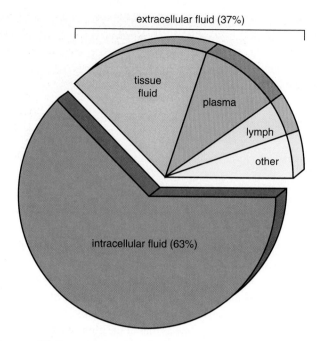

Figure 16.6 Location of fluids in the body. Most of the body's water is inside cells (intracellular fluid), and only about one-third is located outside cells (extracellular fluid).

filtrate are reabsorbed across the plasma membranes of the cells lining the proximal convoluted tubule. But the amount of salt and water reabsorbed is not sufficient to result in a hypertonic urine—one in which the osmolarity is higher than that of blood. How is it, then, that humans produce a hypertonic urine? We now know that the excretion of a hypertonic urine is dependent upon the reabsorption of water from the loop of Henle and the collecting duct.

Loop of Henle and Collecting Duct

A loop of Henle has a descending limb (*descending* through the medulla toward the center of the kidney, in the area near the renal pelvis) and an ascending limb (*ascending* toward the renal cortex). The kidney's longest loops of Henle penetrate deep into the center of the renal medulla (just adjacent to the renal pelvis). The ascending limb has two portions that are structurally different from each other. Salt (NaCl) passively diffuses out of the narrow, first portion

---

TABLE 16.2 Fluid Balance

Water Input	Average ml/day and % of Total	Water Output	Average ml/day and % of Total
In liquids	1,500; 60%	In urine	1,300; 52%
In moist food	750; 30%	In sweat	650; 26%
From metabolism	250; 10%	In exhaled air	450; 18%
		In feces	100; 4%
	Total 2,500; 100%		Total 2,500; 100%

of the ascending loop of Henle because it has a thinner wall. The wide, thick-walled second portion actively transports salt into the tissue of the renal medulla (Fig. 16.7). As the second portion of the loop continues to transport salt, less and less salt is available in the fluid moving through it. Therefore, the concentration of salt is greater in the center of the medulla than in the outer rim of the medulla. It's important to realize that water cannot leave the thick segment of the ascending limb because the ascending limb is impermeable to water.

The large arrow on the left side of Figure 16.7 indicates that the central portion of the medulla (closest to the renal pelvis) has the highest concentration of solutes. You can see that this is due not only to the presence of salt, but also to the presence of urea. Urea is reabsorbed from the lower portion of the collecting duct, and it is this molecule that contributes to the high solute concentration of the lowest portion of the inner medulla.

Because of the osmotic gradient within the renal medulla, water leaves the descending limb along its entire length. There is a

higher concentration of water in the first portion of the descending limb. Thus, less solute is needed for water to be able to diffuse into the renal medulla. The remaining fluid within the descending limb encounters an increasingly greater osmotic concentration of solute as it moves along, so that water continues to leave the descending limb along its entire length. Such a mechanism is called a *counter-current mechanism.*

At the top of the ascending limb, any remaining water enters the distal convoluted tubule. Surprisingly, the fluid inside the nephron is still not hypertonic—the net effect of reabsorption of salt and water so far is the production of a fluid that has the same tonicity as blood plasma. However, the collecting duct also encounters the same osmotic gradient as did the descending limb of the loop of Henle (Fig. 16.7). Therefore, water diffuses out of the collecting duct into the renal medulla, and the urine within the collecting duct becomes hypertonic to blood plasma.

As you can see in Figures 16.4 and 16.5, the renal tubule is surrounded by peritubular capillaries. As blood flows through these capillaries, water will naturally diffuse from high concentration (in the blood) to low concentration (in the renal medulla) because the medulla has high urea and salt concentrations. How is it, then, that the high particle concentration found in the medulla isn't simply "washed away"? A specialized, U-shaped peritubular capillary, the **vasa recta,** parallels the loop of Henle and prevents this from happening. As it passes alongside the loop, the vasa recta collects both salt and water in equal amounts, forming plasma that is quickly transported out of the kidney and back to the circulation. This helps to minimize loss of solutes from the medulla.

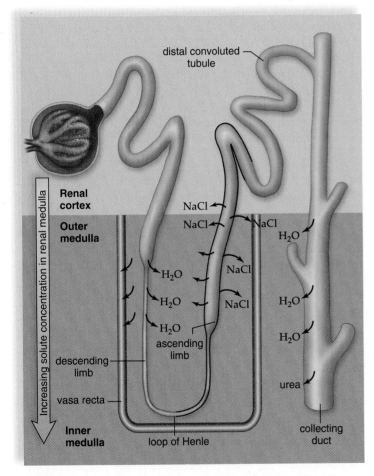

AP|R **Figure 16.7 Reabsorption of water at the loop of Henle and the collecting duct.** A hypertonic environment in the tissues of the medulla of a kidney draws water out of the descending limb and the collecting duct. This water is returned to the cardiovascular system. The thick black line means the ascending limb is impermeable to water. In the distal convoluted tubule and collecting duct, water permeability is hormone dependent.

Antidiuretic Hormone (ADH)

Antidiuretic hormone (ADH) produced by the hypothalamus, and released by the posterior lobe of the pituitary, plays a role in water reabsorption from the distal convoluted tubule and collecting duct. In order to understand the action of this hormone, consider its name. *Diuresis* means flow of urine, and *antidiuresis* means against a flow of urine. In practical terms, if an individual does not drink much water on a given day, the hypothalamus will detect an increase in the osmotic pressure of the blood (i.e., the solutes in the blood will become more concentrated). As an analogy, imagine leaving a glass of salt water on a windowsill for a week or more. The salt will become more concentrated over time because water is lost by evaporation. In the same way, blood solutes become more concentrated if the water lost daily isn't replaced by drinking. In response to increased blood osmolarity, the hypothalamus will produce ADH, which is then released from the posterior pituitary. ADH is transported to the cells of the collecting duct. There, it causes specialized cell membrane channels called *aquaporins* to open. Water diffuses from the collecting duct to the interstitial space of the kidney through these aquaporins. Thus, more water is reabsorbed into the blood, blood volume and blood pressure rise, and less urine is formed. On the other hand, if an individual drinks a large amount of water and does not perspire much, ADH is not released. Collecting duct aquaporins remain closed, more urine forms, and more water is excreted.

Reabsorption of Electrolytes

As previously discussed, the osmolarity of body fluids is dependent upon the concentration of particular electrolytes within the fluids. **Electrolytes** are compounds and molecules that are able to ionize and, thus, carry an electrical current. The kidneys regulate electrolyte excretion and therefore help control blood composition.

The Electrolytes

The most common electrolytes in the plasma are sodium (Na^+), chloride (Cl^-), and bicarbonate (HCO_3^-). The most common electrolytes in the intracellular fluid are potassium (K^+) and phosphate (PO_4^{-3}). Na^+ and K^+ are termed *cations* because they are positively charged. Cl^-, HCO_3^-, and PO_4^{-3} are termed *anions* because they are negatively charged (see Chapter 2).

Sodium You'll remember that the movement of Na^+ across an axon membrane is necessary to the formation of a nerve action potential signal and also to muscle contraction. The concentration of Na^+ in the blood is also the best indicator of the blood's osmolarity.

Potassium The movement of K^+ across an axon membrane is also necessary to the formation of a nerve action potential and muscle contraction. Abnormally low K^+ concentrations in the blood, as might occur if diuretics are abused, can lead to cardiac arrest.

Bicarbonate ion HCO_3^- is the form in which carbon dioxide is carried in the blood. The bicarbonate ion has the very important function of helping maintain the pH of the blood, as will be discussed later in this section.

Other ions The plasma contains many other ions. For example, calcium ions (Ca^{2-}) and monohydrogen phosphate ions (HPO_4^{2-}) are important to bone formation and cellular metabolism. Their absorption from the intestine and excretion by the kidneys is regulated by hormones, as discussed in Chapter 10.

The Kidneys

More than 99% of sodium (Na^+) filtered at the glomerulus is returned to the blood. Most sodium (67%) is reabsorbed at the proximal convoluted tubule, and a sizable amount (25%) is actively transported from the tubule by the ascending limb of the loop of Henle. The rest is reabsorbed from the distal convoluted tubule and collecting duct.

Aldosterone

Hormones control the reabsorption of sodium and water in the distal portion of the renal tubule. **Aldosterone,** a hormone secreted by the adrenal cortex, promotes the excretion of potassium ions (K^+) and the reabsorption of sodium ions (Na^+). Water is subsequently reabsorbed by osmosis. Aldosterone acts on the ascending limb of the loop of Henle, the distal convoluted tubule, and the collecting ducts.

The release of aldosterone is set in motion by the kidneys themselves. The **juxtaglomerular apparatus** is a region of contact between the afferent arteriole and the distal convoluted tubule (Fig. 16.8). When blood volume, and therefore blood pressure, is not sufficient to promote glomerular filtration, the juxtaglomerular apparatus secretes renin. **Renin** is an enzyme that changes angiotensinogen (a large plasma protein produced by the liver) into angiotensin I. Later, angiotensin I is converted in the lungs to angiotensin II, a powerful vasoconstrictor that also stimulates the adrenal cortex to release aldosterone. The reabsorption of sodium ions is followed by the reabsorption of water. Therefore, blood volume and blood pressure increase.

Atrial Natriuretic Hormone (ANH)

Atrial natriuretic hormone (ANH—sometimes referred to as *atrial natriuretic peptide*) is a hormone secreted by the atria of the heart when cardiac cells are stretched due to increased blood volume. ANH inhibits the secretion of renin by the juxtaglomerular apparatus and the secretion of aldosterone by the adrenal cortex. Its effect, therefore, is to promote the excretion of Na^+, called *natriuresis*. When Na^+ is excreted, so is water, and therefore blood volume and blood pressure decrease.

Diuretics

Diuretics are chemicals that increase the flow of urine. Drinking alcohol causes diuresis because it inhibits the secretion of ADH. The dehydration that follows is believed to contribute to the symptoms of a hangover. Caffeine is a diuretic because it increases the glomerular filtration rate and decreases the tubular reabsorption of Na^+. Diuretic drugs developed to counteract high blood pressure cause water reabsorption to diminish. Subsequently, blood pressure decreases because cardiac output is diminished.

Begin Thinking Clinically

Two categories of diuretic drugs are separately classified as *loop diuretics* and *aldosterone antagonists*. Where in the kidney tubule would each of these function? How does each category work? Can you name a trade/generic version for each?

Answers and discussion in Appendix A.

Acid–Base Balance

The pH scale, as discussed in Chapter 2, can be used to indicate the basicity (alkalinity) or the acidity of body fluids. A basic solution has a lower hydrogen ion concentration [H^+] than the neutral pH of 7.0. An acidic solution has a greater [H^+] than neutral pH. The normal pH for body fluids ranges between 7.35 and 7.45. This is the pH at which our proteins, such as cellular enzymes, function properly. If the blood pH rises above 7.45, a person is said to have **alkalosis,** and if the blood pH decreases below 7.35, a person is said to have **acidosis.** If severe, alkalosis and acidosis are conditions that will need medical attention.

The foods we eat add basic or acidic substances to the blood, and so does metabolism. For example, cellular respiration adds carbon dioxide that combines with water to form carbonic acid, and fermentation adds lactic acid. The pH of body fluids stays within the normal range via several mechanisms, primarily acid–base buffer systems, the respiratory center, and the kidneys.

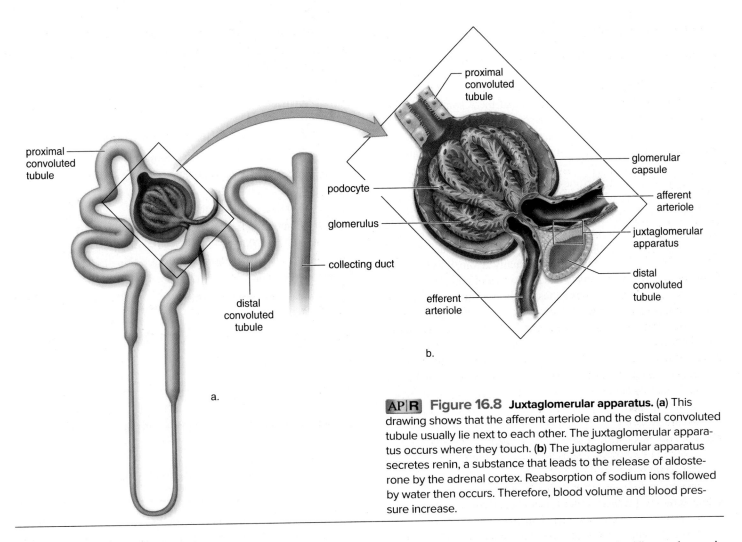

proximal convoluted tubule

distal convoluted tubule

collecting duct

a.

proximal convoluted tubule

podocyte

glomerulus

glomerular capsule

afferent arteriole

juxtaglomerular apparatus

distal convoluted tubule

efferent arteriole

b.

AP|R **Figure 16.8** **Juxtaglomerular apparatus.** (a) This drawing shows that the afferent arteriole and the distal convoluted tubule usually lie next to each other. The juxtaglomerular apparatus occurs where they touch. (b) The juxtaglomerular apparatus secretes renin, a substance that leads to the release of aldosterone by the adrenal cortex. Reabsorption of sodium ions followed by water then occurs. Therefore, blood volume and blood pressure increase.

Acid–Base Buffer Systems

Normally, the pH of the blood stays near 7.4 because the blood is buffered. A **buffer** is a chemical or a combination of chemicals that can take up excess hydrogen ions (H^+) or excess hydroxide ions (OH^-). Proteins are the primary buffers of both the intracellular and extracellular fluid. Their amino acids can combine with or release hydrogen ions as needed. One of the most important buffers in the blood is a combination of carbonic acid (H_2CO_3) and bicarbonate ions (HCO_3^-). Carbonic acid is a weak acid that minimally dissociates and re-forms in the following manner:

$$\underset{\substack{\text{carbonic} \\ \text{acid}}}{H_2CO_3} \underset{\text{re-forms}}{\overset{\text{dissociates}}{\rightleftharpoons}} \underset{\substack{\text{hydrogen} \\ \text{ion}}}{H^+} + \underset{\substack{\text{bicarbonate} \\ \text{ion}}}{HCO_3^-}$$

When hydrogen ions (H^+) are added to blood, the following reaction occurs:

$$H^+ + HCO_3^- \longrightarrow H_2CO_3$$

When hydroxide ions (OH^-) are added to blood, this reaction occurs:

$$OH^- + H_2CO_3 \longrightarrow HCO_3^- + H_2O$$

These reactions temporarily prevent any significant change in blood pH. A blood buffer, however, can be overwhelmed unless some more permanent adjustment is made. The next adjustment to keep the pH of the blood constant occurs at pulmonary capillaries.

Respiratory Regulation of Acid–Base Balance

As discussed in Chapter 14, the respiratory center in the medulla oblongata increases the breathing rate, as well as the depth of respiration, if the hydrogen ion concentration of the blood rises. This change rids the body of hydrogen ions because the following reaction takes place in pulmonary capillaries:

$$H^+ + HCO_3^- \rightleftharpoons H_2CO_3 \rightleftharpoons H_2O + CO_2$$

In other words, when CO_2 is exhaled, H^+ ions recombine with HCO_3^- ions to form carbonic acid. Carbonic acid then dissociates, forming more carbon dioxide to be exhaled. H^+ ions are now no longer free in solution, but tied up harmlessly with water. Increasing respiratory rate provides a rapid way to remove excess free H^+ ions from blood. In contrast, when blood H^+ falls, respiration decreases (hypoventilation), allowing CO_2 to build up. More H^+ is produced and balance is restored.

It is important to have the correct proportion of carbonic acid and bicarbonate ions in the blood. Breathing readjusts this proportion so that this particular acid–base buffer system can continue to absorb both H^+ and OH^- as needed.

Renal Regulation of Acid–Base Balance

As powerful as the acid–base buffer and the respiratory center mechanisms are, only the kidneys can rid the body of a wide range of acidic and basic substances and otherwise adjust the pH. The kidneys are slower acting than the other two mechanisms, but they have a more powerful effect on pH. For the sake of simplicity, we can think of the kidneys as reabsorbing bicarbonate ions and excreting hydrogen ions as needed to maintain the normal pH of the blood (Fig. 16.9). If the blood is acidic, hydrogen ions are excreted and bicarbonate ions are reabsorbed. If the blood is basic, hydrogen ions are not excreted and bicarbonate ions are not reabsorbed. Because the urine is usually acidic, it follows that an excess of hydrogen ions is usually excreted. Ammonia (NH_3) provides a means for buffering these hydrogen ions in urine. Ammonia (whose presence is quite obvious in the diaper pail or kitty litter box) is produced by the deamination of amino acids in both liver and kidney tubule cells. Ammonia is a basic molecule that diffuses easily through tubule cells into the urine. Once there, it combines with hydrogen ions to form ammonium ions (NH_4^+) by this reaction:

$$NH_3 + H^+ \longrightarrow NH_4^+$$

Ammonium ions cannot diffuse back out of the kidney tubules and pass out of the body in the urine. Phosphate provides another means of buffering hydrogen ions in urine.

The importance of the kidneys' ultimate control over the pH of the blood cannot be overemphasized. As mentioned, the enzymes of cells cannot continue to function if the internal environment does not have near-normal pH.

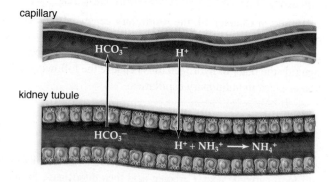

Figure 16.9 **Acid–base balance.** In the kidneys, bicarbonate ions (HCO_3^-) are reabsorbed and hydrogen ions (H^+) are excreted as needed to maintain the pH of the blood. Excess hydrogen ions are buffered, for example, by ammonia (NH_3), which becomes ammonium (NH_4^+). Ammonia is produced in tubule cells by the deamination of amino acids.

7. Which of the following statements about body water is correct?
   a. Males have a higher percentage of body water by weight than females.
   b. Most of the body's water is found inside cells.
   c. Water can be lost from the body in exhaled air.
   d. A portion of daily water intake comes from water produced during metabolism.
   e. All of these statements are correct.

8. Which hormone is produced by the hypothalamus and released from the posterior pituitary?
   a. atrial natriuretic hormone (ANH)
   b. aldosterone
   c. antidiuretic hormone (ADH)
   d. angiotensin

9. Compare and contrast the effects of aldosterone and atrial natriuretic hormone on blood pressure.

Answers in Appendix A.

16.4 Problems with Kidney Function

10. State, in general, the normal composition of urine and the benefits of doing a urinalysis.
11. Discuss the need for hemodialysis and how hemodialysis functions to bring about the normal composition of urine.

The composition of normal urine is given in Table 16.3. Water accounts for almost all of the volume of urine (95%). The remaining 5% consists of electrolytes and various solutes, including nitrogenous end products and substances derived from drugs. Notice that

TABLE 16.3 Composition of Urine

Water		95%
Solids		5%
Organic nitrogenous wastes (per 1,500 ml of urine)		
Urea		30 g
Creatinine		1–2 g
Ammonia		1–2 g
Uric acid		1 g
Electrolytes		25 g
Positive	*Negative*	
Sodium (Na^+)	Chlorides (Cl^-)	
Potassium (K^+)	Sulfates (SO_4^{2-})	
Magnesium (Mg^{2+})	Phosphates (PO_4^{3-})	
Calcium (Ca^{2+})		

urine is typically free of proteins and blood cells because they are not filtered at the glomerulus.

Urinalysis is an examination of the physical, chemical, and microscopic properties of the urine. A urinalysis is done to help determine the state of the body. As discussed in the Focus on Forensics on page 386, the composition of the urine changes if disease has altered body metabolism or if kidney function is abnormal. Abnormal substances in urine and abnormal quantities of normal constituents are both matters of concern.

Many types of illnesses, especially diabetes, hypertension, and inherited conditions, cause progressive renal disease and renal failure. Infections of the urinary tract are fairly common, particularly in females because the urethra is considerably shorter than that of the male. If the infection is localized in the urethra, it is called **urethritis.** If the infection invades the urinary bladder, it is called **cystitis.** Finally, if the kidneys are affected, the infection is called **pyelonephritis.**

Renal calculi are commonly referred to as kidney stones, and most are made from calcium compounds (salts). Most kidney stone sufferers have an inherited condition in which their urine contains excessively high calcium levels. Kidney stones can be extremely painful, and if large enough, can cause permanent kidney damage. If they cannot be safely passed, kidney stones can be broken into smaller bits using sound waves, in a technique called *extracorporeal shock-wave lithotripsy.* In some cases, kidney stones must be surgically removed.

Glomerular damage sometimes leads to blockage of the glomeruli so that glomerular filtration either does not occur or allows large substances to pass through. This is detected when a urinalysis is done. If the glomeruli are too permeable, albumin, white blood cells, or even red blood cells appear in the urine. A trace amount of protein in the urine is usually not a matter of concern.

When glomerular damage is so extensive that more than two-thirds of the nephrons are inoperative, urea and other waste substances accumulate in the blood. This condition is called **uremia.** Although nitrogenous wastes can cause serious damage, the retention of water and salts is of even greater concern. The latter causes edema, fluid accumulation in the body tissues. Imbalance in the ionic composition of body fluids can lead to loss of consciousness and to heart failure.

Hemodialysis

Patients with renal failure can undergo **hemodialysis,** utilizing either an artificial renal dialysis machine or *continuous ambulatory peritoneal dialysis* (CAPD). *Dialysis* is defined as the diffusion of dissolved molecules through a semipermeable natural or synthetic membrane having pore sizes that allow only small molecules to pass through. In an artificial renal dialysis machine (Fig. 16.10), the patient's blood is passed through a membranous tube, which is in contact with a dialysis solution, or **dialysate.** Substances more concentrated in the blood diffuse into the dialysate, and substances more concentrated in the dialysate diffuse into the blood. The dialysate is continuously replaced to maintain favorable concentration gradients. In this way, the artificial kidney can be utilized either to extract substances from blood, including waste products or toxic chemicals and drugs, or to add substances to blood—for example, bicarbonate ions (HCO_3^-) if the blood is acidic. In the course of a three- to six-hour hemodialysis, from 50 to 250 grams of urea can be removed from a patient, which greatly exceeds the amount excreted by normal kidneys. Therefore, a patient needs to undergo treatment only about twice a week.

CAPD is so named because the peritoneal lining of the abdominal cavity is the dialysis membrane. A fresh amount of dialysate is introduced directly into the abdominal cavity from a bag that is temporarily attached to a permanently implanted plastic tube. The dialysate flows into the peritoneal cavity by gravity. Waste and

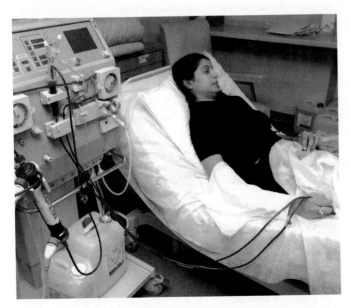
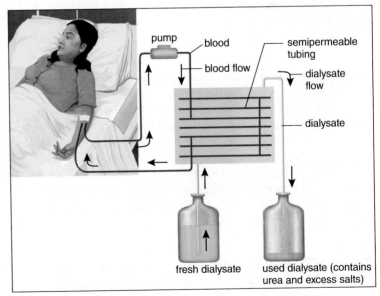

Figure 16.10 An artificial kidney machine. As the patient's blood is pumped through dialysis tubing, it is exposed to a dialysate (dialysis solution). Wastes exit from blood into the solution because of a preestablished concentration gradient. In this way, blood is not only cleansed, but its water–salt and acid–base balances can also be adjusted.

Since ancient times, urinalysis has been used to diagnose disease. As early as 600 B.C., Hindu physicians in India noted that the urine of a diabetic was sweet to the taste. In diabetes mellitus, blood glucose is abnormally high, either because insulin-secreting cells have been destroyed or because cell receptors don't respond to the insulin that is present. Thus, the filtrate level of glucose is extremely high, and the proximal convoluted tubule can't reabsorb it all. Tasting urine to diagnose diabetes persisted through the 1800s (thankfully, modern techniques have made it obsolete). Similarly, the great Greek physician Hippocrates studied and wrote about urinalysis, producing perhaps the first written work about renal failure. Hippocrates noted that shaking the urine of renal failure patients produced frothy bubbles at the surface of the sample. Today, we know that the froth is a sign of *proteinuria,* or protein in the urine. Proteinuria indicates that the glomerulus is more permeable than normal, and is an early indicator of chronic renal failure.

Today, the use of urinalysis has expanded beyond medical applications to include forensic diagnosis of drug abuse. Screening for illegal drug use is now mandated by federal and state agencies as a condition of employment, and most private employers now require it as well. Urinalysis can be court ordered if drug abuse is involved in the commission of a crime. The National Collegiate Athletic Association requires all student athletes to undergo testing, as do many high school athletic programs.

Urinalysis is not used to screen for the drugs themselves, but for drug metabolites—the breakdown products of drugs that are consumed or injected. Once in the body, drugs are metabolized by the liver and filtered by the kidney. Thus, metabolites will be present in the urine of a drug abuser. Two types of techniques can be used for metabolite detection. The first, a screening exam, involves a test strip placed into freshly voided urine. The test strip contains monoclonal antibodies (see pages 306 and 312) specific for metabolites of street drugs. Strips can test for 12 or more different drugs at once, but most screen for five commonly abused drugs: marijuana, amphetamine, PCP, cocaine, and opiates such as heroin. Urine strip testing will give results within minutes, but certain legal, over-the-counter medications can give false-positive results. If a sample tests positive, it can be immediately sent for a sophisticated chemical analysis such as gas chromatography. Tracking long-term drug use may require using hair samples as well as urine samples because drug metabolites can be incorporated into the hair.

The urine specimen must be properly collected to avoid possible tampering by the individual being tested. Specimens can be altered by simple dilution with sink or toilet water, or by contaminating the sample with any number of additives. Bleach, drain cleaner, soft drinks, etc., can be used, as well as products specifically sold for the express purpose of "helping to beat a drug screen." Drug abusers may also attempt to substitute the urine of a "clean" individual for their own. Tampering may be prevented by simply requiring the presence of a witness at all times while the sample is being collected and stored. In addition, proper documentation must accompany any urine sample. Chain-of-custody forms record each step of the handling of a specimen, from collection to disposal. This provides proof of everything that happens to the specimen, and prevents the possibility of the specimen being rejected as evidence in court proceedings.

The goal of any forensic urinalysis testing program should go beyond mere *detection* of illegal drug use. The more important aim of screening should be *intervention.* With proper treatment and counseling, drug abusers can overcome addiction and lead healthier and more productive lives.

---

salt molecules pass from the blood vessels in the abdominal wall into the dialysate before the fluid is collected four or eight hours later. The solution is drained into a bag from the abdominal cavity by gravity, and then it is discarded. One advantage of CAPD over an artificial kidney machine is that the individual can go about his or her normal activities during CAPD. However, CAPD is not appropriate for all patients with renal failure. Because the abdominal cavity has a permanent opening, the patient must be carefully trained and very diligent with sterile technique to avoid causing abdominal infection (peritonitis).

Renal Transplantation

Patients with renal failure sometimes undergo a kidney transplant operation, during which a functioning kidney from a donor is received. As with all organ transplants, there is the possibility of organ rejection. Receiving a kidney from a close relative has the highest chance of success. The current one-year survival rate is 97% if the kidney is received from a relative and 90% if it is received from a nonrelative. Even with a successful surgery, the transplant patient must take antirejection medication for the rest of his or her life.

Content CHECK-UP!

10. Which are not normally found in urine? Choose all that apply.

    a. hydrogen ions   d. urochrome

    b. proteins   e. creatinine

    c. glucose

11. Dialysis is used to treat:

    a. diabetes mellitus.   c. kidney failure.

    b. high blood pressure.   d. kidney stones.

Answers in Appendix A.

MEDICAL FOCUS

Prostate Enlargement and Prostate Cancer

The prostate gland, which is part of the male reproductive system, surrounds the urethra at the point where the urethra leaves the urinary bladder. The prostate gland produces and adds a fluid to semen as semen passes through the urethra within the penis. At about age 50, a man's prostate often begins to enlarge, growing from the size of a walnut to that of a lime or even a lemon. This condition is called **benign prostatic hyperplasia (BPH).** Prostate enlargement is due to a prostate enzyme called 5-alpha reductase. This enzyme acts on the male hormone testosterone, converting it into a substance that promotes prostate growth. That growth is normal during puberty, but continued growth in an adult is undesirable. As it enlarges, the prostate squeezes the urethra, causing urine to back up—first into the bladder, then into the ureters, and finally, perhaps, into the kidneys. If the bladder can't be completely emptied, the patient is at risk for infection and formation of kidney stones.

The treatment for BPH begins by taking dietary supplements or drugs that are expected to shrink the prostate and/or improve urine flow. The first, saw palmetto, is sold in tablet form as an over-the-counter nutrient supplement. It is believed to interfere with the action of 5-alpha reductase. It should not be taken without medical supervision, but it is particularly effective during the early stages of prostate enlargement. The prescription drugs finasteride and dutasteride are more powerful inhibitors of the enzyme, but erectile dysfunction and loss of libido are common side effects. Drugs called alpha-adrenergic blockers can also

be used, and there are several belonging to this category: terazosin, doxazosin, tamsulosin, alfuzosin. All relax muscle tissue in the prostate. However, these drugs do not shrink the prostate. Often, treatment involves using combinations of drugs from each category.

If medication fails to help relieve the symptoms of BPH, a more invasive procedure can reduce the size of the prostate. Prostate tissue can be destroyed by applying microwaves to a specific portion of the prostate, or by using laser or ultrasound. In many cases, however, a physician may decide that prostate tissue should be removed surgically. Rather than performing abdominal surgery, which requires an incision in the abdomen, the physician can often access the prostate via the urethra in an operation called transurethral resection of the prostate (TURP). This commonly used procedure is successful in approximately 97% of patients.

Many men are concerned that BPH may be associated with prostate cancer, but the two conditions are not necessarily related. BPH occurs in the inner zone of the prostate, while cancer tends to develop in the outer area. If prostate cancer is suspected, blood tests and a biopsy, in which a tiny sample of prostate tissue is surgically removed, will confirm the diagnosis.

Although prostate cancer is the second most common cancer in men, it is not a major killer. Typically, the most common form of prostate cancer is so slow growing that the survival rate is about 98% if the condition is detected early. Treatment typically consists of surgery to remove the prostate, along with radiation therapy and/or chemotherapy.

16.5 Effects of Aging

12. Describe the anatomical and physiological changes that occur in the urinary system as we age.

Urinary disorders are significant causes of illness and death among the elderly. Total renal function in an elderly individual may be only 50% of that of the young adult. With increasing age, the kidneys decrease in size and have significantly fewer nephrons. However, vascular changes may play a more significant role in declining renal efficiency than renal tissue loss. Microscopic examination shows many degenerate glomeruli through which blood no longer flows and many other glomeruli that are completely destroyed.

Kidney stones occur more frequently with age, possibly as a result of improper diet, inadequate fluid intake, and kidney infections. Infections of the urethra, bladder, ureters, and kidneys increase in frequency among the elderly. Enlargement of the prostate occurs in males and, as is discussed in the Medical Focus, Prostate Enlargement and Prostate Cancer, this can lead to urine retention and kidney disease. Cancer of the prostate and bladder are the most common cancers of the urogenital system.

The involuntary loss of urine, called **incontinence,** increases with age. The bladder of an elderly person has a capacity of less

than half that of a young adult and often contains residual urine. Therefore, urination is more urgent and frequent.

Content CHECK-UP!

12. Many changes occur in the kidneys as people age. The change that seems to affect renal efficiency most is:

   a. degenerating glomeruli.

   b. improper diet and fluid intake.

   c. kidney stones.

   d. prostate problems.

Answer in Appendix A.

16.6 Homeostasis

13. Describe how the urinary system works with other systems of the body to maintain homeostasis.

The illustration in Human Systems Work Together on page 388 tells how the urinary system works with the other systems of the body to help maintain homeostasis.

Integumentary System

Kidneys compensate for water loss due to sweating; activate vitamin D precursor made by skin.

Skin helps regulate water loss; sweat glands carry on some excretion.

Skeletal System

Kidneys provide active vitamin D for Ca^{2+} absorption and help maintain blood level of Ca^{2+}, needed for bone growth and repair.

Bones provide support and protection.

Muscular System

Kidneys maintain blood levels of Na^+, K^+, and Ca^{2+}, which are needed for muscle contraction, and eliminate creatinine, a muscle waste.

Smooth muscular contraction assists voiding of urine; skeletal muscles support and help protect urinary organs.

Nervous System

Kidneys maintain blood levels of Na^+, K^+, and Ca^{2+}, which are needed for nerve conduction.

Brain controls nerves, which innervate muscles that permit urination.

Endocrine System

Kidneys keep blood values within normal limits so that transport of hormones continues.

ADH, aldosterone, and atrial natriuretic hormone regulate reabsorption and secretion of Na^+ and K^+ by the kidney.

How the Urinary System works with other body systems.

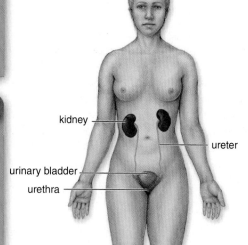

kidney

ureter

urinary bladder

urethra

Reproductive System

Semen is discharged through the urethra in males; kidneys excrete wastes and maintain electrolyte levels for mother and child.

Penis in males contains the urethra and performs urination; prostate enlargement hinders urination.

Cardiovascular System

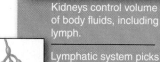

Kidneys filter blood and excrete wastes; maintain blood volume, pressure, and pH; produce renin and erythropoietin.

Blood vessels deliver waste to be excreted; blood pressure aids kidney function; heart produces atrial natriuretic hormone.

Lymphatic System/Immunity

Kidneys control volume of body fluids, including lymph.

Lymphatic system picks up excess tissue fluid, helping to maintain blood pressure for kidneys to function; immune system protects against infections.

Respiratory System

Kidneys compensate for water lost through respiratory tract; work with lungs to maintain blood pH.

Lungs excrete carbon dioxide, provide oxygen, and convert angiotensin I to angiotensin II, leading to kidney regulation.

Digestive System

Kidneys convert vitamin D to active form needed for Ca^{2+} absorption; compensate for any water loss by digestive tract.

Liver synthesizes urea; digestive tract excretes bile pigments from liver and provides nutrients.

Recall that *excretion* means to rid the body of a metabolic waste. Using this definition, it is possible to classify three other organs in addition to the kidneys as excretory organs:

1. The sweat glands in the skin excrete perspiration, which is a solution of water, salt, and some urea. Although perspiration is a form of excretion, we perspire more to cool the body than to rid it of wastes. In times of kidney failure, urea is excreted by the sweat glands and forms a so-called *uremic frost* on the skin.

2. The liver breaks down hemoglobin and excretes bile pigments, which are derived from heme. Bile pigments are incorporated into bile, a substance stored in the gallbladder before it passes into the small intestine by way of ducts. The yellow pigment found in urine, called urochrome, also is derived from the breakdown of heme, but this pigment is deposited in blood and is subsequently excreted by the kidneys. As you know, the liver also produces urea—our main nitrogenous end product—which is excreted by the kidneys.

3. The lungs excrete carbon dioxide. The process of exhalation not only removes carbon dioxide, but also results in the loss of water. The air we exhale contains moisture, as demonstrated by breathing onto a cool mirror.

The kidneys are the primary organs of excretion. They excrete almost all of our nitrogenous wastes—namely urea, creatinine, and uric acid. The liver makes urea, and muscles make creatinine. Urea is the end product of protein metabolism, and creatinine is a breakdown product of creatine phosphate, a molecule that stores energy in muscles. Uric acid is produced by cells when they break down nucleotides. Nitrogenous wastes are carried by the cardiovascular system to the kidneys. In this way, the cardiovascular system and the kidneys work together to clear the blood of nitrogenous end products. The excretion of nitrogenous wastes may not be as critical as maintaining the water–salt and the acid–base balances, but it is still necessary because urea is a toxic substance.

The kidneys are the primary organs of homeostasis because they maintain the water–electrolyte and the acid–base balances of the blood. If blood does not have the usual water–salt balance, blood volume and blood pressure are affected. Without adequate blood pressure, exchange across capillary walls cannot take place, nor is glomerular filtration possible in the kidneys themselves. The kidneys and the endocrine system work together to help maintain blood pressure. The production of renin by the kidneys and subsequently the renin-angiotensin-aldosterone sequence help to ensure that the sodium (Na^+) concentration of the blood, and therefore osmolarity and blood pressure, stay normal. The lymphatic system assists the urinary system because it makes a significant contribution to blood pressure by picking up excess tissue fluid and returning it to the cardiovascular veins.

Aside from producing renin, the kidneys assist the endocrine system and also the cardiovascular system by producing erythropoietin. Erythropoietin stimulates red bone marrow to produce red blood cells. The kidneys assist the skeletal, nervous, and muscular systems by helping to regulate the amount of calcium ions (Ca^{2+}) in the blood. The kidneys convert vitamin D to its active form needed for Ca^{2+} absorption by the digestive tract, and they regulate the excretion of electrolytes, including Ca^{2+}. The kidneys also regulate the sodium (Na^+) and potassium (K^+) content of the blood. These ions are necessary to the contraction of the heart and other muscles in the body, and are also needed for nerve conduction.

We have already described how the kidneys work with the cardiovascular system and the respiratory system to maintain the acid–base balance (pages 382–384). This is a critical function to prevent the occurrence of life-threatening alkalosis or acidosis. This function must be performed by a machine when people undergo hemodialysis. So, while we tend to remember that the kidneys excrete urea, we must also keep in mind all the other functions of the kidneys that are absolutely essential to homeostasis.

Content **CHECK-UP!**

13. List three organs (other than urinary organs) that excrete wastes. What substances do they excrete?

14. How do the kidneys act as endocrine organs?

15. The kidneys maintain homeostatic balance for three ions that are critical for survival. What are they?

Answers in Appendix A.

Selected New Terms

Basic Key Terms

aldosterone (ăl-dŏs′tŭh-rōn), p. 382

antidiuretic hormone (ADH) (ăn″tĭ-dī′ŭ-rĕt′ĭk hŏr′mōn), p. 381

atrial natriuretic hormone (ANH) (ā′trē-ăl nā″trē-yŭ-rĕt′ĭk hŏr′mōn), p. 382

buffer (bŭf′ĕr), p. 383

collecting duct (kŭh-lĕk′tĭng dŭkt), p. 376

creatinine (krē-ăt′ĭ-nēn), p. 373

distal convoluted tubule (dĭs′tăl kŏn′vō-lūt-ĕd tū′byūl), p. 376

electrolytes (ē-lĕk′trŭh-lītz′), p. 382

erythropoietin (ē-rĭth″rō-pōy-ē′tĭn), p. 373

excretion (ĕks-krē′shŭn), p. 373

glomerular capsule (glō-mār′yŭ-lĕr kăp′sūl), p. 375

glomerular filtrate (glō-mār′yŭ-lĕr fĭl′trāt), p. 377

glomerular filtration (glō-mār′yŭ-lĕr fĭl-trā′shŭn), p. 377

glomerulus (glō-mār′yŭ-lŭs), p. 375

hilum (hī′lŭm), p. 374

juxtaglomerular apparatus (jŭks″tŭh-glō-mār′yŭ-lĕr ăp″ŭh-rā′tŭs), p. 382

kidney (kĭd′nē), p. 374

loop of Henle (lūp ŭv Hĕn′lē), p. 376

micturition (mĭk″tū-rĭsh′ŭn), p. 375

nephron (nĕf′rŏn), p. 375

peritubular capillary network (pĕr″ĭ-tū′byū-lĕr kăp′ĭ-lār″ē nĕt′wĕrk), p. 375

prostate gland (prŏs′tāt glănd), p. 374

proximal convoluted tubule (prŏk′sĭ-măl kŏn′vō-lū-tĕd tū′byūl), p. 375

renal artery (rē′nŭl ăr′tĕ-rē), p. 374

renal corpuscle (rē′nŭl kŏr′pŭs-ĕl), p. 375

renal cortex (rē′nŭl kŏr′tĕks), p. 375

renal medulla (rē′nŭl mĕ-dū′lŭh), p. 375

Clinical Key Terms

Summary

16.1 Urinary System
A. The kidneys excrete nitrogenous wastes, including urea, uric acid, and creatinine. They maintain the normal water–salt balance, blood pressure, and acid–base balance of the blood, as well as influencing the secretion of certain hormones.
B. The kidneys produce urine, which is conducted by the ureters to the bladder, where it is stored before being released by way of the urethra.
C. Urine is eliminated from the body via a reflex called micturition. Stretch receptors in the wall of the urinary bladder begin this reflex, which ends in the relaxation of two urinary sphincters: the internal sphincter, made of smooth muscle, and the external sphincter, which is made of skeletal muscle.

16.2 Anatomy of the Kidney and Excretion
Macroscopically, the kidneys are divided into the renal cortex, renal medulla, and renal pelvis.
A. Microscopically, the kidneys contain nephrons. Each nephron has its own blood supply. Blood in the afferent arteriole flows into the glomerulus, a knot of capillaries. Blood in the efferent arteriole flows out of the glomerulus and immediately enters the peritubular capillary network.
B. Each region of the nephron is anatomically suited to its task in urine formation. The spaces between the podocytes of the glomerular capsule allow small molecules to enter the capsule from the glomerulus. The cuboidal epithelial cells of the proximal convoluted tubule have many mitochondria and microvilli to carry out *reabsorption*, which is active transport from the tubule to the blood. In contrast, the cuboidal epithelial cells of the distal convoluted tubule have numerous mitochondria but lack microvilli. They carry out *secretion*, which is active transport from the blood to the tubule.
C. The steps in urine formation are glomerular filtration, tubular reabsorption, and tubular secretion.

16.3 Regulatory Functions of the Kidneys
A. The kidneys regulate the fluid and electrolyte balance of the body. Water comes into the body by drinking, in food, and as a by-product of metabolism. Water leaves the body through urine, sweat, exhalation, and feces.
B. Water is reabsorbed from certain parts of the tubule, and the loop of Henle establishes an osmotic gradient that draws water from the descending loop and also from the collecting duct. The permeability of the collecting duct is under the control of the hormone ADH.
C. The reabsorption of salt increases blood volume and pressure because more water is also reabsorbed. Two other hormones, aldosterone and ANH, control the kidneys' reabsorption of sodium (Na^+) and water and the secretion of potassium (K^+).
D. The kidneys regulate the acid–base balance of the blood. Before the work of the kidneys begins, the acid–base buffer systems of the blood have functioned to keep the pH temporarily under control; also, the respiratory center has regulated the breathing rate to control the excretion of carbon dioxide at pulmonary capillaries. The kidneys largely control acid–base balance by excreting hydrogen ions and reabsorbing bicarbonate ions as needed.

16.4 Problems with Kidney Function
Various types of urinary problems, including repeated urinary infections, can lead to renal failure.
A. Hemodialysis uses machines to mimic the function of the kidneys by removing wastes from the blood. Some patients use CAPD to cleanse their blood.
B. Patients in kidney failure might undergo a kidney transplant. Success rates of this procedure are high, especially when donors and recipients have similar tissue types and immunosuppressant drugs are used.

16.5 Effects of Aging
Kidney function declines with age. Also, kidney stones, infections, and urination problems are more common.

16.6 Homeostasis
The kidneys work with other organs in the body to excrete wastes. The sweat glands of the integumentary system excrete excess water, salts, and urea. The liver excretes products of the breakdown of hemoglobin, while the lungs excrete carbon dioxide. The kidneys are the primary organs of homeostasis: they maintain water–electrolyte balance and acid–base balance. These in turn affect blood volume and blood pressure. The kidneys also act as endocrine organs to secrete renin and erythropoietin.

Study Questions

1. List and explain four functions of the urinary system. (pp. 373–374)
2. Trace the path of urine, and describe the function of each organ mentioned. (p. 374)
3. Explain how urination is controlled. (p. 375)
4. Describe the detailed anatomy of a kidney. (pp. 375–377)
5. Trace the path of blood around the nephron. (pp. 375–379)
6. Name the parts of a nephron, and tell how the structure of the convoluted tubules suits their respective functions. (pp. 375–377)
7. State and describe the three steps of urine formation. (pp. 377–379)
8. Where in particular are salt and water reabsorbed along the length of the nephron? Describe the contribution of the loop of Henle. (pp. 377–381)
9. Describe the action of antidiuretic hormone (ADH), the renin-aldosterone connection, and atrial natriuretic hormone (ANH). (pp. 381–382)
10. How do the kidneys maintain the pH of the blood within normal limits? (pp. 382–384)
11. Explain the two techniques which can be used for renal dialysis. (pp. 385–386)

Learning Outcome Questions

Fill in the blanks.

1. In addition to excreting nitrogenous wastes, the kidneys adjust the _____, _____, and _____ balance of the blood.
2. The primary nitrogenous end product of humans is _____.
3. The accumulation of uric acid crystals in a joint cavity produces a condition called _____.
4. Urine is carried from the kidneys to the urinary bladder by a pair of tubes called _____.
5. Urine leaves the bladder in the _____.
6. The outer granulated layer of the kidney is the renal _____, and the renal _____ is composed of cone-shaped tissue masses called renal _____.
7. The capillary tuft inside the glomerular capsule is called the _____.
8. _____ is a substance that is found in the filtrate, is reabsorbed, and is present in urine.
9. _____ is a substance that is found in the filtrate, is not reabsorbed, and is concentrated in urine.
10. At which parts of the nephron does tubular secretion occur? _____.
11. Reabsorption of water from the collecting duct is regulated by the hormone _____.
12. A _____ is a chemical that can combine with either hydrogen ions or hydroxide ions, depending on the pH of the solution.
13. The lungs are organs of excretion because they rid the body of _____.

Medical Terminology Exercise

After studying this chapter, see if you can derive the definitions for the medical terms listed at right. Many of the prefixes and suffixes used to create these terms can be found throughout the chapter. For additional help, use McGraw-Hill Connect™ at www.mcgrawhillconnect.com and consult Appendix B.

1. hematuria (hĕm″ŭh-tū-rē′ŭh)
2. oliguria (ŏl″ĭ-gū′rē-ŭh)
3. polyuria (pŏl″ē-yū′rē-ŭh)
4. extracorporeal shock wave lithotripsy (ESWL) (ĕks″trŭh-kŏr-pō′rē-ăl lĭth′ō-trĭp″sē)
5. antidiuretic (ăn″tĭ-dī″yū-rĕt′ĭk)
6. cystopyelonephritis (sĭs″tō-pī″ĕ-lō-nĕ-frī′tĭs)
7. nocturia (nŏk-tū′rē-ŭh)
8. glomerulonephritis (glō-măr″yū-lō-nĕ-frī′tĭs)

Online Study Tools

LEARNSMART® **connect**

APR

Anatomy & Physiology REVEALED includes cadaver photos that allow you to peel away layers of the human body to reveal structures beneath the surface. This program also includes animations, radiologic imaging, audio pronunciations, and practice quizzing. To learn more visit www.aprevealed.com

Anatomy & Physiology REVEALED
aprevealed.com

17 The Reproductive System

PART V

" . . . No sooner met but they looked;
No sooner looked but they loved;
No sooner loved but they sighed;
No sooner sighed but they asked one another the reason;
No sooner knew the reason but they sought the remedy;
. . . they are in the very wrath of love . . . clubs cannot part them . . . "

As You Like It, William Shakespeare

Do you believe in love at first sight? There's no doubt that human beings have a particularly acute visual sense, and so poets, authors, and lovers would answer *yes* to that question. But here is another question: Is there "love at first smell," influenced by human pheromones? As you recall from Chapter 10, pheromones are thought to be airborne messenger chemicals that influence human sexual behavior through the sense of smell. While you study the reproductive system in this chapter, the "love at first smell" question is interesting to think about. Pheromones clearly exist for animals, and behavioral studies seem to indicate that we humans have them, too. For example, it's well known that women who live or work together experience menstrual synchrony (menstrual cycles in unison); this effect seems to be caused by a component of axillary sweat, which shortens or lengthens women's cycles until eventually group members menstruate together. Men's axillary secretions have also been shown to influence women's menstrual cycles. Further, a woman's vagina produces fatty-acid compounds called *copulines,* which can raise a man's testosterone (the male sex hormone) level when he is exposed to them. Likewise, women who were first exposed to androstenol, a testosterone-like compound found in male sweat, were subsequently more likely to find a man's photo attractive. However, scientists aren't completely convinced about the influence of human pheromones. The exact compounds responsible for these effects haven't been isolated and identified, nor have their precise receptors in the nose and brain been found. Research into the topic will undoubtedly continue. For now, though, since we don't have all the answers yet, you might not want to count on your pheromone cologne—no matter how good you smell—to bring you true love.

Learning Outcomes

After you have studied this chapter, you should be able to:

17.1 Human Life Cycle
1. Discuss the functions of the reproductive system.
2. Describe the human life cycle.

17.2 Male Reproductive System
3. Trace the path of sperm, from the testes to the urethra.
4. Describe the macroscopic and microscopic anatomy of the testes.
5. Name the glands and describe the secretions that contribute to the composition of semen.
6. Describe the anatomy of the penis and the events preceding and during ejaculation.
7. Discuss hormonal regulation in the male.
8. Outline the actions of testosterone, including its effects on the primary sex organs and secondary sexual characteristics.

17.3 Female Reproductive System
9. Describe the macroscopic and microscopic anatomy of the ovaries.
10. Label a diagram of the external female genitals.
11. Contrast female orgasm with male orgasm.
12. Describe the menstrual cycle.
13. Describe the actions of estrogen and progesterone, including both primary and secondary sexual characteristics.

17.4 Control of Reproduction and Sexually Transmitted Infections
14. List several means of birth control, and describe their effectiveness.
15. Describe the symptoms of genital warts, genital herpes, hepatitis, chlamydia, gonorrhea, and syphilis.

17.5 Effects of Aging
16. Discuss the anatomical and physiological changes that occur in the reproductive system as we age.

17.6 Homeostasis
17. Discuss how the reproductive system works with other systems of the body to maintain homeostasis.

Visual Focus
Anatomy of Ovary and Follicle

Medical Focus
Ovarian Cancer
Breast and Testicular Self-Exams for Cancer
Endocrine-Disrupting Contaminants
Preventing Transmission of STIs

Human Systems Work Together
Reproductive System

Focus on Forensics
Rape

17.1 Human Life Cycle

1. Discuss the functions of the reproductive system.
2. Describe the human life cycle.

Unlike the other systems of the body, which are similar in both genders, the reproductive system is very different in males and females. Development of a child into a sexually mature young adult is a sequence of events called **puberty.** The reproductive system does not begin to fully function until puberty is complete. Sexual maturity occurs between the ages of 11 and 13 in girls, and 14 and 16 in boys. At the completion of puberty, the individual is capable of producing children.

Here's a quick overview of the functions of our reproductive organs, the **genitals:**

1. Both males and females produce reproductive cells called **gametes.** Males produce sperm within testes, and females produce ova (eggs) within ovaries.
2. In males, sperm mature and are then transported in a duct system until they exit through the penis. In females, ova are transported in the uterine tubes to the uterus.
3. The male penis functions to deliver sperm to the female vagina, which can receive sperm. The vagina also transports menstrual fluid to the exterior and is the birth canal.
4. The female uterus allows the fertilized egg to develop within her body. After birth, the female breast provides nourishment in the form of breast milk.
5. The testes and ovaries both produce and are maintained by the sex hormones. These hormones have a profound effect on the body because they bring about masculinization and feminization of various features. In females, ovarian hormones allow a pregnancy to continue.

Meiosis

We studied the type of cell division called mitosis in Chapter 3 and learned that mitosis occurs during growth and repair of the body's tissues. Mitosis is *duplication division* (as an analogy, imagine the cell making exact copies of itself during mitosis, much like a copy machine does with a page of notes). As a result of mitosis, the chromosome number stays constant, and every cell in your body has an identical set of 46 chromosomes. These chromosomes are organized into pairs of homologous chromosomes of approximately the same size and length. As you recall (from Chapter 2), chromosome genes are individual sections of DNA that determine particular traits such as hair or eye color. These specific genes are located in the same position on homologous chromosomes.

It might be helpful to review the cell cycle and mitosis (Chapter 3). The terminology used to describe DNA might seem confusing because the terms sound so similar. As you'll recall, *chromatin* is the term for unorganized DNA found in the nucleus of the cell, and a *chromosome* is formed when the chromatin condenses. There are 46 chromosomes in a normal human cell. A *chromatid* is the term used for each single chromosome while it is attached to its duplicate copy at the centromere (following the S-phase of the cell cycle). In this paired state, the two chromatids

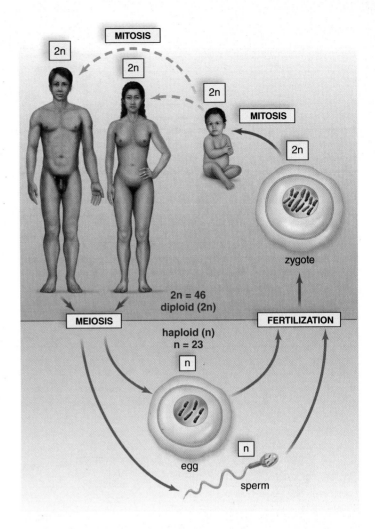

AP|R **Figure 17.1 Human life cycle.** The human life cycle has two types of cell divisions: mitosis, in which the chromosome number stays constant; and meiosis, in which the chromosome number is halved. Meiosis only occurs in the testes of males during the production of sperm and in the ovaries of females during the production of ova (eggs). In human beings, the ovum (egg) and sperm have 23 chromosomes each, called the n number. Following fertilization, the new individual has 46 chromosomes, called the 2n number.

are called *sister chromatids.* Once the chromatids have separated, each is called a chromosome once more.

In addition to mitosis, the human life cycle (Fig. 17.1) includes a type of cell division called **meiosis,** which is *reduction division.* Meiosis only takes place in the testes of males during sperm production and in the ovaries of females during the production of ova. During meiosis, the chromosome number is reduced from the normal 46 chromosomes, called the **diploid** or **2n** number, down to 23 chromosomes, called the **haploid** or **n** number of chromosomes. This occurs in two successive divisions, called meiosis I and meiosis II (Fig. 17.2a). The stages of division in meiosis and mitosis are the same—prophase, metaphase, anaphase, and telophase.

To begin meiosis, the cell (called a primary cell) duplicates each chromatid (individual DNA strand), so there are 92 individual

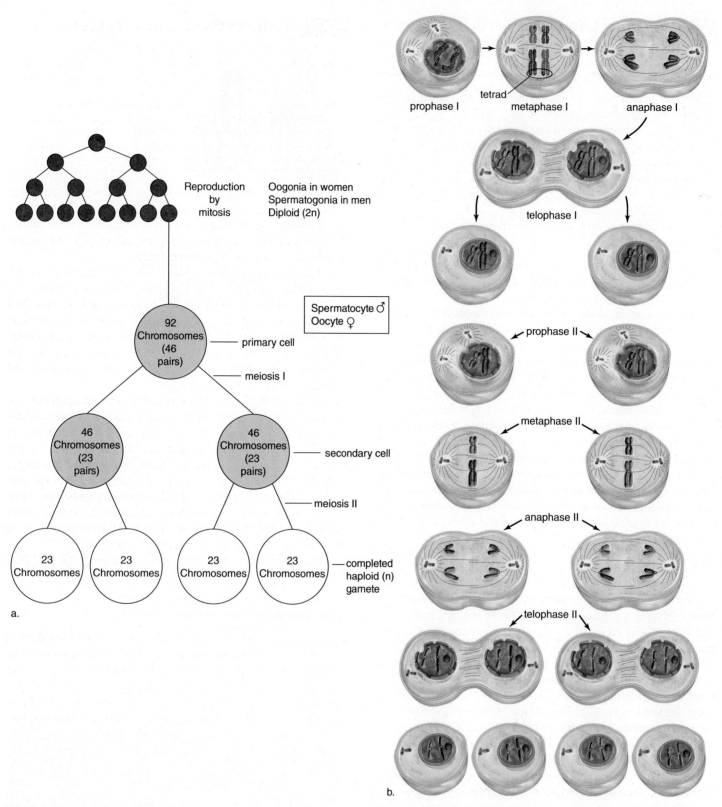

AP|R **Figure 17.2** **Meiosis (a)** Prior to meiosis, the ovaries in females and testes in males produce large quantities of oogonia/spermatogonia by mitosis. Primary cells will divide to form two secondary cells; secondary cells divide further to form four haploid cells. In this way, the chromosome number is reduced from 92, to two sets of 46 chromosomes, to four sets of 23 chromosomes. **(b)** The stages in meiosis are the same as mitosis (prophase, metaphase, anaphase, telophase) except that they occur twice. Note that during prophase I, the paired chromosomes in a tetrad can exchange genetic information with one another.

chromatids, which remain joined together (Fig. 17.2). (This is the same process that occurs in body cells during the S-phase of the cell cycle, which precedes mitosis; see pages 55–59). Next, chromosomes of the same shape and size, called homologous pairs of chromosomes, line up side-by-side in prophase I (Fig. 17.2b, *prophase I*). Each set of four chromatids is called a tetrad. As they line up, each of the chromatids in a tetrad can swap bits and pieces of DNA with the others, and the result is shown in metaphase I (Fig. 17.2b, *metaphase I*). The effect of this "gene-swap" (the technical term is *crossing over*) is that one's gametes (ova or sperm) are *not* exact copies of body cells. Instead, new and different gene combinations are created and passed down to one's children. For humans as a species, forming novel chromosomes by crossing over creates genetic diversity.

Next, in metaphase I, the chromosome pairs line up at the cell's equator, and in anaphase I, spindle fibers draw the pairs apart. In telophase I, two separate cells are formed (Fig. 17.2b, *metaphase I, anaphase I, telophase I*). As meiosis I is completed, two secondary cells are created, each with 46 sister chromatids, which are still joined together. In meiosis II, only one member of each pair goes into the daughter cells (Fig. 17.2b, *telophase II*). To summarize, each primary cell starts with 92 chromatids in 46 pairs, which are divided during meiosis I into two cells with 46 chromatids in 23 pairs, and then divided once more during meiosis II into four cells with 23 chromosomes.

A **zygote,** the first cell of a new human being, forms following *fertilization,* when a sperm joins with an egg. Because the sperm has 23 chromosomes and the egg has 23 chromosomes, the zygote has 23 pairs of homologous chromosomes, or 46 chromosomes altogether. Without meiosis, the chromosome number in each generation of human beings would double, and cells would no longer be able to function. The zygote undergoes mitosis during development to produce the many cells of a newborn, and mitosis also occurs as a child becomes an adult. Of course, mitosis continues throughout adult life as well, every time any body cell divides.

Content CHECK-UP!

1. If you wanted to describe meiosis, which statement(s) could you use?
   a. Meiosis occurs only in the ovaries and testes.
   b. Meiosis begins with 92 chromosomes, in sets called tetrads.
   c. Meiosis results in two identical cells that are haploid.
   d. Statements a and b are correct.
   e. All three statements are correct.

2. At the end of meiosis I in a gamete, there are 46 chromosomes in each cell. At the completion of mitosis, there are also 46 chromosomes in each cell. How are the two cells different?

3. Which type of cell is haploid?
   a. sperm cell   d. sperm and ovum
   b. zygote   e. All of the above.
   c. ovum

Answers in Appendix A.

17.2 Male Reproductive System

3. Trace the path of sperm, from the testes to the urethra.
4. Describe the macroscopic and microscopic anatomy of the testes.
5. Name the glands and describe the secretions that contribute to the composition of semen.
6. Describe the anatomy of the penis and the events preceding and during ejaculation.
7. Discuss hormonal regulation in the male.
8. Outline the actions of testosterone, including its effects on the primary sex organs and secondary sexual characteristics.

The male reproductive system includes the organs shown in Figure 17.3. The *primary sex organs* of a male are the paired testes (sing., **testis**), which are suspended within the sacs of the **scrotum.** The testes are the primary sex organs because they produce sperm and the male sex hormones (**androgens**).

The other organs depicted in Figure 17.3 are the *accessory* (or secondary) *sex organs* of a male. Sperm produced by the testes are stored within the **epididymis** (pl., *epididymides*). Then they enter a **vas deferens** (pl., *vasa deferentia*), which transports them to an **ejaculatory duct.** The ejaculatory ducts enter the **urethra.** (The urethra in males is a part of both the urinary system and the reproductive system.) The urethra passes through the penis and transports sperm to the outside.

Several important secretory structures are also accessory sex organs. The **seminal vesicles** lie lateral to each of the two vas deferentia. The duct of each seminal vesicle joins the duct of the vas deferens beside it. The shared ducts are termed *ejaculatory ducts,* and each empties into the urethra. The **prostate gland** is a single, walnut-sized gland that surrounds the upper portion of the urethra just inferior to the bladder. **Bulbourethral glands** (also called Cowper glands) are pea-sized organs that lie inferior to the prostate on either side of the urethra. At the time of **ejaculation,** sperm leave the penis in a fluid called **semen** (seminal fluid).

Each component of seminal fluid has a particular function; these will be detailed as we discuss each of the accessory organs.

The Testes

The testes lie outside the male's abdominal cavity within the two scrotal sacs, but they were not always outside of the abdomen. The testes begin their development *inside* the abdominal cavity but descend into the scrotal sacs during the last two months of fetal development. If, by chance, the testes do not descend (a condition called **cryptorchidism**), and the male is not operated on to place the testes in the scrotum, sterility—the inability to produce offspring—usually follows. This is because sperm formation requires a temperature approximately 2°C lower than body temperature; thus, the internal temperature of the body is too high to produce viable sperm. A subcutaneous muscle along with an adjoining muscle raise the scrotum during sexual arousal. If necessary, these muscles can also raise the testes closer to the body to warm them.

Anatomy of a Testis

A sagittal section of a testis shows that it is enclosed by a tough, fibrous capsule. The connective tissue of the capsule extends into

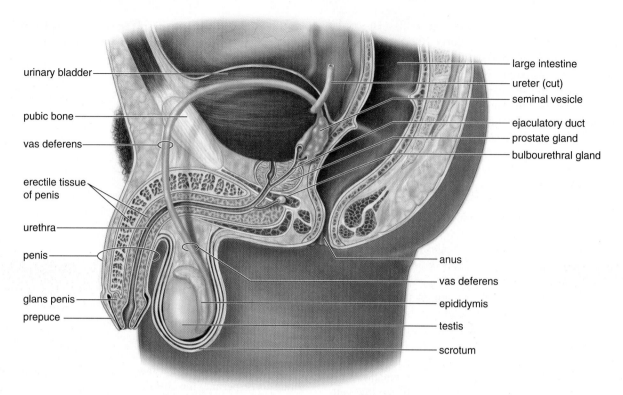

urinary bladder

pubic bone

vas deferens

erectile tissue
of penis

urethra

penis

glans penis

prepuce

large intestine

ureter (cut)

seminal vesicle

ejaculatory duct

prostate gland

bulbourethral gland

anus

vas deferens

epididymis

testis

scrotum

AP|R **Figure 17.3** **Male reproductive system.** The testes produce sperm. The seminal vesicles, the prostate gland, and the bulbourethral glands provide a fluid medium for the sperm, which move from a testis to an epididymis to a vas deferens and through the ejaculatory duct to the urethra in the penis. The foreskin (prepuce) is removed when a penis is circumcised.

the testis, forming *septa,* or walls, that divide the testis into compartments called lobules. Each lobule contains one to three tightly coiled **seminiferous tubules** (Fig. 17.4*a*). Altogether, these tubules have a combined length of approximately 250 m. A microscopic cross section of a seminiferous tubule reveals that it is packed with cells undergoing spermatogenesis (Fig. 17.4*b*), the production of sperm.

Delicate connective tissue surrounds the seminiferous tubules. Cells that secrete the male sex hormones, the androgens, are located here between the seminiferous tubules. Therefore, these endocrine cells are called **interstitial cells** (also called interstitial cells of Leydig or, simply, Leydig cells). The most important of the androgens is **testosterone,** whose functions are discussed later in this section.

Testicular cancer, or cancer of the testes, is one type of cancer that can be detected by self-examination, as explained in the Medical Focus on page 413.

Spermatogenesis

Spermatogenesis, the production of sperm, includes the process of meiosis as the sperm form. Before puberty, the testes, including the seminiferous tubules, are small and nonfunctioning. At the time of puberty, the interstitial cells increase their size and start producing androgens. Then, the seminiferous tubules also enlarge, and they start producing sperm.

The seminiferous tubules contain two types of cells: **germ cells,** which are involved in spermatogenesis, and **sustentacular**

(Sertoli) cells. Sustentacular cells are large; they extend from the capsule to the lumen of the seminiferous tubule. The sustentacular cells support, nourish, and regulate the development of cells undergoing spermatogenesis.

The germ cells near the outer capsule are called **spermatogonia.** The spermatogonia divide, producing more cells by mitosis. Some of these cells remain as spermatogonia, which continue to divide. These provide a constant source of sperm cells throughout a normal man's lifetime. Other spermatogonia replicate their DNA and become **primary spermatocytes,** containing 92 sister chromatids (strands of DNA) held together in 46 pairs (Fig. 17.5). Primary spermatocytes are termed *diploid,* or 2n cells. The primary spermatocytes start the process of meiosis, which requires two divisions. Following meiosis I, two **secondary spermatocytes** are formed, each containing 46 chromatids in 23 pairs. When meiosis II has been completed, there are four spermatids, cells that are termed *haploid* because they possess *half* the normal chromosome number. Spermatids then mature into sperm.

Mature **sperm,** or spermatozoa, have three distinct parts: a head, a middle piece, and a tail (see Fig. 17.4*c*). Mitochondria in the middle piece provide energy for the movement of the tail, which has the structure of a flagellum. The head contains a nucleus covered by a cap called the **acrosome,** which stores enzymes needed to penetrate the ovum. Notice in Figure 17.4*b,* that the sperm are situated so that their tails project into the lumen of the seminiferous tubules.

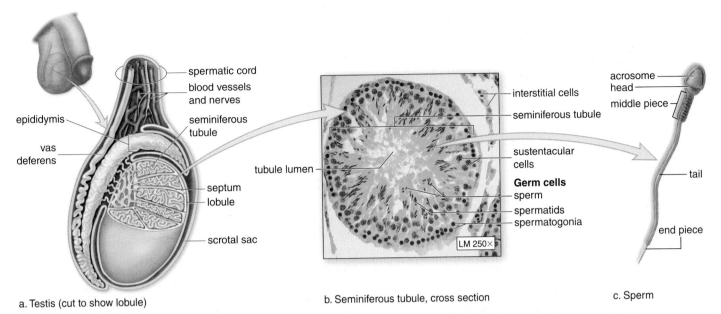

a. Testis (cut to show lobule)

b. Seminiferous tubule, cross section

c. Sperm

AP|R **Figure 17.4** **Testis and sperm. (a)** The lobules of a testis contain seminiferous tubules. **(b)** Micrograph showing a cross section of the seminiferous tubules, where spermatogenesis occurs. **(c)** A sperm has a head, a middle piece, and a tail. The nucleus is in the head, which is capped by the enzyme-containing acrosome.

When formed, the sperm are transported from the seminiferous tubules into a complex network of channels that ultimately form ducts. The ducts join to form the epididymis, which empties into the vas deferens.

The ejaculated semen of a normal human male contains several hundred million sperm, but only one sperm normally enters an egg. Sperm typically survive in the female reproductive tract for approximately 48 hours after sexual intercourse. However, under optimal conditions sperm can survive for 4–6 days after intercourse.

Male Internal Accessory Organs

Table 17.1 lists and Figure 17.6 depicts the internal accessory organs, as well as the other reproductive organs, of the male. Sperm are transported to the urethra by a series of ducts. Along the way, various glands add secretions to seminal fluid.

Epididymides

Each epididymis is a tightly coiled, threadlike tube that would stretch about 6 meters if uncoiled. An epididymis runs posteriorly

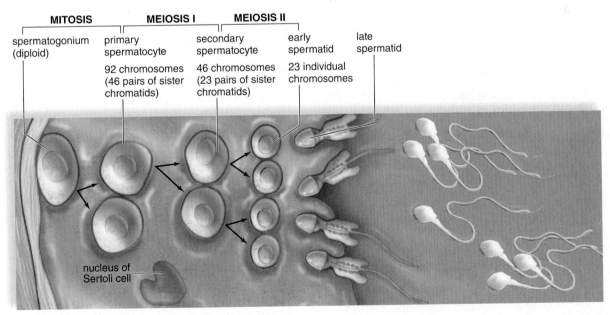

AP|R **Figure 17.5** **Spermatogenesis.** Diploid spermatogonia reproduce by mitosis, and the two phases of meiosis result in spermatids. Spermatids mature into functional sperm cells.

TABLE 17.1	Male Internal Accessory Organs
Organ	**Function**
Epididymides	Ducts where sperm mature and some sperm are stored
Vasa deferentia	Transport and store sperm
Seminal vesicles	Contribute nutrients and fluid to semen
Ejaculatory ducts	Transport sperm
Prostate gland	Contributes basic fluid to semen
Urethra	Transports sperm
Bulbourethral glands	Produce mucoid fluid that neutralizes urine

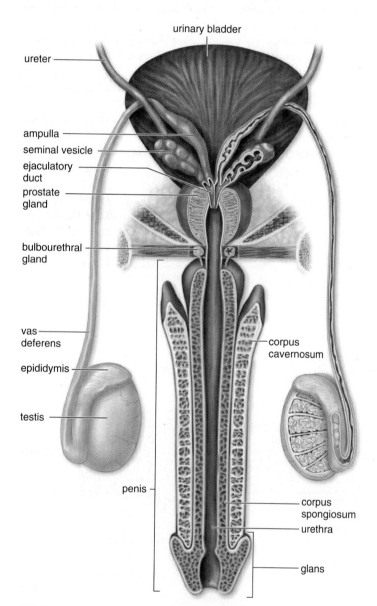

AP|R **Figure 17.6 Male reproductive system, posterior view.** This view shows the duct system that transports sperm from each testis to the urethra, which continues in the penis.

down along each testis and becomes a vas deferens that ascends each testis medially.

The lining of the epididymis consists of pseudostratified columnar epithelium with long cilia. Sperm are stored in the epididymides, and the lining secretes a fluid that supports them. The wall of an epididymis contains a thin layer of smooth muscle. Peristaltic contractions move the sperm along as they mature. By the time the sperm leave the epididymides, they are capable of fertilizing an egg even though they do not "swim" until they enter the vagina.

Vas Deferens

Each vas deferens is a continuation of the epididymis. Like the epididymis, the vas deferens is lined by pseudostratified columnar epithelium that is ciliated at the testicular end of the tube. The vas deferens is contained within a protective structure called the **spermatic cord.** The spermatic cord consists of connective tissue and muscle fibers, and contains the testicular artery, vein, and nervous supply in addition to the vas deferens. As each spermatic cord ascends into the abdomen, it passes through the inguinal canal. This is the passageway by which the testis descended from the abdomen into the scrotum during fetal life. The inguinal canal remains a weak point in the abdominal wall. As such, it is frequently a site of **inguinal hernia.** (A **hernia** is an opening or separation of some part of the abdominal wall through which a portion of an internal organ, usually the intestine, protrudes. Hernias also can occur in the umbilical region, diaphragm, or elsewhere in the abdomen.)

After each vas deferens enters the abdomen, it passes over the bladder's superior surface to reach the posterior side. Each vas deferens widens at its ampulla, located at the posterior base of the urinary bladder, then joins with the duct of a seminal vesicle to form an ejaculatory duct. The ejaculatory ducts pass through the prostate gland to join the urethra.

Seminal Vesicles

The two seminal vesicles lie lateral to each vas deferens on the posterior side of the bladder. They are coiled, membranous pouches about 5 cm long. The glandular lining of the seminal vesicles secretes an alkaline fluid that contains fructose and **prostaglandins** into an ejaculatory duct. The pH of the fluid helps modify the pH of seminal fluid; the fructose provides energy for sperm; and the prostaglandins promote muscular contractions of the female genital tract that help to draw sperm further into the female.

Prostate

The prostate gland encircles the urethra just inferior to the bladder. The walnut-sized gland is about 4 cm across, 2 cm thick, and 3 cm in length. The fibrous connective tissue of its capsule extends inward to divide the gland into lobes, each of which contains about 40 to 50 tubules. The epithelium lining the tubules secretes a fluid that is thin, milky, and alkaline. In addition to adjusting the pH of seminal fluid, prostatic fluid enhances sperm motility. The secretion of the prostate gland enters the urethra when the smooth muscle in its capsular wall contracts.

As discussed in the Medical Focus on page 387, the prostate gland is a frequent site for cancer.

Bulbourethral Glands

The bulbourethral glands (Cowper glands) are two small glands about the size of peas. They are located inferior to the prostate gland and enclosed by fibers of the external urethral sphincter. These glands also contain many tubules that secrete a mucuslike fluid. This fluid lubricates the end of the penis preparatory to sexual intercourse. It also neutralizes any acidic urine that may be in the urethra.

Male External Genitals

The **external genitals** are the sex organs that can be easily observed because they are located outside the body. The **penis** and the scrotum are the external genitals of the male. The penis is the male organ of sexual intercourse by which sperm are introduced into the female reproductive tract.

Penis

The penis has an internal root that anchors it to the body wall, and an external shaft. The tip of the penis is called the *glans penis* (see Fig. 17.6). At the glans penis, the skin folds back on itself to form the prepuce, or foreskin (Fig. 17.7*a*). This is the structure that is removed in the surgical procedure called **circumcision.** The penis contains three cylindrical bodies of erectile tissue. The two paired **corpora cavernosa** (sing., *corpus cavernosum*) are found on either side of the single **corpus spongiosum.** The urethra passes through the corpus spongiosum. These three columns are supported by

fibrous connective tissue, and all are then covered with a thin, loose skin (Fig. 17.7*b*).

Male Sexual Response

A male's sexual response begins with an erection. The erection process is a spinal reflex; thus, it can even occur in male infants in the womb (see Chapter 8 for a review of reflexes). After puberty, sexual excitation in a male can be triggered by touch, smell, visual or emotional stimuli, or a combination of these. The three erectile bodies contain distensible spaces that can fill with blood during the erection process.

During sexual arousal, autonomic nerves release nitric oxide, NO. This stimulus leads to the production of cGMP (cyclic guanosine monophosphate), causing the smooth muscle walls of incoming arteries to relax and the erectile tissue to fill with blood. The veins that take blood away from the penis are compressed, and the penis becomes erect. **Erectile dysfunction** (formerly called impotency) exists when the erectile tissue doesn't expand enough to compress the veins. Medications for treatment of erectile dysfunction inhibit an enzyme that breaks down cGMP, ensuring that a full erection will take place. However, individuals who take these medications may experience vision problems because the same enzyme occurs in the retina.

Orgasm (climax) in males is marked by ejaculation, which has two phases: emission and expulsion. During emission, sperm enter the urethra from each ejaculatory duct, and the prostate, seminal vesicles, and bulbourethral glands contribute secretions to the seminal fluid. Once seminal fluid is in the urethra, rhythmic muscle contractions cause seminal fluid to be expelled from the penis in spurts. During ejaculation, a sphincter closes off the bladder so that no urine enters the

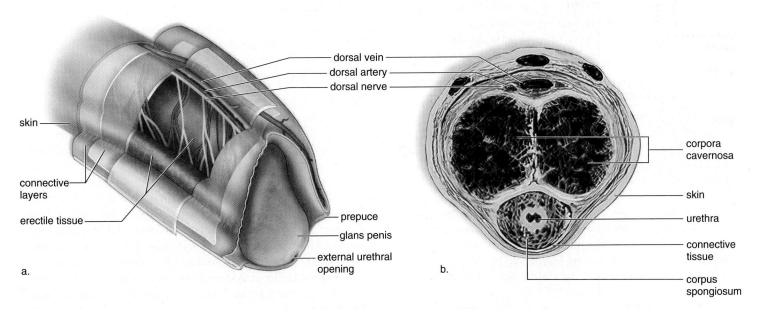

AP|R **Figure 17.7** **Penis anatomy.** (a) Beneath the skin and the connective tissue lies the urethra surrounded by erectile tissue. This tissue expands to form the glans penis, which in uncircumcised males is partially covered by the foreskin (prepuce). (b) Two other columns of erectile tissue in the penis are located posteriorly.

urethra. (Notice that the urethra carries either urine or semen at different times.)

Male orgasm includes expulsion of seminal fluid from the penis but also the physiological and psychological sensations that occur at the climax of sexual stimulation. The psychological sensation of pleasure is centered in the brain, but the physiological reactions involve the genital organs and associated muscles, as well as the entire body. Marked muscular tension is followed by contraction and relaxation.

Following ejaculation and/or loss of sexual arousal, the penis returns to its normal flaccid state. After ejaculation, a male typically experiences a period of time, called the refractory period, during which stimulation does not bring about an erection. The length of the refractory period increases with age.

There may be in excess of 400 million sperm in the 3.5 ml of semen expelled during ejaculation. The sperm count can be much lower than this, however, and fertilization of the egg by a sperm can still take place.

Regulation of Male Hormone Levels

At the time of puberty, the sex organs mature, and then changes occur in the physique of males. The cause of puberty is related to the level of sex hormones in the body, as regulated by the negative feedback system described in Figure 17.8. We now know that this feedback system functions long before puberty, but the level of hormones is low because the hypothalamus is supersensitive to feedback control. At the start of puberty, the hypothalamus becomes less sensitive to feedback control and begins to increase its production of gonadotropin-releasing hormone (GnRH), which stimulates the anterior pituitary to produce the gonadotropic hormones. Two gonadotropic hormones, **FSH (follicle-stimulating hormone)** and **LH (luteinizing hormone),** are named for their function in females but exist in both sexes, stimulating the appropriate organs in each. FSH promotes spermatogenesis in the seminiferous tubules, and LH promotes androgen (e.g., testosterone) production in the interstitial cells. LH in males is also called interstitial cell-stimulating hormone (ICSH).

Negative Feedback Mechanisms

As mentioned, the hypothalamus, anterior pituitary, and testes are involved in a negative feedback system. The system maintains testosterone production at a fairly constant level. When the amount of testosterone in the blood rises to a certain level, it causes the hypothalamus and anterior pituitary to decrease their respective secretion of GnRH and LH. As the level of testosterone begins to fall, the hypothalamus increases its secretion of GnRH, and the anterior pituitary increases its secretion of LH. Thus, stimulation of the interstitial cells occurs. Only minor fluctuations of the testosterone level occur in the male, and the feedback mechanism in this case acts to maintain testosterone at a normal level.

A similar feedback mechanism maintains the continuous production of sperm. The sustentacular cells in the wall of the seminiferous tubules produce a hormone called **inhibin,** which blocks GnRH and FSH secretion when appropriate.

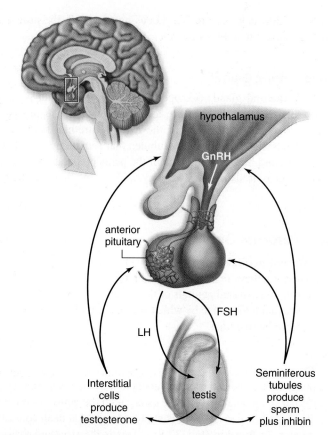

Figure 17.8 Negative feedback. Regulation of testosterone secretion involves negative feedback (reverse arrows) by testosterone on GnRH and LH. Regulation of sperm production involves negative feedback by inhibin on GnRH and FSH.

Testosterone

The male sex hormone, testosterone, has many functions. It is essential for normal development and function of the primary sex organs in males. For example, greatly increased testosterone secretion at puberty stimulates maturation of the penis and testes.

Secondary Sex Characteristics Testosterone also brings about and maintains the male **secondary sex characteristics,** which develop at the time of puberty and visibly distinguish males from females. These characteristics include a pattern of male hair growth, activity of cutaneous glands, deeper pitch to the voice, and muscle strength.

At puberty, males experience growth of a beard, axillary (underarm) hair, and pubic hair. In males, pubic hair tapers toward the navel. A side effect of testosterone activity is baldness. Although genes for baldness are probably inherited by both sexes, the most common inherited form of baldness is androgenetic alopecia, or male-pattern baldness. As you know from Chapter 5, male-pattern baldness is caused by an excess of a form of testosterone called dihydrotestosterone.

Testosterone also causes oil and sweat glands in the skin to increase their secretions, thereby contributing to acne and body odor. The larynx and vocal cords enlarge, causing the voice to change. The "Adam's apple" is a part of the larynx, and it is usually more prominent in males than in females.

Testosterone is responsible for the greater muscular strength of males, which is why some athletes take a supplemental anabolic steroid, such as testosterone or a related chemical. It's important to note that anabolic steroids will build muscle in both men and women. The dangers of anabolic steroid use are discussed in a Medical Focus in Chapter 10, pages 234–235.

Content **CHECK-UP!**

4. Create a list of the ingredients of seminal fluid (semen). What is the function of each?

5. Identify, in order, the structures a sperm cell must pass through as it leaves the testis:

   a. epididymis → vas deferens → ejaculatory duct → urethra → penis

   b. vas deferens → epididymis → ejaculatory duct → urethra → penis

   c. ejaculatory duct → vas deferens → epididymis → urethra → penis

   d. epididymis → vas deferens → urethra → ejaculatory duct → penis

6. Starting at puberty, the hypothalamus secretes _____ hormone, which stimulates the anterior pituitary gland to secrete _____ hormone and _____ hormone. List the effects of these last two hormones.

Answers in Appendix A.

17.3 Female Reproductive System

9. Describe the macroscopic and microscopic anatomy of the ovaries.
10. Label a diagram of the external female genitals.
11. Contrast female orgasm with male orgasm.
12. Describe the menstrual cycle.
13. Describe the actions of estrogen and progesterone, including both primary and secondary sexual characteristics.

The female reproductive system includes the organs depicted in Figure 17.9. The primary sex organs of a female are the paired **ovaries** that lie in shallow depressions, one on each side of the upper pelvic cavity. The ovaries produce **ova** (sing., ovum) and the female sex hormones, estrogen and progesterone.

The other organs depicted in Figure 17.9 are the accessory (or secondary) sex organs of a female. When an ovum leaves an ovary, it is usually swept into a uterine (fallopian) tube by the combined action of the fimbriae (fingerlike projections of a uterine tube) and the beating of cilia that line the uterine tube.

Once in a uterine tube, the ovum is transported toward the uterus. Fertilization, and therefore zygote formation, usually takes place in the uterine tube. The developing embryo normally arrives at the **uterus** several days later, and then **implantation** occurs as the embryo embeds in the uterine lining, which has been prepared to receive it.

Development of the embryo and fetus normally takes place in the uterus. The lining of the uterus, called the **endometrium,** participates in the formation of the placenta (see Chapter 18, pages 432 and 434), which supplies nutrients needed for embryonic and fetal development.

The uterine tubes join the uterus at its upper end, while at its lower end, the **cervix** enters the vagina nearly at a right

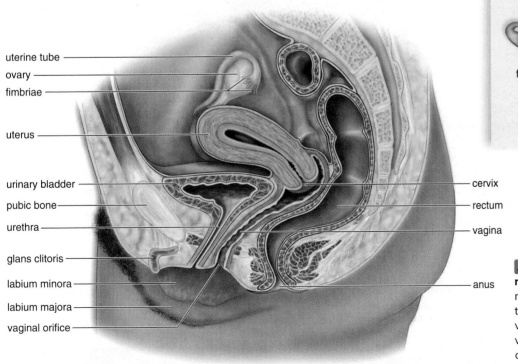

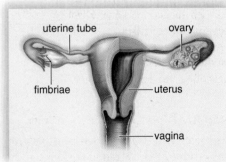

AP|R **Figure 17.9** **The female reproductive system.** The ovaries release one egg per month. Fertilization occurs in the uterine tube and development occurs in the uterus. The vagina is the birth canal as well as the organ of sexual intercourse.

angle. A small opening in the cervix leads from the uterus to the vagina.

The **vagina** is the birth canal and organ of sexual intercourse in females. The vagina also acts as an exit for menstrual flow. If fertilization and implantation of an embryo do not occur, the most superficial layer of the endometrium dies off and is shed during menstruation.

The external genital organs of the female are known collectively as the **vulva.** The vulva consists of the mons pubis, clitoris, labia majora, and labia minora. The area between the labia minora, which is termed the vestibule, contains the openings of the urethra and the vagina.

Notice that the urinary and reproductive systems in the female are entirely *separate*. The urethra carries only urine, and the vagina has functions related only to reproduction: It receives the penis during sexual intercourse and serves as the birth canal.

The Ovary

The ovaries are paired, oval bodies about 3 cm in length by 1 cm in width and less than 1 cm thick. They lie to either side of the uterus on the lateral walls of the pelvic cavity.

Several ligaments hold the ovaries in place (see Fig. 17.12). The largest of these, the broad ligament, is also attached to the uterine tubes and the uterus. The suspensory ligament holds the upper end of the ovary to the pelvic wall, and the ovarian ligament attaches the lower end of the ovary to the uterus.

A sagittal section through an ovary shows that it is made up of an outer cortex and an inner medulla. In the cortex are many **follicles,** each one containing an immature ovum, called an **oocyte** (Fig. 17.10). A female is born with as many as 2 million follicles,

Visual Focus

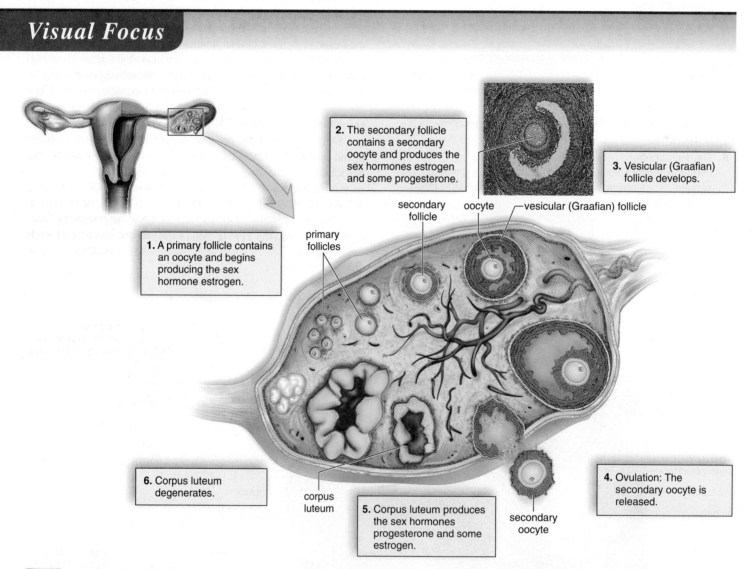

2. The secondary follicle contains a secondary oocyte and produces the sex hormones estrogen and some progesterone.

3. Vesicular (Graafian) follicle develops.

secondary follicle

oocyte

vesicular (Graafian) follicle

primary follicles

1. A primary follicle contains an oocyte and begins producing the sex hormone estrogen.

6. Corpus luteum degenerates.

corpus luteum

5. Corpus luteum produces the sex hormones progesterone and some estrogen.

secondary oocyte

4. Ovulation: The secondary oocyte is released.

AP|R **Figure 17.10** **Anatomy of ovary and follicle.** As a follicle matures, the oocyte enlarges and is surrounded by layers of follicular cells and fluid. The micrograph shows the mature vesicular (Graafian) follicle. Eventually, ovulation occurs, the mature follicle ruptures, and the secondary oocyte is released. A single follicle actually goes through all the stages in one place within the ovary.

but the number is reduced to 300,000–400,000 by the time of puberty. Only a small number of follicles (about 400) ever become completely mature because a female usually produces only one ovum per month during her reproductive years. Because oocytes are present at birth, they age as the woman ages. This may be one reason why older women are more likely to produce children with chromosome defects.

Ovarian cancer (cancer of the ovary), which is discussed in the Medical Focus on page 405, causes more deaths than cervical and uterine cancer.

Oogenesis

Oogenesis, the production of an ovum, includes the process of meiosis. Similar to spermatogenesis, oogenesis begins with a primary oocyte that undergoes meiosis I to become a secondary oocyte having 23 paired sister chromatids. The secondary oocyte undergoes meiosis II, but only if it is first fertilized by a sperm cell. If the secondary oocyte remains unfertilized, it never completes meiosis and will die shortly after being released from the ovary.

Oogenesis begins within a follicle. As the follicle matures, it develops from a primary follicle to a secondary follicle to a **vesicular (Graafian) follicle** (Fig. 17.10). The epithelium of a primary follicle surrounds a primary oocyte. Pools of follicular fluid surround the oocyte in a secondary follicle. In a vesicular follicle, a fluid-filled cavity increases to the point that the follicle wall balloons out on the surface of the ovary.

Figure 17.11 traces the steps of oogenesis. As a follicle matures, the primary oocyte divides, producing two cells. One cell is a secondary oocyte, and the other is a polar body. A **polar body** is formed only during oogenesis, and its function is simply to hold the discarded chromosomes, as a sort of cellular "trash can." Because it is almost completely lacking in cytoplasm, the polar body can never be fertilized or develop further. The vesicular follicle bursts, releasing the secondary oocyte surrounded by a clear membrane and attached follicular cells. This process is referred to as **ovulation.**

The secondary oocyte, often called an egg, enters a uterine tube. If fertilization occurs, a sperm enters the secondary oocyte, which then completes meiosis II. A haploid ovum with 23 chromosomes and a second polar body result. Thus, formation of sperm cells (spermatogenesis) and formation of the ovum (oogenesis) both result in cells that are haploid, containing only 23 chromosomes. However, spermatogenesis produces four sperm cells for every primary cell. In oogenesis, only one haploid ovum is formed for each primary oocyte (the extra chromosomes are contained in two polar bodies). When the sperm nucleus unites with the egg nucleus during conception, a diploid zygote with 46 chromosomes is produced.

A follicle that has lost its egg develops into a **corpus luteum,** a glandlike structure. If implantation does not occur, the corpus luteum begins to degenerate after about 10 days. The remains of a corpus luteum is a white scar called the **corpus albicans.** If implantation does occur, the corpus luteum continues to secrete for about six months. Hormones from the corpus luteum help keep the uterine lining intact and allow a pregnancy to continue.

Although a number of follicles grow during each month, only one reaches full maturity and ruptures to release a secondary oocyte. Scientists believe that the ovaries alternate in producing functional ova. The number of secondary oocytes produced by a female during her lifetime is minuscule compared to the number of sperm produced by a male.

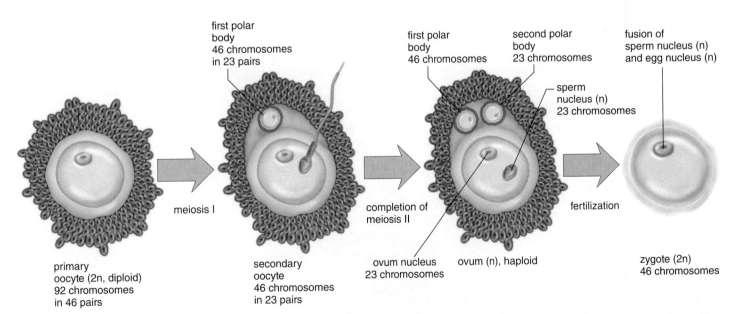

first polar body
46 chromosomes in 23 pairs

first polar body
46 chromosomes

second polar body
23 chromosomes

fusion of sperm nucleus (n) and egg nucleus (n)

sperm nucleus (n)
23 chromosomes

meiosis I

completion of meiosis II

fertilization

primary oocyte (2n, diploid)
92 chromosomes in 46 pairs

secondary oocyte
46 chromosomes in 23 pairs

ovum nucleus
23 chromosomes

ovum (n), haploid

zygote (2n)
46 chromosomes

AP|R **Figure 17.11 Oogenesis in an ovary.** Oogenesis involves meiosis I, during which the chromosome number is reduced, and meiosis II, which results in a single egg. Meiosis II takes place after a sperm enters the secondary oocyte. At the end of oogenesis, there are also at least two polar bodies, nonfunctional cells that later disintegrate.

TABLE 17.2 Female Internal Accessory Organs

Organ	Function
Uterine tubes (fallopian tubes, oviducts)	Transport ovum; location of fertilization
Uterus (womb)	Houses developing fetus
Cervix	Contains opening to uterus
Vagina	Receives penis during sexual intercourse; serves as birth canal and as an exit for menstrual flow

Female Internal Accessory Organs

Table 17.2 lists and Figure 17.12 depicts the internal accessory organs, as well as the other reproductive organs, of a female.

Uterine Tubes

The **uterine tubes,** also called fallopian tubes or oviducts, extend from the uterus to the ovaries. Usually the secondary oocyte enters a uterine tube because the **fimbriae** sweep over the ovary at the time of ovulation, and the beating of the cilia that line uterine tubes creates a suction effect. Once in the uterine tube, the egg is propelled slowly toward the uterus by action of the cilia and by muscular contractions in the wall of the uterine tubes.

Fertilization, the completion of meiosis, and zygote formation normally occur in the upper one-third of a uterine tube. The developing embryo usually does not arrive at the uterus for several days. Once in the uterus, the embryo embeds itself in the uterine lining, which has been prepared to receive it.

Occasionally, the embryo becomes embedded in the wall of a uterine tube, where it begins to develop. Tubular pregnancies cannot succeed and must be surgically removed because the tubes are not anatomically capable of allowing full development to occur. Any pregnancy that occurs outside the uterus is called an **ectopic pregnancy.**

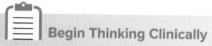

Begin Thinking Clinically

Rarely, an ectopic pregnancy can occur outside of the womb, in the abdominal cavity. How could this occur? On which organs could a fertilized ovum construct a placenta? How would a full-term infant resulting from such a pregnancy be delivered?

Answers and discussion in Appendix A.

Uterus

The *uterus* is a thick-walled, muscular organ about the size and shape of an inverted pear. Normally, it lies above and is tipped over the urinary bladder. The uterus has three sections: The *fundus* is the

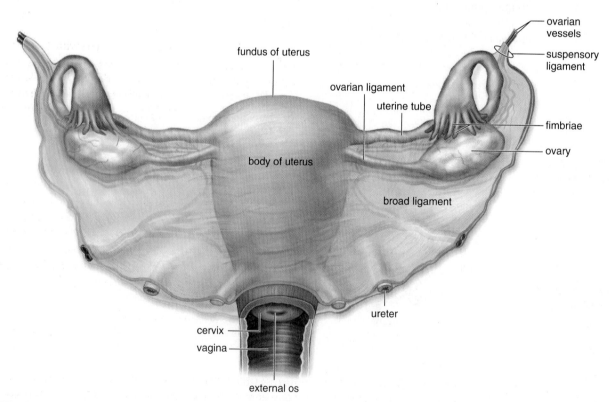

Figure 17.12 Female reproductive system, posterior view.

Ovarian cancer is often "silent," showing no obvious signs or symptoms until late in its development. Rarely is there any abnormal vaginal bleeding, as might be seen with uterine or cervical cancers. The most common sign is enlargement of the abdomen, which is caused by the accumulation of fluid. Vague symptoms may occur: bloating, pelvic or abdominal pain, difficulty eating and/or feeling full quickly, increased urinary frequency, and the urgent need to urinate. When symptoms such as these persist in women over 40 (and if they cannot be explained by any other cause) they may indicate the need for a thorough evaluation for ovarian cancer.

The risk for ovarian cancer increases with age. The highest rates are for women over age 60. Women who have never had children are twice as likely to develop ovarian cancer as those who have. Early age at first pregnancy, early menopause, and the use of oral contraceptives, which reduces ovulation frequency, appear to be protective against ovarian cancer. If a woman has had breast cancer, her chances of developing ovarian cancer double. Certain rare genetic disorders and usage of hormone replacement therapy are also associated with increased risk. The highest incidence rates for ovarian cancer are reported in the more industrialized countries, with the exception of Japan.

Early detection requires periodic, thorough pelvic examinations. The Pap smear, useful in detecting cervical cancer, does not reveal ovarian cancer. Women over age 40 should have a cancer-related check-up every year. Researchers are currently trying to develop a blood test that will allow for consistent diagnosis of early ovarian cancer. Testing for high levels of tumor marker CA-125, a protein antigen, is sometimes used to screen for ovarian cancer. However, CA-125 is unreliable because it gives false positive results for many other disorders, including benign ovarian cysts and pregnancy. A combination of blood testing and ultrasound examination of the ovaries is currently the most accurate way to diagnose ovarian cancer.

Surgery, radiation therapy, and drug therapy are treatment options. Surgery usually includes the removal of one or both ovaries **(oophorectomy),** the uterus **(hysterectomy),** and the uterine tubes **(salpingectomy).** In some very early tumors, only the involved ovary is removed, especially in young women. In advanced disease, an attempt is made to remove all intra-abdominal cancerous tissue to enhance the effect of chemotherapy.

---

region superior to the entrance of the uterine tubes, and the *body* of the uterus is the major region. The *cervix* is the narrow end of the uterus that projects into the vagina. A cervical orifice, or opening, leads to the lumen of the vagina.

Development of the embryo normally takes place in the uterus. This organ, sometimes called the womb, is approximately 5 cm wide in its usual state but is capable of stretching to over 30 cm to accommodate the growing baby. The lining of the uterus, called the *endometrium,* participates in the formation of the placenta (see Chapter 18), which supplies nutrients needed for embryonic and fetal development. In the nonpregnant female, the endometrium varies in thickness during a monthly menstrual cycle, discussed later in this chapter.

Cancer of the cervix is a common form of cancer in women. The Medical Focus Immunization: The Great Protector in Chapter 13 (pages 309–310) describes a vaccine that immunizes against the human papillomavirus that causes the majority of human cervical cancers. Early detection is possible by means of a **Pap smear,** which entails the removal of a few cells from the region of the cervix for microscopic examination. If the cells are cancerous, a hysterectomy (the removal of the uterus) may be recommended. Removal of the ovaries in addition to the uterus is termed an **ovariohysterectomy.** Because the vagina remains intact, the woman can still engage in sexual intercourse.

Vagina

The vagina is a tube that makes a 45° angle with the inferior lumbar region (small of the back). The mucosal lining of the vagina lies in folds that extend when the fibromuscular wall stretches. This capacity to extend is especially important when the vagina serves as the birth canal, and it can also facilitate intercourse, when the vagina receives the penis.

External Genitals

The female external genitals (Fig. 17.13) are known collectively as the vulva. The vulva includes two large, hair-covered folds of skin called the **labia majora** (sing., *labium majus*). They extend posteriorly from the **mons pubis,** a fatty prominence underlying the pubic hair. The **labia minora** (sing., *labium minus*) are two small folds of skin lying just inside the labia majora. They extend forward from the vaginal opening to encircle and form a foreskin for the **clitoris,** an organ that is homologous to the penis. Although quite small, the clitoris has a shaft of erectile tissue and is capped by a pea-shaped glans. The clitoris also has sensory receptors that allow it to function as a sexually sensitive organ.

The **vestibule,** a cleft between the labia minora, contains the orifices of the urethra and the vagina. The vagina can be partially closed by a ring of tissue called the hymen. The hymen

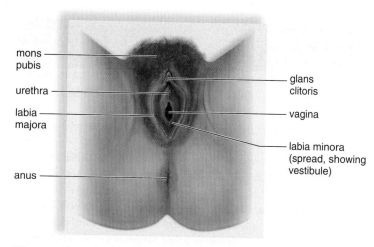

Figure 17.13 **External genitals of the female.** At birth, the opening of the vagina is partially blocked by a membrane called the hymen. Physical activities and sexual intercourse disrupt the hymen.

ordinarily is ruptured by initial sexual intercourse; however, it can also be disrupted by other types of physical activities. If the hymen persists after sexual intercourse, it can be surgically ruptured.

Female Sexual Response

Upon sexual stimulation, the labia minora, the vaginal wall, and the clitoris become engorged with blood. The breasts also swell, and the nipples become erect.

The vagina expands and elongates. Blood vessels in the vaginal wall release small droplets of fluid that seep into the vagina and lubricate it. Mucus-secreting glands called **Bartholin glands,** beneath the labia minora on either side of the vagina, also provide lubrication for entry of the penis into the vagina. (Bartholin glands are similar to male bulbourethral glands). Although the vagina is the organ of sexual intercourse in females, the clitoris plays a significant role in the female sexual response. The extremely sensitive clitoris can swell to two or three times its usual size. The thrusting of the penis and the pressure of the pubic symphyses of the partners act to stimulate the clitoris.

Orgasm occurs at the height of the sexual response. Blood pressure and pulse rate rise, breathing quickens, and the walls of the uterus and uterine tubes contract rhythmically. A sensation of intense pleasure is followed by relaxation when organs return to their normal size. Females have no refractory period, and multiple orgasms can occur during a single sexual experience.

Regulation of Female Hormone Levels

At the time of puberty in females, the hypothalamus increases its secretion of GnRH, and the anterior pituitary releases larger amounts of the gonadotropins, FSH and LH. These hormones stimulate the ovaries to produce and release ova. FSH and LH also cause elevated estrogen and progesterone production by the ovaries (Fig. 17.14).

Estrogen and Progesterone

Estrogen stimulates the growth of the uterus and vagina, and is also necessary for ovum maturation. Development of the secondary sex characteristics in the female, including the onset of the menstrual cycle, is caused by increased estrogen production during puberty. Along with sexual development, girls going through puberty experience an early increase in long bone growth that leads to increased height. However, estrogen also leads to early fusion of the growth plates in long bones, so that most young women have achieved their full adult height by about age 16.

Secondary Sex Characteristics These characteristics include the female pattern of body hair and fat distribution. In general, females have a more rounded appearance than males because of a greater accumulation of fat beneath the skin. Also, the pelvic girdle enlarges in females so that the pelvic cavity has a larger relative size compared to that of males; this means that females have wider hips. Both estrogen and **progesterone** are required for breast development, which is discussed on pages 407 and 409.

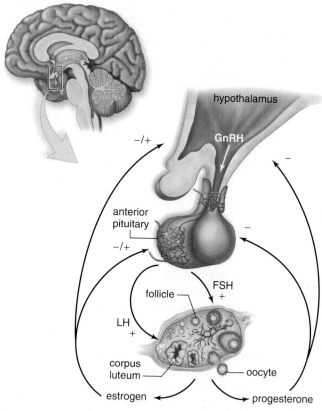

Figure 17.14 **Hormonal control of ovary.** Hypothalamic gonadotropin-releasing hormone (GnRH) causes pituitary secretion of follicle-stimulating hormone (FSH) and luteinizing hormone (LH). FSH primarily stimulates ovarian estrogen secretion, while LH primarily causes corpus luteum progesterone secretion after ovulation. Surges of both LH and FSH cause ovulation. Estrogen and progesterone maintain the female reproductive organs and secondary sexual characteristics, and they exert feedback control over the hypothalamus and pituitary.

Menstrual Cycle

The **menstrual cycle** is a monthly series of events that involve the ovaries and uterus plus the hormones already mentioned. The cycle is about 28 days long, but it can be as short as 18 days or as long as 40 days (Fig. 17.15).

Pre-Ovulation Events

Under the influence of follicle-stimulating hormone (FSH) from the anterior pituitary, several follicles begin developing in the ovary. Therefore, this period of time (days 1–14) is called the *follicular phase* of the ovary (Fig. 17.15). Although several follicles begin growing, only one follicle continues developing, and it secretes increasing amounts of estrogen. This particular follicle becomes more and more sensitive to FSH and then LH. Eventually, the very high level of estrogen exerts *positive feedback control* over the hypothalamus so that it secretes ever greater amounts of GnRH. GnRH induces a surge in FSH and LH secretion by the pituitary. The LH level rises to a greater extent than does the FSH level. Ovulation is triggered by the very high level of hormonal stimulation, and in particular the spike in LH, called the LH surge. (It is interesting to note that over-the-counter ovulation prediction test kits work by detecting high levels of LH metabolites in the urine, combining the metabolites with monoclonal antibodies to produce a color change.)

While the ovary is experiencing its follicular phase, first menstruation and then the proliferative phase occur in the uterus. During menstruation (days 1–5), a low level of female sex hormones in the body causes the endometrial tissue to disintegrate and its blood vessels to rupture. A flow of blood and tissues, known as the **menses,** passes out of the vagina during menstruation, also called the menstrual period.

Under the influence of estrogen released by the new follicle, the endometrium thickens and becomes vascular and glandular. This is the *proliferative phase* of the uterus, which ends when ovulation occurs.

Post-Ovulation Events

Under the influence of LH from the anterior pituitary, the emptied ovarian follicle tissue becomes a hormone-secreting tissue called the corpus luteum. Therefore, this period of time (days 15–28) is known as the *luteal phase* of the ovary (Fig. 17.15). The corpus luteum secretes progesterone and some estrogen. As the blood level of progesterone rises, it exerts *negative feedback control* over the anterior pituitary's secretion of LH so that the corpus luteum in the ovary begins to degenerate. If fertilization of the egg does occur, the corpus luteum persists for reasons that will be discussed shortly.

Under the influence of progesterone secreted by the corpus luteum, a secretory phase (days 15–28) begins in the uterus. During the *secretory phase* of the uterus, the endometrium of the uterus doubles or even triples in thickness (from 1 mm to 2–3 mm), and the uterine glands mature, producing a thick, mucoid secretion. The endometrium is now prepared to receive the embryo. If implantation of an embryo does not take place,

the corpus luteum disintegrates, and menstruation occurs. The disintegrated corpus luteum in the ovary becomes a patch of white scar tissue called the corpus albicans. With age, more of a woman's ovarian tissue becomes nonfunctional corpus albicans. This is why ovulation and menstruation eventually stop during menopause.

If fertilization occurs and is followed by implantation, the developing placenta produces **human chorionic gonadotropin (HCG),** which maintains the corpus luteum in the ovary until the placenta begins to produce its own progesterone and estrogen. The placental hormones shut down the anterior pituitary so that no new ovarian follicle matures. Placental hormones also maintain the endometrium so that the corpus luteum in the ovary is no longer needed. Usually, no menstruation occurs during pregnancy. After human chorionic gonadotropin is metabolized and excreted in the urine, it can be detected using prepared monoclonal antibodies. Over-the-counter home pregnancy test kits function in this fashion, as discussed in Chapter 18.

Menopause

Menopause, the period in a woman's life during which the menstrual cycle ceases, is likely to occur between ages 45 and 55. The ovaries are no longer responsive to the gonadotropic hormones produced by the anterior pituitary, and the ovaries secrete very low amounts of estrogen and progesterone. At the onset of menopause, the uterine cycle becomes irregular, but as long as menstruation occurs, it is still possible for a woman to conceive. Therefore, a woman is usually not considered to have completed menopause until menstruation has been absent for a year.

The hormonal changes during menopause often produce physical symptoms, such as "hot flashes" (caused by circulatory irregularities), dizziness, headaches, insomnia, sleepiness, and depression. These symptoms may be mild or even absent. If they are severe, medical attention should be sought. Women sometimes report an increased sex drive following menopause. It has been suggested that this may be due to androgen production by the adrenal cortex.

Female Breast and Lactation

Early growth of the female breasts during puberty is referred to as *budding* of the breasts. Budding is followed by development of lobes, the functional portions of the breast, and deposition of adipose tissue, which gives breasts their adult shape.

A breast contains 15 to 25 lobules, each with a milk duct that begins at the nipple. The nipple is surrounded by a pigmented area called the **areola.** Within the areola, glands are present that secrete an oily, saliva-resisting lubricant to protect the nipples, particularly during nursing. Smooth muscle fibers in the region of the areola may cause the nipple to become erect in response to sexual stimulation or cold.

Within each lobe, the mammary duct divides into numerous alveolar ducts that end in blind sacs called alveoli (Fig. 17.16). The alveoli are made up of the cells that can produce milk. Estrogen and progesterone are required for lobe development. It is believed that

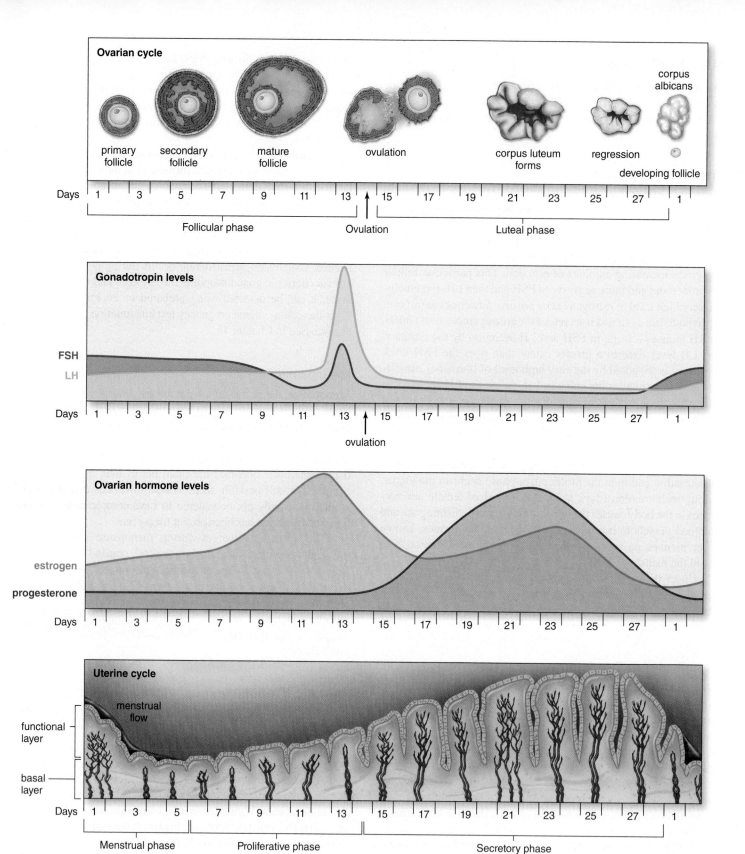

AP|R **Figure 17.15** **Events of the menstrual cycle.** During the menstrual cycle, FSH and LH are released by the anterior pituitary. FSH promotes the maturation of a follicle in the ovary. The follicle produces increasing levels of estrogen, which cause the endometrium to thicken during the proliferative phase in the uterus. An LH surge causes ovulation. After ovulation, LH promotes the development of the corpus luteum. This structure produces increasing levels of progesterone, which causes the endometrial lining to become secretory. Menses due to the breakdown of the endometrium begins when progesterone production declines to a low level due to corpus luteum disintegration.

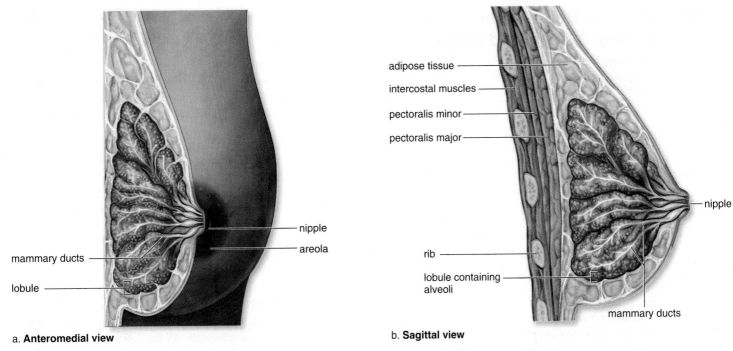

a. **Anteromedial view**

b. **Sagittal view**

adipose tissue

intercostal muscles

pectoralis minor

pectoralis major

nipple

rib

lobule containing alveoli

mammary ducts

nipple

areola

mammary ducts

lobule

AP|R Figure 17.16 **Structure of the female breast and mammary glands. (a)** Anteromedial view. **(b)** Sagittal section.

estrogen causes proliferation of ducts and that both estrogen and progesterone bring about alveolar development. The abundance of these hormones during pregnancy means that the alveoli proliferate at this time. A breast which does not produce milk has ducts but few alveoli, while a lactating (milk-producing) breast has many ducts and alveoli.

During pregnancy, the breasts enlarge as the ducts and alveoli increase in number and size. The same hormones that affect the mother's breasts can also affect the child's. Some newborns, including males, even secrete a small amount of milk for a few days.

Usually, no milk is produced during pregnancy. The hormone prolactin is needed for **lactation** (the process of milk production) to begin. Prolactin formation is suppressed during pregnancy, because negative feedback control by high levels of estrogen and progesterone shuts down prolactin secretion by the anterior pituitary. Once the baby is delivered, however, the pituitary begins secreting prolactin. It takes a couple of days for milk production to begin, and in the meantime, the breasts produce **colostrum,** a thin, yellow, milky fluid rich in protein, including antibodies.

The continued production of milk requires a suckling child. When a breast is suckled, the nerve endings in the areola are stimulated, and a nerve impulse travels along neural pathways from the nipples to the hypothalamus, which directs the posterior pituitary gland to release the hormone oxytocin. When this hormone arrives at the breast, it causes the lobules to contract so that milk flows into the ducts (called milk letdown), where it may be drawn out of the nipple by the suckling child.

Breast cancer is one of the few types of cancer that can be detected by the female herself. The Medical Focus on page 413 tells how to do a monthly check for breast cancer.

<hr/>

Content **CHECK-UP!**

7. Which ovarian structure eventually releases the secondary oocyte? Which produces progesterone? Which is scar tissue? How are they related?

8. The second stage of meiosis in an ovum will only be completed if:

a. it occurs before the ovum is released at ovulation.

b. it occurs after the ovum is released and travels through the uterine tube.

c. the ovum is fertilized by a sperm cell.

d. it occurs just after the ovum is released from the ovary, but before it enters the uterine tube.

9. During which stage of the monthly cycle of the uterus does the endometrium thicken and become glandular and vascular?

a. menstrual phase

b. secretory phase

c. proliferative phase

d. ovulation phase

Answers in Appendix A.

Control of Reproduction and Sexually Transmitted Infections

14. List several means of birth control, and describe their effectiveness.
15. Describe the symptoms of genital warts, genital herpes, hepatitis, chlamydia, gonorrhea, and syphilis.

Contraception

Today, there are more options than ever before for **contraception,** or the prevention of pregnancy. Although many contraceptives are available without prescription, women should consult a health-care provider before choosing a birth control method. Health-care providers can give patients advice on the best method to use and can also ensure that the patient uses the product correctly. However, the patient herself should consider her options carefully, to choose the method that best suits her lifestyle, desire for future pregnancy, and ethical beliefs. Table 17.3 lists the means of birth control used in the United States and rates their effectiveness. The statistics cited are for typical use; i.e., allowing for some users who do not follow directions or use the method consistently.

Oral contraceptives (OCs) come in two basic forms: combinations of a synthetic estrogen and progesterone, or a synthetic progesterone alone. Combination hormone methods can be in the form of pills taken orally, and a new regimen of combined pill allows a woman to have only three or four menstrual cycles a year. In addition, combination hormones can be introduced into the blood via patches that allow hormone absorption through the skin, or by a plastic ring placed in the vagina. The ring slowly releases hormones into the bloodstream. Regardless of delivery method, the estrogen and progesterone in the combined hormone form effectively shut down the pituitary's production of both FSH and LH so that no follicle begins to develop in the ovary. Because ovulation does not occur, pregnancy cannot take place. These contraceptives also change the endometrial lining of the uterus and may prevent implantation of the embryo.

A second hormonal method of contraception involves use of only a synthetic progesterone, without estrogen. The oral form is commonly referred to as the "mini-pill" (because it contains only one hormone) and is taken daily. An injection, Depo-Provera, can be taken once every three months. Another form of the same hormone is an implantable rod that is inserted into the skin under the arm. All three work by inhibiting ovulation and making the mucus of the uterine cervix thick and sticky. Sperm cells cannot swim through the thick mucus, thus preventing fertilization of the ovum. If fertilization does occur and an embryo is created, the uterine lining is changed by this medication so that the embryo will not be able to implant.

All hormonal forms of contraception have potential side effects and health risks, and there are some women who must not take them. Anyone who has had any form of disorder associated with excessive blood clotting (or a family history of abnormal clotting) must never use hormonal contraceptives because of the increased risk of clot formation. A clot that has formed in a blood vessel is a *thrombus* or *thromboembolism* (see Chapter 11), and can cause heart attack or stroke. Likewise, women who smoke are discouraged from taking hormonal contraceptives, because they too have an increased risk of spontaneous clot formation. Women with uncontrolled high blood pressure, liver or kidney disease, or certain forms of hormone-sensitive female cancers must use other forms of contraception as well.

An intrauterine device (IUD) is a small piece of molded plastic that is inserted into the uterus by a physician. IUDs alter the environment of the uterus and uterine tubes, causing inflammation so that fertilization probably will not occur—but if fertilization does occur, implantation cannot take place. One type of IUD has copper wire wrapped around the plastic. Other forms will slowly release progesterone over time into the bloodstream. IUDs are very effective. Their possible side effects include cramping, bleeding, pelvic inflammatory disease, increased risk of ectopic pregnancy, infertility, and a slight risk of uterine perforation.

The diaphragm is a soft latex cup with a flexible rim that lodges behind the pubic bone and fits over the cervix. The diaphragm can be inserted into the vagina no more than two hours before sexual relations. Also, it must be used with spermicidal jelly or cream and should be left in place at least six hours after sexual relations. A cervical shield is smaller than a diaphragm, and the cervical cap is even smaller still. All three devices must be fitted by a physician or other health-care provider. Each carries a slight risk of urinary tract infection in the user. Rarely, incidents of a very serious illness—toxic shock syndrome—have been reported in users of these methods. Use of proper hygiene methods, including handwashing before inserting the device, and prompt removal of the device, can minimize the risk of serious illness.

Traditional barrier methods of birth control remain fairly popular because they are inexpensive, available without prescription, and also offer some protection against sexually transmitted infections. A female condom consists of a large polyurethane tube with a flexible ring that fits onto the cervix. The open end of the tube has a ring that covers the external genitals. A male condom is most often a latex or polyurethane sheath that fits over the erect penis. The semen is trapped inside the sheath and, thus, does not enter the vagina. When used in conjunction with a spermicide, protection is better than with the condom alone. So-called "natural" condoms made of lambskin or other material do not provide any protection against sexually transmitted infection.

People can be surgically sterilized as a form of contraception. In men, a vasectomy procedure will cut or crush the vasa deferentia, preventing sperm from entering the semen. Uterine tubal ligation in women prevents sperm from reaching the ovum in the uterine tube. An implantable metal device (Essure) can be placed at the entrance to the tube without the need for surgery. Once in place, the implant causes scar tissue to form, blocking the uterine tube. However, those who elect to use these techniques must be aware that they are most likely irreversible.

Women who are concerned about environmental pollution caused by hormonal contraceptives (see Medical Focus: Endocrine-Disrupting Contaminants on page 414) as well as those who have ethical concerns about hormonal contraceptives have revived interest in Natural Family Planning. There are several techniques, and all involve abstaining from intercourse on the days of the woman's cycle just prior to when the ovum is released and for three days afterward. On those days, conception is most likely to occur. The most effective

TABLE 17.3 Common Methods of Contraception

Name	Procedure	How Does It Work?	How Effective Is It?	Side Effects and Other Health Risks
NATURAL METHODS				
Abstinence	Refrain from sexual intercourse	No sperm in vagina	100%	None; also protects against sexually transmitted infections
Natural family planning	Determine day of ovulation by keeping records; various testing methods	Intercourse is avoided only during the time period while ovum is viable	80–97%	None (requires training for symptothermal method to be effective)
Withdrawal method	Penis withdrawn from vagina just before ejaculation	Ejaculation outside the woman's body; no sperm in vagina	75%	None
Douching	Vagina cleansed after intercourse	Washes sperm out of woman's body	Less than 70%	May cause infection or inflammation
NONPRESCRIPTION METHODS				
Male condom	Sheath of latex, polyurethane, or natural material fitted over erect penis	Traps sperm and prevents entry into vagina; latex and polyurethane forms protect against STIs	85%	With latex: latex allergy; natural material condoms give no protection against STIs
Female condom	Synthetic latex fitted inside vagina	Traps sperm and prevents entry into vagina; some protection against STIs	79%	Possible allergy or irritation; urinary tract infection
Spermicide: jellies, foams, films, creams, tablets	Spermicidal products inserted into the vagina before intercourse	Spermicidal chemical nonoxynol-9 kills large numbers of sperm cells	50–80%	Irritation, inflammation; allergic reaction; urinary tract infection
Contraceptive sponges	Sponge containing spermicide inserted into vagina and placed against cervix	Spermicidal chemical nonoxynol-9 kills large numbers of sperm cells	72–86%	Irritation, inflammation; allergic reaction; urinary tract infection; toxic shock syndrome
PRESCRIPTION HORMONAL METHODS				
Combined hormone, vaginal ring	Flexible plastic ring inserted into vagina; releases hormones that are absorbed into the bloodstream	Suppresses ovulation by the combined actions of the hormones estrogen and progestin; prevents implantation by embryo	92%	Vaginal irritation, inflammation; dizziness, nausea; changes in menstruation, mood, and weight; rarely: cardiovascular disease, including high blood pressure, blood clots, heart attack, and strokes
Progesterone-only injection (Depo-Provera)	Injection of progestin once every three months	Inhibits ovulation; prevents sperm from reaching the egg; prevents implantation by embryo	97%	Irregular bleeding; weight gain, breast tenderness; headaches; osteoporosis if used for extended period
Progesterone-only implantable rod (Implanon)	Single rod implanted into skin under the arm	Inhibits ovulation; prevents sperm from reaching the egg; prevents implantation by embryo	99%	Irregular bleeding, weight gain, increased risk of blood clots, heart attack, stroke
Emergency contraception	Must be taken within 72 hours after unprotected intercourse	Suppresses ovulation by the combined actions of the hormones estrogen and progestin; prevents implantation by embryo	80%	Nausea; vomiting; abdominal pain; fatigue; headache; excessive bleeding
Combined hormone pill	Pills are swallowed daily; chewable form also available	Suppresses ovulation by the combined actions of the hormones estrogen and progestin; may prevent implantation by embryo	92%	Dizziness, nausea; changes in menstruation, mood, and weight; rarely: cardiovascular disease, including high blood pressure, blood clots, heart attack, and strokes

Continued

TABLE 17.3 Continued

Name	Procedure	How Does It Work?	How Effective Is It?	Side Effects and Other Health Risks
PRESCRIPTION HORMONAL METHODS				
Progestin-only mini-pill	Pills are swallowed daily	Thickens cervical mucus, preventing sperm from contacting ovum; may prevent implantation by embryo	92%	Irregular bleeding; weight gain; breast tenderness
Combined hormone 91, daily regimen	Pills are swallowed daily; user has 3–4 menstrual periods a year	Suppresses ovulation by the combined actions of the hormones estrogen and progestin; may prevent implantation by embryo	98%	Dizziness, nausea; changes in menstruation, mood, and weight; rarely: cardiovascular disease, including high blood pressure, blood clots, heart attack, and strokes
Combined hormone patch	Patch is applied to skin and left in place for 1 week; new patch applied	Suppresses ovulation by the combined actions of the hormones estrogen and progestin; may prevent implantation by embryo	98%	Skin irritation, allergy; dizziness, nausea; changes in menstruation, mood, and weight; rarely: cardiovascular disease, including high blood pressure, blood clots, heart attack, and strokes
PRESCRIPTION BARRIER METHODS				
Diaphragm	Latex cup, placed into vagina to cover cervix before intercourse	Blocks entrance of sperm into uterus; spermicide kills sperm	84% with spermicide	Irritation, inflammation; allergic reaction; urinary tract infection; toxic shock syndrome
Cervical cap	Latex cap held over cervix	Blocks entrance of sperm into uterus; spermicide kills sperm	84% with spermicide	Irritation, inflammation; allergic reaction; toxic shock syndrome; abnormal Pap smear
Cervical shield	Latex cap placed in upper vagina, held in place by suction	Blocks entrance of sperm into uterus; spermicide kills sperm	84% with spermicide	Irritation, inflammation; allergic reaction; toxic shock syndrome; urinary tract infection
Intrauterine device, copper T	Placed in uterus	Causes cervical mucus to thicken; fertilized embryo cannot implant	99%	Cramps; bleeding; pelvic inflammatory disease; infertility; perforation of uterus; ectopic pregnancy risk
Intrauterine device, progesterone-releasing type	Placed in uterus	Prevents ovulation; causes cervical mucus to thicken; fertilized embryo cannot implant	99%	Cramps; bleeding; pelvic inflammatory disease; infertility; perforation of uterus; ectopic pregnancy risk
SURGICAL STERILIZATION				
Vasectomy	Vasa deferentia are cut and tied	No sperm in seminal fluid	Almost 100%	Most likely irreversible; minor risk of surgical complications such as infection or reaction to anesthetic
Tubal ligation: transabdominal surgery	Oviducts are cut and tied, or a clip is placed on the oviduct	Sperm cannot enter oviduct; ova cannot pass through oviduct	Almost 100%	Most likely irreversible; pain, bleeding, infection; other post-surgical complications; ectopic (tubal) pregnancy
Sterilization implant (Essure)	Small metallic implant is placed into uterine tubes; inserted through the vagina using a catheter	Causes scar tissue to form, blocking uterine tubes and preventing conception	Almost 100%	Most likely irreversible; pain after placement; ectopic pregnancy

The American Cancer Society urges women to do a breast self-exam and men to do a testicle self-exam every month. Breast cancer and testicular cancer are far more curable if found early, and either test takes less than half an hour.

Breast Self-Exam for the Ladies:

1. Check your breasts for any lumps, knots, or changes about one week after your period. Don't stand up to do the exam—be sure to lie down on a flat surface. This will allow your breast tissue to spread evenly over your chest, so you can feel a change much more easily.

2. Place your right hand behind your head. Use the fingertips of your middle three fingers for the test; they're the most sensitive part of your hand. Move your *left* hand over the right side of your chest in an up-and-down pattern. Begin at the edge of your armpit, tracing overlapping dime-sized circles all the way to the bottom of your rib cage. Stroke up and down across your entire chest from the armpit to the sternum (breastbone). As you move your fingers across your chest, slide them from the top of your clavicle (collarbone) down to the bottom of your rib cage. Start with a very gentle pressure, so you can feel any lumps that might lie just under the skin. Gradually increase the pressure until you're stroking firmly on the breast. Use each level of pressure before you move onto another area (Fig. 17A).

3. Now place your left hand behind your head and check your *left* breast with your *right* hand.

4. Then, check your breasts while standing in front of a mirror (Fig. 17B). First, put your hands firmly on your hips. Then, raise your arms above your head and stand up very straight so you're tightening your chest muscles and skin. Look for any changes in the way your breasts look: changes in size or shape, dimpling of the skin, changes in the nipple, redness, swelling, or abnormal discharge.

5. If you find any changes during your self check-ups, try not to panic. Finding that there's a change doesn't necessarily mean you've got cancer. But see your physician or other health-care provider right away, just to be sure. Remember—if you find it early, there's an excellent chance that breast cancer can be cured.

You should know that the best check for breast cancer is a mammogram, which is a low-dose X ray of the breast tissue. When your doctor checks your breasts, ask about getting a mammogram. Yearly mammograms are recommended for women over 40 years of age. Women with a personal or family history of breast cancer (mother, sister, or aunt who have or had the disease) may be advised to start such screening exams at an earlier age, and to have the exams more frequently. Additional exams such as ultrasound or MRI of the breast may also be advised.

Testicle Self-Exam for the Guys:

1. Check your testicles once a month, beginning right after puberty. It takes only a few minutes.

2. Do your check right after a warm bath or shower, so that the skin on your scrotum is relaxed. Hold your penis out of the way, and check one testicle at a time. Roll each testicle between your thumb and finger as shown in Figure 17C. Feel for hard lumps or bumps. A normal epididymis often feels like a lump on the outside of your testis, so get used to locating it.

3. If you notice a change or have aches or lumps, try not to be too alarmed. There are many normal structures found in a testicle that may feel "lumpy," including blood vessels and supporting tissues. But be sure to tell your doctor or other health-care professional right away if you're concerned about any differences that you notice.

Testicular cancer is the most common solid tissue cancer in young men between the ages of 15 and 35, so doing a self-exam is very important. Cancer of the testicles can be cured if you find it early. You should also know that prostate cancer is the second most common cancer in men (second only to skin cancer). Men over age 50 should have an annual health checkup that includes a prostate examination.

Breast Self Examination
While you're lying down, examine up to the collarbone, out to armpit, in to middle of chest, and down to bottom of rib cage.

Figure 17A Breast Self-Examination.

Figure 17B Mirror check for breast cancer.

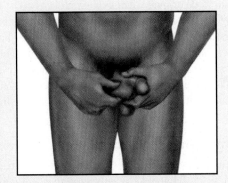

Figure 17C After-shower check for testicular cancer.

MEDICAL FOCUS

Endocrine-Disrupting Contaminants

One of the first synthetic estrogens to be produced in a laboratory was diethylstilbestrol, or DES, created in 1938. In the ensuing years, DES was used to treat a variety of conditions such as prostate cancer in men and menopausal symptoms in women. For a short time, it was even used as an emergency contraceptive. Tragically, DES was also given to many pregnant women to prevent early miscarriage. Years later, those same women's children are suffering from its effects on their reproductive systems. Daughters of the original patients (called *DES daughters*) often have birth defects involving reproductive system organs, including the ovaries, uterus, and vagina. Further, these same women are 2.5 times as likely to have breast cancer after age 40. DES daughters are also more likely to suffer from cancers of the vagina and cervix. DES sons have an increased risk of genital abnormalities, and possibly prostate and testicular cancers as well.

The consequences of DES use have raised concern about the many chemicals created after 1938 that might also affect the endocrine system. Because endocrine hormones influence nearly all aspects of physiology and behavior in animals, synthetic compounds called *endocrine-disrupting contaminants* (EDCs) that show hormonal effects are being studied extensively. Environmental biologists first became aware of EDCs when wildlife—especially fish, amphibians, reptiles, and birds in polluted areas—began to show abnormalities in tissue differentiation, sexual development, and reproduction. Human studies suggest that these same pollutants lower sperm counts, reduce male and female fertility, and increase the rates of certain cancers (breast, ovarian, testicular, and prostate). Further, EDCs are thought to be responsible for early onset of menses in young girls, some of whom begin their periods as early as age 10. Additionally, some studies seem to indicate that EDCs contribute to learning deficits and behavioral problems in children.

Many EDCs are chemicals used as pesticides and herbicides in agriculture, and some are associated with the manufacture of various other synthetic organic compounds such as PCBs (polychlorinated biphenyls). Oral contraceptives and other hormonal pharmaceuticals are introduced into the environment via human sewage. Other chemicals shown to influence hormones are found in plastics, food additives, and personal hygiene products. Baby bottles, plastic cups, and toys recently have been discontinued from the market because compounds used for hardening their plastic are suspected EDCs. In mice, some plastic components affect neonatal development when present in extremely low concentrations (parts per *trillion*). EDCs have been found in nature at levels 1,000 times greater than this—even in amounts comparable to functional hormone levels in the human body. It's also important to recognize that hormones and hormonelike contaminants are concentrated in adipose tissue. Obesity, which is epidemic in the United States, increases the risk that EDCs might be found at toxic levels in humans.

Scientists and those representing industrial manufacturers continue to debate whether EDCs pose a health risk to humans. Currently, the EPA regulates the concentration of 23 inorganic chemicals including mercury, lead, and arsenic, and more than 30 organic chemicals also are regulated. Doubtless other inorganic and organic chemicals will be added to the list as research continues. In the meantime, there are steps you can take to minimize the dangers of EDCs:

- Maintain a healthy weight. Excess adipose tissue can contain high levels of EDCs that you may have been exposed to throughout your life.
- Use less plastic. Substitute reusable cloth bags, washable dinnerware, glass containers, etc., for the plastic products you might otherwise buy. In particular, avoid plastics marked 3, 6, or 7, because these types are the most likely to contain EDCs.
- Recycle plastic, so that EDCs that might be contained in it don't leak into the groundwater.
- Don't heat or reheat food in plastic containers.
- If you eat a lot of fish, be aware of those that may contain high levels of EDCs and other contaminants. The EPA maintains a database that can help you find out which fish in your area are safe to eat: http://www.epa.gov/waterscience/fish/

form, called the *symptothermal method,* uses two separate sets of data to determine when ovulation occurs. The first involves recording basal body temperature, the temperature taken when the woman first awakens. At the time of ovulation, her body temperature will increase abruptly. Simultaneously, the consistency of the mucus discharge from her cervix changes from thick, dense and sticky to elastic (like raw egg whites). When used correctly by trained and motivated couples, the method is 98% effective and free from any side effect.

Emergency Contraception

Emergency contraception, sometimes referred to as "morning-after pills," refers to a medication that will prevent pregnancy after unprotected intercourse. The name is a misnomer because the medication does not have to be taken the next morning—the woman can begin the medication one to several days after having unprotected intercourse.

One type of emergency contraception contains four synthetic estrogen-progesterone pills; a second type contains progesterone-only pills. Two pills are taken up to 72 hours after unprotected intercourse, and two more are taken 12 hours later. The medication upsets the normal uterine cycle, preventing an embryo from implanting in the endometrium. The medication should not be used for regular contraception. Side effects may include nausea, vomiting, abdominal pain, fatigue, and headache. Other, more serious complications include hypertension, cardiovascular disease, and stroke.

Mifepristone, better known as RU-486, does not prevent conception. Rather, this medication is presently used to cause the loss of an implanted embryo by blocking the progesterone receptor proteins of endometrial cells. Without functioning receptors for progesterone, the endometrium sloughs off, carrying the embryo with it. Thus, the drug induces a medical abortion. When taken in conjunction with a prostaglandin to induce uterine contractions, RU-486 is 95% effective in ending an early pregnancy. However, cases of blood-borne infection and severe hemorrhage have occurred in women using the drug, and it has caused fatalities in Europe and the United States. Thus, patients using RU-486 must be under the constant supervision of a qualified physician.

Anyone considering sexual activity must understand that only condoms can prevent a sexually transmitted infection, and even they are not completely reliable. Further, it's important to know that all forms of contraception carry the risk of failure. The only 100% effective method of birth control is complete abstinence, that is, not engaging in sexual intercourse. This form of birth control has the added advantage of preventing transmission of a sexually transmitted infection.

Infertility

Infertility is the failure of a couple to achieve pregnancy after one year of regular, unprotected intercourse. The American Medical Association estimates that 15% of all couples are infertile. The cause of infertility can be attributed to the male (40%), the female (40%), or both (20%).

Causes of Infertility

The most frequent cause of infertility in males is low sperm count and/or a large proportion of abnormal sperm. There are numerous causes for low sperm count. Often, a low sperm count is caused by a chromosome abnormality in the male. Exposure to ionizing radiation can also impair sperm formation. Herbicides and pesticides may cause feminizing effects in the man's body, as discussed in the previous Medical Focus. Testicular infection by the mumps virus, gonorrhea, or chlamydia may destroy the sperm-forming cells. Chemicals used on the job, such as paints, varnishes, and degreasers, may cause infertility as well. Smoking, alcohol, and drug abuse decrease the numbers and quality of a man's sperm cells. In particular, abuse of anabolic (body-building) steroids can shrink the testes and decrease sperm count. Exposing the testicles to overheating impairs sperm production and lowers sperm count. Common causes for overheating include prolonged usage of a sauna or hot tub. Wearing extremely tight clothing, sitting for prolonged periods, and long periods of using a very warm laptop computer may also decrease sperm count in some men.

Body weight appears to be the most significant factor in causing female infertility. Both obese and underweight women may experience difficulty in becoming pregnant. In women of normal weight, fat cells produce a hormone called leptin that stimulates the hypothalamus to release GnRH. In overweight women, excess leptin impairs ovarian function. The ovaries often contain many small follicles, and the woman fails to ovulate. A woman who is too thin may not produce enough GnRH, FSH, and LH, and will have estrogen deficiency as a result. She too will fail to ovulate. Other causes of infertility in females are blocked uterine tubes due to pelvic inflammatory disease (see page 417) and endometriosis. **Endometriosis** is the presence of uterine tissue outside the uterus, particularly in the uterine tubes and on the abdominal organs. Backward flow of menstrual fluid allows living uterine cells to establish themselves in the abdominal cavity, where they go through the usual uterine cycle, causing pain and structural abnormalities that make it more difficult for a woman to conceive.

Training the couple to recognize the signs of a woman's ovulation (using the symptothermal method mentioned previously) may enable them to effectively time their intercourse and become pregnant. Sometimes the causes of infertility can be corrected by medical intervention so that couples can have children. If no obstruction is apparent and body weight is normal, it is possible to give women fertility drugs, which are gonadotropic hormones that stimulate the ovaries and bring about ovulation. However, these types of hormone treatments may cause multiple ovulations and multiple births.

When all attempts to get pregnant fail, many couples adopt a child. Others sometimes try one of the assisted reproductive technologies discussed next.

Assisted Reproductive Technologies

Assisted reproductive technologies (ARTs) consist of techniques used to increase the chances of pregnancy. Often, sperm and/or ova are retrieved from the testes and ovaries, and fertilization takes place in a clinical or laboratory setting.

Artificial Insemination During **artificial insemination,** sperm are placed in the vagina by a physician. Sometimes a woman is artificially inseminated by her partner's sperm. This is especially helpful if the partner has a low sperm count, because the sperm can be collected over a period of time and concentrated so that the sperm count is sufficient to result in fertilization. Often, however, a woman is inseminated by sperm acquired from a donor who is a complete stranger to her. At times, a combination of partner and donor sperm is used.

During *intrauterine insemination (IUI),* fertility drugs are given to stimulate the ovaries, and then the donor's sperm is placed in the uterus rather than in the vagina.

If the prospective parents wish, sperm can be sorted into those believed to be X-bearing or Y-bearing to increase the chances of having a child of the desired sex.

In Vitro Fertilization During **in vitro fertilization (IVF),** conception occurs in the laboratory. Ultrasound machines can now spot follicles in the ovaries that hold immature ova; therefore, the latest method is to forgo the administration of fertility drugs and retrieve immature ova by using a needle. The immature ova are then brought to maturity in the laboratory before concentrated sperm are added. After about two to four days, the embryos are ready to be transferred to the uterus of the woman, who is now in the secretory phase of her uterine cycle. If desired, preimplantation genetic analysis can be done and only embryos found to be free of genetic disorders are used. If implantation is successful, development is normal and continues to term.

Intracytoplasmic Sperm Injection In **intracytoplasmic sperm injection (ICSI),** a highly sophisticated procedure, a single sperm is injected into an egg. This method is used effectively when a man has severe infertility problems.

Sexually Transmitted Infections

Sexually transmitted infections (STIs) are caused by organisms ranging from viruses to arthropods; however, we will discuss only certain STIs caused by viruses and bacteria. Unfortunately, for unknown reasons, humans cannot develop effective immunity to any STIs. Therefore, any person exposed to an STI should seek prompt medical treatment. To prevent the spread of STIs, a latex or polyurethane condom can be used; the concomitant use of a spermicide containing nonoxynol-9 gives added protection.

Among those STIs caused by viruses, treatment is available for AIDS and genital herpes. However, it is important to note that treatment for HIV/AIDS and genital herpes cannot presently eliminate the virus from the person's body. Drugs used for treatment can merely slow replication of the viruses. Thus, neither viral disease is presently curable. Further, antiviral drugs have serious, debilitating side effects on the body.

Only STIs caused by bacteria (e.g., chlamydia, gonorrhea, and syphilis) are curable with antibiotics. Bacteria that acquire antibiotic resistance may necessitate treatment with extremely strong drugs for an extended period to achieve a cure.

Genital Warts

Genital warts are caused by the human papillomaviruses (HPVs). Many times, carriers either do not have any sign of warts or merely have flat lesions. When present, the warts commonly are seen on the penis and foreskin of men and near the vaginal opening in women. A newborn can become infected while passing through the birth canal.

Individuals who are currently infected with visible growths may have those growths removed by surgery, freezing, or burning with lasers or acids. However, visible warts that are removed may recur. You'll recall (from Chapter 13) that a vaccine for the human papillomaviruses that most commonly cause genital warts is now available. The vaccine is an extremely important step in the prevention of cancer, as well as the warts themselves. Genital warts are associated with cancer of the cervix, as well as tumors of the vulva, vagina, anus, and penis. Researchers believe that these viruses may be involved in up to 95% of all cases of cancer of the cervix. Vaccination might make such cancers a thing of the past.

Genital Herpes

Genital herpes is caused by herpes simplex virus. Type 1 usually causes cold sores and fever blisters, while type 2 more often causes genital herpes.

Persons usually get infected with herpes simplex virus type 2 when they are adults. Some people exhibit no symptoms; others may experience a tingling or itching sensation before blisters appear on the genitals (Fig. 17.17*a, b*). Once the blisters rupture, they leave painful ulcers that may take as long as three weeks or as little as five days to heal. The blisters may be accompanied by fever, pain on urination, swollen lymph nodes in the groin, and in women, a copious discharge. At this time, the individual has an increased risk of acquiring an HIV infection.

After the ulcers heal, the disease is only latent, and blisters can recur, although usually at less frequent intervals and with milder

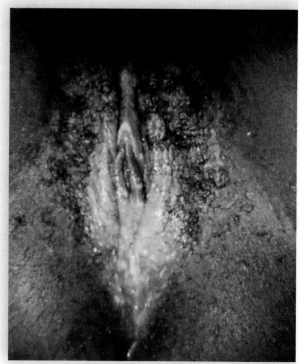

a.

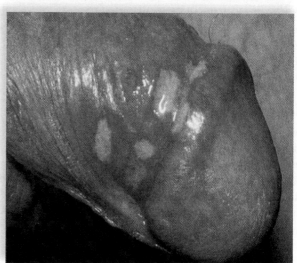

b.

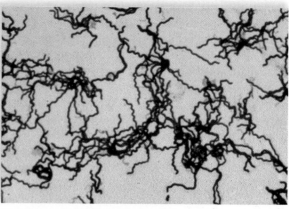

c.

Figure 17.17 Sexually transmitted infections. (a) Herpes simplex virus infection on female labia. **(b)** Herpes simplex virus infection on the penis. **(c)** *Treponema pallidum,* the bacterium that causes syphilis.

symptoms. Fever, stress, sunlight, and menstruation are associated with recurrence of symptoms. Exposure to herpes in the birth canal can cause an infection in the newborn, which leads to neurological disorders and even death. Birth by cesarean section prevents this possibility.

Hepatitis

Hepatitis infects the liver and can lead to liver failure, liver cancer, and death. The type of hepatitis and the virus that causes it are designated by the same letter. There are five known viruses that cause hepatitis, designated A-B-C-D-E. Hepatitis A is usually acquired from sewage-contaminated food or drinking water, but this infection can also be transmitted sexually through oral/anal contact. Hepatitis B is spread through sexual contact and by blood-borne transmission (accidental needle stick on the job, receiving a contaminated blood transfusion, a drug abuser sharing infected needles while injecting drugs, from mother to fetus, etc.). Simultaneous infection with hepatitis B and HIV is common because both share the same routes of transmission. Fortunately, a combined vaccine is available for hepatitis A and B; it is currently recommended that all children receive the vaccine to prevent infection (see pages 309–310). Hepatitis C is a blood-borne illness that causes most cases of post-transfusion hepatitis. Hepatitis D is sexually transmitted, and hepatitis E is acquired from contaminated water. Screening of blood and blood products can prevent transmission of hepatitis viruses during a transfusion. Proper water treatment techniques can prevent contamination of drinking water.

Chlamydia

Chlamydia is named for the tiny bacterium that causes it (*Chlamydia trachomatis*). The incidence of new chlamydia infections has steadily increased since 1984.

Chlamydia infections of the lower reproductive tract are usually mild or asymptomatic, especially in women. About 18 to 21 days after infection, men may experience a mild burning sensation on urination and a mucoid discharge. Women may have a vaginal discharge along with the symptoms of a urinary tract infection. Chlamydia also causes cervical ulcerations, which increase the risk of acquiring HIV.

If the infection is misdiagnosed or if a woman does not seek medical help, there is a particular risk of the infection spreading from the cervix to the uterine tubes so that **pelvic inflammatory disease (PID)** results. This very painful condition can result in blockage of the uterine tubes with the possibility of sterility and infertility. If a baby comes in contact with chlamydia during birth, inflammation of the eyes or pneumonia can result.

Gonorrhea

Gonorrhea is caused by the bacterium *Neisseria gonorrhoeae*. Diagnosis in the male is not difficult, since typical symptoms are pain upon urination and a thick, greenish-yellow urethral discharge. In males and females, a latent infection leads to pelvic inflammatory disease (PID), which can also cause sterility in males. If a baby is exposed during birth, an eye infection leading to blindness can result. All newborns are given eyedrops to prevent this possibility.

Gonorrhea proctitis, an infection of the anus characterized by anal pain and blood or pus in the feces, also occurs in patients. Oral/genital contact can cause infection of the mouth, throat, and tonsils. Gonorrhea can spread to internal parts of the body, causing heart damage or arthritis. If, by chance, the person touches infected genitals and then touches his or her eyes, a severe eye infection can result. Up to now, gonorrhea was curable by antibiotic therapy, but resistance to antibiotics is becoming more and more common, and 40% of all strains are now known to be resistant to therapy.

Syphilis

Syphilis is caused by a bacterium called *Treponema pallidum* (see Fig. 17.17c). As with many other bacterial diseases, penicillin is an effective antibiotic. Syphilis has three stages, often separated by latent periods, during which the bacteria are resting before multiplying again. During the *primary stage,* a hard **chancre** (ulcerated sore with hard edges) indicates the site of infection. The chancre usually heals spontaneously, leaving little scarring. During the *secondary stage,* the victim breaks out in a rash that does not itch and is seen even on the palms of the hands and the soles of the feet. Hair loss and infectious gray patches on the mucous membranes may also occur. These symptoms disappear of their own accord.

During the *tertiary stage,* which lasts until the patient dies, syphilis may affect the cardiovascular system by causing aneurysms, particularly in the aorta. In other instances, the disease may affect the nervous system, resulting in psychological disturbances. Also **gummas,** large destructive ulcers, may develop on the skin or within the internal organs.

Congenital syphilis is caused by syphilitic bacteria crossing the placenta. The child is born blind and/or with numerous anatomical malformations. Control of syphilis depends on prompt and adequate treatment of all new cases; therefore, it is crucial for all sexual contacts to be traced so they can be treated. Diagnosis of syphilis can be made by blood tests or by microscopic examination of fluids from lesions.

Content **CHECK-UP!**

10. A combined hormonal method of contraception is available through the following delivery method(s):

    a. pills taken orally.

    b. patches applied to the skin.

    c. injections received monthly.

    d. All of the above.

11. Name three bacterial forms of sexually transmitted infection.

12. Which of the following is a characteristic of tertiary syphilis?

    a. large, destructive ulcers called gummas

    b. formation of a hard chancre

    c. a skin rash that does not itch

    d. hair loss

Answers in Appendix A.

MEDICAL FOCUS

Preventing Transmission of STIs

It is extremely important to protect yourself from getting a sexually transmitted infection (STI). Some of the STIs, such as gonorrhea, syphilis, and chlamydia, can be cured by taking an antibiotic. Viral STIs, such as hepatitis, herpes, and HIV/AIDS, are incurable, though treatment is available for HIV and herpes. In any case, it is best to prevent the passage of STIs from person to person so that you never need treatment at all.

Sexual Activities Transmit STIs (Fig. 17D)

Abstain from sexual intercourse or develop a long-term monogamous (always the same partner) *sexual relationship* with a partner who is free of STIs.

Refrain from multiple sex partners or having relations with someone who has multiple sex partners. Think about it: if you have sex with two other people and each of these people has sex with two people and so forth, the number of people who might spread an STI dramatically increases.

Remember that, although the prevalence of AIDS is presently higher among homosexuals and bisexuals, the highest rate of increase in infection is now occurring among heterosexuals. The outermost lining of the uterus is quite thin, and it does allow infected cells from a sexual partner to enter.

Be aware that having relations with an intravenous drug user is risky because the behavior of this group risks hepatitis and an HIV infection. Recognize that anyone who already has another active STI is more susceptible to an HIV infection.

Uncircumcised males are more likely to become infected with an STI than circumcised males because vaginal secretions can remain under the foreskin for a long period of time.

Avoid anal-rectal intercourse (in which the penis is inserted into the rectum) because the lining of the rectum is thin and cells infected with HIV can easily enter the body there.

Unsafe Sexual Practices Transmit STIs

Always use a latex condom during sexual intercourse if you don't know for sure that your partner has been free of STIs for some time. Be sure to follow the directions supplied by the manufacturer. Use of a water-based spermicide containing nonoxynol-9 in addition to the condom can offer further protection because nonoxynol-9 immobilizes viruses and virus-infected cells.

Avoid fellatio (kissing and insertion of the penis into a partner's mouth) *and cunnilingus* (kissing and insertion of the tongue into the vagina) because they may be a means of transmission. The mouth and gums often have cuts and sores that facilitate the entrance of infected cells.

Be very careful about using alcohol or any drug that may prevent you from being able to control your behavior. In a social setting where strangers are present, be sure to watch over your food and drinks. Date-rape drugs can be slipped into either one, and they're usually odorless, tasteless, and invisible.

Drug Use Transmits Hepatitis and HIV (Fig. 17E)

Don't ever inject drugs into your veins! Be aware that hepatitis and HIV can be spread by blood-to-blood contact.

Always use a new, sterile needle for injection or one that has been cleaned in bleach if you are a drug user and can't stop your habit. Never share used needles with anyone else!

Figure 17D Sexual activities transmit STIs.

Figure 17E Sharing needles transmits STIs.

17.5 Effects of Aging

16. Discuss the anatomical and physiological changes that occur in the reproductive system as we age.

Sex hormone levels decline with age in both men and women. Menopause, the period in a woman's life during which menstruation ceases, is likely to occur between the ages of 45 and 55. The ovaries are no longer responsive to pituitary gonadotropic hormones because after years of ovulating, much of functional ovarian tissue has been replaced by corpus albicans. After menopause, the ovaries stop producing ova and produce only minimal amounts of estrogen and progesterone. At the onset of menopause, the menstrual cycle becomes irregular and then eventually ceases. Hormonal imbalance often produces physical symptoms, such as dizziness, headaches, insomnia, sleepiness, depression, and "hot flashes" that are caused by circulatory irregularities. Menopausal symptoms vary greatly among women, and some symptoms may be absent altogether.

Following menopause, atrophy of the uterus, vagina, breasts, and external genitals is likely. The lack of estrogen also promotes changes in the skin (e.g., wrinkling; see Chapter 5) and in the skeleton (e.g., osteoporosis; see Chapter 6).

In men, testosterone production diminishes steadily after age 50, which may be responsible for the enlargement of the prostate gland. Sperm production declines with age, yet men can remain fertile well into old age. With age, however, the chance of erectile dysfunction due to degenerative vascular changes in the penis increases.

Sexual desire and activity need not decline with age, and many older men and women enjoy sexual relationships. Men are likely to experience reduced erection until close to ejaculation, and prescription medications for erectile dysfunction (such as Viagra® and Cialis®) can help to restore normal erection. Women may experience a drier vagina, and many over-the-counter lubricants can provide extra moisture.

17.6 Homeostasis

17. Discuss how the reproductive system works with other systems of the body to maintain homeostasis.

Regulation of sex hormone blood level is an example of homeostatic control. Figure 17.8 shows how the blood level of testosterone is maintained within normal limits. Negative feedback results in a self-regulatory mechanism that maintains the appropriate level of these hormones in the blood.

The illustration in Human Systems Work Together on page 420 shows how the reproductive system works with the other systems of the body to maintain homeostasis. Usually we stress that the function of sex hormones produced not only by the gonads but also by the adrenal glands is to promote development of the reproductive organs and to maintain the secondary sex characteristics.

However, these functions of sex hormones don't really pertain to homeostasis, which involves maintaining the internal environment of cells. Other activities of the sex hormones do affect the internal environment, though. For example, estrogen promotes fat deposition, which serves as a source of energy for cells and also helps the body maintain a normal temperature because of its insulating effect.

In recent years, it's been discovered that the sex hormones perform other activities that affect homeostasis even more directly. We are just now beginning to discover the role that estrogen and androgens play in the metabolism of cells and therefore their role in homeostasis in general. Estrogen induces the liver to produce many types of proteins that transport substances in the blood. These include proteins that bind iron and copper and lipoproteins that transport cholesterol. Iron and copper are enzyme cofactors necessary to cellular metabolism. Although we associate cholesterol with cardiovascular diseases, in fact, it is also a substance that contributes to the functioning of the plasma membrane.

Estrogen also induces synthesis of bone matrix proteins and counteracts the loss of bone mass. At menopause, when the rate of estrogen secretion is drastically reduced, osteoporosis (decrease in bone density) may develop. Nevertheless, long-term estrogen therapy (called hormone replacement therapy, or HRT) is no longer recommended. The National Institutes of Health conducted a study of 16,608 healthy women who were taking both estrogen and progesterone or a placebo. The study was halted after 5.2 years because physicians concluded that the risks for the group on HRT outweighed the benefits. Though the benefits of HRT included lower risk of hip fractures and colon cancer, the women on HRT had a small but significant increased risk for breast cancer, coronary heart disease, stroke, and blood clots.

Similarly, besides the action of androgens (e.g., testosterone) on the sexual organs and functions of males, androgens play a metabolic role in cells. They stimulate the synthesis of structural proteins in skeletal muscles and bone, and they also affect the activity of various enzymes in the liver and kidneys. For example, in the kidney, androgens stimulate the synthesis of erythropoietin, the protein that signals the bone marrow to increase the production of red blood cells.

Content CHECK-UP!

13. The stage of life after which females can no longer reproduce is called _____. Do males experience this? Explain.

14. Female hormone production drastically decreases after menopause. What are some benefits and risks of replacing these hormones medically?

Answers in Appendix A.

Human Systems Work Together

Integumentary System

Androgens activate oil gland; sex hormones stimulate fat deposition, affect hair distribution in males and females.

Skin receptors respond to touch; modified sweat glands produce milk; skin stretches to accommodate growing fetus.

Skeletal System

Sex hormones influence bone growth and density in males and females.

Bones provide support and protection of reproductive organs.

Muscular System

Androgens promote growth of skeletal muscle.

Muscle contraction occurs during orgasm and moves gametes; abdominal and uterine muscle contractions occur during childbirth.

Nervous System

Sex hormones masculinize or feminize the brain, exert feedback control over the hypothalamus, and influence sexual behavior.

Brain controls onset of puberty; nerves are involved in erection of penis and clitoris, movement of gametes along ducts, and contraction of uterus.

Endocrine System

Gonads produce the sex hormones.

Hypothalamic, pituitary, and sex hormones control sex characteristics and regulate reproductive processes.

How the Reproductive System works with other body systems.

Cardiovascular System

Sex hormones influence cardiovascular health; sexual activities stimulate cardiovascular system.

Blood vessels transport sex hormones; vasodilation causes genitals to become erect; blood services the reproductive organs.

Lymphatic System/Immunity

Sex hormones influence immune functioning; acidity of vagina helps prevent pathogen invasion of body; milk passes antibodies to newborn.

Immune system does not typically attack sperm or fetus, even though they are foreign to the body.

Respiratory System

Sexual activity increases breathing; pregnancy causes breathing rate and vital capacity to increase.

Gas exchange increases during sexual activity.

Digestive System

Pregnancy crowds digestive organs and promotes heartburn and constipation.

Digestive tract provides nutrients for growth and repair of organs and for development of fetus.

Urinary System

Penis in males contains the urethra and performs urination; prostate enlargement hinders urination.

Semen is discharged through the urethra in males; kidneys excrete wastes and maintain electrolyte levels for mother and child.

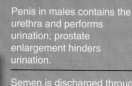

seminal vesicle
ductus deferens
prostate gland
urethra
epididymis
penis
testis
scrotum

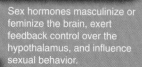

mammary glands
uterine tube
ovary
vagina
external genitalia (clitoris, labia)
uterus

FOCUS on FORENSICS

Rape

Rape is a crime of violence, and not a crime of sexuality. It happens to all kinds of victims: women, men, and children; heterosexual and homosexual; even babies and the elderly. It is a crime that reportedly occurs once every two minutes—but that statistic may be much lower than the number of actual incidents. An estimated 25% of reported cases are committed by a complete stranger. Family members carry out approximately 12% of rapes, but the highest numbers of such crimes—63%—are committed by an acquaintance of the victim. Many rapes go unreported, their victims too afraid or too ashamed to seek medical and legal assistance. When rape victims do seek emergency medical care, it is imperative that their needs for physical, psychological, and legal assistance are met promptly and compassionately.

Many communities have formed Sexual Assault Response Teams (SARTs). A key member of such a team is a trained sexual assault nurse examiner (SANE). Forensic nursing is a subspecialty of nursing recognized by the American Nursing Association, which provides training for nurses desiring SANE certification. The team also includes EMS personnel, physicians, psychologists, social workers, and law enforcement officials.

The first step in care for the rape victim is taking a thorough medical history. The patient should be made as comfortable as possible, and may prefer to remain clothed during the initial assessment. A physical exam will follow during which a standardized rape kit may be used to collect evidence. The patient should disrobe while standing on a sheet of exam-table paper in order to catch any hairs from the suspect that might remain on the patient's body. All clothing should be placed in paper evidence bags, and any fluid on the clothing (blood, urine, semen, etc.) should be allowed to air-dry before the paper bags are sealed. This will prevent mold growth, which might destroy evidence. Bruises and abrasions on the patient's body must be documented and photographed. Vaginal exams of females and oral-rectal exams for either gender may also give evidence of injury. The pubic hair is combed through with a fine-tooth comb to capture any pubic hair from the perpetrator.

Fluids found on the patient's body should be collected on swabs for DNA and other chemical testing, and their presence recorded. The victim's blood, hair, and urine will also be sampled for comparison. A high-frequency ultraviolet lamp called a Wood's lamp can detect the presence of semen on the patient's body. Seminal fluid will fluoresce at specific wavelengths of ultraviolet light. Other chemical testing for the components of semen can be done as well. However, it is important to note that absence of semen does not mean that rape did not occur. The perpetrator may have had a vasectomy. Further, a study of assailants documented that 50% reported erectile dysfunction and inability to ejaculate while attempting a rape.

Medical treatment for the victim will include testing for sexually transmitted infection, including gonorrhea, syphilis, and HIV. Antibiotics and anti-retroviral drugs can be prescribed. If the patient is not immunized for hepatitis B, he or she should receive the vaccine. Testing for HIV and hepatitis should be repeated approximately 6 months after the incident. Female victims should also be tested for a preexisting pregnancy. Nonpregnant women may elect to receive emergency contraception to prevent pregnancy. However, this treatment can damage an existing pregnancy and should not be given to a pregnant woman. The patient should also receive a urinalysis, especially if he or she does not recall details of the rape. Testing can show evidence of so-called "date-rape" drugs, which induce short-term memory loss.

Psychological counseling is essential if a rape victim is to recover. A rape victim may initially feel anger, fear, or anxiety—although he or she may express no emotion at all. Many victims will internalize blame for the incident, believing that some past incident may have invited the assault. Flashbacks—painful, detailed memories of the incident—are common after rape. Anxiety disorders and post-traumatic stress disorder may occur years after the incident, and also require psychological counseling.

By properly documenting history and evidence, EMS, nursing, and medical professionals can enable successful prosecution of the rapist. Detailed evidence is particularly important in cases of non-stranger sexual assault, such as child, spousal, or elder abuse. Evidence from health-care professionals in these cases can support a victim's claims of lack of consent to sexual activity. Just as important, careful medical and psychological care can enable the victim to recover from his or her ordeal.

Selected New Terms

Basic Key Terms

acrosome (ăk′rō-sōm), p. 396

androgens (ăn′drŭh-jĕn), p. 395

areola (ă-rē′ō-lă), p. 407

Bartholin glands (Băr′tō-lĭn glăndz), p. 406

bulbourethral gland (bŭl″bō-yū-rē′thrăl glănd), p. 395

cervix (sĕr′vĭks), p. 401

clitoris (klĭ′tō-rĭs), p. 405

colostrum (kō-lŏs′trŭm), p. 409

corpora cavernosa (kōr′pōr-ă kăv-ĕr-nōs′-ŭh), p. 399

corpus albicans (kōr′pŭs ăl′bĭ-kănz), p. 403

corpus luteum (kōr′pŭs lū′tē-ŭm), p. 403

corpus spongiosum (kōr′pŭs spŭn-gē-ō′sŭm), p. 399

diploid (dĭp′lōyd), p. 393

ejaculation (ē-jăk″yū-lā′shŭn), p. 395

ejaculatory duct (ē-jăk′yū-lŭh-tō″rē dŭkt), p. 395

endometrium (ĕn-dō-mē′trē-ŭm), p. 401

epididymis (ĕp″ĭ-dĭd′ĭ-mĭs), p. 395

estrogen (ĕs′trō-jĕn), p. 406

external genitals (ĕks-tĕr′năl jĕn′ĭ-tălz), p. 399

fimbriae (fĭm′brē-ē), p. 404

follicle (fŏl'ĭ-kŭl), p. 402

FSH (follicle-stimulating hormone) (fŏl'ĭ-kŭl stĭm'yū-lā-tĭng hŏr'mōn), p. 400

gametes (găm'ēt'), p. 393

genitals (jĕn'ĭ-tălz), p. 393

germ cells (jŭrm sĕl), p. 396

haploid (hăp'lōyd), p. 393

human chorionic gonadotropin (hyū'măn kō"rē-ŏn'ĭk gō"năd-ō-trō'pĭn), p. 407

implantation (ĭm"plăn-tā'shŭn), p. 401

inhibin (ĭn-hĭb'ĭn), p. 400

interstitial cell (ĭn"tĕr-stĭsh'ŭl sĕl), p. 396

labia majora (lā'bē-ŭh mŭh-jŏr'ŭh), p. 405

labia minora (lā'bē-ŭh mĭ-nŏr'ŭh), p. 405

lactation (lăk-tā'-shŭn), p. 409

LH (luteinizing hormone) (lū'tē-ĭ-nīz-ĭng hŏr'mōn), p. 400

meiosis (mī-ō'sĭs), p. 393

menopause (mĕn'ō-păwz), p. 407

menses (mĕn'sēz'), p. 407

menstrual cycle (mĕn"strū-ăl sī'kĕl), p. 407

mons pubis (mŏnz pyū'bĭs), p. 405

oocyte (ō'ō-sīt), p. 402

oogenesis (ō"ō-jĕn'ĕ-sĭs), p. 403

orgasm (ŏr'găzm), p. 399

ova (ō'vŭh), p. 401

ovaries (ō'vŭh-rēs), p. 401

ovulation (ŏv"yū-lā'shŭn), p. 403

Pap smear (păp smēr), p. 405

penis (pē'nĭs), p. 399

polar body (pō'lĕr bŏd-ē), p. 403

primary spermatocyte (prī'mā-rē spĕrm-ă'tō-sīt), p. 396

progesterone (prō-jĕs'tŭh-rōn'), p. 406

prostaglandins (prŏs'tŭh-glăn'dĭn), p. 398

prostate gland (prŏs'tāt glănd), p. 395

puberty (pyū'bĕr-tē), p. 393

scrotum (skrō'tŭm), p. 395

secondary sex characteristics (sĕk'ŭn-dār-ē sĕks kār"ăk-tĕr-ĭs'tĭks), p. 400

secondary spermatocyte (sĕk'ŭn-dār-ē spĕrm-ă'tō-sīt), p. 396

semen (sē'mĕn), p. 395

seminal vesicle (sĕm'ĭ-năl vĕs'ĭ-kŭl), p. 395

seminiferous tubule (sē"mĭ-nĭf'ĕr-ŭs tū'byūl), p. 396

sperm (spĕrm), p. 396

spermatic cord (spĕrm-ă'-tĭk kōrd), p. 398

spermatogenesis (spĕr"mŭh-tō-jĕn'ē-sĭs), p. 396

spermatogonia (spĕr"mă-tō-gō'nē-ŭh), p. 396

sustentacular (Sertoli) cells (sŭs"tĕn-tăk'yū-lĕr sĕlls), p. 396

testis (tĕs'tĭs), p. 395

testosterone (tĕs-tŏs'tŭh-rōn"), p. 396

urethra (yū-rē'thrŭh), p. 395

uterine tube (yū'tĕr-ĭn tūb), p. 404

uterus (yū'tĕr-ŭs), p. 401

vagina (vŭh-jī'nŭh), p. 402

vas deferens (văs dĕf'ĕr-ĕns), p. 395

vesicular (Graafian) follicle (vĕs-ĭk'yū-lĕr [grăf'ē-ŭn] fŏl'ĭ-kŭl), p. 403

vestibule (vĕs'tŭh-byūl), p. 405

vulva (vŭl'vŭh), p. 402

zygote (zī'gōt), p. 395

Clinical Key Terms

artificial insemination (ăr"tĭ-fĭ'shŭl ĭn-sĕm-ĭ-nā'shŭn), p. 415

assisted reproductive technologies (ŭh-sĭs'tĕd rē"prō-dŭk'tĭv tĕk-năh'lō-jēz), p. 415

chancre (shăn'kĕr), p. 417

chlamydia (klŭh-mĭ'dē-ŭh), p. 417

circumcision (sĕr"kŭm-sĭzh'ŭn), p. 399

contraception (kŏn-tră-sĕp'shŭn), p. 410

cryptorchidism (krĭp-tŏr'kĭ-dĭzm), p. 395

ectopic pregnancy (ĕk-tŏp'ĭk prĕg'nŭn-sē), p. 404

endometriosis (ĕn-dō-mē"trē-ō'sĭs), p. 415

erectile dysfunction (ē-rĕk'tĭl dĭs-fŭnk'shŭn), p. 399

genital herpes (jĕn'ĭ-tăl hĕr'pēz), p. 416

genital warts (jĕn'ĭ-tăl wŏrts), p. 416

gonorrhea (gŏn-ō-rē'ŭh), p. 417

gumma (gŭm'ŭh), p. 417

hepatitis (hĕp'ŭh-tī'tĭs), p. 417

hernia (hĕr'nē-ŭh), p. 398

hysterectomy (hĭs"tĕr-ĕk'tō-mē), p. 405

infertility (ĭn-fĕr-tĭl'ĭ-tē), p. 415

inguinal hernia (ĭng'gwŭh-nĕl hŭr'nē-ŭh), p. 398

intracytoplasmic sperm injection (ĭn'trŭh-sĭt-ō-plăs-mĭk spĕrm ĭn-jĕk'shŭn), p. 415

in vitro fertilization (ĭn vē'trō fŭr"tĭl-ĭ-zā'shŭn), p. 415

oophorectomy (ō-ŏf-ŏr-ĕk'tō-mē), p. 405

ovarian cancer (ō-vār'ē-ŭn căn'sĕr), p. 405

ovariohysterectomy (ō-vār"ē-ō-hĭs"tĕr-ĕk'tō-mē), p. 405

pelvic inflammatory disease (pĕl'vĭk ĭn-flăm'ŭh-tŏr-ē dĭz-ēz'), p. 417

salpingectomy (săl-pĭn-jĕk'tō-mē), p. 405

sexually transmitted infection (sĕk'shū-ăh-lē trănz-mĭt'ĕd ĭn'fĕk-shŭn), p. 416

syphilis (sĭf-ŭh-lĭs), p. 417

Summary

17.1 Human Life Cycle

The functions of the reproductive system include the production and transportation of gametes by the ovaries and testes, nourishing the developing embryo and fetus, nourishing a newborn via the production and delivery of breast milk, and supplying the body with sex hormones.

A. The life cycle of humans requires two types of nuclear division: meiosis and mitosis. Meiosis occurs in sperm production in males and ova production in females. It results in haploid cells with 23 chromosomes. When fertilization occurs, the zygote has the diploid number of 46 chromosomes. Mitosis is used in the growth of the embryo, fetus, and child.

17.2 Male Reproductive System

The primary male reproductive organs are the testes. The accessory organs include the epididymis, vas deferens, ejaculatory duct, seminal vesicles, prostate gland, bulbourethral glands, and urethra.

A. Spermatogenesis occurs in the seminiferous tubules of the testes.

B. Sperm mature and are stored in the epididymides. Sperm may also be stored in the vasa deferentia before entering the urethra along with secretions produced by the seminal vesicles, prostate, and bulbourethral glands. Sperm and these secretions are called semen, or seminal fluid.

C. The external genitals of males are the penis, the organ of sexual intercourse, and the scrotum, which contains the testes.

D. Erection is a spinal reflex that involves the autonomic nervous system. Orgasm is a physical and emotional climax that results in ejaculation of semen from the penis.

E. Hormone regulation, involving secretions from the hypothalamus, the anterior pituitary, and the testes, maintains testosterone at a fairly constant level. FSH from the anterior pituitary promotes spermatogenesis in the seminiferous tubules, and LH promotes testosterone production by the interstitial cells.

17.3 Female Reproductive System
The primary female reproductive organs are the ovaries. The accessory organs include the uterine tubes, uterus, vagina, mons pubis, clitoris, labia majora, and labia minora.

A. Oogenesis occurs within the ovaries. Typically, one follicle matures each month. This follicle balloons out of the ovary and bursts, releasing a secondary oocyte in a process called ovulation.

B. The ovulated secondary oocyte enters a uterine tube, which leads to the uterus. This is where implantation and development occur if the secondary oocyte is fertilized to become a zygote.

C. The external genitalia include the vaginal opening, mons pubis, clitoris, the labia majora, and labia minora. The vagina receives the penis during intercourse, and allows passage of menstrual fluid and the fetus to the outside.

D. The external genitalia, especially the clitoris, play an active role in orgasm, which culminates in uterine and uterine tube contractions.

E. Hormonal regulation in females, like in males, begins with the secretion of GnRH from the hypothalamus. GnRH promotes the secretion of FSH and LH from the anterior pituitary gland. These hormones stimulate the ovaries to produce estrogen and progesterone.

F. The menstrual cycle can be divided into pre-ovulation and post-ovulation events. Before ovulation, FSH causes an estrogen-producing follicle to begin developing in the ovary. Meanwhile, in the uterus, the menstruation phase occurs before the proliferative phase. During the proliferative phase, estrogen causes the uterine lining to thicken. Ovulation is caused by a positive feedback cycle in which an abundance of estrogen brings about a surge in FSH and LH. After ovulation, the corpus luteum in the ovary secretes primarily progesterone, which causes the uterine lining to become secretory. Unless a fertilized ovum implants into the endometrium, the corpus luteum gradually deteriorates to form the corpus albicans. Progesterone production decreases, and the endometrium is shed once again during menstruation.

G. If fertilization takes place, the embryo implants itself in the thickened endometrium. When this occurs, the developing placenta produces HCG, which maintains the corpus luteum. The corpus luteum continues to secrete progesterone, which prevents menstruation from occurring.

H. Estrogen and progesterone are needed for mammary gland development. During pregnancy, the breasts enlarge and prepare for lactation. Prolactin stimulates breast milk production, and oxytocin causes the release of milk from the breasts.

17.4 Control of Reproduction and Sexually Transmitted Infections

A. Numerous contraceptive methods and devices, such as the birth control pill, diaphragm, condom, surgery, and natural family planning, are available for those who wish to prevent pregnancy. Emergency contraception can be taken after sexual intercourse but before there is any indication of pregnancy. Mifepristone, for example, induces a medical abortion by blocking progesterone receptors, causing an implanted embryo to be shed with the uterine lining. However, the only 100% effective way to prevent pregnancy and sexually transmitted disease is abstinence.

B. Some couples are infertile and may use assisted reproductive technologies in order to have a child. Artificial insemination and in vitro fertilization have been followed by more sophisticated techniques such as intracytoplasmic sperm injection.

C. Sexually transmitted infections can be caused by viruses, bacteria, and parasites. Humans cannot develop effective immunity for these diseases. Viral STIs include HIV/AIDS, herpes, hepatitis, and genital warts. Human papillomavirus (HPV) causes genital warts as well as most cervical cancers. Viral STIs have no cure. Bacterial diseases include syphilis, gonorrhea, and chlamydia. These infections can be cured with antibiotics unless the bacteria have become resistant to these drugs.

17.5 Effects of Aging
Menopause occurs between the ages of 45 and 55 in women. Following menopause, atrophy of the genitals is likely. In men, testosterone production decreases after age 50, and the incidence of erectile dysfunction increases.

17.6 Homeostasis
Maintaining secondary sexual characteristics doesn't directly affect homeostasis of the body's internal environment. But sex hormones have many other effects in the body that do. For example, estrogen maintains bone mass, and after menopause, the decreased level of estrogen in women causes bone weakening and fractures. It also increases the risk of cardiovascular diseases. One of the many functions of androgens, especially testosterone, is to stimulate the synthesis of structural proteins in skeletal muscle and bone.

Study Questions

1. Explain the process of meiosis. How is a haploid cell produced from a diploid cell? (pp. 393–395)

2. Outline the path of sperm. What glands contribute fluids to semen? (pp. 395–397)

3. Discuss the anatomy and physiology of the testes. Describe the structure of sperm. (pp. 395–397)

4. List five male internal accessory organs. What role does each play in forming semen? (pp. 397–399)
5. Describe the structure of the male penis, and what occurs with sexual arousal. (pp. 399–400)
6. Name the endocrine glands involved in maintaining the sex characteristics of males and the hormones produced by each. (pp. 400–401)
7. Describe the organs of the female genital tract. Where do fertilization and implantation occur? Name two functions of the vagina. (pp. 401–402
8. Discuss the anatomy and the physiology of the ovaries, and describe the process of oogenesis. (pp. 402–403)

9. List the four female internal accessory organs, and be able to describe the role of each one. (pp. 404–405)
10. Name and describe the external genitals in females. Outline the steps in the female sexual response. (pp. 405–406)
11. How are female hormones regulated? What endocrine glands are involved? (pp. 406–408)
12. Describe the pre- and post-ovulation events of the menstrual cycle. How is menstruation prevented if pregnancy occurs? (pp. 407–408)
13. Name three functions of the female sex hormones. (pp. 406–409)
14. Outline the process of lactation. (pp. 407, 409)

15. Discuss the various means of birth control and their relative effectiveness in preventing pregnancy. (pp. 410–412, 414–415)
16. Discuss reasons for infertility, and describe three methods of assisted reproductive technology. (p. 415)
17. Discuss the incidence, transmission, and mechanisms for prevention of sexually transmitted infection. Which forms are bacterial/viral/parasitic? Which can be controlled by antibiotics? (pp. 416–418)
18. Discuss the anatomical and physiological changes that occur in the reproductive system as we age. (p. 419)

Learning Outcome Questions

Fill in the blanks.
1. Human haploid cells contain _____ chromosomes. Which cells are haploid?
2. Spermatogenesis occurs within the _____ of the testes.
3. Androgens are secreted by the _____ cells that lie between seminiferous tubules.
4. The primary male sex hormone is _____.
5. In tracing the path of sperm, the structure that follows the epididymis is the _____.
6. The prostate gland, the bulbourethral glands, and the _____ all contribute to seminal fluid.

7. An erection is caused by the entrance of _____ into erectile tissue of the penis.
8. In the female reproductive system, the uterus lies between the uterine tubes and the _____.
9. Which structures are considered parts of the vulva? _____
10. The female sex hormones are _____ and _____.
11. In the menstrual cycle, once each month a(n) _____ produces an ovum and the _____ is prepared to receive the embryo.
12. The release of an ovum from a vesicular follicle is called _____.

13. Once a vesicular follicle has released an egg, it develops into a glandlike structure called a(n) _____.
14. What two hormones are directly involved in lactation? _____
15. The most frequent causes of male infertility are _____ and _____.
16. Whereas AIDS and genital herpes are caused by _____, gonorrhea and chlamydia are caused by _____.

Medical Terminology Exercise

After studying this chapter, see if you can derive the definitions for the medical terms listed at right. Many of the prefixes and suffixes used to create these terms can be found throughout the chapter. For additional help, use McGraw-Hill Connect™ at www.mcgrawhillconnect.com and consult Appendix B.

1. orchidopexy (ŏr″kĭ-dō-pĕk′sē)
2. transurethral resection of prostate (TURP) (trăns″yū-rē′thrăl rē-sĕk′shŭn ŭv prŏs′tāt)
3. gonadotropic (gō″năd-ō-trōp′ĭk)
4. gynecomastia (gīn″ē-kō-măs′tē-ŭh)
5. hysterosalpingo-oophorectomy (hĭs″tĕr-ō-săl-pĭng′gō-ō″ŏf-ŏr-ĕk′tō-mē)

6. multipara (mŭl-tĭp′ŭh-rŭh)
7. seminoma (sĕm″ĭ-nō′mŭh)
8. genitourinary (jĕn″ĭ-tō-yū′rĭ-năr′ē)
9. prostatic hypertrophy (prŏs-tăt′ĭk hy′pĕr-trō″fē)
10. azoospermia (ā-zō′ō-spĕr′mē-ŭh)

Online Study Tools

LEARNSMART®   connect

APR

Anatomy & Physiology REVEALED includes cadaver photos that allow you to peel away layers of the human body to reveal structures beneath the surface. This program also includes animations, radiologic imaging, audio pronunciations, and practice quizzing. To learn more visit www.aprevealed.com

18 Human Development and Birth

The town of Mount Airy, North Carolina, is most famous for its connections to the fictional town of "Mayberry," from the 1960s-era television show, *The Andy Griffith Show.* Mount Airy is the birthplace of Andy Griffith, the show's star and namesake, and Mayberry was modeled after Mount Airy. (If you've never seen *The Andy Griffith Show,* check it out on YouTube®.) Andy Griffith was not the town's only famous resident, however. There were two other Mount Airy citizens who were famous long before *The Andy Griffith Show* ever began—Chang and Eng Bunker. The Bunkers were the original "Siamese twins," a pair of conjoined brothers born in Siam (now modern-day Thailand) in 1811. As you'll discover in this chapter, conjoined twins like Chang and Eng are formed when the developing zygote begins separating but doesn't completely divide. Two individuals are formed but remain connected somewhere along their bodies. The two Bunker brothers were united at the chest by a tube of flesh, and modern-day surgeons speculate that the twins could easily have been separated with today's medical knowledge and surgical procedures. Separation was a medical impossibility then, however, and the two earned a handsome living for a time as human oddities, giving lectures and demonstrations throughout the United States and Europe. In 1839, they settled on a farm in Mount Airy, then married sisters Adelaide and Sarah Yates, and between them, fathered 21 children. The twins died within hours of each other in 1874, at the age of 63.

Learning Outcomes

After you have studied this chapter, you should be able to:

18.1 Fertilization

1. Explain the events of fertilization and the conversion of the ovum into a zygote.

18.2 Development

2. Discuss the processes of development.
3. Name the four extraembryonic membranes, and give a function for each.

4. Describe the events that occur during pre-embryonic and embryonic development.
5. Describe the structures and functions of the placenta and the umbilical cord.
6. Describe the events that occur during fetal development, and compare and contrast development of male and female sex organs.

18.3 Birth

7. Describe the three stages of birth.
8. In general, describe the physical and physiological changes in the mother during pregnancy.

Medical Focus

Therapeutic Cloning
Premature Babies
Preventing Birth Defects

18.1 Fertilization

1. Explain the events of fertilization and the conversion of the ovum into a zygote.

Fertilization of an ovum at conception requires that a single sperm penetrate the ovum. When the genetic materials from ovum and sperm are combined, the **zygote** is formed. This single cell is the foundation for a new, unique individual. Figure 18.1 shows the manner in which an ovum is fertilized by a sperm in humans.

Sperm and Ovum Anatomy

As you'll recall from Chapter 17, a sperm has three distinct parts: a head, a middle piece, and a tail. Very little cytoplasm is found in a sperm cell. The tail is a flagellum, which allows the sperm to swim through the female reproductive tract toward the ovum. The middle piece contains energy-producing mitochondria, and the head contains the sperm nucleus, which encloses the 23 chromosomes from the male. Capping the sperm head is a membrane-bound **acrosome,** which contains digestive enzymes. Notice that only the nucleus from the sperm head fuses with the ovum nucleus. Scientists believe that the few mitochondria in the middle piece of the sperm are rapidly

destroyed by enzymes in the ovum (with only vary rare recorded exceptions). Therefore, the zygote receives cytoplasm, organelles, nutrients, and structural components only from the mother.

The plasma membrane of the ovum is surrounded by a nonliving extracellular matrix termed the **zona pellucida.** In turn, the zona pellucida is surrounded by a few layers of adhering follicular cells, collectively called the **corona radiata.** These cells nourished the ovum when it was in a follicle of the ovary.

Steps of Fertilization

During fertilization, (1) several sperm penetrate the corona radiata, (2) several sperm attempt to penetrate the zona pellucida, and (3) one sperm enters the ovum. The acrosome plays a role in allowing sperm to penetrate the zona pellucida. After a sperm head binds tightly to the zona pellucida, the acrosome releases digestive enzymes that forge a pathway for the sperm through the zona pellucida. When a sperm binds to the ovum, their plasma membranes fuse, and this sperm (the head, the middle piece, and usually the tail) enters the ovum. Fusion of the sperm nucleus and the ovum nucleus follows.

To ensure proper development, only one sperm should enter an ovum. Accidental entry of more than one sperm cell will halt any

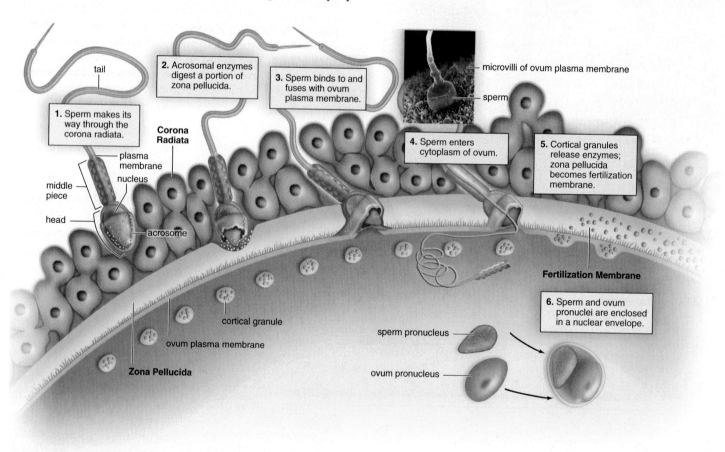

1. Sperm makes its way through the corona radiata.

2. Acrosomal enzymes digest a portion of zona pellucida.

3. Sperm binds to and fuses with ovum plasma membrane.

4. Sperm enters cytoplasm of ovum.

5. Cortical granules release enzymes; zona pellucida becomes fertilization membrane.

6. Sperm and ovum pronuclei are enclosed in a nuclear envelope.

tail

Corona Radiata

plasma membrane

nucleus

middle piece

head

acrosome

microvilli of ovum plasma membrane

sperm

Fertilization Membrane

cortical granule

ovum plasma membrane

Zona Pellucida

sperm pronucleus

ovum pronucleus

AP|R **Figure 18.1 Fertilization.** Fertilization is accomplished in a series of steps: **1.** sperm cells break aside the cells of the corona radiata, until a single sperm is able to penetrate and adhere to the zona pellucida; **2.** enzymes in the sperm's acrosome digest the zona pellucida; **3.** sperm plasma membrane binds to the ovum plasma membrane; **4.** entire sperm enters the ovum, and its nucleus is released; **5.** fertilization membrane is formed, polyspermy is prevented; and **6.** sperm and ovum pronuclei fuse together in a nuclear envelope.

further development by the zygote. Cell division cannot take place because additional chromosomes in the zygote will prevent the pairing of chromosomes essential for mitosis. Prevention of *polyspermy* (entrance of more than one sperm) depends on changes in the ovum's plasma membrane and in the zona pellucida. As soon as the sperm cell membrane fuses with the ovum cell membrane, the ovum's plasma membrane depolarizes (from −65 mV to 10 mV). Then, the ovum releases small vesicles called *cortical oocyte granules,* which contain lysosomal enzymes. In response, the zona pellucida alters its structure and composition, becoming an impenetrable fertilization membrane. As a result, the ovum cell membrane and the zona pellucida cooperate to prevent entry of any additional sperm cells.

Once a zygote has been formed, it must implant into the endometrial layer of the uterus in order to survive. If implantation is unsuccessful, the zygote survives for only a very short period. Following a successful implantation, the mother is clinically pregnant. She will soon have a positive result on a blood or urine pregnancy test because of the presence of human chorionic gonadotropin (HCG), which is described in the following section.

Content CHECK-UP!

1. The nonliving extracellular matrix material surrounding the ovum is called the:

   a. corona radiata.   c. corpus albicans.

   b. zona pellucida.   d. acrosome.

2. Put the following events of fertilization in the correct order as they occur:

   1. Sperm nucleus and ovum nucleus fuse.

   2. Several sperm penetrate the corona radiata.

   3. One sperm penetrates the ovum.

   4. Barriers are set up to prevent polyspermy.

   a. 2-3-4-1   c. 3-2-4-1

   b. 2-4-3-1   d. 3-4-2-1

Answers in Appendix A.

18.2 Development

2. Discuss the processes of development.
3. Name the four extraembryonic membranes, and give a function for each.
4. Describe the events that occur during pre-embryonic and embryonic development.
5. Describe the structures and functions of the placenta and the umbilical cord.
6. Describe the events that occur during fetal development, and compare and contrast development of male and female sex organs.

Before we discuss the stages of development, you'll want to become familiar with the processes of development and the names and functions of the extraembryonic membranes.

Processes of Development

Here's a quick introduction to the stages of a human being's development:

Cleavage. Immediately after fertilization, the zygote begins to divide so that there are first 2, then 4, 8, 16, 32 cells, and so forth. During these first divisions, the ball of cells that is formed does not increase in size; it remains the same size as the zygote (see Fig. 18.3). Cell division during cleavage is mitotic, and each cell receives a full complement of chromosomes and genes.

Growth. During embryonic development, cell division is accompanied by an increase in the size of the daughter cells.

Morphogenesis. Morphogenesis refers to the shaping of the embryo and is first evident when certain cells are seen to migrate to new positions relative to other cells. As these movements continue in a very orderly sequence, the embryo begins to take on a defined shape.

Differentiation. Differentiation occurs as each group of cells takes on a specific structure and function. The first system to become visibly differentiated is the nervous system.

Extraembryonic Membranes

The **extraembryonic membranes** are not part of the embryo and fetus; instead, as implied by their name, they are outside the embryo (Fig. 18.2). The names of the extraembryonic membranes in humans are strange to us because they are named for their function in shelled animals—birds and reptiles! In shelled animals, the chorion lies next to the shell and carries on gas exchange. The amnion contains the protective amniotic fluid, which bathes the developing embryo. The allantois collects nitrogenous wastes, and the yolk sac surrounds the yolk, which provides nourishment.

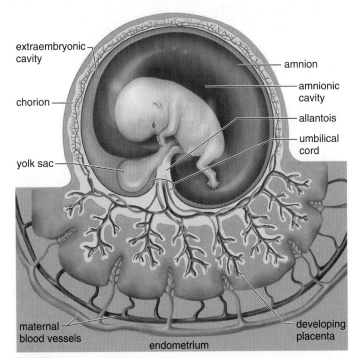

extraembryonic cavity

chorion

yolk sac

amnion

amnionic cavity

allantois

umbilical cord

maternal blood vessels

endometrium

developing placenta

AP|R **Figure 18.2 Extraembryonic membranes.** The chorion and amnion surround the embryo. The other two extraembryonic membranes, the yolk sac and allantois, contribute to the umbilical cord.

The functions of the extraembryonic membranes are different in humans because we humans develop inside the uterus. The extraembryonic membranes have these functions in humans:

1. **Chorion.** The chorion develops into the fetal half of the **placenta,** the organ that provides the embryo/fetus with nourishment and oxygen and takes away its waste.
2. **Yolk sac.** The yolk sac has little yolk and is the first site of blood cell formation.
3. **Allantois.** The allantois blood vessels become the umbilical blood vessels.
4. **Amnion.** The amnion contains fluid to cushion and protect the embryo, which develops into a fetus.

Stages of Development

Development encompasses the events that occur from fertilization to birth. In humans, this **gestation** period is usually calculated by adding 280 days to the start of the last menstruation, a date that is usually known. However, only about 5% of babies actually arrive on the predicted date.

Pre-Embryonic Development

Table 18.1 shows that we can subdivide development into pre-embryonic, embryonic, and fetal development. **Pre-embryonic development** encompasses the events of the first week, as shown in Figure 18.3.

TABLE 18.1 Human Development		
Time	Events for Mother	Events for Baby
PRE-EMBRYONIC DEVELOPMENT		
First week	Ovulation occurs.	Fertilization occurs. Cell division begins and continues. Chorion appears.
EMBRYONIC DEVELOPMENT		
Second week	Symptoms of early pregnancy (nausea, breast swelling and tenderness, fatigue) are present. Blood pregnancy test is positive.	Implantation occurs. Amnion and yolk sac appear. Embryo has tissues. Placenta begins to form.
Third week	First menstruation is missed. Urine pregnancy test is positive. Symptoms of early pregnancy continue.	Nervous system begins to develop. Allantois and blood vessels are present. Placenta is well formed.
Fourth week		Limb buds form. Heart is noticeable and beating. Nervous system is prominent. Embryo has tail. Other systems form.
Fifth week	Uterus is the size of a hen's egg. Mother feels frequent need to urinate due to pressure of growing uterus on bladder.	Embryo is curved. Head is large. Limb buds show divisions. Nose, eyes, and ears are noticeable.
Sixth week	Uterus is the size of an orange.	Fingers and toes are present. Skeleton is cartilaginous.
Two months	Uterus can be felt above the pubic bone.	All systems are developing. Bone is replacing cartilage. Facial features are becoming refined. Embryo is about 38 mm (1½ in.) long.
FETAL DEVELOPMENT		
Third month	Uterus is the size of a grapefruit.	Gender can be distinguished by ultrasound. Fingernails appear.
Fourth month	Fetal movement is felt by a mother who has previously been pregnant.	Skeleton is visible. Hair begins to appear. Fetus is about 150 mm (6 in.) long and weighs about 170 g (6 oz).
Fifth month	Fetal movement is felt by a mother who has not previously been pregnant. Uterus reaches up to level of umbilicus, and pregnancy is obvious.	Protective cheesy coating, called vernix caseosa, begins to be deposited. Heartbeat can be heard.
Sixth month	Doctor can tell where baby's head, back, and limbs are. Breasts have enlarged, nipples and areolae are darkly pigmented, and colostrum is produced.	Body is covered with fine hair called lanugo. Skin is wrinkled and reddish.
Seventh month	Uterus reaches halfway between umbilicus and rib cage.	Testes descend into scrotum. Eyes are open. Fetus is about 300 mm (12 in.) long and weighs about 1,350 g (3 lb).
Eighth month	Weight gain is averaging about a pound a week. Standing and walking are difficult because center of gravity is thrown forward.	Body hair begins to disappear. Subcutaneous fat begins to be deposited.
Ninth month	Uterus is up to rib cage, causing shortness of breath and heartburn. Sleeping becomes difficult.	Fetus is ready for birth. It is about 530 mm (20–21 inches) long and weighs about 3,400 g (7.5 lb).

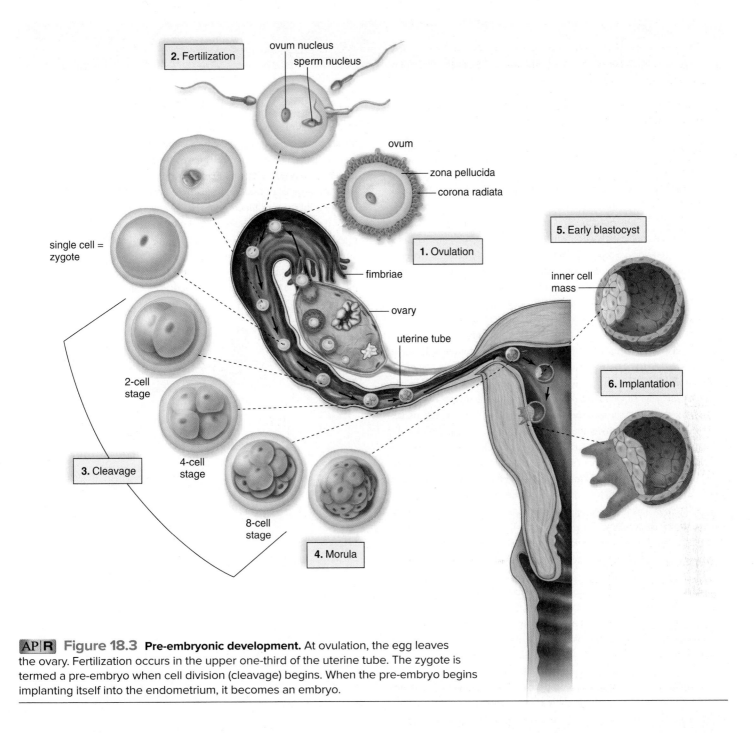

2. Fertilization
ovum nucleus
sperm nucleus

ovum
zona pellucida
corona radiata

1. Ovulation

single cell = zygote

fimbriae

ovary

uterine tube

2-cell stage

3. Cleavage

4-cell stage

8-cell stage

4. Morula

5. Early blastocyst
inner cell mass

6. Implantation

AP|R **Figure 18.3** **Pre-embryonic development.** At ovulation, the egg leaves the ovary. Fertilization occurs in the upper one-third of the uterine tube. The zygote is termed a pre-embryo when cell division (cleavage) begins. When the pre-embryo begins implanting itself into the endometrium, it becomes an embryo.

Immediately after fertilization, the zygote divides repeatedly as it passes down the uterine tube to the uterus. A **morula** is a compact ball of embryonic cells that becomes a **blastocyst.** The many cells of the blastocyst arrange themselves so that there is an **inner cell mass** surrounded by a layer of cells, the **trophoblast.** The trophoblast will become the *chorion.* The early appearance of the chorion emphasizes the complete dependence of the developing embryo on this extraembryonic membrane. The inner cell mass will become the embryo.

Each cell within the morula and the inner cell mass of the blastocyst has the genetic capability of becoming any tissue. This recognition has recently led to a new procedure called therapeutic cloning, as discussed in the Medical Focus reading on page 430. Sometimes during development, either the cells of the morula separate, or the inner cell mass splits, and two embryos are present rather than one. If each of the two embryos is able to complete development, the two babies formed will be *identical twins* because they have inherited exactly the same chromosomes. Should separation of the cell mass be incomplete, the identical twins formed are called *conjoined twins* (formerly referred to as Siamese twins). Conjoined twins can be joined at any part of the body. *Fraternal twins,* who arise when two different ova are fertilized by two different sperm, do not have identical chromosomes.

The term *cloning* means making exact multiple copies of genes, a cell, or an organism. For example, identical twins are clones of a single zygote. Theoretically, cloning of human beings may someday be possible, though certainly ethically objectionable. The procedure might begin as described in Figure 18A. The person being cloned need not contribute sperm or an egg to the process. Instead, in a process called somatic-cell nuclear transfer (SCNT), a 2n (diploid) nucleus from, say, a fibroblast, can be placed in an enucleated ovum, which then begins developing. The pre-embryo (blastocyst) would be implanted in the uterus of a surrogate mother where it would develop until birth. Although subject to environmental influences, the clone would be expected to closely resemble its "parent." (For example, Dolly the sheep, cloned in 1996, was the first mammal ever to be produced through SCNT.)

Therapeutic cloning is not the same as cloning a human being because the cells of the created pre-embryo are never introduced into the uterus. Instead, they are separated and treated to become particular tissues, which can then be used to treat the patient. The separated cells of a pre-embryo are called stem cells because they divide repeatedly and can become various tissues, as shown in Figure 18A. Theoretically, tissues resulting from this procedure will not be subject to rejection by the patient because they will bear the same surface molecules as the patient's cells. However, it's important to remember that an ovum contains mitochondrial DNA (mtDNA) from the mother. Researchers are uncertain what, if any, effect mtDNA could have on the development of rejection.

Researchers hope that one day therapeutic clone cells may provide insulin-secreting cells for diabetics, nerve cells for stroke patients or those with Parkinson's disease, cardiac cells for those with heart disease, and so forth. However, so far therapeutic cloning is experimental and has not been perfected. To date, the research is extremely expensive to carry out and has a very high failure rate. Further, questions remain about what the new cells will do once implanted in the person's body—will they spread like cancer cells do? Moreover, it is important to recognize that ethical concerns about therapeutic cloning will always remain—after all, any pre-embryo is potentially an entire, living human being.

Anticipating intense interest in therapeutic cloning, the U.S. National Academy of Sciences proposed strict new guidelines for federally funded research in 2005. As a result, Embryonic Stem Cell Research Oversight, or ESCRO, committees were created to approve any research before it could begin. An ESCRO committee review is carried out in addition to the current review always required by an Institutional Review Board, or IRB. ESCRO committees include bioethicists and legal experts, as well as members of the general public.

ESCRO rules require that prior to beginning any research, informed consent from the donors of ova or sperm must be obtained. Further, stem cells created for therapeutic cloning must never be used for reproductive purposes. Additional restrictions require that the embryos created cannot be grown in culture for longer than 14 days. At that point, the primitive streak of the developing nervous system begins to form. The Academy formed a national agency charged with regularly reviewing all guidelines on stem cell research, and these reviews were conducted in 2008 and 2010.

Problems with rejection of embryonic stem cells, as well as ethical issues surrounding their use, have led researchers to pursue other sources of stem cells. Numerous exciting studies have shown that the adult body has plenty of stem cells: blood, skin, liver, bone, skeletal muscle, and neural stem cells have all been found. Once scientists can effectively get adult stem cells to reproduce in cell culture, they could be a source of tissue cells that won't ever be rejected. It has even been possible to coax blood stem cells and neural stem cells to become other types of human tissues in the body. Another potential source of blood stem cells is a baby's umbilical cord, and umbilical blood can now be indefinitely stored for future use. As technology improves, these types of abundant, easily obtainable stem cells might just hold the promise of a cure for many human diseases and disorders.

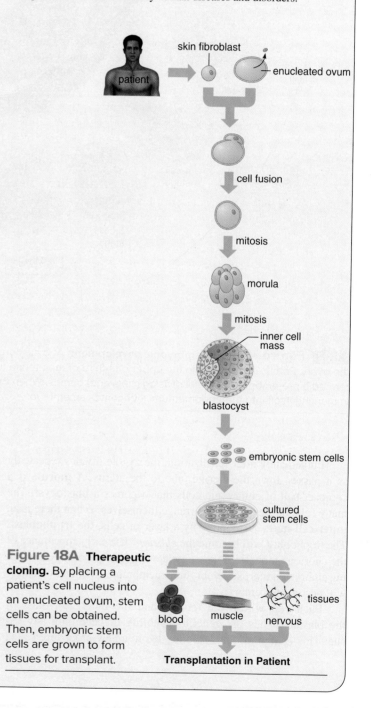

Figure 18A Therapeutic cloning. By placing a patient's cell nucleus into an enucleated ovum, stem cells can be obtained. Then, embryonic stem cells are grown to form tissues for transplant.

Embryonic Development

Embryonic development begins with the second week and lasts until the end of the second month of development. During this stage, the developing human is called an **embryo.**

Second Week At the end of the first week, the embryo usually begins the process of implanting itself in the wall of the uterus. If **implantation** is successful, the woman is clinically pregnant. On occasion, it happens that the embryo implants itself in a location other than the uterus—most likely, the uterine tube. Implantation of the embryo in any location outside the uterus is called *extra-uterine* or *ectopic* pregnancy (see Chapter 17, p. 404). An ectopic pregnancy in the uterine tube must be surgically removed, because the tube is too narrow to allow the continuing growth of the embryo. Should the pregnancy continue, the uterine tube will eventually rupture, causing a fatal hemorrhage to both mother and embryo. In rare circumstances, ectopic pregnancies elsewhere in the abdominal cavity (for example, on the outside of the uterine wall or on the wall of the urinary bladder) can continue to term, producing a live infant. Of course, birth is by *cesarean section* in this circumstance. A cesarean section is carried out by making an incision to open the abdominal and uterine walls and withdrawing the baby from the uterus.

During implantation, the trophoblast secretes enzymes to digest away some of the tissue and blood vessels of the endometrium of the uterus. The trophoblast also begins to secrete **human chorionic gonadotropin (HCG),** the hormone that is the basis for both blood and urine pregnancy tests. HCG acts like luteinizing hormone (LH) in that it serves to maintain the corpus luteum past the time it normally disintegrates. Because it is being stimulated, the corpus luteum secretes progesterone, the endometrium is maintained, and the expected menstruation does not occur.

The embryo is now about the size of the period at the end of this sentence. As the week progresses, further growth of cells below the inner cell mass causes it to separate from the trophoblast. The inner cell mass now becomes the **embryonic disk,** and two more extraembryonic membranes form. The yolk sac is the first site of blood cell formation. The amniotic cavity surrounds the embryo (and then the fetus) as it develops. In humans, amniotic fluid acts as an insulator against cold and heat and also absorbs shock, such as that caused by the mother exercising.

Primary Germ Layers With the start of the major event called **gastrulation,** the inner cell mass becomes the embryonic disk. Gastrulation is an example of morphogenesis (see page 427) during which cells move or migrate, in this case to become tissue layers called the **primary germ layers.** By the time gastrulation is complete, the embryonic disk has become an embryo with three primary germ layers: **ectoderm**, **mesoderm,** and **endoderm.** Figure 18.4 shows the significance of the primary germ layers—all the organs of an individual can be traced back to one of the primary germ layers. Notice also that the original trophoblast layer and the mesoderm layer together form the chorion, the fetal half of the placenta.

Third Week Two important organ systems make their appearance during the third week (Fig. 18.5). The nervous system is the first organ system to be visually evident. Growth of the nervous system begins when the ectoderm layer on the posterior surface of the embryonic disk becomes thickened. This thickened area is termed the **neural plate.** A visible trench, called the **neural groove,** next begins to form in the center of the neural plate. As the neural groove deepens, two ridges of tissue become evident on either side; these are the **neural folds.** When the neural folds meet in the middle, the hollow **neural tube** is formed. This tube will later develop into the brain and spinal cord. Complete closure of the neural tube is critical for normal development. Should this fail to occur for whatever reason, a **neural tube defect** will be the result. Failure of the inferior spinal cord area to close causes **spina bifida,** which can sometimes be surgically repaired. Failure of the superior neural tube to close causes a fatal condition called **anencephaly** (literally, "without brain"), resulting in a fetus whose brain and cranium never complete development. Multivitamin supplements containing the vitamin folic acid (folate) have been shown to help to prevent the occurrence of neural tube defects (see the Medical Focus on pages 440–441).

Development of the heart begins in the third week and continues into the fourth week. At first, there are right and left heart tubes; when these fuse, the heart begins pumping blood, even though the chambers of the heart are not fully formed. The veins enter posteriorly, and the arteries exit anteriorly from this largely tubular heart, but later the heart twists so that all major blood vessels are located anteriorly.

Fourth and Fifth Weeks At four weeks, the embryo is barely larger than the height of this print. A body stalk connects the caudal (tail) end of the embryo with the chorion. Branching projections of the chorion called **chorionic villi** extend into the endometrial wall of the uterus (Fig. 18.5). Within each chorionic villus is a capillary network. Exchange of gases, nutrients, and wastes from mother to fetus can take place across this capillary network. The fourth extraembryonic membrane, the allantois, lies within the body stalk, and its blood vessels become the umbilical blood vessels. The head and the tail then lift up, and the body stalk moves anteriorly by constriction. Once this process is complete, the **umbilical cord,** which connects the developing embryo to the placenta, is fully formed (Fig. 18.5c).

Paddle-shaped tissue stalks called limb buds appear (Fig. 18.6); later, the arms and the legs develop from the limb buds, and even the hands and the feet become apparent. At the same time—during the fifth week—the head enlarges and the sense organs become more prominent. It is possible to make out the developing eyes and ears, and even the nose.

Sixth Through Eighth Weeks During the sixth through eighth weeks of development, the embryo changes to a form that is easily recognized as a human being. Concurrent with brain development, the head achieves its normal relationship with the body as a neck region develops. The nervous system is developed well enough to permit reflex actions, such as a startle response to touch. At the end of this period, the embryo is about 38 mm (1.5 in.) long and weighs no more than an aspirin tablet, even though all organ systems have been established.

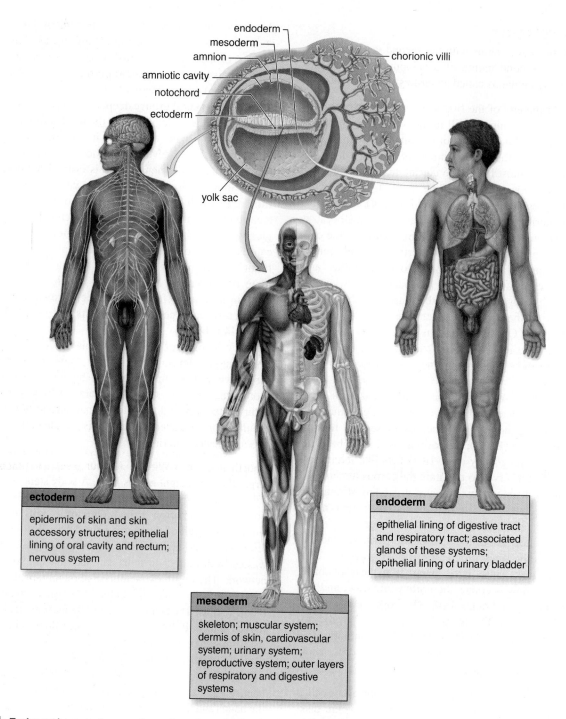

endoderm —
mesoderm —
amnion —
amniotic cavity —
notochord —
ectoderm —

chorionic villi

yolk sac

ectoderm

epidermis of skin and skin accessory structures; epithelial lining of oral cavity and rectum; nervous system

endoderm

epithelial lining of digestive tract and respiratory tract; associated glands of these systems; epithelial lining of urinary bladder

mesoderm

skeleton; muscular system; dermis of skin, cardiovascular system; urinary system; reproductive system; outer layers of respiratory and digestive systems

Figure 18.4 Embryonic germ layers. An embryo has three primary germ layers: ectoderm, mesoderm, and endoderm. Organs and tissues can be traced back to a particular germ layer as indicated in this illustration.

Placenta

The placenta is shaped like a pancake, measuring 15 to 20 cm in diameter and 2.5 cm thick. The placenta is normally fully formed and functional by the end of the embryonic period and before the fetal period begins. The placenta is expelled as the **afterbirth** following the birth of an infant.

The placenta has two portions, a fetal portion composed of chorionic tissue and a maternal portion composed of uterine tissue. Chorionic villi cover the entire surface of the chorion until about the eighth week when they begin to disappear, except in one area. These villi are surrounded by maternal blood, and it is here that exchanges of materials take place across the placental membrane. The **placental membrane** consists of the epithelial wall of an embryonic capillary and the epithelial wall of a chorionic villus. Maternal blood rarely mingles with fetal blood. Instead, oxygen and nutrient molecules, such as glucose and amino acids, diffuse from maternal blood across the placental membrane into fetal blood, and carbon dioxide and other wastes, such as urea, diffuse out of fetal blood into maternal blood.

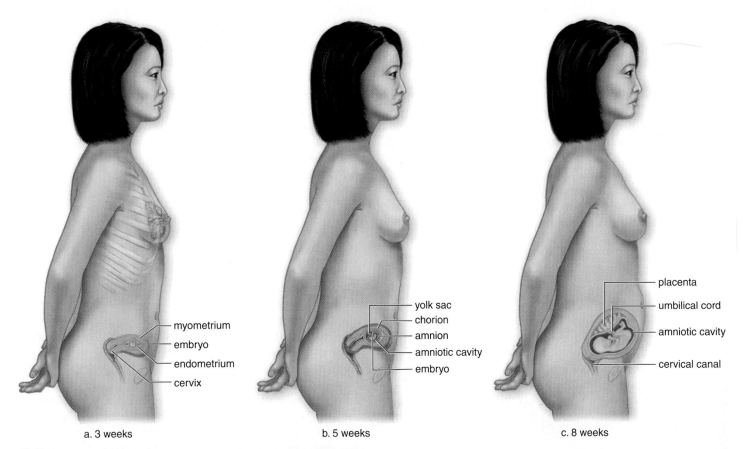

myometrium
embryo
endometrium
cervix

a. 3 weeks

yolk sac
chorion
amnion
amniotic cavity
embryo

b. 5 weeks

placenta
umbilical cord
amniotic cavity
cervical canal

c. 8 weeks

AP|R **Figure 18.5** **Embryonic development within the uterus. (a)** Three weeks after fertilization. **(b)** Five weeks after fertilization, amnion and chorion are present, and the uterus is about the size of a hen's egg. **(c)** Two months after fertilization, the placenta and umbilical cord are well formed.

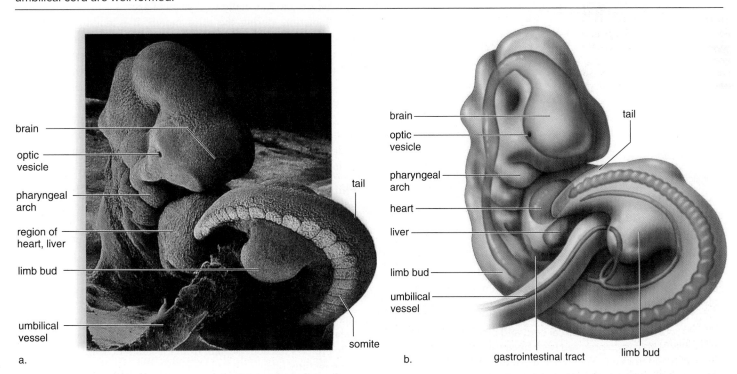

brain
optic vesicle
pharyngeal arch
region of heart, liver
limb bud
umbilical vessel

a.

tail

somite

brain
optic vesicle
pharyngeal arch
heart
liver
limb bud
umbilical vessel

tail

gastrointestinal tract

limb bud

b.

Figure 18.6 **Human embryo at beginning of fifth week. (a)** Scanning electron micrograph. **(b)** The embryo is curled so that the head touches the region of the heart and liver. The organs of the gastrointestinal tract are forming, and the arms and the legs develop from the bulges called limb buds. The tail is an evolutionary remnant; its bones regress and become those of the coccyx (tailbone).

Because of the placenta, the digestive system, lungs, and kidneys do not function in the fetus (though they continue in their development). The functions of these organs are not needed because the placenta supplies the fetus with its nutritional and excretory needs.

The umbilical cord transports fetal blood to and from the placenta (Fig. 18.7; see Fig. 12.20). The umbilical cord is the fetal lifeline because it contains the two umbilical arteries and a single umbilical vein, which transport waste molecules (carbon dioxide and urea) to the placenta for disposal and oxygen and nutrient molecules from the placenta to the rest of the fetal circulatory system.

As mentioned, the trophoblast cells of the chorion and then the placenta produce HCG, the hormone detected by a pregnancy test. The test uses monoclonal antibodies (see Chapter 13, pages 306 and 312) that are specific for HCG. When the antibodies bind to HCG, activated

dye molecules create the lines or symbols that indicate a positive test. In the woman's body, HCG prevents the normal degeneration of the corpus luteum of the ovary. Instead, HCG stimulates the corpus luteum to secrete even larger quantities of progesterone. Later, the placenta begins to produce progesterone and estrogen, and the corpus luteum degenerates—it is no longer needed. Placental estrogen and progesterone maintain the endometrium and have a negative feedback effect on the anterior pituitary so that it ceases to produce gonadotropic hormones during pregnancy. Menstruation ordinarily does not occur during the length of pregnancy.

Fetal Development

Fetal development includes the third through the ninth months of development. At this time, the fetus definitely looks human (Fig. 18.8).

Third and Fourth Months At the beginning of the third month, the fetal head is still very large, the nose is flat, the eyes are far apart, and the ears are well formed. Head growth now begins to slow down as the rest of the body increases in length. Epidermal refinements, such as fingernails, nipples, eyelashes, eyebrows, and hair on the head, appear.

Cartilage begins to be replaced by bone as ossification centers appear in most of the bones. Cartilage remains at the ends of the long bones, and ossification is not complete until age 18 or 20 years. The skull has six large membranous areas called **fontanels,** which permit a certain amount of flexibility as the head passes through the birth canal and allow rapid growth of the brain during infancy. Progressive fusion of the skull bones causes the fontanels to close, usually by 2 years of age.

Sometime during the third month, it is possible to distinguish males from females using ultrasound. Researchers have discovered a series of genes on the X and Y chromosomes that cause the differentiation of gonads into testes and ovaries. Once these have differentiated, they produce the sex hormones that influence the differentiation of the genital tract.

During the fourth month, the fetal heartbeat is loud enough to be heard when a physician applies a stethoscope to the mother's abdomen. By the end of this month, the fetus is about 152 mm (6 in.) in length and weighs about 171 g (6 oz).

Fifth Through Seventh Months During the fifth through seventh months (Fig. 18.8), the mother begins to feel movement. At first, she experiences only a fluttering sensation, but as the fetal legs grow and develop, she feels kicks and jabs. The fetus remains tightly curled in the uterus, with the head bent down and in contact with the flexed knees. This posture is aptly termed the fetal position.

The wrinkled, translucent, pink-colored skin is covered by a fine downy hair called **lanugo.** This in turn is coated with a white, greasy, cheeselike substance called **vernix caseosa,** which probably protects the delicate skin from the amniotic fluid. The eyelids are now fully open.

At the end of this period, the fetus's length has increased to about 300 mm (12 in.), and it weighs about 1,380 g (3 lb). With improved technology, infants born during this period can survive:

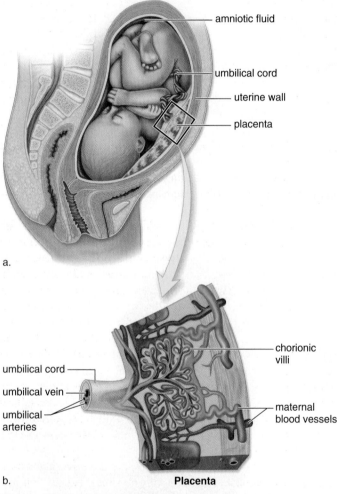

a.

b.   **Placenta**

Figure 18.7 The placenta. (a) Fetus with umbilical cord and placenta. Blood vessels within the umbilical cord lead to the placenta, where exchange takes place between fetal blood and maternal blood. **(b)** Detail of umbilical cord and placenta. Note that the umbilical vein is colored red here because it is bringing oxygenated blood from the placenta to the fetus. Likewise, the umbilical arteries are colored blue to reflect the deoxygenated blood flowing from the fetus to the placenta for gas exchange with the maternal blood vessels.

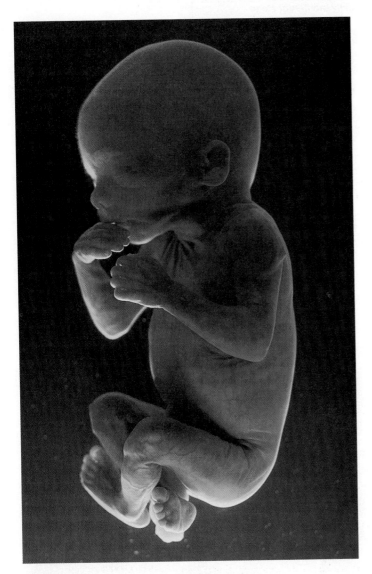

Figure 18.8 Five- to seven-month-old fetus.

An infant born at 26 weeks gestation, weighing between 1,000 and 1,250 g, has a 90% chance of survival (though he or she may have serious medical issues: see the Medical Focus on page 439).

Eighth Through Ninth Months As the end of development approaches, the fetus usually rotates so that the head is pointed toward the cervix. However, if the fetus does not turn, a **breech birth** (rump first) is likely. It is very difficult for the cervix to expand enough to accommodate this form of birth, and asphyxiation of the baby is more likely to occur. Thus, a **cesarean section** may be prescribed for delivery of the fetus (incision through the abdominal and uterine walls).

At the end of nine months, the fetus is about 530 mm (20½ in.) long and weighs about 3,400 g (7½ lb). Weight gain is due largely to an accumulation of fat beneath the skin. Full-term babies have the best chance of survival; as discussed in the Medical Focus on page 439, premature babies are subject to various challenges.

Development of Male and Female Sex Organs

The gender of an individual is determined at the moment of fertilization. Both males and females have 23 pairs of chromosomes; in males, one of these pairs is an X and Y, while females have two X chromosomes. During the first several weeks of development, it is impossible to tell by external inspection whether the unborn child is a boy or a girl. Gonads don't start developing until the seventh week of development. The tissue that gives rise to the gonads is called *indifferent* because it can become testes or ovaries depending on the action and concentration of hormones. Genes on the Y chromosome cause the production of testosterone, and then the indifferent tissue becomes testes.

In Figure 18.9*a*, notice that at six weeks both males and females have the same types of ducts, called **Wolffian ducts** and **Müllerian ducts.** During this indifferent stage, an embryo has the potential to develop into a male or a female. If a Y chromosome is present, testosterone stimulates the Wolffian ducts to become male genital ducts. The Wolffian ducts enter the urethra, which belongs to both the urinary and reproductive systems in males. The testes secrete an anti-Müllerian hormone that causes the Müllerian ducts to regress. In both genders, the probladder will ultimately become the urinary bladder as development continues.

In the absence of a Y chromosome, ovaries develop instead of testes from the same indifferent tissue. Now the Wolffian ducts regress, and the Müllerian ducts develop into the uterus and uterine tubes. A developing vagina also extends from the uterus. As you know, there is no connection between the urinary and genital systems in females (see Chapter 17).

At 14 weeks, both the primitive testes and ovaries are located deep inside the abdominal cavity. An inspection of the interior of the testes would show that sperm are even now starting to develop, and similarly, the ovaries already contain large numbers of tiny follicles, each having an ovum. Later, in the last trimester of fetal development, both testes and ovaries descend into the pelvis. As you know from Chapter 17, the ovaries remain in the pelvis. However, the testes descend out of the pelvis into the scrotal sacs (scrotum). Because the testes are normally found in the scrotal sacs, testicular temperature is cooler than body temperature, and this allows sperm cells to develop. As you'll recall from Chapter 17, sometimes the testes fail to descend, a condition called cryptorchidism. In that case, surgery will be done later to place them in their proper location.

Figure 18.9*b* shows the development of the external genitals. These tissues are also indifferent at first—they can develop into either male or female genitals. At six weeks, a small bud appears between the legs; this can develop into the male penis or the female clitoris, depending on the presence or absence of the Y chromosome and testosterone. At nine weeks, a urogenital groove bordered by two swellings appears. By 14 weeks, this groove has disappeared in males, and the scrotum has formed from the original swellings. In females, the groove persists and becomes the vaginal opening. Labia majora and labia minora are present instead of a scrotum.

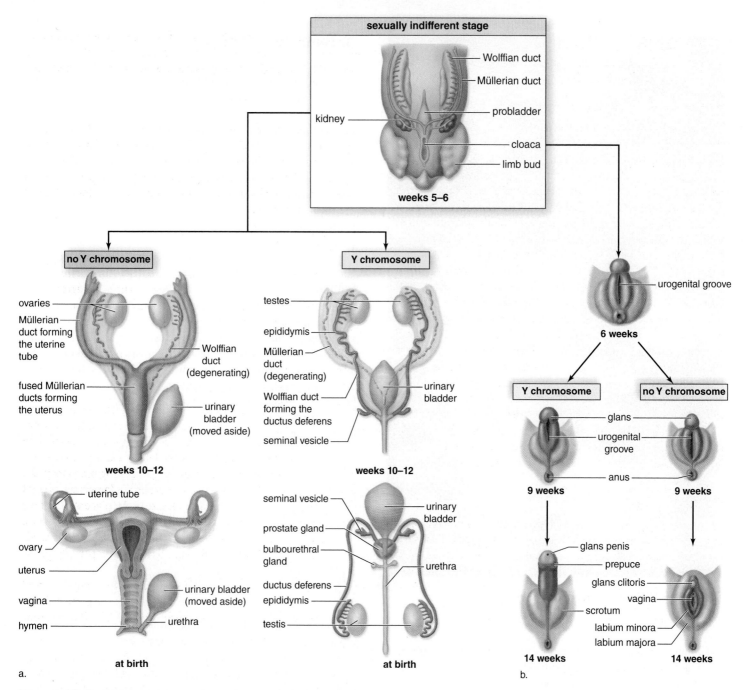

Figure 18.9 **Male and female organs.** (a) Development of gonads and ducts. (b) Development of external genitals.

Content **CHECK-UP!**

3. The site of red blood cell production in an embryo is the:
   a. bone marrow.
   b. chorion.
   c. yolk sac.
   d. allantois.

4. The umbilical cord is the embryo/fetus's lifeline, connecting it to the mother. Describe its structure. In which vessel(s) would you expect the oxygen content to be highest?

5. The inner cell mass of the blastocyst becomes the _____, while the trophoblast becomes part of the _____.

6. If an embryo has a Y chromosome, which of the following will happen?
   a. The embryo will develop into a male.
   b. The Wolffian duct will develop fully, whereas the Müllerian duct regresses.
   c. The Müllerian duct will develop fully, whereas the Wolffian duct regresses.
   d. a and b
   e. a and c

Answers in Appendix A.

18.3 Birth

7. Describe the three stages of birth.
8. In general, describe the physical and physiological changes in the mother during pregnancy.

The uterus has contractions throughout pregnancy. At first, these are light, lasting about 20–30 seconds and occurring every 15–20 minutes. Near the end of pregnancy, the contractions may become stronger and more frequent so that a woman might think she is in labor. "False-labor" contractions are called **Braxton-Hicks** contractions. However, the onset of true labor is marked by uterine contractions that occur regularly every 15–20 minutes (and more frequently in labor's final stages) and last for 40 seconds or longer.

A positive feedback mechanism can explain the onset and continuation of labor. Uterine contractions are induced by a stretching of the cervix, which also brings about the release of oxytocin, which is produced by the hypothalamus and released from the posterior pituitary. Oxytocin stimulates the uterine muscles, both directly and through the action of prostaglandins. Uterine contractions push the fetus downward, and the cervix stretches even more. This cycle keeps repeating itself until birth occurs (see Chapter 1, Fig. 1.10).

Prior to or at the first stage of **parturition,** which is the process of giving birth to an offspring, there can be a "bloody show." This is caused by expulsion of the mucus plug from the cervix. During pregnancy, this thick, sticky mucus effectively sealed the uterine cervix, preventing bacteria and sperm from entering the uterus. Loss of the plug is often a first sign that the baby's birth is imminent.

Stage 1

During the first stage of parturition, the uterine contractions of labor occur in such a way that the cervical canal slowly disappears as the lower part of the uterus is pulled upward toward the baby's head. This process is called effacement, or "taking up the cervix." With further contractions, the baby's head acts as a wedge to assist cervical dilation (Fig. 18.10b). If the amniotic membrane has not already ruptured, it is apt to do so during this stage, releasing the amniotic fluid, which leaks out the vagina (an event sometimes referred to as "breaking water"). The first stage of parturition ends once the cervix is dilated completely (to about 10 cm.).

Stage 2

During the second stage of parturition, the uterine contractions occur every 1–2 minutes and last about one minute each. They are accompanied by a desire to push, or bear down. As the baby's head gradually descends into the vagina, the desire to push becomes greater. When the baby's head reaches the exterior, it turns so that the back of the head is uppermost (Fig. 18.10c). Because the vaginal orifice may not expand enough to allow passage of the head, an **episiotomy** is often performed to prevent the mother's tissues from tearing. This incision, which enlarges the opening, is sewn together later. As soon as the head is delivered, the baby's shoulders rotate so that the baby faces either to the right or the left. At this time, the physician may hold the head and guide it downward, while one shoulder and then the other emerges. The rest of the baby follows easily.

Once the baby is breathing normally, the umbilical cord is cut and tied, severing the child from the placenta. The stump of the cord shrivels and falls off to create a scar, which is the umbilicus.

Stage 3

The placenta, or afterbirth, is delivered during the third stage of parturition (Fig. 18.10d). About 15 minutes after delivery of the baby, uterine muscular contractions shrink the uterus and dislodge the placenta. The placenta then is expelled into the vagina. As soon as the placenta and its membranes are delivered, the third stage of parturition is complete.

Begin Thinking Clinically

During an actual delivery, a pregnant woman's uterine contractions will sometimes begin, seemingly strengthen, and then stop abruptly. What might cause this to occur? What should be done to correct this?

Answers and discussion in Appendix A.

Effects of Pregnancy on the Mother

Major changes take place in the mother's body during pregnancy. When first pregnant, the mother may experience nausea and vomiting, loss of appetite, and fatigue. These symptoms subside, and some mothers report increased energy levels and a general sense of well-being despite an increase in weight. During pregnancy, the mother gains weight due to breast and uterine enlargement, weight of the fetus, amount of amniotic fluid, size of the placenta, her own increase in total body fluid, and an increase in storage of proteins, fats, and minerals. The increased weight can lead to lordosis ("swayback," see Fig. 6.9) and lower back pain, especially in the third trimester.

Aside from an increase in weight, many of the physiological changes in the mother are due to the presence of the placental hormones that support fetal development (Table 18.2). Progesterone

TABLE 18.2	Effect of Placental Hormones on Mother
Hormone	**Chief Effects**
Progesterone	Relaxation of smooth muscle; reduced uterine motility; reduced maternal immune response to fetus
Estrogen	Increased uterine blood flow; increased renin-angiotensin-aldosterone activity; increased protein biosynthesis by the liver
Peptide hormones	Increased insulin resistance

Source: Moore, Thomas R., *Gestation Encyclopedia of Human Biology*, Vol. 7, 7th edition. Copyright © 1997 Academic Press.

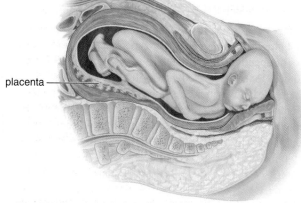

a. 9-month-old fetus

pubic symphysis
urethra
urinary bladder
vagina
cervix
rectum

placenta

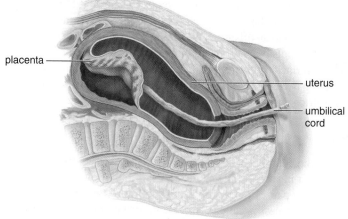

b. First stage of birth: cervix dilates

ruptured
amniotic
sac

placenta

c. Second stage of birth: baby emerges

placenta

uterus

umbilical
cord

d. Third stage of birth: afterbirth is expelled

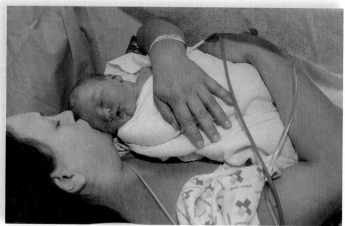

Baby has arrived

Figure 18.10 Three stages of parturition (birth). (a) Position of fetus just before birth begins. **(b)** Dilation of cervix. **(c)** Birth of baby.
(d) Expulsion of afterbirth.

MEDICAL FOCUS

Premature Babies

In Memoriam: Brandon Nelson, 1982–1983

Each and every day in the United States, approximately 1,300 babies will be born too soon. **Premature birth** is defined as birth prior to 37 weeks gestation (normal gestation is approximately 40 weeks). Many other babies, although full-term, have a low birth weight (less than 5.5 pounds at birth). These babies face weeks to months in the neonatal intensive care unit, or NICU. Fortunately, advances in the care of these tiniest patients allow most of them to continue to grow and develop while slowly gaining weight. Most eventually go home. However, "premies" and term babies with low birth weight face many serious challenges to their survival:

Respiratory distress syndrome (hyaline membrane disease) The lungs do not produce enough of the chemical surfactant that helps the alveoli stay open (see Chapter 14, page 324). Therefore, the lungs tend to collapse, instead of expanding to be filled with air.

Retinopathy of prematurity The high level of oxygen needed to ensure adequate gas exchange by the immature lungs can lead to proliferation of blood vessels within the eyes, with ensuing blindness.

Intracranial hemorrhage The delicate blood vessels in the brain are apt to break, causing swelling and inflammation of the brain. If not fatal, this can lead to brain damage.

Jaundice The immature liver fails to excrete the waste product bilirubin, which instead builds up in the blood, possibly causing brain damage.

Infections The level of antibodies in the body is low, and the various medical procedures performed could possibly introduce pathogens. Also, bowel infection is common, along with perforation, bleeding, and shock.

Circulatory disorders Fetal circulation, discussed in Chapter 12, has two features: (1) the oval opening between the atria, and (2) the arterial duct that allows blood to bypass the lungs. If these features persist in the newborn, oxygen-rich blood will mix with oxygen-poor blood. Blood circulation will be impaired, perhaps leading to the delivery of a "blue baby"—that is, a baby with cyanosis, a bluish cast to the skin. Heart failure can also result from these conditions.

Risk for permanent disability Many premies grow up to be normal and healthy, but many others face permanent disability. Developmental delays, cerebral palsy, learning disorders, chronic lung disease, blindness, and deafness are all potential consequences of premature birth. Further, pediatricians estimate that fully half of all neurological disabilities in children are related to premature birth.

Why do women deliver prematurely? Obstetricians (physicians who specialize in the treatment of women during and after pregnancy) have come up with four general causes for preterm labor and delivery. The placenta may be disrupted by a bacterial infection; this triggers uterine contractions. Bleeding from the placenta or from the uterus itself may trigger labor. The uterus may be stretched excessively, as often occurs in multiple births. Finally, maternal or fetal stress may trigger a hormone cascade that leads to uterine contraction and preterm delivery. Pregnant women who have already had a previous preterm delivery are known to be at risk for subsequent preterm delivery, as are women pregnant with twins, triplets, or more. Structural abnormalities of the uterus or the cervix may also lead to delivery of a premature baby.

Unfortunately, despite increased research and improvements in medical care, the rate of premature birth has increased by 31% in the past 20 years. Obviously, more needs to be done to prevent premature birth, and you can read more about it in the Medical Focus on pages 440–441. The March of Dimes supports educational efforts and research into the causes of birth defects and prematurity. More information can be obtained at www.marchofdimes.com.

---

decreases uterine motility by relaxing smooth muscle, including the smooth muscle in the walls of arteries. The arteries expand, and this leads to a low blood pressure. Hypotension causes the kidneys to release their enzyme, renin. The renin-angiotensin-aldosterone mechanism is thus activated, and is promoted by estrogen. Aldosterone activity promotes sodium and water retention, and blood volume increases until it reaches its peak sometime during weeks 28–32 of pregnancy. Altogether, blood volume increases from 5 liters to 7 liters—a 40% rise. An increase in the number of red blood cells follows. With the rise in blood volume, cardiac output increases by 20–30%. Blood flow to the kidneys, placenta, skin, and breasts rises significantly. Smooth muscle relaxation also explains the common gastrointestinal effects of pregnancy. The heartburn experienced by many is due to relaxation of the esophageal

sphincter and reflux of stomach contents into the esophagus. Constipation is caused by a decrease in intestinal tract motility.

Pregnancy results in respiratory system changes caused both by progesterone and by the increasing size of the uterus. Under the influence of progesterone, the bronchial and bronchiolar smooth muscle relaxes. The brain is signaled to lower CO_2, which it does by increasing respiratory rate and depth. Compared to nonpregnant values, the maternal blood oxygen level changes little, but blood carbon dioxide levels fall by 20%, creating a concentration gradient that favors the flow of carbon dioxide from fetal blood to maternal blood at the placenta. The increasing size of the uterus from a nonpregnant weight of 60–80 g to 900–1,200 g means that it will occupy most of the abdominal cavity, reaching nearly to the xiphoid process of the sternum. The pregnant uterus not only

Before the nursery room gets painted pink or blue, before the baby registry is completed at the department store, before deciding whether the baby gets named after Great-Aunt Adelaide or Grandma Gertrude, a couple that has decided to start (or add to) a family must take precautions to prevent **congenital,** or **birth, defects.** Some birth defects are unavoidable—those caused by chromosomal or genetic abnormalities, for example. Tragically, however, many other birth defects are completely preventable.

For decades, the placenta was believed to be an effective barrier between mother and fetus, but continuing research has shown instead that the placenta is really a *filter* through which many substances can enter the baby's bloodstream from the mother's circulation. Thus, almost anything that mom eats, drinks, smokes, or injects—good and bad—is potentially capable of winding up in the baby's body. Chemicals, bacteria, viruses, parasites, antibodies from the mother's immune system (both harmful and beneficial) can all act as **teratogens**—agents capable of causing birth defects.

Ideally, preventing birth defects, is a process that should begin before a woman ever becomes pregnant. At a complete physical exam before pregnancy, she can receive any necessary immunization boosters, or "catch up" with immunizations she may have missed as a child (see pages 309–310). It is especially important to immunize against **rubella (German measles)** because it can cause blindness, deafness, mental retardation, heart malformations, and other serious problems in an unborn child. Antibiotic treatment can cure bacterial STIs (syphilis, gonorrhea, chlamydia, etc.) and prevent serious harm to the embryo. Although there are no cures for viral STIs, it is still important to know if a woman is infected. As you know, HIV can cross the placenta, but highly active anti-retroviral therapy (HAART) can often prevent fetal infection (see pages 307–308). Health-care providers will also recommend prenatal vitamin/mineral supplements with high levels of vitamin B_{12} and folate, before and after conception, to minimize the risk of neural tube defects such as spina bifida.

Couples trying to conceive should also adapt their lifestyle, if necessary. If either or both partners smoke, there will never be a more perfect time to quit (see pages 333–334 for helpful tips). Males who smoke are more likely to have erectile dysfunction than nonsmokers. Further, it will later be essential for both the pregnant mother and the fetus to be in a smoke-free environment at all times.

Because the placenta functions as a filter, if mom smokes or encounters secondhand smoke, her baby "smokes," too. Cleft lip/cleft palate may occur because of exposure to toxins in cigarette smoke. Children of smoking mothers are more likely to be stillborn, or die shortly after birth. Death from sudden infant death syndrome (SIDS—unexplained death of an otherwise healthy baby) is more common in babies of smoking mothers. The occurrence of preterm babies with low birth weight is doubled for women who smoke.

Alcohol easily crosses the placenta, and there is no safe dosage during pregnancy. Even one drink a day appears to increase the chances of miscarriage. The more alcohol consumed, the greater the baby's chance of physical abnormalities. **Fetal alcohol syndrome (FAS) is** the term given to the collection of physical, mental, and behavioral abnormalities seen in babies born to heavy drinkers. FAS babies have decreased weight, height, and head size, with malformation of the head and face. Later, developmental delay is common, as are numerous other physical malformations. Babies born to heavy drinkers are apt to undergo an extremely painful withdrawal, called **delirium tremens,** after birth—shaking, vomiting, and extreme irritability.

Any use of recreational drugs must end before attempting to conceive, because all will decrease a man's sperm count. Certainly, illegal drugs such as marijuana, cocaine, and heroin are teratogens that must be completely avoided during pregnancy. All increase the probability of pregnancy complications, including stillbirth. In particular, cocaine use causes severe fluctuations in a mother's blood pressure that temporarily deprive the developing fetus's brain of oxygen. Approximately 60% of drug-affected babies are cocaine babies; if they survive, they can have visual problems, lack coordination, and are often developmentally delayed.

Once pregnancy occurs, prenatal medical care and good health habits must continue. A healthy pregnancy diet should contain moderate increases in calories and protein to meet the demands of both fetus and mother. The pregnant woman of normal weight should

---

pushes the intestines, liver, stomach, and diaphragm superiorly, but it also widens the thoracic cavity. As a result, a pregnant woman cannot breathe quite as deeply, nor exhale as forcefully.

The enlargement of the uterus does result in additional problems. In the pelvic cavity, compression of the ureters and urinary bladder can result in *stress incontinence,* during which the pregnant woman involuntarily urinates. Compression of the inferior vena cava, especially when lying down, decreases venous return, and the result is edema and varicose veins.

Aside from the steroid hormones progesterone and estrogen, the placenta also produces some peptide hormones. One of these

makes cells resistant to insulin, and the result can be *gestational* (pregnancy-induced) diabetes. Some of the integumentary changes observed during pregnancy are also due to placental hormones. **Striae gravidarum,** commonly called "stretch marks," typically form over the abdomen and lower breasts in response to increased steroid hormone levels rather than stretching of the skin. Melanocyte activity also increases during pregnancy. Darkening of the areolae, skin in the line from the navel to the pubis, areas of the face and neck, and vulva is common.

Changes in breast anatomy and the development of lactation are discussed in Chapter 17, pages 407 and 409.

expect to gain 25–35 pounds during her pregnancy. Without adequate weight gain for mom, the infant may have a low birth weight, which increases the risk of complications (such as respiratory problems) immediately after birth. Furthermore, growing numbers of studies confirm that low birth weight newborns (less than 5 pounds, 8 ounces at birth) are more likely to develop certain chronic diseases, such as diabetes and high blood pressure, when they become adults.

Along with a balanced diet, those prenatal vitamin/mineral supplements must be taken throughout pregnancy. They have calcium for bone growth, iron for red blood cell formation, and vitamin B_6 for proper metabolism. Most important, they have that vitally important folate (folic acid) to meet the increased rate of cell division and DNA synthesis in both mother and fetus. Perhaps as many as 75% of neural tube defects could be avoided by adequate folate intake. Consuming fortified breakfast cereals is a good way to meet folate needs because they contain a more absorbable form of folate.

Certain foods must be avoided during pregnancy, however. Game fish species, such as shark, mackerel, swordfish, and salmon, should be passed up because of potentially high levels of mercury. Mom should also turn down unpasteurized milk, along with raw or undercooked shellfish, poultry, pork, beef, or eggs. These foods may contain bacteria or parasites that could sicken her and/or infect the unborn baby.

Mom has to be careful at home and on the job as well. Pesticides and many organic industrial chemicals, such as vinyl chloride, formaldehyde, asbestos, and benzenes, are teratogens and can cross the placenta. Cleaning the kitty litter box should become someone else's job during the mom's entire pregnancy—cat feces can cause her to contract **toxoplasmosis,** a dangerous parasitic infection. Lead (from old-fashioned, lead-based paint or pipes) can cause severe developmental delays. If mom's a health-care worker (hospital, doctor, dentist office, etc.), she must avoid accidental X-ray exposure. X rays and other ionizing radiation can cause mutations in a developing embryo or fetus and increase the baby's risk of developing leukemia after birth. If a pregnant woman must have diagnostic X rays (if she has broken a bone, for example), she should inform her health-care provider that she is pregnant. Shielding can minimize or eliminate danger to the fetus.

The medicine cabinet can be a pretty dangerous spot if you're pregnant. The most notorious teratogen incident in recent history occurred in Europe (and to a lesser extent, the United States) during the 1950s and 1960s when the drug thalidomide was taken to prevent nausea in pregnant women. Tragically, it arrested the development of arms and legs in some children and also damaged heart and blood vessels, ears, and the digestive tract. Some mothers of affected children reported taking the drug for only a few days. Today, many currently used medications, including herbal supplements, nonprescription medications, and prescription drugs, can also be teratogenic. For example, the isotretinoins, a group of drugs prescribed for severe acne, are known to cause severe birth defects. Before receiving these drugs, a woman must have a negative pregnancy test and sign a contract agreeing to simultaneously use two forms of birth control. Because of these situations and others, physicians are generally very cautious about prescribing drugs during pregnancy. No pregnant woman should take any drug—including ordinary cold remedies and herbal supplements—without checking first with her health-care provider.

As her big day draws nearer, mom must continue to be careful to prevent birth defects and illness in her newborn. When a mother has HIV, herpes, gonorrhea, or chlamydia, newborns can become infected as they pass through the birth canal. Cesarean section birth can prevent the blindness and other physical and mental defects that these infections might cause. As you know from Chapter 11, an Rh-negative woman who has given birth to an Rh-positive child should receive a Rho-Gam injection within 72 hours after birth. Rho-Gam will help to prevent her body from producing Rh antibodies if her child's red blood cells entered her bloodstream during the birth process. The first Rh-positive baby is not usually affected by anti-Rh antibodies. However, in subsequent pregnancies, antibodies created at the first birth cross the placenta and destroy the fetal red blood cells, causing severe anemia and other complications in the baby.

With proper prenatal care, a woman's chances of having a normal pregnancy and healthy baby are better now than ever before in human history. Now that health-care providers and laypeople are aware of the various ways birth defects can be prevented, perhaps the incidence of birth defects will decrease in the future. For more information on all aspects of pregnancy, consult www.marchofdimes.com.

Content CHECK-UP!

7. Oxytocin is an essential hormone both for labor and delivery and the months after birth as well. What are its functions during this entire time? What are the consequences if oxytocin production does not occur?

8. Effacement is the technical term for:
   a. leakage of amniotic fluid.
   b. false labor contractions.
   c. the process of gradually stretching the cervix until it slowly disappears.
   d. loss of the mucous plug that blocked the cervical canal.

9. Which of the following is an effect of placental estrogen?
   a. reduced uterine motility
   b. relaxation of smooth muscle
   c. increased insulin resistance
   d. increased uterine blood flow

Answers in Appendix A.

Selected New Terms

Basic Key Terms

acrosome (ăk′rō-sōm), p. 426

afterbirth (ăf′tĕr-bĕrth), p. 432

allantois (ŭh-lăn′tō-ĭs), p. 428

amnion (ăm′nē-ŏn), p. 428

blastocyst (blăs′tō-sĭst), p. 429

chorion (kō′rē-ŏn), p. 428

chorionic villi (kō″rē-ŏn′ĭk vĭl′lī), p. 431

cleavage (klēv′ĭj), p. 427

corona radiata (kō-rō′nă rā-dē-ă′tŭh), p. 426

differentiation (dĭf′ĕr-ĕn″shē-ā′shŭn), p. 427

ectoderm (ĕk′tō-dĕrm), p. 431

embryo (ĕm′brē-ō), p. 431

embryonic development (ĕm″brē-ŏn′ĭk dĕ-vĕl′ŏp-mĕnt), p. 431

embryonic disk (ĕm″brē-ŏn′ĭk dĭsk), p. 431

endoderm (ĕn′dō-dĕrm), p. 431

extraembryonic membrane (ĕks″trŭh-ĕm″brē-ŏn′ĭk mĕm′brăn), p. 427

fertilization (fĕr′tĭ-lĭ-zā′shŭn), p. 426

fetal development (fē′tŭhl dē-vĕl′ŏp-mĕnt), p. 434

fontanels (fŏn″tŭn-ĕlz′), p. 434

gastrulation (găs″trŭ-lā′shŭn), p. 431

gestation (jĕs-tā′shŭn), p. 428

growth (grōth), p. 427

human chorionic gonadotropin (HCG) (hyū′măn kōr″ē-ŏn′ĭk gō-năd″ō-trō′pĭn), p. 431

implantation (ĭm″plăn-tā′shŭn), p. 431

inner cell mass (ĭn′ĕr sĕl măs), p. 429

lanugo (lŭh-nū′gō), p. 434

mesoderm (mĕz′ō-dĕrm), p. 431

morphogenesis (mŏrf″ō-jĕn′ĕ-sĭs), p. 427

morula (mŏr′ū-lŭh), p. 429

Müllerian duct (myū′lĕr-ē′ŭn dŭkt) p. 435

neural fold (nū′răl fōld), p. 431

neural groove (nū′răl grūv), p. 431

neural plate (nū′răl plāt), p. 431

neural tube (nū′răl tūb), p. 431

placenta (plŭh-sĕn′tŭh), p. 428

pre-embryonic development (prē′-ĕm′brē-ŏn″ĭkdĕ-vĕl′ŏp-mĕnt), p. 428

primary germ layers (prī′mă-rē gĕrm lā′yĕrz), p. 431

trophoblast (trōf′ō-blăst), p. 429

umbilical cord (ŭm-bĭl′ĭ-kŭl kōrd), p. 431

vernix caseosa (vĕr′nĭks ka″sē-ō′sŭh), p. 434

Wolffian duct (wŭlf′ē-ŭn dŭkt) p. 435

yolk sac (yōk săk), p. 428

zona pellucida (zō′nă pĕl-lū′sĭ-dŭh), p. 426

zygote (zī′gōt), p. 426

Clinical Key Terms

anencephaly (ăn′ĕn-sĕf′ŭh-lē), p. 431

Braxton-Hicks contraction (brăks′tŏn hĭks cŏn-trăk′shŭn), p. 437

breech birth (brēch bĕrth), p. 435

cesarean section (sĭ-zār′ē-ŭn sĕk′shŭn), p. 435

congenital (birth) defect (kŏn-jĕn′ĭ-tŭl [bŭrth] dē′fĕkt), p. 440

delirium tremens (dē-lēr′ē-ŭm trē′mĕns), p. 440

episiotomy (ĕ-pĭz″ē-ŏt′ō-mē), p. 437

fetal alcohol syndrome (fē′tŭhl ăl′cŭh-hōl sĭn′drōm), p. 440

neural tube defect (nū′răl tūb dē′fĕkt), p. 431

parturition (păr″tū-rĭsh′ŭn), p. 437

premature birth (prē-măh-tyūr′ bŭrth), p. 439

rubella (German measles) (rū-bĕl′ŭh), p. 440

spina bifida (spī′nŭh bĭf′ĭ-dŭh), p. 431

striae gravidarum (strī′ē grăv-ĭ-dăr′ŭm), p. 440

teratogen (tĕr-ăh′tō-jĕn), p. 440

toxoplasmosis (tŏk″sō-plăz-mō′sĭs) p. 441

Summary

18.1 Fertilization

A. A sperm consists of a head, a middle piece, and a flagellum tail. The ovum is surrounded by a non-living matrix called the zona pellucida and an outermost cellular corona radiata.

B. During fertilization, the sperm's acrosomal enzymes help to digest the layers around the ovum so a single sperm can penetrate it. There, the sperm nucleus fuses with the ovum nucleus to create a zygote with a unique set of chromosomes. Following successful implantation by the embryo, a positive blood or urine test for HCG will confirm the pregnancy.

18.2 Development

A. Cleavage of the zygote and early embryo forms a ball of cells that then differentiate. The first organ system to appear is the nervous system.

B. The extraembryonic membranes, placenta, and umbilical cord allow humans to develop within the uterus. These structures protect the embryo and allow it to exchange wastes, nutrients, and oxygen with the mother's blood.

C. At the end of the embryonic period, all organ systems are established, and there is a mature and functioning placenta. Fetal development extends from the third through the ninth months. During the third and fourth months, the skeleton is becoming ossified and the sex of the fetus becomes distinguishable. During the fifth through ninth months, the fetus continues to grow and to gain weight. Babies born after six or seven months may survive, but full-term babies have a better chance of survival.

D. The gonads start to develop during the seventh week. Testosterone induces the gonads to form into testes; without testosterone, they will form into ovaries.

18.3 Birth

A. During stage 1 of parturition, the cervix effaces and dilates.

B. During stage 2, the child is born.

C. During stage 3, the afterbirth is expelled.

D. During pregnancy, the mother's uterus enlarges greatly, resulting in weight gain, standing and walking difficulties, and general discomfort. It comes to occupy most of the abdominal cavity with resultant annoyances such as incontinence. Many of the complaints of pregnancy, such as constipation, heartburn, darkening of certain skin areas, and diabetes of pregnancy are due to the presence of placental hormones.

Study Questions

1. Describe the process of fertilization and the events immediately following it. (pp. 426–427)
2. Name the four extraembryonic membranes, and give a function for each. (pp. 427–428)
3. What is the basis of the pregnancy test? (p. 427)
4. Specifically, what events normally occur during embryonic development? What events normally occur during fetal development? (pp. 428–429, 431–436)
5. Describe the structure and function of the umbilical cord. (p. 434)
6. Describe the structure and function of the placenta. (pp. 432, 434)
7. What are the three stages of birth? Describe the events of each stage. (pp. 437–438)
8. In general, describe the physical changes in the mother during pregnancy. (pp. 437, 439–440)
9. Label the diagram at right to review the events that occur during pre-embryonic development. (p. 429)

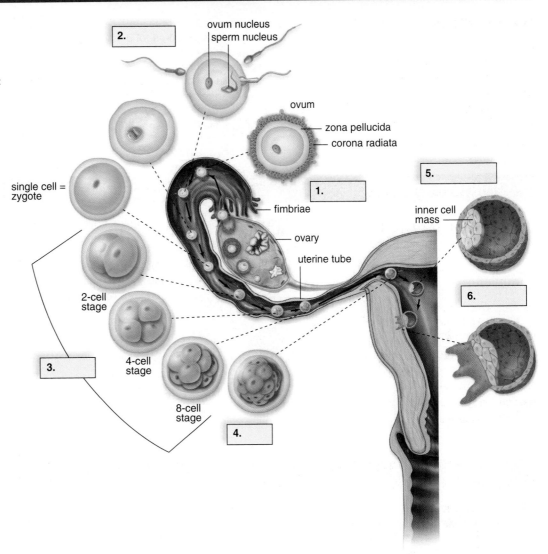

Learning Outcome Questions

Fill in the blanks.

1. Fertilization occurs when the _____ nucleus fuses with the _____ nucleus.
2. Once the blastocyst arrives at the uterus, it begins to _____ itself in the endometrium.
3. The zygote divides as it passes down a uterine tube. This process is called _____.
4. When cells take on a specific structure and function, _____ occurs.
5. During development, the nutrient needs of the developing embryo (fetus) are served by the _____.
6. The _____ membranes include the chorion, the _____, the yolk sac, and the allantois.
7. During embryonic development, all major _____ form.
8. Fetal development begins at the end of the _____ month.
9. In most deliveries, the _____ appear(s) before the rest of the body.
10. During pregnancy, constipation, darkening of certain areas of the skin, and diabetes are all due to _____ produced by the placenta.

Medical Terminology Exercise

After studying this chapter, see if you can derive the definitions for the medical terms listed at right. Many of the prefixes and suffixes used to create these terms can be found throughout the chapter. For additional help, use McGraw-Hill Connect™ at www.mcgrawhillconnect.com and consult Appendix B.

1. morphogenesis (mŏr″fō-jĕn′ĕ-sĭs)
2. neonatologist (nē″ō-nā-tŏl′ō-jĭst)
3. prenatal (prē-nā′tŭhl)
4. hyperemesis gravidarum (hī″pĕr-ĕm′ĕ-sĭs grăv-ĭd-ăr′ŭm)
5. dysmenorrhea (dĭs″mĕn-ō-rē′ŭh)
6. primigravida (prī″mĭ-grăv′ ĭ-dŭh)
7. oligospermia (ŏl″ĭ-gō-spĕr′mē-ŭh)
8. perineorrhaphy (pĕr″ĭ-nē-ŏr′ŭh-fē)
9. abruptio placentae (ăb-rŭp′shē-ō plŭh-sĕn′tē)
10. polyhydramnios (pŏl″ē-hī-drăm′nē-ŏs)
11. amniorrhea (ăm″nē-ō-rē′ŭh)
12. placenta previa (plŭh-sĕn′tŭh prē′vē-ŭh)

Online Study Tools

■■LEARNSMART®   Mc Graw Hill Education connect®

APR

Anatomy & Physiology REVEALED includes cadaver photos that allow you to peel away layers of the human body to reveal structures beneath the surface. This program also includes animations, radiologic imaging, audio pronunciations, and practice quizzing. To learn more visit www.aprevealed.com

Anatomy & Physiology | REVEALED®
aprevealed.com

19 Human Genetics

Learning Outcomes

As you begin to study the topic of human genetics in this chapter, you might be interested to know that genetic chance was, in part, responsible for changing European history. In the early 1900s, the last Russian monarch, Tsar Nicholas Romanov, and his wife, Tsarina Alexandra, had four daughters but were desperate for a male heir for the throne. When their fifth child, crown prince Alexei, was born in 1904, the couple was heartbroken to discover that their son suffered from hemophilia, a sex-linked genetic disorder that causes uncontrolled bleeding in affected boys. When physicians of the time couldn't help their son, the Romanovs turned to Grigory Rasputin, a Russian mystic often called the "mad monk." Historians aren't sure how Rasputin did it, but he was able to alleviate Alexei's suffering. In so doing, Rasputin won the trust of the monarchs, and gained great influence over their decisions. On Rasputin's advice, Tsar Nicholas took control of the Russian army from his generals during World War I, a decision that scholars agree helped to contribute to Russia's defeat. Would Nicholas have been a more effective ruler (and might Europe look different today?) if his only son hadn't suffered from a genetic disorder? No one will ever know. Nicholas abdicated the throne on March 9, 1917, setting the stage for the Russian revolution, and the entire Romanov family was assassinated in July 1918.

After you have studied this chapter, you should be able to:

19.1 Chromosomal Inheritance

1. Explain the normal chromosomal inheritance of humans.
2. Describe how a karyotype is prepared and two ways to obtain fetal chromosomes.
3. Explain how nondisjunction results in inheritance of an abnormal chromosome number.
4. Describe Down syndrome and various syndromes that result from the inheritance of an abnormal sex chromosome number.

19.2 Genetic Inheritance

5. Explain autosomal dominant and recessive allele inheritance.
6. Explain X-linked inheritance and why males have more X-linked disorders than females.
7. Relate the inheritance of an allele to protein synthesis.
8. Describe how to construct and interpret a Punnett square.
9. Tell how a genetic counselor could help a couple who are carriers for cystic fibrosis.

19.3 DNA Technology

10. Explain how gene therapy is being used to treat genetic disorders.
11. Discuss genomics, including how genomics might lead to better treatments for illnesses.

Medical Focus
Preimplantation Genetic Studies

What's New
A Profound Dilemma: Bioengineered Babies

Focus on Forensics
The Innocence Project

19.1 Chromosomal Inheritance

1. Explain the normal chromosomal inheritance of humans.
2. Describe how a karyotype is prepared and two ways to obtain fetal chromosomes.
3. Explain how nondisjunction results in inheritance of an abnormal chromosome number.
4. Describe Down syndrome and various syndromes that result from the inheritance of an abnormal sex chromosome number.

Normally, both males and females have 23 pairs of chromosomes, for a total of 46 chromosomes. The first 22 pairs, or chromosomes 1–44, are called **autosomes** (literally, "self" chromosomes). The final pair, chromosomes 45 and 46, are the **sex chromosomes.** These chromosomes determine the gender of the individual. In humans, males have a **Y chromosome** and an X chromosome. Females have two **X chromosomes.**

Various human disorders result from the inheritance of an abnormal chromosome number, or from inheriting chromosomes that are abnormal in size or shape. Such a disorder may be a **syndrome,** which is a group of symptoms that always occur together. Table 19.1 lists several syndromes that are due to an abnormal chromosome number. It is possible to view an individual's chromosomes by constructing a **karyotype,** a display of the chromosomes arranged by size, shape, and banding pattern. Karyotypes reveal whether an individual has inherited an abnormal number of chromosomes (too many or too few). Karyotyping can also often reveal if a chromosome has an abnormal length (too long or too short) or is otherwise abnormal in shape.

Karyotyping

Any cell in the body except red blood cells, which lack a nucleus, can be a source of chromosomes for karyotyping. In adults, it is easiest to use white blood cells separated from a blood sample for this purpose. Cell-free fetal DNA can be found in the mother's bloodstream beginning at about five weeks of pregnancy. These fetal chromosomes can be studied for certain types of abnormalities, but results can be unreliable. The chromosomes of embryos or fetuses are usually obtained by either amniocentesis or chorionic villi sampling (CVS). Because each carries a risk of spontaneous abortion (miscarriage), these procedures are not routinely performed for every pregnancy, but only when maternal history and/or prenatal testing indicate it is necessary. For example, a pregnant woman who is over the age of 40 may elect to have CVS or amniocentesis because older women are at increased risk for chromosomal abnormalities in the embryo. Blood testing for pregnant women of any age may indicate the need for follow-up chromosomal studies. A blood test routinely offered to pregnant women measures the levels of four chemicals present in the blood: alpha-fetoprotein, produced by the fetal liver; estriol, an estrogen-like hormone produced by the placenta; human chorionic gonadotropin, the pregnancy hormone, and inhibin-A, a hormone produced by the fetus and placenta. Abnormally high or low blood levels of these chemicals in the blood are linked to certain chromosomal and structural abnormalities in the embryo and fetus. Amniocentesis or CVS may be recommended if abnormal levels are detected. It is important to note that this blood test does not *confirm* any abnormality; it just *suggests* that possibility.

During **amniocentesis,** a sample of amniotic fluid is withdrawn from the uterus of a pregnant woman. Blood tests and the age of the mother are used to determine whether the procedure should be done. Amniocentesis is not usually performed until about the fourteenth to the seventeenth week of pregnancy. A long needle is passed through the abdominal and uterine walls to obtain a small amount of fluid, which also contains fetal cells (Fig. 19.1a). Testing the cells and karyotyping the chromosomes may be delayed as long as four weeks so that the cells can be cultured to increase their number. As many as 400 chromosomal and biochemical problems can be detected by testing the cells and the amniotic fluid.

The risk of spontaneous abortion increases by about 0.3% due to amniocentesis, and doctors only use the procedure if it is medically warranted.

Chorionic villi sampling (CVS) is a procedure for obtaining chorionic cells in the region where the placenta will develop. This procedure can be done as early as the fifth week of pregnancy. A long, thin catheter tube is inserted through the vagina into the uterus (Fig. 19.1b). Ultrasound, which gives a picture of the uterine contents, is used to place the tube between the uterine lining and the chorionic villi. Remember that the chorion is the most superficial membrane surrounding the embryo, and it is formed from trophoblast cells of the embryo (see Chapter 18, page 429). Thus, the chromosomes in the cells of chorionic villi are the same as those of the embryo, although chorionic villi cells do not contribute to forming the embryo itself. Once the catheter is in place, a

TABLE 19.1 Syndromes from Abnormal Chromosome Numbers

Syndrome	Sex	Disorder	Chromosome Number	Frequency	
				SPONTANEOUS ABORTIONS	LIVE BIRTHS
Down	M or F	Trisomy 21	47	1/40	1/800
Poly-X	F	XXX (or XXXX)	47 or 48	0	1/1,500
Klinefelter	M	XXY (or XXXY)	47 or 48	1/300	1/800
Jacobs	M	XYY	47	?	1/1,000
Turner	F	XO	45	1/18	1/2,500

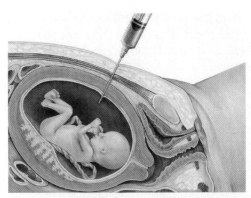

a. During amniocentesis, a long needle is used to withdraw amniotic fluid containing fetal cells.

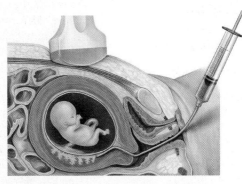

b. During chorionic villi sampling, a suction tube is used to remove cells from the chorion, where the placenta will develop.

c. Cells are microscopically examined and photographed.

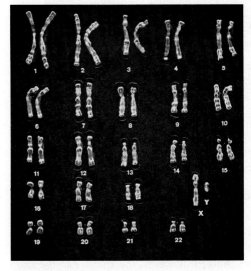

d. Normal male karyotype with 46 chromosomes

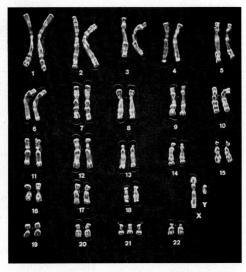

e. Down syndrome karyotype with an extra chromosome 21

AP|R **Figure 19.1** **Human karyotype preparation.** A karyotype is an arrangement of an individual's chromosomes into numbered pairs according to their size, shape, and banding pattern. (**a**) Amniocentesis and (**b**) chorionic villi sampling provide cells for karyotyping to determine if the unborn child has a chromosomal abnormality. (**c**) After cells are treated as described in the text, the karyotype can be constructed. (**d**) Karyotype of a normal male. (**e**) Karyotype of a male with Down syndrome. A Down syndrome karyotype has three number 21 chromosomes.

tiny sample of chorionic cells is withdrawn by suction. The cells do not have to be cultured, and karyotyping can be done immediately. However, this sampling procedure does not gather any amniotic fluid, so the biochemical tests done on the amniotic fluid following amniocentesis are not possible. CVS also carries a greater risk of spontaneous abortion than amniocentesis—0.8% compared to 0.3%. Further, CVS has been suggested as the cause behind certain limb and facial congenital abnormalities. The advantage of CVS is that the results of karyotyping are available at an earlier date.

Preparing the Karyotype

After a sample of cells has been obtained, the cells are stimulated to divide in a culture medium. A chemical is used to stop mitosis during metaphase when chromosomes are the most highly compacted and condensed. The cells are then spread on a microscope slide and dried. Stains are applied to the slides, and the cells are photographed. Staining produces dark and light cross-bands of varying widths, and these can be used in addition to size and shape to help pair up the chromosomes. Today, a computer may be used to arrange the chromosomes in pairs. It is possible to photograph the

nucleus of a cell that is about to divide (the chromosomes are more visible then), so that a picture of the chromosomes is obtained. The picture may be entered into a computer, and the chromosomes electronically arranged by pairs (Fig. 19.1c). The resulting display of chromosomes is the karyotype. Figure 19.1d,e compares a normal karyotype with that of a person who has Down syndrome, the most common autosomal abnormality.

Nondisjunction

An abnormal chromosomal makeup in an individual can be due to nondisjunction. **Nondisjunction** occurs during meiosis I when both members of a homologous pair go into the same daughter cell or during meiosis II when the sister chromatids fail to separate and both daughter chromosomes go into the same gamete (Fig. 19.2). If an ovum with 24 chromosomes is fertilized by a normal sperm cell with 23 chromosomes, the zygote formed has a **trisomy:** One type of chromosome is present in three copies instead of the normal two, and the total chromosome count is 47 instead of the normal 46. Conversely, if an ovum with 22 chromosomes is fertilized with a normal sperm, the zygote formed carries a **monosomy:** One type

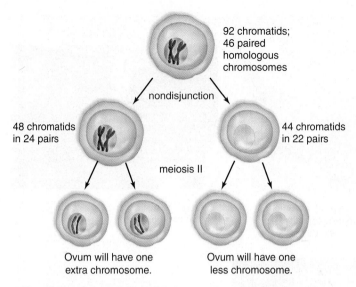

92 chromatids;
46 paired
homologous
chromosomes

nondisjunction

48 chromatids
in 24 pairs

44 chromatids
in 22 pairs

meiosis II

Ovum will have one
extra chromosome.

Ovum will have one
less chromosome.

a. **Nondisjunction during meiosis I**

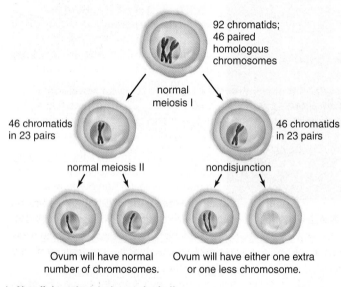

92 chromatids;
46 paired
homologous
chromosomes

normal
meiosis I

46 chromatids
in 23 pairs

46 chromatids
in 23 pairs

normal meiosis II

nondisjunction

Ovum will have normal
number of chromosomes.

Ovum will have either one extra
or one less chromosome.

b. **Nondisjunction during meiosis II**

Figure 19.2 Nondisjunction during oogenesis. Nondisjunction can occur (**a**) during meiosis I if homologous chromosomes fail to separate or (**b**) during meiosis II if the sister chromatids fail to separate completely. In either case, abnormal gametes have an extra chromosome or lack a chromosome.

of chromosome is present in a single copy, and the zygote has only 45 chromosomes. It is important to note that nondisjunction can occur during the formation of gametes in either gender. Thus, although Figure 19.2 shows nondisjunction during the formation of ova, it could also happen during sperm formation.

Researchers believe that when the zygote carries either a monosomy or a trisomy, most nondisjunctions produce embryos that cannot survive and develop normally. The result is an early spontaneous abortion (miscarriage), which may occur before a woman even realizes that she is pregnant. Note in Table 19.1 the high rate of miscarriages occurring from monosomy or trisomy.

Figure 19.3 Down syndrome. Down syndrome occurs when the egg or the sperm has an extra chromosome 21 due to nondisjunction in either meiosis I or meiosis II. Characteristics include a wide, rounded face and narrow, slanting eyelids. Developmental delay to varying degrees is usually present.

Down Syndrome

Down syndrome is also called trisomy 21 because the individual usually has three copies of chromosome 21. In most instances, the ovum had two copies instead of one of this chromosome. (In 23% of the cases studied, however, the sperm had the extra chromosome 21.)

Individuals who have Down syndrome share characteristic physical features (Fig. 19.3). Their heads are rounded; they have a short, webbed neck, and eyes that slant upward and have an inner eyelid fold. Facial features are typically small: a small, flattened nose; small ears that may fold over at the top; and a small mouth that often makes the tongue seem large and protruding. The Down syndrome individual is short statured, with small, short hands that have a large, single palmar crease (the so-called simian line). Mild to moderate developmental delays are typical of the syndrome. With proper assistance, the Down syndrome individual can often function well in school and can live and work independently. In rare cases, developmental delays can be severe.

The chance of a woman having a Down syndrome child increases rapidly with age, starting at about age 40. The frequency of Down syndrome is 1 in 800 births for mothers under 40 years of age and 1 in 80 for mothers over 40 years of age. Most Down syndrome babies are born to women younger than age 40, however, because this is the age group having the most babies. Maternal blood testing can screen for abnormal levels of chemicals associated with Down syndrome, but this type of test is not a positive confirmation. Amniocentesis followed by karyotyping is needed to confirm the diagnosis of Down syndrome in the fetus.

It is known that the genes that cause Down syndrome are located on the bottom third of chromosome 21, and extensive investigative work has been directed toward discovering the specific genes responsible for the characteristics of the syndrome. One day it might be possible to control the expression of these genes even before birth, so that the symptoms of Down syndrome do not appear.

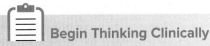

Begin Thinking Clinically

Individuals with Down syndrome often have a shorter-than-normal life span. What are some of the reasons for this?

Answer and discussion in Appendix A.

Sex Chromosome Inheritance

As stated, the sex chromosomes in humans are called X and Y. Because women are XX, an ovum always bears an X, but since males are XY, a sperm can bear an X or a Y. Therefore, the gender of the newborn child is determined by the father. If a Y-bearing sperm fertilizes the egg, then the XY combination results in a male. On the other hand, if an X-bearing sperm fertilizes the egg, the XX combination results in a female. All factors being equal, there is a 50% chance of having a girl or a boy.

♀ \ ♂	X	Y
X	XX	XY

Nondisjunction also occurs with regard to the sex chromosomes. Ova or sperm with too many or too few sex chromosomes can occur. Therefore, nondisjunction accounts for the birth of individuals with too few or too many sex chromosomes.

Too Many/Too Few Sex Chromosomes

From birth, an XO individual with **Turner syndrome** has only one sex chromosome, an X; the O signifies the absence of a second sex chromosome. Turner females are short, with a broad chest and a webbed neck. The ovaries, uterine tubes, and uterus are very small and nonfunctional. Turner females do not undergo puberty or menstruate, and their breasts do not develop (Fig. 19.4a). They are usually of normal intelligence and can lead fairly normal lives. Though Turner females have occasionally given birth following in vitro fertilization using donor ova, there are significant risks for both the mother and baby. Pregnant Turner syndrome women are at high risk of developing pregnancy-related high blood pressure, a condition called *pre-eclampsia*. If untreated, pre-eclampsia can lead to heart attack and stroke. Further, almost 40% of babies of Turner mothers are born prematurely.

A male with **Klinefelter syndrome** has two or more X chromosomes in addition to a Y chromosome, and is sterile. The testes

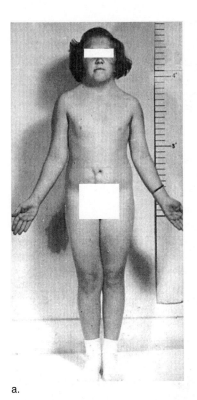 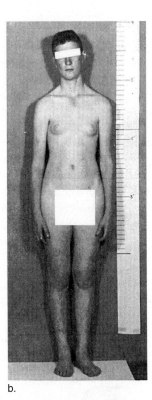

a.   b.

Figure 19.4 Syndromes due to an abnormal sex chromosome number. (a) A female with Turner (XO) syndrome has a short, thick neck, short stature, and lack of breast development. **(b)** A male with Klinefelter (XXY) syndrome has a feminine body shape and some development of the breasts.

and prostate gland are underdeveloped, and the individual has no facial hair. Also, some breast development may occur (Fig. 19.4b). Affected individuals have large hands and feet and very long arms and legs. They may have developmental delays, particularly if they inherit more than two X chromosomes.

A **poly-X female** has more than two X chromosomes. Females with three X chromosomes have no distinctive physical appearance, aside from a tendency to be tall and thin. Some exhibit delayed motor and language development, though most are normal. Some may have menstrual difficulties, but many menstruate regularly and are fertile. Their children usually have a normal karyotype.

Females with more than three X chromosomes occur rarely. Unlike XXX females, XXXX females are usually tall and severely developmentally delayed. Various physical abnormalities are seen, but they may menstruate normally.

XYY males with **Jacobs syndrome** can only result from nondisjunction during spermatogenesis. This is because only males have a Y chromosome. Affected males are usually taller than average, suffer from persistent acne, and tend to have speech and reading problems. At one time, it was suggested that these men were likely to be criminally aggressive, but it has since been shown that the incidence of such behavior among them may be no greater than among XY males.

Notice that there are no YO males. This shows that at least one X chromosome is needed for survival. However, XXY individuals are males, not females.

19.2 Genetic Inheritance

5. Explain autosomal dominant and recessive allele inheritance.
6. Explain X-linked inheritance and why males have more X-linked disorders than females.
7. Relate the inheritance of an allele to protein synthesis.
8. Describe how to construct and interpret a Punnett square.
9. Tell how a genetic counselor could help a couple who are carriers for cystic fibrosis.

Genes, which are sections of chromosomes, control body traits, such as height, eye color, or type of earlobe. Genes can also control the production of specific proteins, such as structural proteins, cell membrane proteins, and enzymes. Recall that as described in Chapter 17, page 393, each cell contains homologous pairs of chromosomes. Alternate forms of a gene having the same position (called the **locus**), on a pair of homologous chromosomes and affecting the same trait are called **alleles.** It is customary to designate an allele by a letter, which represents the specific characteristic it controls. A **dominant allele** "dominates" over a **recessive allele;** that is, the dominant trait will always be expressed. Thus, the dominant allele is assigned an uppercase (capital) letter, whereas a recessive allele is given the same letter but in lowercase. In humans, for example, unattached (free) earlobes are dominant over attached earlobes, so we could use the letter *E* for unattached earlobes and *e* for attached earlobes.

Inheritance of Genes on Autosomal Chromosomes

An individual normally has two alleles for an autosomal trait. Just as one member of each pair of chromosomes is inherited from each parent, so too is one of each pair of alleles inherited from each parent. If the two alleles are identical, the pair is termed **homozygous;** if not, the pair is **heterozygous.**

The term **genotype** refers to the genes of the individual. Figure 19.5 shows four possible fertilizations and the resulting genotype of the individual for earlobe attachment. In the first instance, the chromosomes of both the sperm and the ovum carry an *E.* Consequently, the zygote and subsequent individual have the alleles *EE,* which can be called a **homozygous dominant** genotype. A person with genotype *EE* obviously has unattached earlobes. The physical appearance of the individual—in this case, unattached earlobes—is called the **phenotype.**

In the second fertilization, the zygote has received two recessive alleles (*ee*), and the genotype is called **homozygous recessive.** An individual with this genotype has the recessive phenotype, which is attached earlobes. In both the third and fourth examples of fertilization, the resulting individual has the alleles *Ee,* which is a heterozygous genotype. A heterozygote shows the dominant characteristic; therefore, the phenotype of this individual is unattached earlobes.

How many dominant alleles does an individual need to inherit to have a dominant phenotype? These examples show that a

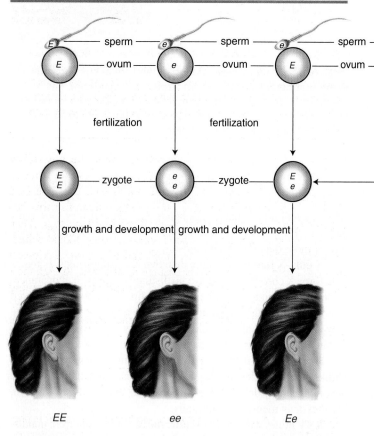

Figure 19.5 Genetic inheritance. Individuals inherit a minimum of two alleles for every characteristic of their anatomy and physiology. The inheritance of a single dominant allele (*E*) causes an individual to have unattached earlobes; two recessive alleles (*ee*) cause an individual to have attached earlobes. Notice that each individual receives one allele from the father (by way of a sperm) and one allele from the mother (by way of an ovum).

dominant allele contributed from only one parent causes expression of a particular dominant phenotype. How many recessive alleles does an individual need to inherit to have the recessive phenotype? A recessive allele must be received from both parents to bring about the recessive phenotype.

Sex-Linked Inheritance

The sex chromosomes contain genes just as the autosomal chromosomes do. Some of these genes determine whether the individual is a male or a female. Investigators have now discovered a series of genes on the Y chromosome that determine the development of male genitals, and at least one on the X chromosome that seems to be necessary for the development of female genitals.

Some traits controlled by alleles on the sex chromosomes have nothing to do with the gender of the individual. These traits are said to be **sex-linked.** An allele that is only on the X chromosome is **X-linked** and an allele that is only on the Y chromosome is **Y-linked.** Most sex-linked traits are controlled by alleles on the larger X chromosome. The very small Y chromosome does not have the corresponding allele, simply because it is so short. Thus, the sex chromosomes in males—X and Y—are not truly homologous chromosomes.

As you know, a female has two X chromosomes, and thus her ova will contain either of her two X chromosomes. A male always receives a sex-linked recessive condition from his mother, from whom he inherited his single X chromosome. The Y chromosome from his father does not carry a corresponding allele for the trait. Therefore, males need receive only one recessive allele to have the X-linked disorder. Consider red-green color blindness, for example (Fig. 19.6a). This genetic disorder results in an individual who cannot see red or green because the cone cells of the eye are abnormal. It is caused by a recessive gene on the X chromosome. Note that although the mother has one recessive gene on an X chromosome, she has normal vision due to the dominant gene on her other X chromosome. She is referred to as a **carrier** for the trait. A carrier is an individual who is heterozygous for a recessive genetic disorder and therefore has no symptoms. If the carrier's ova are fertilized by sperm from a male with normal vision, any male children she bears will have a 50% chance of inheriting the disorder. Similarly, any female children will have a 50% chance of being a carrier for the disorder.

Would it ever be possible for a female to have red-green color blindness? It is possible, although unlikely. A female must receive two recessive X-linked alleles, one from each parent, before she has an X-linked recessive condition. The inheritance of a dominant allele from either her mother or her father can offset the inheritance of a recessive X-linked allele. A female with red-green color blindness will have inherited a recessive X chromosome from both her carrier mother and her color-blind father (Fig. 19.6b).

Punnett Square to Determine Probability

A Punnett square is a "short-cut" method originally devised by a mathematician/geneticist named Reginald Punnett that allows one to predict the *probability* of inheritance of a particular trait. It can be used to answer questions about both autosomal and sex-linked inheritance. Consider this example: An

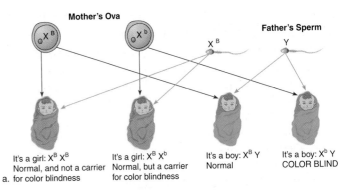

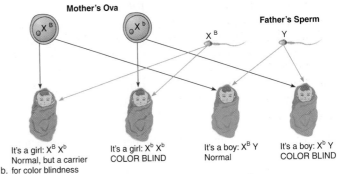

Figure 19.6 **Sex-linked inheritance.** (a) A typical pattern of sex-linked inheritance for the color blindness trait. The mother is a carrier for the trait, and her sons have a 50% chance of inheriting the trait. (b) A sex-linked trait can be inherited by a daughter only if her father is affected and her mother is a carrier or color blind also.

African-American couple is concerned about the possibility of conceiving a child with sickle-cell anemia. As you'll recall from Chapter 11, sickle-cell anemia is a disease characterized by abnormal hemoglobin. Red blood cells containing this hemoglobin are very fragile and easily destroyed (see page 249). Sickle-cell anemia is an autosomal recessive trait, like attached/detached earlobes. Both the husband and wife have sickle-cell trait (i.e., both carry the recessive gene for the trait, though their phenotypes are normal). What is their probability of conceiving a child with the disease?

To solve this problem, first draw a Punnett square. It's just like a tic-tac-toe board:

Then, put the father's genotype in the top boxes. The father is phenotypically normal but carries a recessive gene, so he is heterozygous, Aa. Any symbols can be used to represent the trait, so long as the uppercase/lowercase letters look different (to avoid confusion).

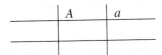

Next, put the mother's genotype in the side boxes. Again, she is heterozygous like the father, since she carries a recessive gene, but is phenotypically normal herself. It makes no difference if the father's symbols are on top, with the mother's on the side, or vice versa—the results will be the same.

	A	a
A		
a		

Then, bring the father's genes down, as though he was passing them on to his offspring:

	A	a
A	A	a
a	A	a

And finally, bring the mother's genes across, as though she was passing them on to her offspring.

	A	a
A	AA	aA
a	Aa	aa

The four filled boxes give the possibilities for this couple's future children, and allow probability to be determined. The first possibility is 1:4, or 25% probability, of producing a normal child who does not carry the recessive trait, as in the starred box:

	A	a
A	AA*	aA
a	Aa	aa

The second possibility is 2:4, or 50% probability, of producing a child who is phenotypically normal, but carries the recessive gene, just like his/her parents, as in the starred boxes. It's important to note that whether the dominant or recessive trait is written first, the outcome will be the same:

	A	a
A	AA	aA*
a	Aa*	aa

The third possibility is 1:4, or 25% probability, of producing a child who has sickle-cell anemia. Remember that sickle-cell anemia is an autosomal recessive disorder—thus, for the child to be affected, he or she must be homozygous for the recessive gene. That would be the case in the starred box below:

	A	a
A	AA	aA
a	Aa	aa*

Keep in mind that dominant traits are not always "good" or beneficial. Consider the inheritance of achondroplasia, a disorder that results in individuals with very short stature. The torso is normal in size, but the limbs are abnormally short. In this disorder, the normal trait is actually a recessive gene. If two heterozygous adults with achondroplasia have children, the Punnett square would be identical to the one above. However, in this case, there would be 75% probability of an affected child and only 25% probability of having a nonaffected child. It's also important to recognize that Punnett squares only allow one to determine the probability of a single genetic event. If the parents with sickle-cell trait have a child with sickle-cell anemia, there will still be a 25% probability that any subsequent children will have sickle-cell anemia. As an analogy, consider tossing a coin—with every independent coin toss, there is an equal probability of "heads" or "tails."

The Punnett square method can be used to predict probability in sex-linked inheritance as well. Consider hemophilia, the sex-linked disease inherited by Alexei Romanov, whose history was briefly described in the chapter introduction. Alexei's mother, Alexandra, was the carrier, though she was a phenotypically normal female. Her genotype can be expressed as $X^H X^h$, where X^H is the dominant trait and X^h is the recessive trait. Tsar Nicholas Romanov, Alexei's father, was normal: $X^H Y$. The Punnett square used for sex-linked inheritance is similar to one used for autosomal inheritance; the only difference is the addition of the X and Y chromosomes. The completed square should look like this:

	X^H	X^h
X^H	$X^H X^H$	$X^H X^h$
Y	$X^H Y$	$X^h Y$

The results of this Punnett square show that for every Romanov child conceived, there was a 25% probability of a phenotypically normal noncarrier female ($X^H X^H$), a 25% probability of a phenotypically normal carrier female ($X^H X^h$), a 25% chance of a normal male ($X^H Y$), and a 25% chance of an affected male such as Alexei ($X^h Y$).

As a historical footnote, you might be interested to know that the original carrier female for the hemophilia trait was Queen Victoria of England (Tsarina Alexandra was Queen Victoria's granddaughter). Because her descendants were royalty themselves, the hemophilia trait also occurred in the royal households of England, Spain, Russia, and Prussia. Eventually, the trait was no longer passed down to Victoria's descendants, so the illness is no longer found among European royalty.

Genetic Counseling

Couples who believe that they may be at risk for having a child with a chromosomal or genetic disorder may choose to seek the advice of a genetic counselor. Genetic counselors are professionals with special training and certification. They help families to understand chromosomal and genetic inheritance and to make informed decisions about pregnancy and childbearing. In order to understand genetic counseling, we will consider the inheritance of cystic fibrosis. **Cystic fibrosis (CF)** is the most common lethal genetic disorder among Caucasians in the United States. About one in 20 Caucasians is a carrier for CF, and about one in 3,000 newborns has the disorder. In children with CF, the mucus in the bronchi, bronchioles, and pancreatic ducts is particularly thick and viscous, interfering with the function of the lungs and pancreas. Affected individuals suffer from recurring lung infections, which are often fatal. Because pancreatic enzymes cannot digest food, a CF patient may be malnourished as well. However, in the past few years, new treatments have raised the average life expectancy for CF patients to approximately 37 years of age.

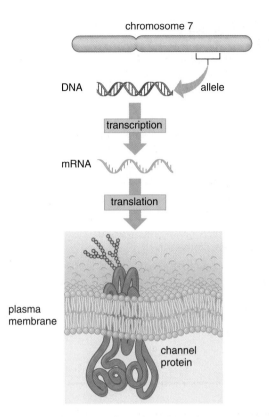

chromosome 7

DNA

allele

transcription

mRNA

translation

plasma
membrane

channel
protein

Figure 19.7 Cystic fibrosis. A person with cystic fibrosis has an abnormal allele, and the result is an abnormal channel protein in the plasma membrane. Chloride ions cannot exit the cell, resulting in a very thick mucus, particularly in the bronchial tubes and pancreatic duct.

Normal individuals have at least one dominant allele for a plasma membrane channel protein. Figure 19.7 shows the relationship between inheritance of a normal CF allele on a chromosome (chromosome 7) and development of the channel protein. The illustration emphasizes that alleles are actually a segment of DNA and that alleles cause the production of certain proteins in a cell.

As shown in Figure 19.8*a*, if a genetic counselor knows the genotype of the potential parents, the counselor can predict the chances that a couple will have a child having a recessive autosomal disorder such as CF. If the parents are both carriers, each offspring has a 25% chance of receiving two recessive alleles and having CF. This pattern of inheritance is the same for all autosomal recessive disorders.

If the counselor does not know the inheritance pattern of a disorder, it is sometimes possible to deduce it by studying a pedigree. A **pedigree** is a diagram that depicts the inheritance of a particular trait: Circles are females, and squares are males; shaded-in symbols represent those who have a trait; half-shaded symbols represent carriers; and Roman numerals indicate generations. Notice that the pedigree in Figure 19.8*b* has to be for a recessive disorder because unaffected parents have a child with the disorder. This can only happen when a condition is recessive. If the condition were dominant, one of the parents would definitely be affected.

Prenatal Testing for Genetic Disorders

Commonly inherited genetic disorders are listed in Table 19.2. Several of these disorders do not appear unless multiple

TABLE 19.2 Inheritance of Some Genetic Disorders

Dominant	Recessive	X-Linked	Multiple Genes
Examples of dominantly inherited disorders include:	Examples of recessive inherited disorders include:	Examples of X-linked disorders include:	Examples of multifactorial inheritance include:
• Neurofibromatosis—benign tumors in skin or deeper	• Cystic fibrosis—disorder affecting function of mucous and sweat glands	• Agammaglobulinemia—lack of immunity to infections	• Cleft lip and/or palate
• Achondroplasia—a form of dwarfism	• Galactosemia—inability to metabolize milk sugar	• Color blindness—inability to distinguish certain colors	• Clubfoot
• Chronic simple glaucoma (some forms)—a major cause of blindness if untreated	• Phenylketonuria—essential liver enzyme deficiency	• Hemophilia (some forms)—defect in blood-clotting mechanisms	• Congenital dislocation of the hip
• Huntington disease—progressive nervous system degeneration	• Sickle-cell disease—blood disorder primarily affecting blacks	• Muscular dystrophy (some forms)—progressive wasting of muscles	• Spina bifida—open spine
• Familial hypercholesterolemia—high blood cholesterol levels, propensity to heart disease	• Thalassemia—blood disorder primarily affecting persons of Mediterranean ancestry	• Spinal ataxia (some forms)—spinal cord degeneration	• Hydrocephalus (with spina bifida)—water on the brain
• Polydactyly—extra fingers or toes	• Tay-Sachs disease—lysosomal storage disease leading to nervous system destruction		• Pyloric stenosis—narrowed or obstructed opening from stomach into small intestine
			• Breast cancer
			• Diabetes mellitus
			• Coronary artery disease

Source: Data from the National Foundation/March of Dimes.

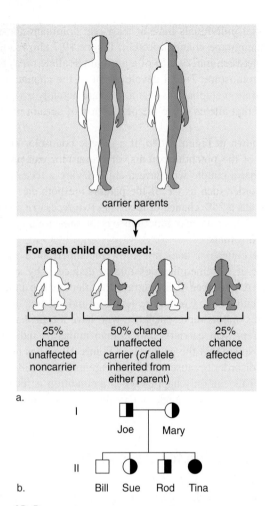

carrier parents

For each child conceived:

25% chance unaffected noncarrier

50% chance unaffected carrier (*cf* allele inherited from either parent)

25% chance affected

a.

I   Joe   Mary

II   Bill   Sue   Rod   Tina

b.

Figure 19.8 Inheritance pattern for CF, an autosomal recessive disorder. (a) The figures below the parents show four possible combinations of inherited alleles. Therefore, each offspring has a 25% chance of inheriting two recessive alleles and having CF. **(b)** A sample pedigree for carrier parents and their children. Note that Sue and Rod are carriers like their parents Joe and Mary, and Tina is affected with CF.

Content **CHECK-UP!**

4. The outward, or physical expression of a person's genes for a trait is the:

   a. allele.   c. genotype.

   b. phenotype.   d. recessive gene.

5. A couple has just had a child diagnosed with sickle-cell anemia, which is an autosomal recessive trait. Neither parent has the disease. What are the parents' genotypes? (Use the letter *A* to represent the dominant trait, and the letter *a* to represent the recessive trait.)

   a. father *Aa*, mother *aa*

   b. father *aa*, mother *aa*

   c. father *aa*, mother *Aa*

   d. father *Aa*, mother *Aa*

6. Imagine you're a genetic counselor who is advising grieving parents who just lost their 2-year-old son to Tay-Sachs disease. They want to know if there is a chance that they could have other children with the disease. What will you tell them?

Answers in Appendix A.

19.3 DNA Technology

10. Explain how gene therapy is being used to treat genetic disorders.

11. Discuss genomics, including how genomics might lead to better treatments for illnesses.

Previously, you studied the structure of DNA and how it replicates and carries out protein synthesis. **DNA technology** includes our ability to work directly with DNA to determine the relatedness of individuals, to assist forensics in determining whether a person has committed a crime, and to develop new treatments for human illnesses, called gene therapy.

Gene Therapy

Gene therapy is the insertion of genetic material into human cells for the treatment of a disorder. It includes procedures that give a patient healthy genes to make up for faulty genes and also includes the use of genes to treat various other human illnesses, such as cancer and cardiovascular disease. Currently, approximately 1,700 patients are enrolled in over 300 approved gene therapy trials in the United States.

How can a molecule as large as a gene be introduced, intact, into a cell? The most common approach used by scientists is to employ a *vector*—a carrier whose job is to transport the DNA directly into the cell. Viruses are nature's own vectors because they replicate themselves by inserting their own genetic material into a host cell. Scientists studying gene therapy techniques have adapted viruses to function as vectors. The viruses are first modified: Viral genetic material is removed or altered so that it does not cause disease, and human genes are inserted into the virus instead. Next, the virus is allowed to infect human tissue, where it inserts the

abnormal genes are inherited. The inheritance of these conditions, listed in the "Multiple Genes" column in Table 19.2, is complex.

Often parents want to improve their chances of having a child who is free of a particular genetic disorder that runs in their families. As you have learned, testing of cell-free fetal DNA in the maternal bloodstream can be followed by chorionic villus sampling or amniocentesis. The Medical Focus reading on page 455 describes techniques for preimplantation genetic diagnosis (PGD). Using PGD, the ovum can be analyzed to ensure that it does not carry an abnormal allele prior to in vitro fertilization, or the embryo itself is tested for genetic defects before being introduced into its mother's uterus. In the future, it might be possible to use gene therapy to cure any genetic defects found in an ovum or embryo.

Preimplantation Genetic Studies

We have already seen that in vitro fertilization can lead to therapeutic cloning (see page 430) for the production of various tissues. Now, we'll consider the medical implications of the fact that each cell of the pre-embryo is *totipotent,* meaning that each cell can become a complete embryo because the cells have not yet become specialized. Not surprisingly, then, if a single cell is removed from an eight-celled cleavage embryo, the seven-celled embryo will still go on and develop into a normal newborn (Fig. 19A). Researchers are using this knowledge to allow them to carry out preimplantation genetic analysis.

Consider that one partner (or both) of a couple may have one of the genetic disorders listed in Table 19.2. As potential parents, wouldn't the couple want the assurance that the embryo is free of genetic disorders and will develop into a healthy adult? In some instances, it's possible to perform in vitro fertilization and then preimplantation genetic diagnosis (PGD) in order to determine the genotype of the embryo with regard to particular genetic disorders. Then, only a healthy embryo is implanted into the female. Currently, PGD can be used to screen for over 30 chromosomal and genetic abnormalities, including Down syndrome, cystic fibrosis, hemophilia, and sickle-cell anemia. Proponents cite its advantages: By selecting only healthy embryos for implantation, a couple is less likely to suffer a miscarriage that results from chromosomal abnormality. Multiple embryos will not have to be implanted, and multiple pregnancy, with its increased risk of premature birth, can be avoided. Furthermore, the couple need not go through chorionic villus sampling or amniocentesis if pregnancy is achieved, and won't have to face the painful decision to terminate an existing pregnancy because of a genetic or chromosomal disorder. Although the PGD procedure is expensive, proponents argue that lifetime medical care for a child with a genetic or chromosomal disorder costs far more. In the future, it's possible that PGD might be coupled with gene therapy so that any embryo would be suitable for implantation.

Recently, the effectiveness of single-cell PGD has been called into question. Researchers note that fewer than half of the embryo's chromosomes are currently tested, and other defects may be present

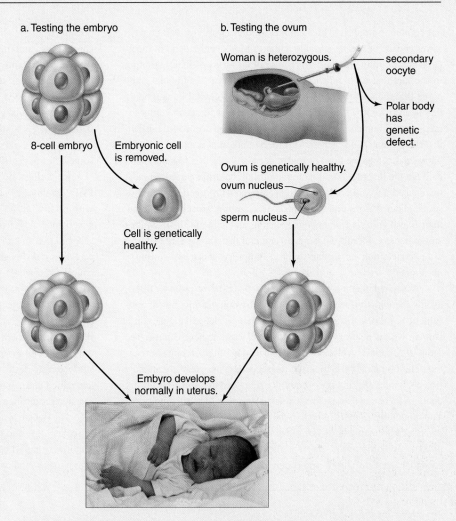

Figure 19A **Preimplantation genetic diagnosis.** **(a)** Following in vitro fertilization, a cell is removed from an 8-cell embryo and its chromosomes are tested for genetic defects. If genetically healthy, the embryo is implanted into the uterus. **(b)** Before in vitro fertilization, the chromosomes of the polar body accompanying a secondary oocyte are tested for genetic defects. If the polar body contains the abnormal gene, the ovum is assumed to be genetically healthy. Once fertilized with sperm, it is implanted into the uterus.

on the remaining chromosomes. Newer techniques are being tested on embryos in the blastocyst stage, when 100 cells or more are present. In this later stage, five cells can be removed for testing instead of a single cell, and there is less risk of damaging the embryo. Then, a new screening method called comparative genomic hybridization (CGH) allows all chromosomes in the embryonic cells to be studied.

However, all of the testing takes time, and many IVF embryos die before the testing can be completed, while those that remain may

—Continued

Continued—

not be able to implant. Thus, couples who elect to have PGD have a lower chance of having a viable embryo. Moreover, experts in medical ethics continue to question the use of all forms of PGD. Pointing to the fact that each cell in an eight-cell embryo is totipotent, they argue that the single cell used for testing is a potential human being that is destroyed by the testing process. Likewise, defective embryos are potential human beings that will ultimately be destroyed. Further, it's known that many genetic disorders don't appear in the affected individual before the third or fourth decade of life. It's feasible that a cure for the disorder could be found during that time period. Finally, medical ethicists stress that although children with genetic or chromosomal abnormalities may have medical issues that must be overcome, their lives can have tremendous meaning and value to their families and to society. Moral philosophers contend that ending a life shortly after conception is an act whose morality must be seriously questioned.

A second method of PGD, called polar body biopsy (PBB), avoids the ethical issues surrounding embryonic testing, but its usefulness is limited to detecting the presence of defective genes in a woman who is heterozygous for a particular single gene disease (see Table 19.2). PBB involves removal and testing of the chromosomes in the polar body that lies alongside the secondary oocyte. As you know (from Chapter 18), the polar body is like a "garbage can," because it contains 23 pairs of sister chromatids that are discarded after meiosis I. The remaining 23 pairs remain in the secondary oocyte. For example, consider a woman who is heterozygous for Tay-Sachs disease. If her polar body chromosome pair has the Tay-Sachs recessive gene, the secondary oocyte chromosome pair has the dominant gene and is presumed to be disease-free. Likewise, if the polar body sister chromatids contain the Huntington's disease dominant gene in a woman with a family history of that disease, the secondary oocyte has the normal recessive gene. In either case, normal secondary oocytes are then fertilized and implanted into the woman's uterus, and defective secondary oocytes are destroyed. (But the technique is not perfect; there's a slight chance that crossing-over during prophase I may have put the defective gene into the secondary oocyte's chromosomes. See pages 393–395 to review crossing-over.)

Regardless of the technique that is used, ethical issues about PGD continue to surface. In several instances worldwide, families with a child suffering from a serious blood disorder (leukemia or genetic anemias) have already used PGD to create a "savior baby"—a second child free of the genetic disorder whose umbilical cord blood was then transfused into the ill sibling. This practice creates further concern that PGD could be used to create babies for their "spare parts." Perhaps most disturbing is the notion that PGD could eventually be used to select only for embryos with a couple's desired characteristics: gender, intelligence, physical appearance, or athletic ability. PGD is already being used by couples who want a male child to select only male embryos for implantation—and any female embryos are subsequently destroyed. In societies worldwide where male children are valued more than females, practitioners of PGD find a very lucrative market for their skills.

The debate will undoubtedly continue over the uses of this type of technology, and others like it that will develop in the future. It is important that all of us are well informed and understand the scientific issues so that we can make informed decisions.

---

healthy human genes into tissue cells. Four classes of viruses have been employed in gene therapy research: (1) retroviruses, like the virus that causes AIDS; (2) adenoviruses, which are a form of cold virus; (3) herpes viruses, which cause cold sores on the mouth; and (4) the most commonly used virus vector, the adeno-associated virus, a very small virus that can have all of its own genetic material removed and still be capable of infecting cells. Scientists have seen some early successes in the treatment of several immune system deficiency diseases, as well as inherited diseases affecting metabolism.

Other mechanisms to insert genes into cells are also being investigated. One way is to introduce DNA directly into cells. A second mechanism involves attaching the gene to a liposome. Liposomes are microscopic vesicles that form spontaneously when lipoproteins are put into a solution. Still other scientists attach the DNA to molecules that have specific receptors on the cell membrane, then allowing the cell to incorporate the DNA by phagocytosis.

Among the many gene therapy trials, one is for the treatment of familial hypercholesterolemia, a condition that develops when liver cells lack a receptor for removing cholesterol from the blood. The high levels of blood cholesterol make the patient subject to fatal heart attacks at a young age. In a newly developed procedure, a small portion of the liver is surgically excised and infected with a virus containing a normal gene for the receptor. Several patients have experienced lowered serum cholesterol levels following this procedure.

Cystic fibrosis patients have an abnormal gene for a transmembrane carrier of the chloride ion. Patients often die due to numerous infections of the respiratory tract. In a newly developed procedure, liposomes have been coated with the gene needed to cure cystic fibrosis. The liposomes have then been delivered to the lower respiratory tract.

Genes are also being used to treat medical conditions other than the known genetic disorders. VEGF (vascular endothelial growth factor) can cause the growth of blood vessels. The gene that

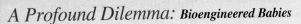

What's New

A Profound Dilemma: **Bioengineered Babies**

Consider for a moment that you're a young woman who is going blind, and the doctors tell you that you've inherited a condition called Leber hereditary optic neuropathy. After researching the condition on the Internet[1], you discover that this disease is not inherited in the way you might have expected (by mutation of the chromosomes in your cells' nuclei). Rather, this mutation took place in the DNA of your mitochondria. As you know from Chapter 18, human mitochondria come from the ovum that formed the zygote, not the sperm. That means your children's mitochondria will all have this mutation, though they may or may not display disease symptoms. But what if you first replace your mitochondria with a donor's healthy ones?

That's the basic premise behind a technique called mitochondrial donation, which was approved for human experimentation in the United Kingdom in February 2015. There are two methods for the procedure. In *maternal spindle transfer,* the pronucleus in an ovum donated by a healthy woman (i.e., one who is free of mitochondrial disease) is destroyed and replaced with the pronucleus from the mother's own defective ovum. This reengineered ovum is then fertilized with the husband's sperm. With the second method, *pronuclear transfer,* two embryos are first created by fertilizing both the donor ovum and the mother's ovum with the father's sperm. The nucleus of the healthy donor embryo is removed and destroyed, then replaced with the nucleus from the mother's embryo. In both techniques, the resulting embryos are theoretically free of the mother's defective mitochondria and can then be introduced into the mother's uterus. If the embryo implants and is carried to term, the baby will have three sets of DNA: mitochondrial DNA from the ovum donor and nuclear DNA from the mother and father. Though neither of these techniques has been attempted in humans, experimentation in mice and monkeys has successfully produced offspring that seem healthy and disease-free.

Though mitochondrial donation seems an "easy" cure for a number of serious illnesses, the entire process still has serious short-term and long-term technical problems. Scientists know that nuclear and mitochondrial DNA interact, but these interactions are currently very poorly understood. Thus, no one can predict what will happen when an artificial combination is created. Moreover, one scientist has compared the procedure to the cellular equivalent of a heart transplant. What if the technique isn't perfect and maternal mitochondria are accidentally transferred along with the mother's nucleus? What will happen to the donor ovum's other organelles? How will the reengineered DNA combination ultimately affect the mother? What if the donor's mitochondrial DNA passes into the mother's bloodstream? Could it cause an immune response, or even cancer? And the most fundamentally important question of all: is there any guarantee that the resulting child will be healthy and normal?

Other countries, including the United States, are carefully considering Britain's decision to permit mitochondrial donation. Bioethicists everywhere fear that mitochondrial donation is unethical for *all* individuals involved. The ovum donor and the mother must take hormones to make multiple ova develop within their ovaries and then undergo laparoscopic surgery to have the ova removed. Both the hormones and the surgery involve serious potential (and potentially fatal) health risks. Should an ovum donor be subject to these risks, and should she be paid? Will her DNA contribution show up in the child's physical, intellectual, or psychological makeup? Does she have any legal rights at all regarding the child created from her ovum? Furthermore, should a healthy embryo be destroyed in pronuclear transfer so that scientists can create the new "hybrid" one? What if the resulting child has a physical or intellectual disability or a shortened life span?

Even more concerning is a new technology called clustered regularly interspaced short palindromic repeat technology (CRISPR). Using this technique, scientists can change the actual chromosomes of a cell by adding or deleting snippets of DNA. If the process is only used to correct mutations in somatic stem cells (the body's own stem cells, which produce muscle cells, nerve cells, etc.), it holds the promise of correcting and curing genetic disorders such as cystic fibrosis or Tay-Sachs disease. Altered somatic cell chromosomes aren't passed to a person's children. But what if the procedure is used to alter the chromosomes in embryos? If the embryo implants and completes development, its transformed DNA will theoretically pass to future generations. In theory, this means that the next generation's future children could inherit undesirable traits, such as a tendency for uncontrolled rage. Moreover, what if the technology is used to attempt to create a "perfect race?" Novels such as Aldous Huxley's *Brave New World* explore the fictional consequences of such a decision, and historical examples—most notably, the Holocaust— demonstrate the devastation that results when a small group of individuals assumes the power to make value judgments based on genetics and race.

And perhaps the most important question of all—and one with profound implications for society—do scientists ever have the right to bioengineer a child?

[1]You can find out more about this condition at: http://ghr.nlm.nih.gov /condition/leber-hereditary-optic-neuropathy

---

codes for this growth factor can be injected alone or within a virus into the heart to stimulate branching of coronary blood vessels. These additional branches improve coronary circulation, providing additional oxygen and nutrients to the myocardium. Coronary patients report that they have less chest pain and can run longer on a treadmill.

Some of the newest applications of gene therapy are approaches to correct the errors in transcription and translation that result from a genetic mutation. As you know from Chapter 3, DNA serves as the template for translating messenger RNA, and messenger RNA codes for a protein. If DNA has mutated, it will not correctly translate into mRNA, and the protein product can't be formed. These novel medications are proteins themselves, which don't repair the genetic mutation. Instead, they fix the "manufacturing process," enabling the correct protein to be translated.

Gene therapy is increasingly used as part of cancer therapy. Genes are being used to make healthy cells more tolerant of, and tumors more vulnerable to, chemotherapy. The gene *p53* brings about the death of cells, and there is much interest in introducing it into cancer cells, and in that way killing them off.

Genomics

Genomics is the molecular analysis of a **genome,** which is all the genetic information in all the chromosomes of an individual. Recall from Chapter 2, Figure 2.14, that the two DNA strands of a double helix are held together by bonding between their bases, designated as A, T, G, and C. Base A is always paired with base T, and base G is always paired with base C. In a worldwide effort known as the Human Genome Project, researchers set out to map the human chromosomes in two ways: (1) by constructing a map that shows the sequence of base pairs along all the human chromosomes, and

(2) by constructing a map that shows the sequence of genes along the human chromosomes.

The Base Sequence Map

Researchers achieved the first goal in 2003, thanks to an international effort. It took some 15 years for scientists to complete this monumental task. It is now known that human genetic material contains 3.16 billion base pairs, and the order of 99.9% of those base pairs is exactly the same in all people. To date, no one knows how much of the DNA consists of truly functional genes that actually code for proteins. It is estimated that 50% or more is so-called "junk DNA," which does not code for protein at all. The total number of truly functional genes is estimated to be between 30,000 and 50,000—a much lower number than originally projected, and less than 2% of the entire genome.

The Genetic Map

The genetic map tells the location of genes along each chromosome. Sixteen of the 46 human chromosomes have been mapped, including chromosome 22 (the first human chromosome to be completely mapped) and chromosome 4 (the most recently completed). Completing the genetic map should accelerate now that the base sequence map is done. Researchers need only know a short sequence of bases in a gene of interest

FOCUS on FORENSICS

The Innocence Project

Imagine that you're a young man who's been charged with murder. At your trial, a truck driver who gives "eyewitness" testimony states that you were the hitchhiker he picked up near the crime scene—but he's blind in one eye and admits it was very dark that night. A jailhouse snitch tells the judge that you confessed to the murder—and then, rape charges against him are later dropped by the prosecution. A former girlfriend testifies against you—but then recants her testimony two weeks later. A scent-dog handler tells the judge and jury that his dog identified your scent on a bloody T-shirt found near the scene—but the man's credentials and the dog's tracking ability are both later called into question. You maintain that you were five miles away from the scene at the time the murder occurred, and others confirm your alibi. However, five days later, you're convicted of first-degree murder, sentenced to life in prison, and incarcerated. This was the nightmarish scenario faced by William Dillon in 1981.

Now fast-forward to 2007: With help from the attorneys of the Innocence Project, Dillon's repeated requests for an investigation into his conviction were finally granted by a Florida judge. The Innocence Project—an advocacy group composed of attorneys, law students, and volunteers—uses modern DNA testing in its mission of re-examining

past criminal convictions in which evidence is highly suspect. As you remember from Chapter 3, DNA analysis is often a vital tool for criminal identification. However, DNA evidence can also be used to review court findings and potentially overturn a wrongful conviction. In Dillon's case, although nearly all of the evidence from his original murder trial had been lost or destroyed, the blood-stained T-shirt from the crime was fortunately still available. DNA testing was performed on the shirt, and the testing showed that the blood belonged to the victim and to another unidentified man. There was no trace of Dillon's DNA on the shirt. With this new evidence, William Dillon was finally proven innocent, after spending 26 years in prison for a crime that someone else committed. He was cleared of all charges by the prosecution, and was freed from prison on December 10, 2008. In June 2011, 30 years after the original murder, cold-case detectives named four new suspects.

William Dillon's case is one of a multitude of decades-old criminal cases that have been reopened through the work of the Innocence Project. To date, the Project's efforts have exonerated more than 329 individuals using DNA technology, including 18 who had been serving time on death row.

in order for computers to search the genome for a match. Then, computers can help to pinpoint where this gene is located. However, decades—if not centuries—of research remain to be completed about the genome. Many questions remain unanswered: What does each gene do? How is the gene turned on and off? What is all the junk DNA for? How do genes interact to make each unique individual? What gene, or combination of genes, causes disease? For each question answered, many others will continue to arise.

Many ethical questions also arise regarding how we should use our knowledge of the human genome. Who owns genetic information? How should it be used and accessed? Should humans attempt to control and/or manipulate human genes and, if so, how? As you know from the What's New reading on page 457, manipulation of human genes is already underway, in laboratories around the world. It's imperative that you be knowledgeable about the human genome so that you can help decide these issues.

Content CHECK-UP!

7. Name the four classes of viruses that are being studied as possible gene vectors. How are these viruses used in gene therapy?

8. An abnormal gene for the carrier that transports chloride across the plasma membrane results in:

   a. familial hypercholesterolemia.

   b. cystic fibrosis.

   c. hemophilia.

   d. sickle-cell anemia.

9. Which percentage of DNA base pairs is the same for all human beings?

   a. 99.9%   b. 75%   c. less than 2%   d. 50%

Answers in Appendix A.

Selected New Terms

Basic Key Terms

allele (ă-lēl′), p. 450

amniocentesis (ăm″nē-ō-sĕn-tē′sĭs), p. 446

autosome (ăw′tō-sōm), p. 446

carrier (kār′ē-ĕr), p. 451

chorionic villi sampling (kō-rē-ŏn′ĭk vĭl′ī săm′plĭng), p. 446

dominant allele (dŏm′ĭ-nănt ă-lēl′), p. 450

genes (jēnz), p. 450

genome (jē′nōm), p. 458

genomics (jē′nō-mĭks), p. 458

genotype (jē′-nō-tīp), p. 450

heterozygous (hĕt″ĕr-ō-zī′gŭs), p. 450

homozygous (hō-mō-zī′gŭs), p. 450

homozygous dominant (hō-mō-zī′gŭs dŏm′ĭ-nănt), p. 450

homozygous recessive (hō-mō-zī′gŭs rē-sĕs′ĭv), p. 450

karyotype (kār′ē-ō-tīp), p. 446

locus (lō′kŭs), p. 450

monosomy (mŏ′nō-sō-mē), p. 447

nondisjunction (nŏn″dĭs-jŭnk′shŭn), p. 447

pedigree (pĕd′ĭ-grē), p. 453

phenotype (fē′nō-tīp), p. 450

recessive allele (rē-sĕs′ĭv ă-lēl′), p. 450

sex chromosome (sĕks krō′mō-sōm), p. 446

sex-linked (sĕks-lĭnkt), p. 451

syndrome (sĭn′drōm), p. 446

trisomy (trī′sō-mē), p. 447

X chromosome (x krō′mō-sōm), p. 446

X-linked (x-lĭnkt), p. 451

Y chromosome (y krō′mō-sōm), p. 446

Y-linked (y-lĭnkt), p. 451

Clinical Key Terms

cystic fibrosis (sĭs′tĭk′ fī-brō′sĭs), p. 452

DNA technology (DNA tĕk-năwl′ō-jē), p. 454

Down syndrome (dŏwn sĭn′drōm), p. 448

gene therapy (jēn thĕr′ŭh-pē), p. 454

Jacobs syndrome (jā′kŭbs sĭn′drōm), p. 449

Klinefelter syndrome (klĭn′fĕl-tĕr sĭn′drōm), p. 449

poly-X female (pŏ′lē x fē′māl), p. 449

Turner syndrome (tŭr′nĕr sĭn′drōm), p. 449

XYY male (xyy māl), p. 449

Summary

19.1 Chromosomal Inheritance

A. Normally, an individual inherits 22 pairs of autosomal chromosomes and one pair of sex chromosomes. Females are XX and males are XY.

B. Amniocentesis and chorionic villi sampling are used to provide cell samples for karyotyping fetal chromosomes.

C. Nondisjunction during oogenesis or spermatogenesis explains the inheritance of an abnormal number of chromosomes.

D. The most common autosomal abnormality is Down syndrome, which is due to the inheritance of an extra chromosome 21.

E. Abnormal combinations of sex chromosomes include XO (Turner syndrome), XXX (poly-X), XXY (Klinefelter syndrome), and XYY (Jacobs syndrome).

19.2 Genetic Inheritance

A. Genes control human traits. Uppercase letters designate dominant alleles; lowercase letters designate recessive alleles.

B. The genotype represents the genes of an individual, and the phenotype refers to outward expression of the gene. A homozygous dominant individual inherited two dominant alleles and has the dominant phenotype; a heterozygous individual inherited one dominant and one recessive allele and has the dominant phenotype; a homozygous recessive individual inherited two recessive alleles and has the recessive phenotype. A Punnett square can be used to predict the probability of gene inheritance.

C. The inheritance of X-linked alleles differs in males and females. Males require only one recessive allele to have an X-linked trait; females require two recessive alleles. This means that males are more likely to inherit an X-linked disorder.

D. Genetic counselors help couples determine the chances of having children with a genetic disorder, such as cystic fibrosis. They can also determine the pattern of inheritance from studying a family's pedigree.

19.3 DNA Technology
A. Gene therapy, which involves replacing defective genes with healthy genes, is now a reality. Researchers are envisioning various applications aimed at curing human genetic disorders, as well as many other types of illnesses.

B. Genomics is an actively growing field. Because the sequence of base pairs along the human chromosomes has now been determined, new treatments for genetic disorders are expected.

Study Questions

1. What is the normal chromosome inheritance of humans? (p. 446)
2. How is a karyotype prepared? What are the possible sources for cell samples in an adult? In the fetus? (pp. 446–447)
3. What is nondisjunction, and when does nondisjunction occur during meiosis? (pp. 447–448)
4. What are the characteristics of a person with Down syndrome? (pp. 448–449)
5. What are the characteristics of the most common human conditions resulting from inheritance of abnormal numbers of sex chromosomes? (p. 449)
6. Explain autosomal dominant and recessive genetic inheritance. (pp. 450–451)
7. Explain X-linked allele inheritance in humans. (p. 451)
8. Describe how to construct and interpret a Punnett square. (pp. 451–452)
9. What type of information does a genetic counselor give parents who might pass on a genetic disorder? (pp. 452–453)
10. Give examples of dominant, recessive, and X-linked genetic disorders in humans. (p. 453)
11. Describe the function of an allele using the CF allele as an example. (pp. 452–453)
12. What is gene therapy, and what types of genetic disorders have been treated thus far? (pp. 454, 456–458)
13. What is a genome? What is genomics, and what might be the benefits of genomics in the future? (pp. 458–459)

Learning Objective Questions

For questions 1-9, fill in the blanks. For questions 10 and 11, choose the one best answer.

1. Genes are found on the _____.
2. A person with Down syndrome has inherited _____ copies of chromosome 21.
3. The sex chromosomes of a male are labeled _____.
4. A person with Klinefelter syndrome has the chromosomes _____.
5. A dominant autosomal genetic disorder only requires the inheritance of _____ (one or two) abnormal gene(s).
6. Someone who is heterozygous for a recessive disorder is called a(n) _____ of that trait. He or she doesn't display symptoms of this disorder.
7. If a person inherits an autosomal genetic disorder and both parents are unaffected by the disorder, the disorder is _____.
8. Replacing defective genes with healthy genes is the goal of _____.
9. The molecular analysis of a genome is called _____.
10. Karyotyping reveals:
    a. what alleles a person has.
    b. the entire sequence of genes of a person.
    c. the number and sizes of chromosomes of a person.
    d. a person's phenotype.
11. A male with an X-linked recessive disorder inherited this disorder from his:
    a. mother.
    b. father.

Medical Terminology Exercise

After studying this chapter, see if you can derive the definitions for the medical terms listed at right. Many of the prefixes and suffixes used to create these terms can be found throughout the chapter. For additional help, use McGraw-Hill Connect™ at www.mcgrawhillconnect.com and consult Appendix B.

1. neogenesis (nē″ō-jĕn′ĕ-sĭs)
2. regeneration (rē-jĕn″ĕr-ā′shŭn)
3. fetoscope (fē′tō-skōp)
4. polydysplasia (pŏl″ē-dĭs-plā′zē-ŭh)
5. congenital (kŏn-jĕn′ĭ-tăl)

Online Study Tools

LEARNSMART® **connect®**

APR

Anatomy & Physiology REVEALED includes cadaver photos that allow you to peel away layers of the human body to reveal structures beneath the surface. This program also includes animations, radiologic imaging, audio pronunciations, and practice quizzing. To learn more visit www.aprevealed.com

Chapter 1

Content Check-Up!
1. structural organization; 2. organ systems; 3. organelles; 4. superior; 5. buccal; 6. horizontal; 7. parietal pleura—b, visceral pericardium—a, visceral peritoneum—c; 8. chest pain, particularly when a person inhales; 9. endocrine and nervous; 10. digestive; 11. a; 12. digestive and urinary; 13. d; 14. b.

Begin Thinking Clinically
Page 8: Because the child is pointing to the umbilical region, the source of his pain is most likely the small intestine. Two common reasons for intestinal pain are stretching the small intestine (as when the organ overfills with gas) or irritating the intestinal lining (as in bacterial food poisoning).

Learning Outcome Questions
1. c; 2. f; 3. g; 4. a; 5. d; 6. c; 7. e; 8. b; 9. a; 10. d; 11. g; 12. f; 13. d; 14. c; 15. e; 16. a; 17. d; 18. b; 19. g; 20. c; 21. d; 22. e; 23. f; 24. a; 25. b; 26. anatomy; 27. organ; 28. midsagittal or median; 29. homeostasis; 30. negative feedback.

Medical Terminology Exercise
1. above; 2. below; 3. stomach; 4. abdomen; 5. head; 6. thorax; 7. sides; 8. eye; 9. back; 10. study; 11. a; 12. a; 13. blood; 14. d; 15. a.

Chapter 2

Content Check-Up!
1. 20; 2. F=F; F_2; 3. 1×10^{-4} moles per liter, or 0.0001 moles per liter; 4. a. urine at pH 5 is more acidic; b. there are 1000 times more free hydrogen ions in pH 5 urine; 5. carbohydrates, lipids, proteins and nucleic acids; 6. Digestion of food macromolecules occurs by hydrolysis, or splitting with water. A constant supply of water is needed for the process to continue; 7. a—monosaccharide, b—monosaccharide, c—disaccharide, d—polysaccharide, e—polysaccharide, f—polysaccharide; 8. cellulose; 9. saturated, unsaturated; 10. Their polar phosphate heads are water-soluble, but their lipid tails are oil-soluble; 11. cholesterol; 12. a; 13. The protein's three-dimensional structure is changed when the protein is denatured. Biologically active proteins may no longer function; 14. b; 15. a. RNA, b. RNA, c. both, d. DNA, e. DNA, f. RNA; 16. This data doesn't make sense because of complimentary base pairing. The percentage of adenine must match that of thymine. Likewise, the percentage of cytosine and guanine must be the same; 17. AATCG in DNA; AAUCG in RNA; 18. adenosine diphosphate.

Begin Thinking Clinically
Page 24: Because strontium is similar to calcium in structure, it could replace bone calcium and damage bone.

Begin Thinking Clinically
Page 37: When DNA is defective, the structure of the messenger RNA transcribed from it will also be abnormal. In turn, the amino acid sequence in the protein translated from messenger RNA will be altered as well. Just a single base substitution in DNA causes sickle-cell hemoglobin formation and all of the symptoms of the disease. If untreated, sickle cell anemia is often fatal. You can read more about sickle-cell disease in Chapter 11.

Learning Outcome Questions
1. atoms; 2. neutrons; 3. ionic, covalent; 4. hydrogen; 5. hydrogen, lower; 6. glucose, energy; 7. glycerol, fatty acid; 8. amino acids, coil, structure or configuration; 9. enzymes; 10. DNA, nucleotide.

Chapter 3

Content Check-Up!
1. a-2, b-3, c-1; 2. Nuclear envelope, Golgi apparatus, lysosomes, vesicles are the components of the endomembrane system. Endoplasmic reticulum forms proteins and lipids, the Golgi apparatus forms vesicles, lysosomes destroy cells or cell parts, and the nuclear membrane contains the nucleus; 3. a-2, b-4, c-1, d-3; 4. c; 5. The bacteria were taken into the cell by phagocytosis; 6. a; 7. c; 8. b; 9. d.

Begin Thinking Clinically
Page 52: When brain tissue is damaged, fluid escapes into the spaces between the nerve cells. This swelling can put pressure on the brain (because it's enclosed by the skull) and cause further brain damage. If a hypertonic solution flows through the blood vessels supplying the brain, osmosis will cause excess fluid to be returned to the bloodstream.

Begin Thinking Clinically
Page 53: Excess blood glucose is filtered by the kidneys, traveling from the blood into the fluid that will ultimately become urine. Because it can't be completely reabsorbed, the excess glucose is eliminated in the urine. Glucosuria (the technical term for glucose in the urine) causes increased urinary output (called polyuria) and increased thirst (polydipsia). These three are common symptoms of uncontrolled diabetes.

Learning Outcome Questions
1. c; 2. e; 3. a; 4. d; 5. b; 6. cytoplasm, organelles; 7. protein, phospholipid; 8. ribosomes; 9. centrioles; 10. hypotonic; 11. carrier molecule, ATP; 12. endocytosis; 13. 46; 14. d; 15. b; 16. a; 17. c; 18. d; 19. f; 20. a; 21. c; 22. b; 23. e.

Chapter 4

Content Check-Up!
1. d; 2. d; 3. Microvilli indicate that the cells are specialized for absorption; 4. fibrous connective tissue, tendons, ligaments and aponeuroses; 5. chondrocytes, lacunae; 6. c; 7. c; 8. a-3, b-1, c-2; 9. cardiac and smooth muscle; 10. intercalated disks; 11. b; 12. Schwann cells and oligodendrocytes; 13. a-2, b-3, c-1; 14. Exocrine glands secrete their products through ducts. Salivary glands are exocrine glands. Endocrine glands secrete their products directly into the blood. The thyroid gland is an endocrine gland; 15. a-3, b-1, c-2.

Begin Thinking Clinically
Page 68: Smoking causes constant "wear-and-tear" on the delicate pseudostratified ciliated columnar epithelium of the airways. Over time, this

tissue will gradually be replaced by stratified squamous epithelium, which is much sturdier (though unable to move mucus and clear airway pollutants).

Learning Outcome Questions

1. tissues; 2. stratified, cilia, columnar; 3. matrix, fibers; 4. connective; 5. a; 6. epithelial, loose connective; 7. c; 8. a; 9. b; 10. d; 11. c; 12. a; 13. b; 14. neuron, neuroglia; 15. desmosomes, tight junctions and gap junctions; 16. endocrine, exocrine; 17. mucous.

Chapter 5

Content Check-Up!

1. c; 2. b; 3. The deepest layer is the stratum basale, and the most superficial layer is stratum corneum; 4. a; 5. Hirsutism (excess body hair in women) may result from testosterone excess. Alopecia may be genetic or caused by an immune disorder; 6. b; 7. Sweat contains salts and urea, which are also excreted by the kidneys; 8. constrict. The process keeps blood in the body's torso, retaining heat in the body; 9. basal cell carcinoma, melanoma, ultraviolet light; 10. fibroblasts; 11. Adipose tissue concentration in the dermis decreases, decreasing the amount of insulation for the body; 12. Vitamin D, produced by skin cells, is converted to calcitriol. Calcitriol enables calcium absorption in the digestive tract.

Begin Thinking Clinically

Page 88: In psoriasis, overgrowth of the epidermis causes the skin to have a reddened, crusted appearance. Patches of affected skin are covered with scales, and may itch. Psoriasis is not serious unless it is part of an autoimmune disorder that affects other body tissues (see Chapter 13 for further discussion of autoimmune disorders).

Learning Outcome Questions

1. both b and c; 2. c; 3. a; 4. b; 5. b; 6. temperature; 7. hair follicles, sebum; 8. mechanical trauma, pathogen invasion, water gain or loss; 9. D; 10. melanoma, basal cell carcinoma; 11. depth of burn and extent of burn.

Chapter 6

Content Check-Up!

1. b; 2. compact; 3. d; 4. epiphyseal growth plate; 5. foramen; 6. d; 7. atlas and axis; 8. vertebral foramen; 9. c; 10. Glenoid cavity and head of humerus. The joint is unstable because the glenoid cavity is shallow; 11. a; 12. c; 13. The sutures of the skull and the epiphyseal growth plates. If these close prematurely, no further brain or long bone growth will occur; 14. a; 15. osteoarthritis, articular cartilage; 16. flexion and extension of the leg; 17. integumentary, skeletal, digestive and endocrine.

Begin Thinking Clinically

Page 113: Depending on the severity of the blow, he may have fractured the bones of the anterior orbit, or eye socket. These include the frontal, lacrimal, maxilla, and zygomatic bones. He may also have a fracture of the nasal bone. If he lost consciousness after the blow, he must be carefully monitored for a possible brain injury (see the In Case of Emergency reading in Chapter 8 for a discussion of traumatic brain injury).

Learning Outcome Questions

1. c; 2. g; 3. a; 4. d; 5. f; 6. e; 7. d; 8. c; 9. g; 10. a; 11. f; 12. b; 13. e; 14. longer; 15. spongy; 16. sinuses; 17. axial skeleton, vertebral column and pelvic girdle, axial skeleton and rib cage; 18. support and flexibility, support and stability; 19. hand and foot; 20. hinge; 21. c; 22. e; 23. b; 24. d; 25. a.

Chapter 7

Content Check-Up!

1. d; 2. b; 3. Epimysium (superficial); perimysium; endomysium (deep); 4. c; 5. Actin, troponin, tropomyosin; 6. c; 7. d; 8. b; 9. Intermediate twitch; used for walking, jogging, or biking because these activities require moderate strength for shorter periods; 10. a; 11. d; 12. You might use zygomaticus, levator labii superioris, and levator anguli oris; 13. The cells themselves are fewer and smaller, with fewer mitochondria. 14. Skeletal muscle compresses cardiovascular veins, thus forcing blood back to the heart.

Begin Thinking Clinically

Page 135: Compartment syndrome usually occurs in the muscles of the thigh or leg. It can result from severe bruising of the muscle or from bone fracture. Compartment syndrome is treated by elevating the affected limb and using anti-inflammatory drugs to reduce swelling. In severe cases, the fascia that creates the compartment must be cut open, a procedure called fasciotomy.

Begin Thinking Clinically

Page 154: The muscles that lift the foot are on the anterior leg, commonly called the shin. The two primary muscles are tibialis anterior and extensor digitorum longus.

Learning Outcome Questions

1. smooth; 2. endomysium; 3. actin; 4. ATP; 5. muscle tone; 6. prime mover, synergists; 7. biceps brachii; 8. lower spine, humerus; 9. temporalis; 10. deltoid; 11. triceps brachii; 12. gluteus maximus; 13. h; 14. e; 15. c; 16. f; 17. b; 18. a.

Chapter 8

Content Check-Up!

1. d; 2. a; 3. a; 4. The muscle would be more likely to contract, because the receptors for inhibitory neurotransmitters would be blocked; 5. c; 6. central canal, ventricles; 7. thalamus, temporal lobe; 8. d; 9. thalamus, hypothalamus, epithalamus; 10. a. cerebellum—motor coordination, b. limbic system—emotional response, c. brain stem—visual and auditory reflexes; 11. d; 12. c; 13. A sensory receptor for pain has an action potential, and the sensory neuron transmits its action potential to the spinal cord, where it synapses with an interneuron. In turn, the interneuron synapses with a motor neuron. The motor neuron activates skeletal muscle to contract; 14. Maintaining cardiovascular health through diet and exercise is essential. As a person ages, s(he) should continue to remain mentally active as well. A well-developed social network can help to prevent depression, which can impair memory; 15. If the stroke sufferer loses the ability to move upper limb muscles, those muscles will decrease in size and density. Because the bones are then no longer exercised, they will lose density as well.

Begin Thinking Clinically

Page 175: The sensory (receiving) and association (memory) areas for vision are located in the occipital lobe. The person who strikes his or her head in a fall may lose vision or visual memories. Depending on the severity of the injury, the loss may be permanent.

Learning Outcome Questions

1. brain, spinal cord, cranial nerves, spinal nerves; 2. axon; 3. effectors (muscle cells, glands, organs); 4. sodium, inside; 5. synaptic cleft; 6. acetylcholinesterase; 7. meninges; 8. interneuron; 9. EEG or electroencephalogram; 10. learning, memory; 11. cerebellum; 12. reticular formation; 13. cranial

nerve, somatic and parasympathetic motor functions. It also transmits somatic sensory information; 14. sympathetic, parasympathetic; 15. a. sensory neuron, b. interneuron, c. motor neuron, d. sensory receptor, e. neuron cell body, f. dendrites, g. axon, h. nucleus of Schwann cell (part of myelin sheath), i. node of Ranvier, j. axon terminals.

Chapter 9
Content Check-Up!
1. Your touch receptors would react to the apple's texture, your visual receptors will respond to its color, chemoreceptors for smell and taste will carry sensory information as you bite and chew it, and pressure receptors in the mouth will transmit information about the force your jaw generates; 2. a; 3. d; 4. e; 5. Alkaline compounds will taste bitter, so bitter taste receptors will have to be the most sensitive; 6. a; 7. d; 8. b; 9. The image travels on the optic nerves to the optic chiasma. From there, optic tracts route the image to the thalamus, which directs the information via the optic radiations to the occipital lobe; 10. Sound waves strike the tympanic membrane, causing its vibrations and those of the malleus, incus and stapes in the inner ear. The stapes vibrates against the oval window, creating fluid movement in the cochlear duct; 11. b; 12. VIII (8), vestibulocochlear nerve, temporal; 13. a; 14. static or gravitational, dynamic or rotational; 15. presbyopia.

Begin Thinking Clinically
Page 197: An uncinate fit is a form of epileptic seizure. Epilepsy results from excessive neuron activity in a particular brain area. In an uncinate fit, the temporal lobe areas controlling smell may be damaged, and the seizure results.

Learning Outcome Questions
1. mechanoreceptors; 2. chemoreceptors; 3. rods, cones, retina; 4. color, bright; 5. accommodates (becomes thicker); 6. distant objects far away, concave; 7. malleus, incus, stapes; 8. inner, hair; 9. cochlear duct in the cochlear; inner ear; 10. dynamic or rotational; 11. brain.

Chapter 10
Content Check-Up!
1. a; 2. c; 3. You would produce ADH anytime you're dehydrated; for example, after heavy exercise; 4. follicle-stimulating hormone (FSH), luteinizing hormone (LH), thyroid hormone (thyrotropin or TH), and/or adrenocorticotropic hormone (corticotropin or ACTH); 5. a; 6. d; 7. a; 8. One's diet must contain adequate iodine to synthesize thyroid hormone. Populations living in inland areas where there is little soil iodine are at risk for hypothyroidism; 9. c; 10. e; 11. c; 12. You would produce aldosterone anytime you're dehydrated; for example, after heavy exercise; 13. d; 14. a; 15. Insulin is produced when blood glucose is high (after a meal, for example). Glucagon is produced whenever blood glucose is low (when we're fasting between meals); 16. Female sex hormones cause long bone growth and secondary sexual characteristic development, menstruation, ovulation and maintenance of a pregnancy; 17. b; 18. d; 19. a-2, b-3, c-1; 20. thyroid; 21. a-3, b-2, c-1.

Begin Thinking Clinically
Page 224: A prolactin-producing adenoma in a man might cause his mammary glands to swell and produce milk. These adenomas also cause impotence, or inability to achieve erection of the penis.

Learning Outcome Questions
1. steroid, protein or amine; 2. negative feedback; 3. ADH, oxytocin; 4. releasing and inhibiting hormones; 5. anterior; 6. too little thyroid hormone, too little iodine; 7. calcium; 8. medulla, cortex, ACTH (corticotropin); 9. cortex; 10. Cushing's syndrome; 11. pancreas, insulin receptors; 12. blood.

Chapter 11
Content Check-Up!
1. The formed elements are red blood cells (erythrocytes), white blood cells (leukocytes), and platelets (thrombocytes). From top to bottom of the test tube, the three layers are: plasma layer, buffy coat (containing both leukocytes and thrombocytes) and red blood cell layer; 2. d; 3. by absorbing body heat from muscles and transporting it around the body, by maintaining salt-water and protein balance, by buffering acids and bases; 4. The fluid would be replaced through the actions of ADH and aldosterone (see Chapter 10). The kidneys would respond to lower blood oxygen levels by producing erythropoietin, which would stimulate red blood cell production by the red bone marrow; 5. d; 6. b; 7. d; 8. c; 9. d; 10. b; 11. c; 12. c; 13. Because of a poor diet, the elderly may have protein, iron or vitamin B_{12} deficiency. Any or all of these will cause anemia, because all three are necessary for red blood cell production.

Begin Thinking Clinically
Page 247: Polycythemia can occur whenever the tissues (and the kidney in particular) have a lowered oxygen supply. Thus, heart disease or lung disease can result in polycythemia. Polycythemia also results from unknown causes. When the polycythemia is caused by heart or lung disease, treating the underlying condition will alleviate polycythemia. Removing blood through a vein can also help to alleviate symptoms.

Learning Outcome Questions
1. plasma; 2. oxygen and carbon dioxide, function in defense; 3. oxyhemoglobin; 4. nucleus, 120; 5. neutrophil; 6. antibodies, destroy; 7. fibrin; 8. A, B, neither; 9. Rh negative, Rh positive; 10. c; 11. b; 12. a; 13. d.

Chapter 12
Content Check-Up!
1. The three layers are fibrous pericardium, parietal serous pericardium, and visceral pericardium (also called the epicardium). The fibrous pericardium protects the heart from injury and protects it from overfilling, which would reduce its contraction strength; 2. d; 3. a; 4. b; 5. a; 6. The trained athlete's heart will have increased its size and contraction strength, and his cardiac output will increase. Without training, the second athlete's cardiac output will remain unchanged. The trained athlete will have the advantage in the race; 7. c; 8. d; 9. b; 10. c; 11. a; 12. Cardiac output increases during exercise because heart rate and stroke volume both increase. Stroke volume increases due to increased venous return caused by muscle contraction. Increased cardiac output causes increased blood pressure during exercise; 13. a; 14. c; 15. Because blood will be diverted away from the pulmonary trunk, the pulmonary circuit will fail to develop; 16. The left ventricle. The chamber gradually enlarges because it must pump harder to eject blood through the stiff aortic valve; 17. The integumentary system helps to rid the body of excess heat by sweating.

Begin Thinking Clinically
Page 282: Figure 12.17 shows other ways for blood to return from the foot and leg to the heart. For example, blood could travel from the anterior tibial vein or small saphenous vein to the popliteal vein. However, the clot would cause severe pain, and could break free to form a thromboembolism (see Chapter 11). To prevent blood clotting, drink lots of water and stay hydrated, stretch your legs, and walk around as much as possible.

1. lungs; 2. bicuspid; 3. aorta; 4. sinoatrial; 5. away from; 6. mean arterial pressure or blood pressure; 7. systolic, diastolic; 8. cardiac output, peripheral resistance; 9. the pumping force of the ventricles, skeletal muscle contraction, the respiratory pump, and pressure difference between veins and right atrium; 10. subclavian, common iliac and femoral; 11. coronary; 12. digestive organs, liver; 13. cerebral arterial circle (Circle of Willis); 14. foramen ovale, ductus arteriosus.

Chapter 13

Content Check-Up!

1. d; 2. Lymphatic capillaries will drain to progressively larger lymphatic vessels, to the right lymphatic duct. This duct will drain to the right subclavian vein. The subclavian vein drains to the right brachiocephalic vein, which empties to the superior vena cava. 3. c; 4. The thymus produces thymosins (thymic hormones), and is the site where T lymphocytes mature; 5. d; 6. a; 7. b; 8. a; 9. b; 10. Interferon from virus-infected cells binds to non-infected cells, causing them to prepare for virus attack by producing antiviral substances; 11. c; 12. Passive immunity is obtained when antibodies are transferred from one person to another. Natural transfer occurs across the placenta from a mother to her fetus. Breast milk also contains antibodies that pass from the nursing mother to her baby. Artificial passive immunity involves injecting antibodies from one person into another. Neither form of passive immunity is as effective or long-lasting as active immunity, because the active immune response allows the person to make his/her own antibodies. 13. d; 14. The thymus gland gradually shrinks with age, decreasing the number of T cells. This decreases the strength of the immune response; 15. B and T cells create a specific immune response, whereas the other white blood cells are non-specific.

Begin Thinking Clinically

Page 297: If the cancerous tumor has begun to spread throughout the body, a lymph node will often be the first place for cancer cells to lodge. A sentinel lymph node is the first node to receive lymph from a malignant tumor. If this node is free of cancer cells, others are likely to be free as well.

Learning Outcome Questions

1. valves; 2. B lymphocytes, T lymphocytes; 3. thymus; 4. lymph, blood; 5. non-specific; 6. neutrophils, eosinophils, monocytes/macrophages; 7. complement; 8. plasma, memory; 9. antibody; 10. cell; 11. vaccine; 12. passive, active; 13. histamine.

Chapter 14

Content Check-Up!

1. a; 2. trachea, primary bronchi, secondary bronchi, tertiary bronchi, bronchioles, alveoli; 3. b 4. external intercostals, diaphragm, accessory muscles; 5. b; 6. You might start by asking the patient to close his/her nose, and then to become familiar with the apparatus by breathing normally into it. A minute or two of normal quiet breathing determines tidal volume. You would then prompt your patient to take the deepest breath possible, and then to exhale as forcefully as possible. Your patient's deepest breath is inspiratory reserve volume, and his or her strongest exhalation is expiratory reserve volume; 7. d; 8. b; 9. As carbonic acid/bicarbonate ion, dissolved, attached to Hb; 10. Bacterial infections can be treated with antibiotics, but they are not effective against viral infections. Using antibiotics unnecessarily can cause bacteria to become antibiotic resistant, and disrupt the person's normal bacterial populations; 11. d; 12. b; 13. b; 14. a.

Begin Thinking Clinically

Page 321: The tissues of the soft palate and tonsils are often enlarged in someone with OSA. The condition is usually treated using a continuous positive airway pressure (CPAP) device, which constantly blows air into the patient's airways to keep enlarged tissues from collapsing and blocking airways.

Learning Outcome Questions

1. larynx; 2. epiglottis; 3. alveoli; 4. alveoli, pulmonary capillaries; 5. expanded, opened up; 6. tidal volume; 7. vital capacity; 8. carbon dioxide; 9. diffusion; 10. enters; 11. enters; 12. bicarbonate; 13. smoking; 14. primary bronchi; 15. a. nasal cavity, b. nostril, c. pharynx, d. epiglottis, e. glottis, f. larynx, g. trachea/cartilage, h. bronchus, i. bronchiole.

Chapter 15

Content Check-Up!

1. c; 2. b; 3. Lipids diff use through the phospholipid cell membrane, and can be directly absorbed into the lymphatic lacteals. As you recall from Chapter 13, all lymphatic capillaries empty into larger and larger lymphatic vessels. The largest of these—right lymphatic duct and thoracic duct—empty into the right and left subclavian veins, respectively; 4. d; 5. The liver produces bile, which emulsifies fats, allowing them to be effectively digested by lipase enzymes. As intestinal blood passes through its hepatic portal system, the liver removes and stores nutrients (such as glucose) and vitamins and minerals (such as iron and the fat soluble vitamins A, D, E, and K) Regulation of blood cholesterol and forming urea from amino acids are other digestive functions of the liver. By making plasma proteins, the liver regulates blood osmotic pressure. Finally, liver macrophages help to remove pathogens from the blood, while liver enzymes detoxify poisonous substances; 6. a; 7. b; 8. d; 9. Physical digestion will begin in your mouth, as you bite and chew the burger. In the stomach, hydrochloric acid and pepsin will start the process of chemical digestion. Three pancreatic enzymes—trypsin, chymotrypsin, and carboxypeptidase—will continue chemical digestion, breaking down the proteins into smaller peptides and free amino acids. Finally, the intestinal peptidases found on intestinal villi will complete the process, breaking down large peptides into dipeptides and free amino acids. Both are molecules that can be absorbed into the blood; 10. malnutrition; 11. endocrine; 12. essential; 13. b.

Begin Thinking Clinically

Page 353: If normal flora bacteria are destroyed by antibiotic abuse, pathogenic microbes can grow in the colon instead. The first symptom caused by these harmful organisms is typically diarrhea. Long-term antibiotic abuse can result in vitamin deficiency.

Learning Outcome Questions

1. amylase; 2. pepsin; 3. esophagus, protein; 4. acidic; 5. cecum, ascending, transverse, descending and sigmoid portions, rectum, anus; 6. duodenum; 7. glycogen; 8. bile, emulsifies; 9. trypsin, chymotrypsin, and carboxypeptidase, lipase, amylase; 10. villi; 11. a.

Chapter 16

Content Check-Up!

1. a; 2. In response to decreased blood pressure, the kidneys produce the enzyme renin. Renin causes the first step in activation of angiotensinogen (a blood plasma protein, produced by the liver) to angiotensin. As it travels through the lungs, angiotensin is completely activated. Angiotensin causes blood vessels to constrict, increasing blood pressure; 3. d; 4. c; 5. First you would find the sodium ion in Bowman's capsule, the structure that collects tubular fluid as it forms. Next, the sodium ion passes into the proximal

convoluted tubule, where it can be reabsorbed and returned to the blood. If it continues into the loop of Henle, it may be actively transported by ascending limb cells into the renal medulla. However, if it travels into the distal convoluted tubule and the proximal collecting duct, its fate depends on two hormones. Aldosterone will cause the ion to be reabsorbed and returned to the blood, and atrial natriuretic hormone will allow it to be eliminated in the urine; 6. blood cells and large plasma proteins; 7. e; 8. c; 9. Each causes its effect by changing the amounts of sodium and water that are reabsorbed from the distal convoluted tubule and proximal collecting duct. Aldosterone increases sodium and water reabsorption, thus raising blood pressure. Atrial natriuretic hormone decreases reabsorption and blood pressure falls as a result; 10. Both b and c; 11. c; 12. a; 13. The sweat glands excrete water, salt and urea. The liver excretes bile pigments. The lungs excrete carbon dioxide and water; 14. They produce erythropoietin; 15. sodium, potassium, and calcium.

Begin Thinking Clinically

Page 382: Loop diuretics work in the ascending loop of Henle. They inhibit secretion of sodium into the medulla, so sodium is gradually washed away and the sodium concentration gradient disappears. Less water is reabsorbed in the kidneys, and the volume of urine increases as a result. One generic version is furosemide. Aldosterone antagonist diuretics work primarily in the distal convoluted tubule. By interfering with aldosterone, they inhibit the reabsorption of sodium and water. Once again, more urine is formed. Drugs called spironolactones are aldosterone antagonist diuretics.

Learning Outcome Questions

1. sodium, water, pH; 2. urea; 3. gout; 4. ureters; 5. urethra; 6. cortex, medulla, pyramids; 7. glomerus; 8. sodium; 9. creatinine and/or urochrome; 10. loop of Henle, distal convoluted tubule, collecting duct; 11. ADH; 12. buffer; 13. carbon dioxide and water.

Chapter 17

Content Check-Up!

1. d; 2. The chromosomes in the two cells formed by mitosis are identical. Because of crossing over and the random distribution of sister chromatids, the chromosomes of the two cells formed by meiosis are not identical; 3. d; 4. Semen contains sperm cells from the testes, in the alkaline fluid from the seminal vesicles and prostate. The seminal vesicles also add fructose and prostaglandins. Fructose serves as an energy source for sperm; prostaglandins produce contractions of the female genital tract that help to move sperm closer to the ovum; 5. a; 6. gonadotropin-releasing, follicle-stimulating, luteinizing. Follicle-stimulating hormone stimulates development of sperm and ova, and luteinizing hormone stimulates testosterone production by the male testes. In females, LH stimulates ovulation and progesterone secretion; 7. vesicular (Graafian) follicle, corpus luteum, corpus albicans. The three occur in sequence from a single primary follicle; 8. c; 9. c; 10. d; 11. syphilis, gonorrhea, chlamydia; 12. a; 13. Menopause. Males do not stop producing the sex hormones, so they don't experience menopause. 14. The benefits include relief from menopause symptoms and increased bone density. Risks include increased rates of cancer, heart disease, stroke and blood clots.

Begin Thinking Clinically

Page 404: An abdominal ectopic pregnancy occurs when an ovum is fertilized outside the uterine tube, and doesn't successfully enter the uterine tube to travel to the uterus. The organs most likely to be able to support a placenta (and allow the fetus to grow) would be the external uterus, urinary bladder, and the intestines. The infant would have to be delivered through the abdominal wall by Caesarian section.

Learning Outcome Questions

1. 23, sperm and ova; 2. seminiferous tubules; 3. interstitial; 4. testosterone; 5. vas deferens; 6. seminal vesicle; 7. blood; 8. vagina; 9. labia majora, mons pubis, labia minora, clitoris, vestibule, orifices of the urethra and vagina; 10. estrogen, progesterone; 11. vesicular follicle of an ovary, uterus; 12. ovulation; 13. corpus luteum; 14. prolactin and oxytocin; 15. low sperm count, abnormal sperm; 16. viruses, bacteria.

Chapter 18

Content Check-Up!

1. b; 2. a; 3. c; 4. The umbilical cord is a tube that contains two umbilical arteries and a single umbilical vein. Surrounding them is a complex polysaccharide that acts like a pillow and keeps these blood vessels from being compressed. Because the placenta is the organ for fetal blood oxygenation, oxygen content would be highest in the blood vessel that drains the placenta: the umbilical vein; 5. embryo, chorion; 6. d; 7. During labor, oxytocin promotes uterine contractions that help to deliver the fetus. Without it, labor would not progress. After delivery of the placenta, oxytocin causes the uterine smooth muscle to continue contracting. This helps to seal the uterine surface that was previously covered by the placenta, preventing a possible fatal hemorrhage. Oxytocin causes the milk let-down reflex in a nursing mother, which allows her milk to be emptied into the breast ducts and delivered to her baby. Without the hormone, her milk would not be expelled, resulting in breast swelling (and a hungry baby!); 8. c; 9. d

Begin Thinking Clinically

Page 437: Precise mechanisms are unclear, but research suggests that in this circumstance, cervical stimulation fails to cause adequate release of oxytocin. In this circumstance, oxytocin can be administered through an intravenous solution.

Learning Outcome Questions

1. ovum, sperm; 2. implant; 3. pre-embryonic development; 4. differentiation; 5. placenta; 6. extra-embryonic, amnion; 7. organ systems; 8. second; 9. head; 10. placental hormones.

Chapter 19

Content Check-Up!

1. d; 2. Each is a trisomy, where three chromosomes are found instead of the normal two. Edwards syndrome is trisomy 18. Patau syndrome is trisomy 13. Both cause severe congenital (birth) defects, and affected infants usually die within their first year of life; 3. d; 4. b; 5. d; 6. Unfortunately, as in all autosomal recessive diseases, any time the couple conceives a child, there is a 25% chance that the child will inherit the disorder; 7. retroviruses, adenoviruses, herpes viruses, adeno-associated viruses; 8. b; 9. a.

Begin Thinking Clinically

Page 449: Down syndrome individuals can be born with heart and lung abnormalities that can be fatal (though many can be surgically corrected). Further, Down syndrome adults are more susceptible to a certain form of leukemia, or cancer of the white blood cells. Alzheimer disease occurs in Down's adults at an earlier age, and the person is more susceptible to infection throughout life. The average age at death is 49.

Learning Outcome Questions

1. chromosomes; 2. c; 3. three; 4. XY; 5. XXY; 6. one; 7. a. mother; 8. carrier; 9. recessive; 10. gene therapy; 11. genomics.

Photo Credits

Photo and Text Art

Design Elements

Ambulance: Jeremy Hoare/Life File/Photodisc/Getty Images RF; Stethoscope: TRBfoto/Photodisc/Getty Images RF; Doctor and microscope: Digital Vision RF; Thumbprint: Steve Allen/Brand X Pictures RF.

Front Matter

Photo of Sue Longenbaker: Courtesy of Sue Longenbaker; TOC 1: © Jim Connely; TOC 2a: © Science Source; TOC 2b: © John Bell/Getty Images RF; TOC 2c: © Edelcio Muscat/Getty Images RF; TOC 2d: ©Samuel Betkowski/Getty Images RF; TOC 3: © SPL/Science Source; TOC 4: © Obstetrics & Gynaecology/Science Source; TOC 5:© Al Telser/ The McGraw-Hill Higher Education Inc.; TOC 5a: © Buero Monaco/Getty Images RF; TOC 5b: © Jordan Siemens/Getty Images RF; TOC 5c: © sam74100/Getty Images RF; TOC 5d: © Indeed/Getty Images RF; TOC 5e: ©Westend61/Getty Images RF; TOC 6: © SuperStock RF; TOC 7: © RapidEye/Getty Images RF; TOC 7b: © Don Hammond/DesignPics RF; TOC 7c: © Jose Luis Pelaez, Inc/Getty Images RF; TOC 7d: © galinast/Getty Images RF; TOC 7e: © baona/Getty Images RF; TOC 8: © VladTeodor/Getty Images RF; TOC 9: © DRB Images, LLC/Getty Images RF; TOC 10: © ROBERT KNUDSEN/Office of the Naval Aide to the President/EPA/Newscom; TOC 11: © Frank Conlon/Star Ledger/Corbis News/Corbis; TOC 12a: Courtesy of nasaimages.org/NASA; TOC 12b: Courtesy of nasaimages.org/NASA; TOC 13: ©Photodisc Collection/Getty Images RF; TOC 14: © Stockbyte/Getty Images RF; TOC 15: © Bettmann/Corbis; TOC 16a: © Dick Darrell/Toronto Star/Getty Images; TOC 16b: © Bettmann/Corbis; TOC 17: © Bruce Grenville Matthews/Photodisc/Getty Images RF; TOC 18: © Hulton-Deutsch Collection/Historical/Corbis; TOC 19: © FPG/The Image Bank/Getty Images.

Chapter 1

Opener: © Jim Connely; 1.2: © Eric Wise/Wise Anatomy Collection/McGraw-Hill Higher Education; 1.4: © Joe De Grandis/McGraw-Hill Education; 1.4a-c: © Joe De Grandis/McGraw-Hill Education Inc.; 1.7a-b: © McGraw-Hill Education/MCOF Enterprises, Ltd.; 1Ba: Larry McKee/Stock Connection Distribution/Alamy; 1Bb: © Allen Bell/Spirit/Corbis RF; 1Bc: Philippe Psaila/Science Source; 1Bd: Mazzlota et al/Science Source.

Chapter 2

Opener: © Science Source; CO 2b: © John Bell/Getty Images RF; CO 2c: © Edelcio Muscat/Getty Images RF; CO 2d: © Samuel Betkowski/Getty Images RF; 2.2a-b: © Evelyn Jo Johnson/McGraw-Hill Education Inc.; 2.7: © Jeremy Burgess/SPL/Science Source; 2.8: © Don W. Fawcett/Science Source; 2.11a: © Jon Kopaloff/Getty Images; 2.11b: © Mark Sullivan/Getty Images.

Chapter 3

Opener: © SPL/Science Source; 3.1a: © Alfred Pasieka/Science Source; 3.3: © Dennis Kunkel/Newscom; 3.4: © Don W. Fawcett/Jessica Wilson/Science Source; 3.5a: © Don W. Fawcett/Science Source; 3.5b: Courtesy Tim Wakefield, John Brown University; 3.6a: © CNRI/Getty Images; 3.7a: © Oliver Meckes/Science Source; 3.7b: © J. Walsh/Science Source; 3.8a: © Don W. Fawcett/Science Source; 3.8b: © Biology Pics/Science Source; 3.10a: © David M. Phillips/Science Source; 3.10b: © Custom Medical Stock Photo/Newscom; 3.10c: © Custom Medical Stock Photo/Newscom; 3.17a-d: © Michael Abbey/Science Source; 3B: Courtesy Dr. Michael Baird, Lifecodes Corporation.

Chapter 4

Opener: © Obstetrics & Gynaecology/Science Source; 4.1, 4.2a: © Ed Reschke; 4.2b: © Dr. Alvin Telser/McGraw-Hill Education; 4.3-4.8: © Ed Reschke; 4.9: © Dr. Alvin Telser/McGraw-Hill Education; 4.10a-b: © Ed Reschke; 4.10c: © Dr. Alvin Telser/McGraw-Hill Education; 4.11a: © Ed Reschke; 4.11b: © Victor Eroschenko; 4.12: © Dr. Alvin Telser/McGraw-Hill Education; 4.13: © Ed Reschke; 4.14: © Dennis Strete/McGraw-Hill Education; 4.15, 4.16: © Ed Reschke; p. 76 : Courtesy Forbes Library/Calvin Coolidge Presidential Library & Museum.

Chapter 5

Opener: © Al Telser/McGraw-Hill Education Inc.; CO 5a: © Buero Monaco/Getty Images RF; CO 5b: © Jordan Siemens/Getty Images RF; CO 5c: © sam74100/Getty Images RF; CO 5d: ©Indeed/Getty Images RF; CO 5e: ©Westend61/Getty Images RF; 5.2b: © Al Telser/McGraw-Hill Education; 5.3: © McGraw-Hill Education/MCOF Enterprises, Ltd., photographer; 5.5a: © Living Art Enterprises, LLC/Science Source; 5.5b: Courtesy Dr. George Bogumill; 5.6a: Mediscan/Alamy; 5.6b: Dr. P. Marazzi/SPL/Science Source; 5.6c: © James Stevenson/SPL/Science Source; 5A: Courtesy Kent Van De Graaff, photo by James M. Clayton.

Chapter 6

Opener: © SuperStock RF; 6.1a-e: © McGraw-Hill Companies, Inc./MCOF Enterprises, Ltd., photographer; 6.2c: © Cultura Science/Alvin Telser, PhD/Getty Images; 6.5: © McGraw-Hill Companies, Inc./MCOF Enterprises, Ltd., photographer; 6Bb: © Robert Marien/Spirit/Corbis; 6.6a-b, 6.7a-b, 6.8, 6.10b, 6.11, 6.12a-c, 6.13a-b, 6.14,6.15, 6.16c, 6.17a-b, 6.18, 6.19: © McGraw-Hill Education/MCOF Enterprises, Ltd., photographer; 6B: © Robert Marien/Spirit/Corbis RF.

Chapter 7

Opener: © RapidEye/Getty Images RF; CO 7b: © Don Hammond/Design-Pics RF; CO 7c: © Jose Luis Pelaez, Inc/Getty Images RF; CO 7d: © galinast/Getty Images RF; CO 7e: © baona/Getty Images RF; 7.1a-c: © Ed Reschke; 7.2c: © Dennis Strete/McGraw-Hill Education; 7.3b: © Biology Pics/Science Source; 7.4b: © Dr. Thomas Caceci/Tom Caceci RF; 7.8a-b: © Lyn Wolf.

Chapter 8

Opener: © VladTeodor/Getty Images RF; 8.2a: © EM Research Services, Newcastle University RF; 8.2b: © David McCarthy/Science Source; 8.5b: © Manfred Kage/Science Source; 8.7a: © Karl E. Deckart/Phototake; 8.7d: © Rebecca Gray/Don Kincaid/McGraw-Hill Education; 8.9b: © Colin Chumbley/Science Source; 8.11b: © Steve Gschmeissner/Science Source; p. 177: © S. Granitz/Getty Images.

Chapter 9

Opener: © DRB Images, LLC/Getty Images RF; 9.3a: © Coneyl Jay/Science Source; 9.3c.1-2: © Science Source; 9.3d: © David Gregory/Debbie Marshall/Wellcome Images; 9.9a: © Science Photo Library/Getty Images RF; 9.10b: © Biophoto Associates/Science Source; 9.13: © P. Motta/SPL/Science Source; 9B.a-b: Courtesy National Eye Institute.

Chapter 10

Opener: © ROBERT KNUDSEN/Office of the Naval Aide to the President/EPA/Newscom; 10.5: © General Photographic Agency/Hulton Archive/Getty Images; 10.6a-b: © Photo by Incredible Features/Barcroft Media/Getty Images; 10.7: © Biophoto Associates/SPL/Science Source; 10.8: © Mediscan/Alamy; 10.12a-b: © BSIP/Science Source; 10.13a-b: Courtesy Shannon Halverson; 10.15: © Rob Melnychuk/Getty Images RF; 10A: Courtesy Robert P. Lanza.

Chapter 11

Opener: © Frank Conlon/Star Ledger/Corbis News/Corbis; 11.1b: © Ed Reschke; 11.7: © Eye of Science/Science Source; 11.9a-b: © Jean Claude Revy - ISM/Phototake; 11A: © Bill Longcore/Science Source; 11B: © Ed Reschke.

Chapter 12

Opener: Courtesy of nasaimages.org/NASA; CO 12b: Courtesy of nasaimages.org/NASA; 12.3b: © McGraw-Hill Education; 12.5b: © Thomas Deerinck, NCMIR/Science Source; 12.8a: © Ed Reschke; 12.8b: ©Biophoto Associates/Science Source; 12.15: © Ariel Skelley/Blend Images/Getty Images RF; 12Ab: Biophoto Associates/Science Source.

Chapter 13

Opener: ©Photodisc Collection/Getty Images RF; 13.1c_a: © Al Telser/McGraw-Hill Education; 13.1c_b: © Ed Reschke/Photolibrary/Getty Images; 13.1c_c: © Al Telser/McGraw-Hill Education; 13.1c_d: © Ed Reschke; 13.2a: © Ed Reschke/Photolibrary/Getty Images; 13.2b-c: © Ed Reschke; 13.2d-e: © Ed Reschke/Photolibrary/Getty Images; 13.4b: Courtesy Dr. Arthur J. Olson, The Scripps Research Institute; 13.7d: © Steve Gschmeissner/Science Source; 13.9: © JGI/Blend Images LLC RF; 13C: © C. S. Goldsmith and A. Balish/Centers for Disease Control and Prevention; p. 310: © Jill Braaten/McGraw-Hill Education.

Chapter 14

Opener: © Stockbyte/Getty Images RF; 14.3c: © CNRI/Phototake; 14.4b: © Ed Reschke; 14.4c: © Nibsc/Getty Images; 14.8: © Burger/Phanie/AGE Fotostock; 14.12a: © Matt Meadows/Photolibrary/Getty Images; 14.12b: © Dennis K. Burns, Md Travis G. Brown, Md Walter L. Kemp, Md/McGraw-Hill Education; 14.12c: © Biophoto Associates/Science Source.

Chapter 15

Opener: © Bettmann/Corbis; 15.4b: © Biophoto Associates/Science Source; 15.5c: © Ed Reschke/Photolibrary/Getty Images; 15.7b: © Victor P. Eroschenko RF; 15.7c: © Steve Gschmeissner/Science Photo Library/Getty Images RF; 15.9d: © Ed Reschke; 15.13: Courtesy USDA, ChooseMyPlate.gov website; 15.14a: © Wellcome Image Library/Custom Medical Stock Photo; 15.14b: © Jack Star/PhotoLink/Photodisc/Getty Images RF.

Chapter 16

Opener: © Dick Darrell/Toronto Star/Getty Images; CO 16b: © Bettmann/Corbis; 16.1: © The McGraw-Hill Companies, Inc./MCOF Enterprises, Ltd., photographer; 16.3b: © McGraw-Hill Education/MCOF Enterprises, Ltd., photographer; 16.4b: © Steve Gschmeissner/Science Photo Library/Getty Images RF; 16.4c-d: © Ed Reschke/Photolibrary/Getty Images; 16.10: HPA-Voisin/Science Source.

Chapter 17

Opener: © Bruce Grenville Matthews/Photodisc/Getty Images RF; 17.4b: © Al Telser/McGraw-Hill Education; 17.5(right): © David M. Phillips/Science Source; 17.7b: © Anatomical Travelogue/Science Source; 17.10: © Ed Reschke/Photolibrary/Getty Images; 17.17a: © Centers for Disease Control and Prevention; 17.17b: © Biophoto Associates/Science Source; 17.17c: © Melba Photo Agency/Alamy RF; 17Da: © Dave and Les Jacobs/Blend Images LLC RF; 17Db: © David Raymer/Comet/Corbis; 17E: © Steve Nagy/Design Pics RF.

Chapter 18

Opener: © Hulton-Deutsch Collection/Historical/Corbis; 18.1: © David M. Phillips/Science Source; 18.6a: © Lennart Nilsson/Scanpix; 18.8: © James Stevenson/SPL/Science Source; 18.10: © Rune Hellestad/Encyclopedia/Corbis.

Chapter 19

Opener: © FPG/The Image Bank/Getty Images; 19.1c: © Ermakoff/Science Source; 19.1d-e: © CNRI/Science Source; 19.3: © Jill Cannefax; 19.4a-b: Courtesy of G. H. Valentine, Earl Plunkett; 19A1: © Brand X/SuperStock RF.

Glossary/Index

Page references followed by *f* and *t* indicate figures and tables, respectively. Bolded terms are defined here.

A

A band Region in the center of a muscle sarcomere that contains overlapping thick and thin filaments, 138

Abdominal cavity Portion of the body between the diaphragm and the pelvis, 6*f*, 7, 7*t*, 8*f*

Abdominal wall, 149

Abdominopelvic cavity Pertaining to the abdominal and pelvic regions, 80, 374

Abducens nerve, 200

Abduction Movement of a body part away from the midline, 124, 126*f*, 148

ABO blood types, 242, 252–253, 252*t*, 253*f*

Abstinence, 410, 411*t*, 418

Accessory pancreatic duct, 354*f*

Accommodation Ability of the lens to change its focus from distant to near objects; achieved through the action of the ciliary muscles that change the shape of the lens, 201, 202–203, 202*f*

Acetabulum Socket in the lateral surface of the hipbone into which the head of the femur articulates, 120, 121*f*

Acetylcholine (ACh) Neurotransmitter secreted at the ends of many neurons; responsible for the transmission of a nerve impulse across a synaptic cleft, 138, 139*f*, 167, 186

Acetylcholinesterase (AChE) Enzyme in the membrane of postsynaptic cells that breaks down acetylcholine; this enzymatic reaction inactivates the neurotransmitter, 168

ACh. *See* Acetylcholine (ACh)

AChE. *See* Acetylcholinesterase (AChE)

Achondroplasia, 453*t*

Acid Solution in which pH is less than 7; substance that contributes or liberates hydrogen ions in a solution; opposite of *base*, 25–27, 26*f*

Acid-base balance, 373, 382–384

Acid-base buffer systems, 383

Acidosis Excessive accumulation of acids in body fluids, 27, 230, 330, 382

Acinar cells, 353

Acne vulgaris Inflammation of sebaceous glands; the common form of acne, 92

Acromegaly Condition resulting from an increase in growth hormone production after adult height has been achieved, 224, 225*f*

Acromion process, 118, 118*f*

Acrosome Covering on the tip of a sperm cell's nucleus that is believed to contain enzymes necessary for fertilization, 396, 397*f*, 426, 426*f*

ACTH. *See* Adrenocorticotropic hormone (ACTH)

Actin One of the two major proteins of muscle; makes up thin myofilaments in myofibrils of muscle cells, 31. *See also* Myosin

Actin filaments Long, extremely thin fibers that can be isolated from microvilli and cells that undergo movement, 44*f*, 49, 138–139, 140*f*

Action potential Change in potential propagated along the membrane of a neuron; the nerve impulse, 165, 166*f*, 167, 168*f*

Active immunity Resistance to disease due to the immune system's response to a microorganism or a vaccine, 305

Active transport Transfer of a substance into or out of a cell against a concentration gradient by a process that requires a plasma membrane carrier protein and an expenditure of energy, 53–55, 53*f*, 53*t*

Acute bronchitis Infection of the primary and secondary bronchi, 331, 332*f*

Acute disease Sudden in onset and severe, 17

Acute lymphoblastic leukemia (ALL) Cancer of the blood in which immature lymphocytes proliferate in bone marrow, the thymus, and lymph nodes, 249

Adam's apple, 321*f*, 401

Addison's disease Condition resulting from a deficiency of adrenal cortex hormones, 217, 230*f*

Adduction Movement of a body part toward the midline, 124, 126*f*, 148

Adductor group, 152, 152*t*, 153*f*

Adductor longus, 146*f*, 153*f*

Adductor magnus, 146, 146*f*, 153*f*

Adenoids, 321

Adenosine diphosphate (ADP) Molecule produced when the terminal phosphate is lost from a molecule of adenosine triphosphate; ATP, 37, 37*f*, 53*f*, 138–142, 140*f*, 141*f*

Adenosine triphosphate (ATP) Molecule used by cells when energy is needed, 36–37, 37*f*, 49, 140–142

ADH. *See* Antidiuretic hormone (ADH)

Adhesion The molecular attraction exerted between the surfaces of bodies in contact, 25

Adhesion junction Junction between cells in which the adjacent plasma membranes do not touch but are held together by intercellular filaments attached to buttonlike thickenings, 78, 78*f*

Adipose tissue, 70, 70*f*, 235

ADP. *See* Adenosine diphosphate (ADP)

Adrenal cortex Outer portion of the adrenal gland, 227–229, 228*f*, 229*f*

Adrenal cortex malfunction, 228–229, 230*f*

Adrenal gland Endocrine gland located on the superior portion of each kidney, 219*f*, 220*t*, 227–229, 228*f*, 229*f*

Adrenaline. *See* Epinephrine

Adrenal medulla Inner portion of the adrenal gland; produces epinephrine and norepinephrine, 227

Adrenocorticotropic hormone (ACTH) Hormone secreted by the anterior lobe of the pituitary gland that stimulates the adrenal cortex to produce cortisol, 220*t*, 223*f*, 224, 227

Afferent arteriole, 376*f*, 378*f*, 383*f*

Afferent system, 180

Afterbirth Placenta and the extraembryonic membranes, which are delivered (expelled) during the third stage of parturition, 432

Agammaglobulinemia, 453*t*

Agglutination Clumping of cells, particularly in reference to red blood cells involved in an antigen-antibody reaction, 252, 301

Agonist, 145

Agranular leukocyte White blood cell with poorly visible cytoplasmic granules, 248

AID. *See* Artificial insemination

AIDS (acquired immunodeficiency syndrome) Disease caused by a retrovirus and transmitted via body fluids; characterized by failure of the immune system, 307–308, 308*f*, 313, 416, 418

Albinism Genetic disorder characterized by a defect in pigment production, 87*f*, 88

Albumin Plasma protein that helps the osmotic concentration of blood, 244, 245*f*

Alcohol, 440

Aldosterone Hormone secreted by the adrenal cortex that functions in regulating sodium and potassium excretion by the kidneys, 219*f*, 228, 382

Alimentary canal Tubular portion of the digestive tract, 343, 343*f*, 347*f*

Alkalosis Excessive accumulation of bases in body fluids, 27, 330, 382

ALL. *See* Acute lymphoblastic leukemia (ALL)

Allantois Extraembryonic membrane that serves as a source of blood vessels for the umbilical cord, 428

Allele Different forms of a gene, 450

Allergen Foreign substance capable of stimulating an allergic reaction, 312

Allergy Immune response to substances that usually are not recognized as foreign, 293, 312

All-or-none law For muscle cells, property of muscle that states that a muscle cell contracts completely or not at all. For neurons, property states that neurons either conduct a nerve signal completely or not at all, 143

Alopecia Loss of hair, 90

Alveolar processes, 111*f*, 113

Alveolus Air sac of a lung (pl., *alveoli*), 320*t*, 323–324, 324*f*

Alzheimer cataract A clouding of the lens of the eye or its surrounding transparent membrane that obstructs the passage of light; currently being investigated as an early signal of Alzheimer disease, 201

Alzheimer disease Brain disorder characterized by a general loss of mental abilities, 170, 178, 201

Amino acid Unit of a protein that takes its name from the fact that it contains an amino group (—NH₂) and an acid group (—COOH), 32, 33*f*, 35, 36, 53*t*, 57*f*, 58, 361–362

Ammonia, 384

Amniocentesis Method of retrieving fetal cells for genetic testing in which a long needle is used to withdraw a sample of amniotic fluid, 446

Amnion One of the extraembryonic membranes; a fluid-filled sac around the embryo, 428, 433*f*

Amniotic cavity, 432*f*, 433*f*

Amphiarthrosis Joint that allows slight movement, 123

Ampulla Expansion at the end of each semicircular canal that contains receptors for rotational equilibrium, 210, 211*f*

Amyotrophic lateral sclerosis Disease caused by gradual death of motor neurons, which leads to loss of all movement, including swallowing and breathing, 157

Anabolic steroid Synthetic steroid that mimics the effect of testosterone, 232, 234–235, 235*f*

Anal canal, 352*f*

Anal intercourse, 418

Anal sphincters, 352*f*

Anaphase Stage in mitosis when replicated chromosomes separate and move to opposite poles of the cell, 56*f*, 58–59, 59*f*, 60*f*, 394*f*, 395

Anaphylactic shock A severe systemic form of anaphylaxis involving bronchiolar constriction, impaired breathing, vasodilation, and a rapid drop in blood pressure with a threat of circulatory failure, 312

Anatomical position Standard body position in which the person is erect, with body, head, arms, palms, and feet facing forward, 3, 3*f*

Anatomical terms, 3–5, 3*f*–5*f*

Anatomy Branch of science dealing with the form and structure of body parts, 2

Androgen Male sex hormone, 232, 395

Androgenetic alopecia Hair loss caused by excessive testosterone formation, 90

Androstenol, 392

Anemia Condition characterized by a deficiency of red blood cells or hemoglobin, 248, 249, 254. *See also* Iron deficiency anemia; Pernicious anemia

Anencephaly Congenital absence of the cranial vault, with cerebral hemispheres completely missing or reduced to small masses attached to the base of the skull, 431

Aneurysm Saclike expansion of a blood vessel wall, 267, 268*f*

Angina pectoris Condition characterized by thoracic pain resulting from occluded coronary arteries; precedes a heart attack, 268

Angioplasty, 268*f*

Angiotensin Plasma protein; in its active state it causes blood vessels to constrict, 337, 373

Anions, 22

Ankle-jerk reflex, 183

Anorexia nervosa Eating disorder caused by the fear of becoming obese; includes loss of appetite and inability to maintain a normal minimum body weight, 365–366, 366*f*

Antagonist Muscle that acts in opposition to a prime mover, 145

Anterior Pertaining to the front; the opposite of *posterior*, 3, 3*f*

Anterior cavity, 6*f*, 7

Anterior cerebral artery, 283*f*

Anterior communicating artery, 283*f*

Anterior pituitary Front lobe of the pituitary gland, 219*f*, 220*t*, 222–224, 223*f*, 224*f*

Anterior superior iliac spine, 120, 121*f*

Anterior tibial artery, 280*f*

Anterior tibial vein, 281*f*

Antibody Protein produced in response to the presence of some foreign substance in the blood or tissues, 81, 250, 301–302, 301*f*, 302*f*, 302*t*, 306, 306*f*, 312

Antibody-mediated immunity Resistance to disease-causing agents resulting from the production of specific antibodies by B lymphocytes; humoral immunity, 301

Antibody titer Amount of antibody present in a sample of blood serum, 305

Anticodon Three contiguous nucleotides of a transfer RNA molecule that are complementary to a specific mRNA codon, 58

Antidiuretic hormone (ADH) Hormone released from the posterior lobe of the pituitary gland that enhances water conservation by the kidneys; sometimes called vasopressin, 219*f*, 220*t*, 222, 223*f*, 237, 276, 381

Antigen Foreign substance, usually a protein, that stimulates the immune system to produce antibodies, 248, 300, 303

Antigenic drift, 311

Antigen-presenting cell (APC) The cell that displays the antigen to the cells of the immune system so they can defend the body against that particular antigen, 303

Antigen receptor Receptor proteins in the plasma membrane of immune system cells whose shape allows them to combine with a specific antigen, 300–303, 302*f*, 302*t*, 303*f*

Antiretroviral therapy, 307–308

Anus Outlet of the alimentary canal, 343, 343*f*, 396*f*, 401*f*, 406*f*

Aorta Major systemic artery that receives blood from the left ventricle, 262*f*, 264*f*, 279*f*, 280*f*, 284*f*

Aortic aneurysm, 268*f*

Aortic arch, 279*t*, 280*f*, 284*f*

Aortic body Receptor in the aortic arch sensitive to oxygen content, carbon dioxide content, and blood pH, 328

Aortic semilunar valve, 263

Aplastic anemia Insufficient number of red blood cells brought on by damage to the red bone marrow due to radiation or chemicals, 249

Apnea Temporary cessation of breathing, 321

Apocrine glands, 91

Aponeurosis A broad, flat sheet of dense regular connective tissue that covers and attaches various muscles (pl. *aponeuroses*), 70, 71*f*

Apoptosis Programmed cell death and destruction, 55, 301

Appendicitis Infected swelling of the appendix, 9, 249, 356

Appendicular portion Pertaining to the upper limbs (arm) and lower limbs (legs), 4

Appendicular skeleton Part of the skeleton forming the upper limbs, pectoral girdle, lower limbs, and pelvic girdle, 110, 118–123, 118*f*–122*f*

Appendix Small, tubular appendage that extends outward from the cecum of the large intestine, 298, 343*f*, 349*f*, 351, 352*f*

Appetite, 235

Appositional growth In bone, growth that leads to increase in diameter of a long bone, 106–107

Aqueous humor Watery fluid that fills the anterior cavity of the eye, 200*f*, 201, 201*t*

Arachnoid membrane Weblike middle covering (one of the three meninges) of the central nervous system, 171, 171*f*

Arch, 279*t*

Areola Dark, circular area surrounding the nipple of the breast, 407, 409*f*

Arm muscles, 151, 151*f*

Arrector pili Smooth muscle in the skin associated with a hair follicle, 90

Arrhythmia Abnormal heart rhythm, 22, 265

Arterial duct. *See* Ductus arteriosus

Arteriole Branch from an artery that leads into a capillary, 271, 272*f*

Arteriosclerosis Thickening and hardening of arterial walls, 267, 267*f*

Arteriovenous shunt, 272*f*

Artery Vessel that takes blood away from the heart; characteristically possesses thick elastic walls, 271, 272*f*, 279, 280*f*

Arthritis Inflammation of a joint, 126, 312

Articular cartilage Hyaline cartilaginous covering over the articulating surface of the bones of synovial joints, 103, 104*f*, 124*f*, 135*f*

Articular process, 115, 115*f*

Articulation Joining together of bones at a joint, 103

Artificial insemination Placement of donated sperm in the vagina so that fertilization followed by pregnancy might occur, 415

Artificial organs, 372

Ascending aorta, 279*t*, 280*f*

Ascending colon Portion of the large intestine that travels superiorly as it extends from the entry of the small intestine to the transverse colon, 351, 352*f*

Ascending lumbar vein, 281*f*

Ascending tracts, 173

Assisted reproductive technologies (ART) Medical techniques, sometimes performed in vitro, that are done to increase the chances of pregnancy, 415

Association area Region of the cerebral cortex related to memory, reasoning, judgment, and emotional feelings, 175, 175*f*

Aster Short, radiating fibers about the centrioles at the poles of a spindle, 58

Asthma Condition in which bronchioles constrict and cause difficulty in breathing, 312, 332*f*, 335, 336

Astigmatism Visual defect due to errors in refraction caused by abnormal curvatures in the surface of the cornea or lens, 205

Astrocytes, 77

Atelectasis, 323

Atherosclerosis Condition in which fatty substances accumulate abnormally beneath the inner linings of the arteries, 267, 267*f*

Atherosclerotic plaque Fatty lesion in the tunica intima layer of the arteries, 267

Athlete's foot Skin disease caused by fungal infection, usually of the toes and soles of the foot, 93

Athletics, 144–145, 234–235, 235*f*

Atlas First cervical vertebra; it supports and balances the head, 115–116, 115*f*

Atom Smallest unit of matter, 2, 2*f*, 21–22, 21*f*

Atomic number Number of protons and electrons an atom of an element has when the element is electrically neutral, 21

ATP. *See* Adenosine triphosphate (ATP)

Atrial diastole, 266, 266*f*

Atrial natriuretic hormone (ANH) Substance secreted by the atria of the heart that accelerates sodium excretion so that blood volume decreases, 228, 276, 276*f*, 382

Atrial systole, 265, 266*f*

Atriopeptide Atrial natriuretic peptide, or (ANP), formerly referred to as atrial natriuretic hormone, 228

Atrioventricular (AV) bundle Part of the cardiac conduction system that extends from the AV node to the bundle branches, 264–265, 265*f*

Atrioventricular (AV) node Small region of neuromuscular tissue located near the septum of the heart that transmits impulses from the SA node to the ventricular walls, 264, 265*f*

Atrioventricular (AV) valve Valve located between the atrium and the ventricle, 262*f*, 263

Atrium Chamber; particularly an upper chamber of the heart that lies above the ventricles (pl., *atria*), 261, 262, 262*f*, 263, 284*f*

Atrophy Wasting away or decrease in size of an organ or tissue, 144

Auditory association area, 175*f*

Auditory canal Tube in the outer ear that leads to the tympanic membrane, 208, 209*f*, 213*t*

Auditory nerve, 209

Auditory tube, 209*f*

Auditory (eustachian) tube Air tube that connects the pharynx to the middle ear, 331

Auricle, 261, 264*f*

Autoimmune disease Disease that results when the immune system mistakenly attacks the body's own tissues, 312, 313, 357

Autoimmune response Disease process in which the patient's own antibodies attack normal healthy tissue, 301

Autologous chondrocyte implantation (ACI), 126

Autonomic motor system Sympathetic and parasympathetic portions of the nervous system that function to control the actions of the visceral organs and skin, 180

Autonomic nervous system (ANS), 183, 185*f*, 186*t*

Autosome Chromosome other than a sex chromosome, 446

Autotransfusion Return of collected blood to the patient's own circulatory system, 255

AV bundle. *See* Atrioventricular (AV) bundle

AV node. *See* Atrioventricular node

AV valve. *See* Atrioventricular (AV) valve

Axial portion Pertaining to the body's axis, 4

Axial skeleton Portion of the skeleton that supports and protects the organs of the head, neck, and trunk, 109–117, 109*f*–112*f*, 114*f*–116*f*

Axillary artery, 277*f*, 280*f*

Axillary lymph nodes, 295*f*

Axillary vein, 281*f*

Axis Second cervical vertebra upon which the atlas rotates, allowing the head to turn, 115–116, 115*f*

Axon Process of a neuron that conducts nerve impulses away from the cell body, 77, 77*f*, 164, 164*f*, 169*f*

Axon terminal, 164, 164*f*, 169*f*

B

Back pain, 117

Ball-and-socket joint The most freely movable type of joint (for example, the shoulder or hip joint), 124, 125*f*

Balloon angioplasty Procedure for treating a blocked coronary artery: A flexible guide wire is pushed into the coronary artery, and a miniature balloon catheter is pushed down the wire to the blockage; repeated inflations of the balloon decrease or relieve the blockage, 268*f*

Bariatric surgery Surgery conducted on the digestive tract (stomach, small intestine, or both) used in the treatment of obesity, 365, 367, 367*f*

Baroreceptors, 270

Barriers, in immune system, 298

Bartholin gland Either of two oval glands on each side of the lower part of the vagina that secrete a lubricating mucus, 406

Basal body Cytoplasmic structure that is located at the base of and may organize cilia or flagella, 50, 51*f*

Basal cell carcinoma Form of skin cancer that begins in the epidermis and rarely metastasizes but has the capacity to invade local tissues, 93

Basal nuclei Mass of gray matter located deep within a cerebral hemisphere of the brain, 176

Base Solution in which pH is more than 7; a substance that contributes or liberates hydroxide ions in a solution; alkaline; opposite of *acid*, 25–27, 26*f*

Baseball, 234

Basement membrane, 66, 66*f*

Basilar artery, 283*f*

Basilar membrane, 209, 210*f*

Basilic vein, 281*f*

Basophil Leukocyte with a granular cytoplasm and that is able to be stained with a basic dye, 245*f*, 246*f*, 248, 296*f*

B cells, 301, 302*f*

Bedsores, 89, 89*f*

Benign prostatic hyperplasia (BPH) Enlargement of the prostate gland, 387

Benign tumor Mass of cells derived from a single mutated cell that has repeatedly undergone cell division but remained at the site of origin, 79

Bicarbonate ion The form in which carbon dioxide is carried in the blood; HCO_3^2, 330, 382, 384*f*

Biceps brachii, 146, 146*f*, 150*t*, 151, 151*f*

Biceps femoris, 153*f*

Bicuspid (tooth) Flat premolar teeth used for grinding food, 344

Bicuspid valve Atrioventricular valve between the left atrium and the left ventricle; also known as the mitral valve, 262*f*, 263

Bile Secretion of the liver that is temporarily stored in the gallbladder before being released into the small intestine, where it emulsifies fat, 30, 349, 354*f*

Bile duct, 355*f*

Bile pigments Group of compounds formed by the liver from globin, then excreted as waste in bile, 248

Bile salts, 358

Bilirubin, 358

Bioengineering, of babies, 457

Biopsy Removal of sample tissue by plungerlike devices to diagnose a disease, 79

Biotin, 362*t*

Birth, 437–441, 437*t*, 438*f*

Birth control pill Oral contraceptive containing estrogen and progesterone, 410

Birth defects, 440–441

Black lung, 332

Bladder, 353*f*, 374, 374*f*, 385, 396*f*, 398*f*, 401*f*, 438*f*

Blastocyst Early stage of embryonic development that consists of a hollow ball of cells, 429, 429*f*

Blind spot Area where the optic nerve passes through the retina and where vision is not possible due to the lack of rod cells and cone cells, 204

Blood Connective tissue composed of cells separated by plasma. *See also* Cardiovascular system
in aging, 254
components of, 243, 243*f*, 244–250, 245*f*–249*f*
as connective tissue, 69*t*, 74, 74*f*
functions of, 243–244
pH of, 337

Blood flow velocity, 274, 274*f*

Blood pressure Force of blood against a blood vessel wall, 267, 270, 272, 274–278, 275*f*, 277*f*, 289, 337, 373

Blood spatter analysis, 256

Blood sugar, 155

Blood transfusion Introduction of whole blood or a blood component directly into the bloodstream, 252–254, 252*t*, 253*f*, 254*f*, 255, 255*f*

Blood types, 242, 252–254, 252*t*, 253*f*, 254*f*

Blood vessels, 271–274, 272*f*, 273*f*

B lymphocyte Type of lymphocyte that is responsible for antibody-mediated immunity, 246*f*, 301

Body art, 98

Body region (stomach) Central and widest portion of the stomach, 348, 348*f*

Body temperature, 93, 97, 244

Body wall, 7*f*

Bolus A soft mass of chewed food, 345*f*, 346

Bond, 22–24, 23*f*, 32

Bonds, Barry, 234

Bone Connective tissue having a hard matrix of calcium salts deposited around protein fibers
anatomy, 103–107, 104*f*, 105*f*
broken, 108, 108*f*
compact, 69*t*, 72, 73*f*, 103, 124*f*
as connective tissue, 69*t*, 72, 73*f*
of cranium, 110–113, 110*f*–113*f*
development, 105–107
of face, 111*f*, 112*f*, 113
growth, 105–107
head of, 107*t*
long, 103–107, 104*f*, 105*f*
remodeling of, 107
spongy, 69*t*, 72, 73*f*, 103–105, 104*f*, 124*f*
surface features of, 107*t*

Bone marrow, 103, 104*f*, 124*f*, 294–296, 295*f*, 296*f*

Bony callus, 108

Bony labyrinth, 208

Bony pelvis, 120

Botulism toxin, 138

Brachial artery, 277*f*, 280*f*

Brachialis, 146*f*, 150*t*, 151, 151*f*

Brachial vein, 281*f*

Brachiocephalic Pertaining to the arm and head, as in the brachiocephalic artery and vein

Brachiocephalic artery, 262*f*, 264*f*, 279, 279*t*, 280*f*

Brachiocephalic vein, 281*f*, 281*t*

Brachioradialis, 146

Bradycardia Slow heart rate, characterized by fewer than 60 heartbeats per minute, 269

Brain, 173, 174*f*, 282–283, 283*f*. *See also* Cerebrum

"Brain freeze," 163

Brain injury, 179

Brain stem Portion of the brain that includes the midbrain, pons, and medulla oblongata, 174*f*, 179–180

Braxton-Hicks contractions Strong, late-term uterine contractions prior to cervical dilation; also called false labor, 437

Breast, female, 407–409, 409*f*

Breast cancer, 409, 413, 413*f*

Breast self-exam, 413, 413*f*

Breathing, 325–328, 326*f*–328*f*

Breech birth Birth in which the baby is positioned rump first, 435

Broad ligament of uterus, 404*f*

Broca's area Brain center associated with the motor control of speech and usually located in the left frontal gyrus, 175*f*, 176

Bronchial thermoplasty, 336

Bronchial tree, 323, 332f

Bronchiole Smaller air passages in the lungs, 319f, 323, 324f

Bronchitis, 331, 332f, 335

Bronchogenic carcinoma Form of cancer originating in the air passages of the lungs, 333

Bronchus One of the two major divisions of the trachea; leads to the lungs (pl., *bronchi*), 319f, 320t, 323

Buccinator, 147, 147f, 148t

Buffer Substance or compound that prevents large changes in the pH of a solution, 27, 383

Buffy coat The layer of white blood cells covering red blood cells in a sample of centrifuged blood, 243

Bulbourethral gland Gland located in the pelvic cavity that adds secretions to seminal fluid within the urethra, 395, 396f, 398f, 398t, 399

Bulimia nervosa Eating disorder characterized by binge eating followed by purging, 365

Buproprion-naltrexone, 365

Burns, 96, 96f

Bursa Saclike, fluid-filled structure, lined with synovial membrane, that occurs near a joint (pl., *bursae*), 123, 124f

Bursitis Inflammation of any of the friction-easing sacs called bursae within the knee joint, 123

C

Calcaneal tendon, 146f, 153f

Calcaneus Heel bone, 122f, 123

Calcitonin Hormone secreted by the thyroid gland that helps regulate the level of blood calcium level, 219f, 220t, 226, 226f

Calcitriol, 226–227, 226f

Calcium, 363t, 382

Canaliculi, 103, 104f

Cancer Rapid, uncontrolled, and disorganized growth of tissue cells, 79
blood, 249, 254
breast, 409, 413, 413f
cervical, 310
chemical signaling and, 236
chemotherapy for, 305
cytokines and, 305
exercise and, 155
lung, 333, 335–336, 335f
lymphatic system, 297
ovarian, 405
prostate, 387
skin, 93–94, 94f
smoking and, 333
T-cells and, 304f
testicular, 413, 413f
treatment, 20, 80–81

Candidiasis, 93

Canines, 344f

Capillary Microscopic vessel located in the tissues connecting arterioles to venules; molecules either exit or enter the blood through the thin walls of capillaries, 262, 271

Capillary bed, 272f

Capillary exchange, 271–273, 273f

Capitulum, of humerus, 119, 119f

Capsid The protein shell of a virus particle that surrounds its nucleic acid, 307

Carbaminohemoglobin Hemoglobin carrying carbon dioxide, 330

Carbohydrate Organic compounds with the general formula $(CH_2O)_n$, including sugars and glycogen, 28–29, 28f, 29f, 364, 364f

Carbon dioxide transport, 330

Carbonic anhydrase Enzyme that promotes the formation of carbonic acid from carbon dioxide and water, 330

Carboxy peptidase, 360t

Carcinogen Any agent that causes cancer, 79

Carcinoma Cancer arising in epithelial tissue, 79

Cardiac cycle Series of myocardial contractions that constitutes a complete heartbeat, 265–266, 266f

Cardiac muscle Heart muscle (myocardium) consisting of striated muscle cells that interlock, 74t, 75–76, 75f, 134, 134f, 140, 154–156, 261f, 265f

Cardiac output (CO), 266–270, 267f–270f, 275

Cardiac region The most superior portion of the stomach, 347–348, 348f

Cardiac vein Blood vessel that returns blood from the venules of the myocardium to the coronary sinus, 263

Cardiopulmonary resuscitation (CPR), 278

Cardioregulatory center Portion of the medulla oblongata that regulates the heartbeat rate, 269–270, 270f

Cardiovascular disease, 267–268, 267f, 268f, 284–285

Cardiovascular system, 10. *See also* Blood
aging of, 286
blood vessels in, 271–274, 272f, 273f
circulation, 274–286, 274f–277f, 280f–284f
heart in, 260–270, 260f–262f, 264f–270f
in homeostasis, 9, 14f, 287–289

Caries Destruction of tooth enamel by oral bacteria, 344

Carotene, 88

Carotid artery Either of two arteries branching off the aortic arch and supplying blood to the head and neck, 260f

Carotid body Structure located at the branching of the carotid arteries that contains chemoreceptors, 328

Carpals Bones of the wrist, 109f, 120f

Carpal tunnel, 120

Carpal tunnel syndrome, 120

Carrier Molecule that combines with a substance and actively transports it through the plasma membrane, 53–55, 53f

Cartilage Type of avascular connective tissue containing a firm, jellylike matrix embedded with protein fibers that provides support & protection, 69t, 72, 72f

Cartilaginous joint Two or more bones joined by cartilage, 123

Cataract Opaqueness of the lens of the eye, making the lens incapable of transmitting light, 201, 206, 213

Cavities, body, 6–8, 6f–7f

CCK. *See* Cholecystokinin (CCK)

Cecum Blind pouch, such as the one below where the small intestine enters the large intestine, 351, 352f

Celiac artery, 279t, 280f

Cell Structural and functional unit of an organism; smallest structure capable of performing all the functions necessary for life, 2, 2f, 43–50, 43t, 44f–46f, 48f–50f, 105

Cell body Portion of a nerve cell that includes a cytoplasmic mass and a nucleus, and from which the nerve fibers extend, 77, 77f, 164, 164f, 169f

Cell cycle Life cycle of a cell consisting of G_1 (growth), S (DNA synthesis), G_2 (growth), and mitosis (division), 55–60, 56f, 57f, 59f, 60f

Cell-mediated immunity Immunological defense provided by killer T cells, which destroy virus-infected cells, foreign cells, and cancer cells, 303–305, 303f, 304f

Cellular respiration Process that releases energy from organic compounds in cells, 49, 141

Cellulose Polysaccharide very abundant in plant tissues that human enzymes cannot break down, 28–29, 29f

Cementum, 344f

Central Situated at the center of the body or an organ, 3f, 4

Central canal Tube within the spinal cord that is continuous with the ventricle of the brain and contains cerebrospinal fluid, 103, 104f, 172, 172f

Central nervous system (CNS) Brain and spinal cord, 163f, 163f, 171–180, 171f, 172f, 174f–176f, 178f. *See also* Nervous system

Central sulcus, 175f

Central white matter, 176

Centriole Short, cylindrical organelle that contains microtubules in a 9 1 0 pattern and is associated with the formation of the spindle during cell division, 43t, 44f, 50

Centromere, 58, 59f

Centrosome, 44f, 49, 59f

Cephalic phase In gastric secretion, the phase that is stimulated by central nervous system stimuli such as the taste, sight, or smell of food, 350

Cephalic vein, 281f

Cerebellum Part of the brain that controls muscular coordination, 174f, 178–179

Cerebral arterial circle A complete ring of arteries at the base of the brain and formed by the cerebral and communicating arteries; more commonly referred to as the Circle of Willis, 283, 283f

Cerebral cortex Outer layer of the cerebrum, 173

Cerebral hemisphere One of the large, paired structures that together constitute the cerebrum of the brain, 173, 175f

Cerebral palsy Spastic weakness of the arms and legs due to damage to the motor areas of the cerebral cortex, 175

Cerebrospinal fluid (CSF) Fluid found within the ventricles of the brain and surrounding the CNS in association with the meninges, 171–172, 171f, 172f

Cerebrovascular accident (CVA) Condition resulting when an arteriole in the brain bursts or becomes blocked by an embolism; stroke, 251, 287

Cerebrum Main portion of the vertebrate brain that is responsible for consciousness, 171f, 173–176, 174f, 174f–176f

Cervical canal, 433f

Cervical cancer, 310

Cervical cap, 412t

Cervical nerves, 183f

Cervical shield, 412t

Cervix Narrow end of the uterus that projects into the vagina, 401–402, 401f, 404t, 433f, 438f

Cesarean section Birth by surgical incision of the abdomen and uterus, 431

Chancre Primary sore or ulcer that is the initial lesion of syphilis, 417

Chemical digestion, 358–360, 359f, 360t

Chemical signals, 236

Chemistry, 21–24, 21f, 23f

Chemoreceptor Sensory receptor that responds to a particular chemical in air, water, or blood, 194, 196

Chemotherapy Use of drugs to kill cancer cells, 305

Chewing tobacco, 334

Cheyne-Stokes respiration Type of respiration characterized by alternate periods of deep, labored breathing and no breathing at all, 328

Chief cells, 348f

Chlamydia Sexually transmitted disease caused by the bacterium *Chlamydia trachomatis*; often causes painful urination and swelling of the testes in men; is usually symptomless in women but can cause inflammation of the cervix or uterine tubes, 417

Chloride, 363t

Chloride shift Movement of chloride ions from the blood plasma into red blood cells as bicarbonate ions diffuse out of the blood cells into the plasma, 330

Cholecystokinin (CCK) Hormone secreted by the small intestine that stimulates the release of pancreatic juice from the pancreas and bile from the gallbladder, 350, 351, 351f

Cholesterol, 31

Chordae tendinae Tough bands of connective tissue that attach the papillary muscles to the atrioventricular valves within the heart, 262, 262f

Chorion Extraembryonic membrane that forms an outer covering around the embryo and contributes to the formation of the placenta, 428

Chorionic villi Projections from the chorion that appear during implantation and that in one area contribute to the development of the placenta, 431, 432f, 434f

Chorionic villi sampling (CVS) Method of retrieving fetal cells for genetic testing in which a long, thin tube is passed through the vagina into the uterus, and suction is used to obtain a sample of chorionic villi cells, 446–447, 447f

Choroid Vascular, pigmented middle layer of the wall of the eye, 200f, 201t

Choroid plexus, 172, 174f

Chromatids Two identical parts of a chromosome following replication of DNA, 56, 59f, 393, 395, 397f, 448f

Chromatin Threadlike network in the nucleus that condenses to become the chromosomes just before cell division, 44f, 46, 46f

Chromium, 363t

Chromosomal inheritance, 446–449, 446t, 447f–449f

Chromosome Rod-shaped body in the nucleus, particularly during cell division, that contains the hereditary units, or genes, 46, 56, 58, 59f, 393, 394f, 395, 397f, 403f, 446, 450–451

Chronic bronchitis Obstructive pulmonary disorder that tends to recur, marked by inflamed airways filled with mucus, and degenerative changes in the bronchi, including loss of cilia, 332f, 335

Chronic disease Long and continued but not acute, 17

Chronic obstructive pulmonary disease (COPD) Continued interference with airflow in the lungs due to chronic bronchitis or emphysema, 332–335, 332f

Chronic simple glaucoma, 453t

Chyme Semifluid food mass leaving the stomach, 343f, 346t, 349

Chymotrypsin, 360t

Cilia Membrane-bounded microtubular structures that project from a cell, and in multicellular animals facilitate the flow of materials over the cell surface

Ciliary body Structure associated with the choroid layer of the eye that secretes aqueous humor and contains the ciliary muscle, 200, 200f, 201t, 202f

Ciliary muscle Muscle that controls the curvature of the lens of the eye, 200, 201t, 202f

Cilium Short, hairlike projection from the plasma membrane, occurring usually in large numbers, 42, 43t, 50, 50f, 319

Circadian rhythm Pattern of repeated behavior associated with the cycles of night and day, 234, 235f

Circle of Willis Arterial ring located on the ventral surface of the brain, 283, 283f

Circular fold Permanent transverse folds of the luminal surface of the small intestine, involving the mucosa and the submucosa, 349

Circulation Movement of the blood through the heart and blood vessels, 274–286, 274f–277f, 280f–284f

Circulatory routes, 278–286, 279t, 280f–284f, 281t

Circumcision Removal of the prepuce (foreskin) of the penis, 399, 418

Circumduction Conelike movement of a body part, such that the distal end moves in a circle, while the proximal portion remains relatively stable, 124, 126f

Circumflex artery, 264f

Cirrhosis Chronic, irreversible injury to liver tissue; commonly caused by frequent alcohol consumption, 357

Clavicle Bone extending from the sternum to the scapula, 109f, 118, 118f

Cleavage Early, successive divisions of the blastocyst cells into smaller and smaller cells, 427, 429f

Cleavage furrow Site of cell division of the fertilized egg that is unaccompanied by growth. Numerous small cells result, 59, 59f

Clitoris Small, erectile, female organ located in the vulva; homologous to the penis, 406f

Clonal selection theory States that the antigen selects which lymphocyte will undergo clonal expansion and produce more lymphocytes bearing the same type of receptor, 301

Clone An individual grown from a single body cell of its parent and genetically identical to it; in the immune system, a single lymphocyte and all of the identical cells that derive from it, 301, 430, 430f

Clustered regularly interspaced short palindromic repeat technology (CRISPR), 457

CNS. *See* Central nervous system (CNS)

Coagulation Blood clotting, 250, 250f, 251

Coal-dust pneumoconiosus, 332

Coccygeal nerve, 183f

Coccyx Caudal end of the vertebral column formed by the fusion of four vertebrae; tailbone, 109f, 114f, 116

Cochlea Portion of the inner ear that contains the receptors for hearing, 208, 209f, 210f, 211f, 213t

Cochlear canal Canal within the cochlea that bears small hair cells that function as hearing receptors, 209, 210f

Cochlear implant Prosthetic device used to help persons with severe hearing impairment; the device converts sound to an electrical impulse that directly stimulates the auditory nerve, 212

Cochlear nerve Either of two cranial nerves that carry nerve impulses from the spiral organ to the brain; auditory nerve, 209f, 210f

Codon Set of three nucleotides of a messenger RNA molecule corresponding to a particular amino acid, 57f, 58

Cofactor A metal ion, vitamin, or other substance that acts with an enzyme to bring about certain effects; also called a *coenzyme,* 32

Cohesion The force of attraction by which the molecules of a solid or liquid tend to remain together, 25

Colitis, 312, 357

Collagen, 31

Collagen fibers Very strong and flexible fibers of the protein collagen, 69, 70f, 71f, 88

Collecting duct Tube that receives urine from several distal convoluted tubules, 376, 376f, 378f, 380–381, 381f, 383f

Colon Large intestine, 351, 352f

Color blindness Genetic disorder in which the affected individual lacks particular cone cells and cannot distinguish their corresponding colors, 203, 453t

Color vision Ability to detect the color of an object, dependent on three kinds of cone cells, 201t, 203

Colostomy Attachment of a shortened colon to a surgical opening in the abdominal wall

Colostrum First secretion of a woman's mammary glands after she gives birth, 409

Columnar Column-shaped; cells that are taller than they are wide, 66, 66t, 68, 68f, 322, 322f

Coma A state of profound unconsciousness caused by disease, injury, or ingestion of poison, 180, 232

Common bile duct, 354f

Common carotid artery, 262f, 264f, 277f, 279, 279t, 280f

Common hepatic duct, 354f, 355f, 358

Common iliac artery, 260f, 279, 279t, 280f, 284f

Common iliac vein, 260f, 280, 281f, 281t

Compact bone Hard bone consisting of osteons cemented together, 69t, 72, 73f, 103, 124f

Compartment syndrome A painful condition resulting from the expansion or overgrowth of a leg muscle within its connective tissue sheath, producing pressure that interferes with circulation and adversely affects the function and health of the muscle, 135

Complement system Group of proteins in plasma that aid the general defense of the body by destroying bacteria; often called complement, 300

Complex carbohydrates, 28–29, 29f

Complimentary base pairing In DNA, temporary hydrogen bonds formed between complimentary nucleotides A—T and C—G; in RNA, A—U and C—G, 36

Compound Chemical substance having two or more different elements in fixed ratio, 22–23, 23*f*

Computed tomography (CT), 16, 16*f*

Concha Shell-shaped structure, such as that seen in the bones of the nasal cavity (pl., *conchae*), 111*f*, 113, 320, 320*f*

Condom For males, a latex sheath used to cover the penis during sexual intercourse; for females, a large polyurethane tube with a flexible ring that fits onto the cervix. Both male and female condoms function as contraceptives and help minimize the risk of transmitting infection, 410, 411*t*

Conduction deafness Hearing loss or impairment resulting from interference with sound wave transmission to the inner ear, 212

Conduction system of the heart Neuromuscular tissue and fibers that control the cardiac cycle; includes the SA node, the AV node, the AV bundle and its branches, and the Purkinje fibers, 264–265, 265*f*

Condyle Large, rounded surface at the end of a bone, 107*t*

Condyloid joint Bone with an oval-shaped projection at one end joined with a bone processing a complementary elliptical cavity, 124, 125*f*

Cone cell Color receptor located in the retina of the eye, 201, 201*t*, 203, 204

Congenital defect Body abnormality arising from birth and due to hereditary factors, 440–441

Congenital hypothyroidism Abnormal condition marked by physical stunting and severe developmental delay, caused by severe thyroid deficiency, 225, 226*f*

Congenital valve defect, 263

Congestive heart failure Inability of the heart to maintain adequate circulation, especially of the venous blood returned to it, 286

Conjoined twins, 425, 429

Conjunctiva The mucous membrane that lines the inner surface of the eyelids and is continued over the anterior surface of the eyeball, except over the surface of the cornea, 199, 199*f*, 200*f*

Connective tissue Type of tissue characterized by cells separated by a matrix; often contains fibers, 69–74, 69*t*, 70*f*–74*f*, 135, 135*f*

Constipation Infrequent, difficult defecation caused by insufficient water in the feces, 356

Continuous ambulatory peritoneal dialysis (CAPD), 385–386

Contraception Deliberate prevention of pregnancy, 410, 411*t*–412*t*, 414–415

Contralateral Structures located on opposite sides of the body, 3*f*, 4

Convulsion An abnormal, violent, and involuntary contraction or series of contractions of the muscles, 157

COPD. *See* Chronic obstructive pulmonary disease (COPD)

Copper, 363*t*

Coracoid process, 118, 118*f*

Cornea Transparent, anterior portion of the outer layer of the eyeball, 199*f*, 200, 200*f*, 201*t*

Coronal suture Line of junction of the frontal bone with the two parietal bones, 110*f*, 111*f*

Corona radiata Group of small follicular cells immediately surrounding the ovum before and after ovulation, 426, 426*f*, 429*f*

Coronary artery Artery that supplies blood to the wall of the heart (myocardium), 263, 264*f*, 279*t*

Coronary artery disease Condition caused by atherosclerosis that reduces blood flow through coronary arteries to heart muscle; typically results in chest pain or myocardial infarction (heart attack), 267–268, 267*f*

Coronary bypass operation Therapy for blocked coronary arteries in which part of a blood vessel from another part of the body is grafted around the obstructed artery, 268

Coronary circulation, 263

Coronary sinus A large vessel on the posterior surface of the heart into which the cardiac veins drain, 262

Coronoid fossa, of humerus, 119, 119*f*

Corpora cavernosa Two bundles of erectile tissue found in the penis or clitoris that distend with blood during sexual arousal to cause an erection (sing, *corpus cavernosum*), 399, 399*f*

Corpus albicans White, fibrous tissue that replaces the regressing corpus luteum in the ovary in the latter half of pregnancy, 403, 408*f*

Corpus callosum Mass of white matter within the brain, composed of nerve fibers connecting the right and left cerebral hemispheres, 173, 174*f*, 398*f*

Corpus cavernosum, 398*f*, 399, 399*f*

Corpus luteum Structure that forms from the tissues of a ruptured ovarian follicle and functions to secrete female hormones, 402*f*, 403, 406*f*, 408*f*

Corpus spongiosum The center column of erectile tissue of the penis that contains the urethra and is anterior to the corpora cavernosa, 399, 399*f*

Corrective lenses, 205, 205*f*

Corticotropin Hormone manufactured by the anterior pituitary gland that stimulates the adrenal cortex; called also *adrenocorticotropic hormone*, 220*t*, 223*f*, 224, 227

Cortisol Glucocorticoid secreted by the adrenal cortex, 219*f*, 227

Cortisone, 227

Costal cartilage, 109*f*, 117, 118*f*

Costal facets, 115

Covalent bond Chemical bond created by the sharing of electrons between atoms, 22–23, 23*f*, 24

Coxal bone Bone of the pelvic girdle, 109*f*, 120

Cramp A painful, involuntary muscle spasm, 157

Cranial cavity Hollow space in the cranium containing the brain, 6, 6*f*, 7*t*

Cranial nerve Nerve that arises from the brain, 182*f*

Cranium, 110–113, 110*f*–113*f*

C-reactive protein (CRP), 267

Creatine phosphate Muscle biochemical that stores energy, 141, 141*f*

Creatinine Nitrogenous waste, the end product of creatine phosphate metabolism, 373

Crenation Shrinking of red blood cells often caused by osmotic conditions, 52

Crest (bone), 107*t*

Cretinism Condition resulting from a lack of thyroid hormone in an infant, 225, 226*f*

Creutzfeldt-Jakob disease, 34

Cribriform plate, 112*f*, 113

Criminology, 128, 256, 345

Crista galli, 110*f*, 112*f*, 113

Crohn's disease, 312, 357

Cryptorchidism A condition in which one or both testes fail to descend normally into the scrotal sacs, 395

CSF. *See* Cerebrospinal fluid (CSF)

Cuboidal Cube-shaped; cells that are approximately the same width and height, 66, 66*t*, 67, 67*f*

Cuboid bone, 122*f*

Cuneiform bones, 122*f*

Cunnilingus, 418

Cupula, 210, 211*f*

Curvature, spinal, 114, 114*f*

Cushing's syndrome Condition characterized by thin arms and legs and a "moon face;" accompanied by high blood glucose and sodium levels due to hypersecretion of cortical hormones, 229, 230*f*

Cuspid A canine tooth, 344

Cutaneous membrane Pertaining to the skin, 82, 87

Cutaneous receptors, 194–196, 195*f*

Cuticle, 90, 91*f*

Cyanosis Bluish cast to the skin due to an increased amount of deoxyhemoglobin in the blood; sometimes due to a defective atrial septum, which incompletely closes the foramen ovale after birth, 284

Cyclic AMP Derivative of ATP that responds to messages entering a cell and triggers the cell's response; also known as cAMP, 218

Cystic duct, 354*f*, 355*f*, 358

Cystic fibrosis (CF) Generalized, autosomal recessive disorder of infants and children, in which there is widespread dysfunction of the exocrine glands, 55, 452, 453*f*, 453*t*, 454*f*, 456, 457

Cystitis Inflammation of the urinary bladder, 385

Cytokine Type of protein secreted by a T lymphocyte that attacks viruses, virally infected cells, and cancer cells, 301, 305

Cytokinesis Division of the cytoplasm following mitosis and meiosis, 56*f*, 58, 59

Cytoplasm Ground substance of cells located between the nucleus and the plasma membrane, 43, 44*f*

Cytoskeleton System of microtubules and filaments that reinforces a cell's three-dimensional form; maintains cell shape and allows movement of cell and its contents, 43*t*, 44*f*, 45, 49–50

Cytotoxic T cell T lymphocyte that attacks and kills antigen-bearing cells, 304, 304*f*

D

Dandruff Skin disorder characterized by flaking, itchy scalp; caused by accelerated keratinization of the scalp, 93

Daughter cell Cell that arises from a parent cell by mitosis or meiosis, 58, 59*f*

Dead-space, 327

Deafness, 212, 212*t*

Decubitus ulcer Skin sore due to restricted blood flow to the area in bedridden patients; also called a bedsore, 89, 89*f*

Deep Located away from the surface of the body or an organ, 3*f*, 4

Deep brachial artery, 280*f*

Deep-brain stimulation In treatment of Parkinson's disease, deep-brain stimulation (DBS) uses a surgically implanted neurostimulator to apply targeted electrical shock to brain areas that control movement, blocking the abnormal nerve signals that cause tremor and disease symptoms, 189

Deep fascia, 135, 135*f*, 137*f*

Deep femoral artery, 280*f*

Defecation Discharge of feces from the rectum through the anus, 351–352

Degradation-decomposition reaction Chemical reaction in which larger molecules are broken down into smaller and simpler molecules, 32

Dehydration Loss of water, 54, 54f

Dehydration reaction Anabolic process that joins small molecules and releases water as a by-product of the reaction; synthesis reaction, 27f, 28

Delayed allergic response Allergic response initiated at the site of the allergen by sensitized T cells, involving macrophages and regulated by cytokines, 312

Delirium tremens Alcohol withdrawal, 440

Deltoid, 146, 146f, 149f, 150f, 150t, 151, 151f

Deltoid tuberosity, 119, 119f

Denaturation Loss of normal shape by an enzyme so that it no longer functions; caused by a less than optimal pH or temperature, 32

Dendrite Process of a neuron, typically branched, that conducts nerve impulses toward the cell body, 77, 77f, 164, 164f, 169f

Dens, 115–116

Dense connective tissue Type of tissue containing many collagen fibers packed together; found in tendons and ligaments, for example, 69t, 70, 71f

Dental care, 286

Dentin, 344, 344f

Dentition, 345

Deoxyhemoglobin Hemoglobin not carrying oxygen, 330

Deoxyribonucleic acid (DNA) Nucleic acid; the genetic material found in the nucleus of a cell, 35
  cancer and, 80–81
  fingerprinting, 61, 61f
  in meiosis, 393–395, 394f
  in protein synthesis, 56, 57f
  replication, 56, 57f
  RNA vs., 35t
  structure, 36f
  technology, 454–459, 455f

Depolarization Loss in polarization, as when a nerve impulse occurs, 165, 166f, 168f

Depression Movement of a synovial joint that lowers a body part, 125

Depressor anguli oris, 147, 147f, 148t

Depressor labii inferioris, 147, 147f, 148t

Dermis Thick skin layer that lies beneath the epidermis, 87f, 88

Descending aorta, 279f, 280f, 352f

Descending colon That portion of the large intestine that travels inferiorly as it extends from the transverse colon to the sigmoid colon, 351, 352f

Descending tracts, 173

Dessication, 67

Detrusor muscle, 374

Development, 427–436, 427f, 428t, 429f, 430f, 432f–436f

Diabetes insipidus Condition characterized by an abnormally large production of urine, due to a deficiency of antidiuretic hormone, 222

Diabetes mellitus Condition characterized by a high blood glucose level and the appearance of glucose in the urine, due to a deficiency of insulin, 230–231, 233, 237

Diabetic coma A state of profound unconsciousness caused by insulin insufficiency that leads to excessive blood glucose, 232

Diabetic ketoacidosis (DKA), 232

Diabetic retinopathy Noninflammatory retinal disorder caused by chronic, uncontrolled diabetes mellitus; retinal damage can cause blindness, 206

Diagnosis Decision based on an examination to determine the nature of a diseased condition, 79

Dialysate Material that passes through the membrane in dialysis, 385

Dialysis, 372, 385–386, 385f

Diaphragm Sheet of muscle that separates the thoracic cavity from the abdominopelvic cavity; also, a birth control device inserted in front of the cervix in females, 149, 319f, 343f, 353f, 410, 412t

Diaphysis Shaft of a long bone, 103, 104f

Diarrhea Frequent, watery defecation, often caused by digestive infection or stress, 356

Diarthrosis Freely movable joint, 123

Diastole Relaxation of heart chambers, 265, 266f

Diastolic pressure Arterial blood pressure during the diastolic phase of the cardiac cycle, 277–278

Diencephalon Portion of the brain in the region of the third ventricle that includes the thalamus and hypothalamus, 174f, 176

Diet, 285, 361–366, 361f, 362t–363t, 364f, 366f

Dietary anemias, 249

Diethylstilbestrol (DES), 414

Differential white blood cell count Microscopic examination of a blood sample in which each type of white blood cell is counted, 249

Differentiation Process by which a cell becomes specialized for a particular function, 427

Diffusion Passive movement of molecules from an area of greater concentration to an area of lesser concentration, 51–52, 52f, 53t

Digestive system, 10
  aging of, 360
  anatomy of, 343–353, 343f–345f, 346t, 347f–353f
  chemical digestion in, 358–360, 359f, 360t
  disorders, 356–357
  functions of, 343
  in homeostasis, 13, 15f, 361
  nutrition and, 361–366, 361f, 362t–363t, 364f, 366f

Dipeptide, 32

Diploid Having the haploid chromosome number of 23 doubled; having 46 chromosomes, 393

Disaccharase Enzyme that digests a disaccharide such as sucrose, 359

Disaccharide, 28

Disease Any abnormal condition considered harmful to the body; an illness or disorder, 17

Distal Further from the midline or origin; opposite of *proximal*, 3f, 4

Distal convoluted tubule Highly coiled region of a nephron that is distant from the glomerular capsule, 376, 376f, 377f, 378f, 381f, 383f

Diuresis, 381

Diuretic Drug used to counteract hypertension by inhibiting Na^1 reabsorption so that less water is reabsorbed in the nephron, 382

Diverticulosis Presence of diverticula, or saclike pouches, in the colon, 357

DNA. *See* Deoxyribonucleic acid (DNA)

Dominant allele Hereditary factor that expresses itself even when there is only one copy in the genotype, 450

Doping, 234–235, 235f

Dorsal Pertaining to the back or posterior portion of a body part; opposite of *ventral*, 3f

Dorsal artery of penis, 399f

Dorsal cavity, 6, 6f

Dorsalis pedis artery, 277f, 280f

Dorsal nerve of penis, 399f

Dorsal vein of penis, 399f

Double bonds, 23–24, 23f

Douching, in contraception, 411t

Down syndrome Human congenital disorder associated with an extra chromosome 21, 446t, 448–449, 448f

Drinking, 440

Drug abuse, 285, 418, 418f

dual X-ray absorptiometry (DEXA), 106

Duchenne muscular dystrophy, 157

Ductus arteriosus Fetal connection between the pulmonary artery and the aorta; venous artery, 283, 284f

Ductus venosus Fetal connection between the umbilical vein and the inferior vena cava; also called venous duct, 283, 284f

Duodenum First portion of the small intestine into which ducts from the gallbladder and pancreas enter, 348f, 349, 349f, 353f, 354f

Duplicated chromosome Chromosome having two sister chromatids held together by a centromere, 56

Duplication division, 393

Dural venous sinuses, 171

Dura mater Tough outer layer of the meninges; membranes that protect the brain and spinal cord, 171, 171f

Dust cells, 324

Dwarfism, 224, 224f

Dynamic equilibrium, 210, 211f

E

Ear, anatomy and physiology of, 208–210, 209f–211f

Ear infection, 331

Earworm, 193

Eating disorders, 365

Eccrine glands, 91

ECG. *See* Electrocardiogram

E-cigarettes, 334

Ectoderm The outermost of the three primary germ layers of an embryo, 431, 432f

Ectopic pacemaker A group of cells in the heart that temporarily serves to direct heart rate and contraction, 265

Ectopic pregnancy, 404

Ectopic pregnancy Implantation of the embryo in a location other than the uterus, most often in a uterine tube, 404

Eczema Form of noncontagious dermatitis that begins with itchy red patches that thicken and crust over, 93

Edema Swelling due to tissue fluid accumulation in the intercellular spaces, 294

Effectors, 163

Efferent arteriole, 376f, 378f, 383f

Eggs, 393

Ejaculation Ejection of seminal fluid, 395, 399–400

Ejaculatory duct Tube, formed by the joining of the vas deferens and the tube from the seminal vesicle, that transports sperm to the urethra, 395, 396f, 398f, 398t

EKG. *See* Electrocardiogram; Electrocardiogram

Elastic cartilage Cartilage composed of elastic fibers, allowing greater flexibility, 69t, 72, 72f

Elastic fibers Flexible connective tissue fibers that contain elastin, 69, 70f, 88

Electrocardiogram (ECG) Recording of the electrical activity that accompanies the cardiac cycle, 265, 269, 269f

Electroencephalogram (EEG) Graphic recording of the brain's electrical activity, 173

Electrolyte Any substance that ionizes and conducts electricity; electrolytes are present in the body fluids and tissues, 27, 382

Electrolyte balance, 380

Electrolyte reabsorption, 382

Electron Small, negatively charged particle that revolves around the nucleus of an atom, 21

Electronegative In a covalent molecule, the atom(s) that tend to retain electrons and thus possess a partial negative charge, 24

Electronic cigarettes, 334

Element The simplest of substances, consisting of only one type of atom (for example, carbon, hydrogen, oxygen), 21–22, 21f

Elephantiasis Swelling of the arms, legs, or external genitalia due to failure of the lymphatic system to remove excess fluid, 297

Elevation Movement of a synovial joint that raises a body part, 125

Embolic stroke, 287

Embolus Moving blood clot that is carried through the bloodstream, 251

Embryo Organism in its early stages of development; in humans, the organism in its second week to two months of development, 431, 433f

Embryonic development Period of development from the second through eighth weeks, 428t, 431, 432f

Embryonic disk Flattened area between the amniotic cavity and the yolk sac from which the embryo arises, 431

Emergency contraception, 411t, 414–415

Emphysema Lung impairment caused by deterioration of the bronchioles, which traps air in alveoli, 332f, 335

Emulsification Breaking up of fat globules into smaller droplets by the action of bile salts, 30, 349

Emulsifiers, 30

Enamel, 344f

Endocardium Inner layer of the heart wall, 261, 261f

Endochondral ossification Ossification that begins as hyaline cartilage that is subsequently replaced by bone tissue, 105, 105f

Endocrine cells, 354f

Endocrine-disrupting contaminants (EDCs), 414

Endocrine gland Gland that secretes hormones directly into the bloodstream or body fluids, 218–221, 219f, 220t, 221f

Endocrine system, 10
  adrenal glands in, 227–229, 228f, 229f
  aging of, 237
  chemical signals in, 236
  glands in, 218–221, 219f, 220t, 221f
  gonads in, 232, 234
  in homeostasis, 13, 14f, 237–239
  hypothalamus in, 222–224, 223f–225f
  pancreas in, 229–231, 231f
  parathyroid gland in, 225–227, 225f, 226f
  pineal gland in, 234, 235f
  pituitary gland in, 222–224, 223f–225f
  thymus gland in, 234
  thyroid in, 225–227, 225f, 226f

Endocytosis Process in which a vesicle is formed at the plasma membrane to bring a substance into the cell, 44f, 53t, 55

Endoderm The innermost of the three primary germ layers of an embryo, 431, 432f

Endolymph, 208, 211f

Endomembrane system Collection of membranous structures involved in transport within the cell, 47, 48f

Endometriosis Implantation of uterine endometrial tissue in the abdominal cavity, possibly as a result of irregular menstrual flow, 415

Endometrium Lining of the uterus that becomes thickened and vascular during the menstrual cycle, 401, 433f

Endomysium, 135, 135f, 137f

Endoneurium, 182f

Endoplasmic reticulum (ER) Complex system of tubules, vesicles, and sacs in cells; sometimes has attached ribosomes, 43t, 44f, 47

Endothelium, 261, 261f

Enteric nervous system Nerve network that is built into the wall of the organs of the digestive tract, 350

Enterokinase Intestinal enzyme that activates trypsinogen to trypsin, 354

Entry barriers, in immune system, 298

Enzymatic reactions, 32, 34, 35f

Enzyme Protein catalyst that speeds up a specific reaction or a specific type of reaction, 32

Eosinophil Granular leukocyte capable of being stained with the dye eosin, 245f, 246f, 248, 296f

Ependymal cells, 77

Epicardium Visceral portion of the pericardium on the surface of the heart, 261

Epicondyle, 107t

Epidermal growth factor, 236

Epidermis Organism's outer layer of cells, 87–88, 87f

Epididymis Coiled tubules next to the testes where sperm mature and may be stored for a short time, 395, 396f, 397–398, 397f, 398f, 398t

Epidural hematoma Bleeding between the dura mater and the bone, as a result of a head injury, 179

Epidural stimulation, 177

Epiglottis Structure that covers the glottis during the process of swallowing, 197f, 321f, 322, 345f

Epimysium, 135f, 137f

Epinephrine Hormone produced by the adrenal medulla that stimulates "fight-or-flight" reactions; also called adrenaline, 219f, 220t, 227

Epiphyseal plate Cartilaginous layer within the epiphysis of a long bone that functions as a growing region, 105f, 106

Epiphysis End segment of a long bone, separated from the diaphysis early in life by an epiphyseal plate, but later becoming part of the larger bone, 103, 104f

Episiotomy Surgical procedure performed during childbirth in which the opening of the vagina is enlarged to avoid tearing, 437

Epithelial tissue Type of tissue that lines the body's internal cavities and covers the body's external surface, 66–69, 66f–68f, 66t

Equilibrium, 210–213, 211f, 212t, 213t

Erectile dysfunction Failure of the penis to achieve erection, 399

Erectile tissue, 399f

Erythroblasts, 246f

Erythrocyte Nonnucleated, hemoglobin-containing blood cell capable of carrying oxygen; red blood cell, 74, 74f, 244–248, 245f–247f, 246f, 249, 249f

Erythropoietin (EPO) Kidney hormone that promotes red blood cell formation, 247, 373

Esophageal sphincter, 357

Esophagus Tube that transports food from the mouth to the stomach, 320f, 343f, 345f, 346t, 347, 348f, 360

Esophagus disorders, 356

Essential amino acid Amino acid that is necessary in the diet because the body is unable to manufacture it, 361–362

Essential fatty acid Fatty acid that is necessary in the diet because the body is unable to manufacture it, 361

Essential hypertension, 278

Estradiol, 31f

Estrogen Female sex hormone secreted by the ovaries that, along with progesterone, promotes the development and maintenance of the primary and secondary female sex characteristics, 219f, 220t, 232, 402f, 406, 406f, 414, 419, 437t

Ethmoid bone, 110f, 111f, 112f, 113

Eupnea Easy or normal respiration, 327

Eversion Movement of the foot in which the sole is turned outward, 125, 126f

Excretion Elimination of metabolic wastes, 373

Exercise, 144, 155, 156t, 286

Exocrine gland Particular glands with ducts, such as salivary glands, whose secretions are deposited into cavities, 354f

Exocrine glands, 78, 78f, 80

Exocytosis Process in which an intracellular vesicle fuses with the plasma membrane so that the vesicle's contents are released outside the cell, 44f, 48f, 53t, 55

Exophthalmic goiter Enlargement of the thyroid gland, accompanied by an abnormal protrusion of the eyes, 225

Expiration Process of expelling air from the lungs; exhalation, 319, 325–326, 326f

Expiratory capacity, 327, 327f

Expiratory reserve volume Volume of air that can be forcibly exhaled after normal exhalation, 327, 327f

Extension Movement that increases the angle between parts at a joint, 124, 126f, 148

Extensor carpi group, 146f, 150t, 151, 151f

Extensor digitorum, 146, 150t, 151, 151f

Extensor digitorum longus, 146f, 152t, 153, 154f

External auditory meatus Opening through the temporal bone that connects with the tympanum and the middle ear chamber and through which sound vibrations pass, 110

External carotid artery, 280f

External genitals Sex organs that occur outside the body, 399, 399f, 405–406, 406f

External iliac artery, 260f, 279, 280f

External iliac vein, 260f, 280, 281f

External intercostals, 149, 149f, 149t

External jugular vein, 279, 281f

External obliques, 146, 146f, 149, 149f, 149t, 150f

Gallbladder Saclike organ associated with the liver that stores and concentrates bile, 343*f*, 354*f*, 355*f*, 357, 358

Gamete Sex cell; sperm cells in males and ova in females, 393. *See also* Ovum; Sperm

Gamma globulin Most common form of plasma protein; a protein fraction of blood rich in antibodies, administered especially for passive immunity against measles, German measles, hepatitis A, or poliomyelitis, 306

Ganglion Collection of neuron cell bodies outside the central nervous system (pl., *ganglia*), 180

Gap junction Junction between cells formed by the joining of two adjacent plasma membranes; it lends strength and allows ions, sugars, and small molecules to pass between cells, 78, 78*f*, 265

Gas exchange, 324*f*, 328–330, 329*f*, 337

Gas transport, 329–330, 329*f*

Gastric bypass, 367, 367*f*

Gastric gland Gland within the stomach wall that secretes gastric juice, 348, 348*f*

Gastric inhibitory polypeptide Hormone produced by small intestine cells that inhibits stomach movement and acid secretion; also promotes release of insulin in response to glucose consumption; also referred to as *glucose-dependent insulinotropic peptide,* 350

Gastric juice Thin, watery digestive fluid secreted by the gastric glands in the stomach and containing hydrochloric acid and enzymes, such as pepsin, 348

Gastric phase The time period of gastric juice secretion that is stimulated by gastrin, caused by the presence of food in the stomach and by stomach stretching, 350

Gastric pits Openings into gastric glands, 348, 348*f*

Gastric ulcer Open sore in the lining of the stomach; frequently caused by bacterial infection, 356

Gastric veins, 282*f*

Gastrin, 348, 351*f*

Gastrocnemius, 146*f*, 152*t*, 153

Gastroesophageal reflux disease (GERD), 356

Gastrointestinal tract, 343, 343*f*

Gastrulation Formation of a gastrula from a blastula; characterized by an invagination of the cell layers to form a caplike structure, 431

Gene Unit of heredity located on a chromosome, 35, 395, 450–451

General interpretative area, 175*f*, 176

Gene therapy Method of replacing a defective gene with a healthy gene, 454–458

Genetic counseling, 452–453, 453*t*

Genetic inheritance, 450–454, 450*f*, 451*f*, 453*f*, 454*f*

Genetic testing, 453–454

Genital Pertaining to the genitalia (internal and external organs of reproduction)

Genital herpes Sexually transmitted disease caused by herpes simplex virus and sometimes accompanied by painful ulcers on the genitals, 416–417, 416*f*

Genitals, 393, 399, 399*f*

Genital wart Raised growth on the genitals due to a sexually transmitted disease caused by human papillomavirus, 416

Genome All of the DNA in a cell of an organism, 458

Genomics, 458–459

Genotype Combination of genes present within a zygote or within the cells of an individual, 450

German measles, 440

Germ cell An ovum or sperm cell, or a cell that gives rise to either ova or sperm, 396, 397*f*

Germ layers, 431, 432*f*

Gestation Period of development, from the start of the last menstrual cycle until birth; in humans, typically 280 days, 428

GH. *See* Growth hormone (GH)

Ghrelin Hormone manufactured primarily by stomach cells that causes growth hormone secretion, and stimulates fat storage and food intake, 235, 351

Giantism, 224*f*

Gingivitis Inflammation of the gums, 344

Gland Epithelial cell or group of epithelial cells that are specialized to secrete a substance, 78, 78*f*, 80, 88, 91–92, 91*f*

  apocrine, 91

  eccrine, 91

Glans clitoris, 406*f*

Glans penis, 396*f*, 398*f*, 399*f*

Glasses, 205, 205*f*

Glaucoma Increasing loss of field of vision, caused by blockage of the ducts that drain the aqueous humor, creating pressure buildup and nerve damage, 206, 213, 453*t*

Glenoid cavity, 118, 118*f*

Gliding joint Two bones with nearly flat surfaces joined together, 124, 125*f*

Globulin Type of protein in blood plasma. There are alpha, beta, and gamma globulins, 244, 245*f*

Glomerular capsule Double-walled capsule surrounding the glomerulus of a nephron through which glomerular filtrate passes to the proximal convoluted tubule, 375, 376*f*, 378*f*, 383*f*

Glomerular filtrate Liquid that passes out of the glomerular capillaries in the kidney into the glomerular capsules, 377

Glomerular filtration Process whereby blood pressure forces liquid through the glomerular capillaries in the kidney into the glomerular capsule, 377–379, 378*f*

Glomerulus Cluster of capillaries surrounded by the glomerular capsule in a kidney nephron, 375, 376*f*, 378*f*, 383*f*

Glottis Slitlike opening between the vocal cords, 319*f*, 320*t*, 321, 345*f*

Glucagon Hormone secreted by the pancreatic islets that causes the release of glucose from glycogen, 219*f*, 220*t*, 230

Glucocorticoid Any one of a group of hormones secreted by the adrenal cortex that influences carbohydrate, fat, and protein metabolism, 219*f*, 220*t*, 227

Glucocorticoid therapy, 227

Glucose Blood sugar that is broken down in cells to acquire energy for ATP production, 28

Gluteus maximus, 146, 146*f*, 152, 152*t*, 153*f*

Gluteus medius, 146*f*, 152, 152*t*, 153*f*

Glycerol Three-carbon molecule that joins with fatty acids to form fat, 30

Glycogen Polysaccharide that is the principal storage compound for sugar in animals, 28, 29*f*, 136*t*, 231*f*

Glycolipids, 45

Glycoproteins, 45

Glycosuria Presence of glucose in the urine, typically indicative of a kidney disease, diabetes mellitus, or other endocrine disorder, 230

Goiter, 225, 225*f*

Golgi apparatus Organelle that consists of concentrically folded membranes and that functions in the packaging and secretion of cellular products, 43*f*, 44*f*, 47, 48*f*

Golgi tendon organs, 194

Gonad Organ that produces sex cells: the ovary, which produces eggs, and the testis, which produces sperm, 219*f*, 220*t*, 232, 234

Gonadal arteries, 279*t*, 280*f*

Gonadal vein, 281*f*, 281*t*

Gonadotropic hormone Type of hormone that regulates the activity of the ovaries and testes; principally, follicle-stimulating hormone and luteinizing hormone, 219*f*, 220*t*, 223*f*, 224

Gonadotropin, 408*f*

Gonadotropin-releasing hormone (GnRH), 400*f*, 406, 406*f*, 407

Gonorrhea Sexually transmitted disease caused by the bacterium *Neisseria gonorrhoeae* that causes painful urination and swollen testes in men and is usually symptomless in women, but can cause inflammation of the cervix and uterine tubes, 417

Gout Joint inflammation caused by accumulation of uric acid, 373

Gracilis, 146*f*, 153*f*

Graded potential In a neuron, each of many small signals at a synapse; together they can be summed to bring the axon of the neuron to threshold and cause an action potential, 167

Granular leukocyte White blood cell with prominent granules in the cytoplasm, 248

Granulocyte and macrophage colony-stimulating factor (GM-CSF), 236

Graves' disease Autoimmune disease with swollen throat due to an enlarged, hyperactive thyroid gland; patients often have protruding eyes and are underweight, hyperactive, and irritable, 225

Gravitational equilibrium Maintenance of balance when the head and body are motionless, 210, 211*f*, 213, 213*t*

Gray matter Nonmyelinated nerve fibers in the central nervous system, 171*f*, 172*f*

Greater omentum, 353, 353*f*

Greater sciatic notch Indentation in the posterior coxal bone through which pass the blood vessels and the large sciatic nerve to the lower leg, 120

Greater trochanter, of femur, 121, 122*f*

Greater tubercle, of humerus, 119, 119*f*

Great saphenous vein, 281*f*

Ground substance, 70, 70*f*

Growth Increase in the number of cells and/or the size of these cells, 105–107, 427

Growth factor Chemical signal that regulates mitosis and differentiation of cells that have receptors for it; important in such processes as fetal development, tissue maintenance and repair, and hematopoiesis; sometimes a contributing factor in cancer, 236

Growth hormone (GH) Hormone released by the anterior lobe of the pituitary gland that promotes the growth of the organism; also known as somatotropin, 219*f*, 220*t*, 223*f*, 224, 224*f*

Gumma Tumor of rubbery consistency that is characteristic of the tertiary stage of syphilis, 417

Gustatory regions, 196

Gyrus Convoluted elevation or ridge (pl., *gyri*), 173

H

H1N1 flu, 311

Hair Consists of a cylindrical shaft and a root, which is contained in a flasklike depression (hair follicle) in the dermis and subcutaneous tissue. The base of the root is expanded into the hair bulb, which rests upon and encloses the hair papilla, 86, 89

Hair cell Mechanoreceptor in the inner ear that lies between the basilar membrane and the tectorial membrane and triggers action potentials in fibers of the auditory nerve, 208, 209, 210f

Hair follicle Tubelike depression in the skin in which a hair develops, 86, 90, 90f

Hair matrix, 90, 90f

Hamstring group, 152t, 153, 153f

Hand, 120, 120f

Haploid Having the number of chromosomes of a gamete; half the number characteristic of body cells, 393, 394f, 403f

Hard palate Anterior portion of the roof of the mouth that contains several bones, 113, 344f, 345f

Haustra One of the pouches or small sacs into which the large intestine is divided, 351, 352f

Hay fever Seasonal variety of allergic reaction to a specific allergen. Characterized by sudden attacks of sneezing, swelling of nasal mucosa, and often asthmatic symptoms, 312

hCG. *See* Human chorionic gonadotropin (HCG)

Head Pertaining to the skeleton, an enlargement on the end of a bone, 147, 147f, 148t

Healing, wound, 94, 95f

Hearing, 208–210, 209f–211f

Hearing damage, 212, 212t

Heart Muscular organ located in the thoracic cavity that is responsible for maintenance of blood circulation
aging of, 286
anatomy of, 260–263, 260f–262f
conduction system of, 264–265, 265f
coronary circulation in, 263
nodal tissue in, 264–265, 265f
physiology of, 264–270, 265f–270f
sounds, 263

Heart attack. *See* Myocardial infarction

Heart block Impairment of conduction of an impulse in heart excitation, 265

Heartburn Burning pain in the chest occurring when part of the stomach contents escapes into the esophagus, 356

Heart disease, 267–268, 267f, 268f, 284–285

Heart murmur Clicking or swishing sounds often due to leaky valves, 263

Heart rate, 266, 269–270, 270f

Heart valve Valve found between the chambers of the heart or between a chamber and a vessel leaving the heart, 262f, 263

Heat of vaporization Amount of energy needed to turn water into steam, 25

Heat stroke, 93

HeLa cells, 65

Heliobacter pylori, 267

Helminthic therapy, 313

Helper T cell Secretes lymphokines, which stimulate all kinds of immune cells, 304. *See also* T lymphocyte

Hematocrit Percent of the volume of whole blood that is composed of red blood cells, 243

Hematoma, 108

Hematopoiesis Formation of blood cells in red bone marrow, 103, 244, 246f

Heme Iron-containing portion of a hemoglobin molecule, 247f, 248, 256

Hemispheres, cerebral, 173, 175f

Hemodialysis Mechanical way to remove nitrogenous wastes and to regulate blood pH when the kidneys are unable to perform these functions, 372, 385–386, 385f

Hemoglobin Pigment of red blood cells responsible for oxygen transport, 244, 246–247, 247f, 329–330

Hemolysis Lysis of red blood cells that releases hemoglobin, 52, 52f, 248

Hemolytic anemia Insufficient number of red blood cells caused by an increased rate of red blood cell destruction, 249

Hemolytic disease of the newborn Destruction of a fetus's red blood cells by the mother's immune system, caused by differing Rh factors between mother and fetus, 249, 253–254, 254f

Hemophilia Most common of the severe clotting disorders caused by the absence of a blood clotting factor, 251, 453t

Hemorrhage Excessive blood loss from the body, 252

Hemorrhagic anemia Condition in which the blood is deficient in red blood cells, in hemoglobin, or in total volume because of a hemorrhage, 249

Hemorrhagic bleeding Escape of blood from blood vessels, 251

Hemorrhagic stroke, 287

Hemorrhoids Abnormally dilated blood vessels of the rectum, 273

Hemostasis Stoppage of blood flow, 250–251, 250f

Hepatic artery, 355f

Hepatic lobule One of the small masses of tissue of which the liver is made, 354, 355f

Hepatic portal system Portal system that begins at the villi of the small intestine and ends at the liver, 282, 282f, 354

Hepatic portal vein Vein leading to the liver and formed by the merging blood vessels of the small intestine, 260f, 280, 282, 282f, 355f

Hepatic vein, 260f, 280, 281f, 281t, 282, 282f

Hepatitis Inflammation of the liver; often due to a serious infection by any of a number of viruses, 81, 357, 417, 418

Hepatitis A, 310

Hepatitis B, 310

Hering-Brewer reflex, 328

Hernia Protrusion of an organ through an abnormal opening, such as the intestine through the abdominal wall near the scrotum (inguinal hernia) or the stomach through the diaphragm (hiatal hernia), 398

Herniated disk Fibrous ring of cartilage between two vertebrae that has ruptured, 114, 117, 173

Herpes, 416–417, 416f

Heterozygous Different alleles in a gene pair, 450

Hexose Six-carbon sugar, 28

Hiccups, 318

Highly active antiretroviral therapy (HAART), 307–308

Hinge joint Type of joint characterized by a convex surface of one bone fitting into a concave surface of another so that movement is confined to one place, such as in the knee or interphalangeal joint, 124, 125f

Hip, muscles of, 151–153, 152t, 153f, 154f

Hippocampus An important part of the limbic system; involved in forming, storing, and processing memory, 178

Hirsutism Excessive body and facial hair in women, 89–90

Histamine Substance produced by basophil-derived mast cells in connective tissue that causes capillaries to dilate; causes many of the symptoms of allergy, 289f, 298

HLA (human leukocyte-associated) antigen Protein in a plasma membrane that identifies the cell as belonging to a particular individual and acts as an antigen in other organisms, 303

Hodgkin disease Cancer of the lymph glands that is normally localized in the neck region, 297

Homeostasis Constancy of conditions, particularly the environment of body cells: constant temperature, blood pressure, pH, and other body conditions, 10
in body maintenance, 10–13, 11f–15f
cardiovascular system in, 287–289
digestive system in, 361
endocrine system in, 237–239
immune system in, 315
integumentary system in, 97
muscular system in, 154–157, 156t
nervous system in, 187
reproductive system in, 419
respiratory system in, 337
skeletal system in, 127–128
urinary system in, 387–389

Homozygous Having two identical genes on paired homologous chromosomes, 450

Homozygous dominant Possessing two identical alleles, such as *AA*, for a particular trait, 450

Homozygous recessive Possessing two identical alleles, such as *aa*, for a particular trait, 450

Hormone Substance secreted by an endocrine gland that is transmitted in the blood or body fluids, 218–221, 219f, 220t, 221f, 276
in contraception, 410, 411t, 412t
in digestion, 351f
female, 406, 406f
male, 400–401, 400f

Human chorionic gonadotropin (HCG) Hormone produced by the placenta that helps maintain pregnancy and is the basis for the pregnancy test, 407, 431

Human development, 427–436, 427f, 428t, 429f, 430f, 432f–436f

Human immunodeficiency virus (HIV) Virus responsible for AIDS, 307–308, 308f, 313, 416, 418

Human life cycle, 393–395, 394f, 395f

Human papillomavirus (HPV), 310, 416

Humerus Heavy bone that extends from the scapula to the elbow, 109f, 119, 119f

Hunger, 235

Huntington disease Genetic disease marked by progressive deterioration of the nervous system due to deficiency of a neuron-transmitter, 176, 453t

Hyaline cartilage Cartilage composed of very fine collagenous fibers and a matrix of a glassy, white, opaque appearance, 69t, 72, 72f

Hyaline membrane disease, 439

Hydrocephalus Enlargement of the brain due to abnormal accumulation of cerebrospinal fluid, 172

Hydrogen bond Weak attraction between a partially positive hydrogen and a partially negative oxygen or nitrogen some distance away; found in proteins and nucleic acids, 24, 24*f*

Hydrolysis reaction Splitting of a bond by the addition of water, 27*f,* 28

Hydrolytic enzyme Enzyme that catalyzes a reaction in which the substrate is broken down by the addition of water, 358

Hydrophilic Type of molecule that interacts with water by dissolving in water and/or forming hydrogen bonds with water molecules, 25, 45

Hydrophobic Type of molecule that does not interact with water because it is nonpolar, 25, 45

"Hygiene hypothesis," 313

Hymen, 405–406

Hyoid bone, 109*f,* 113, 321*f*

Hypercalcemia, 227

Hyperglycemia Excessive glucose in the blood, 230

Hyperopia Inability to see nearby objects, 205

Hyperpnea Deep and labored breathing, 328

Hypersensitivity reactions, 312

Hypertension Elevated blood pressure, particularly the diastolic pressure, 22, 267, 278, 285, 286

Hyperthermia Abnormally high body temperature, 93

Hypertonic solution Solution that has a higher concentration of solute and a lower concentration of water than the cell, 52*f,* 53

Hypertrophy Increase in the size of an organ, usually by an increase in the size of its cells, 144

Hypocalcemia, 227

Hypodermic needle Slender, hollow instrument for introducing material into or removing material from or below the skin, 88

Hypodermis Mainly composed of fat, this loose layer is directly beneath the dermis; subcutaneous, 87, 87*f,* 88–89

Hypoglycemia Insufficient amount of glucose in the blood, 232

Hypotension, 275, 278

Hypothalamic-inhibiting hormone One of many hormones produced by the hypothalamus that inhibits the secretion of an anterior pituitary hormone, 224

Hypothalamic-releasing hormone One of many hormones produced by the hypothalamus that stimulates the secretion of an anterior pituitary hormone, 224

Hypothalamus Region of the brain; the floor of the third ventricle that helps maintain homeostasis, 176, 219*f,* 220*t,* 222–224, 223*f*–225*f*

Hypothalamus-pituitary portal system, 282

Hypothermia Abnormally low body temperature, 93

Hypothyroidism, congenital, 225, 226*f*

Hypotonic solution Solution that has a lower concentration of solute and a higher concentration of water than the cell, 52, 52*f*

Hypovolemic shock Form of shock caused by excessive blood, tissue fluid, or plasma loss, 252

Hysterectomy Surgical removal of the uterus, 405

H zone In a sarcomere, the region in the center of an A band where only myosin filaments are found, 138

I

I band In a sarcomere, the region that is light colored and attached to the Z line; comprised of only thin filaments, 138

Ileocecal valve, 349, 349*f,* 352*f*

Ileum Lower portion of the small intestine, 349, 349*f,* 352*f,* 353*f*

Iliac crest, 120, 121*f*

Iliacus, 153*f*

Iliopsoas, 152, 152*t,* 153*f*

Ilium One of the bones of a coxal bone or hipbone, 120, 121*f*

Imaging, 16, 16*f*

Immediate allergic response Allergic response that occurs within seconds of contact with an allergen; caused by the attachment of the allergen to IgE antibodies, 312

Immune complex, 301–302

Immune deficiency, 313

Immune response, 305–313, 305*f,* 306*f,* 308*f*–311*f*

Immune system All the cells in the body that protect the body against foreign organisms and substances and also against cancerous cells, 294. *See also* Antibody
  active immunity in, 305
  aging of, 313
  barriers in, 298
  cell-mediated immunity in, 303–305, 303*f,* 304*f*
  complement system in, 300
  in homeostasis, 315
  inflammatory reaction in, 298–300, 299*f*
  lymphatic organs in, 294–298, 296*f,* 297*f*
  lymphatic system in, 294
  nonspecific defenses in, 298–300, 299*f,* 300*t*
  passive immunity in, 305–306, 306*f*
  proteins in, 300
  specific defenses in, 300–305, 300*t,* 301*f*–304*f,* 302*t*

Immunity Resistance to disease-causing organisms, 294

Immunization Strategy for achieving artificial immunity to the effects of specific disease-causing agents, 305, 305*f,* 309–310, 309*f*

Immunoglobulin (Ig) Globular plasma proteins that function as antibodies, 302, 302*t*

Immunoglobulin E-mediated allergic response, 312

Immunosuppressive Inactivating the immune system to prevent organ rejection, usually via a drug, 312

Immunotherapy, 81

Impetigo Contagious skin disease caused by bacteria in which vesicles erupt and crust over, 93

Implantation, 401, 429*f*

Implantation Attachment and penetration of the embryo to the lining (endometrium) of the uterus, 431

Impotence, 399

Incisors, 344*f*

Incontinence Involuntary loss of urine, 387, 440

Incus The middle of three ossicles of the ear; serves with the malleus and the stapes to conduct vibrations from the tympanic membrane to the oval window of the inner ear, 208, 209*f*

Induced pluripotent stem cells (iPSCs), 255

Infant respiratory distress syndrome Condition in newborns, especially premature ones, in which the lungs collapse because of a lack of surfactant lining the alveoli, 324

Infarct An area of tissue death resulting from obstruction of blood flow, 268

Infection
  opportunistic, 307
  sexually transmitted, 416–417, 417*f,* 418
  upper respiratory, 330–331

Inferior Situated below something else; pertaining to the lower surface of a part, 3, 3*f*

Inferior mesenteric artery, 279*t,* 280*f*

Inferior mesenteric vein, 282*f*

Inferior nasal concha, 111*f,* 113, 320*f*

Inferior oblique, 199, 200*f*

Inferior rectus, 199, 199*f,* 200*f*

Inferior vena cava Large vein that enters the right atrium from below and carries blood from the trunk and lower extremities, 260*f,* 262*f,* 264*f,* 279, 281*f,* 281*t,* 284*f*

Infertility Inability to have as many children as desired, 415

Inflammatory reaction Tissue response to injury that is characterized by dilation of blood vessels and accumulation of fluid in the affected region, 298–300, 299*f*

Influenza, 311, 311*f*

Infrahyoid muscles, 148

Infraspinatus, 150*f,* 150*t,* 151*f*

Inguinal hernia A hernia in which part of the intestine protrudes into the inguinal canal, 398

Inguinal lymph nodes, 295*f*

Inheritance
  chromosomal, 446–449, 446*t,* 447*f*–449*f*
  genetic, 450–454, 450*f,* 451*f,* 453*f,* 454*f*
  sex chromosome, 449, 449*f,* 450*f*
  sex-linked, 451, 451*f*

Inhibin Hormone secreted by seminiferous tubules that inhibits the release of follicle-stimulating hormone from the anterior pituitary, 400, 400*f*

Inner cell mass An aggregation of cells at one pole of the blastocyte, which is destined to form the embryo proper, 429

Inner ear Portion of the ear, consisting of a vestibule, semicircular canals, and the cochlea, where balance is maintained and sound is transmitted, 181*t,* 208, 209*f*

Innocence Project, 458

Inorganic molecule Type of molecule that is not an organic molecule; not derived from a living organism, 24

Insertion End of a muscle that is attached to a movable part, 145, 145*f*

Inspiration The act of breathing in; inhalation, 319, 325–326, 326*f*

Inspiratory capacity, 327, 327*f*

Inspiratory reserve volume Volume of air that can be forcibly inhaled after normal inhalation, 327, 327*f*

Insulin Hormone produced by the pancreas that regulates glucose storage in the liver and glucose uptake by cells, 218, 219*f,* 220*t,* 229, 231*f*

Insulin-dependent diabetes mellitus (IDDM) Type 1 diabetes mellitus characterized by abrupt onset of symptoms, dependence on exogenous insulin, and a tendency to develop ketoacidosis, 231, 233, 237

Insulin shock Severe hypoglycemia caused by excessive insulin; left untreated, can result in convulsions and progressive development of coma, 232

Integration Summing up of excitatory and inhibitory signals by a neuron or by some part of the brain, 163

Integument Pertaining to the skin, 87

Integumentary system Pertaining to the skin and accessory organs, 8, 9, 12, 14*f,* 87–89, 87*f. See also* Skin

Interatrial septum Wall between the atria of the heart, 261

Intercalated disk Membranous boundary between adjacent cardiac muscle cells, 76

Intercostal arteries, 279*t*, 280*f*

Intercostal muscles, 409*f*

Intercostal nerves, 182*t*

Interferon Protein formed by a cell infected with a virus that can increase the resistance of other cells to the virus, 300

Interleukin Class of immune system chemicals (cytokines) having varied effects on the body, 305

Intermediate filaments Tough protein strands that form cell-to-cell junctions, 44*f*, 49

Intermediate-twitch fibers, 144

Internal carotid artery, 283*f*

Internal iliac artery, 260*f*, 279, 280*f*, 284*f*

Internal iliac vein, 260*f*, 280, 281*f*

Internal intercostals, 149, 149*f*, 149*t*

Internal jugular vein, 279, 281*f*

Internal obliques, 149, 149*f*, 149*t*

Internal respiration Exchange of oxygen and carbon dioxide between blood and tissue fluid, 329

Interneuron Neuron found within the central nervous system that takes nerve impulses from one portion of the system to another, 164*f*, 165

Interphase General term for that period of the cell cycle in which cell division is not occurring, 55–58, 56*f*, 57*f*

Interstitial cell Hormone-secreting cell located between the seminiferous tubules of the testes, 396, 397*f*

Intertubercular groove, of humerus, 119, 119*f*

Interventricular septum, 261

Intervertebral disk Layer of cartilage located between adjacent vertebrae, 113, 114, 114*f*, 173

Intervertebral foramina, 115

Intestinal phase The time period of gastric juice secretion that is stimulated by the presence of food in the intestine and by intestinal stretching, 350

Intestinal polyps, 356–357

Intracellular fluid, 380, 380*f*

Intracranial hemorrhage, 439

Intracytoplasmic sperm injection (ICSI), 415

Intradiscal electrothermal therapy (IDET), 117

Intramembranous ossification Bone that forms from membranelike layers of primitive connective tissue, 105, 105*f*

Intraoperative blood salvage, 255

Intrapleural pressure Pressure found in the space between the parietal pleura and the visceral pleura in the thoracic cavity, 325

Intrauterine device Solid object placed in the uterine cavity for purposes of contraception; IUD, 410, 412*t*

Intrauterine insemination (IUI) Process of achieving pregnancy in which donated sperm are deposited in the uterus, 415

Intrinsic factor Protein produced by the normal gastrointestinal mucosa that facilitates absorption of vitamin B$_{12}$, 348

Intrinsic mechanism, 251

Inversion Movement of the foot so that the sole is turned inward, 125, 126*f*

In vitro fertilization (IVF) Process of achieving pregnancy in which eggs retrieved from an ovary are fertilized in a laboratory; viable

embryos are then placed into the woman's uterus, 415

Iodine, 363*t*

Ion A charged atom, 21

Ionic bond Chemical attraction between a positive ion and a negative ion, 22, 23*f*

Ionic lattice, 22

Ionizing radiation, 38

Ipsilateral Structures located on the same side of the body, 3*f*, 4

Iris Muscular ring that surrounds the pupil and regulates the passage of light through this opening, 199*f*, 200, 200*f*, 201*t*

Iron, 363*t*

Iron deficiency anemia Abnormally low amount of red blood cells or hemoglobin, due to a lack of iron in the diet, 249, 254

Ischemic heart disease Insufficient oxygen delivery to the heart, usually caused by partially blocked coronary arteries, 268

Ischial spine Projection of the coxal bone into the pelvic cavity, 120

Ischial tuberosity, 120, 121*f*

Ischium Most inferior of the three bones that comprise the pelvic bone, 120, 121*f*

Islets of Langerhans, 229, 233, 233*f*

Isotonic solution Solution that contains the same concentration of solutes and water as does the cell, 52, 52*f*

Isotope One of two or more atoms with the same atomic number that differs in the number of neutrons and, therefore, in weight, 22

IUI. *See* Intrauterine insemination (IUI)

IVF. *See* In vitro fertilization (IVF)

J

Jacobs syndrome Abnormal condition in a male characterized by a single X chromosome and two Y chromosomes, 446*t*, 449

Jaundice Yellowish tint to the skin caused by an abnormal amount of bilirubin in the blood, indicating liver malfunction, 357, 439

Jejunum Middle portion of the small intestine, 349, 349*f*

Joint Union of two or more bones; an articulation, 123–127, 124*f*–127*f*

cartilaginous, 123

damage, 126

fibrous, 123

repair, 126

replacement, 126, 127*f*

synovial, 102, 123–125, 124*f*, 125*f*

Joint capsule, 124*f*

Joint cavity, 124*f*

Jugular Any of four veins that drain blood from the head and neck, 260*f*

Juxtaglomerular apparatus Structure located in the walls of arterioles near the glomerulus that regulates renal blood flow, 382, 383*f*

K

Kaposi's sarcoma, 94

Karyotype Arrangement of all the chromosomes from a nucleus by pairs in a fixed order, 446–447, 447*f*

Keratin Insoluble protein present in the epidermis and in epidermal derivatives, such as hair and nails, 87*f*, 88

Keratinocytes, 87*f*, 88

Ketonuria Abnormal presence of acidic molecules called ketones in the urine, 230

Kidney Organ in the urinary system that forms, concentrates, and excretes urine, 373–374, 374*f*, 375–386, 376*f*–378*f*, 379*t*, 380*f*, 380*t*, 381*f*, 383*f*–385*f*, 384*t*. *See also* Urinary system

Kidney stones, 385, 387

Kidney transplantation, 386

Kinocilium, 211*f*, 213

Klinefelter syndrome Condition caused by the inheritance of XXY chromosomes, 446*t*, 449

Knee-jerk reflex Automatic, involuntary response initiated by tapping the ligaments just below the patella (kneecap), 183

Knee joint, 124*f*

Knuckle cracking, 102

Korotkoff sounds, 277

Krause end bulbs, 195, 195*f*

Kyphoplasty, 117

Kyphosis Increased roundness in the thoracic curvature of the spine; also called *hunchback*, 114, 114*f*

L

Labels, nutrition, 364, 364*f*

Labia majora Two large, hairy folds of skin of the female external genitalia, 401*f*, 405, 406*f*

Labia minora Two small folds of skin inside the labia majora and encircling the clitoris, 401*f*, 405, 406*f*

Lacrimal apparatus Structures that provide tears to wash the eye, consisting of the lacrimal gland and the lacrimal sac with its ducts, 199

Lacrimal bone, 111*f*, 113

Lacrimal gland, 199*f*, 320, 320*f*

Lacrimal sac, 199*f*

Lactase, 360*t*

Lactate, 142

Lactation Production and secretion of milk by the mammary glands, 407–409, 409*f*

Lacteal Lymph vessel in a villus of the wall of the small intestine, 294

Lactose, 28

Lacuna Small pit or hollow cavity, as in bone or cartilage, where a cell or cells are located (pl., *lacunae*), 72, 72*f*

Lambdoidal suture Line of junction between the occipital and parietal bones, 110*f*, 111*f*, 112*f*

Lamellae, 103, 104*f*

Laminae, 115, 115*f*

Laminectomy, 117

Langerhans cells Specialized epidermal cells that assist the immune system, 87*f*, 88

Lanugo Short, fine hair that is present during the later portion of fetal development, 434

Large intestine Portion of the digestive tract that extends from the small intestine to the anus, 343*f*, 346*t*, 349*f*, 351–353, 352*f*, 396*f*

Large intestine disorders, 356–357

Laryngitis Inflammation of the larynx, 331

Laryngopharynx Lower portion of the pharynx near the opening to the larynx, 320*f*, 346

Larynx Structure that contains the vocal cords; also known as the voice box, 319*f*, 320*t*, 321–322, 321*f*, 345*f*

Lateral Pertaining to the side, 3*f*, 4

Lateral condyle, of tibia, 121, 122*f*

Lateral epicondyle, of femur, 121

Lateral malleolus Rounded protuberance on the lateral surface of the ankle joint, 122, 122*f*

Lateral rectus, 199, 200*f*

Lateral sulcus, 175*f*

Lateral ventricle, 174*f*

Latissimus dorsi, 146*f*, 150*f*, 150*t*, 151, 151*f*

Left ventricular assist device (LVAD), 286

Lens Clear, membranelike structure that is found in the eye behind the iris and that brings objects into focus, 199*f*, 200*f*, 201, 201*t*, 202–203, 202*f*

Leptin Peptide hormone produced by fat cells; plays a role in body weight regulation by acting on the hypothalamus to suppress appetite and burn fat stored in adipose tissue, 235, 351

Lesser omentum, 353, 353*f*

Lesser trochanter, of femur, 121, 122*f*

Lesser tubercle, of humerus, 119, 119*f*

Leukemia Form of cancer characterized by uncontrolled production of leukocytes in red bone marrow, 79, 249, 254

Leukocytes Several types of colorless, nucleated blood cells that, among other functions, resist infection; white blood cells, 74, 74*f*, 234, 245*f*, 248–250, 248*f*

Leukocytosis Abnormally large increase in the number of white blood cells, 249

Leukopenia Abnormally low number of leukocytes in the blood, 249

Levator anguli oris, 147, 147*f*, 148*t*

Levator labii superioris, 147, 147*f*, 148*t*

Levator palpebrae superioris, 199, 199*f*, 200*f*

Leydig cells, 396

LH. *See* Luteinizing hormone (LH)

Life cycle, human, 393–395, 394*f*, 395*f*

Ligament Strong connective tissue that joins bone to bone, 70, 71*f*, 123, 124*f*

Ligamentum arteriosum, 284*f*

Ligamentum teres, 284*f*

Ligamentum venosum, 284*f*

Limb bud, 433*f*

Limbic system System that involves many different centers of the brain and that is concerned with visceral functioning and emotional responses, 178

Linea alba, 149*f*

Linea aspera, 121, 122*f*

Lingual tonsils, 297

Lipase Enzyme secreted by the pancreas that digests or breaks down fats, 354, 359, 359*f*

Lipid Group of organic compounds that are insoluble in water—notably, fats, oils, and steroids, 29–30, 30*f*

Lipoproteins Any of a large class of proteins composed of protein and lipid, 244

Liver Largest organ in the body, located in the abdominal cavity below the diaphragm; performs many vital functions that maintain homeostasis of blood, 343*f*, 353*f*, 354–355, 354*f*, 355*f*, 357, 358, 360

Local disease Disease that is confined to a particular area of the body, 17

Locus The position in a chromosome of a particular gene or allele, 450

Loop of Henle Portion of a nephron between the proximal and distal convoluted tubules; functions in water reabsorption, 376, 376*f*, 378*f*, 380–381, 381*f*

Loose connective tissue Tissue that is composed mainly of fibroblasts separated by collagenous and elastin fibers and that is found beneath epithelium, 69*t*, 70, 70*f*

Lorcaserin, 365

Lordosis Exaggerated lumbar curvature of the spine; also called "swayback.," 114, 114*f*

Love, 392

Lower gastroesophageal sphincter, 348*f*

Lower limb, 121–123, 122*f*, 151–153, 152*t*, 153*f*, 154*f*

Lower respiratory infection, 331–332

Low sperm count, 415

Lumbar nerves, 183*f*

Lumen Space within a tubular structure such as a blood vessel or intestine, 346

Lung Internal respiratory organ containing moist surfaces for gas exchange, 260*f*, 262, 319*f*, 320*t*, 323–325, 324*f*

Lung cancer Cancer neoplasm found in the airways or lobes of the lungs, 333, 335–336, 335*f*

Lung capacity, 327, 327*f*

Lung collapse, 323

Lunula Pale, half-moon–shaped area at the base of nails, 90

Luteinizing hormone (LH) Hormone produced by the anterior pituitary that stimulates the development of the corpus luteum in females and the production of testosterone in males, 219*f*, 220*t*, 223*f*, 224, 400, 400*f*, 406, 406*f*, 407

Lymph Fluid having the same composition as tissue fluid; carried in lymph vessels, 294, 380*f*

Lymphadenitis Infection of the lymph nodes, 297

Lymphangitis Infection of the lymphatic vessels, 297

Lymphatic nodules, 297–298

Lymphatic organ Organ other than a lymphatic vessel that is part of the lymphatic system; includes lymph nodes, tonsils, spleen, thymus gland, and bone marrow, 294–298, 295*f*, 296*f*

Lymphatic system Vascular system that takes up excess tissue fluid and transports it to the bloodstream, 10, 13, 15*f*, 97, 294–298, 296*f*, 297, 297*f*

Lymphatic tissue Reticular connective tissue found in lymphatic organs such as lymph nodes and spleen, 70

Lymphatic vessel Vessel that carries lymph, 294

Lymphedema Edema due to faulty lymphatic drainage, 297

Lymph node Mass of lymphatic tissue located along the course of a lymphatic vessel, 295*f*, 297

Lymphoblasts, 246*f*

Lymphocyte Type of white blood cell characterized by agranular cytoplasm; lymphocytes usually constitute 20–25% of the white cell count, 234, 245*f*, 248, 296*f*

Lymphoma Cancer of lymphatic tissue (reticular connective tissue), 79, 297

Lysis To split or divide, 52

Lysosome Organelle involved in intracellular digestion; contains powerful digestive enzymes, 43*t*, 44*f*, 47, 48*f*

Lysozyme, 319

M

Macromolecule Large molecule composed of smaller molecules, 2, 2*f*, 27*f*

Macrophage Enlarged monocyte that ingests foreign material and cellular debris, 267, 298

Macular degeneration Disruption of the macula lutea, a central part of the retina, causing blurred vision, 206, 206*f*

Mad cow disease, 34

Magnesium, 363*t*

Magnetic resonance imaging (MRI), 16, 16*f*

Major histocompatibility complex A group of genes that function especially in determining the tissue compatibility antigen molecules found on cell surfaces; abbreviated *MHC*, 303

Male external genitalia, 399, 399*f*

Male reproductive system, 395–401, 396*f*–400*f*

Male sex organ development, 435, 436*f*

Male sexual response, 399–400

Malignant The power to threaten life; cancerous, 79

Malleolus Rounded projection from a bone, 107*t*

Malleus First of three ossicles of the ear; serves with the incus and stapes to conduct vibrations from the tympanic membrane to the oval window of the inner ear, 209*f*

Maltase Enzyme that catalyzes conversion of maltose into glucose, 360*t*

Maltose, 28

Mammary ducts, 409*f*

Mammary gland Milk-secreting gland that develops within the breast in pregnancy and lactation; only minimally developed in the breast of a nonpregnant or nonlactating woman, 92

Mandible, 109*f*, 111*f*, 113

Mandibular angle, 111*f*

Mandibular condyle, 111*f*, 113

Mandibular fossa, 110, 112*f*

Manganese, 363*t*

Manubrium, 116–117, 116*f*

Masseter, 146*f*, 147, 147*f*, 148*t*

Mass number, 21

Mast cell Cell to which antibodies, formed in response to allergens, attach, bursting the cell and releasing allergy mediators, which cause symptoms, 298

Mastication, 147

Mastoiditis Inflammation of the mastoid sinuses of the skull, 110

Mastoid process, 111*f*, 112*f*, 113

Maternal spindle transfer, 457

Matrix Secreted basic material or medium of biological structures, such as the matrix of cartilage or bone, 69

Maxilla, 109*f*, 110*f*, 111*f*, 113

Mean arterial blood pressure (MABP), 274–275

Measles, 309–310

Meatus, 107*t*

Mechanoreceptor A tactile or stretch receptor that responds to a mechanical stimulus such as touch or stretch, 194, 210*f*

Medial Toward or near the midline, 3, 3*f*

Medial condyle, of tibia, 121, 122*f*

Medial epicondyle, of femur, 121

Medial malleolus, of tibia, 122, 122*f*

Medial rectus, 199, 200*f*

Median cubital vein, 281*f*

Median nerve, 182*t*

Mediastinum Tissue mass located between the lungs, 6*f*, 7

Medulla oblongata Lowest portion of the brain; concerned with the control of internal organs, 174*f*, 179

Medullary cavity Within the diaphysis of a long bone, cavity occupied by yellow marrow, 103, 104*f*

Megakaryoblasts, 246*f*

Megakaryocyte Large bone marrow cell that gives rise to blood platelets, 74, 74*f*, 246*f*, 250

Meiosis Type of cell division in which the daughter cells have 23 chromosomes; occurs during spermatogenesis and oogenesis, 60, 393–395, 394*f*, 397*f*, 403*f*, 448*f*

Meissner corpuscles, 195*f*

Melanin Pigment found in the skin and hair of humans that is responsible for coloration, 87*f*, 88

Melanocyte Melanin-producing cell, 87*f*, 88

Melanocyte-stimulating hormone (MSH) Substance that causes melanocytes to secrete melanin in lower vertebrates, 220*t*

Melanoma Deadly form of skin cancer that begins in the melanocytes, pigment cells present in the epidermis, 93–94, 94*f*

Melatonin Hormone, secreted by the pineal gland, that is involved in biorhythms, 219*f*, 220*t*, 234, 235*f*

Membranes
 body, 6–8, 6*f*–7*f*
 cutaneous, 82, 87
 extraembryonic, 427–428, 427*f*
 mucous, 80
 periodontal, 344*f*
 placental, 432
 respiratory, 324–325, 324*f*
 serous, 80
 synovial, 82, 123, 124*f*
Membranous labyrinth, 208

Memory B cell Cells derived from B lymphocytes that remain within the body for some time and account for active immunity, 301, 302*f*

Memory T cells, 304–305

Meninges Protective membranous coverings around the brain and spinal cord (sing., *meninx*), 6, 6*f*, 7*t*, 82, 171–172, 171*f*, 172*f*, 174*f*

Meningitis Inflammation of the meninges around the brain and spinal cord usually caused by bacterial infection, 9, 9*f*

Meniscus Piece of fibrocartilage that separates the surfaces of bones in the knee (pl., *menisci*), 123, 124*f*

Menopause Termination of the menstrual cycle in older women, 407, 419

Menses (menstruation) Loss of blood and tissue from the uterus, 407

Menstrual cycle Female reproductive cycle characterized by regularly occurring changes in the uterine lining, 407

Menstruation, 236, 392

Merkel cells Cells that signal the brain that an object has touched the skin; also called *tactile cells*, 87*f*, 88

Merkel disks, 194, 195*f*

Mesentery Fold of peritoneal membrane that attaches an abdominal organ to the abdominal wall, 80, 352–353, 353*f*

Mesocolon, 353*f*

Mesoderm The middle of the three primary germ layers of an embryo; source of bone, muscle, connective tissue, and dermis, 431, 432*f*

Messenger RNA (mRNA) Nucleic acid (ribonucleic acid) complementary to genetic DNA; has codons that direct cell protein synthesis at the ribosomes, 46, 56, 57*f*, 58, 221*f*, 458

Metabolic waste, 373

Metabolism All of the chemical changes that occur within cells, 32

Metacarpal Bone of the hand between the wrist and the finger bones, 109*f*, 120*f*

Metaphase Stage in mitosis when chromosomes align in the center of the cell, 56*f*, 58, 59*f*, 60*f*, 394*f*, 395

Metastasis Mechanism of cancer spread in which cancer cells break off from the initial tumor, enter the blood vessels or lymphatic vessels, and start new tumors elsewhere in the body, 79, 335

Metatarsal bones Bones found in the foot between the ankle and the toes, 122*f*, 123

Microglia, 77

Microtubule Hollow rod of the protein tubulin in the cytoplasm, 44*f*, 49

Microvillus Cylindrical process that extends from some epithelial cell membranes and increases the membrane surface area (pl., *microvilli*), 49–50, 68, 78*f*, 197*f*, 349

Micturition Emptying of the bladder; urination, 375, 375*f*

Midbrain Small region of the brain stem located between the forebrain and the hindbrain; contains tracts that conduct impulses to and from the higher parts of the brain, 174*f*, 179

Middle ear Portion of the ear consisting of the tympanic membrane, the oval and round windows, and the ossicles, where sound is amplified, 209*f*

Middle nasal concha, 111*f*, 113, 320*f*

Mifepristone, 415

Mineral Inorganic substance; certain minerals must be in the diet for normal metabolic functioning of cells, 363–365, 363*t*, 364, 364*f*

Mineralocorticoid Hormones the adrenal cortex secretes that influence the concentrations of electrolytes in body fluids, 219*f*, 220*t*, 227, 228

Mitochondrial donation, 457

Mitochondrion Organelle in which cellular respiration produces the energy molecule ATP, 43*t*, 44*f*, 49, 49*f*

Mitosis Type of cell division in which two daughter cells receive 46 chromosomes; occurs during growth and repair, 58–60, 59*f*, 60*f*, 393*f*, 394*f*, 397*f*

Mitotic spindle, 58

Mitotic stage, 55

Mitral valve Valve in the heart that controls blood flow between the left atrium and the left ventricle; prevents the blood in the ventricle from returning to the atrium (also called *bicuspid valve*), 263

Mixed nerve Nerve that contains both the long dendrites of sensory neurons and the long axons of motor neurons, 180, 181

Molar (tooth) Three posterior pairs of teeth in each human jaw with a rounded or flattened surface adapted for grinding; found behind the incisors and canines, 344, 344*f*

Mole Raised growth on the skin due to an overgrowth of melanocytes, 22

Molecule Smallest quantity of a substance that retains its chemical properties, 2, 2*f*
 in basic chemistry, 22–23, 23*f*
 inorganic, 24
 in levels of organization, 2, 2*f*
 of life, 27–28, 27*f*
 organic, 24
 polar, 24

Molybdenum, 363*t*

Monoblasts, 246*f*

Monoclonal antibody Antibody of one type that is produced by cells derived from a lymphocyte that has fused with a cancer cell, 306, 306*f*, 312

Monocyte Type of white blood cell that functions as a phagocyte, 245*f*, 246*f*, 250, 296*f*

Monohydrogen phosphate, 382

Mononucleosis Viral disease characterized by an increase in atypical lymphocytes in the blood, 249

Monosaccharide Simple sugar; a carbohydrate that cannot be decomposed by hydrolysis, 28

Monosomy Having one less than the diploid number of chromosomes, 447

Mons pubis The rounded fleshy prominence over the pubic symphysis, 405, 406*f*

Morning-after pill, 411*t*, 414–415

Morphogenesis Establishment of shape and structure in an organism, 427

Morula Early stage in development in which the embryo consists of a mass of cells, often spherical, 429, 429*f*

Motor neuron Neuron that takes nerve impulses from the central nervous system to an effector; also known as an efferent neuron, 138, 164*f*, 165

Motor speech area, 175*f*, 176

Motor unit Motor neuron and all the muscle fibers it innervates, 138, 143, 144

Mouth Opening through which food enters the body, 343*f*, 344

MS. *See* Multiple sclerosis (MS)

Mucociliary escalator Protective mechanism involving cilia of the mucous membranes of the respiratory system; transports mucus containing inhaled pollutants to the pharynx where the mucus can be swallowed or expectorated, 320, 322, 322*f*

Mucosa A membrane rich in mucous glands that lines body passages and cavities that communicate directly or indirectly with the exterior, such as the digestive, respiratory, and genitourinary tracts, 346, 348*f*

Mucous membrane Membrane lining a cavity or tube that opens to the outside of the body; also called mucosa, 80

Müllerian ducts, 435

Multiple sclerosis (MS) Disease in which the outer, myelin layer of nerve fiber insulation becomes scarred, interfering with normal conduction of nerve impulses, 312

Multipotent stem cell An unspecialized parent cell that gives rise to differentiated cells, 244, 246*f*

Muscle compartment, 135

Muscle fiber Muscle cell, 134, 134*f*, 136–138, 136*t*, 137*f*

Muscle spindles, 194, 195*f*

Muscle tone, 132–144, 194

Muscle twitch Contraction of a whole muscle in response to a single stimulus, 142

Muscular dystrophy Progressive muscle weakness and atrophy caused by deficient dystrophin protein, 157, 453*t*

Muscularis The smooth muscular layer of the wall of contractile organs, such as the digestive organs and the bladder, 346, 348*f*

Muscular system, 10. *See also* Skeletal muscle

Muscular tissue Major type of tissue that is adapted to contract; the three kinds of muscle are cardiac, smooth, and skeletal, 74–76, 74*t*, 75*f*

Musculocutaneous nerve, 182*t*

Mutation Alteration of the DNA in a tissue cell, 79

Myalgia Pain in a muscle or muscles, 157

Myasthenia gravis Muscle weakness due to an inability to respond to the neurotransmitter acetylcholine, 157, 199, 312

Myelin sheath Fatty plasma membranes of Schwann cells that cover long neuron fibers and give them a white, glistening appearance, 77–78, 164, 164*f*, 168*f*, 182*f*

Myeloblasts, 246*f*

Myocardial infarction Damage to the myocardium due to blocked circulation in the coronary arteries; a heart attack, 251, 268, 286

Myocardium Heart (cardiac) muscle consisting of striated muscle cells that interlock, 261*f*

Myofibril Contractile portion of muscle fibers, 136, 136*t*, 137–138, 137*f*

Myofilament, 136*t*, 137*f*, 138

Myoglobin Pigmented compound in muscle tissue that stores oxygen, 136, 136*t*

Myography, 143*f*

Myometrium, 433*f*

Myopia Inability to see distant objects clearly, 205

Myosin Thick myofilament in myofibrils that is made of protein and is capable of breaking down ATP, 31, 137. *See also* Actin

Myosin filaments, 138–139, 140*f*

Myxedema Condition resulting from a deficiency of thyroid hormone in an adult, 225

N

Nail body, 90, 91*f*

Nail root, 90

Nails, 90

Nasal bone, 110*f*, 111*f*, 113

Nasal cavity Space within the nose, 198*f*, 319*f*, 320*t*

Nasal concha, 111*f*, 113, 320, 320*f*

Nasolacrimal duct, 199*f*

Nasopharynx Portion of the pharynx associated with the nasal cavity, 345*f*, 346

Natriuresis, 276

Natural Family Planning, 410, 411*t*

Natural killer cell (NK) Lymphocyte that causes an infected or cancerous cell to burst, 245*f*, 248, 300

Neck, muscles in, 147–148, 147*f*, 148*t*

Necrotizing fasciitis, 76

Negative feedback Mechanism that is activated by a surplus imbalance and acts to correct it by stopping the process that brought about the surplus, 11–12, 11*f*, 12*f*, 400

Neoplasm New tissue form that may be harmless, or a precursor to cancer, 79. *See also* Cancer

Nephron Anatomical and functional unit of the kidney; kidney tubule, 375–377, 376*f*, 377*f*

Nerve Bundle of long nerve fibers that run to and/or from the central nervous system, 77, 77*f*, 164

Nerve deafness Hearing impairment that usually occurs when the cilia on the sensory receptors within the cochlea have worn away, 212

Nerve fiber Thin process of a neuron (i.e., axon, dendrite), 77

Nerve impulse Change in polarity that flows along the membrane of a nerve fiber, 77

Nerve signal conduction, 165–167, 166*f*

Nervous system, 10. *See also* Central nervous system (CNS); Peripheral nervous system (PNS)
    aging of, 186–187
    brain stem in, 174*f*, 179–180
    in breathing, 328, 328*f*
    cerebellum in, 174*f*, 178–179
    cerebrospinal fluid in, 171–172, 171*f*, 172*f*
    cerebrum in, 173–176, 174*f*–176*f*
    divisions of, 163–164, 163*f*

    enteric, 350
    in homeostasis, 12, 14*f*, 187
    limbic system in, 178
    meninges in, 171–172, 171*f*, 172*f*
    parasympathetic, 184–186, 185*f*, 186*f*, 186*t*
    spinal cord in, 172–173, 172*f*
    sympathetic, 184, 185*f*, 186*f*, 186*t*

Nervous tissue Tissue of the nervous system, consisting significantly of neurons and neuroglia, 77–78, 77*f*, 164–165, 164*f*

Neural folds, 431

Neural groove, 431

Neural plate, 431

Neural tube, 431

Neural tube defect, 431

Neurofibromatosis, 453*t*

Neuroglia Nonconducting nerve cells that are intimately associated with neurons and function in a supportive capacity, 77–78, 164

Neurolemmocyte Type of neuroglial cell that forms a myelin sheath around axons; also called a Schwann cell, 164

Neuromuscular diseases, 157

Neuromuscular junction Junction between a neuron and a muscle fiber, 138, 139*f*

Neuron Nerve cell that characteristically has three parts: dendrite, cell body, and axon, 77, 77*f*, 164–165

Neurotransmitter Chemical made at the ends of axons that is responsible for transmission across a synapse, 138, 139*f*, 167–168, 168*f*, 169*f*

Neutral fat A triglyceride, 30

Neutron Electrically neutral particle in an atomic nucleus, 21

Neutrophil Phagocytic white blood cell that normally constitutes 60–70% of the white blood cell count, 245*f*, 246*f*, 248, 296*f*

Niacin, 362*t*

Nipple, 409*f*

Nociceptor Pain receptor, 196

Nodal tissue, 264–265, 265*f*

Node of Ranvier Gap in the myelin sheath of a nerve fiber, 77, 164, 164*f*, 168*f*

Nondisjunction Failure of the chromosomes (or chromatids) to separate during meiosis, 447–449, 448*f*

Noninsulin-dependent diabetes mellitus NIDDM Type 2 diabetes mellitus, usually characterized by gradual onset with minimal or no symptoms of metabolic disturbance and no required exogenous insulin to prevent ketonuria and ketoacidosis, 231

Nonpolar covalent bond Bond created when electrons shared by adjacent atoms are shared equally, 24

Norepinephrine Hormone secreted by the adrenal medulla to help initiate the "fight-or-flight" reaction, 167, 184, 219*f*, 220*t*, 227

Nose Specialized structure on the face that serves as the sense organ of smell and as part of the respiratory system, 320–321, 320*f*

Nostril One of the external orifices of the nose, 320, 320*f*

Notochord, 432*f*

Nuclear envelope Membrane surrounding the cell nucleus and separating it from the cytoplasm, 44*f*, 46, 46*f*

Nuclear pore Opening in the nuclear envelope, 47

Nuclease, 360*t*

Nucleic acid Large organic molecule found in the nucleus (DNA and RNA) and in the cytoplasm (RNA), 34–37, 35*f*–37*f*, 35*t*

Nucleolus Organelle found inside the nucleus and composed largely of RNA for ribosome formation (pl., *nucleoli*), 43, 43*t*, 44*f*, 46*f*

Nucleosidase, 360*t*

Nucleotide Building block of a nucleic acid molecule, consisting of a sugar, a nitrogen-containing base, and a phosphate group, 34–35

Nucleus Large organelle that contains the chromosomes and acts as a cell control center, 43, 43*t*, 44*f*, 46–47, 46*f*

Nutrient absorption, 349–350

Nutrition, 361–366, 361*f*, 362*t*–363*t*, 364*f*, 366*f*

Nutrition labels, 364, 364*f*

O

Obesity Excess adipose tissue; exceeding desirable weight by more than 20%, 285, 365, 367, 415

Obstructive sleep apnea Sleep pattern characterized by interruption of breathing; caused by obstruction of the upper airway by weakened or excessive tissues of the pharynx, 321

Obturator foramen, 120, 121*f*

Occipital bone, 109*f*, 110, 110*f*, 111*f*, 112*f*

Occipital condyle One of two processes on the lateral portions of the occipital bone; for articulation with the atlas, 110, 110*f*, 112*f*

Occipitalis, 146*f*

Occipital lobe Area of the cerebrum responsible for vision, visual images, and other sensory experiences, 175*f*, 207*f*

Occluded coronary arteries Blocked blood vessels that serve the needs of the heart, 268

Oculomotor nerve, 200

Odontoid process, 115–116, 115*f*

Oil Substance, usually of plant origin and liquid at room temperature, formed when a glycerol molecule reacts with three fatty acid molecules, 29–30, 30*f*

Olecranon fossa, 119, 119*f*

Olecranon process, 119, 119*f*

Olfactory bulb, 198*f*

Olfactory cell Cell located high in the nasal cavity that bears receptor sites on cilia for various chemicals and whose stimulation results in smell, 197, 198*f*

Olfactory epithelium, 198*f*

Oligodendrocytes, 77, 164

Oocyte Developing female gamete, 402–403, 402*f*, 403*f*

Oogenesis Production of eggs in females by the process of meiosis and maturation, 402*f*, 403, 403*f*, 406*f*, 448*f*

Oophorectomy Surgical removal of one or both ovaries, 405

Opportunistic infection Disease that arises in the presence of a severely impaired immune system, 307

Optic chiasma Structure found on the inferior surface of the brain where optic nerves meet and cross, 204, 207*f*, 223*f*

Optic disk, 200*f*, 201

Optic nerve Nerve composed of the ganglion cell fibers that form the innermost layer of the retina, 199*f*, 200*f*, 201, 201*t*, 207*f*

Optic radiations, 207*f*

Optic tracts, 204, 207*f*

Optic vesicle, 433*f*

Oral cavity, 346*t*

Oral contraceptives (OCs), 410

Orbicularis oculi, 146, 146*f*, 147, 148*t*, 199, 199*f*, 200*f*

Orbicularis oris, 146*f*, 147, 147*f*, 148*t*

Organ Structure consisting of a group of tissues that perform a specialized function; a component of an organ system, 2

Organelle Part of a cell that performs a specialized function, 2, 2*f*
in cellular structure, 43, 43*t*
in levels of organization, 2, 2*f*

Organic molecule Carbon-containing molecule, 24

Organism Individual living thing, 2, 2*f*

Organization levels, 2, 2*f*

Organ of Corti. *See* Spiral organ

Organ system Group of related organs working together, 2, 2*f*, 8–10

Organ transplantation Replacement of a diseased or defective organ with a healthy one, 386

Orgasm Physical and emotional climax during sexual intercourse; results in ejaculation in the male, 399–400, 406

Origin End of a muscle that is attached to a relatively immovable part, 145, 145*f*

Orlistat, 365

Oropharynx Portion of the pharynx in the posterior part of the mouth, 320*f*, 346

Orthostatic hypotension, 278

Osmosis Movement of water from an area of greater concentration to an area of lesser concentration across the plasma membrane, 52, 52*f*, 53*f*

Osmotic pressure The amount of pressure needed to stop osmosis; the potential pressure of a solution caused by nondiffusible solute particles in the solution, 52, 244, 272

Ossicles Tiny bones in the middle ear; malleus (hammer), incus (anvil), and stapes (stirrup), 208, 209*f*, 210*f*, 213*t*

Ossification Formation of bone, 105, 105*f*

Osteoarthritis Disintegration of the cartilage between bones at a synovial joint, 126

Osteoblast Bone-forming cell, 105

Osteoclast Cell that causes the erosion of bone, 105

Osteocyte Mature bone cell, 105

Osteon Cylinder-shaped units that comprise compact bone, 103, 104*f*

Osteoporosis Weakening of bones due to decreased bone mass, 106, 107

Osteoprogenitor cells Cells found on or near all of the free surfaces of bone, which undergo division and transform into osteoblasts, 105

Otitis media Inflammation of the middle ear, 331

Otolith Granule that lies above, and whose movement stimulates, ciliated cells in the utricle and saccule, 211*f*

Otolithic membrane, 211*f*, 213

Otosclerosis Overgrowth of bone that causes the stapes to adhere to the oval window, resulting in conductive deafness, 214

Ototoxic Damaging to any of the elements of hearing or balance, 212

Outer ear Portion of the ear consisting of the pinna and the auditory canal, 209*f*

Oval window Membrane-covered opening between the stapes and the inner ear, 208, 209*f*, 210*f*

Ovarian cancer Cancer of an ovary, 405

Ovarian cycle, 408*f*

Ovarian ligament, 404*f*

Ovariohysterectomy Surgical removal of the ovaries and uterus, 405

Ovary Female gonad; the organ that produces eggs, estrogen, and progesterone, 219*f*, 220*t*, 232, 234, 393, 401*f*, 402–403, 402*f*, 403*f*, 404*f*, 429*f*

Ovulation Discharge of a mature egg from the follicle within the ovary, 402*f*, 403, 407, 408*f*, 428*t*

Ovum (pleural, ova) Female gamete, contributing most of the cytoplasm of the zygote, 426, 426*f*, 429*f*

Oxygen debt Amount of oxygen required after anaerobic exercise to replace that used during the exercise, 142

Oxygen deficit Amount of oxygen needed to metabolize the lactic acid that accumulates during vigorous exercise, 142

Oxygen transport, 329–330, 329*f*

Oxyhemoglobin Hemoglobin bound to oxygen in a loose, reversible way, 329–330

Oxytocin Hormone released by the posterior pituitary that causes contraction of uterus and milk letdown, 219*f*, 220*t*, 222, 223*f*

P

Pacemaker Small region of neuromuscular tissue that initiates the heartbeat; also called the *SA node*, 265

Pacinian corpuscles, 195, 195*f*

Pain, 117, 196

Pain receptors Receptors that respond to tissue damage or oxygen deprivation, 194, 196

Palatine bone, 110*f*, 112*f*, 113

Palatine process, 112*f*, 113

Palatine tonsil Either of two small, almond-shaped masses located on either side of the oropharynx, composed mainly of lymphatic tissue; believed to act as sources of bacteria-killing phagocytes, 297

Pallidotomy Surgical destruction of brain tissue for the treatment of involuntary movements in Parkinson's disease, 189

Pancreas Endocrine organ located near the stomach that secretes digestive enzymes into the duodenum and produces hormones, notably insulin, 219*f*, 220*t*, 229–231, 231*f*, 343*f*, 353–354, 353*f*, 354*f*, 357

Pancreatic acinar cells, 353

Pancreatic amylase Enzyme that digests starch to maltose, 354, 360*t*

Pancreatic duct, 354*f*

Pancreatic islets (of Langerhans) Distinctive groups of cells within the pancreas that secrete insulin and glucagon, 229, 233, 233*f*, 354*f*

Pancreatic juice, 354, 354*f*

Pancreatic lipase, 360*t*

Pancreatitis, 357

Pandemic, 311

Pantothenic acid, 362*t*

Papillae, 197*f*

Papillary muscle Muscle that extends inward from the ventricular walls of the heart and to which the chordae tendinae attach, 262*f*

Pap smear Sample of cells removed from the tip of the cervix and then stained and examined microscopically, 79, 405

Paranasal sinus One of several air-filled cavities in the maxillary, frontal, sphenoid, and ethmoid bones that is lined with mucous membrane and drains into the nasal cavity, 320, 320*f*

Paraplegia Paralysis of the lower body and legs, due to injury to the spinal cord between vertebrae T1 and L2, 177

Parasympathetic division Portion of the autonomic nervous system that usually promotes those activities associated with a normal state, 184–186, 185*f*, 186*f*, 186*t*

Parathyroid gland One of four small endocrine glands embedded in the posterior portion of the thyroid gland, 219*f*, 220*t*, 225–227, 225*f*, 226*f*

Parathyroid hormone (PTH) Hormone secreted by the parathyroid glands that raises the blood calcium level primarily by stimulating reabsorption of bone, 218, 219*f*, 220*t*, 226

Parental cell Cell that divides so as to form daughter cells, 58, 59*f*

Parietal Pertaining to the wall of a cavity

Parietal bone, 109*f*, 110, 110*f*, 111*f*, 112*f*

Parietal lobe Area of the cerebrum responsible for sensations involving temperature, touch, pressure, pain, and speech, 175*f*

Parietal pericardium Outer layer of the two layers of the serous pericardium, lining the fibrous pericardium, 6*f*, 7, 352

Parietal peritoneum Lines the abdominal and pelvic walls and the inferior surface of the thoracic diaphragm, 6*f*, 7, 7*t*, 352, 353*f*, 374

Parietal pleura Membrane that lines the inner wall of the thoracic cavity, 6*f*, 7, 7*t*

Parietal serous membrane Serous membrane that covers the inner body wall, 7, 7*f*

Parietal serous pericardium, 261*f*

Parkinson's disease Progressive deterioration of the central nervous system due to a deficiency in the neurotransmitter dopamine; also called paralysis agitans, 189

Parotid glands, 343*f*

Partial pressure, 329

Parturition Processes that lead to and include the birth of a human and the expulsion of the extraembryonic membranes through the terminal portion of the female reproductive tract, 437, 438*f*

Passive immunity Protection against infection acquired by transfer of antibodies to a susceptible individual, 305–306, 306*f*

Passive smoking, 333

Patella Bone of the kneecap, 109*f*, 153*f*, 154*f*

Patellar ligament, 124*f*, 153*f*, 154*f*

Patellar surface, 121

Pathogen Disease-causing agents, such as bacteria and viruses, 243

Pathologist Person trained in knowledge of diseases and their symptoms, allowing for diagnosis of disease, 79

Peanut allergy, 293

Pectineus, 146*f*, 153*f*

Pectoral girdle Portion of the skeleton that provides support and attachment for the upper limbs, 109*f*, 118, 118*f*

Pectoralis major, 146*f*, 149*f*, 150*t*, 151, 151*f*, 409*f*

Pectoralis minor, 409*f*

Pedicles, 115, 115*f*

Pedigree Chart showing the relationships of relatives and which ones have a particular trait, 453

Pelvic brim, 121*f*

Pelvic cavity Hollow place within the ring formed by the sacrum and coxal bones, 6*f*, 7, 7*t*

Pelvic girdle Portion of the skeleton to which the lower limbs are attached, 109f, 120–121, 121f

Pelvic inflammatory disease (PID) Latent infection of gonorrhea or chlamydia in the vasa deferentia or uterine tubes, 410, 412t, 417

Pelvis, 120, 121f

Penis Male excretory and copulatory organ, 393, 396f, 398f, 399, 399f

Pentose Five-carbon sugar; deoxyribose is the pentose sugar found in DNA; ribose is a pentose sugar found in RNA, 28

Pepsin Protein-digesting enzyme produced by the stomach, 348, 360t

Peptidase Enzyme that catalyzes the breakdown of polypeptides, 359, 360t

Peptide bond Bond that joins two amino acids, 32

Peptide hormone Type of hormone that is a protein, a peptide, or derived from an amino acid, 218, 221f, 437t

Perforating canals, 103, 104f

Perforin Protein released by cytotoxic T cells that attaches to an antigen, 304

Pericardial cavity Cavity around the heart created by the pericardial sac, 6f, 7, 7t, 261f

Pericardial fluid Fluid found in small amounts in the potential space between the parietal and visceral laminae of the serous pericardium, 7, 261

Pericarditis Inflammation of the pericardium, 9

Pericardium Protective serous membrane that surrounds the heart, 6f, 7, 80, 261f

Perilymph, 208

Perimysium, 135, 135f

Perineurium, 182f

Periodontal disease, 286

Periodontal membrane, 344f

Periodontitis Inflammation of the periodontal membrane that lines tooth sockets, causing loss of bone and loosening of teeth, 344, 360

Periosteum Fibrous connective tissue covering the surface of bone, 103, 104f, 124f, 135f

Peripheral Situated away from the center of the body or an organ, 3f, 4

Peripheral nervous system (PNS) Nerves and ganglia of the nervous system that lie outside the brain and spinal cord, 163, 163f, 180–186, 181t, 182f–186f, 182t, 186t. See also Nervous system

Peripheral resistance, 276, 276f

Peristalsis Rhythmical contraction that moves the contents along in tubular organs, such as the digestive tract, 345f, 347, 360

Peritoneal cavity, 353f

Peritoneum Serous membrane that lines the abdominopelvic cavity and encloses the abdominal viscera, 80, 352–353, 353f

Peritonitis Generalized infection of the lining of the abdominal cavity, 9, 356

Peritubular capillary network Capillary network that surrounds a nephron and functions in reabsorption during urine formation, 375, 378f

Permeability, 51

Pernicious anemia Insufficiency of mature red blood cells, due to poor absorption of vitamin B₁₂, 249, 254

Peroxisomes Small vesicles within a cell which contain enzymes for the breakdown of fatty acids and hydrogen peroxide, 43t, 47

Perpendicular plate, 110f, 111f, 113

Peyer patches Lymphatic organs located in small intestine, 298

Phagocytosis Taking in of bacteria and/or debris by engulfing; also called cell eating, 53t, 55

Phalanges Bones of the fingers and thumb in the hand and of the toes in the foot (sing., *phalanx*), 109f, 120, 120f, 122f, 123

Pharyngeal arch, 433f

Pharyngeal tonsil Diffuse lymphatic tissue and follicles in the roof and posterior wall of the nasopharynx, 297, 320f, 321

Pharynx Common passageway for both food intake and air movement; the throat, 319f, 320f, 320t, 321, 343f, 346, 346t

Phenotype Physical manifestation of a trait that results from the action of a particular set of genes, 450

Phenteramine-topiramate, 365

Pheromones, 236, 392

Phlebitis Inflammation of a vein, 274

Phosphatase, 360t

Phospholipid Lipid that contains two fatty acid molecules and a phosphate group combined with a glycerol molecule, 30, 31f

Phosphorus, 363t

Photoreceptors Cells in the retina of the eye that can respond to light stimuli, 194, 198–199, 203, 203f

Phrenic nerve, 182t

pH scale Measure of the hydrogen ion concentration; any pH below 7 is acidic, and any pH above 7 is basic, 25–27, 26f, 337, 373

Physiology Branch of science dealing with the study of body functions, 2

Pia mater Innermost meningeal layer that is in direct contact with the brain and spinal cord, 171f, 172

Piercings, 98

Pineal gland Small endocrine gland, located in the third ventricle of the brain, that secretes melatonin and is involved in biorhythms, 174f, 176, 219f, 220t, 234, 235f

Pinna Outer, funnel-like structure of the ear that picks up sound waves, 208, 209f, 213t

Pinocytosis Form of endocytosis in which the cell takes in liquids, 53t, 55

Pituitary dwarfism Condition in which a person has normal proportions but small stature; caused by inadequate growth hormone, 224, 224f

Pituitary gland (hypophysis) Endocrine gland attached to the base of the brain that consists of anterior and posterior lobes, 218, 219f, 220t, 222–224, 223f–225f, 283f

Pivot joint The end of a bone moving within a ring formed by another bone and connective tissue, 124, 125f

Placenta Structure formed from the chorion and uterine tissue, through which nutrient and waste exchange occurs for the embryo and later the fetus, 284f, 428, 432, 434, 434f, 438f

Placental membrane Semipermeable membrane that separates the fetal from the maternal blood in the placenta, 432

Planes, body, 5, 5f

Plaque Accumulation of soft masses of fatty material, particularly cholesterol, beneath the inner linings of arteries, 267

Plasma Liquid portion of blood, 243, 244, 245f, 380f

Plasma cell Cell derived from a B lymphocyte that is specialized to mass-produce antibodies, 301

Plasma membrane Membrane that surrounds the cytoplasm of cells and regulates the passage of molecules into and out of the cell, 43, 43t, 44f, 45–46, 45f, 51–55, 52f–55f, 53t

Plasmin, 251

Platelet Cell-like disks formed from fragmentation of megakaryocytes that initiate blood clotting, 74, 74f, 245f, 250–251, 250f

Platelet-derived growth factor, 236

Platelet plug platelets that stick and cling to each other in order to seal a break in a blood vessel wall, 250

Pleura (pl., *pleurae*) Serous membrane that covers the lungs and lines the walls of the chest and the diaphragm, 6f, 7, 7t, 80, 323, 325

Pleurisy Inflammation of the pleura, 9

Pneumonectomy Surgical removal of all or part of a lung, 336

Pneumonia Infection of the lungs that causes alveoli to fill with mucus and pus, 331, 332f.

Pneumothorax, 323

PNS. *See* Peripheral nervous system (PNS)

Podocyte, 383f

Poisoning, 26

Polar body Small, nonfunctional cell that is a product of meiosis in the female, 403, 403f, 455f

Polar covalent bond Bond created when electrons shared by adjacent atoms are shared unequally, 24

Polar molecule Combination of atoms in which the electrical charge is not distributed symmetrically, 24

Pole During cell division, the elongated side of a cell, 58

Polycythemia Abnormally high number of red blood cells in the blood, 247

Polydactyly, 453t

Polydipsia Chronic, excessive intake of water, 230

Polyp Small, abnormal growth on any mucous membrane, such as in the large intestine, 356–357

Polypeptide A compound formed by the union of many amino acid molecules, 32, 33f

Polyphagia Excessive eating, 230

Polyribosome String of ribosomes simultaneously translating regions of the same mRNA strand during protein synthesis, 44f, 47

Polysaccharide Carbohydrate composed of many bonded glucose units—for example, glycogen, 28–29

Polyuria Excessive output of urine, 230

Poly-X female Female who has more than two X chromosomes, 446t, 449

Pons Portion of the brain stem above the medulla oblongata and below the midbrain; assists the medulla oblongata in regulating the breathing rate, 174f, 179

Popliteal artery, 277f, 280f

Popliteal vein, 281f

Portal system A system of veins that begins and ends in capillaries, 222, 223f

Portal triad Grouping of the tributaries of the hepatic artery, vein, and bile duct at the angles of the lobules of the liver, 354–355, 355f

Positive feedback Process by which changes cause more changes of a similar type, producing unstable conditions, 12, 13f, 222

Posterior Toward the back; opposite of *anterior*, 3, 3f

Posterior body cavity, 6, 6f

Posterior cerebral artery, 283f

Posterior communicating artery, 283*f*

Posterior pituitary (neurohypophysis) Portion of the pituitary gland connected by a stalk to the hypothalamus, 219*f*, 220*t*, 222, 223*f*, 381, 409, 437

Posterior (dorsal)-root ganglion Mass of sensory neuron cell bodies located in the dorsal root of a spinal nerve, 172*f*, 173

Posterior superior iliac spine, 120, 121*f*

Posterior tibial artery, 277*f*, 280*f*

Posterior tibial vein, 281*f*

Postganglionic fiber In the autonomic nervous system, the axon that leaves, rather than goes to, a ganglion, 183, 184, 186*f*

Potassium, 363*t*, 382

Power stroke During muscle contraction, the action that pulls thin filaments to the center of the sarcomere, 138

Prediabetes, 231

Pre-embryonic development, 428–429, 428*t*, 429*f*

Prefrontal area Association area in the frontal lobe that receives information from other association areas and uses it to reason and plan actions, 175, 175*f*

Preganglionic fiber In the autonomic nervous system, the axon that goes to, rather than leaves, a ganglion, 184, 185*f*

Preimplantation genetic diagnosis (PGD), 454, 455–456, 455*f*

Premature birth Child born before full term and weighing 5 pounds, 8 ounces, or less, 439

Premolars, 344*f*

Premotor area, 175, 175*f*

Prenatal care, 440–441

Prepuce, 396*f*, 399*f*

Presbycusis Loss of hearing that accompanies old age, 213

Presbyopia A visual condition that occurs with age and is caused by loss of lens elasticity; results in defective accommodation and inability to focus sharply for near vision, 213

Primary auditory area, 175*f*

Primary germ layers Three layers (endoderm, mesoderm, and ectoderm) of embryonic cells that develop into specific tissues and organs, 431, 432*f*

Primary motor area Area in the frontal lobe where voluntary commands begin; each section controls a part of the body, 173–174, 175*f*, 176*f*

Primary olfactory area, 175*f*

Primary ossification center, 105

Primary respiratory center Group of neurons in the medulla oblongata that regulates respiration, 327

Primary somatosensory area Area posterior to the central sulcus where sensory information arrives from the skin and skeletal muscles, 174, 175*f*, 176*f*

Primary spermatocyte Cell dividing into two secondary spermatocytes, 396, 397*f*

Primary structure, 32

Primary taste area, 174, 175*f*

Primary visual area, 174–175, 175*f*, 207*f*

Prime mover For muscle action, that muscle or group of muscles that carries out the primary action; for example, the biceps brachii is the prime mover for elbow flexion, 145

Prion Proteinaceous infectious particles; thought to cause neurological disease, 34

PRL. *See* Prolactin (PRL)

Process, 107*t*

Processing centers, 175–176

Progesterone Female sex hormone secreted by the ovaries that, along with estrogen, promotes the development and maintenance of the primary and secondary female sex characteristics, 219*f*, 220*t*, 232, 402*f*, 406, 406*f*, 411*t*, 414, 437*t*

Progestin, 412*t*

Prolactin (PRL) Hormone secreted by the anterior pituitary that stimulates milk production in the mammary glands; also known as lactogenic hormone, 219*f*, 220*t*, 223*f*, 224

Prolactin-inhibiting hormone (PIH), 224

Promoter Any influence that causes a mutated cancer cell to start growing in an uncontrolled manner, 79

Pronation Rotation of the forearm so that the palm faces backward, 125, 126*f*

Prophase Stage of mitosis when chromosomes become visible, 56*f*, 58, 59*f*, 60*f*, 394*f*, 395

Proprioceptor Sensory receptor that assists the brain in knowing the position of the limbs, 194

Prostaglandins Hormones that have various and powerful effects, often within the cells that produce them, 236, 398

Prostate cancer, 387

Prostate enlargement, 387

Prostate gland Gland situated around the base of the male urethra; secretes a viscous alkaline fluid that is a major constituent of the ejaculatory fluid, 374, 395, 396*f*, 398, 398*f*, 398*t*

Protease Enzyme that digests protein, 307

Protective proteins, 300

Protein Macromolecule composed of amino acids, 31–34, 33*f*, 300, 364, 364*f*

Protein synthesis, 56, 57*f*

Prothrombin Plasma protein made by the liver that must be present in blood before clotting can occur, 250, 250*f*

Prothrombin activator Enzyme that catalyzes the transformation of the precursor prothrombin to the active enzyme thrombin, 250*f*, 251

Proton Positively charged particle in an atomic nucleus, 21

Proximal Closer to the midline or origin; opposite of *distal*, 3*f*, 4

Proximal convoluted tubule Highly coiled region of a nephron near the glomerular capsule, 375–376, 376*f*, 377*f*, 383*f*

Pseudostratified columnar Appearance of layering in some epithelial cells when, actually, each cell touches a baseline and true layers do not exist, 66*t*, 68, 68*f*, 322, 322*f*

Psoas major, 153*f*

Psoriasis Common chronic, inherited skin disease in which red patches are covered with scales; occurs most often on the elbows, knees, scalp, and trunk, 88, 333

PTH. *See* Parathyroid hormone (PTH)

Puberty Stage of development in which the reproductive organs become functional, 232, 393, 400–401

Pubic arch, 121*f*

Pubic bone, 396*f*, 401*f*

Pubic symphysis Slightly movable cartilaginous joint between the anterior surfaces of the hip bones, 120, 121*f*, 438*f*

Pubis Anterior-most bone of the three bones that comprise the pelvic bone, 120, 121*f*

Pulmonary arteriole, 324*f*

Pulmonary artery Blood vessel that takes blood away from the heart to the lungs, 260*f*, 262*f*, 264*f*, 279, 284*f*, 324*f*

Pulmonary capillaries Capillaries found between pulmonary arterioles and pulmonary venules, 262, 320*t*, 323, 324*f*

Pulmonary circuit Path of blood through vessels that take O_2-poor blood to and O_2-rich blood away from the lungs, 260, 260*f*

Pulmonary edema Excessive fluid in the lungs caused by congestive heart failure, 286, 297

Pulmonary embolism Blockage of a pulmonary artery by a blood clot that commonly originates in a vein of the lower legs, 274

Pulmonary fibrosis Accumulation of fibrous connective tissue in the lungs; caused by inhaling irritating particles, such as silica, coal dust, or asbestos, 332, 332*f*

Pulmonary semilunar valve, 262, 262*f*

Pulmonary trunk, 262, 262*f*, 264*f*

Pulmonary tuberculosis Tuberculosis of the lungs, caused by the tubercle bacillus, 331–332, 332*f*

Pulmonary vein Blood vessel that takes blood away from the lungs to the heart, 260*f*, 262*f*, 264*f*, 279, 284*f*, 324*f*

Pulmonary venules, 262, 324*f*

Pulp, 344, 344*f*

Pulse Vibration felt in arterial walls due to expansion of the aorta following ventricular contraction, 276–277, 277*f*

Punnett square, 451–452

Pupil Opening in the center of the iris that controls the amount of light entering the eye, 199*f*, 200, 200*f*, 201*f*

Purkinje fiber Specialized muscle fiber that conducts the cardiac impulse from the AV bundle into the ventricular walls, 264–265, 265*f*

Pus Thick, yellowish fluid composed of dead phagocytes, dead tissue, and bacteria, 300

Pyelonephritis Inflammation of the kidney; due to bacterial infection, 385

Pyloric region Distal-most portion of the stomach; joins the duodenum of the small intestine, 348, 348*f*

Pyloric sphincter, 348*f*

Pylorus, 348*f*

Q

Quadrants, abdominopelvic, 8*f*

Quadriceps femoris, 146, 153*f*

Quadriceps femoris group, 152*t*, 153

Quadriceps tendon, 153*f*

Quadriplegia Paralysis of the entire body and all four limbs, due to injury to the spinal cord between vertebrae C4 and T1, 177

Quaternary structure, 32

R

Radial artery, 277*f*, 280*f*

Radial nerve, 182*t*

Radial notch, 119*f*, 120

Radial tuberosity, 119, 119*f*

Radial vein, 281*f*

Radioactive isotope Atom whose nucleus undergoes degeneration and in the process gives off radiation, 22

Radius Elongated bone located on the thumb side of the lower arm, 109f, 119, 119f, 120f

Rape, 421

Receptor potential In sensory receptors, graded potentials that can be summed together to signal other neurons to create an action potential, 167, 194

Recessive allele Hereditary factor that expresses itself only when two copies are present in the genotype, 450

Recruitment Increase in the number of motor units activated as intensity of stimulation increases, 143

Rectum Terminal portion of the intestine, 343f, 351, 352f, 353f, 401f, 438f

Rectus abdominis, 146, 146f, 149, 149f, 149t

Rectus femoris, 146f, 153f

Red blood cell. See Erythrocyte

Red bone marrow Blood cell-forming tissue located in spaces within certain bones, 103, 104f, 294–296, 295f, 296f

Reduced hemoglobin (HHb) Hemoglobin that is carrying hydrogen ions, 273, 330

Reduction To align the ends of a fractured bone so that proper healing can occur, 108

Referred pain Pain perceived as having come from a site other than that of its actual origin, 196

Reflex Automatic, involuntary response of an organism to a stimulus, 183, 184f

Refractory period During an action potential, the time during which a second stimulus won't cause a second action potential, 167

Renal artery Vessel that originates from the aorta and delivers blood to the kidney, 260f, 279t, 280f, 376f, 378f

Renal calculi, 385, 387

Renal corpuscle The part of a nephron that consists of the glomerulus and surrounding Bowman's capsule, 375

Renal cortex Outer, primarily vascular portion of the kidney, 375, 376f, 381f

Renal medulla Inner portion of the kidney, including the renal pyramids, 375, 376f, 381f

Renal pelvis Inner cavity of the kidney formed by the expanded ureter and into which the collecting ducts open, 375, 376f

Renal pyramid, 376f

Renal transplantation, 386

Renal tubule The part of a nephron that leads away from a glomerulus; is made up of a proximal convoluted tubule, loop of Henle, and distal convoluted tubule, and empties into a collecting tubule, 375–377, 377f

Renal vein Vessel that takes blood from the kidney to the inferior vena cava, 260f, 281f, 376f, 378f

Renin Secretion from the kidney that activates angiotensinogen to angiotensin I, 228, 276, 382

Replacement reactions Chemical reaction involving both decomposition and synthesis, in which two new compounds are formed, 34

Replication Production of an exact copy of a DNA sequence, 56, 57f

Repolarization Recovery of a neuron's polarity to the resting potential after the neuron ceases transmitting impulses, 165, 166f, 168f

Reproduction, 10

Reproductive system, 10
 aging of, 419
 female, 401–409, 401f–404f, 406f, 408f, 409f

in homeostasis, 13, 15f, 419
 human life cycle and, 393–395, 394f, 395f
 male, 395–401, 396f–400f

Residual volume Volume of air that remains in the lungs after normal exhalation, 327, 327f

Respiration Transport and exchange of gases between the atmosphere and the cells via the lungs and blood vessels, 328–329, 330–336, 332f, 335f. See also Respiratory system

Respiratory capacities, 326–327, 326f

Respiratory distress syndrome, 439

Respiratory membrane Alveolar wall plus the capillary wall, across which gas exchange occurs, 324–325, 324f

Respiratory pump, 275–276

Respiratory system, 10
 acid-base balance and, 383–384
 aging of, 336
 breathing in, 325–328, 326f–328f
 functions of, 319
 gas exchange in, 324f, 328–330, 329f
 in homeostasis, 13, 15f, 337
 infections, 330–332
 lungs in, 323–325, 324f

Respiratory tract, 319–323, 319f–322f, 320t

Respiratory volumes, 326–327, 326f

Resting potential Potential energy of a resting neuron, created by separating unlike charges across the neuron cell membrane, 165, 166f

Restrictive pulmonary disorders, 332

Reticular connective tissue Collagen fibers that form the framework for lymphoid organs, 70

Reticular fibers Thin, highly branched, collagenous fibers that form delicate supporting networks, 69

Reticular formation Nerve cells and fibers situated primarily in the brain stem and functioning upon stimulation, especially in arousal, 179–180

Reticulocytes, 246f, 247

Retina Innermost layer of the eyeball that contains the rod cells and cones cells, 200f, 201, 201t, 203–204, 204f

Retinal A form of vitamin A, 203, 203f

Retinal hemorrhage, 208

Retinopathy of prematurity, 439

Retrovirus Any of the family of single-stranded RNA viruses, 307

Rheumatoid arthritis Persistent inflammation of synovial joints, often causing cartilage destruction, bone erosion, and joint deformities, 126, 312

Rh factor Type of antigen on red blood cells, 253–254, 254f

Rhodopsin Light-sensitive biochemical in the rod cells of the retina; visual purple, 203

Rib cage Bony framework of the thoracic cavity; created by the thoracic vertebrae, ribs, costal cartilages, and sternum, 109f, 116–117, 116f

Riboflavin, 362t

Ribonucleic acid (RNA) Nucleic acid that helps DNA in protein synthesis, 35, 35t, 58

Ribosomal RNA (rRNA) RNA (ribonucleic acid) occurring in ribosomes, structures involved in protein synthesis, 58

Ribosome Minute particle, found attached to the endoplasmic reticulum or loose in the cytoplasm, that is the site of protein synthesis, 43t, 44f, 46f, 47, 48f

Ribs, 109f, 116, 116f

Rickets Defective mineralization of the skeleton, usually due to inadequate vitamin D in the body, 22, 92, 92f

Right lymphatic duct, 294, 295f

Rigor mortis, 144

RNA. See Ribonucleic acid (RNA)

Rod cell Dim-light receptor in the retina of the eye that detects motion but not color, 201, 201t, 203, 204

Romanov family, 445

Root canal, 344f

Root ganglia, 172f, 173

Root hair plexus, 194, 195f

Rotation Movement of a bone around its own longitudinal axis, 125, 126f, 148

Rotational equilibrium Maintenance of balance when the head and body are suddenly moved or rotated, 210, 211f, 213, 213t

Rotator cuff, 150f, 150t, 151, 151f

Rough ER Endoplasmic reticulum that is studded with ribosomes on the side of the membrane that faces the cytoplasm, 43t, 44f, 46f, 48f. See also Smooth ER

Round ligament of liver, 355f

Round window Membrane-covered opening between the inner ear and the middle ear, 208, 209f, 210f

Roux-en-Y gastric bypass, 367, 367f

RU-486, 415

Rubella An acute, infectious disease affecting the respiratory tract in children and nonimmune young adults; characterized by a slight cold, sore throat, and fever, and the appearance of a fine, pink rash, 440

Ruffini endings, 195, 195f

Rugae Deep folds, as in the wall of the stomach, 347, 348f

S

Saccule Saclike cavity of the inner ear that contains receptors for gravitational equilibrium, 211f, 213, 213t

Sacral nerves, 183f

Sacroiliac joint Connection between the coxal bone and the sacrum, 120, 121f

Sacrum Bone consisting of five fused vertebrae that form the posterior wall of the pelvic girdle, 109f, 114f, 116, 121f

Saddle joint Two bones joined, having convex and concave surfaces that are complementary, 124, 125f

Sagittal plane Plane or section that divides a structure into right and left portions, 5, 5f

Sagittal suture Line of junction between the two parietal bones in the cranium, 123

Salivary amylase In humans, enzyme in saliva that digests starch to maltose, 345–346, 360t

Salivary gland Gland associated with the mouth, secretes saliva, 343f, 344–346

Salpingectomy Surgical removal of the uterine tubes, 405

Salt Compound produced by a reaction between an acid and a base, 22

Saltatory conduction Conduction of an action potential down the length of a myelinated axon, by jumping from one node of Ranvier to the next, 167

Salt balance, 373

SA node. See Sinoatrial (SA) node

Smoking, 42, 285, 333–334

Smooth ER Synthesizes the phospholipids that occur in membranes, among other functions, depending on the particular cell, 43t, 44f, 46f, 48f

Smooth muscle Contractile tissue that comprises the muscles in the walls of internal organs; also called visceral muscle, 74t, 75, 75f, 134–135, 134f, 140, 156

Snellen chart, 205

Sodium intake, 285, 363t, 382

Sodium-potassium pump, 165, 166f, 168f

Soft palate Entirely muscular posterior portion of the roof of the mouth, 344, 344f, 345f

Solute The substance dissolved in a solution, 52

Solvent Major component of a solution, usually liquid or gas, 24

Somatic system Portion of the peripheral nervous system containing motor neurons that control skeletal muscles, 180

Somatosensory association area, 175, 175f

Somatostatin, 220t, 229, 230

Somatotropin. *See* Growth hormone (GH)

Somite, 433f

Spasm Sudden, violent, involuntary contraction of a muscle or a group of muscles, 157

muscle, 157

vascular, 250

Specific heat capacity Amount of energy needed to change an object's temperature by exactly 18C, 25

Sperm Male gamete having a haploid number of chromosomes and the ability to fertilize an egg, the female gamete, 393, 396, 397f, 426, 426f

Spermatic cord Cord that suspends the testis within the scrotum; contains the vas deferens and vessels and nerves of the testis, 397f, 398

Spermatid Intermediate stage in the formation of sperm cells, 397f

Spermatocytes, 396, 397f

Spermatogenesis Sperm production in males by the process of meiosis and maturation, 396–397, 397f

Spermatogonia, 396, 397f

Spermatozoa Developing male gametes, 396

Sperm count, 415

Spermicide, 411t

S phase, 56, 56f

Sphenoid bone, 111f, 112f, 113

Sphenoid sinus, 320f

Sphincter Muscle that surrounds a tube and closes or opens the tube by contracting and relaxing, 347

Sphygmomanometer, 277, 277f

Spina bifida Disorder caused by failure of the neural tube to close, 431

Spinal ataxia, 453t

Spinal cord Portion of the central nervous system extending downward from the brain stem through the vertebral canal, 172–173, 172f

Spinal cord injury, epidural stimulation in, 177

Spinal meningitis Inflammation of the meninges of the spinal cord, 82

Spinal nerve Nerve that arises from the spinal cord, 172f, 182f, 183f

Spinal reflex, 183

Spindle Apparatus composed of microtubules to which the chromosomes are attached during cell division, 49, 58, 59f

Spine (of bone), 107t

Spinous process, 114f, 115, 115f

Spiral organ Structure in the vertebrate inner ear that contains auditory receptors; also called organ of Corti, 209

Spirometer Instrument for measuring the air entering and leaving the lungs, 326–327, 326f

Spleen Large, glandular organ located in the upper left region of the abdomen that stores and purifies blood, 295f, 296

Splenic vein, 282f

Sponge, contraceptive, 411t

Spongy bone Bone found at the ends of long bones; consists of bars and plates separated by irregular spaces, 69t, 72, 73f, 103–105, 104f, 124f

Sports, 234–235, 235f

Sprain Joint injury in which some of the fibers of a supporting ligament are ruptured, but the continuity of the ligament remains intact, 157

Squamosal suture Type of suture formed by overlapping of the broad, beveled edges of the participating bones, 111f

Squamous cell carcinoma Type of skin cancer that occurs in the epidermis; has warty appearance, bleeds, and/or forms a scab, 93, 94f

Squamous epithelium, 66, 66t

Squamous tissue Tissue formed from groups of flattened cells, 110f

Stapes The last of three ossicles of the ear; serves with the malleus and incus to conduct vibrations from the tympanic membrane to the oval window of the inner ear, 208, 209f, 210f

Staphylococcus aureus, 310

Starch Polysaccharide that is common in foods of plant origin, 28, 29f

Static equilibrium, 210, 211f

Stem cell A precursor cell, 244, 246f, 255

Stent Device used to prop open or reinforce a blood vessel, 268

Stereoscopic vision, 207

Sterilization, 410, 412t

Sternocleidomastoid, 146, 146f, 147f, 148, 148t, 149f

Sternum Breastbone to which the ribs are ventrally attached, 109f, 116–117, 116f

Steroid Lipid-soluble, biologically active molecules having four interlocking rings; examples are cholesterol, progesterone, and testosterone, 29, 30, 31f, 234–235, 235f

Steroid hormone Type of hormone that has the same complex of four-carbon rings, but each one has different side chains, 218, 221f

Stimulus In nerve physiology, energy needed to cause a neuron to respond, 165

Stomach Saclike, expandable digestive organ located between the esophagus and the small intestine, 343f, 346t, 347–349, 348f, 353f

Stomach disorders, 356

Strain An overstretching or overexertion of some muscles, 157

Strand, 35

Stratified Layered, as in stratified epithelium, which contains several layers of cells, 66t, 68, 68f

Stratified cuboidal epithelium Tissue created by layers of cube-shaped cells; found in gland ducts, 66t, 67, 67f

Stratified squamous epithelium Covering of the internal or external surfaces of the body composed of layered, flattened, platelike cells, 66, 66t, 67, 67f

Stratum basale Deepest layer of the epidermis, where cell division occurs, 87–88, 87f

Stratum corneum Uppermost keratinized layer of the epidermis, 87f, 88

Stratum granulosum, 87f, 88

Stratum lucidum Layer of epidermis that is found only in thick skin (i.e., the soles of feet), 87f, 88

Stratum spinosum, 87f, 88

Strep throat, 331

Stress incontinence, 440

Striae gravidarum Stretch marks of pregnancy; caused by weakening of collagen in the skin, 440

Striations, 134, 134f

Stroke. *See* Cerebrovascular accident (CVA)

Stroke volume (SV), 266, 270

Sty Inflammation of a sebaceous gland, 199

Subarachnoid space, 172

Subclavian artery Either of two arteries branching off the aortic arch and supplying the arms, 262f, 264f, 279, 279t, 280f

Subclavian vein, 280, 281f, 295f

Subcutaneous injection Introduction of a substance beneath the skin, using a syringe, 88

Subcutaneous tissue Tissue beneath the dermis that tends to contain fat cells, 87, 87f

Subdural hematoma Accumulation of blood between the dura mater and the brain, 179

Sublingual glands, 343f

Submandibular glands, 343f

Submucosa, 346

Subscapularis, 150t

Substrate The reactant in an enzymatic reaction, 32

Sucrase, 360t

Sucrose, 28

Summation In muscle contraction, increased contraction until maximal sustained contraction is achieved, 143, 143f

Superficial Near the surface, 3f, 4

Superficial fascia, 135

Superficial temporal artery, 277f, 280f

Superior Toward the upper part of a structure or toward the head, 3, 3f

Superior mesenteric artery, 279t, 280f

Superior mesenteric vein, 282f

Superior nasal concha, 113, 320f

Superior oblique, 199, 200f

Superior rectus, 199, 199f, 200f

Superior vena cava Large vein that enters the right atrium from above and carries blood from the head, thorax, and upper limbs to the heart, 262f, 264f, 279, 281f, 281t, 284f

Supination Rotation of the forearm so that the palm faces forward when in the anatomical position, 125, 126f

Suprahyoid muscles, 148

Supraspinatus, 150f, 150t, 151f

Surface tension Force that holds moist membranes together when water molecules attract, 323

Surfactant Agent that reduces the surface tension of water; in the lungs, a surfactant prevents the alveoli from collapsing, 324, 324f

Suspensory ligament of eyeball, 200f, 201t, 202f

Suspensory ligament of uterus, 404f

Sustentacular cell In the testis, a supporting epithelial cell; also called a *Sertoli cell*, 396, 397f

Suture Type of immovable joint articulation found between bones of the skull, 110

as bone feature, 107t

cranial, 110, 110f

Swallowing, 148, 345f, 346

Sweat gland Skin gland that secretes a fluid substance for evaporative cooling; also called sudoriferous gland, 91–92, 91f

Swine flu, 311

Sympathetic division Part of the autonomic nervous system whose effects are generally associated with emergency situations, 184, 185f, 186f, 186t

Tonsil Partly encapsulated lymph nodule located in the pharynx, 197f, 295f, 297, 320f, 321, 331, 344f

Tonsillectomy Surgical removal of the tonsils, 331

Tonsillitis Inflammation of the tonsils, 331

Tooth decay, 286, 344

Total lung capacity, 327, 327f

Toxoplasmosis Disease caused by toxoplasma parasites that invades the tissues and can seriously damage the central nervous system, especially of infants, 441

Trabeculae Branching bony plate that separates irregular spaces within spongy bone, 72, 73f

Tracer Substance having an attached radioactive isotope that allows a researcher to track its whereabouts in a biological system, 22

Trachea Windpipe; serves as a passageway for air, 319f, 320t, 321f, 322–323, 322f

Tracheal cartilage, 321f

Tracheostomy Creation of an artificial airway by incision of the trachea and insertion of a tube, 322–323

Tract Bundle of neurons forming a transmission pathway through the brain and spinal cord, 77, 164

Transcription Manufacturing RNA from DNA, 56, 57f, 58

Transfer RNA (tRNA) Molecule of RNA (ribonucleic acid) that carries an amino acid to a ribosome engaged in the process of protein synthesis, 58

Transfusion, blood, 252–254, 252t, 253f, 254f, 255, 255f

Transient ischemic attack Brief episode of decreased blood flow to the brain; usually characterized by temporary blurring of vision, slurring of speech, numbness, paralysis, or fainting; often predictive of a serious stroke; abbreviated *TIA;* also called *mini-stroke,* 287

Transitional epithelium Tissue that forms the lining of the ureters, urinary bladder, and urethra, 66t, 68

Translation Assembly of an amino acid chain according to the sequence of base triplets in a molecule of mRNA, 56, 57f, 58

Transmissible spongiform encephalopathies (TSEs), 34

Transplantation, renal, 386

Transport, 53–55, 53f

Transport vesicle, 48f

Transverse colon Portion of the large intestine that travels transversely as it extends from the ascending colon to the descending colon, 351, 352f, 353f

Transverse ligament, spinal, 115f

Transverse plane Plane or section that divides a structure horizontally to give a cross section, 5, 5f

Transverse process, 114f, 115, 115f

Transverse tubules In muscle, structures formed from the sarcolemma; they extend into the sarcoplasm and contact but do not touch the sarcoplasmic reticulum, 136, 137f

Transversus abdominis, 149, 149f, 149t

Trapezius, 146f, 147f, 148, 148t, 149f, 150f

Traumatic brain injury (TBI), 179

Triceps brachii, 150t, 151, 151f

Trichuris suis ova (TSO), 313

Tricuspid valve Atrioventricular valve between the right atrium and the right ventricle, 262, 262f

Trigeminal nerve, 163

Triglyceride Lipid composed of three fatty acids combined with a glycerol molecule, 30

Triiodothyronine (T₃), 225

Triple bonds, 23–24, 23f

Triplet code Three-nucleotide base unit coding for a particular amino acid during protein synthesis, 56

Trisomy State of having an extra chromosome—three instead of the normal two, 447

Trochanter, 107t

Trochlea, of humerus, 119, 119f

Trochlear nerve, 200

Trochlear notch, 119f, 120

Trophoblast Outer cells of a blastocyst that help form the placenta and other extra-embryonic membranes, 429

Tropomyosin One of three proteins that comprise the thin filaments in a sarcomere, 138

Troponin One of three proteins that comprise the thin filaments in a sarcomere, 138

Troposin, 138

Trunk, muscles of, 148–149, 149f, 149t

Trypsin Protein-digesting enzyme produced by the pancreas, 360t

TSH. *See* Thyroid-stimulating hormone (TSH)

T (transverse) tubule Membranous channel that extends inward from a muscle fiber membrane and passes through the fiber, 136, 137f

Tubal ligation Method for preventing pregnancy in which the uterine tubes are cut and sealed, 410, 412t

Tuberculosis, 331–332, 332f

Tuberosity, 107t

Tubular reabsorption Process that transports substances out of the renal tubule into the interstitial fluid from which the substances diffuse into peritubular capillaries, 378f, 379

Tubular secretion Process of substances moving out of the peritubular capillaries into the renal tubule, 378f, 379

Tumor Abnormal growth of tissue that serves no useful purpose, 79. *See also* Cancer

Tumor angiogenesis factor, 236

"tumor paint," 20

Tunica externa The outer layer that makes up a blood vessel; composed of collagenous and elastic fibers, 271, 272f

Tunica intima The innermost lining of a blood vessel, consisting of an endothelial layer backed by connective and elastic tissue, 271, 272f

Tunica media Middle layer of the walls of arteries and veins; composed of smooth muscle, 271, 272f

Turner syndrome Condition caused by the inheritance of a single X chromosome, 446t, 449, 449f

Twins, conjoined, 425, 429

Tympanic canal, 210f

Tympanic duct, 210f

Tympanic membrane Membrane located between the external and middle ear; the eardrum, 209f, 213t

Tympanostomy tubes, 331

U

Ulcerative colitis, 312, 357

Ulna Elongated bone within the lower arm, 109f, 119–120, 119f, 120f

Ulnar artery, 280f

Ulnar nerve, 182t

Ulnar notch, 119, 119f

Ulnar vein, 281f

Ultrasound, 3D, 1

Umami, 196

Umbilical Pertaining to the umbilicus

Umbilical artery One of a pair of fetal arteries that pass through the umbilical cord to the placenta to which they carry the deoxygenated blood from the fetus, 283, 284f, 434f

Umbilical cord Cord through which blood vessels that connect the fetus to the placenta pass, 431, 433f, 434f, 438f

Umbilical vein A vein that passes through the umbilical cord to the fetus and returns the oxygenated and nutrient blood from the placenta to the fetus, 283, 284f, 434f

Uncinate fit Form of temporal lobe epileptic seizure characterized by hallucinations of taste and odor, 197

Unipolar neuron, 164

Universal donor, 252

Universal recipient, 252

Unsaturated fatty acid Organic compound that includes a fatty acid molecule having one or more double bonds between the atoms of its carbon chain, 30

Upper limb, 118–120, 118f–120f

Upper respiratory infection, 330–331

Urea Primary nitrogenous waste of mammals, 355, 372, 373

Uremia High level of urea nitrogen in the blood, 385

Ureters Tubes that take urine from the kidneys to the bladder, 374, 374f, 376f, 396f, 398f, 404f

Urethra Tube that takes urine from the bladder to the outside of the body, 374, 396f, 398f, 398t, 399f, 401f, 406f, 438f

Urethral sphincters, 374

Urethritis Inflammation of the urethra, 385

Uric acid Product of nucleic acid metabolism in the body, 373, 389

Urinalysis Examination of a urine sample to determine its chemical, physical, and microscopic aspects, 385, 386

Urinary bladder Organ where urine is stored before being discharged by way of the urethra, 353f, 374, 374f, 385, 396f, 398f, 401f, 438f

Urinary incontinence, 387, 440

Urinary system, 10
  aging of, 387
  functions of, 373–374
  in homeostasis, 13, 15f, 387–389
  kidneys in, 375–386, 376f–378f, 379t, 380f, 380t, 381f, 383f–385f, 384t
  organs, 374, 374f, 375f
  urination in, 375, 375f

Urination, 375, 375f

Urine composition, 384t

Urine formation, 377–379, 378f, 379t

Urticaria Skin eruption characterized by the development of welts as a result of capillary dilation, 93

Uterine cycle, 408f

Uterine tube Tube that extends from the uterus on each side toward an ovary and transports sex cells; also called fallopian tube or oviduct, 401, 401f, 404, 404t, 429f

Uterus Female organ in which the fetus develops, 393, 401, 401f, 404–405, 404f, 404t, 438f, 439–440

Utricle Saclike cavity of the inner ear that contains receptors for static equilibrium, 211f, 213, 213t

Uvula A fleshy portion of the soft palate that hangs down above the root of the tongue, 344, 344f

V

Vaccine Treated antigens that can promote active immunity when administered, 305, 305f, 309–310, 309f

Vacuole, 43t, 47, 49